Water
Pollution

Water Pollution

WK BERRY

CBS

CBS Publishers & Distributors Pvt Ltd

New Delhi • Bengaluru • Chennai • Kochi • Kolkata • Mumbai • Pune
Hyderabad • Nagpur • Patna • Vijayawada

Water Pollution

ISBN: 978-81-239-2838-8

First Edition: 2016
Reprint: 2017

Published by Satish Kumar Jain and produced by Varun Jain for

CBS Publishers & Distributors Pvt Ltd
4819/XI Prahlad Street, 24 Ansari Road, Daryaganj, New Delhi 110 002, India.
Ph: 23289259, 23266861, 23266867   Website: www.cbspd.com
Fax: 011-23243014   e-mail: delhi@cbspd.com; cbspubs@airtelmail.in.
Corporate Office: 204 FIE, Industrial Area, Patparganj, Delhi 110 092
Ph: 4934 4934   Fax: 4934 4935   e-mail: publishing@cbspd.com; publicity@cbspd.com

Branches

- **Bengaluru:** Seema House 2975, 17th Cross, K.R. Road,
  Banasankari 2nd Stage, Bengaluru 560 070, Karnataka
  Ph: +91-80-26771678/79   Fax: +91-80-26771680   e-mail: bangalore@cbspd.com
- **Chennai:** 7, Subbaraya Street, Shenoy Nagar, Chennai 600 030, Tamil Nadu
  Ph: +91-44-26680620, 26681266   Fax: +91-44-42032115   e-mail: chennai@cbspd.com
- **Kochi:** Ashana House, No. 39/1904, AM Thomas Road, Valanjambalam,
  Ernakulam 682 016, Kochi, Kerala
  Ph: +91-484-4059061-65   Fax: +91-484-4059065   e-mail: kochi@cbspd.com
- **Kolkata:** 6/B, Ground Floor, Rameswar Shaw Road, Kolkata-700 014, West Bengal
  Ph: +91-33-22891126, 22891127, 22891128   e-mail: kolkata@cbspd.com
- **Mumbai:** 83-C, Dr E Moses Road, Worli, Mumbai-400018, Maharashtra
  Ph: +91-22-24902340/41   Fax: +91-22-24902342   e-mail: mumbai@cbspd.com

Representatives

• **Hyderabad**	0-9885175004	• **Jharkhand**	0-9811541605	• **Nagpur**	0-9021734563
• **Patna**	0-9334159340	• **Pune**	0-9623451994	• **Uttarakhand**	0-9716462459

Printed at India Binding House, Noida, UP

Preface

Water is the nature's free gift to the human race. It is available in various forms such as rivers, lakes, streams, etc. The importance of water in human life is so much that the development of any city of the world has practically taken place near some sources of water supply. It may also further be noted that the water is available in solid, liquid and gaseous forms. The occurrence of water in all these three forms is basically important for human beings for comfort, luxury and various other necessities of life.

Water is one of the abundantly available substances in nature. It is an essential constituent of all animal and vegetable matter and forms about 75 per cent of the matter of earth's crust. It is also an essential ingredient of animal and plant life. Water is distributed in nature in different forms, such as rain water, river water, spring water and mineral water. Rain water is the purest form of naturally occurring water. It evaporates from sea as a result of extensive heat. The water vapours thus rising from the surface are drifted by the winds onwards. On rising to high altitudes, they undergo condensation and form small droplets. These droplets move in the sky in the form of clouds and aggregate continuously till they become heavy enough unable to support their weight. As a result, they fall down in the form of rain. Since rain water is produced by a process of distillation, it is considered to be the purest form of water. The rain water, however, is associated with dissolved gases such as CO_2, SO_2, NH_3, etc. from the atmosphere.

This reference textbook is divided into five sections. Section I deals with characteristics and perspective of water. Chapter 1 is devoted to sources of water. Chapters 2 and 3 deal with physical, chemical and biological water quality parameters.

Section II concentrates on engineering systems for water purification. Chapter 4 acquaints the readers with basic concepts of engineering systems for water purification. Chapter 5 brings to light water conditioning and softening and discusses the various softening methods. Chapter 6 deals with aeration which occupies a significant place in waste-water quality management and is an important factor in the purification of polluted water. Chapter 7 focuses on coagulation, flocculation and filtration which convert non-settleable turbidity particles into settleable form for their effective removal by gravity. Chapter 8 deals with ion exchange and carbon adsorption which are physical operations as they affect physical composition of the water. Chapter 9 is devoted to water sterilisation technologies and provides a very basic overview of the principles and technologies associated with water purification or more specifically, sterilisation.

Section III focuses on engineering systems for waste-water treatment, disposal and analysis. Chapter 10 concentrates on basic concepts for waste-water treatment and disposal. Chapter 11 brings to light primary and secondary treatment. Various primary and secondary methods for waste-water treatment are discussed. Chapter 12 focuses on advance waste-water treatment which refers to the methods and processes that remove more contaminants from waste-water than are taken out by convectional biological treatment. Chapter 13 focuses on waste-water treatment by rootzone technology. Various rootzone technologies along with process of waste-water treatment have been discussed. Membrane separation systems and membrane reactors are available today for reducing pollution control cycles. Chapter 14 deals with membrane technology for production of waste-waters. Chapter 15 concentrates on treating the sludge. Sludge or solid waste is unavoidably produced in the treatment of water containing suspended solids. Various methods of treatment of sludge are discussed in detail. Chapter 16 focuses on analysis of water. Various methods for water analysis are discussed in detail.

Section IV attempts to study recycling and reuse of waste-water from various industries. Chapter 17 concentrates on chemical and allied industries. Chapter 18 is devoted to coal and metal mining. Chapter 19 deals with iron and steel industry in which waste-water is generated from Blast furnaces and casting operations. Various operations of reduction and recycling of waste-water have been discussed. Chapter 20 focuses on coal based thermal power plants. In thermal pollution water is used for cooling the working steam which is led back to the source at an appreciable higher temperature, affecting aquatic organisms. Various waste-water treatment schemes have been discussed. Chapter 21 concentrates on cement and ceramic industry and discusses the characteristics, sources and recycling of waste-water. Chapter 22 focuses on electrical and electronic industries. Chapter 23 is devoted to nuclear and radioactive industry. The menace of radioactive pollution enters into the environment in waste streams and stack gases from various operations. Chapter 24 discusses about common effluent treatment plant. The chemical process industry has attracted a great deal of opprobrium on account of its poor record in the management of liquid and solid wastes. The quality and quantity of raw effluents vary not only from industry to industry but from the same type of industry. For example, it is no surprise that effluents from the cotton textile industry differ from those of synthetic textiles but it is surprising that effluents from one nylon textile factory may not be the same in quantity and quality as another nylon factory. In the light of this quality and quantity fluctuation, the design of an effluents treatment plant (ETP) has to be tailored to suit a particular industry.

Section V discusses case studies related to water pollution. Chapter 25 discusses the case studies of Hindustan Petroleum Corporation Limited (HPCL), JK Rayon and Synthetics, and Radioactive waste treatment of nuclear plant.

Glossary and index have been provided at the end for quick reference. Figures and tables supplement the text. All topics have been covered in a cogent and lucid style to help the reader grasp the information quickly and easily.

It may not be wrong to hold that the present reference textbook of *Water Pollution* is a complete treatise on water pollution and its control. It is essential reading for all students and teachers of environment, engineering and life sciences. In addition, researchers in water management, environmental and allied fields will also find it highly useful and informative.

The book also caters to the requirement of the syllabus prescribed by various Indian universities for undergraduate student pursuing engineering, life sciences, environment and allied courses. It has been prepared with meticulous care, aiming at making the book error-free. Constructive suggestions are always welcome from users of this book.

WK Berry

Contents at a Glance

SECTION V

Contents

SECTION II

SECTION III

SECTION IV

SECTION V

SECTION I

Water: Characteristics and Parameters of Quality

Chapter 1

Sources of Water

INTRODUCTION

Water is one of the abundantly available substances in nature. It is an essential constituent of all animal and vegetable matter and forms about 75 per cent of the matter of earth's crust. It is also an essential ingredient of animal and plant life. Water is distributed in nature in different forms, such as rain water, river water, spring water and mineral water. Rain water is the purest form of naturally occurring water. It evaporates from sea as a result of extensive heat. The water vapours thus rising from the surface are drifted by the winds onwards. On rising to high altitudes, they undergo condensation and form small droplets. These droplets move in the sky in the form of clouds and aggregate continuously till they become heavy enough unable to support their weight. As a result, they fall down in the form of rain. Since rain water is produced by a process of distillation, it is considered to be the purest form of water. The rain water, however, is associated with dissolved gases such as CO_2, SO_2, NH_3, etc. from the atmosphere. India, being a vast country, with an area of about 806 million acres, the rainfall constitutes one of the most important and largest source of water. The rain water in the hilly districts and that from the snows that melt in the mountain regions, flows in the form of rivers. The original river water is very pure, but it takes up suspended impurities as it flows through the plains. In this process, it also dissolves CO_2 from the atmosphere, which enables water to dissolve carbonates as it passes over the beds. Some of the rain water percolates underneath the surface of earth till it reaches an impervious strata which prevents its further penetration. This exudes in the from of springs. Water from some springs contains dissolved sulphur compounds. Such water is helpful in the cure of some skin diseases. If water from the springs contains some salts such as $MgCl_2$ and $MgSO_4$, it is also known as saline water.

Water is an essential ingredient of animal and plant life. Generally the municipal water is used for drinking purposes and other domestic purposes in cities and towns and hence water conditioning and waste water treatment have long been essential practical functions of municipalities. The importance of suitably preparing water for chemical industry is now well recognised and various processes for purification or conditioning of water obtained from natural sources have actually been proposed. The treatment of water to which it is subjected, depends on the purpose for which the treated water has to be applied. The choice generally depends on the use, whether for power generation, heating, cooling or actual incorporation in a product or its manufacturing process. The quality or quantity of surface as well as ground water are very important in the location of the chemical plant. Rivers, lakes, streams, rains, etc. are the various natural sources of water which supply abundant water containing large number of impurities in most cases. The impurities present in water vary greatly from one place to another. Treatment of water obtained from different sources is, therefore, essential so that a part or whole of the treated water may be used safely for municipal purpose, laundry purpose, boiler purpose or other industrial purpose.

RAIN WATER

The chief source of all water supply schemes at present is rainfall. As time may pass, it may become necessary to find out substitutes for rainfall as sources of water supply schemes. The scientists have already started experiments in this line and attempts are being made to find out feasibilities of converting ocean water and sewage effluent into potable water.

Rain water harvesting traces its origin back to Europe. It refers to the deliberate collection of rain water from a surface (catchments) and its storage to provide a supply of water.

The uses of rainwater harvesting are:
1. To increase the groundwater table.
2. To minimise the decreasing of water table.
3. To ameliorate the quality of groundwater.
4. To avoid the entry of sea water into the freshwater of ground level, in coastal regions.
5. To avoid flood damages.
6. To decrease soil erosion.
7. To bring more wasteland to productive lands.
8. To be used as drinking water.

Rain Water Harvesting for Drinking Purpose

Rain water harvesting is being increasingly followed for meeting the drinking water needs of the rural areas particularly during the period of drought. Recently, it has been made compulsory for all the existing buildings as well as new buildings by the Government of Tamil Nadu. This effort is found to be fruitful, as the water table increased soon after the monsoon.

The following steps are commonly followed in rain water harvesting from roofs:
1. Collection of rain water.
2. Separation of first rain flush.
3. Filtration of rain water.
4. Storage of rain water.
5. Distribution of water.

A system normally used for the concrete buildings and thatched houses and its application and the expenses are given in Figs 1.1 and 1.2.

Rain Water Harvesting for Agriculture

Rain water harvesting is more important because the country loses 50–60 per cent of rain water resulting in acute soil moisture deficit. In low rainfall areas, it is not sufficient even to meet the crop sowing moisture requirements. Under these circumstances water harvesting techniques are the principal means of water conservation to enhance agricultural production.

Surface Runoff

The rainfall on an area is expressed as so many millimeters over the entire area for a certain fixed interval of time, i.e. day, month, season or year. Thus the quantity of water obtained from rainfall during a certain interval of time can be easily worked out by the multiplication of the area and depth of rainfall.

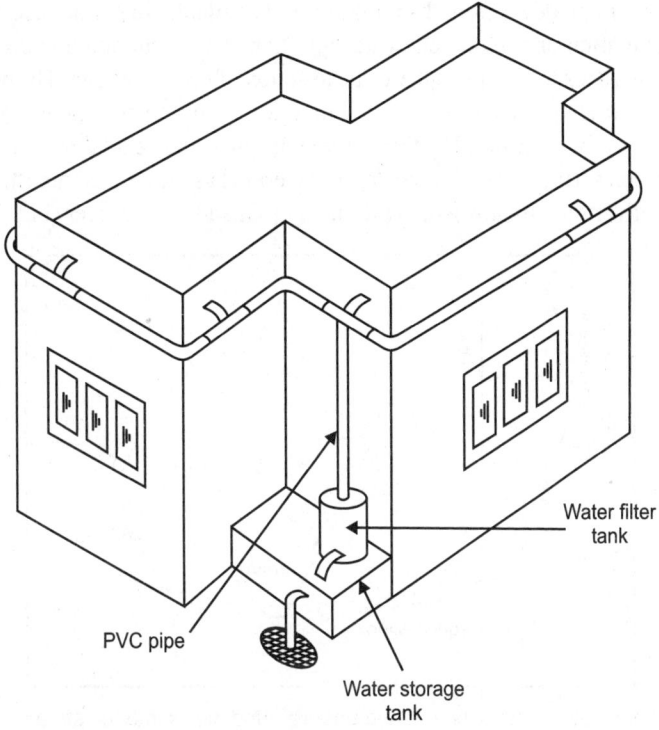

Fig. 1.1. Rain water harvest in a concrete house.

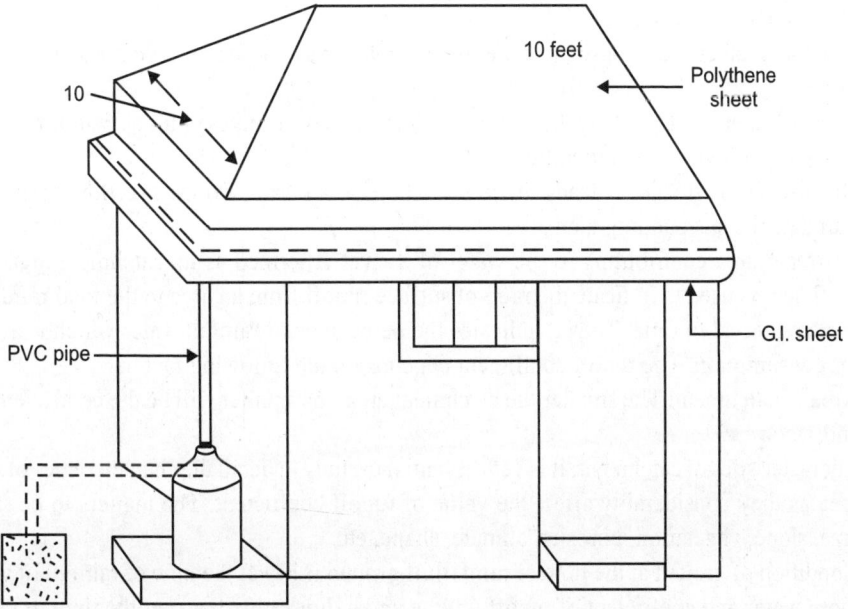

Fig. 1.2. Rain water harvest in thatched house.

But all the water coming down from the rainfall is not available for further use. Some quantity of it is lost either in evaporation or percolation or transpiration. The evaporation is the loss of water from land and water surfaces back to the atmosphere due to action of heat of the sun. The percolation indicates the loss of water penetrated into the soil and it may join some underground source of water. The transpiration is the loss of water caused by the leaves of the growing vegetation. The net quantity of rain water which remains on surface after all these losses is termed as the surface runoff. This surface runoff is seen in the form of various streams which ultimately join and form a river (Fig. 1.3.).

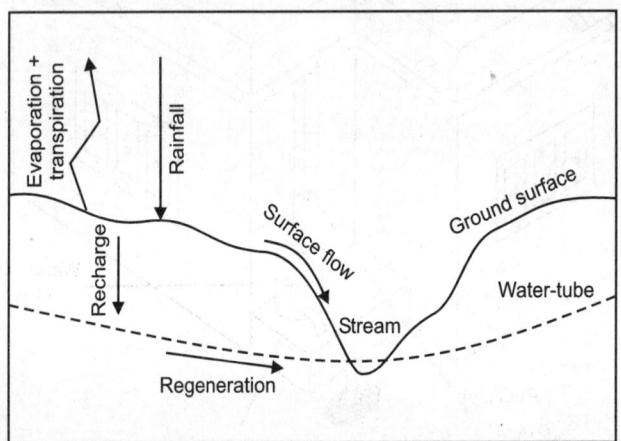

Fig. 1.3. Much of our water, both overland and underground, flows into the streams, then to the sea.

The surface runoff is harmful because of the following reasons:
1. Economic use: If surface runoff is to be used economically, it requires costly reservoirs or land improvement schemes.
2. Erosion: It takes away topsoil and the soil erosion due to surface runoff causes serious economic losses.
3. Loss of water: It takes away the water which might have been used for agriculture. It thus leads to the loss of water for agriculture.
4. Occurrence of floods: It leads to floods. The rivers during floods overflow their banks and inundate the surrounding land area.

The upstream area contributing to the water of a river is termed as its catchment area. The term runoff coefficient is used to indicate the ratio of surface runoff from an area to the total rainfall on that area in a fixed interval of time. Thus it indicates the percentage of rainfall water which is available on surface for consumption. The runoff coefficient depends on the following factors:
1. Area of catchment: The smaller the catchment area, the smaller will be the coefficient of runoff and *vice versa*.
2. Characteristics of catchment: It is very essential to study in detail the characteristics of catchment area as they considerably affect the value of runoff coefficient. The matters to be studied are size, slope, vegetation, porosity, climate, shape, etc.
3. Condition of ground at the time of rainfall: If ground is dry at the time of rainfall, it will absorb more water and coefficient of runoff will be small. For ground, wet at the time of rainfall, the reverse will be the case.

4. Intensity of rainfall: If it rains heavily in short duration of time, the soil does not get opportunity to absorb all water. It thus increases surface flow and consequently, the coefficient of runoff is also increased.

5. Interval between successive showers: The smaller the interval between successive rainfall showers, the greater will be the coefficient of runoff and *vice versa*.

6. Season of rainfall: The rainfall during hot season gives less surface flow than that during cold season.

7. Yearly rainfall: The greater the annual rainfall, the greater is the runoff coefficient and *vice versa*.

Precipitation

The term precipitation is used to indicate the water which returns to the surface of earth in various form like rain, snow, etc. The major part of precipitation occurs in the form of rain and only a small portion of it occurs in the form of snow, etc.

The usual process by which the atmospheric air is cooled to cause precipitation is the lifting of the air mass. There are three methods by which the air mass gets lifted and accordingly the precipitations are of the following three types:

1. Convective precipitation: This type of precipitation occurs in tropics where on a hot day, the ground surface gets heated in an unequal manner. The warmer air lifts up and its place is taken by the cool air. The vertical currents of air thus set up possess tremendous velocities and are found harmful to the movements of aircraft also. The precipitation takes the form of showers of high intensity of short duration.

2. Cyclonic precipitation: The cyclone is a very large mass of air having diameter of about 800 km to 1600 km and moving with a velocity of 50 km p.h. The pressure in the central portion of cyclone is low and it acts like a chimney through which the air gets lifted. The air thus lifted expands, cools and ultimately gets condensed which causes cyclonic precipitation. It takes the form of drizzle intermittent rain or steady rain. In cyclonic precipitation, if cold air replaces the warm air, it is known as the warm front and on the other hand, if warm air replaces the cold air, it is known as the cold front. The precipitation caused by the warm front is more continuous and that which is caused by the cold front is very intense and of short duration.

3. Orographic precipitation: This type of precipitation is caused by air masses which strike some natural topographic barriers or obstructions like mountains and when they cannot move up, they rise up leading to condensation and precipitation. The greatest amount of precipitation falls on the windward side and the leeward side often has very little precipitation.

The rainfall due to orographic precipitation is composed of showers and steady rainfall. This type of precipitation is responsible for most of the heavy rains in our country. The rainfall is measured by standard instruments which are known as the rain gauges.

Rainfall

Following three terms will be discussed in connection with the rainfall of a locality:

1. Average annual rainfall: The annual rainfall at a given rain gauge station is recorded for a number of years and the mean of annual rainfall from the records of 35 years or so is worked out. This is known as the average annual rainfall at the given rain gauge station. It is to be remembered that whenever the rainfall of a particular locality is mentioned, it indicates the average annual rainfall of that place.

2. Index of wetness: The ratio of the actual rainfall in a given particular year at a given place to its average annual rainfall is known as the index of wetness. Thus,

$$\text{Index of wetness} = \frac{\text{Actual rainfall in a particular year}}{\text{Average annual rainfall}}$$

The index of wetness thus gives an idea about the wetness of the year and if it is less than 100 per cent, it indicates the deficiency of rain. For instance, if the index of wetness is 60 per cent, it means that there is a rain deficiency of 40 per cent.

If the deficiency is about 30 per cent to 45 per cent, it is known as large deficiency; if it is about 45 per cent to 60 per cent, it is known as serious deficiency; and if it exceeds 60 per cent, it is referred to as disastrous deficiency.

3. Minimum annual rainfall: The term bad year or dry year or sub-normal year is used to mean the year in which the rainfall is less than the average annual rainfall. The study of rainfall records of the locality is made say for 35 years or so and the minimum of all the bad years is obtained. This is known as the minimum annual rainfall of the locality and in rare cases only, there are three successive bad years indicating minimum annual rainfall. The provision of water in storage reservoirs is therefore usually made for two or three successive bad or dry years.

TYPES OF SOURCES

The sources from which water is available for water supply schemes can conveniently be classified into the following two categories according to their proximity to the ground surface:
1. Surface sources.
2. Underground sources.

We will now discuss various forms of surface sources and underground sources. But it will be necessary to consider the following important factors while making choice of source of water supply for a particular town or city:

1. Cost: The selection of source should be such that the overall cost of the water supply project is brought down to the minimum.
2. Elevation: The source of water supply should be at a higher level so that it becomes possible to supply water by the gravity flow only. If water source is at a lower level, it will involve huge expenditure on the operational and maintenance costs of pumping.
3. Location: The source whether surface or underground should be situated as near to the town or city as possible because such a location will require less lengths of pipes and few associated appurtenances.
4. Quality of water: The source should contain water which is free from pollution or other undesirable impurities and capable of being easily and cheaply treated.
5. Quantity of water: The source should be able to supply enough quantity of water to meet the demands of town or city for various purposes like domestic, industrial, fire fighting, etc. In some cases, part of the available source may be used to meet with the present demand and additional units may be brought into use as demand increases with passage of time.

Surface Sources

In this type of source, the surface runoff is available for water supply schemes. The usual forms of surface sources are as follows:

Lakes and streams

A natural lake represents a large body of water within land with impervious bed. Hence it may be used as a source of water supply scheme for nearby localities. The quantity of runoff that goes to the lake should be accurately determined and it should be seen that it is at least equal to the expected demand of locality. Similar is the case with streams which are formed by the surface runoff.

It is found that the flow of water in streams is quite ample in rainy season. But it becomes less and less in hot season and sometimes the stream may even become absolutely dry.

The catchment area of lakes and streams is very small and hence the quantity of water available from them is also very low. Hence the lakes and streams are not considered as principal sources of water supply schemes for large cities. But they can be adopted as sources of water supply schemes for hilly areas and small towns. The water which is available from lakes and streams is generally free from undesirable impurities and can, therefore, be safely used for drinking purposes.

Ponds

A pond is a man-made body of standing water smaller than a lake. Thus the ponds are formed due to excessive digging of ground for the construction of roads, houses, etc. and they are filled up with water in rainy season. The quantity of water in pond is very small and it contains many impurities.

A pond cannot be adopted as a source of water supply and its water can only be used for washing of clothes or for animals only.

Rivers

It is observed that rivers are studied more thoroughly than other sources of water. Since the dawn of civilisation, the ancient man settled on the banks of river, drank river water, ate fish caught from river water and sailed down rivers to find out unknown lands. Even the occurrence of floods did not disappoint the man and he tried to study the regularities of floods and make use of flood waters for irrigation of his fields. As a matter of fact, many ancient civilisations such as India, Egypt, etc. were inseparably bound up with rivers.

The large rivers constitute the principal source of water supply schemes for many cities. Some rivers are perennial while others are non-perennial. The former rivers are snowfed and hence the water flows in such rivers for all the seasons. The latter type of rivers dries in summer either wholly or partly and in monsoon, the heavy flood visits them.

For such types of rivers, it is desirable to store the excess water of flood in monsoons by constructing dams across such rivers. This stored water may then be used in summer.

The principal uses of a river can be summarised as follows:

1. It can be developed as the chief source of water supply for a town or a city.
2. It can be used for navigation.
3. It can be used to supply water for irrigation purposes.
4. It can serve as an agent of purification of wastes.
5. It can serve as a center of recreational activities such as bathing, boating, fishing, fountains, etc.

In order to ascertain the quantity of water available from the river, the discharges at various periods of the year are taken and recorded. The observations over a number of years serve as a good guide for estimating the quantity of water available from the river in any particular period of the year.

Generally the quantity of water available from non-perennial rivers is variable throughout the year and it is likely to fall down in hot season when demand of water is maximum. It becomes, therefore, essential to augment such source of water supply by some other sources so as to make the water supply scheme successful.

The quality of surface water obtained from rivers is not reliable. It contains silt and suspended impurities. When completely or partly treated sewage is being discharged into the river at some upstream point, the river water is to be suspected for high contamination. The river water requires to be properly analysed as regards to the contents of disease bacteria, harmful impurities, etc. The presence of all such undesirable elements in river water requires an exhaustive treatment of water before it can be made fit for drinking purposes.

It should, however, be noted that the quality of river water is subject to the widest variations because it depends on various uncertain factors such as character of the catchment area, discharges of sewage and industrial wastes, climatic conditions, season of the year, etc. The character of the water differs not only with each individual river, but also at many points along the course of the same river. It is usually found that the quality of river water at its head is good, but it goes on deteriorating as the river proceeds along its course. The main reasons why the river pollution is undesirable are as follows:

1. Contamination of water supplies resulting in additional load on treatment plants.
2. Creation of nuisances in the form of appearance and odour.
3. Detrimental effect on fish life.
4. Hindrance to the navigation by banks of deposited solids.
5. Restriction of recreational use.

The chief points to be considered in investigating a river supply of water are as follows:

1. Adequacy of storage of purified water so as not to disturb the distribution system during periods of flood when the river water is turbid.
2. Efficiency of the subsequent stages of purification system adopted.
3. General nature of river, the rate of flow and the distance between the sources of pollution and the intake of the water.
4. Relative proportions of the polluting matter and the flow of river when at its minimum.

Storage reservoirs

An artificial lake formed by the construction of dam across a valley is termed as a storage reservoir. Whatever may be the size or use of a reservoir, the main object or function of a reservoir is to store water and thus it stabilises the flow of water. The most important physical characteristic of a reservoir is, therefore, its storage capacity. The topographic survey of the dam site is carried out and a contour map is prepared. The capacity of reservoir is then worked out with the help of the contour map. A schematic diagram of underground reservoir is shown in Fig. 1.4.

A storage reservoir essentially consists of the following three parts: (i) a dam to hold water; (ii) a spillway to allow the excess water to flow; and (iii) a gate chamber containing necessary valves for regulating the flow of water.

At present, this is rather the chief source of water supply schemes for very big cities. The multi-purpose reservoirs also make provisions for other uses in addition to water supply such as irrigation and power generation. The subject of reservoir design is a topic by itself. Its salient features in brief are discussed below.

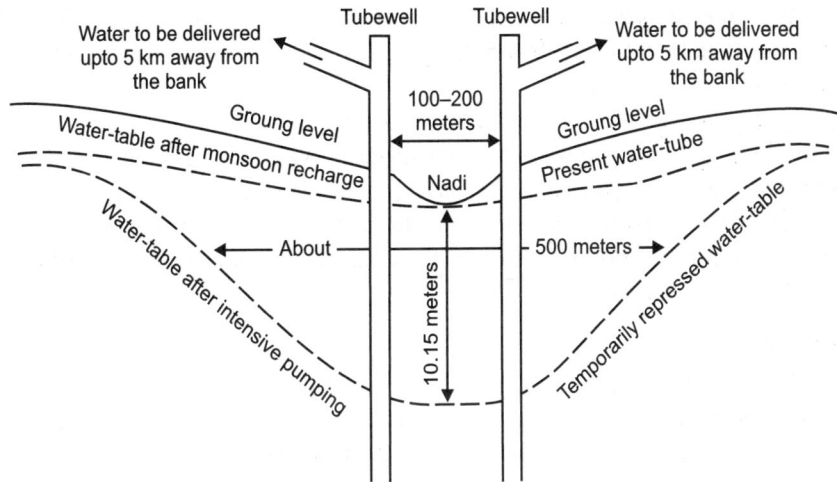

Fig. 1.4. A schematic diagram of underground reservoir.

Selection of site

Following are the factors, which are to be taken into consideration, while selecting the site for a storage reservoir:

1. Area of land to be submerged by the construction of reservoir.
2. Availability of construction materials and possibilities of using local materials for the construction of dam.
3. Availability of good foundation bed for dam.
4. Availability of skilled labour for the construction of dam.
5. Chances of biological troubles.
6. Characteristics of catchment area.
7. Density of population over the catchment area.
8. Distance between the proposed site and the point of distribution.
9. Elevation of reservoir level.
10. Facilities of transport for men and materials.
11. Geological conditions of basin of storage area.
12. Nature of land to be acquired.
13. Possibilities of earthquake occurrences due to the storage of water.
14. Quality of water available.
15. Quality of water likely to come to the reservoir site.
16. Water-tightness of the reservoir area.

Storage capacity of the reservoir

An analysis of demand and supply of water per month of the year is made. Following procedure is adopted:

1. The average monthly rainfall for every month of year is determined.
2. The average coefficient of runoff for different months of year is worked out by suitable method.
3. The multiplication of (1) and (2) indicates the total surface flow in the stream for different months of the year.

4. From the available surface runoff, the quantity representing various losses such as evaporation loss, penetration loss, etc. is subtracted. This gives the net supply of water from the stream for different months of the year.
5. Now the demand of water for every month of year is worked out.
6. The surplus or deficiency of water for each month is obtained by manipulation of above results. When supply is more, it indicates surplus and when supply is less, it indicates deficiency.
7. The total deficiency during successive months gives the storage capacity of reservoir.
8. If provision is to be made for two or three successive dry years, the capacity obtained in (7) above is increased accordingly.

Fixing height of dam

The height of dam for the creation of storage reservoir is decided as follows:
1. The site of storage reservoir is surveyed and the contours at suitable intervals are drawn.
2. The area of each contour is calculated by using a planimeter.
3. Starting from the lowest contour, the volume between successive contours is then worked out and a statement showing the cumulative volume against the elevation of contour is prepared. A graph is then drawn as shown in Fig. 1.5. It indicates the volume of water stored for different heights of dam.
4. The gross storage required for the reservoir is worked out by adding suitable allowances for various losses that would occur due to the storage of water.
5. The height of dam corresponding to the gross storage required is then read out from the reservoir capacity curve as shown in Fig. 1.5.
6. The height of dam as obtained above is the Full Supply Level (FSL) of reservoir. To get the overall height of dam, the provisions are made for the Highest Flood Level (HFL) freeboard; Dead Storage Level (DSL) and depth of foundation as shown in Fig. 1.6.

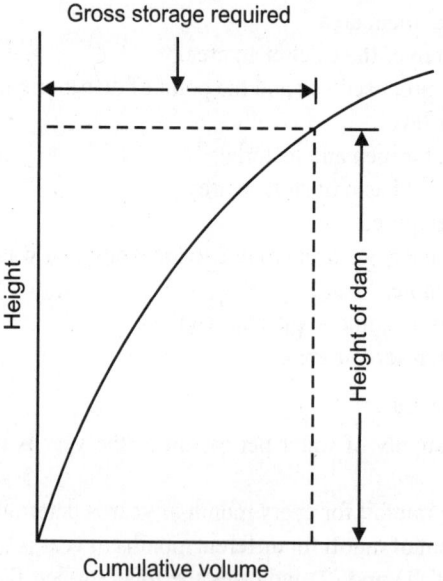

Fig. 1.5. Reservoir capacity curve.

Reservoir losses

Following are the three important losses which occur in the stored water of a reservoir:

1. Absorption losses: These losses depend on the nature of soil forming the reservoir site. They are significant in the initial stage but get reduced as the pores become saturated. These losses as such do not play any important role in the planning of a reservoir.

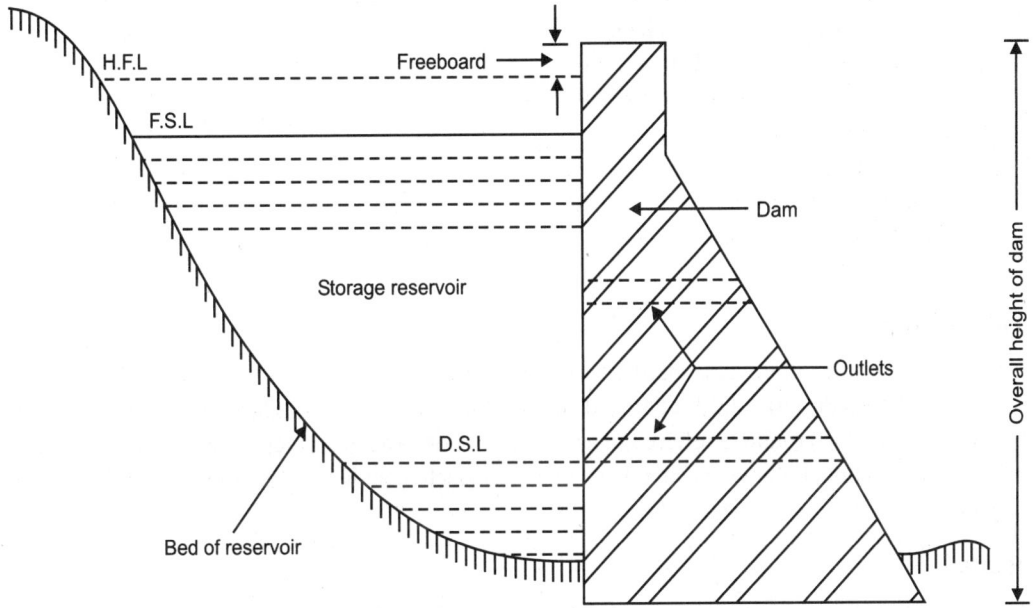

Fig. 1.6. Storage reservoir with dam.

2. Evaporation losses: These losses mainly depend upon the exposed surface area of the reservoir and influenced by various other factors like wind velocity, relative humidity, temperature, etc. These losses are sometimes enormous and to reduce their magnitude, certain chemical compounds are sometimes spread on the reservoir surface at regular intervals.

3. Percolation losses: The leakage which occurs through the banks of reservoir is sometimes more due to fractured rocks, etc. The cracks in banks should, therefore, be sealed by the pressure grouting technique to bring down the amount of percolation losses.

Reservoir clearance

The process of removal of trees, bushes and other vegetation from the reservoir area is known as the reservoir clearance and if it is not carried out properly, it leads to the following difficulties:

1. The decay of organic material will grant objectionable odour and taste to the stored water.
2. The floating bushes, trees, etc. will create debris problem at the dam site.
3. The trees projecting above the water surface will exhibit undesirable appearance for the reservoir to be used as picnic spot.

Economic height of dam

The cost of the dam which is minimum per unit of storage is known as the economic height of dam.

Choice of type of dam

There are various types of dams such as earth dams, masonry dams, R.C.C. dams, steel dams, arch dams, etc. Each type of dam requires special considerations for its design and utility.

However, the following are the general considerations which are to be taken into account while deciding the particular type of dam at a given site for water supply schemes:

1. Availability of materials of construction and type of labour.
2. Available natural foundation bed.
3. Cost of maintenance of dam section.
4. Cross-section of river or stream banks at site.
5. Importance of stored water.
6. Length and height of dam.
7. Nature of river or stream banks at site.
8. Overall cost of dam section.
9. Site for spillway.

Underground Sources

In this type of source, the water that has percolated into the ground is brought on the surface. The difference between the terms infiltration and percolation should be noted. The entrance of rain water or melted snow into the ground is referred to as the infiltration. The movement of water after entrance is called the percolation. Following general remarks in connection with the underground sources are worth to be noted:

Aquifers

It is observed that the surface of earth consists of alternate courses of pervious and impervious strata. The pervious layers are those through which water can easily pass while it is not possible for water to go through an impervious layer.

The pervious layers are known as the aquifers or water-bearing strata. If aquifer consists of sand and gravel strata, it gives good supply of drinking water. The aquifer of limestone strata can supply good amount of drinking water, provided there is presence of cracks or fissures in it.

Movement of groundwater and Its velocity

The underground water moves due to actions of gravity and molecular attraction of surface tension. The velocity of flow depends on three factors:

1. Slope of groundwater surface.
2. Hydraulic properties of soil through which it flows.
3. Temperature of water.

Permeability

The ability of a rock or unconsolidated sediment to transmit or pass water through itself is known as the permeability. The capability of the entire soil of full width and depth (i.e. area = width × depth) is represented by permeability. The coefficient of permeability is used to measure the permeability and is defined as the rate of flow of water through a unit cross-sectional area of the water bearing material under a unit hydraulic gradient and at a temperature of 20°C.

Porosity

The term porosity is used to indicate the ratio of the volume of openings or pores or voids in the material to its total volume, expressed in percentage. Thus the expression for porosity will be as follows:

$$P = \frac{V_1}{V_2} \times 100$$

where,

P = Porosity of the soil

V_1 = Openings or pores or voids in the material

V_2 = Total volume of the material.

The terms high porosity, medium porosity and small porosity are used to indicate porosity above 20 per cent, between 5 per cent to 20 per cent and below 5 per cent respectively.

The sand and gravel possess high porosity and hence they form the most important aquifers for public water supply schemes. The clay is highly porous but it is practically impervious because of its very fine particles. The limestone possesses small or medium porosity and it can permit the movement of water only if it contains cracks, fissures or faults.

Quality of water

The process of natural filtration takes place when rainfall water percolates through the ground. Hence the underground waters are generally clear and free from impurities. They are also likely to contain very small quantity of bacteria. However, it is advisable to protect them from possible sources of contamination. The groundwaters, however, absorb various salts during the process of percolation. The amount and character of salts present in ground waters will depend upon the chemical composition of the strata and the length of underground travel. The usual minerals present are calcium, iron, magnesium, manganese, potassium and sodium. It may become necessary to give treatment to the water for removing these minerals.

In general, the rocky strata will give more or less pure water. The water through clayey soils will possess high turbidity. Some aquifers like limestones permit long travel of underground passage and hence they are likely to permit polluted water to travel for long distances. The chances of such danger are, however, less in case of sandy soils since the passage of underground water is limited to about 30 meters or so in such soils.

Transmissibility

The term transmissibility is used to indicate the same physical meaning, of the permeability, but differing only mathematically. The capability of the soil of unit width and full depth (i.e. width = 1 and hence A = depth) is known as the transmissibility.

Water table

The uppermost layer of soil or topsoil at ground level is generally pervious. The rain water which is directly percolated through this topsoil is contained by it. The upper surface of free water in topsoil is termed as the ground water level or water table.

The level of water table is variable. It rises with increase in percolation in wet season and falls down in dry season. It is generally not horizontal but is found to follow the profile of ground level.

Forms of underground sources

Following are the four forms in which underground sources are found:

Infiltration galleries

An infiltration gallery is a horizontal or nearly horizontal tunnel which is constructed through water bearing strata. It is sometimes referred to as the horizontal well. The gallery is usually constructed of brick walls with slab roof as shown in Fig. 1.7. The gallery obtains its water from water bearing strata by various porous drain pipes. These pipes are covered with gravel, pebble, etc. so as to prevent the entry of very fine material into the pipe.

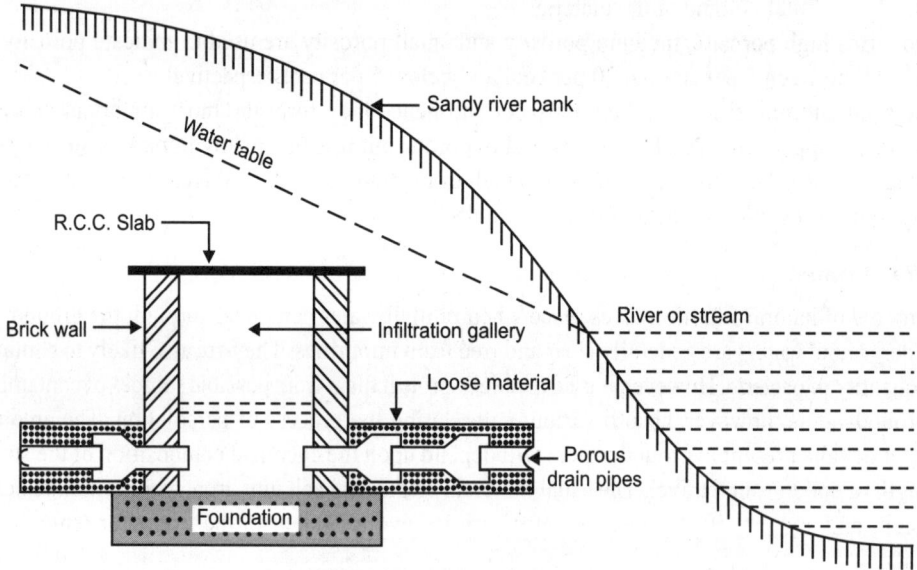

Fig. 1.7. Infiltration gallery.

The gallery is laid at a slope and the water collected in the gallery is led to a sump from where it is pumped and supplied to consumers after proper treatment. The manholes are provided along the infiltration gallery for the purposes of cleaning and inspection.

The infiltration galleries are useful as source of water supply when ground water is available in sufficient quantity just below ground level or so. The galleries are usually constructed at depth of about 5 to 10 meters from the ground level.

In another form, the gallery assumes the form of an outlet pipe having perforations all around its surface. These perforations are covered with gravel, pebble, etc. and they work as filters. Such construction is known as the infiltration pipes and it is adopted when quantity of available ground water is small.

Infiltration wells

In order to obtain large quantities of water, the infiltration wells are sunk in series in the banks of river. The wells are closed at top and open at bottom. They are constructed of brick masonry with open joints as shown in Fig. 1.8.

For the purpose of inspection of well, the manholes are provided in the top cover. The water infiltrates through the bottom of such wells and as it has to pass through sand bed, it gets purified to some extent.

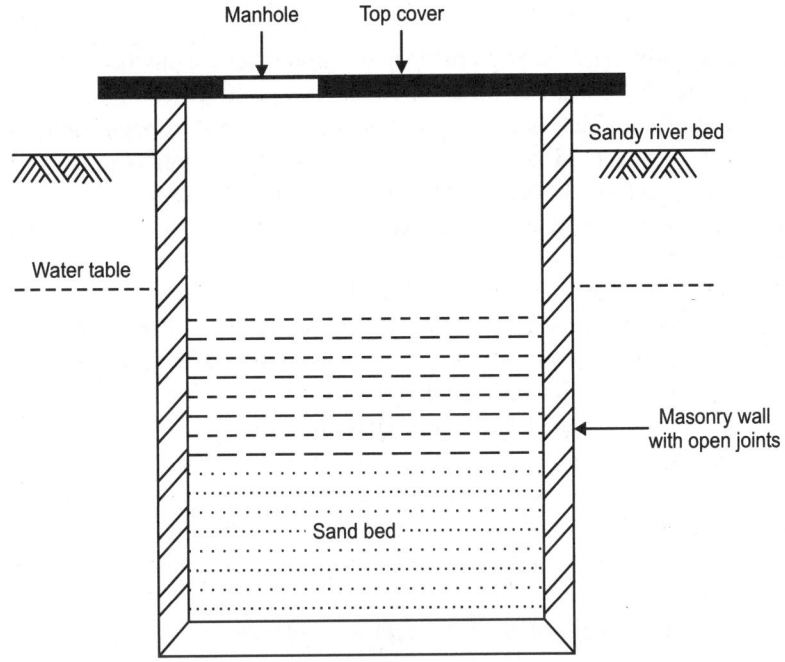

Fig. 1.8. Infiltration well.

The infiltration wells in turn are connected by porous pipes to a collecting sump, known as Jack well and the water thus collected through the infiltration wells then flows by gravity into the jack well. The water from jack well is pumped to the purification plant for treatment.

Springs

When groundwater appears at the surface for any reason, the springs are formed. They serve as source of water supply for small towns, especially near hills or bases of hills. Some springs discharge hot water due to the presence of sulphur and other minerals in their formations. These hot springs cannot be used to supply water for domestic purposes. But the hot water is found to cure some of the human disorders.

A good spring proves to be a sure source of water. But it is difficult to find a good spring for the purpose of water supply scheme. However, when a spring is to be developed as a source of water supply, the following factors should be carefully ascertained:

1. It should be easier, cheaper and surer enough to develop the spring for the locality than to adopt any other source of water supply.
2. The flow of water should be adequate even in dry weather.
3. The spring should be adequately protected from the water pollution sources.
4. The spring should be so located as to have natural gravity flow.
5. The water should be of good quality.

It is found that the quality of spring water depends on geological and topographical conditions and it may be hard or soft, pure or polluted or sometimes saline, sulphurous, etc. Similarly the yield from springs is mostly inadequate, except for small supplies. The spring water which is not disturbed by rainfall is usually attractive in appearance and of good palatability. However, the content of free carbonic acid is sometimes high and the spring water may possess corrosive and plumbo-solvent properties.

Wells

A well is defined as an artificial hole or pit made in the ground for the purpose of tapping water. The holes made for tapping oil are also known as the wells. But in the general sense, a well indicates a source of water. In India, the chief source of water supply for most of its population is wells and it is estimated that 50 to 60 per cent of Indian population has to depend on wells for its water supply.

The three factors which form the basis of theory of wells are as follows:

1. Geological conditions of the earth's surface.
2. Porosity of various layers.
3. Quantity of water which is absorbed and stored in different layers.

The geological conditions of the earth's surface indicate the slope of water bearing strata. If the slope of water bearing layers is towards the well, there will be some quantity of water in the well even during the severe hot season. On the other hand, if the slope of water bearing layers is away from the well, such well will soon get dry and it will only give some quantity of water only in monsoon.

The porosity of aquifers will also play a great role in determining the quantity of water in the well. If the porosity of aquifers is more, the well will easily collect more quantity of water in less time. The capacity of aquifers to absorb and store water will determine the supply rate of water to the well. If the aquifers are capable of storing more water, the well will get more quantity of water and practically at a constant rate.

Following is the general classification of different types of wells:

Shallow wells

The shallow wells are constructed in the uppermost layer of the earth's surface. They obtain their quota of water supply from the ground water table as shown in Fig. 1.9. The diameter of shallow wells varies from 2 to 6 meters. They may be lined or unlined from inside. The lining is also called the steining and its thickness varies from 30 cm to 50 cm.

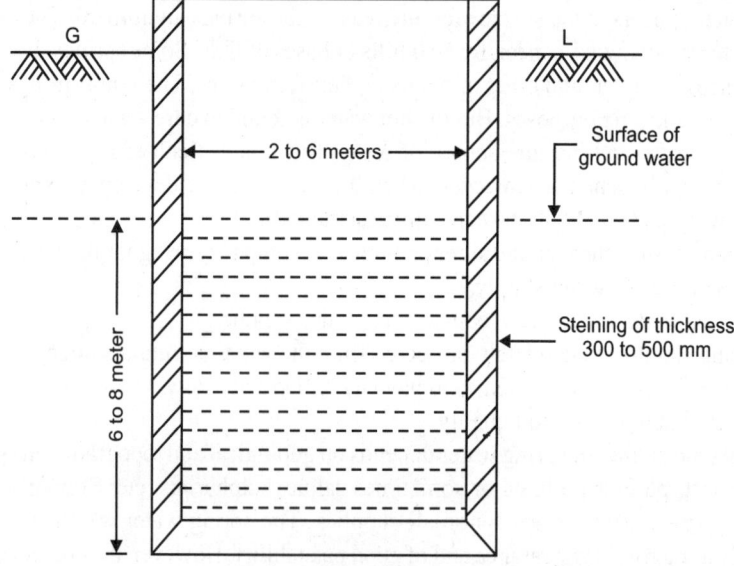

Fig. 1.9. Shallow well.

Figure 1.9 shows a shallow well with steining. The unlined wells are generally constructed upto a maximum depth of about 7 meters or so. But for greater depths, the soil cannot stand vertically and hence the steining becomes essential for such wells. These wells are also sometimes referred to as the draw wells or gravity wells or open wells or dug wells or percolation wells.

The quantity of water available from shallow wells is generally limited as their source of supply is the uppermost layer of earth only. They sometimes even dry up in summer. In order to ensure the supply of water from shallow wells, even in dry years, they are taken much below the surface of groundwater table. The depth below water table is kept as about 6 to 8 meters so that even if the water table falls by 3 to 5 meters, some water will be available from the shallow wells. In any case, the discharge of shallow wells does not exceed 5 liters per second and hence, they are not suitable for public water supply schemes.

The quality of water obtained from shallow wells is better than river water. But it is not reliable and requires purification. The main source of contamination is the effluent from nearby septic tanks, soak wells, etc. It is, therefore, desirable to construct shallow wells away from such possible sources of contamination. It may also be noted that the shallow well water is notoriously liable to intermittent pollution and hence the samples of water should always be collected after heavy rainfall, when marked deterioration in purity may be revealed.

Looking to the uncertain supply of water and bad quality of water, the shallow wells are used as source of water supply for small villages, undeveloped municipal towns, isolated buildings, camps, etc.

Deep wells

The deep wells obtain their quota of water from an aquifer below an impervious layer as shown in Fig. 1.10. The theory of deep well is based on the travel of water from the outcrop to the site of deep well. The outcrop is the place where aquifer is exposed to the atmosphere as shown in Fig. 1.10. The entry of rain water takes place at outcrop and it reaches the site of deep well. During travel, the water gets thoroughly purified. But it dissolves certain salts and may, therefore, become hard. In such cases, some treatment would be necessary to remove the hardness of water.

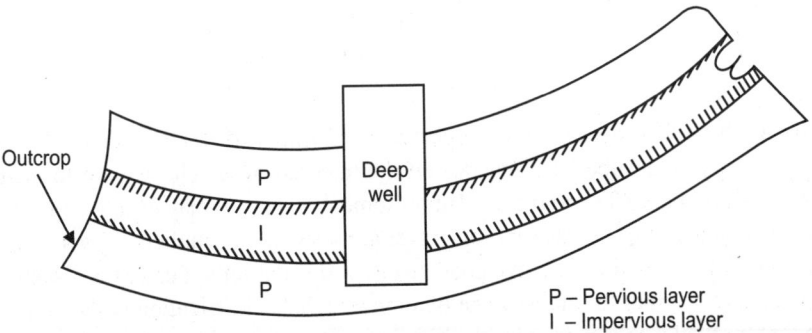

Fig. 1.10. Deep well.

The depth of deep well should be decided in such a way that the location of outcrop is not very near to the site of well. The water of deep wells is contained in lower embedded aquifers and hence it is always available at a pressure greater than the atmospheric pressure. The deep wells are, therefore, referred to as the pressure wells.

Tube wells

A tube well is a deep well having a diameter of about 50 mm to 200 mm and it obtains its quota of water from a number of aquifers as shown in Fig. 1.11. The blind pipes are placed against the impervious layers.

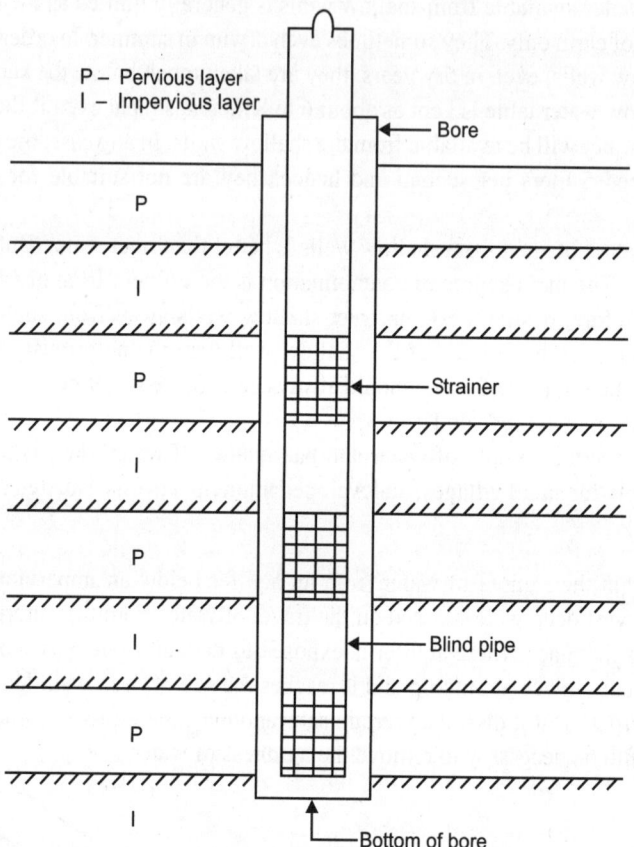

P – Pervious layer
I – Impervious layer

Fig. 1.11. Tube well.

The depth of tube well is decided with respect to the quantity of water required. The discharge of various aquifers composing tube well depends on the material of which they are formed. The usual depth of a tube well is about 30 to 50 meters. But in some dry areas, it may even go upto 300 meters or so. The aquifers composed of coarse sand or gravel are very good suppliers of water while aquifers composed of limestone or marble will give good quantity of water only if cracks or fissures are present in them. A schematic diagram for intensive cultivation by tube well irrigation is shown in Fig. 1.12.

The pipe for tube well is then inserted in the bore hole. It consists of strainers and blind sections. The pumping is then started. It should be done gradually to avoid sticking of fine sand particles on the external surface of the strainers. The strainers can be cleaned through perforations by allowing water from a high tank under pressure or by blowing air under pressure or by reversing the direction of flow. The process of removing the finer particles surrounding the strainers is known as the well development and it grants the following advantages:

1. It increases the specific yield of well.

2. It prevents the entry of fine sand particles in well pipe along with water.
3. The economic life of well is increased.

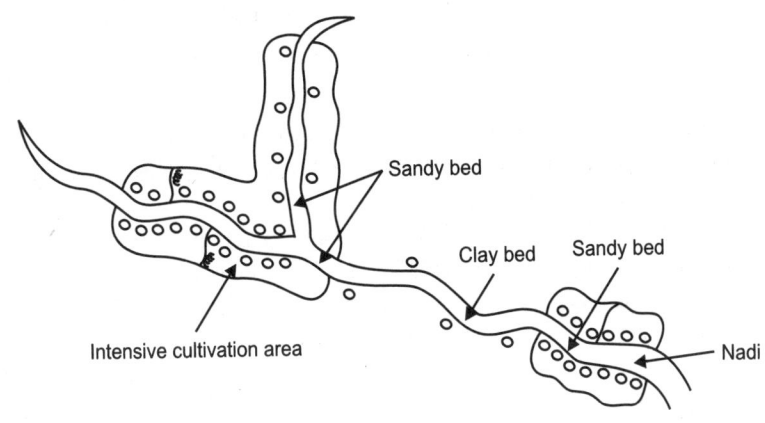

Fig. 1.12. A schematic diagram for intensive cultivation by tube well irrigation.

Quantity of tube well water: The quantity of water available from a tube well is generally sufficient and more or less reliable. The discharge from a tube well does not exceed 40 to 50 liters per second.

Quality of tube well water: The quality of tube well water is generally very good and in many cases, it can be used without any treatment. However, it is found to possess hardness and it may become necessary to remove it.

Use of tube well: The tube wells form the sources of water supply schemes for residential colonies, small towns, isolated portions of cities, big gardens, etc.

Maintenance of tube well

The maintenance work involves the method of cleaning or replacing the strainers or tube well pipe to make it workable for longer period. The following measures are generally taken:

1. If the yield of the well is reduced due to the clogging of strainer, then the strainer may be cleaned by surging.
2. Sometimes compressed air is forced through the well pipe to remove the clogging.
3. Dry ice may be dropped into the well pipe and the well top is tightly capped. The pressure of carbon dioxide vapour forces the soil particles away from the strainer.
4. The internal corrosion of the well pipe or strainer may be removed by sulphuric acid.
5. If muddy water is discharged through the well, then it is an indication of perforation in the well pipe or damaged strainers. In that case, the resinking of the well should be done by removing the affected pipe and changing the wire net of the strainer.

Failure of tube well

A tube well may fail due to the following reasons:

Corrosion

The groundwater contains acids, chlorides, sulphates, etc. which cause the corrosion of pipes. Again, the corrosion damages the strainers.

The following steps may be taken to reduce the corrosion:
1. Always thick pipes should be used.
2. Galvanised pipes or other anti-corrosive coated pipes should be used.
3. Periodical washing of the tube well should be done by sulphuric acid.

Incrustation

The groundwater also contains calcium-bicarbonate, magnesium salts, etc. These compounds get deposited inside the tube well and the diameter of the well is reduced. This is known as incrustation. The incrustation can be arrested by the following steps:
1. The water of the tube well is tested in the laboratory to determine the presence of alkali salts. The salts which are responsible for incrustation may be removed by titration. The titration may be done by forcing the proper doses of acids to the well. The water of the well is then pumped out.
2. The tube well should not be left unused for long period.

Yield of a well

The fundamental principles involved in the yield of a well are as follows:
1. It decreases if there is interference from the neighbouring well.
2. It decreases with the increase in radius of influence.
3. It increases very rapidly with the coarseness of particles of the water-bearing layer of subsoil.
4. It is approximately directly proportional to the depth of penetration of well in a water-bearing layer of subsoil.
5. It is approximately proportional to the depression head or drawdown.
6. It is not increased appreciably by the increase of the diameter of well.

TESTS FOR YIELD OF A WELL

Following are the two tests by which the yield of a well can be determined:
1. Pumping test or constant level test.
2. Recuperation test.

PUMPING TEST OR CONSTANT LEVEL TEST

In this test, the water level of the existing open well is depressed to such an extent which represents the safe working head for subsoil at the site of well. The rate of pumping is so adjusted that the water level in well remains constant which means that whatever water percolates into the well is taken out by the regulated pumping. Thus the rate of pumping indicates the yield of a well. The test may also be carried out for working out the yield of a proposed well. For this purpose, a bore is driven into the ground and by regulated pumping, the yield of well per hour per base area of the bore is calculated and by knowing the diameter of proposed open well, its probable yield can then be calculated.

It is, however, very difficult to maintain the constant level in the well and hence this test is generally not adopted to ascertain the quantity of water available from a well.

RECUPERATION TEST

In this test, the water is first pumped to a certain depressed head which should be less than the safe working depressed head. The pumping is then stopped and time required by water to come to the original level is recorded. From this data, it is possible to work out the recuperative power of well at different drawdowns.

SANITARY PROTECTION OF WELLS

A well should be properly protected from the possible contamination from undesirable water entering into it.

Following are the precautions to be taken for the sanitary protection of wells:

1. Connection of pump: The connection between casing and pumping unit should be watertight.
2. Covered top: The top of well should be properly covered to prevent the entry of ground waters from top. The ground must slope away from the well.
3. Depth of casing: The casing should extend by about 3 meters below the ground water table.
4. Distance from source of contamination: The minimum horizontal distance of the possible source of contamination from the well should be 15 meters and preferably, it should be 90 meters.
5. Drilled wells: In case of drilled wells, the hollow space between the well hole and casing should be filled up by cement grout to a depth of at least 3 meters.
6. Presence of trees: No tree should be allowed to be grown over or near the well. There are chances of well water to be contaminated by the fallen leaves, etc.
7. Priming of pumps: The priming of pumps should not be carried out by unsafe water.
8. Pump house: The pump house should be adequately drained and it should be protected against flooding.
9. Rate of pumping: The pumping rate from the well should be normal and not excessive.
10. Vents: The wells should be provided with enough vents so as to prevent the suction of contaminated water into the wells.
11. Washing of clothes: It is desirable to disallow or prevent the washing of clothes, utensils, etc. at or near the wells.
12. Well pits: The pumping machinery should not be installed below ground in pits.

Table 1.1 shows the comparative study of the surface and under ground sources of water supply with respect to certain features.

Table 1.1. Comparison of surface and underground sources of water.

Item	Surface sources	Underground sources
Forms in which available	Lakes, streams, ponds, rivers and storage reservoirs	Infiltration galleries, infiltration wells, springs and wells
Quality of water	They are sometimes highly polluted and unsafe to consume. They contain inorganic impurities, organic impurities and industrial wastes	They are generally free from impurities because of natural filtration but may contain large amounts of dissolved salts, minerals and gases
Quantity of water	Huge quantity of water is available during monsoon, but is considerably reduced during summer	The quantity of water available is generally limited
Treatment	They are to be suitably tested and a line of treatment is to be decided before they are adopted for public use	They can be supplied to the public with no or minor treatment only
Use	They are useful for big towns and cities. They can be adopted for irrigation facilities also	They are useful for small towns and villages only

Chapter 2

Physical and Chemical Parameters of Water Quality

INTRODUCTION

The availability of a water supply adequate in terms of both quantity and quality is essential to human existence. Early people recognised the importance of water from a quantity viewpoint. Civilisation developed around water bodies that could support agriculture and transportation as well as provide drinking water. Recognition of the importance of water quality developed more slowly. Early humans could judge water quality only through the physical senses of sight, taste and smell. Not until the biological, chemical and medical sciences developed, were methods available to measure water quality and to determine its effects on human health and well-being.

Physical parameters define those characteristics of water that respond to the senses of sight, touch, taste or smell. Suspended solids, turbidity, colour, taste and odour and temperature fall into this category.

The development of the science of water chemistry roughly paralleled that of water microbiology. Many of the chemicals used in industrial processes and agricultrure have been identified in water. However, the effort to identify other chemical compounds which may already be found in trace quantities in many water supplies and to determine their effect on human health was only recently begun. It is likely that new analytical techniques will be developed that will identify compounds not yet known to exist in water and it is conceivable that these meterials will also be linked to human health. Thus, the science of water quality will remain a challenge for engineers and scientists for years to come.

SUSPENDED SOLIDS

Solids can be dispersed in water in both suspended and dissolved forms. Although some dissolved solids may be perceived by the physical senses, they fall more appropriately under the category of chemical parameters and will be discussed more fully in a later section.

Sources

Solids suspended in water may consist of inorganic or organic particles or of immiscible liquids. Inorganic solids such as clay, silt and other soil constituents are common in surface water. Organic material such as plant fibers and biological solids (algal cells, bacteria, etc.) are also common constituents of surface waters. These materials are often natural contaminants resulting from the erosive action of water flowing over surfaces. Because of the filtering capacity of the soil, suspended material is seldom a constituent of groundwater.

Other suspended material may result from human use of the water. Domestic waste-water usually contains large quantities of suspended solids that are mostly organic in nature. Industrial use of water may result in a wide variety of suspended impurities of either organic or inorganic nature. Immiscible liquids such as oils and greases are often constituents of waste-water.

Impacts

Suspended material may be objectionable in water for several reasons. It is aesthetically displeasing and provides adsorption sites for chemical and biological agents. Suspended organic solids may be degraded biologically, resulting in objectionable by-products. Biologically active (live) suspended solids may include disease-causing organisms as well as organisms such as toxin-producing strains of algae.

Measurement

There are several tests available for measuring solids. Most are gravimetric tests involving the mass of residues. The total solids test quantifies all the solids in the water, suspended and dissolved, organic and inorganic. This parameter is measured by evaporating a sample to dryness and weighing the residue. The total quantity of residue is expressed as milligrams per liter (mg/l) on a dry-mass-of-solids basis. A drying temperature slightly above boiling ($104°C$) is sufficient to drive off the liquid and the water adsorbed to the surface of the particles, while a temperature of about $180°C$ is necessary to evaporate the occluded water.

Most suspended solids can be removed from water by filtration. Thus, the suspended fraction of the solids in a water sample can be approximated by filtering the water, drying the residue and filter to a constant weight at $104°C$ ($\pm 1°C$) and determining the mass of the residue retained on the filter. The results of this suspended solids test are also expressed as dry mass per volume (milligrams per liter). The amount of dissolved solids passing through the filters, also expressed as milligrams per liter, is the difference between the total-solids and suspended-solids content of a water sample.

It should be emphasised that filtration of a water sample does not exactly divide the solids into suspended and dissolved fractions according to the definitions presented earlier. Some colloids may pass through the filter and be measured along with the dissolved fraction while some of the dissolved solids adsorb to the filter material. The extent to which this occurs depends on the size and nature of the solids and on the pore size and surface characteristics of the filter material. For this reason, the terms filterable residues and nonfilterable residues are often used. Filterable residues pass through the filter along with the water and relate more closely to dissolved solids, while non-filterable residues are retained on the filter and relate more closely to suspended solids. 'Filterable residues' and 'non-filterable residues' are terms more frequently used in laboratory analysis while the 'dissolved solids' and 'suspended solids' are terms more frequently used in water-quality-management practice. For most practical applications, the distinction between the two is not necessary.

Once samples have been dried and measured, the organic content of both total and suspended solids can be determined by firing the residues at $600°C$ for 1 hour. The organic fraction of the residues will be converted to carbon dioxide, water vapour and other gases and will escape. The remaining material will represent the inorganic or fixed residue. When organic suspended solids are being measured, a filter made of glass fiber or some other material that will not decompose at the elevated temperature must be used.

Use

Suspended solids, where such material is likely to be organic and/or biological in nature, are an important parameter of waste-water. The suspended-solids parameter is used to measure the quality of the waste-water influent, to monitor several treatment processes and to measure the quality of the effluent. EPA has set a maximum suspended-solids standard of 30 mg/l for most treated waste-water discharges.

TURBIDITY

A direct measurement of suspended solids is not usually performed on samples from natural bodies of water or on potable (drinkable) water supplies. The nature of the solids in these waters and the secondary effects they produce are more important than the actual quantity. For such waters a test for turbidity is commonly used.

Turbidity is a measure of the extent to which light is either absorbed or scattered by suspended material in water. Because absorption and scattering are influenced by both size and surface characteristics of the suspended material, turbidity is not a direct quantitative measurement of suspended solids. For example, one small pebble in a glass of water would produce virtually no turbidity. If this pebble were crushed into thousands of particles of colloidal size, a measurable turbidity would result, even though the mass of solids had not changed.

Sources

Most turbidity in surface waters results from the erosion of colloidal material such as clay, silt, rock fragments and metal oxides from the soil. Vegetable fibers and micro-organisms may also contribute to turbidity. Household and industrial waste-waters may contain a wide variety of turbidity-producing material. Soaps, detergents and emulsifying agents produce stable colloids that result in turbidity. Although turbidity measurements are not commonly run on waste-water, discharges of waste-waters may increase the turbidity of natural bodies of water.

Impacts

When turbid water in a small, transparent container, such as a drinking glass, is held up to the light, an aesthetically displeasing opaqueness or 'milky' colouration is apparent. The colloidal material associated with turbidity provides adsorption sites for chemicals that may be harmful or cause undesirable tastes and odours and for biological organisms that may be harmful. Disinfection of turbid waters is difficult because of the adsorptive characteristics of some colloids and because the solids may partially shield organisms from the disinfectant.

In natural water bodies, turbidity may impart a brown or other colour to water, depending on the light-absorbing properties of the solids and may interfere with light penetration and photosynthetic reactions in streams and lakes. Accumulation of turbidity-causing particles in porous streambeds results in sediment deposits that can adversely affect the flora and fauna of the stream.

Measurement

Turbidity is measured photometrically by determining the percentage of light of a given intensity that is either absorbed or scattered. The original measuring apparatus, called a Jackson turbidimeter, was based on light absorption and employed a long tube and standardised candle. The candle was placed beneath the glass tube that was then housed in a black metal sheath so that the light from the candle could only be seen from above the apparatus. The water sample was then poured slowly into the tube until the lighted candle was no longer visible, i.e. complete absorption had occurred. The glass tube was calibrated with readings for turbidity produced by suspensions of silica dioxide (SiO_2), with one Jackson turbidity unit (JTU) being equal to the turbidity produced by 1 mg SiO_2 in 1 liter of distilled water.

In recent years this awkward apparatus has been replaced by a turbidity meter in which a standardised electric bulb produces a light that is then directed through a small sample vial. In the absorption mode, a photometer measures the light intensity on the side of the vial opposite from the light source, while in the scattering mode, a photometer measures the light intensity at a 90° angle from the light source. Although most turbidity meters in use today work on the scattering principle, turbidity caused by dark substances that absorb rather than reflect light should be measured by the absorption technique. Formazin, a chemical compound, provides more reproducible standards than SiO_2 and has replaced it as a reference. Turbidity meter readings are now expressed as formazin turbidity units or FTUs. The term nephelometry turbidity units (NTU) is often used to indicate that the test was run according to the scattering principle.

Use

Turbidity measurements are normally made on 'clean' waters as opposed to waste-waters. Natural waters may have turbidities ranging from a few FTUs to several hundred. EPA drinking-water standards specify a maximum of 1 FTU, while the American Water Works Association has set 0.1 FTU as its goal for drinking water.

COLOUR

Pure water is colourless, but water in nature is often coloured by foreign substances. Water whose colour is partly due to suspended matter is said to have apparent colour. Colour contributed by dissolved solids that remain after removal of suspended matter is known as true colour.

Sources

After contact with organic debris such as leaves, conifer needles, weeds or wood, water picks up tannins, humic acid and humates and takes on yellowish-brown hues. Iron oxides cause reddish water and manganese oxides cause brown or blackish water. Industrial wastes from textile and dyeing operations, pulp and paper production, food processing, chemical production and mining, refining and slaughterhouse operations may add substantial colouration to water in receiving streams.

Impacts

Coloured water is not aesthetically acceptable to the general public. In fact, given a choice, consumers tend to choose clear, non-coloured water of otherwise poorer quality over treated potable water supplies with an objectionable colour. Highly coloured water is unsuitable for laundering, dyeing, papermaking, beverage manufacturing, dairy production and other food processing and textile and plastic production. Thus, the colour of water affects its marketability for both domestic and industrial use.

While true colour is not usually considered unsanitary or unsafe, the organic compounds causing true colour may exert a chlorine demand and thereby seriously reduce the effectiveness of chlorine as a disinfectant. Perhaps more important are the products formed by the combination of chlorine with some colour-producing organics. Phenolic compounds, common constituents of vegetative decay products, produce very objectionable taste and odour compounds with chlorine. Additionally, some compounds of naturally occurring organic acids and chlorine are either known to be or are suspected of being carcinogens (cancer-causing agents).

Measurement

Although several methods of colour measurement are available, methods involving comparison with standardised coloured materials are most often used. Colour-comparison tubes containing a series of standards may be used for direct comparison of water samples that have been filtered to remove apparent colour. Results are expressed in true colour units (TCUs) where one unit is equivalent to the colour produced by 1 mg/l of platinum in the form of chlorplatinate ions. For colours other than yellowish-brown hues, especially for coloured waters originating from industrial waste effluents, special spectrophotometric techniques are usually employed.

In fieldwork, instruments employing coloured glass disks that are calibrated to the colour standards are often used. Because biological and physical changes occurring during storage may affect colour, samples should be tested within 72 hours of collection.

Use

Colour is not a parameter usually included in waste-water analysis. In potable water analysis, the common practice is to measure only the true colour produced by organic acid resulting from decaying vegetation in the water. The resulting value can be taken as an indirect measurement of humic substances in the water.

TASTE AND ODOUR

The terms taste and odour are themselves definitive of this parameter. Because the sensations of taste and smell are closely related and often confused, a wide variety of tastes and odours may be attributed to water by consumers. Substances that produce an odour in water will almost invariably impart a taste as well. The converse is not true, as there are many mineral substances that produce taste but no odour.

Sources

Many substances with which water comes into contact in nature or during human use may impart perceptible taste and odour. These include minerals, metals and salts from the soil, end products from biological reactions and constituents of waste-water. Inorganic substances are more likely to produce tastes unaccompanied by odour. Alkaline material imparts a bitter taste to water, while metallic salts may give a salty or bitter taste.

Organic material, on the other hand, is likely to produce both taste and odour. A multitude of organic chemicals may cause taste and odour problems in water, with petroleum-based products being prime offenders. Biological decomposition of organics may also result in taste and odour-producing liquids and gases in water. Principal among these are the reduced products of sulphur that impart a 'rotten egg' taste and odour. Also, certain species of algae secrete an oily substance that may result in both taste and odour. The combination of two or more substances, neither of which would produce taste or odour by itself, may sometimes result in taste and odour problems. This synergistic effect was noted earlier in the case of organics and chlorine.

Impacts

Consumers find taste and odour aesthetically displeasing for obvious reasons. Because water is thought of as tasteless and odourless, the consumer associates taste and odour with contamination and may prefer to use a tasteless, odourless water that might actually pose more of a health threat. And odours produced by organic substances may pose more than a problem of simple aesthetics, since some of those substances may be carcinogenic.

Measurement

Direct measurement of materials that produce tastes and odours can be made if the causative agents are known. Several types of analysis are available for measuring taste-producing inorganics. Measurement of taste and odour-causing organics can be made using gas or liquid chromatography. Because chromatographic analysis is time-consuming and requires expensive equipment, it is not routinely performed on water samples, but should be done if problem organics are suspected. However, because of the synergism noted earlier, quantifying the sources does not necessarily quantify the nature or intensity of taste and odour.

Quantitative tests that employ the human senses of taste and smell can be used for this purpose. An example is the test for the threshold odour number (TON). Varying amounts of odorous water are poured into containers and diluted with enough odour-free distilled water to make a 200-ml mixture. An assembled panel of five to ten 'noses' is used to determine the mixture in which the odour is just barely detectable to the sense of smell. The TON of that sample is then calculated, using the formula

$$TON = \frac{A+B}{A} \qquad \qquad ...(2.1)$$

where, A is the volume of odorous water (ml) and B is the volume of odour-free water required to produce a 200 ml mixture. Threshold odour numbers corresponding to various sample volumes are shown in Table 2.1. A similar test can be used to quantify taste or the panel can simply rate the water qualitatively on an 'acceptability' scale.

Table 2.1. Threshold odour numbers (TON) corresponding to sample volume diluted to 200 ml.

Sample volume (A), ml	TON
200	1.0
175	1.1
150	1.3
125	1.6
100	2.0
75	2.7
67	3.0
50	4.0
40	5.0
25	8.0
10	20.0
2	100
1	200

Use

Although odours can be a problem with waste-water, the taste and odour parameter is only associated with potable water. EPA does not have a maximum standard for TON. A maximum TON of 3 has been recommended by the Public Health Service and serves as a guideline rather than a legal standard.

TEMPERATURE

Temperature is not used to evaluate directly either potable water or waste-water. It is, however, one of the most important parameters in natural surface-water systems. The temperature of surface waters governs to a large extent the biological species present and their rates of activity. Temperature has an effect on most chemical reactions that occur in natural water systems. Temperature also has a pronounced effect on the solubilities of gases in water.

Sources

The temperature of natural water systems responds to many factors, the ambient temperature (temperature of the surrounding atmosphere) being the most universal. Generally, shallow bodies of water are more affected by ambient temperatures than are deeper bodies. The use of water for dissipation of waste heat in industry and the subsequent discharge of the heated water may result in dramatic, though perhaps localised, temperature changes in receiving streams. Removal of forest canopies and irrigation return flows can also result in increased stream temperature.

Impacts

Cooler waters usually have a wider diversity of biological species. At lower temperatures, the rate of biological activity, i.e. utilisation of food supplies, growth, reproduction, etc. is slower. If the temperature is increased, biological activity increases. An increase of 10°C is usually sufficient to double the biological activity, if essential nutrients are present. At elevated temperatures and increased metabolic rates, organisms that are more efficient at food utilisation and reproduction flourish, while other species decline and are perhaps eliminated altogether. Accelerated growth of algae often occurs in warm water and can become a problem when cells cluster into algae mats. Natural secretion of oils by algae in the mats and the decay products of dead algae cells can result in taste and odour problems. Higher-order species, such as fish, are affected dramatically by temperature and by dissolved oxygen levels, which are a function of temperature. Game fish generally require cooler temperatures and higher dissolved-oxygen levels.

Temperature changes affect the reaction rates and solubility levels of chemicals, a subject more fully explored in later sections of this chapter. Most chemical reactions involving dissolution of solids are accelerated by increased temperatures. The solubility of gases, on the other hand, decreases at elevated temperatures. Because biological oxidation of organics in streams and impoundments is dependent on an adequate supply of dissolved oxygen, decrease in oxygen solubility is undesirable.

Temperature also affects other physical properties of water. The viscosity of water increases with decreasing temperature. The maximum density of water occurs at 4°C and density decreases on either side of that temperature, a unique phenomenon among liquids. Both temperature and density have a subtle effect on planktonic micro-organisms in natural water systems.

CHEMICAL WATER QUALITY PARAMETERS

Water has been called the universal solvent and chemical parameters are related to the solvent capabilities of water. Total dissolved solids, alkalinity, hardness, fluorides, metals, organics and nutrients are chemical parameters of concern in water quality management. The following review of some basic chemistry related to solutions should be helpful in understanding subsequent discussions of chemical parameters.

Chemistry of Solutions

An atom is the smallest unit of each of the elements. Atoms are building blocks from which molecules of elements and compounds are constructed. For instance, two hydrogen atoms combine to form a molecule of hydrogen gas:

$$H + H \rightarrow H_2$$

Adding one atom of oxygen to the hydrogen molecule results in one molecule of the compound water:

$$H_2 + O \rightarrow H_2O$$

A relative mass has been assigned to a single atom of each element based on a mass of 12 for carbon. The sum of the atomic mass of all the atoms in a molecule is the molecular mass of that molecule. The atomic mass of hydrogen is 1 and the atomic mass of oxygen is 16. Thus, the molecular mass of the hydrogen molecule is 2 and the molecular mass of water is 18. A mole of an element or compound is its molecular mass expressed in common mass units, usually grams. A mole of hydrogen is 2 gram, while a mole of water is 18 gram. One mole of a substance dissolved in sufficient water to make one liter of solution is called a one molar solution.

Bonding of elements into compounds is sometimes accomplished by electrical forces resulting from transferred electrons. When these compounds dissociate in water, they produce species with opposite charges. An example is sodium chloride:

$$NaCl \rightleftharpoons Na^+ + Cl^-$$

The charged species are called ions. Positively charged ions are called cations and negatively charged ions are called anions. The number of positive charges must equal the number of negative charges to preserve electrical neutrality in a chemical compound. The number of charges on an ion is referred to as the valence of that ion. Thus, the valence of sodium (Na^+) is 1, while the valence of calcium (Ca^{2+}) is 2. Some compounds, called radicals, also possess charges. An example of a cationic radical is ammonium (NH_4^+), while carbonate (CO_3^{2-}) is an anionic radical.

When ions or radicals react with each other to form new compounds, the reactions may not always proceed on a one-to-one basis as was the case for sodium chloride. They do, however, proceed on an equivalence basis that can be related to electroneutrality. Technically, the equivalence of an element or radical is defined as the number of hydrogen atoms that element or radical can hold in combination or can replace in a reaction. In most cases, the equivalence of an ion is the same as the absolute value of its valence. An equivalent of an element or radical is its gram molecular mass divided by its equivalence. A milli-equivalent is the molecular mass expressed in milligrams divided by the equivalence and is often more useful in water chemistry because concentrations of dissolved substances are more often in the milligrams per liter range. Compounds are formed by the combination of elements or radicals on a one-to-one equivalent basis.

Equivalents are very important in water chemistry. In addition to being useful in calculating chemical quantities for desired reactions in water and waste-water treatment, equivalents also provide a means of expressing various constituents of dissolved solids in a common term. An equivalent of one substance is chemically equal to an equivalent of any other substance. Therefore, the concentration of substance A can be expressed as an equivalent concentration of substrate B by the following method:

$$\frac{(g/l)A}{(g/equiv)A} \times (g/equiv)B = (g/l)A \text{ expressed as } B \qquad \text{... (2.2)}$$

Historically, constituents of dissolved solids have been reported in terms of equivalent calcium carbonate concentrations.

Many solid substances, particularly those with crystalline structure, ionise readily in water. Water may or may not be a chemical reactant in the process. In (Eq. 2.3), water is a reactant, while in (Eq. 2.4) it is not.

$$CaO + H_2O \rightarrow Ca^{2+} + 2OH^- \qquad \text{... (2.3)}$$

$$NaCl + H_2O \rightleftharpoons Na^+ + Cl^- + H_2O \qquad \qquad ... (2.4)$$

When water is not a reactant, it is customary to omit it from the equation.

TOTAL DISSOLVED SOLIDS

The material remaining in the water after filtration for the suspended-solids analysis is considered to be dissolved. This material is left as a solid residue upon evaporation of the water and constitutes a part of total solids.

Sources

Dissolved material results from the solvent action of water on solids, liquids and gases. Like suspended material, dissolved substances may be organic or inorganic in nature. Inorganic substances which may be dissolved in water include minerals, metals and gases. Water may come in contact with these substances in the atmosphere, on surfaces and within the soil. Materials from the decay products of vegetation, from organic chemicals and from the organic gases are common organic dissolved constituents of water. The solvent capability of water makes it an ideal means by which waste products can be carried away from industrial sites and homes.

Impacts

Many dissolved substances are undesirable in water. Dissolved minerals, gases and organic constituents may produce aesthetically displeasing colour, tastes and odours. Some chemicals may be toxic and some of the dissolved organic constituents have been shown to be carcinogenic. Quite often, two or more dissolved substances—especially organic substances and members of the halogen group—will combine to form a compound whose characteristics are more objectionable than those of either of the original materials.

Not all dissolved substances are undesirable in water. For example, essentially pure, distilled water has a flat taste. Additionally, water has an equilibrium state with respect to dissolved constituents. An under-saturated water will be 'aggressive' and will more readily dissolve materials with which it comes in contact. Readily dissolvable material is sometimes added to a relatively pure water to reduce its tendency to dissolve pipes and plumbing.

Measurement

A direct measurement of total dissolved solids can be made by evaporating to dryness a sample of water which has been filtered to remove the suspended solids. The remaining residue is weighed and represents the total dissolved solids (TDS) in the water. The TDS is expressed as milligrams per liter on a dry-mass basis. The organic and inorganic fractions can be determined by firing the residue at 600°C.

An approximate analysis for TDS is often made by determining the electrical conductivity of the water. The ability of a water to conduct electricity, known as the specific conductance, is a function of its ionic strength. Specific conductance is measured by a conductivity meter employing the Wheatstone bridge principle. The standard procedure is to measure the conductivity in a cubic-centimeter field at 25°C and express the results in millisiemens per meter (mS/m).

Unfortunately, specific conductance and concentration of TDS are not related on a one-to-one basis. Only ionised substances contribute to specific conductance. Organic molecules and compounds that dissolve without ionising are not measured. Additionally, the magnitude of the specific conductance is influenced by the valence of the ions in solution, their mobility and relative numbers.

The temperature also has an important effect, with specific conductance, increasing as the water temperature increases. Conversion of units to milligrams per liter or milli-equivalents per liter must be made by use of an appropriate constant. A multiplier ranging from 0.055 to 0.09 is used to convert millisiemens to milligrams per liter. To use specific conductance as a quantitative test, sufficient analysis for filterable residue must be run to determine the conversion factor. For this reason, specific conductance is most often used in a qualitative sense to monitor changes in TDS occurring in natural streams or treatment processes.

Use

Because no distinction among the constituents is made, the TDS parameter is included in the analysis of water and waste-water only as a gross measurement of the dissolved material. While this is often sufficient for waste-waters, it is frequently desirable to know more about the composition of the solids in water that is intended for use in potable supplies, agriculture and some industrial processes. When this is the case, tests for several of the ionic constituents of TDS are made.

Ion Balance

The ions usually accounting for the vast majority of TDS in natural waters are listed in Table 2.2. Those listed under major constituents are often sufficient to characterise the dissolved-solids content of water. These are called common ions and are often measured individually and summed on an equivalent basis to represent the approximate TDS. As a check, the sum of the anions should equal the sum of the cations because electroneutrality must be preserved. A significant imbalance suggests that additional constituents are present or that an error has been made in the analysis of one or more of the ions.

Table 2.2. Common ions in natural waters.

Major constituents, 1.0–1000 mg/l	Secondary constituents, 0.01–10.0 mg/l
Sodium	Iron
Calcium	Strontium
Magnesium	Potassium
Bicarbonate	Carbonate
Sulphate	Nitrate
Chloride	Fluoride
	Boron
	Silica

It is important to arrange the cations and anions in the order shown for convenience in determining types of hardness and the quantities of chemicals needed for softening.

Several of the constituents of dissolved solids have properties that necessitate special attention. These constituents include alkalinity, hardness, fluoride, metals, organics and nutrients.

ALKALINITY

Alkalinity is defined as the quantity of ions in water that will react to neutralise hydrogen ions. Alkalinity is thus a measure of the ability of water to neutralise acids.

Sources

Constituents of alkalinity in natural water systems include CO_3^{2-}, HCO_3^-, OH^-, $HSiO_3^-$, $H_2BO_3^-$, HPO_4^{2-}, $H_2PO_4^-$, HS^- and NH_3^0. These compounds result from the dissolution of mineral substances in the soil and atmosphere. Phosphates may also originate from detergents in waste-water discharges and from fertilisers and insecticides from agricultural land. Hydrogen sulphide and ammonia may be products of microbial decomposition of organic material.

By far the most common constituents of alkalinity are bicarbonate (HCO_3^-), carbonate (CO_3^{2-}) and hydroxide (OH^-). In addition to their mineral origin, these substances can originate from carbon dioxide, a constituent of the atmosphere and a product of microbial decomposition of organic material. These reactions are as follows:

$$CO_2 + H_2O \rightleftharpoons H_2CO_3^* \qquad \text{(dissolved } CO_2 \text{ and carbonic acid)} \qquad ... (2.5)$$

$$H_2CO_3^* \rightleftharpoons H^+ + HCO_3^- \qquad \text{(bicarbonate)} \qquad ... (2.6)$$

$$HCO_3^- \rightleftharpoons H^+ + CO_3^{2-} \qquad \text{(carbonate)} \qquad ... (2.7)$$

$$CO_3^{2-} + H_2O \rightleftharpoons HCO_3^- + OH^- \qquad \text{(hydroxide)} \qquad ... (2.8)$$

The reaction represented by (Eq. 2.8) is a weak reaction chemically. However, utilisation of the bicarbonate ion as a carbon source by algae can drive the reaction to the right and result in substantial accumulation of OH^-. Water with heavy algal growths often has pH values as high as 9 to 10. Because the reactions represented by the above equations involve hydrogen or hydroxide ions, the relative quantities of the alkalinity species are pH dependent. These relationships are shown graphically in Fig. 2.1.

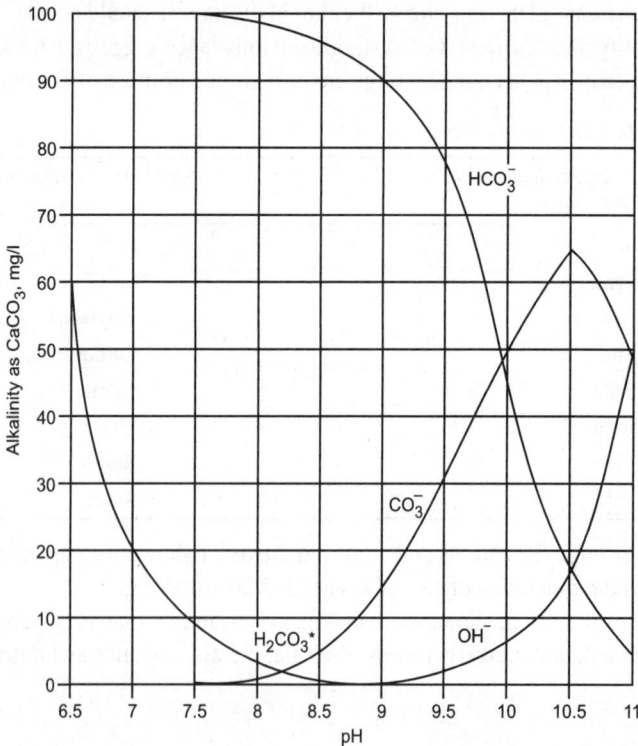

Fig. 2.1. Alkalinity species vs. pH. Values are calculated for water at 25°C containing a total alkalinity of 100 mg/l as $CaCO_3$.

Impacts

In large quantities, alkalinity imparts a bitter taste to water. The principal objection to alkaline water, however, is the reactions that can occur between alkalinity and certain cations in the water. The resultant precipitate can foul pipes and other water-systems appurtenances.

Measurement

Alkalinity measurements are made by titrating the water with an acid and determining the hydrogen equivalent. Alkalinity is then expressed as milligrams per liter of $CaCO_3$. If 0.02 N H_2SO_4 is used in the titration, then 1 ml of the acid will neutralise 1 mg of alkalinity as $CaCO_3$. Hydrogen ions from the acid react with the alkalinity according to the following equations:

$$H^+ + OH^- \rightleftharpoons H_2O \qquad \qquad \text{... (2.9)}$$

$$CO_3^{2-} + H^+ \rightleftharpoons HCO_3^- \qquad \qquad \text{... (2.10)}$$

$$HCO_3^- + H^+ \rightleftharpoons H_2CO_3 \qquad \qquad \text{... (2.11)}$$

If acid is added slowly to water and the pH is recorded for each addition, a titration curve similar to that shown in Fig. 2.2 is obtained. Of particular significance are the inflection points in the curve that occur at approximately pH 8.3 and pH 4.5. The conversion of carbonate to bicarbonate Eq. 2.10 is essentially complete at pH 8.3. However, because bicarbonate is also an alkalinity species, an equal amount of acid must be added to complete the neutralisation. Thus, the neutralisation of carbonate is only one-half complete at pH 8.3. Because the conversion of hydroxide to water is virtually complete at pH 8.3 (Fig. 2.1), all of the hydroxide and one-half of the carbonate have been measured at pH 8.3. At pH 4.5 all of the bicarbonate has been converted to carbonic acid Eq. 2.11 including the bicarbonate resulting from the reaction of the acid and carbonate Eq. 2.10. Thus, the amount of acid required to titrate a sample to pH 4.5 is equivalent to the total alkalinity of the water.

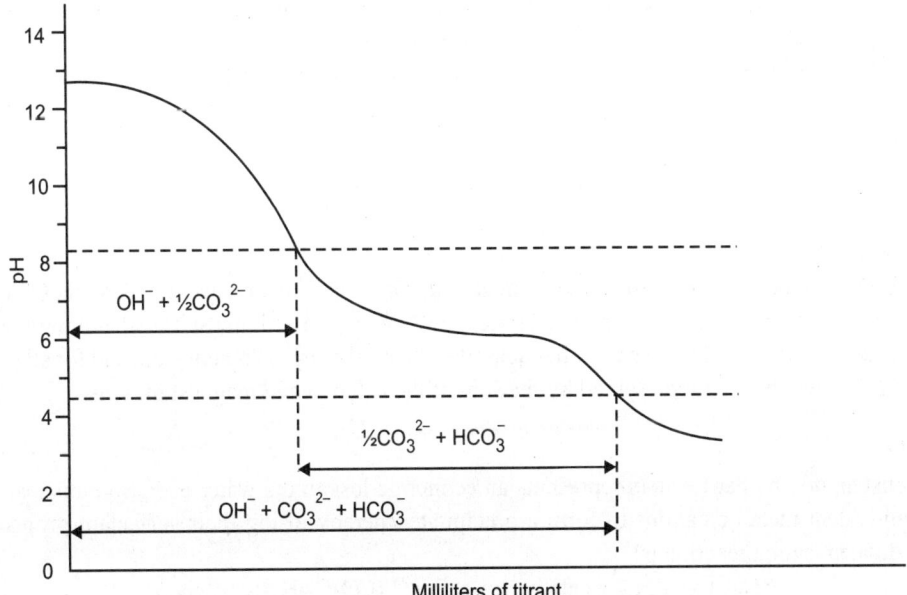

Fig. 2.2. Alkalinity titration curve.

If the volume of acid needed to reach the 8.3 end point is known, the species of alkalinity can also be determined. Because all of the hydroxide and one-half of the carbonate have been neutralised at pH 8.3, the acid required to lower the pH from 8.3 to 4.5 must measure the other one-half of the carbonate, plus all of the original bicarbonate. If P is the amount of acid required to reach pH 8.3 and M is the total quantity of acid required to reach 4.5, the following generalisations concerning the forms of alkalinity can be made:

if P = M, all alkalinity is OH$^-$

$\qquad$ P = M/2, all alkalinity is CO$_3^{2-}$

$\qquad$ P = 0 (i.e. initial pH is below 8.3), all alkalinity is HCO$_3^-$

$\qquad$ P < M/2, predominant species are CO$_3^{2-}$ and HCO$_3^-$

$\qquad$ P > M/2, predominant species are OH$^-$ and CO$_3^{2-}$

In observing the pH dependency of the species in Fig. 2.1, it is noted that the quantity of OH$^-$ becomes significant at pH less than about 9.0. Without introducing significant error, it can be assumed that the OH$^-$ of samples with pH less than 9.0 is insignificant. The CO$_3^{2-}$ would then be measured by 2P and the HCO$_3^-$ would be measured by the remainder (M – 2P).

Use

Alkalinity measurements are often included in the analysis of natural waters to determine their buffering capacity. It is also used frequently as a process control variable in water and waste-water treatment. Maximum levels of alkalinity have not been set by EPA for drinking water or for waste-water discharges.

HARDNESS

Hardness is defined as the concentration of multivalent metallic cations in solution. At supersaturated conditions, the hardness cations will react with anions in the water to form a solid precipitate. Hardness is classified as carbonate hardness and noncarbonate hardness, depending upon the anion with which it associates. The hardness that is equivalent to the alkalinity is termed carbonate hardness, with any remaining hardness being called noncarbonate hardness.

Carbonate hardness is sensitive to heat and precipitates readily at high temperatures.

$$Ca(HCO_3)_2 \xrightarrow{\Delta} CaCO_3 + CO_2 + H_2O \qquad \text{... (2.12)}$$

$$Mg(HCO_3)_2 \xrightarrow{\Delta} Mg(OH)_2 + 2CO_2 \qquad \text{... (2.13)}$$

Sources

The multivalent metallic ions most abundant in natural waters are calcium and magnesium. Others may include iron and manganese in their reduced states (Fe^{2+}, Mn^{2+}), strontium (Sr^{2+}) and aluminium (Al^{3+}). The latter are usually found in much smaller quantities than calcium and magnesium and for all practical purposes, hardness may be represented by the sum of the calcium and magnesium ions.

Impacts

Soap consumption by hard waters represents an economic loss to the water user. Sodium soaps react with multivalent metallic cations to form a precipitate, thereby losing their surfactant properties. A typical divalent cation reaction is:

$$2NaCO_2C_{17}H_{33} + cation^{2+} \rightarrow cation^{2+} (CO_2C_{17}H_{33})_2 + 2Na^+ \qquad \text{... (2.14)}$$
$$\underset{\text{Soap}}{} \qquad\qquad\qquad \underset{\text{Precipitate}}{}$$

Lathering does not occur until all of the hardness ions are precipitated, at which point the water has been 'softened' by the soap. The precipitate formed by hardness and soap adheres to surfaces of tubs, sinks and dishwashers and may stain clothing, dishes and other items. Residues of the hardness-soap precipitate may remain in the pores, so that skin may feel rough and uncomfortable. In recent years these problems have been largely alleviated by the development of soaps and detergents that do not react with hardness.

Boiler scale, the result of the carbonate hardness precipitate may cause considerable economic loss through fouling of water heaters and hot-water pipes. Changes in pH in the water distribution systems may also result in deposits of precipitates. Bicarbonates begin to convert to the less soluble carbonates at pH values above 9.0.

Magnesium hardness, particularly associated with the sulphate ion, has a laxative effect on persons unaccustomed to it. Magnesium concentrations of less than 50 mg/l are desirable in potable waters, although many public water supplies exceed this amount. Calcium hardness presents no public health problem. In fact, hard water is apparently beneficial to the human cardiovascular system.

Measurement

Hardness can be measured by using spectrophotometric techniques or chemical titration to determine the quantity of calcium and magnesium ions in a given sample. Hardness can be measured directly by titration with ethylenediamine tetraacetic acid (EDTA) using eriochrome black T (EBT) as an indicator. The EBT reacts with the divalent metallic cations, forming a complex that is red in colour. The EDTA replaces the EBT in the complex and when the replacement is complete, the solution changes from red to blue. If 0.01 M EDTA is used, 1.0 ml of the titrant measures 1.0 mg of hardness as $CaCO_3$.

Use

Analysis for hardness is commonly made on natural waters and on waters intended for potable supplies and for certain industrial uses. Hardness may range from practically zero to several hundred or even several thousand, parts per million. Although acceptability levels vary according to a consumer's acclimation to hardness, a generally accepted classification is as follows:

Soft	< 50 mg/l as $CaCO_3$
Moderately hard	50–150 mg/l as $CaCO_3$
Hard	150–300 mg/l as $CaCO_3$
Very hard	> 300 mg/l as $CaCO_3$

The Public Health Service standards recommend a maximum of 500 mg/l of hardness in drinking water.

FLUORIDE

Generally associated in nature with a few types of sedimentary or igneous rocks, fluoride is seldom found in appreciable quantities in surface waters and appears in groundwater in only a few geographical regions. Fluoride is toxic to humans and other animals in large quantities, while small concentrations can be beneficial. Concentrations of approximately 1.0 mg/l in drinking water help to prevent dental cavities in children. During formation of permanent teeth, fluoride combines chemically with tooth enamel, resulting in harder, stronger teeth that are more resistant to decay. Fluoride is often added to drinking water supplies if sufficient quantities for good dental formation are not naturally present.

Excessive intakes of fluoride can result in discolouration of teeth. Noticeable discolouration, called mottling, is relatively common when fluoride concentrations in drinking water exceed 2.0 mg/l, but is rare when concentrations are less than 1.5 mg/l. Adult teeth are not affected by fluoride, although both the benefits and liabilities of fluoride during tooth-formation years carry over into adulthood. Excessive dosages of fluoride can also result in bone fluorosis and other skeletal abnormalities. Concentrations of less than 5 mg/l in drinking water are not likely to cause bone fluorosis or related problems and some water supplies are known to have somewhat higher fluoride concentrations with no discernible problem other than severe mottling of teeth. On the assumption that people drink more water in warmer climates, EPA drinking-water standards base upper limits for fluoride on ambient temperatures.

METALS

All metals are soluble to some extent in water. While excessive amounts of any metal may present health hazards, only those metals that are harmful in relatively small amounts are commonly labelled toxic; other metals fall into the nontoxic group. Sources of metals in natural waters include dissolution from natural deposits and discharges of domestic, industrial or agricultural waste-waters. Measurement of metals in water is usually made by atomic absorption spectrophotometry.

Nontoxic Metals

In addition to the hardness ions, calcium and magnesium, other nontoxic metals commonly found in water include sodium, iron, manganese, aluminium, copper and zinc. Sodium, by far the most common nontoxic metal found in natural waters, is abundant in the earth's crust and is highly reactive with other elements. The salts of sodium are soluble in water. Excessive concentrations cause a bitter taste in water and are a health hazard to cardiac and kidney patients. Sodium is also corrosive to metal surfaces and, in large concentrations, is toxic to plants.

Iron and manganese quite frequently occur together and present no health hazards at concentrations normally found in natural waters. As already discussed, iron and manganese in very small quantities may cause colour problems. Iron concentrations of 0.3 mg/l and manganese concentrations as low as 0.05 mg/l can cause colour problems. Additionally, some bacteria use iron and manganese compounds for an energy source and the resulting slime growth may produce taste and odour problems.

When significant quantities of iron are encountered in natural water systems, it is usually associated with chloride ($FeCl_2$), bicarbonate [$Fe(HCO_3)_2$] or sulphate [$Fe(SO_4)$] anions and exists in a reduced state. In the presence of oxygen, the ferrous (Fe^{2+}) ion is oxidised to the ferric (Fe^{3+}) ion and forms an insoluble compound with hydroxide [$Fe(OH)_3$]. Thus, significant quantities of iron will usually be found only in systems devoid of oxygen such as groundwaters or perhaps the bottom layers of stratified lakes. Similarly, manganese ions (Mn^{2+} and Mn^{4+}) associated with chloride, nitrates and sulphates are soluble, while oxidised compounds (Mn^{3+} and Mn^{5+}) are virtually insoluble. It is possible, however, for organic acids derived from decomposing vegetation to chelate iron and manganese and prevent their oxidation and subsequent precipitation in natural waters.

The other nontoxic metals are generally found in very small quantities in natural water systems and most would cause taste problems long before toxic levels were reached. However, copper and zinc are synergetic and when both are present, even in small quantities, may be toxic to many biological species.

Toxic Metals

As noted earlier, toxic metals are harmful to humans and other organisms in small quantities. Toxic metals that may be dissolved in water include arsenic, barium, cadmium, chromium, lead, mercury and silver. Cumulative toxins such as arsenic, cadmium, lead and mercury are particularly hazardous. These metals are concentrated by the food chain, thereby posing the greatest danger to organisms near the top of the chain. Fortunately, toxic metals are present in only minute quantities in most natural water systems. Although natural sources of all the toxic metals exist, significant concentration in water can usually be traced to mining, industrial or agricultural sources.

ORGANICS

Many organic materials are soluble in water. Organics in natural water systems may come from natural sources or may result from human activities. Most natural organics consist of the decay products of organic solids, while synthetic organics are usually the result of waste-water discharges or agricultural practices. Dissolved organics in water are usually divided into two broad categories: biodegradable and nonbiodegradable (refractory).

Biodegradable Organics

Biodegradable material consists of organics that can be utilised for food by naturally occurring micro-organisms within a reasonable length of time. In dissolved form, these materials usually consist of starches, fats, proteins, alcohols, acids, aldehydes and esters. They may be the end product of the initial microbial decomposition of plant or animal tissue or they may result from domestic or industrial waste-water discharges. Although some of these materials can cause colour, taste and odour problems, the principal problem associated with biodegradable organics is a secondary effect resulting from the action of micro-organisms on these substances.

Microbial utilisation of dissolved organics can be accompanied by oxidation (addition of oxygen to or the deletion of hydrogen from elements of the organic molecule) or by reduction (addition of hydrogen to or deletion of oxygen from elements of the organic molecule). Although it is possible for the two processes to occur simultaneously, the oxidation process is by far more efficient and is predominant when oxygen is available. In aerobic (oxygen-present) environments, the end products of microbial decomposition of organics are stable and acceptable compounds. Anaerobic (oxygen-absent) decomposition results in unstable and objectionable end products. Should oxygen later become available, anaerobic end products will be oxidised to aerobic end products. The oxygen-demanding nature of biodegradable organics is of utmost importance in natural water systems. When oxygen utilisation occurs more rapidly than oxygen can be replenished by transfer from the atmosphere, anaerobic conditions that severely affect the ecology of the system will result.

The amount of oxygen consumed during microbial utilisation of organics is called the biochemical oxygen demand (BOD). The BOD is measured by determining the oxygen consumed from a sample placed in an air-tight container and kept in a controlled environment for a pre-selected period of time. In the standard test, a 300 ml BOD bottle is used and the sample is incubated at 20°C for 5 days. Light must be excluded from the incubator to prevent algal growth that may produce oxygen in the bottle. Because the saturation concentration for oxygen in water at 20°C is approximately 9 mg/l, dilution of the sample with BOD-free, oxygen-saturated water is necessary to measure BOD values greater than just a few milligrams per liter.

The BOD of a diluted sample is calculated by

$$BOD = \frac{DO_I - DO_F}{P} \qquad \qquad ...(2.15)$$

where DO_I and DO_F are the initial and final dissolved-oxygen concentrations (mg/l) and P is the decimal fraction of the sample in the 300 ml bottle.

Ranges of BOD covered by various dilutions are shown in Table 2.3. These values assume an initial dissolved-oxygen concentration of 9 mg/l in the mixture, with a minimum of 2 and a maximum of 7 mg/l of O_2 being consumed. Calculations of BOD_5 from this testing procedure are illustrated in the following example.

Table 2.3. Ranges of BOD values covered by various dilutions.

By using per cent mixture		By direct pipetting into 300 ml bottles	
% mixture	Range of BOD	ml	Range of BOD
0.01	20000–70000	0.02	30000–1,05,000
0.02	10000–35000	0.05	12000–42000
0.05	4000–14000	0.10	6000–21000
0.1	2000–7000	0.20	3000–10500
0.2	1000–3500	0.50	1200–4200
0.5	400–1400	1.0	600–2100
1.0	200–700	2.0	300–1050
2.0	100–350	5.0	120–420
5.0	40–140	10.0	60–210
10.0	20–70	20.0	30–105
20.0	10–35	50.0	12–42
50.0	4–14	100.0	6–21
100.0	0–7	300.0	0–7

Most natural water and municipal waste-waters will have a population of micro-organisms that will consume the organics. In sterile waters, micro-organisms must be added and the BOD of the material containing the organisms must be determined and subtracted from the total BOD of the mixture. The presence of toxic materials in the water will invalidate the BOD results.

The BOD_5 only represents the oxygen consumed in 5 days. The total BOD or BOD for any other time period, can be determined provided additional information is known or obtained. The rate at which organics are utilised by micro-organisms is assumed to be a first-order reaction; that is, the rate at which organics utilised is proportional to the amount available.

Nonbiodegradable Organics

Some organic materials are resistant to biological degradation. Tannic and lignic acids, cellulose and phenols are often found in natural water systems. These constituents of woody plants biodegrade so slowly that they are usually considered refractory. Molecules with exceptionally strong bonds (some of the polysaccharides) and ringed structures (benzene) are essentially nonbiodegradable. An example is the detergent compound alkyl benzene sulphonate (ABS) which, with its benzene ring, does not biodegrade. Being a surfactant, ABS causes frothing and foaming in waste-water treatment plants and increases turbidity by stabilising colloidal suspensions. This problem was largely alleviated when detergent manufacturers switched to a linear alkyl sulphonate (LAS) compound, which is biodegradable. Many of the organics associated with petroleum and with its refining and processing also contain benzene and are essentially nonbiodegradable.

Some organics are nonbiodegradable because they are toxic to organisms. These include the organic pesticides, some industrial chemicals and hydrocarbon compounds that have combined with chlorine.

Pesticides, including insecticides and herbicides, have found widespread use in modern society in both urban and agricultural settings. Poor application practices and subsequent washoff by rainfall and runoff may result in contamination of surface streams. Organic insecticides are usually chlorinated hydrocarbons (i.e. aldrin, dieldrin, endrin and lindane), while herbicides are usually chlorophenoxys (e.g. 2,4-dichlorophenoxyacetic acid and 2,4,5-trichlorophenoxy-propionic acid). Many of the pesticides are cumulative toxins and cause severe problems at the higher end of the food chain. An example is the near-extinction of the brown pelican that feeds on fish and other macroaquatic species by the insecticide DDT, the use of which is now banned in the United States.

Measurement of nonbiodegradable organics is usually by the chemical oxygen demand (COD) test. Nonbiodegradable organics may also be estimated from a total organic carbon (TOC) analysis. Both COD and TOC measure the biodegradable fraction of the organics, so the BOD must be subtracted from the COD or TOC to quantify the nonbiodegradable organics. Specific organic compounds can be identified and quantified through analysis by gas chromatography.

NUTRIENTS

Nutrients are elements essential to the growth and reproduction of plants and animals and aquatic species depend on the surrounding water to provide their nutrients. Although a wide variety of minerals and trace elements can be classified as nutrients, those required in most abundance by aquatic species are carbon, nitrogen and phosphorus. Carbon is readily available from many sources. Carbon dioxide from the atmosphere, alkalinity and decay products of organic matter all supply carbon to the aquatic system. In most cases, nitrogen and phosphorus are the nutrients that are the limiting factors in aquatic plant growth.

NITROGEN

Nitrogen gas (N_2) is the primary component of the earth's atmosphere and is extremely stable. It will react with oxygen under high-energy conditions (electrical discharges or flame incineration) to form nitrogen oxides. Although a few biological species are able to oxidise nitrogen gas, nitrogen in the aquatic environment is derived primarily from sources other than atmospheric nitrogen.

Nitrogen is a constituent of proteins, chlorophyll and many other biological compounds. Upon the death of plants or animals, complex organic matter is broken down to simple forms by bacterial decomposition. Proteins, for instance, are converted to amino acids and further reduced to ammonia (NH_3). If oxygen is present, the ammonia is oxidised to nitrite (NO_2^-) and then to nitrate (NO_3^-). The nitrate can then be reconstituted into living organic matter by photosynthetic plants.

Other sources of nitrogen in aquatic systems include animal wastes, chemical (particularly chemical fertilisers) and waste-water discharges. Nitrogen from these sources may be discharged directly into streams or may enter waterways through surface runoff or groundwater discharge. Nitrogen compounds can be oxidised to nitrate by soil bacteria and may be carried into the groundwater by percolating water. Once in the aquifer, nitrates move freely with the groundwater flow. Groundwater contamination by nitrogen from animal feedlots and septic-tank drain fields has been recorded in numerous instances.

In addition to the over-enrichment problems alluded to earlier, nitrogen can have other serious consequences. Ammonia is a gas at temperatures and pressures normally found in natural water systems. The gas (NH_3) exists in equilibrium with the aqueous ionic form called ammonium (NH_4^+).

$$NH_3 + H_2O \rightleftharpoons NH_4^+ + OH^- \qquad \text{... (2.16)}$$

The hydroxyl ion concentration of the water and thus the pH, controls the relative abundance of each species. Oxidation of NH_3 and NH_4^+ to nitrate and on to nitrate by aquatic microbes results in an additional biochemical oxygen demand as discussed in the preceding section.

Nitrate poisoning in infant animals, including humans, can cause serious problems and even death. Apparently, the lower acidity in an infant's intestinal tract permits growth of nitrate-reducing bacteria that convert the nitrate to nitrite, which is then absorbed into the bloodstream. Nitrite has a greater affinity for haemoglobin than does oxygen and thus replaces oxygen in the blood complex. The body is denied essential oxygen and, in extreme cases, the victim suffocates. Because oxygen starvation results in a bluish discolouration of the body, nitrate poisoning has been referred to as the 'blue baby' syndrome, although the correct term is *methemoglobinemia*. Once the flora of the intestinal tract has fully developed, usually after the age of 6 months, nitrate conversion to nitrite and subsequent methemoglobinemia from drinking water is seldom a problem. Fortunately, the natural oxidation of nitrite to nitrate occurs quickly so that significant quantities of nitrites are not found in natural water.

Tests for nitrogen forms in water commonly include analysis for ammonia (including both ammonia and ammonium), nitrate and organic nitrogen. The results of the analyses are usually expressed as milligrams per liter of the particular species as nitrogen. Tests for ammonium and organic nitrogen are more common on waste-water and other polluted waters, while the test for nitrate is the most common on clean-water samples and treated waste-waters.

Phosphorus

Phosphorus appears exclusively as phosphate (PO_4^{3-}) in aquatic environments. There are several forms of phosphate, however, including orthophosphate, condensed phosphates (pyro, meta and polyphosphates) and organically bound phosphates. These may be insoluble or particulate form or may be constituents of plant or animal tissue. Like nitrogen, phosphates pass through the cycles of decomposition and photosynthesis. Phosphate is a constituent of soils and is used extensively in fertiliser to replace and/or supplement natural quantities on agricultural lands. Phosphate is also a constituent of animal waste and may become incorporated into the soil in grazing and feeding areas. Runoff from agricultural areas is a major contributor to phosphate in surface waters. The tendency for phosphate to adsorb to soil particles limits its movement in soil moisture and groundwater, but results in its transport into surface waters by erosion. Municipal waste-water is another major source of phosphate in surface water. Condensed phosphates are used extensively as builders in detergents and organic phosphates are constituents of body waste and food residue. Other sources include industrial waste in which phosphate compounds are used for such purposes as boiler-water conditioning.

While phosphates are not toxic and do not represent a direct health threat to human or other organisms, they do represent a serious indirect threat to water quality. As noted earlier, phosphate is often the limiting nutrient in surface waters. When the available supply is increased, rapid growth of aquatic plants usually results, with severe consequences. Phosphate can also interfere with water-treatment processes. Concentrations as low as 0.2 mg/l interfere with the chemical coagulation of turbidity.

Phosphates are measured colorimetrically. Orthophosphates can be measured directly, while condensed forms must be converted to orthophosphate by acid hydrolysation and organic phosphates must be converted to orthophosphates by acid digestion. Results of the analysis are reported as milligrams per liter of phosphate as phosphorus. Careful handling of samples prior to analysis is crucial. For example, acid-washed glass bottles should be used for sampling, as bottles washed in phosphate detergent may contaminate samples.

Chapter 3

Biological Parameters of Water Quality

INTRODUCTION

Water may serve as a medium in which literally thousands of biological species spend part, if not all, of their life cycles. Aquatic organisms range in size and complexity from the smallest single-cell micro-organism to the largest fish. All members of the biological community are, to some extent, water quality parameters, because their presence or absence may indicate in general terms the characteristics of a given body of water. As an example, the general quality of water in a trout stream would be expected to exceed that of a stream in which the pre-dominant species of fish is carp. Similarly, abundant algal populations are associated with a water rich in nutrients.

Biologists often use a species-diversity index (related to the number of species and the relative abundance of organisms in each species) as a qualitative parameter for streams and lakes. A body of water hosting large numbers of species with well-balanced numbers of individuals is considered to be a healthy system. Based on their known tolerance for a given pollutant, certain organisms can be used as indicators of the presence of pollutants.

PATHOGENS

From the perspective of human use and consumption, the most important biological organisms in water are pathogens, those organisms capable of infecting, or of transmitting diseases to humans. These organisms are not native to aquatic systems and usually require an animal host for growth and reproduction. They can, however, be transported by natural water systems, thus becoming a temporary member of the aquatic community. Many species of pathogens are able to survive in water and maintain their infectious capabilities for significant periods of time. These waterborne pathogens include species of bacteria, viruses, protozoa and helminths (parasitic worms). The characteristics of the primary waterborne pathogens are listed in Table 3.1.

Bacteria

Bacteria are considered to be single-celled plants because of their cell structure and the way they take in food. They utilise soluble food taken in through a rigid cell wall. But unlike green plants that use photosynthesis, bacteria do not produce their own food.

Bacteria are very small, typically about 2 μm in size and can be seen only with the aid of a microscope. They occur in three basic cell shapes: rod shaped or *bacillus*, sphere shaped or coccus and spiral shaped or *spirellus*. In some cases, the individual cells grow together in larger groups or chains. *Sphaerotilus natans* is an example of a species of bacteria that grows in a chain or filament enclosed within a long sheath or tube. Excessive growth of these filamentous organisms is known to be one of the causes of reduced treatment efficiency in biological sewage treatment plants.

Table 3.1. Common waterborne pathogens.

Organism	Disease
Bacteria	
Francisella tularensis	Tularemia (deer fly fever)
Leptospirae	Leptospirosis (Weil's disease, swineherd's disease, hemorrhagic jaundice)
Salmonella paratyphi (A,B,C)	Paratyphoid (enteric fever)
Salmonella typhi	Typhoid fever, enteric fever
Shigella (S. flexneri, S. sonnei, S. dysenteriae, S. boydii)	Shigellosis (bacillary dysentery)
Vibrio comma (Vibrio cholerae)	Cholera (Asiatic, Indian, El Tor)
Viruses	
Enteric cytopathogenic human orphan (ECHO)	Aseptic meningitis, epidemic exanthem, infantile diarrhoea
Poliomyelitis (3 types)	Acute anterior poliomyelitis, infantile paralysis
Unknown viruses	Infectious hepatitis
Protozoa	
Entamoeba histolytica	Amebiasis (amebic dysentery, amebic enteritis, amebic colitis)
Giardia lamblia	Giardiasis (Giardia enteritis, lambliasis)
Helminths (parasitic worms)	
Dracunculus medinensis	Dracontiasis (dracunculiasis; dracunculosis; medina; serpent, dragon or guinea-worm infection)
Echinococcus	Echinococcosis (hydatidosis; granulosus; dog tapeworm)
Schistosoma (S. mansoni, S. japonicum, S. haematobium)	Schistosomiasis (bilharziasis or 'Bill Harris' or 'blood fluke' disease)

In less than 30 minutes, a single bacterial cell can mature and divide into two new cells. This process of reproduction is called binary fission. Under favourable conditions of food supply, temperature and pH, bacteria can reproduce so rapidly that a bacterial culture may contain as many as 20 million individual cells per milliliter after just one day of growth. This rapid growth of visible colonies of bacteria on a suitable nutrient medium makes it possible to detect and count the number of bacteria in water. This is discussed in more detail in the section on coliform bacteria.

There are several distinctions among the various species of bacteria. One depends on how they metabolise their food. Bacteria that require oxygen for their metabolism are called aerobic bacteria or aerobes.

Those that live only in an oxygen-free environment are called anaerobic bacteria or anaerobes. The distinction between aerobes and anaerobes is of great significance in water pollution and waste-water treatment. (Some species, called facultative bacteria, can live in either the absence or presence of oxygen.)

Another distinction among species of bacteria is a function of the type of food that they require. Those that utilise simple inorganic compounds for nourishment are called autotrophic bacteria; those that require complex organic substances are called heterotrophic bacteria. The nitrifying bacteria, for example, which use ammonia as food and convert it to nitrate, are among the autotrophs. Other examples of autotrophs include the iron bacteria and the sulphur bacteria. Iron bacteria thrive in some water pipelines and often cause taste and odour problems in drinking water. The sulphur bacteria, which are also anaerobes, are active in sewers and speed the deterioration of concrete pipes by converting hydrogen sulphide gas to sulphuric acid.

One of the most important factors affecting the growth and reproduction of bacteria is temperature. At low temperatures, bacteria grow and reproduce slowly. As the temperature increases, the rate of growth and reproduction just about doubles for every additional 10°C (up to the optimum temperature for the species). The majority of species of bacteria are classified as *mesophilic*, having an optimum temperature of about 35°C. Those that do best at elevated temperatures of about 60°C are called *thermophilic* bacteria. Bacteria with an optimum growth temperature between 0°C and 20°C are called *psycrophilic* bacteria.

Algae

Algae are microscopic plants that contain photosynthetic pigments, such as chlorophyll. They are autotrophic organisms that support themselves by converting inorganic materials into organic matter using energy from the sun. During the process of photosynthesis, they take in carbon dioxide from the air and give off oxygen.

A basic characteristic of these simple plants is their lack of roots, stems and leaves. Free-floating algae are also called phytoplankton. (*Plankton* are tiny floating plants or animals that live in either fresh or salt waters. Over 90 per cent of atmospheric oxygen is produced by salt water or marine *phytoplankton*, by the process of photosynthesis.) Even though most species of algae are microscopic, they can be easily noticed when their numbers proliferate in the water. Excessive growths of algae, called algal blooms, are often unsightly. Some algal species are multicellular, growing as filaments that sometimes appear as a green slime in the water.

Common species include the blue-green algae such as *Anabaena*, green algae such as *Spirogyra*, yellow-green algae such as *Botrydium* and red algae such as *Gelidium*. Another important group of algae, called *diatoms*, produce hard shells of silica. Deposits of these shells, from dead diatoms, that have accumulated over many hundreds of years form *diatomaceous earth*, a material sometimes used for filtering water.

Algae play a role in the ageing or eutrophication of lakes. They also are important in waste-water treatment stabilisation ponds. Algae are generally nuisance organisms in public water supplies because of the taste and odour problems that they cause and because of the extra expense required to filter them out of the water.

Protozoa

Protozoa are the simplest of animal species. These single-celled microscopic animals consume solid or ganic particles, bacteria and algae for food. They are, in turn, ingested as food by higher-level multicellular animals. Floating freely in water, these zooplankton, as they are sometimes called, are a vital part of the natural aquatic food chain. They are also of significance in biological waste-water treatment systems.

Amoebae are protozoa that move by projecting sections of their bodies; this mobile protoplasm of the amoebae is also used to surround and engulf food particles. Amoebae are commonly found in slimes formed in certain types of sewage treatment processes.

A group of protozoa called *flagellates* move around in water by means of a long threadlike strand, called a *flagella*, that propels them with its whiplike action. One such organism, *Giardia lamblia*, is an intestinal parasite that causes a form of dysentery in humans. Another type of protozoa has hundreds of short hairs called *cilia* that propel the organism through the water and that serve to direct food particles into its digestive system. The *paramecia*, for example, are ciliated protozoa commonly found in freshwater ponds and lakes.

A species of protozoa called *Cryptosporidium* has been found to be the cause of recent waterborne gastrointestinal disease outbreaks in the United States. These pathogens are frequently found in lakes and streams and are very resistant to disinfection by chlorination (although they can be controlled by ozonation). The 1996 amendments to the Safe Drinking Water Act call for enhanced surface water treatment rules to prevent such outbreaks. Several types of protozoa, as well as some common forms of algae and bacteria, are illustrated in Fig. 3.1.

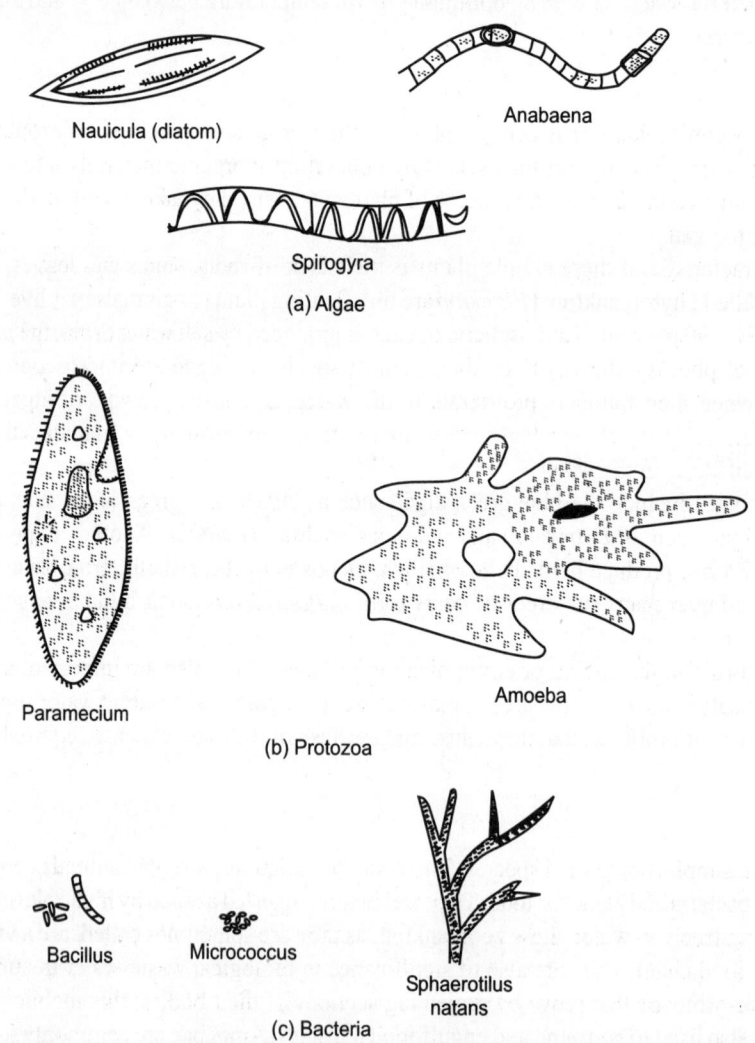

Fig. 3.1. Sketches of some typical micro-organisms found in water and/or sewage.

Viruses

Viruses are the smallest biological structures known to contain all the genetic information necessary for their own reproduction. So small that they can only be 'seen' with the aid of an electron microscope, viruses are obligate parasites that require a host in which to live. Symptoms associated with waterborne viral infections usually involve disorders of the nervous system rather than of the gastrointestinal tract. Waterborne viral pathogens are known to cause poliomyelitis and infectious hepatitis and several other viruses are known to be or suspected of being, waterborne.

Immunisation of individuals has reduced the incidence of polio to a few isolated cases each year in developed nations. Outbreaks of hepatitis are common throughout the world. Most of the hepatitis cases result from persons eating shellfish contaminated by viruses from polluted waters, although an occasional outbreak will occur at campgrounds or other facilities where crowds gather and where water-supply protection and sanitary facilities are poor.

Although standard disinfection practices are known to kill viruses, confirmation of effective viral disinfection is difficult, owing to the small size of the organism and the lack of quick and conclusive tests for viable virus organisms. The uncertainty of viral disinfection is a major obstacle to direct recycling of waste-water and is a cause of concern regarding the increasing practice of land application of waste-water (Table 3.1).

HELMINTHS

The life cycles of helminths, or parasitic worms, often involve two or more animal hosts, one of which can be human and water contamination may result from human or animal waste that contains helminths. Contamination may also be via aquatic species of other hosts, such as snails or insects. While aquatic systems can be the vehicle for transmitting helminthal pathogens, modern water-treatment methods are very effective in destroying these organisms. Thus, helminths pose hazards primarily to those persons who come into direct contact with untreated water. Sewage plant operators, swimmers in recreational lakes polluted by sewage or storm water runoff from cattle feedlots and farm labourers employed in agricultural irrigation operations are at particular risk.

PATHOGEN INDICATORS

Analysis of water for all the known pathogens would be a very time-consuming and expensive proposition. Tests for specific pathogens are usually made only when there is a reason to suspect that those particular organisms are present. At other times, the purity of water is checked using indicator organisms.

An indicator organism is one whose presence presumes that contamination has occurred and suggests the nature and extent of the contaminant(s). The ideal pathogen indicator would: (i) be applicable to all types of water, (ii) always be present when pathogens are present, (iii) always be absent when pathogens are absent, (iv) lend itself to routine quantitative testing procedures without interference from or confusion of results because of extraneous organisms, and (v) for the safety of laboratory personnel, not be a pathogen itself.

Most of the waterborne pathogens are introduced through fecal contamination of water. Thus, any organism native to the intestinal tract of humans and meeting the above criteria would be a good indicator organism. The organisms most nearly meeting these requirements belong to the fecal coliform group. Composed of several strains of bacteria, principal of which is *Escherichia coli*, these organisms are found exclusively in the intestinal tract of warm-blooded animals and are excreted in large numbers with feces. Fecal coliform organisms are non-pathogenic and are believed to have a longer survival time outside the animal body than do most pathogens. Because the die-off rate of fecal coliforms is logarithmic, the number of surviving organisms may be an indication of the time lapse since contamination.

There are other coliform groups which flourish outside the intestinal tract of animals. These organisms are native to the soil and decaying vegetation and are often found in water that was in recent contact with these materials. Because the life cycles of some pathogens (particularly helminths) may include periods in the soil, this group of coliform organisms also serves as an indicator of pathogens.

It is the usual practice in the United States to use the total coliform group (those of both fecal and non-fecal origin) as indicators of the sanitary quality of drinking water, while the indicator of choice for waste-water effluents is the fecal coliform group. Relatively simple tests have been devised to determine the presence of coliform bacteria in water and to enumerate the quantity. The tests for total coliform organisms employ slightly different culture media and lower incubation temperatures than those used to identify fecal coliform organisms.

The membrane-filter technique, a technique popular with environmental engineers, gives a direct count of coliform bacteria. In this test, a portion of the sample is filtered through a membrane, the pores of which do not exceed 0.45 μm. Bacteria are retained on the filter that is then placed on selective media to promote growth of coliform bacteria while inhibiting growth of other species. The membrane and media are incubated at the appropriate temperature for 24 hours, allowing coliform bacteria to grow into visible colonies that are then counted. The results are reported in number of organisms per 100 ml of water.

An alternative method often preferred by microbiologists is the multiple-tube fermentation test. Coliform organisms are known to ferment lactose, with one of the end products being a gas. A broth containing lactose and other substances which inhibit non-coliform organisms is placed in a series of test tubes which are then inoculated with a decimal fraction of 1 ml (100, 10, 1.0, 0.1, 0.01, etc.). These tubes are incubated at the appropriate temperature and inspected for development of gas. This first stage of the procedure is called the presumptive test and tubes with gas development are presumed to have coliforms present. A similar test, called the confirmed test is then set up to confirm the presence of coliform organisms.

Sampling techniques and subsequent handling of the samples are extremely important because samples can easily be contaminated.

It should be emphasised again that pathogens are not identified by the coliform test. The presence of coliform organisms in water does, however, indicate that some portion of the water has recently contacted soil or decaying vegetation or has been through the intestinal tract of a warm-blooded animal. The assumption must then be made that pathogens may have accompanied the coliform bacteria.

SECTION II

Engineering Systems for Water Purification

Chapter 4

Basic Concepts of Engineering Systems for Water Purification

INTRODUCTION

An adequate supply of pure water is absolutely essential to human existence. The consequences of a contaminated water supply can be illustrated by conditions prevalent during the industrial revolution in Europe when large numbers of peasants were attracted to the cities where they crowded together with little or no sanitary facilities. Human waste or 'night soil' as it was called, was tossed into the streets or emptied into pits in common courtyards, often near the shallow wells that served as the neighbourhood water supply. Seepage into these wells and runoff into nearby streams provided a direct link in the infection cycle and once an outbreak of disease occurred it usually spread rapidly through the community.

The development of effective water-treatment methods has virtually eliminated major waterborne epidemics in developed countries. This is not to suggest, however, that the problem of waterborne diseases has been eliminated. Developing nations, where treated water is not available to all the population, still experience occasional epidemics of cholera and typhoid, as well as many outbreaks of less severe disease. Even highly developed countries, including the United States, where public water supplies are almost universally treated, are not totally immune from an occasional outbreak of gastrointestinal illnesses traceable to biologically contaminated water supplies.

Chemical contamination of water supplies has become a concern in more recent times. Industrial facilities in developed countries produce and use literally thousands of chemical compounds. Along with an abundant array of household and agricultural chemicals, these materials often find their way into water supplies. While some of these chemical compounds are known toxicants, mutagents or carcinogens, the health effects of many others are not presently known. It is ironic that the high standard of living that allows industrialised nations to provide biologically pure water to the majority of then population also results in the discharge of chemical waste that may eventually have more deleterious effects on human health than the domestic waste that helped spread the plagues of past centuries.

The treatment of water intended for human consumption is a very old practice. Baker reports references in Sanskrit literature dating back to 2000 BC to such practices as the boiling and filtering of drinking water. Wick siphons that transferred water from one vessel to another, filtering out the suspended impurities in the process, were pictured in Egyptian drawings of the thirteenth century BC and were referred to in early Greek and Roman literature. The fact that these practices were recorded in the medical documents of the times indicates that the connection between water and health had been observed. In fact, Hippocrates (460–354 BC), considered to be the father of modern medicine, wrote that 'whosoever wishes to investigate medicine properly should—consider the water that the inhabitants use—for water contributes much to health.'

These early water-treatment devices were used in individual households; there is no indication of community water supplies being treated until around the first century. Some of the Roman aqueducts had settling basins at the headworks and incorporated 'pebble catchers' in the aqueduct channel. These aqueducts supplied a few private taps and provided fountains or reservoirs for use by the general public. The city of Venice, situated on islands with no freshwater resource, channelled rain water from roofs and courtyards into elaborate cisterns through sand filters surrounding the reservoir. The first of these cisterns was built around the fifth century AD and provided private and public water supplies for about 13 centuries.

Water-treatment practice apparently lagged during the Middle Ages, with a renewed interest emerging in the eighteenth century. Several patents were issued for filtering devices, primarily in France and England. As in ancient times, however, these devices were for use in private households, institutions, ships, etc. It was not until the beginning of the nineteenth century that the treatment of public water supplies was attempted on a large scale. The city of Paisley, Scotland, is generally credited with being the first city with a treated water supply. That system consisted of settling operations followed by filtration and was put in service in 1804. This practice slowly spread through Europe and by the end of the century, most major municipal water supplies were filtered. These filters were the 'slow sand' type.

The development of water treatment in America lagged behind the European practice. The first attempt at filtration was made at Richmond, Virginia, in 1932. This project was a failure and several years intervened before another significant effort was made. After the Civil War, other attempts were made to follow the sand filtration practice of Europe, few of which were successful. Apparently the nature of the suspended solids in American streams was significantly different from that of the solids in European streams and the slow sand process was not as effective. The development of the hydraulically cleaned rapid sand filter during the latter part of the nineteenth century provided a more workable process and by the end of the century its use was widespread.

During the first two-thirds of the nineteenth century, filtration was practiced to improve the aesthetic quality of the drinking water. An unknown benefit was the removal of micro-organisms, including pathogens, which made the water more wholesome as well. The acceptance of this fact in the last quarter of the century spurred the construction of the filter plants throughout Europe and America. At the turn of the century, filtration was the primary defense against waterborne disease.

Acceptance of the germ theory of disease transmittal led to the disinfection of public water supplies. First used on a temporary basis, disinfection with bleach powders and hypochlorites was used in isolated cases in the eighteen-nineties. The first permanent installation for chlorinating water was made in Belgium in 1902. The production of liquid chlorine began in 1909 and was first used for water disinfection in Philadelphia in 1913. Other means of disinfection, notably ozonation, were developed simultaneously but did not find widespread use. The drastic reduction in deaths due to waterborne diseases as a result of disinfection led to the widespread chlorination of public water supplies.

Other water-treatment processes developed more slowly and less dramatically. Coagulation as an adjunct to settling was developed along with the rapid sand filter in America. Softening of hard waters was demonstrated in Europe during the nineteenth century but did not find widespread use in public water supplies until well into the twentieth century. The capacity of charcoal to remove dissolved organics was observed by early experimenters in filtration but did not find application in public water supplies. The improvement of this material into activated carbon and its use in water-treatment plants is a recent occurrence, as is the use of synthetic membranes for hyperfiltration to remove dissolved inorganic material.

More progress has been made in water purification in the last century than in all of the previously recorded history. With few exceptions, treatment processes developed in the absence of scientific knowledge concerning the basic principles upon which they operate and often with little means to quantitatively assess their effectiveness. Only within the last 30 to 40 years has the body of scientific knowledge caught up with the practice of water purification. It is interesting to note that the development of a theory base has resulted in few changes in the basic processes of water purification. Understanding of scientific principles has, however, led to refinements of processes, development of better equipment and an overall increase in operating efficiencies in water treatment. The following section gives an overview of modern water treatment processes, while the remaining sections of the chapter contain a detailed description of the individual processes.

WATER TREATMENT PROCESSES

Past practices in US and other developing countries have often been to obtain the purest possible source, even at the expense of transporting water over long distances and to deliver it to the consumer with little or no treatment. Some cities still own large tracts of land near the headwaters of stream and restrict activities on these watersheds to minimise contamination. Although the benefits of source protection are recognised as a 'first line of defense' in preserving water quality, all natural waters will require some degree of treatment in order to meet modern drinking-water standards. The nature and extent of treatment will, of course, depend upon the nature and extent of impurities.

The processes selected for the treatment of potable water depend on the quality of the raw water supply. Most ground waters are clear and pathogen-free and do not contain significant amounts of organic materials. Such waters may often be used in potable systems with a minimal dose of chlorine to prevent contamination in the distribution system. Other groundwaters may contain large quantities of dissolved solids or gases. When these include excessive amounts of iron, manganese or hardness, chemical and physical treatment processes may be required. Treatment systems commonly used to prepare potable water from groundwater are shown in Fig. 4.1.

Surface waters often contain a wider variety of contaminants than groundwater and treatment processes may be more complex. Most surface waters contain turbidity in excess of drinking-water standards. Although fast-moving streams may carry larger material in suspension, most of the solids will be colloidal in size and will require chemical coagulation for removal. Depending on the geology of the watershed, hardness may or may not be a problem in surface waters. If low levels of colour and other organic material are present, adsorption onto surface-active material, a process not significant in natural water systems, may be necessary. A wide variety of micro-organisms, some of which may be pathogenic, are also common constituents of surface waters. Treatment systems commonly used in treating surface waters are shown in Fig. 4.2.

Water Treatment Processes: Theory and Application

It is generally convenient to group human use of water into two broad categories depending upon the location of the use relative to the source. In-place use of water includes navigation, recreation, wildlife propagation and the dilution, assimilation and transportation of waste-water. Although hydroelectric power generation requires brief diversion of water through turbine penstocks, this use is also considered an in-place use.

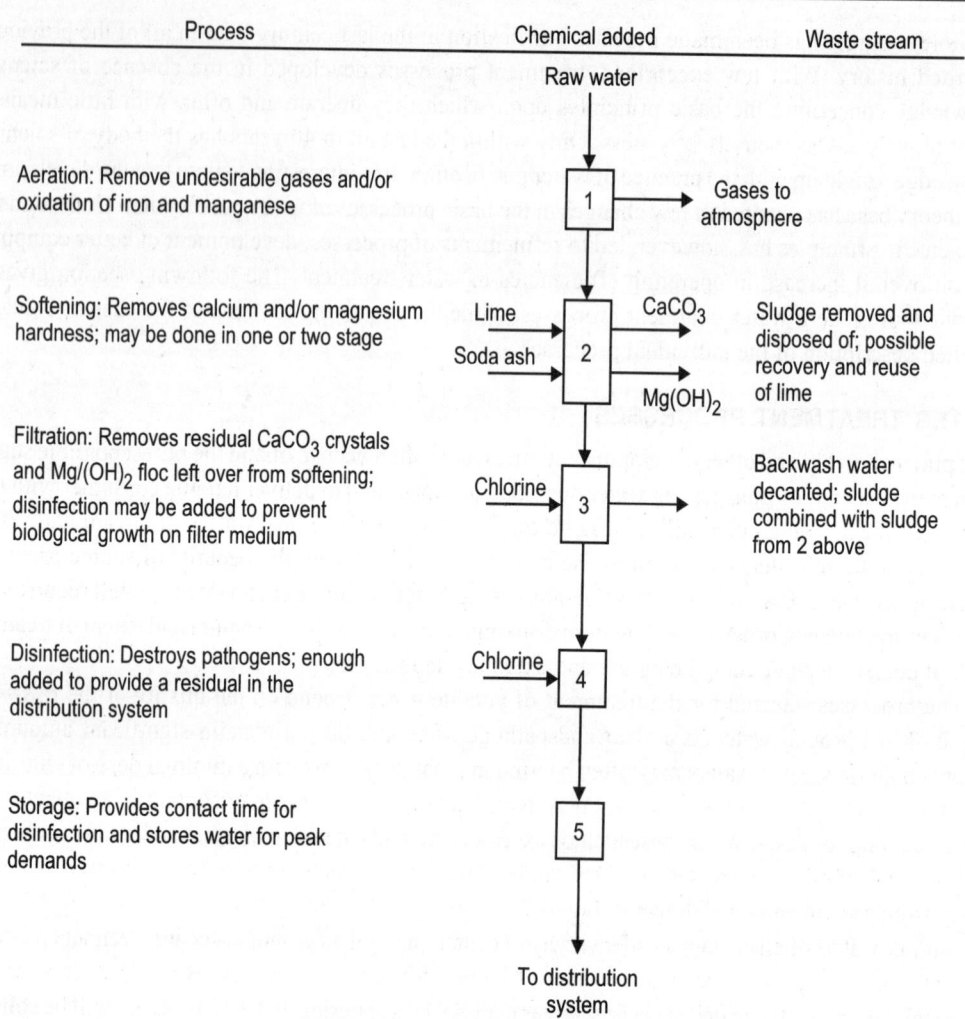

Process	Chemical added	Waste stream

Raw water

Aeration: Remove undesirable gases and/or oxidation of iron and manganese

Gases to atmosphere

Softening: Removes calcium and/or magnesium hardness; may be done in one or two stage

Lime

Soda ash

$CaCO_3$

$Mg(OH)_2$

Sludge removed and disposed of; possible recovery and reuse of lime

Filtration: Removes residual $CaCO_3$ crystals and $Mg/(OH)_2$ floc left over from softening; disinfection may be added to prevent biological growth on filter medium

Chlorine

Backwash water decanted; sludge combined with sludge from 2 above

Disinfection: Destroys pathogens; enough added to provide a residual in the distribution system

Chlorine

Storage: Provides contact time for disinfection and stores water for peak demands

To distribution system

Fig. 4.1. Typical plant treating hard groundwater.

For irrigation and industrial use and for individual and public domestic supplies, water must be withdrawn from streams, lakes or aquifers in the natural hydrologic cycle. The pollutants most deleterious to crops (inorganic salts and metals) are difficult and expensive to remove. The vast quantity of irrigation water used and the low margin of profit associated with farming virtually preclude any treatment of this water. Water not suited for irrigation is simply abandoned and available capital is used instead to secure an alternate source of acceptable quality. Many industries with needs for small amounts of essentially potable water obtain their supplies from public systems. Some industrial water supplies, such as boiler feed water, may require a chemical purity an order of magnitude greater than potable water. Engineering design for treatment of other types of industrial water supplies may also be necessary. Cooling water, particularly that used only once and discharged back to nature, has few quality constraints. Individual domestic supplies are usually drawn from wells or springs of acceptable quality and serve individual homes or farmsteads. Such systems are seldom engineered but are installed and operated by the home owners, perhaps with the advice of the well-driller and the distributor of home water-treatment units.

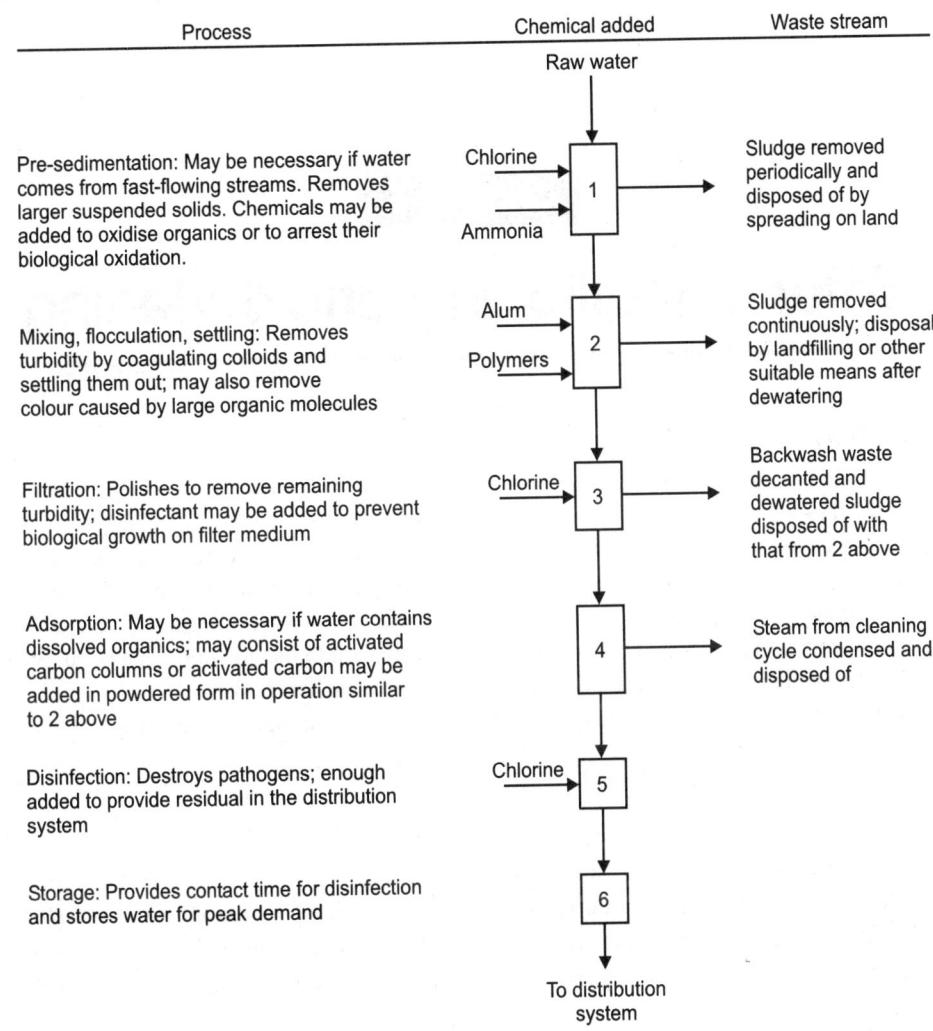

Process	Chemical added	Waste stream

Raw water

Pre-sedimentation: May be necessary if water comes from fast-flowing streams. Removes larger suspended solids. Chemicals may be added to oxidise organics or to arrest their biological oxidation.

Chlorine / Ammonia → **1**

Sludge removed periodically and disposed of by spreading on land

Mixing, flocculation, settling: Removes turbidity by coagulating colloids and settling them out; may also remove colour caused by large organic molecules

Alum / Polymers → **2**

Sludge removed continuously; disposal by landfilling or other suitable means after dewatering

Filtration: Polishes to remove remaining turbidity; disinfectant may be added to prevent biological growth on filter medium

Chlorine → **3**

Backwash waste decanted and dewatered sludge disposed of with that from 2 above

Adsorption: May be necessary if water contains dissolved organics; may consist of activated carbon columns or activated carbon may be added in powdered form in operation similar to 2 above

4

Steam from cleaning cycle condensed and disposed of

Disinfection: Destroys pathogens; enough added to provide residual in the distribution system

Chlorine → **5**

Storage: Provides contact time for disinfection and stores water for peak demand

6

To distribution system

Fig. 4.2. Typical plant treating turbid surface water with organics.

Public water supplies, while only a fraction of the total water use, require by far the largest amount of effort expended by environmental engineers in the water-treatment field.

Chapter 5

Water Conditioning and Softening

INTRODUCTION

Softening of water is the removal of bivalent calcium and magnesium ions (Ca^{+2}, Mg^{+2}). These ions come from the dissolved compounds of calcium and magnesium. Their presence is known as hardness of water. Hardness is defined as the concentration of multivalent metallic cations in solution. At supersaturated conditions, the hardness cations will react with anions in the water to form a solid precipitate. Hardness is classified as carbonate hardness and noncarbonate hardness, depending upon the anion with which it associates. The hardness that is equivalent to the alkalinity is termed carbonate hardness with any remaining hardness being called non-carbonate hardness.

Carbonate hardness is sensitive to heat and precipitates readily at high temperatures.

$$Ca(HCO_3)_2 \xrightarrow{\Delta} CaCO_3 + CO_2 + H_2O \qquad \text{... (5.1)}$$

$$Mg(HCO_3)_2 \xrightarrow{\Delta} Mg(OH)_2 + 2CO_2 \qquad \text{... (5.2)}$$

The multivalent metallic ions most abundant in natural waters are calcium and magnesium. Others may include iron and manganese in their reduced states (Fe^{2+}, Mn^{2+}), strontium (Sr^{2+}) and aluminum (Al^{3+}). The latter are usually found in much smaller quantities than calcium and magnesium and for all practical purposes, hardness may be represented by the sum of the calcium and magnesium ions.

SOFTENING

The reduction of hardness or softening, is a process commonly practised in water treatment. Softening may be done by the water utility at the treatment plant or by the consumer at the point of use, depending on the economics of the situation and the public desire for soft water. Generally, softening of moderately hard water (50 to 150 mg/l hardness) is best left to the consumer, while harder water should be softened at the water-treatment plant. Softening processes commonly used are chemical precipitation and ion exchange, either of which may be employed at the utility-owned treatment plant. Home-use softeners are almost exclusively ion-exchange units.

TYPES OF HARDNESS

Carbonate or temporary hardness is mainly due to bicarbonates (HCO_3^-) of calcium and magnesium. It is called temporary because it can be removed by boiling the water, which converts some of the bicarbonates into insoluble carbonates.

$$Ca(HCO_3)_2 \rightarrow CaCO_3\downarrow + H_2O + CO_2\uparrow$$

Noncarbonate or permanent hardness is caused by soluble compounds other than bicarbonates, such as sulphates, nitrates and chlorides of calcium and magnesium. These compounds are more stable than bicarbonates; they are not removed by boiling the water. Calcium sulphate (gypsum) and magnesium sulphate (epsom) are the common causes of noncarbonate hardness.

Problems Caused by Hardness

Hardness is undesireable for several reasons. For example, hardness is:

1. A nuisance in laundering, due to wastage of soap and collection of dirty precipitate on fibers.
2. A nuisance in bathing.
3. A source of a dirty ring in the tubs and sinks.
4. Responsible for a residue on washed objects like cars and utensils.
5. Responsible for deposits on faucets and shower heads.
6. Responsible for forming a carbonate scale inside the steam boilers.

Impacts

Soap consumption by hard waters represents an economic loss to the water user. Sodium soaps react with multivalent metallic cations to form a precipitate, thereby losing their surfactant properties. A typical divalent cation reaction is:

$$2NaCO_2C_{17}H_{33} + cation^{2+} \rightarrow cation^{2+}(CO_2C_{17}H_{33})_2 + 2Na^+$$

$$\text{Soap} \qquad\qquad\qquad \text{Precipitate}$$

... (5.3)

Lathering does not occur until all of the hardness ions are precipitated, at which point the water has been 'softened' by the soap. The precipitate formed by hardness and soap adheres to surfaces of tubs, sinks and dishwashers and may stain clothing, dishes and other items. Residues of the hardness-soap precipitate may remain in the pores, so that skin may feel rough and uncomfortable. In recent years these problems have been largely alleviated by the development of soaps and detergents that do not react with hardness.

Boiler scale, the result of the carbonate hardness precipitate may cause considerable economic loss through fouling of water heaters and hot-water pipes. Changes in pH in the water distribution systems may also result in deposits of precipitates. Bicarbonates begin to convert to the less soluble carbonates at pH values above 9.0.

Magnesium hardness, particularly associated with the sulphate ion, has a laxative effect on persons unaccustomed to it. Magnesium concentrations of less than 50 mg/l are desirable in potable waters, although many public water supplies exceed this amount. Calcium hardness presents no public health problem. In fact, hard water is apparently beneficial to the human cardiovascular system.

Measurement

Hardness can be measured by using spectrophotometric techniques or chemical titration to determine the quantity of calcium and magnesium ions in a given sample. Hardness can be measured directly by titration with ethylenediamine tetraacetic acid (EDTA) using eriochrome black T (EBT) as an indicator. The EBT reacts with the divalent metallic cations, forming a complex that is red in colour. The EDTA replaces the EBT in the complex and when the replacement is complete, the solution changes from red to blue. If 0.01 M EDTA is used. 1.0 ml of the titrant measures 1.0 mg of hardness as $CaCO_3$.

Use

Analysis for hardness is commonly made on natural waters and on waters intended for potable supplies and for certain industrial uses. Hardness may range from practically zero to several hundred, or even several thousand, parts per million. Although acceptability levels vary according to a consumer's acclimation to hardness, a generally accepted classification is as follows:

Soft   < 50 mg/l as $CaCO_3$
Moderately hard   50–150 mg/l as $CaCO_3$
Hard   150–300 mg/l as $CaCO_3$
Very hard   > 300 mg/l as $CaCO_3$

SOFTENING METHODS

The water-softening methods can be classified as chemical precipitation and nonchemical precipitation methods.

Chemical Precipitation Methods

Lime or lime-soda ash softening method

In this method, soluble calcium and magnesium compounds are converted to insoluble calcium carbonate ($CaCO_3$) and magnesium hydroxide [$Mg(OH)_2$], respectively. For this purpose, lime [quick lime, CaO or slaked lime, $Ca(OH)_2$] and soda ash/sodium carbonate (Na_2CO_3) are added to the water in the coagulation or flocculation basins. Lime is added after the alum and soda ash is applied after the lime. This sequence is important for proper reactions. Alum needs to react with turbidity and precipitate it out before reacting with lime. After the alum and lime reaction, soda ash is added to react with permanent hardness (to prevent its reaction with alum or lime). Lime and soda ash should never be feed through a common line because they will react and plug up the line. Lime–soda ash reactions occur during the flocculation to form a part of the floc. Insoluble calcium carbonate and magnesium hydroxide settle out in the sedimentation basins along with the turbidity.

Lime removes all the carbonate hardness and noncarbonate magnesium hardness. It forms insoluble $CaCO_3$ and $Mg(OH)_2$. The following chemical reactions show the removal of these hardnesses:

$$Ca(HCO_3)_2 + Ca(OH)_2 \rightarrow CaCO_3\downarrow + 2H_2O$$

$$Mg(HCO_3)_2 + 2Ca(OH)_2 \rightarrow Mg(OH)_2\downarrow + 2CaCO_3\downarrow + 2H_2O$$

$$MgSO_4 + Ca(OH)_2 \rightarrow Mg(OH)_2\downarrow + CaSO_4$$

These reactions show that removal of magnesium carbonate hardness, requires twice the amount of lime than the removal of calcium carbonate hardness and magnesium noncarbonate hardness removal produces the equivalent amount of calcium noncarbonate hardness, which needs soda ash for its removal. Soda ash removes the noncarbonate calcium hardness.

$$CaSO_4 + Na_2CO_3 \rightarrow CaCO_3\downarrow + Na_2SO_4$$

Removal of calcium chlorides and nitrates is similar, except that the products are sodium chloride and sodium nitrate. Due to a high amount of lime use, magnesium hardness, removal needs pH above 10.6, whereas calcium is removed above pH 9.4.

These equations are used to determine the lime and soda ash doses corresponding to the degree of hardness removal.

Terminology of lime and lime-soda ash softening treatment based on the removal of various degrees of hardness

1. Partial lime softening uses a small amount of lime to remove the desired amount of calcium carbonate hardness.
2. Lime softening is done with lime only. It is applied when water has only high carbonate hardness. This process requires pH of 9.6 to 9.8.
3. Excess lime softening is the use of excess lime to remove high magnesium and calcium hardness. It needs pH above 10.6.
4. Lime-soda ash softening is the use of lime and some soda ash. This process is used for waters with high calcium carbonate hardness, low magnesium hardness and only some of the noncarbonate calcium hardness.
5. Excess lime-soda ash softening uses excess lime and soda ash to remove high calcium and magnesium carbonate hardness and high noncarbonate hardness.

Dose calculation

It is important to know how much of each chemical is needed to remove the desired amount of hardness. To calculate the hardness removal, determine the total alkalinity, total hardness, calcium hardness and possibly the CO_2 contents of the water. As discussed before, total alkalinity is equal to carbonate hardness and noncarbonate hardness is the difference between total hardness and alkalinity.

Lime also reacts with coagulants, carbon dioxide, iron and manganese. Therefore, an excess amount of lime is needed. This can be determined by jar testing that takes into consideration all the reactants. The jar test simplifies the calculations.

Table 5.1 can be used to calculate the softening chemicals needed to remove hardness. For example, suppose, river water has 300 mg/l total hardness, 200 mg/l total alkalinity and 200 mg/l calcium hardness. How much quick lime and soda ash will be required to remove 100 mg/l of carbonate hardness, 50 mg/l of noncarbonate calcium hardness and 50 mg/l of magnesium carbonate hardness for lime-soda ash softening? All these values are in mg/l as calcium carbonate.

Carbonate hardness – Total alkalinity – 200 mg/l
Noncarbonate hardness = 300 mg/l – 200 mg/l = 100 mg/l
Calcium hardness = 200 mg/l
Thus, magnesium hardness = 300 mg/l – 200 mg/l = 100 mg/l. 100 mg/l of calcium carbonate hardness needs,
(100 × 0.56) mg/l of CaO = 56 mg/l of CaO
50 mg/l magnesium carbonate hardness needs,
(50 × 1.12) mg/l = 56 mg/l of CaO.
Excess lime for magnesium removal and other reactants is an estimated 50 mg/l to raise the pH to 10.6.
Total quick lime required = (56 + 56 + 50) mg/l = 162 mg/l
50 mg/l of calcium noncarbonated needs,
(50 × 1.06) mg/l of Na_2CO_3 = 53 mg/l of $Na_2 CO_3$
Thus, we need 162 mg/l of quick lime and 53 mg/l of soda ash doses. First, run a jar test on this water by using 162 mg/l of lime. After 10 to 15 minutes, add 53 mg/l of soda ash to the same sample and determine the pH (which needs to be above 10.6) and actual hardness removal. Then adjust the dose as required. Apply the dose at plant-scale level and make the adjustments as needed to have the desired results.

Table 5.1. Softening chemicals required to remove hardness.

To reduce 1 mg/l of:	Requires the following amount of (mg/l)		
	Quick lime	*Slacked lime*	*Soda ash*
Noncarbonate hardness			1.06
Carbonate hardness, Ca	0.56	0.74	
Carbonate hardness, Mg	1.12	1.48	
Carbon dioxide	1.27	1.68	

Quick lime = CaO

Slaked lime = Ca (OH)$_2$

Calculated by stoichiometry from the softening equations

It is cheaper to remove calcium hardness than magnesium hardness and cheaper to remove carbonate hardness than noncarbonate; soda ash is a relatively expensive chemical when compared to lime.

Due to the lack of a required detention time and other unknown factors, it is difficult to produce waters with less than 50 mg/l hardness by lime-soda ash softening. In waters softened by this method, there is generally 50 to 80 mg/l total hardness with 30 to 50 mg/l of calcium. Mostly, municipal waters in the regions with high hardness are treated to have 100 ± 20 mg/l total hardness, which is cost effective and practical. Lime-soda ash softening produces large amounts of solids. They are disposed of into a sludge lagoon, into a landfill, into the river, or onto the land.

Water pH, after lime or lime-soda ash softening, is high and the water is very depositing. pH is adjusted by adding CO_2 into the final sedimentation basin. This process is known as recarbonation. pH is adjusted to below 9.3 to avoid any carryover calcium carbonate to precipitate in the filter media.

Permutit or Zeolite Process

In this process, water is softened through a natural or artificial zeolite. Permutit is an artificial zeolite, called as hydrate of sodium aluminium orthosilicate and can be obtained in the form of a coarse sand by fusing together sodium carbonate (Na_2CO_3), alumina (Al_2O_3) and silica (SiO_2). Zeolites, known as green sand are used for water softening, but artificial zeolite, known as permutit is more common and it has general formula.

$Na_2O \cdot Al_2O_3 \cdot n\ SiO_2 \cdot xH_2O$ (n = 5–13, x = 3–4). Permutit or zeolite is insoluble in water, but can act as base exchanger when brought in contact with water containing cations. The zeoloite or permutit is placed in a suitable column as shown in the Fig. 5.1 and hard water containing Ca^{++} and Mg^{++} ions is allowed to percolate through it. This process removes both temporary and permanent hardness.

$$Na_2Z + Ca(HCO_3)_2 \rightarrow 2NaHCO_3 + CaZ \qquad Na_2Z + MgSO_4 \rightarrow Na_2SO_4 + MgZ$$

$$Na_2Z + Mg(HCO_3)_2 \rightarrow 2NaHCO_3 + MgZ \qquad Na_2Z + CaCl_2 \rightarrow 2NaCl + CaZ$$

$$Na_2Z + CaSO_4 \rightarrow Na_2SO_4 + CaZ \qquad Na_2Z + MgCl_2 \rightarrow 2NaCl + MgZ$$

The permutit or zeolite is represented as Na_2Z and after base exchange it is converted into CaZ and MgZ. Sodium present in the zeolite is replaced by divalent Ca^{++} and Mg^{++} ions present in hard water. Water is thus softened as sodium salts do not cause hardness. After some time of use, the whole of zeolite gets exhausted. The zeolite may be regenerated by treatment for some hours with 10 per cent solution of sodium chloride, when the sodium salt of zeolite is formed. The soluble chloride of Ca and Mg passing into the solution can thus be washed away.

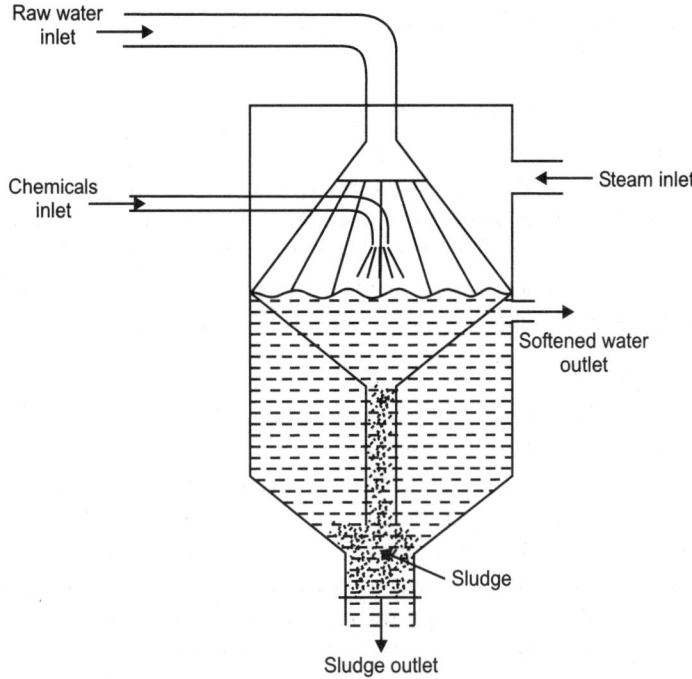

Fig. 5.1. Continuous type hot lime soda softener.

$$CaZ + 2NaCl \rightarrow CaCl_2 + Na_2Z$$

The permutit or zeolite process has become a commercial success, because calcium and magnesium zeolite formed by passing hard water through the bed of sodium zeolite can be easily recovered into sodium zeolite as discussed above. Soft water obtained by this method is used mostly for laundry purpose. Water softened by this process cannot be used for boiler purpose.

Nonprecipitation methods

Membrane softening

Membrane filtration, such as reverse osmosis, removes hardness without producing any residual solids. Currently, membrane filtration is used only for small operations. In the future, it may replace lime-soda ash softening treatment. This is discussed in detail in Chapter 7.

Ion exchange softening

This method uses zeolites (Z) or ion exchange resins that exchange calcium and magnesium ions for sodium or hydrogen ions, receptively. If hard water is allowed to stand in the sodium zeolite Ca^{++} and Mg^{++} ions replace Na + ions and the water is softened. This discussed in detail in Chapter 8.

Iron and Manganese in Water

The presence of other metal ions such as iron (Fe^{+2}) and manganese (Mn^{+2}) also cause hardness, but the amount is insignificant. They are commonly present in water with low pH and no dissolved oxygen. Both iron and manganese can be present in all groundwaters. Water containing more than 0.3 mg/l of iron and 0.05 mg/l of manganese is aesthetically objectionable. Soluble forms of iron and manganese are ferrous (Fe^{+2}) and manganous (Mn^{+2}) compounds. They are removed as insoluble ferric (Fe^{+3}) and manganic (Mn^{+3}) compounds.

Problems caused by iron and manganese

1. Iron and manganese stain clothes and enamel yellow and black, respectively.
2. They are undesirable for bottling, laundries, paper mills, tanning and ice manufacturing.
3. Iron deposits in the distribution systems cause red water complaints and inaccurate meter readings.

These metals can be removed by precipitation and non-precipitation methods similar to calcium and magnesium removal. In all the precipitation methods, they form ferric oxide (Fe_2O_3) and manganic oxide (Mn_2O_3). Ferric oxide is commonly called rust.

Chemical precipitation

1. Lime-soda ash treatment removes iron and manganese while removing calcium and magnesium. They are removed above pH 9.4. They precipitate out as Fe_2O_3 and Mn_2O_3. When hydrated (wet), they are ferric hydroxide [$Fe(OH)_3$] and manganic hydroxide [$Mn(OH)_3$].
2. Aeration is the adding of oxygen into water. It is done by passing air through the water in aeration towers, by cascading, or mechanical aerating. Oxygen reacts with iron and manganese compounds to form ferric and manganic oxides. For effective iron removal, pH should be around 7.5. Manganese oxidises slower than iron and requires a higher pH.
3. Chlorination is the process by which chlorine also removes iron and manganese.
4. Chlorine dioxide and ozone treatment use the disinfectants chlorine dioxide and ozone to remove iron and manganese.

Nonprecipitation method

The zeolite treatment is a non-precipitation treatment. Removal of iron and manganese by the cation exchanger method is similar to and simultaneous with the removal of calcium and magnesium. Water should not be aerated before the zeolite treatment to avoid any accumulation of ferric hydroxide in the zeolite bed.

Stabilisation

Complete removal of hardness cannot be accomplished by chemical precipitation. Under conditions normally prevailing in water-treatment plants, up to 40 mg/l $CaCO_3$ and 10 mg/l $Mg(OH)_2$ usually remain in the softened water. Precipitation of the supersaturated solution of $CaCO_3$ will continue slowly, however, resulting in deposits in water lines and storage facilities. It is, therefore, necessary to 'stabilise' the water by converting the supersaturated $CaCO_3$ back to the soluble form. $Ca^{2+} + 2\ (HCO_3)^-$. Stabilisation can be accomplished by the addition of anyone of several acids. Using sulphuric acid as an example:

$$2CaCO_3 + H_2SO_4 \rightarrow 2Ca^{2+} + 2(HCO_3)^- + SO_4^{2-} \qquad ... (5.4)$$

$$Mg(OH)_2 + H_2SO_4 \rightarrow Mg^{2+}\ SO_4^{2-} + 2H_2O \qquad ... (5.5)$$

The most common practice, however, is to make the conversion with carbon dioxide:

$$CaCO_3 + CO_2 + H_2O \rightarrow Ca^{2+} + 2(HCO_3)^- \qquad ... (5.6)$$

$$Mg(OH)_2 + 2CO_2 \rightarrow Mg^{2+} + 2(HCO_3) \qquad ... (5.7)$$

This process is generally called recarbonation.

If the pH has been raised to facilitate the precipitation of magnesium, it will be necessary to neutralise the excess hydroxyl ions prior to stabilisation. This necessitates a two-stage treatment process. Typical reactions are:

With sulphuric acid

$$Ca^{2+} + 2OH^- + H_2SO_4 \rightarrow Ca^{2+} + SO_4^{2-} + 2H_2O \qquad \qquad \text{... (5.8)}$$

$$2Na^+ + 2OH^- + H_2SO_4 \rightarrow 2Na^+ + SO_4^{2-} + 2H_2O \qquad \qquad \text{... (5.9)}$$

With carbon dioxide

$$Ca^{2+} + 2OH^- + 2CO_2 \rightarrow CaCO_3s\downarrow + H_2O \qquad \qquad \text{... (5.10)}$$

$$2Na^+ + 2OH^{2-} + CO_2 \rightarrow 2Na^+ + CO_3^{2-} + H_2O \qquad \qquad \text{... (5.11)}$$

The pH must be lowered to approximately 9.5 before significant stabilisation occurs.

Softening operations

Softening operations consist of several steps and may be carried out in one or two stages. The operations include mixing of the chemicals with the water, flocculation to aid in precipitate growth, settling of precipitate and stabilisation. Softening systems operate in much the same manner as the systems for coagulating and removing turbidity. Design, criteria, however, are slightly different and are summarised in Table 5.2.

Table 5.2. Typical design criteria for softening systems.

Parameter	Mixer	Flocculator	Settling basin	Solids-contact basin
Detention time*	5 min	30–50 minute	2–4 hr.	1–4 hrs.
Velocity gradient, s⁻¹	700	10–100	NA	†
Flow-through velocity, ft/s	NA	0.15–0.45	0.15–0.45	NA
Overflow rate, gal/min/ft²	NA	NA	0.85–1.71	4.27‡

* This should be confirmed by pilot-plant analysis for each water.

† Velocity gradient in mixer and flocculator component should be approximately the same as in flow-through units.

‡ At slurry blanket-clarifier water interface.

Water with high magnesium hardness is often softened by a process called split treatment. This process bypasses the first-stage softening unit with a part of the incoming water. Excess lime is added to facilitate the removal of magnesium in the first stage and, instead of being neutralised thereafter, is used to precipitate the calcium hardness in the bypassed water in the second stage. Since no magnesium is removed in the bypassed water, the initial magnesium hardness and the allowable magnesium hardness in the finished water govern the quantity that may be bypassed:

$$Q_x = \frac{Mg_f - Mg_1}{Mg_r - Mg_1} \qquad \qquad \text{... (5.12)}$$

where,

Q_x = Fraction of the total flow bypassed

Mg_f = Magnesium concentration in the finished water, 40–50 mg/l (as $CaCO_3$) usually acceptable

Mg_r = Magnesium concentration in the raw water, mg/l

Mg_1 = Magnesium concentration remaining in the fraction of the water receiving first-stage treatment. [As previously stated, practical limits are 10 mg/l $Mg(OH)_2$ (as $CaCO_3$).]

A typical split-treatment system for removing magnesium is shown in Fig. 5.2.

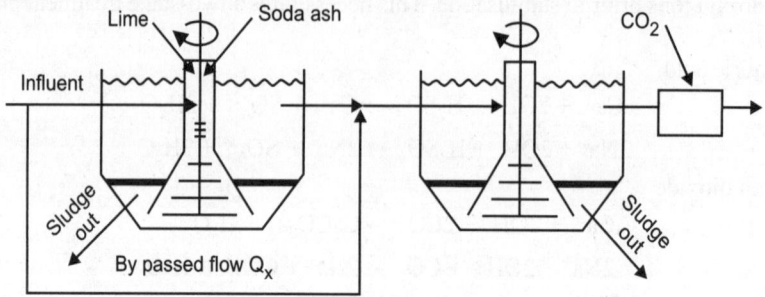

Fig. 5.2. Flow diagram for softening by split treatment.

SOFTENING PROBLEMS AND POSSIBLE SOLUTIONS

Refer to Table 5.3 for some common problems and possible solutions.

Table 5.3. Softening problems and their solutions.

Problems	Possible causes	Possible solutions
Soda ash does not effectively remove hardness	Hardness is gone up in the raw by water	Run hardness test; determine right dose jar test; apply it
	Lime and soda ash react together before they react with the hardness	Check feed lines because they might be feeding too closely. Feed lime first and soda ash later
	Feeder does not feed correctly	Check feeder belt speed; look for any obstruction in feeding system and correct it
Soda ash line plugs up	Mixing of lime slurry with soda ash solution	Check lines and any other possible mixing of these two chemicals. They react and produce calcium carbonate. Calcium carbonate will deposit in the lines
Soda ash solution is milky and gritty	In some way, there is a mixing of some lime with soda ash in the storage bin	Check soda ash bin for any contamination with lime
Lime is not effectively softening	Lime dose is too high. It will be indicated by the light and flakey floc and slightly milky water due to undissolved calcium hydroxide which causes high calcium hardness	Check lime dose by jar testing and reduce it as required
	Coagulant (alum) dose is too low	Run jar test; determine right dose of coagulant; and feed correct dose. Alum will react with lime and reduce hardness
	Quick lime (CaO) is not slaking properly. It is indicated by grit still slaking in the grit drum, which means an improper grade of lime	Ask supplier to provide proper grade of lime

(Contd...)

Problems	Possible causes	Possible solutions
During winter, softening sludge is heavy, gritty, and like white sand	An insufficient alum dose. Especially during winter, an improper dose of alum can cause sandlike calcium carbonate sludge formation that separates from the rest of the lighter sludge. It is easily visible as a bottom layer in a graduated cylinder while running the settlability test. If not removed, this sludge can lock up scrapers	Run jar test and determine optimum alum dose until all the sludge has a uniform density. It does not stratify in the jar test

Chapter 6

Aeration

INTRODUCTION

Aeration occupies a significant place in waste-water quality management and is an important factor in the purification of polluted water. Gas transfer is a physical phenomena in which gas molecules are exchanged between a liquid and a gas at a gas-liquid interface. This physical phenomenon of gas molecules exchanged between the liquid and gas at the liquid-gas interface may also be accompanied by biological, biochemical, biophysical and chemical action. These results are often the primary purpose of the gas transfer operation and methods of achieving the desired results may vary. Principal objectives of aeration, however, usually add or remove gases or volatile substances to water or carry out both objectives simultaneously. In the biological process, aerators function to transfer the required oxygen and include sufficient mixing to maintain uniform dispersed oxygen throughout the basin and keep biological solids in suspension in aerobic basins and the activated sludge process. For high-rate organic loadings, the power required may be determined by oxygen transfer requirements rather than mixing.

AERATION

Aeration is one of the important unit operation of gas transfer. The aim of the aeration is to create extensive, new and self-renewing interfaces between air and water, to keep interfacial films from building up in thickness.

Objectives

Aeration of water is done to accomplish the following objectives:
1. It removes tastes and odours caused by gases due to organic decomposition.
2. It increases the dissolved oxygen content of the water.
3. It removes hydrogen sulphide and hence odour due to this is also removed.
4. It decreases the carbon dioxide content of water and thereby reduces its corrosiveness and raises its pH value.
5. It converts iron and manganese from their soluble states to their insoluble states, so that these can be precipitated and removed.
6. Due to agitation of water during aeration, bacteria may be killed to some extent.
7. It is also used for mixing chemicals with water, as in the Aeromix process and in the use of diffused compressed air.

Types of Aerators

Aeration is done by the following main types of aerators:
1. Free fall aerators or gravity aerators:
    (a) Cascade aerators.
    (b) Inclined apron aerators.
    (c) Slat tray aerators.
    (d) Gravel bed aerators (trickling beds).
2. Spray aerators.
3. Air diffuser basins.

Cascade aerators

Cascade aerators are the simplest of the free fall aerator. Weirs and waterfalls of any kind are cascade aerators (Fig. 6.1a). A simple cascade consists of a series of three or four steps—of concrete or metal. Water is allowed to fall through a height of 1 to 3 meters and due to this it comes into close contact with air. The cascades can be either in open air or may be in a room which has plenty of louvred air inlet. The reduction of CO_2 is usually in the range of 50 to 60 per cent.

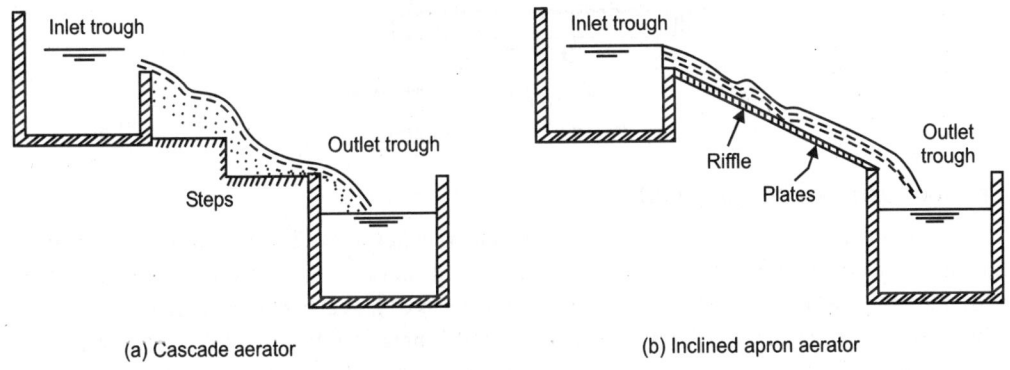

(a) Cascade aerator   (b) Inclined apron aerator

Fig. 6.1. Gravity aerators.

Inclined apron aerator with riffle plates

In this type of aerator, water is allowed to fall along an inclined plane/apron which is usually studded with riffle plates in herring bone fashion. The breaking-up to the sheet of water will cause agitation of water and consequent aeration (Fig. 6.1b).

Slat tray aerators

This is most commonly used. It consists of a closed round or square structure containing a series of closely-stacked superimposed wood-slat trays (Fig. 6.2). Water enters the top of the aerator and is evenly distributed over the topmost tray. The slats in the trays are staggered so that the films of water raining over the edges of the slat in one tray fall on the centers of the slats in the tray just below. Air is supplied to the bottom of the aerator with the help of a blower, which blows it upward. A ventilator is provided at the top, which discharges air and gases to the atmosphere. Water is collected in the collector pan at the bottom, from where it flows to a catch basin or reservoir.

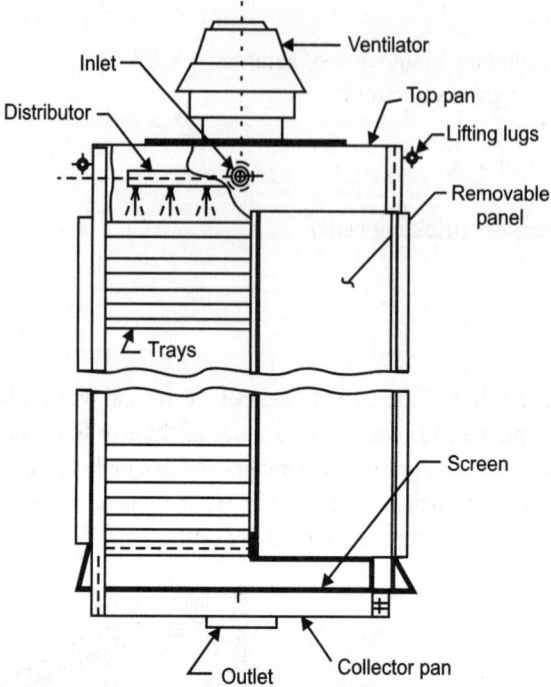

Fig. 6.2. Slat tray aerator.

Gravel bed aerators (trickling beds)

Cascading through beds of coke, limestone or anthracite is believed to have more efficient CO_2 removal than other methods. Figure 6.3 shows a typical gravel bed aerator in which water is applied at the top and trickles down while air is blown upwards. The thickness of gravel bed may be from 1 to 1.5 meters.

In another form, commonly known as trickling beds, three or four trays filled with coke, slag or stone are used. The thickness of bed in each tray is kept about 0.5 to 0.6 meter, while the vertical distance between the bed is kept about 0.5 meter. Water is applied from top through a perforated distribution pipe as shown in Fig. 6.4. During the trickling process, aeration takes place.

Spray aerators

Spray aerators divide the water flow into fine streams and small droplets which come into intimate contact with the air in their trajectory. Water is sprinkled in fine jets through nozzles. It requires considerable head (0.75 to 1.5 kg/cm^2), but it reduces carbon dioxide by 70 to 90 per cent or more.

Air diffusion

In this method, perforated pipe network is installed at the bottom of the aeration tank and compressed air is blown through these pipes. The air bubbles travel upward through water, thus causing aeration. Air diffuser basins have a retention period of about 15 minutes and a depth of 3 to 5 meters.

Alternatively, compressed air may be injected into the flow of water in a pipe or air at atmospheric pressure may be drawn in the pipe where a constriction, such as the throat of a venturi tube reduces the water pressure below atmospheric. However, aeration under pressure does not remove CO_2.

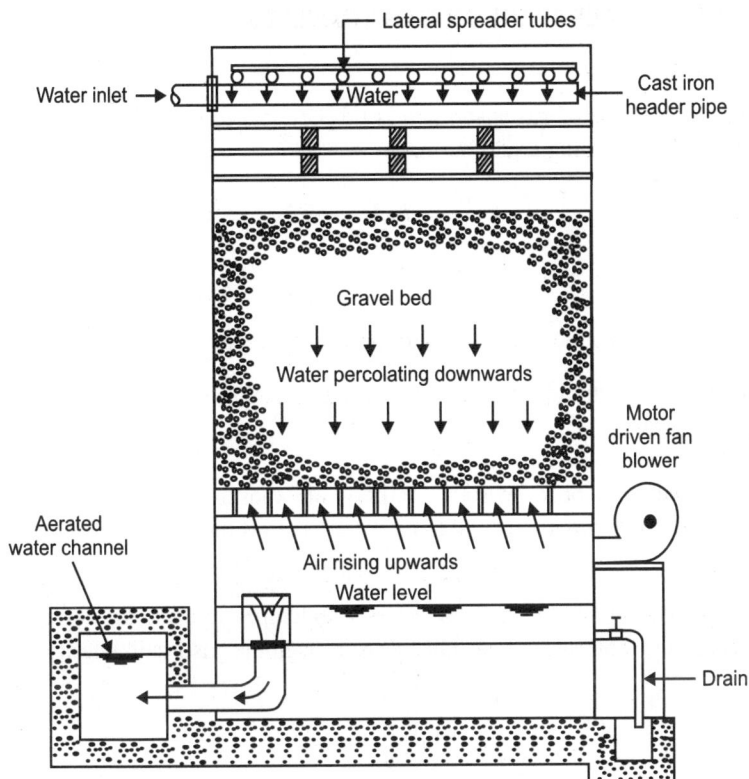

Fig. 6.3. Gravel bed aerator.

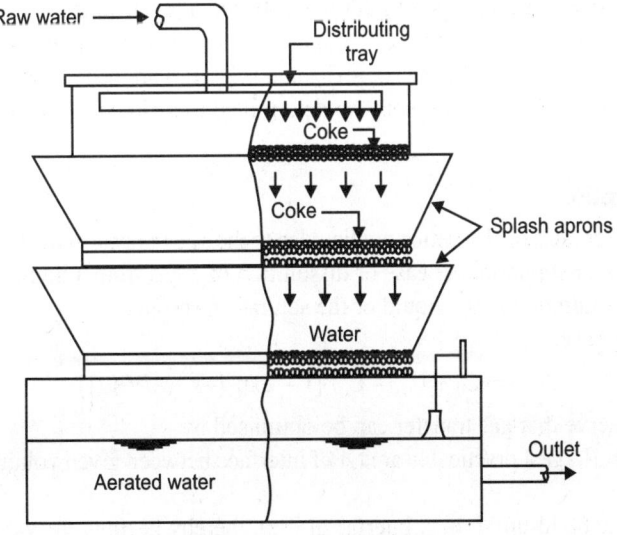

Fig. 6.4. Trickling bed aerator.

FACTORS GOVERNING AERATION OR GAS TRANSFER

Rate of Gas Absorption

If it is postulated that the rate of gas absorption is proportional to the degree of under saturation (or saturation deficit) in the absorbing liquid, we have

$$\frac{dC}{dt} = K_g(C_s - C_t) \qquad \qquad ...(6.1)$$

where,

$\dfrac{dC}{dt}$ = Change in concentration or rate of absorption, transport or transfer at time t.

C_s = Saturation concentration of gas in water, at a given temperature (i.e. maximum concentration in absorption).

C_t = Concentration at time t.

K_g = Proportionality factor for existing condition of exposure.

Integrating between the limits C_0 at $t = 0$ to C_t at $t = t$, we get the basic equation.

$$C_t - C_0 = (C_s - C_0)\,[1 - \exp(-K_g t)] \qquad \qquad ...(6.2)$$

In Eq. 6.2, the proportionality factor K_g increases with temperature and the degree of mixing of the gas with liquid. Since the molecules of gas must pass through the gas-liquid interface, K_g is a function of A/V ratio.

Therefore,
$$K_g = k_g \frac{A}{V} \qquad \qquad ...(6.3)$$

where,

k_g = Gas transfer coefficient, having dimension of velocity.

A/V = Area of interface per unit volume of liquid.

A = Largest practicable area of interface between a given water volume V and air (or gas).

For absorption $C_0 < C_t < C_s$ and both $(C_t - C_0)$ and $(C_s - C_0)$ are positive.

For desorption, $C_s < C_t < C_0$ and both $(C_t - C_0)$ and $(C_s - C_0)$ are negative.

The value of K_g $(= k_g A/V)$ must normally be determined experimentally and verified in plant-scale tests.

Rate of Gas Desorption

Equation 6.2 derived for rates of absorption applies also to the rate of desorption. In contrast to absorption, the rate of desorption, precipitation, release or dissolution of a gas from a liquid becomes proportional to its degree of over-saturation in the liquid or the saturation surplus.

From Eq. 6.2, we have:

$$C_t = C_0 + (C_s - C_0)\,[1 - \exp\{-k_g\,(A/V)\,t\}]$$

From this, we observe that gas transfer can be optimised by:

1. Generating the largest practicable area A of interface between given volume (V) of water and air or gas.
2. Preventing the build-up of thick interfacial film, thereby keeping the value of k_g higher.
3. Inducing longer time (t) of exposure.

4. Ventilating the aerator and its components.

Hence to ensure proper aeration it is necessary:

1. To increase the area of water in contact with air i.e. if the water is sprayed, the smaller the droplets produced, the greater will be the area available. Similarly, if the water is being made to fall as a film over packing material in a tower, the smaller the size of the packing material, the greater will be the area available.

2. To keep the surface of the liquid constantly agitated so as to reduce the thickness of the liquid film which would govern the resistance offered to the rate of exchange of the gas.

3. To increase the time of contact of water droplets with air or increase the time of flow which can be achieved by increasing the height of jet in spray aerator and increasing the height of tower in the case of packed madia.

Introducing factor $f = \left(\dfrac{A}{V}\right) t$, let us compare the efficiencies of spherical bubbles of air, rising up through water, with that of spherical droplets of water falling under gravity.

Spherical droplets of water

In the case of spray aerators, let the spherical droplets of water fall down after attaining a height h. Then, we have

$$h = \tfrac{1}{2} g\, t^2$$

Therefore

$$t = \sqrt{\frac{2h}{g}}$$

therefore

$$\text{Factor } f_s = \left[\left(\frac{A}{V}\right) t\right]_{\text{spherical}} = \frac{6}{d}\sqrt{\frac{2h}{g}} \qquad \ldots (6.4)$$

(since ratio A/V for a sphere is equal to $6/d$ where d is the diameter of spherical droplets).

Spherical bubbles of air

Let the bubbles rise through water column of height h, with a velocity v.

Therefore

$$\text{Time taken, } t = \frac{h}{v}$$

therefore

$$\text{Factor } f_b = \left[\left(\frac{A}{V}\right) t\right]_{\text{bubble}} = \frac{6}{d} \cdot \frac{h}{v} \qquad \ldots (6.5)$$

Hence, ratio

$$\frac{f_b}{f_d} = \left(\frac{6}{d}\frac{h}{v}\right) \div \left(\frac{6}{d}\sqrt{\frac{2h}{g}}\right)$$

or

$$\frac{f_b}{f_d} = \frac{1}{v}\sqrt{\frac{gh}{2}} \qquad \ldots (6.6)$$

For illustration purpose, let the rising velocity of air bubbles be 0.3 m/sec, rising through a height $h = 3$ meters.

therefore
$$\frac{f_b}{f_d} = \frac{1}{0.3}\sqrt{\frac{9.18 \times 3}{2}} \approx 12.4$$

This shows that bubble aeration has an advantage over falling droplets of water.

DESIGN OF GRAVITY AERATORS

In the case of gravity aerators, water is made to fall through the available head h, which can be put to use either in single descent or in multiple descent.

At a given instant, $\dfrac{dh}{dt} = v = g \cdot t$

therefore
$$\int_0^h dh = g\int_0^t t\, dt$$

or
$$h = \tfrac{1}{2}g\, t^2 \qquad \qquad ... (6.7)$$

where,

$\qquad$ h = height of fall, in meters.

$\qquad$ t = time of fall, in seconds.

$\qquad$ v = velocity in m/sec.

$\qquad$ g = acceleration due to gravity, in m/sec^2.

In a single descent through height h,

$$\text{Elapsed time} \qquad t_1 = \sqrt{\frac{2h}{g}} \qquad \qquad ... (6.8a)$$

In n descents through the same total height h,

$$t_n = n\sqrt{\frac{2(h/n)}{g}} = \sqrt{\frac{2nh}{g}} \qquad \qquad ... (6.8b)$$

Hence, we have $\qquad t_n = t_1 \sqrt{n} \qquad \qquad ... (6.8c)$

It should be noted that the quality of exposure is poorer in multiple descents because droplets do not necessarily break away from jets of falling water as soon as they strike the air.

DESIGN OF FIXED SPRAY AERATORS

The following three factors affect the hydraulic performance of fixed-spray pressure aerators:

1. Orifice and nozzle behaviour.
2. Wind effects.
3. Pipe friction associated with multiple take-offs.

Orifice and Nozzle Behaviour

Spray aerators are normally composed of perforated or nozzle pipes which create a spray pattern. The initial spray velocity (v) is given by:

$$v = C_v\sqrt{2gh} \qquad \qquad ... (6.9)$$

where,

C_v = velocity coefficient (≈ 0.95).

h = orifice head or driving head.

For a pipe having multiple openings,

therefore
$$Q = C(Sa)\sqrt{2gh} \qquad \qquad ...\,(6.10)$$

Q = rate of discharge.

C = discharge coefficient.

= 0.8 for rounded openings.

= 0.85 to 0.92 for nozzles.

= 0.6 for sharp edged openings.

Sa = total area of openings

If there are n openings, each of equal area a,
$$\Sigma a = na \qquad \qquad ...\,(6.11)$$

The water rises either vertically or at an angle (α) and falls onto a collecting apron, after moving along a trajectory (Fig. 6.5).

Let h = driving head.

t_r = time of rise of spray.

t = total time of exposure = $2\,t_r$.

α = inclination of jet.

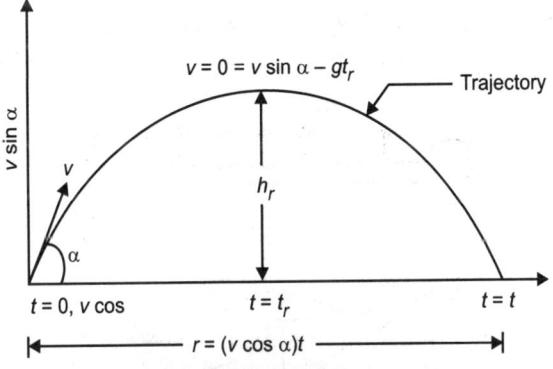

Fig. 6.5. Water rises at an angle α.

At the time of rise t_r, $v = 0$.

Therefore
$$v = 0 = v\sin\alpha - g\,t_r$$

therefore
$$t_r = \tfrac{1}{2}t = \frac{v\sin\alpha}{g}$$

or
$$t_r = \frac{C_v\sqrt{2gh}\,\sin\alpha}{g} = C_v\sqrt{\frac{2h}{g}}\cdot\sin\alpha \qquad \qquad ...\,(6.12)$$

The driving head h, carrying the spray to the top of the trajectory is:

$$h = \frac{g\,t_r^2}{2C_v^2\sin^2\alpha} \qquad \qquad ...\,(6.13)$$

The horizontal carry or radius of circle of spray is given by

$$r = (v \cos \alpha)t = 2v \cdot t_r \cos \alpha$$

$$= 2\{C_v \sqrt{2gh}\}\left\{C_v \sqrt{\frac{2h}{g}} \sin \alpha\right\} \cos \alpha$$

or
$$r = 2 C_v^2 h \sin^2 \alpha \qquad \qquad \text{... (6.14)}$$

The vertical rise h_r of the spray is given by:

$$h_r = \tfrac{1}{2} g \, t_r^2 = C_v^2 \cdot h \sin^2 \alpha \qquad \qquad \text{... (6.15)}$$

Wind Effects

The above analysis is based on the assumption that wind does not affect the trajectory. Though air resistance does not reduce the height or time of rise appreciably, the wind does affect performance of the spray. The distance l by which the spray droplets are carried forward by the wind depends upon the wind velocity (v_w) and the time of exposure and its magnitude is given by:

$$l_w = 2C_D \cdot v_w \cdot t_r \qquad \qquad \text{... (6.16)}$$

where, C_D = coefficient of drag ≈ 0.6.

Friction in Perforated Pipes

Figure 6.6a shows a perforated or nozzled pipe in which the flow decreases stepwise at each opening. Figure 6.6b shows the variation of discharge Q and unit frictional-resistance(s) along the length.

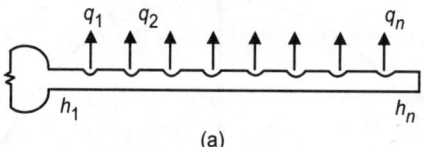

(a)

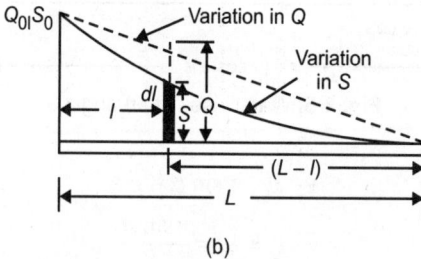

(b)

Fig. 6.6. Uniformly decreasing flow.

For an idealised flow in a slotted pipe,

$$s = kQ^n \qquad \qquad \text{... (6.17)}$$

Also, at distance $(L - 1)$ from the

$$Q = \frac{Q_e(L - l)}{L} \qquad \qquad \text{... (6.18)}$$

where subscript e denotes the entrant flow and resistance to it.

The total head loss h_f is given by:

$$h_f = \int_0^l s \, dl = k\left(\frac{Q_e}{L}\right)^n \int_0^l (L-l)^n dl$$

or

$$= \frac{S_e}{L^n} \int_0^l (L-l)^n dl$$

or

$$h_f = \frac{S_e}{n+1}\left[L - \frac{(L-1)^{n+1}}{L^n}\right]$$ ... (6.19)

Taking $l = L$ and; $n = 2$ for the Chezy formula, we get

$$h_f = \frac{S_e}{2+1}[L] = \tfrac{1}{3} s_e \cdot L$$ ... (6.20)

This shows that the resistance to flow within the perforated section closely equals the resistance that would be offered to the full entrant floor by one-third the length of the perforated pipe. Equation 6.20 is also evident from the fact that area outside the parabola is one-third the area of the pertinent rectangle.

AERATION UNITS

The aeration facilities are designed to meet the calculated oxygen demand of the process while maintaining in the aeration tank a minimum DO of about 1–2 mg/l which is necessary for proper development of biological sludge. In addition to supplying dissolved oxygen, the aeration devices have also to provide adequate mixing and agitation so that the mixed liquor suspended solids do not settle down. This way, aeration increases the contact opportunity between the floc and sewage. To summarise, aeration serves the following three functions: (i) oxygenation of the mixed liquor, (ii) flocculation of the colloids in sewage influent, and (iii) suspension of activated sludge floc.

Following are the three methods which are employed for the purpose of aeration in activated sludge process: (i) diffused air aeration, (ii) mechanical aeration, (iii) combined diffused air and mechanical aeration.

Diffused Air System

In this system compressed air is blown through sewage in aeration tanks. Diffused air aeration involves the introduction of compressed air into the sewage through submerged diffusers or nozzles. The aerators may be of fine bubble or coarse bubble type. In the former, compressed air is released at or near the bottom of the aeration tank through porous tubes or plates made of aluminium oxide or silicon oxide grains cemented together in a ceramic matrix.

The permeability of a diffuser plate is defined as the volume of free air in m^3/min that will pass through 1 m^2 of diffuser at 50 mm differential pressure under dry conditions at temperature of 21°C. Recommended permeabilities lie between 12 and 24, larger rating tending to give uneven distribution.

Standard ceramic plate diffusers have dimensions 0.3 m × 0.3 m × 25 mm with pores of about 0.3 mm diameter. Such plates, under water, pass 1.2 m^3 of air/min/m^2 with pressure losses between about 100 mm and 200 mm of water. For the best uniformity of distribution and prevention of clogging problems, a minimum of 0.6 m^3 of air/min/m^2 of plate surface under water is advisable and should not exceed

2.5 m³ because of pressure losses, poor air economy and more rapid clogging due to corrosion. The air supply, should be at least 0.25 m³ per minute per meter length of channel. Spacing of the diffusers should be 0.6 meter and preferably 1 meter apart between centers to avoid rising streams of bubbles.

Other types of diffusers using porous material having the form of long tubes or mushroom shaped domes installed at a level of 0.3 to 0.6 meter above the floor of the aeration tank are preferred sometimes to the diffuser plates which are placed at the floor level, because of the ease of cleaning and replacement with the unit under operation. Tube diffusers are generally 0.6 meter long with internal diameter as 75 mm and thickness of wall equal to 15 mm.

The 180 mm dome diffusers pass about 0.85–1.4 m³ of air per hour through each dome. Porous diffusers are liable to clogging from the inside by the dust carried in the air and also clogging from the outside by the suspended solids in the sewage. The air supplied must, therefore, be free of dust, by providing air filters.

Coarse bubble aerators consist of proprietary devices such as Monosparj, Deflectofuser, Discfuser, etc. They have slightly lower aeration efficiency than fine bubbles aerators, but are cheaper in first cost and are less liable to clogging and do not require filtration of air. Air diffusers are generally placed along one side of the aeration tank, helping to set up a spiral flow in the tank (Fig. 6.8) which improves mixing and prevents the solids from settling. They are located 0.3 to 0.6 meter above tank floor to aid in tank cleaning and reduce clogging during shutdown.

The quantity of oxygen to be delivered through the diffuser system depends on the oxygen demand of the sewage and the efficiency of oxygen transfer of the diffusers with the latter being controlled by the size of the air bubble produced and the depth of submersion of the diffusers. The oxygen transfer efficiency at 1 to 2 mg/l of DO in aeration tank varies from 5 to 15 per cent for most diffusers with 8 per cent being common for fine bubble diffusers and 6 per cent for coarse bubble diffusers.

The quantity of air to be delivered through the diffuser system can be worked out from the quantity of oxygen to be delivered assuming 23.2 per cent oxygen in air and air density of 1.43 kg/m³ under standard conditions. The air requirements for different types of activated sludge system are given in Table 6.1.

The air delivery systems are designed to deliver 1.5 times the normal air requirements and compressors are installed in multiple units do enable increase or decrease of the air supply. The pressure developed by the air compressors should equal the depth of sewage to the diffuser units plus losses in diffuser plus about 25 per cent extra for losses in transmission or about 0.4 to 0.65 kgf/cm² total. Air-pipings are designed for velocities of 6 to 30 m/s for pipe diameters of 25 mm to 1500 mm. Air header pipes should be located above the tank water level to avoid back siphonage when the compressors trip.

The advantages of flexibility in the aeration arrangements are given particular attention for large plants. Greater flexibility requires costlier air supply piping systems — additional fittings, valves, air flow meters, perhaps less efficient compressor operation, etc. — but the added costs may be justified by improved purification efficiencies. Automatic dissolved oxygen or redox potential measurements may also be provided.

Types of aeration tanks

Following are two types of aeration tanks generally used in the diffused air aeration: (i) ridge and furrow type tank, and (ii) spiral flow type tank.

Table 6.1. Characteristics and design parameters of different activated sludge systems.

Process type	Flow regime	MLSS (mg/l)	MLVSS/ MLSS	F/M	HRT (hr.)	Volumetric loading (kg BOD$_5$/m^3)	SRT (day)	$r = \dfrac{Qr}{Q}$	BOD removal (%)	kg O$_2$ per kg BOD$_5$ removal	Air required per kg BOD$_5$ (m^3)
Conventional	Plug	1500–3000	0.8	0.4–0.2	4–8	0.3–0.7	5–15	0.25–0.5	85–95	0.8–1.1	40–100
Tapered aeration	Plug	1500–3000	0.8	0.4–0.2	4–8	0.3–0.8	5–15	0.25–0.5	85–95	0.7–1.0	50–75
Step aeration	Plug	2000–3000	0.8	0.4–0.2	3–5	0.7–1.0	5–15	0.25–0.75	85–95	0.7–1.0	50–75
Contact stabilisation	Plug	1000–3000* 3000–6000**	0.8	0.5–0.2	0.5–1.5* 3–6**	1.0–1.2	5–15	0.25–1	85–95	0.7–1.0	50–75***
Complete mix.	Complete mix.	3000–6000	0.8	0.6–0.2	3–5	0.8–2.0	5–15	0.25–1.0	85–95	07–1.0	50–75
Modified aeration	Plug	300–800	0.8	5–1.5	1.5–3	1.2–2.4	0.2–0.5	0.05–0.15	60–75	0.4–0.6	25–50
Extended aeration	Complete mix.	3000–8000	0.5–0.6	0.15–0.05	18–36	0.2–0.4	20–30	0.35–1.5	90–98	1.0–1.2	100–125

* In contact aeration tank; ** In sludge reaeration tank; *** Divided equally between contact aeration tank and sludge reaeration tank.

Ridge and furrow type tank

Figure 6.7 shows the ridge and furrow type aeration tank, used with fine bubble aeration through diffuser plates placed in the furrows or depressions. These tanks are in the form of narrow rectangular channels, 30 to 120 meters long, 4.5 to 9 meters wide and 3 to 4.5 meters deep, laid parallel to each other. The diffuser tiles are fixed in the furrow portion by cement or bituminous compounds and are made air-tight by rubber rings. Air is supplied to the diffuser by header pipe through air distributing pipes. The diffuser plates are provided at right angles to the direction of flow. A free board of 60 cm is provided at the top.

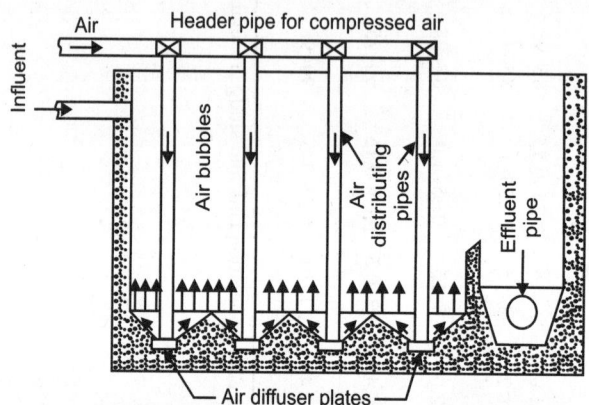

Fig. 6.7. Ridge and furrow type aeration tank.

Spiral flow type tank

In this type of tank, diffused air may be supplied either through plate diffusers placed at the bottom along one side of the tank, or through tube diffusers kept suspended from the top, though tube diffusers are more commonly used. Figure 6.8a and 6.8b show the details of spiral flow type tanks. In the spiral flow tank, the corners are chamfered and the diffusers are placed to one side only.

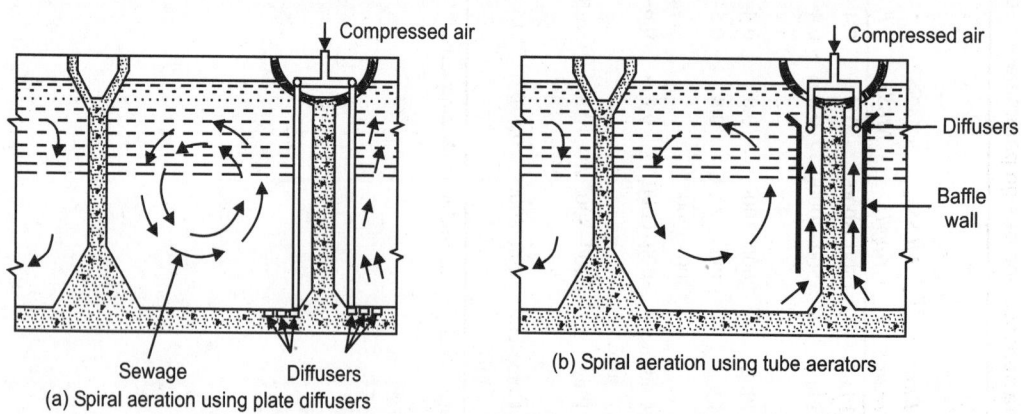

(a) Spiral aeration using plate diffusers

(b) Spiral aeration using tube aerators

Fig. 6.8. Spiral aeration tank using diffusers.

As the bubbles rise, they are deflected by the chamfered corners of the top, creating a spiral motion. This, combined with the logitudinal motion of the sewage flow, causes a helical track and therefore a longer travel. This results in a saving of about 25 per cent of diffusers and compressed air. Also, spiral flow tanks are less costly to build than the ridge and furrow type. Deposition of solids on the bottom is prevented by maintaining a transverse velocity across the bottom of about 0.4 to 0.5 m/s.

Aeration period

The aeration period is the detention time of the raw waste-water flow in the aeration tank, expressed in hours. Period of aeration depends upon the following: (i) strength of sewage and MLSS concentration, (ii) desired degree of purification in term of BOD removal, (iii) rate of aeration, and (iv) proportion of returned activated sludge.

A number of empirical formulae and charts are available to determine the aeration period. The following are two available empirical formulae:

1. American public health association formula

$$T = \frac{L_a}{20} - 1 \qquad\qquad ...(6.21)$$

where, T = aeration time (hours) and
          L_a = BOD of the aeration tank sewage influent (mg/liter) to be removed.

2. M/s Ames Crosta Mills and Co. Ltd. (England) formula

$$T = \left(\frac{L_a}{10}\right)^{3/4} \qquad\qquad ...(6.22)$$

For complete treatment (such that the effluent is fairly stabilised with the presence of nitrates and some dissolved oxygen) a period of 4 to 6 hours is required in America, 10 to 12 hours in Britain and 6 to 10 hours in India. A noteworthy feature of the activated sludge process is the rapidity with which organic matter is oxidised when the sewage is first brought in contact with the active sludge. About 60 per cent of the organic matter is oxidised during the first hour and only 30 to 35 per cent in the next 5 or 6 hours. Generally, aeration time varies from 4 to 8 hours, the common value being 4½ hours.

Volume of returned activated sludge

The volume of returned activated sludge from the secondary clarifier to the aeration tank (or contactor) mainly depends upon the extent of BOD desired to be removed. Table 6.2 gives the volume of activated sludge to be added to remove the desired BOD.

Table 6.2. Volume of returned activated sludge.

Desired BOD removal (ppm or mg/l)	Percentage of returned activated sludge
150	25
250	30
300	35
400	40
500	48
600	53

Figure 6.9 gives an empirical chart to determine the aeration time as well as returned activated sludge, both of which are primarily dependant on the desired BOD removal.

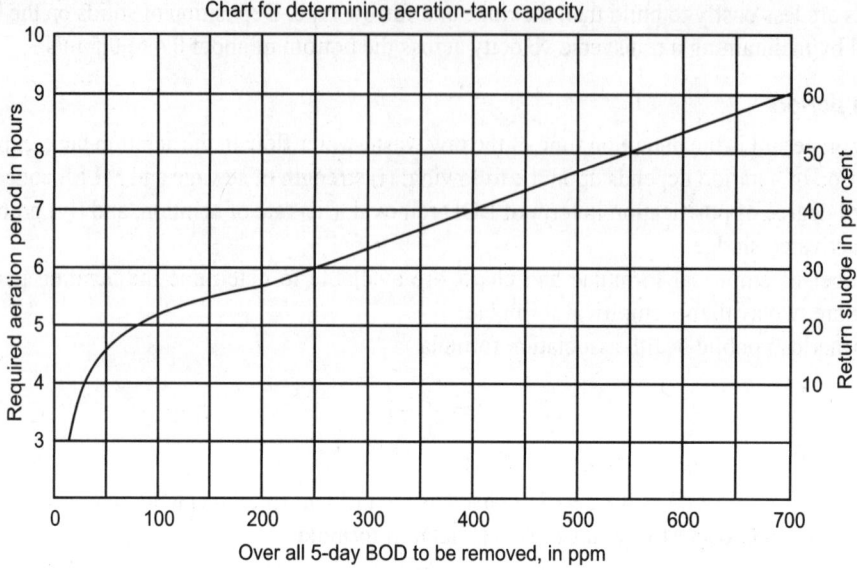

Fig. 6.9. Chart for determination of aeration tank capacity.

Capacity of aeration tank

The capacity (C) of aeration tank depends upon the following three factors: (i) aeration period, (ii) volume of returned sludge, and (iii) volume of flow of sewage.

Out of these, the first two can be determined from the chart of Fig. 6.9. Thus we have,

$$V = (Q + Q_s)\frac{T}{24} \qquad \qquad ... (6.23)$$

where, V = Capacity of aeration tank (m³).
   Q = Volume of flow of sewage (m³/day).
   Q_s = Volume of returned sludge (m³/day).
   T = Aeration period (hours).

Mechanical Aeration Systems

In the diffused air system, the air which is actually utilised in oxidation is only about 5 per cent, while the remainder is required for the purpose of agitating the liquid to meet the mixing requirements. A need was, therefore, led to the development of some cheaper system to achieve the purpose. This has led to many mechanical devices. With the development of efficient mechanical surface aerators requiring very little operational attention, diffused air aeration is falling out of use. Mechanical aerators were linked to small, installations in the past but with recent improvements in their design, they are being increasingly used for large plants in preference to diffused air aeration systems. Some of their advantages are: (i) higher oxygen transfer capacity, (ii) absence of air piping and air filter, and (iii) simplicity of operation and maintenance.

Mechanical aerators generally consist of large diameter impeller plates revolving on vertical shaft at the surface of the liquid with or without draft tubes. A hydraulic jump is created by the impellers at the surface causing air entrainment in the sewage. The impellers also induce mixing. The speed of rotation of impellers is usually 70–100 rpm. The agitator-sparjer is a special mechanical system involving the release of compressed air at the bottom of the aeration tank in large bubbles and breaking up of the bubbles into fine bubbles by submerged turbine rotors located above the air outlets. The turbine rotors also provide mixing.

Mechanical aerators are rated based on the amount of oxygen they can transfer to tap water under standard conditions of 20°C, 760 mm kg barometric pressure and zero DO. The oxygen transfer capacity under field conditions can be calculated from the standard oxygen transfer capacity by the following formula:

$$N = N_S \cdot \frac{C_S - C_L}{9.17} \times 1.024^{T-20} \times \alpha \qquad \qquad \text{... (6.24)}$$

where, N = oxygen transfer under field conditions (kg O_2/h).
N_S = oxygen transfer capacity under standard conditions.
C_S = dissolved oxygen saturation value for sewage at operating temperature.
C_L = operating DO level in aeration tank (usually 1 to 2 mg/l).
T = temperature °C.
α = correction factor for oxygen transfer of sewage, usually 0.8 to 0.85.

The oxygen transfer capacity of mechanical aerators under standard conditions is about 1.9 kg/kwh compared to 1.3 kg/kwh for sparjer system, 1.5 kg/kwh for diffused air fine bubble aerators and 0.9 kg/kwh for diffused air coarse bubble aerators. Oxygen transfer capacities of surface aerators should be supported by actual test data for the model and size offered.

Mixing requirements: The aeration equipment has also to provide adequate mixing in the aeration tank to keep the solids in suspension. Mixing considerations require that the power input in mechanical aerators should not be less than 0.015 – 0.026 kw/m³ of tank volume. The power input of mechanical aerators derived from oxygenation considerations should be checked to satisfy the mixing requirements and increased where required.

Systems of mechanical aeration

Following are some of the patented systems of mechanical aeration:
1. Haworth paddle or Sheffield aeration system.
2. Hartley paddle or Birmingham bio-flocculation system.
3. Simplex aeration system.
4. Link belt aeration system.
5. Kessner brush aeration system.

Haworth paddle or Sheffield aeration system

In the Haworth system, adopted in Sheffield (England), the aeration tank is divided with a number of long and relatively narrow parallel channels by means of thin dividing walls. Two rows of paddles mounted on horizontal shafts, provided about midway between ends of each channel, rotate at a speed of 15 rpm. The rotation of paddles arranged in staggered fashion, cause spiral motion of sewage required for the aeration. The inlet and outlet positions are arranged as shown in Fig. 6.10 and they are cross connected to achieve the recirculation of a part of the flow through the tank. The depth of the tank is kept about 1.2 m and the width of each channel is kept between 1.2 to 2 meters.

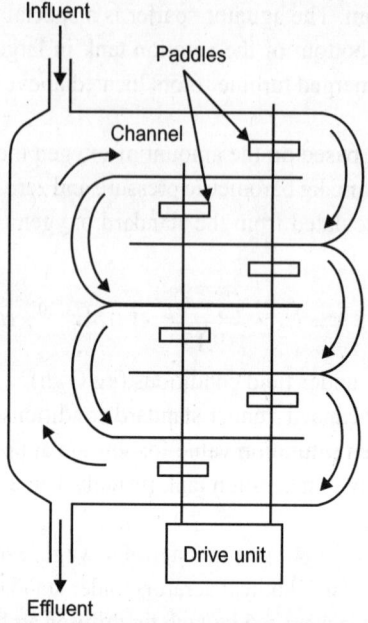

Fig. 6.10. Haworth paddle system.

Hartley paddle system or Birmingham bio-flocculation system

In the Hartley system, adopted at Birmingham (England) the tank is divided into channels and at one of the ends of each bend, partially submerged paddles are provided, inclined at a small angle from the vertical, as shown in Fig. 6.11. These paddles give forward movement to the sewage and at the same time set up a wave action which brings the sewage in intimate contact with the atmospheric oxygen. Diagonal flat baffle plates are provided in each channel, at regular intervals to reduce the velocity of sewage and to maintain spiral flow. This prevents sedimentation of sludge and also increases the efficiency of aeration.

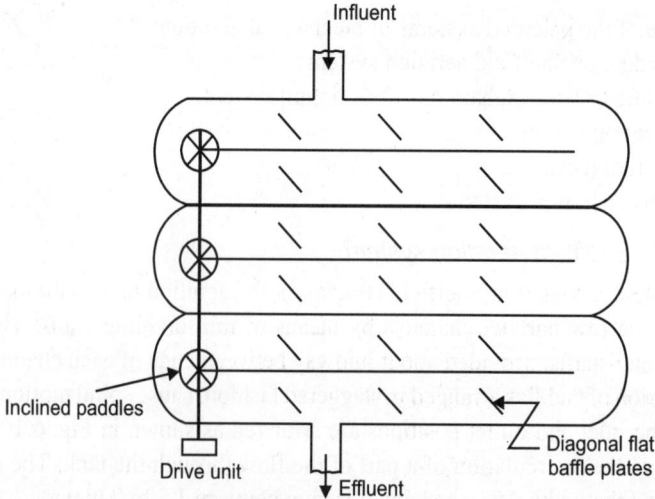

Fig. 6.11. Hartley system.

Symplex system

The symplex system, also known as bioaeration system, consists of deep hopper bottom vertical flow tanks, generally of square shape. At the center of each tank, a hollow uptake or draft tube is suspended from the top, keeping it about 15 cm above the tank floor. There is a steel cone with vanes, which, by rotating by means of gear at the top end of the tube, draws the mixed liquor up through the tube, by the suction effect and sprays it over the surface. The spray absorbs oxygen from the atmosphere and keeps the surface in motion. The speed of rotation is kept about 60 rpm, by means of motor located at the top (Fig. 6.12).

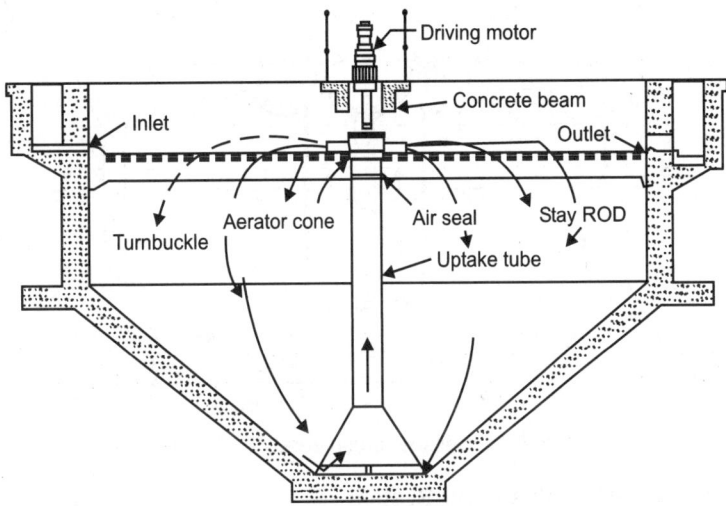

Fig. 6.12. Simplex system.

Link belt aeration system

In this system, the aeration tank is 2.5 to 3 meters deep and 3 to 4 meters wide. The stirring and circulation of mixed liquor are obtained by means of a 75 cm diameter steel paddle wheel suspended near the top, partially submerged and extending the full length of the sides, with blades in the form of narrow ribbons. The speed of rotation is about 48 rpm. A longitudinal vertical baffle wall is provided at a distance of about 45 cm from the wall supporting the above mechanism for aeration, with an opening at the bottom, thus forming a sort of a lift channel. The baffle partition wall carries a narrow trough at right angles at the top (Fig. 6.13). When the paddle wheel is rotated, it pushes the liquid down in the tank which then rises through the opening at the bottom of the partition and up behind the latter into the trough and is forced across the surface producing waves and bringing fresh surfaces in contact with air.

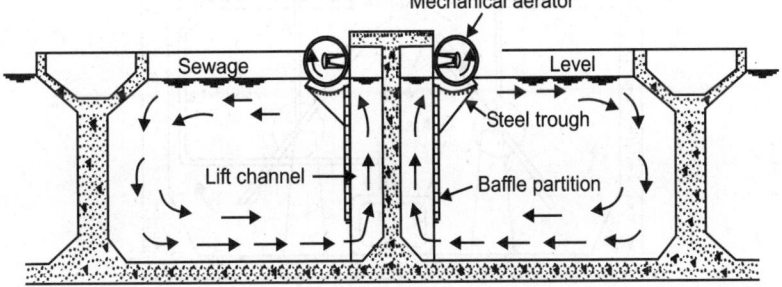

Fig. 6.13. Link-belt aeration system.

Kessner brush aeration system

This process, developed for small installations in Holland, consists of a long tank with an agitating device in the form of partially submerged horizontal brush located at one side of the tank, as shown in Fig. 6.14. Originally the brushes were ordinary street cleaning brushes, but they are now made in the form of stainless steel comb. The brush may be partially submerged to a depth of about 5 mm to 40 mm and is rotated at a speed of about 40 to 50 rpm. The rotation of the brush cause wave action in the sewage to bring about the necessary aeration.

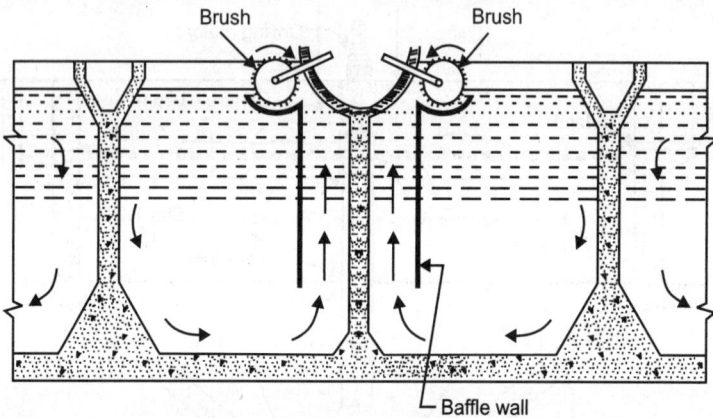

Fig. 6.14. Kessner brush aeration system.

Combined Mechanical and Diffused Air System

In this system, diffused air aeration and mechanical aeration are combined in a single unit, so as to achieve both the efficiency of diffused air system and economy of the mechanical aeration system simultaneously. This reduces the cost of construction and maintenance. In the combined system, aeration of sewage is done by compressed air, using air diffusers placed at the bottom of the tank and the agitation of sewage is carried out by means of mechanical paddles. An example of this is found in Imhoff type aeration tank patented by Dorr Oliver Co. (America), under the name 'Dorr Aerator', shown in Fig. 6.15.

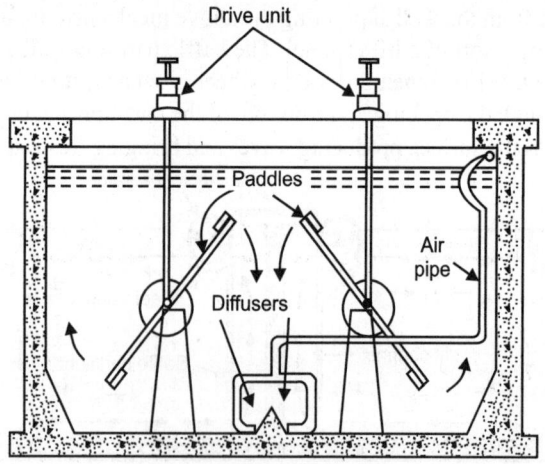

Fig. 6.15. Dorr aerator.

Dorr aerator consists of a tank 3.5 to 4.5 meters deep and about 7.5 meters square, having two rows of air diffusers fixed in the bottom. These diffusers induce upward flow of sewage. There are paddle wheels with two blades each mounted on horizontal shafts, making 10 to 12 rpm in a direction opposite to that of the rising air bubbles. The detention period is 5 hours. For low flows at night and when the sewage is weak, only the paddle wheel may be kept working to prevent the sludge deposition and the diffused air may be completely stopped, with the sole aim of achieving of saving in power.

AERATION TANK LOADING CRITERIA

The loading rates of aeration tank is based on the following four criteria:
1. Hydraulic retention time (HRT).
2. Volumetric BOD loading.
3. Organic loading based on food to micro-organisms ratio (F/M ratio).
4. Solids retention time (SRT) or Mean cell residence time (MCRT) or sludge age.

Hydraulic Retention Time (HRT) or Aeration Period

The aeration period or loading rate expresses the rate at which sewage is applied in the aeration tank. A loading parameter that has been developed empirically over the years is the hydraulic retention time (HRT) which is expressed as follows:

$$\text{HRT (hours)} = \frac{V}{Q \times 1000} \times 24 \qquad \qquad \text{... (6.25)}$$

where, V = Volume of aeration tank (m^3).
Q = Sewage inflow, mLd (excluding sludge recycle).

Volumetric BOD$_5$ Loading

Another empirical loading parameter is volumetric loading which is defined as the BOD$_5$ applied per unit volume of aeration tank, expressed as under:

$$\text{Volumetric load (kg BOD}_5/\text{m}^3) = \frac{Q \times L_a}{V} \qquad \qquad \text{... (6.26)}$$

where, L_a = influent BOD$_5$ to aeration tank, (mg/l).

Organic Loading Based on F/M Ratio

It is an important organic loading criterion in which BOD loading (representing food F to the micro-organisms) is expressed with regard to microbial mass M (represented by MLSS in the aeration tank). The organic loading rate is defined as the ratio of kg BOD$_5$ applied per day (representing microbial food) to kg MLSS in aeration tank (representing micro-organisms), expressed as under:

$$\text{F/M} = \frac{Q \times L_a}{(V/1000)x_t} \qquad \qquad \text{... (6.27)}$$

where, x_t = mixed liquor suspended solids (MLSS), mg/l.

The F/M ratio is the main factor controlling BOD removal. Lower the F/M value, the higher will be the BOD removal in the plant. The F/M ratio can be varied by varying the MLSS concentration in the aeration tank.

Sludge Age: Solids Retention Time (SRT)

Another parameter that could be used for checking the design of activated sludge systems is the solids retention time (SRT), also known as mean cell residence time (MCRT) or sludge age (q_c) defined by the expression:

$$\theta_c = \frac{X}{(\Delta X / \Delta t)} \qquad \qquad ...(6.28)$$

where, X = total microbial mass in a reactor.

$\Delta X / \Delta t$ = total quantity of solids withdrawn daily, including solids deliberately wasted and those in the effluent.

Thus, the mean cell residence time (q_c) or sludge age may be defined as average time for which the mass of suspended solids (or the biological solids) remain under aeration. Though the hydraulic retention time may be only few hours, the residence time of biological solids is much greater and while the sewage (liquid) passes through the aeration tank only once, within the hydraulic retention time (HRT), the resultant biological growth and the extracted organic solids are repeatedly recycled from the secondary settling tank back to the aeration tank thereby increasing the retention time of solids known as solids retention time (SRT) or mean cell residence time (MCRT).

Completely mixed-cellular recycle system

Figure 6.16 shows a completely mixed process of solids recycle applicable to activated sludge systems of any configuration (see Eq. 6.28). The model was specifically developed for completely mixed systems, but is conservative in its prediction of effluent quality for plug-flow systems. In the model, it is assumed that there are no micro-organisms in the influent and that the system contains a unit in which the cells from the reactor are settled and then returned to the reactor. Because of the presence of settling unit, following two additional assumptions are made: (i) waste utilisation by the micro-organisms occurs only in the reactor unit. This assumption leads to a conservative model, since in some systems, there may be some waste utilisation in the settling unit, and (ii) the volume used in calculating the mean cell residence time for the system includes only the volume of the reactor unit (i.e. aeration unit).

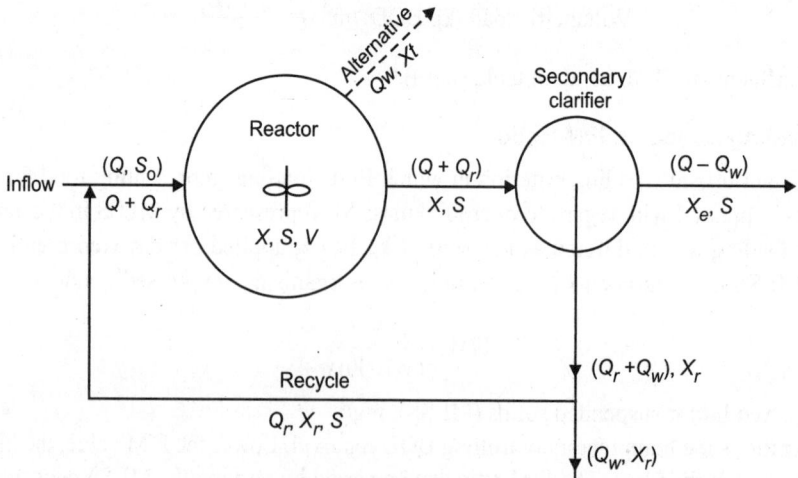

Fig. 6.16. Completely mixed process of solids recycle.

Hence the mean cell residence time (q_c) or the sludge age (days) for this system by definition (Eq. 6.28) is given by:

$$\theta_c = \frac{\text{mass of suspended solids in the system}}{\text{mass of solids/leaving the system per day}} = \frac{X}{\Delta X / \Delta t}$$

Let, Q = the waste flow rate (or volume of flow per day).

X = microbial mass concentration=mixed liquor volatile suspended soils concentration, MLVSS or simply, VSS.

V = volume of the reactor (or aerator).

Q_w = volume of wasted sludge per day.

x_r = clarifier underflow suspended solids concentration (i.e. concentration of solids in the returned sewage or in the wasted sludge, both being equal).

x_e = concentration of solids in the effluent.

OXYGEN REQUIREMENT AND CONTROL

Oxygen is required in the activated sludge process for the oxidation of a part of the influent organic matter and also for the endogenous respiration of the micro-organisms in the system. The former is a function of BOD removed, while the latter is the function of MLVSS in the aeration tank. However, a part of BOD is removed from the system, without being oxidised, in the form of wastage of excess sludge (synthesised biomass) from the system. Hence, the oxygen requirement will be equal to the amount that would be required if all the BOD is to be removed by oxidation only (i.e. the BOD) less a credit for the fraction of BOD removed by sludge wasting.

It has been found from Stoichiometry of cell oxidation that 1.42 kg of O_2 is required per kg of cells.

$$O_2 \text{ required, kg/d} = \left(\frac{\text{total mass of BOD}_L}{\text{utilised, kg/d}}\right) - 1.42 \left(\frac{\text{mass of organisms}}{\text{wasted, kg/d}}\right) \qquad \text{... (6.29)}$$

Also, taking a BOD rate constant = 0.23/day at 20°C

$$\frac{BOD_L}{BOD_5} = \frac{1}{1 - e^{-0.23 \times 5}} = 1.47$$

which gives $\qquad\qquad BOD_L = 1.47 \, BOD_5 \qquad\qquad$ ... (6.30a)

or $\qquad\qquad\qquad BOD_5 = 0.68 \, BOD_L \qquad\qquad$ ... (6.30b)

Hence from Eq. 6.27, we obtain

$\qquad O_2$ required/day = 1.47 (BOD_5 removed/day) – 1.42 (excess sludge waste/day)

or $\qquad O_2$ required/day = $1.47 Q (S_0 - S) - 1.42 V (x/q_c)$ $\qquad$ ... (6.31)

or $\qquad O_2$ required/day = $1.47 Q (S_0 - S) - 1.42 P_x$ $\qquad$ ... (6.31a)

where, S_0, S = substrate concentration at the inlet and outlet of the system respectively, BOD_5 in mg/l.

$\qquad V$ = volume of the reactor.

$\qquad x$ = micro-organism concentration in the reactor, which may be taken equal to MLVSS, mg/l

$\qquad q_c$ = mean cell residence time.

Eckenfelder of O' Connor gave a similar expression for the O_2 requirement, as given below:

$$O_2 \text{ required/day} = a' \, Q \,(S_0 - S) + b' \, V \cdot X \qquad \qquad ... (6.32)$$

where, a' and b' are constants determined experimentally. For municiple sewage, the values of a and b have been reported to be 0.4 to 0.65 and 0.1 to 0.3 respectively.

The formula does not allow for nitrification but allows only for carbonaceous BOD removal. The extra oxygen requirements for nitrification is 4.56 kg O_2/per kg NH_3–N oxidised to NO_3–N.

The total oxygen requirements per kg BOD_5 removed for different activated sludge processes are given in Table 6.2. The amount of oxygen required for a particular process will increase within the range shown in the table as the F/M value decreases.

After having known the total O_2 requirement, the actual quantity of air to be supplied is found by considering the fraction of oxygen in air and the oxygen transfer efficiency of aerators. The specific weight of air, at mean sea level is 1.2 kg/m^3 at 20°C (and 1.16 kg/m^3 at 30°C), while fraction of oxygen in air is 23.2 per cent.

$$\text{Hence theoretical } Q_{air} = \frac{O_2 \text{ demand}}{0.232 \,(1.20)} = \frac{O_2 \text{ demand}}{0.278} \qquad \qquad ... (6.33)$$

For porous tube diffusers, used in conventional activated sludge units, oxygen transfer efficiency is 8 per cent.

Therefore $$\text{Actual } Q_{air} = \frac{O_2 \text{ demand}}{0.278 \,(0.08)} \qquad \qquad ... (6.34a)$$

The oxygen transfer efficiency for coarse bubble diffusers is about 6 per cent.

Therefore $$\text{Actual } Q_{air} = \frac{O_2 \text{ demand}}{0.278 \,(0.06)} \qquad \qquad ... (6.34b)$$

The air supply must be adequate to: (i) satisfy the BOD of the waste, (ii) satisfy the endogenous respiration by the sludge organisms, (iii) provide adequate mixing, and (iv) maintain of minimum dissolved-oxygen concentration of 1 to 2 mg/l throughout the aeration tank.

The following information about air supply is noteworthy:
1. For F/M ratios greater than 0.3, air requirements for conventional process amount to 30 to 55 m^3/kg of BOD_5 removed.
2. For F/M ratios lesser than 0.3, endogenous respiration, nitrification and prolonged aeration periods increase air use to 75 to 115 m^3/kg of BOD_5 removed.
3. For diffused air aeration, the amount of air used has commonly ranged from 3.75 to 15.0 m^3/m^3 at different plants, with 7.5 m^3/m^3 an early rule-of-thumb design factor.
4. The Ten States Standards required the air diffusion system to be capable of delivering 150 per cent of normal requirements, which are assumed to be 62 m^3/kg of BOD in the waste-water applied to the aeration tanks.

AERATION TANK DESIGN CONSIDERATIONS

Capacity

The aeration tank capacity is determined from the F/M and MLSS values selected for the plant. The F/M and MLSS levels generally employed in different types of activated sludge process are given in

Table 6.2 along with their corresponding BOD removal efficiencies. The lower F/M values are recommended also when winter operating temperatures are low and near freezing point. The tank capacity is determined on the basis of organic loading formula. The capacity obtained should be checked against the empirical design criteria for HRT and volumetric loading.

Configuration

Except in the case of extended aeration plants and complete mix plants, the aeration tanks are designed as narrow channels. This configuration is achieved by provision of round-the-end baffles in small plants when only one or two tank units are proposed and by construction as long and narrow rectangular tanks with common intermediate walls in large plants when several units are proposed. In extended aeration plants other than oxidation ditches and in complete mix plants, the tank shape may be circular or square when the plant capacity is small or rectangular with several side inlets and equal number of side outlets, when the plant capacity is large.

Dimensions

The width and depth of aeration channel depends on the type of aeration equipment employed. The depth controls the aeration efficiency usually ranges from 3 to 4.5 meters, the latter depth being formed to be more economical for installations treating more than 50 mLd. Beyond 70 mLd, duplicate units are preferred. The width controls the mixing and is usually kept between 5 and 10 meters. Width-depth ratio should be adjusted to be between 1.2 to 2.2. The length should not be less than 30 m or not ordinarily longer than 100 meters in a single section length before doubling back. The horizontal velocity should be around 1.5 m/min. Excessive width may lead to settlement of solids in the tank. Triangular baffles and fillets are used to eliminate dead spots and induce spiral flow in the tanks. Tank free board is generally 0.5 meter. Due consideration must be given in the design of aeration tanks to the need for emptying them for maintenance and repair of the aeration equipment. Intermediate walls should be designed for empty conditions on either side. The method of dewatering should be considered in the design and provided during construction.

Inlet and Outlet Arrangements

The inlet and outlet channels of the aeration tank should be designed to maintain a minimum velocity of 0.2 m/s to avoid deposition of solids. The channels or conduits and their appurtenances should be sized to carry the maximum hydraulic load to the remaining aeration tank units when anyone unit is out of operation.

The inlet should provide for free fall into the aeration tank when more than one tank unit or more than one inlet is proposed. The free fall will enable positive control of the flows through the different inlets. Outlets usually consist of free-fall weirs. The weir length should be sufficient to maintain a reasonably constant water level in the tank. When multiple inlets or multiple tanks are involved, the inlets should be provided with valves, gates or stop planks to enable regulation of flow through each inlet.

Characteristics and typical applications of air-water contact systems that fall into one of these four groups are summarised in Table 6.3. Some of these systems may be used to contact water with gases other than air and while these uses are listed in Table 6.3.

Table 6.3. Characteristics of gas-liquid contacting systems.

Type of contacting device	Process description	Method of gas introduction	Typical applications	Oxygen transfer rate, kg O$_2$/kWh	Number of transfer units (NTU)	Hydraulic head required, m (ft)	Loading factor
Spray aerator	Water to be treated is sprayed through nozzles to form disperse droplets; typically a fountain configuration. Nozzle diameters usually range from 2.5 to 4 cm (1–1.6 inch) to minimise clogging	Natural aeration through convection	H$_2$S, CO$_2$ and marginal VOC removal; taste and odour control, oxygenation	–	0.5–0.7	1.5–7.6 (5–25)	Surface area of 0.10–0.30 m^2·s/l
Spray tower	Water to be treated is sprayed downward through nozzles to form disperse droplets in a tower configuration; air-water ratio is controlled; typically countercurrent flow	Forced-draft aeration	H$_2$S, CO$_2$ and VOC removal; taste and odour control	–	1–1.5	1.5–7.6 (5–25)	Surface area of 0.10–0.30 m^2·s/l
Packed tower	Water to be treated is sprayed onto high-surface area packing to produce a thin-film flow; process configuration typically countercurrent	Forced-draft aeration	H$_2$S, CO$_2$ and VOC removal; taste and odour control	–	1–4	3–12 (10–40)	
Cascade	Water to be treated flows over the side of sequential pans, creating a waterfall effect to promote droplet-type aeration	Aeration primarily by natural convection	CO$_2$ removal, taste and odour control, aesthetic value, oxygenation	–	0.5–0.7	0.9–3 (3–10)	
Multiple tray	Water to be treated trickles by gravity through trays containing media [layers 0.1–0.15 m (4–6 inch) deep] to produce thin-film flow. Typical media used include coarse stone or coke [50–150 mm (2–6 inch) in diameter] or wood salts	Natural or forced-draft aeration	H$_2$S, CO$_2$ removal, taste and odour control	–	<1	1.5–3 (5–10)	0.007– 0.014 m/s (10 to 20 gpm/ft^2)

(Contd...)

Type of contacting device	Process description	Method of gas introduction	Typical applications	Oxygen transfer rate, kg O_2/kWh	Number of transfer units (NTU)	Hydraulic head required, m (ft)	Loading factor
Low profile (sieve tray)	Water flows from entry at the top of the tower horizontally across series of perforated trays. Large air flow rates are used, causing frothing upon air-water contact, which provides large surface area for mass transfer. Units are typically less than 3 m (10 ft) high.	Air introduced at bottom of tower	VOC removal	–	–	–	Water flow rates less than 0.065 m^3/s (1000 gpm)
Diffuser	Fine bubbles are supplied through porous diffusers submerged in the water to be treated; tank depth is typically restricted to 4.5 m (15 ft)	Compressed air or ozone	Fe and Mn removal, CO_2 removal, taste and odour control, oxygenation, ozonation	0.5	0.5–1.5	–	0.1–1 L air/l water
Dispersed air	Compressed air is supplied through a stationary sparger orifice-type dispersion apparatus located directly below a submerged high-speed turbine	Compressed air or ozone	Ozonation especially when high concentrations of Fe and Mn are present due to clogging of porous diffusers	1.5	1–2		
Hydraulic aspirator	A gas stream is educted into the liquid stream with a venturi-type device	Compressed ozone, CO_2, Cl_2	Ozonation, CO_2 addition, Cl_2 disinfection	1.5–3.5	–	3–6 (10–20)	
Mechanical aspirator	A hollow-blade impeller rotates at a speed sufficient to aspirate and discharge a gas stream into the water	Compressed air or ozone	Ozonation, CO_2 addition	0.7			
Mechanical aerator	Surface aerators (brush or turbine types) and aeration pumps are primary types of mechanical aerators	Mechanical agitation of water into surrounding air	O_2 absorption, VOC removal when < 90% required	1.5–4.5 (turbine) 2.5 (brush)			

LIMITATIONS OF AERATION

Aeration has following limitations:

1. It is not an efficient method of removal or reduction of tastes and odours caused by relatively nonvolatile substances such as oils of algae etc.
2. Odour removal is 50 per cent only when symura was causative organism.
3. Tastes and odours caused by chemicals due to industrial wastes discharged into receiving waters are not satisfactorily reduced.
4. Aeration may add more oxygen in water making it more corrosive while removing iron and manganese.
5. Iron and manganese can be precipitated by aeration only when organic matter is not present.
6. Possibility of air borne contamination in water is there.
7. Additional lime may be required to neutralise the CO_2 that would be removed by aeration.
8. Aeration is economical only in warmer months.

Chapter 7

Coagulation, Flocculation and Filtration

INTRODUCTION

Coagulation and flocculation convert non-settlable turbidity particles into settlable form for their effective removal by gravity.

After pre-sedimentation, these particles are mostly colloidal type. Colloidal turbidity particles are too small (1–100 nm) to settle by gravity. They stay suspended and cause turbidity. Mostly, they are negatively charged. Their removal is accomplished by using substances that make them clump together to form large and heavy particles known as floc that will settle. These substances are known as coagulants. A coagulant is an electrolyte that provides cations (positively charged ions) to precipitate out the negatively charged colloidal turbidity particles. As a rule, the higher the charge on the cation, the more effective is the coagulant. Therefore, commonly used coagulants are aluminium and ferric compounds that provide Al^{+3} and Fe^{+3} cations, respectively. Some other substances are often used to facilitate the coagulants; those are known as coagulant aids. This treatment phase is the second barrier to remove turbidity, waterborne pathogens and other contaminants. It consists of three parts: rapid mixing, coagulation and flocculation.

RAPID MIXING

Rapid mixing is the fast and thorough mixing—flash mixing—of the various chemicals, such as coagulants and coagulant aids, with water for their proper chemical reactions. It has only 30 ± 15 seconds detention time in a small tank. Rapid mixing disperses the chemicals immediately to reach their targets and start their precipitation. Precipitation (separation of a solid from a liquid) in water treatment is known as coagulation. It is the start of the removal of colloidal particles. Rapid mixing is followed by coagulation.

COAGULATION

Coagulation is the precipitation of the colloidal turbidity particles, coagulants and coagulant aids.

Steps of Coagulation

Coagulation occurs in three steps. First, Al^{+3} or Fe^{+3} ions attract a considerable number of negative colloidal turbidity particles. Second, due to aggregation, they form small clumps, called micro-floc. Third, micro-floc, due to its positive charge, still attracts negative ions such as alkalinity (OH^- from lime) and floc compounds precipitate due to their low solubility as shown in Fig. 7.1.

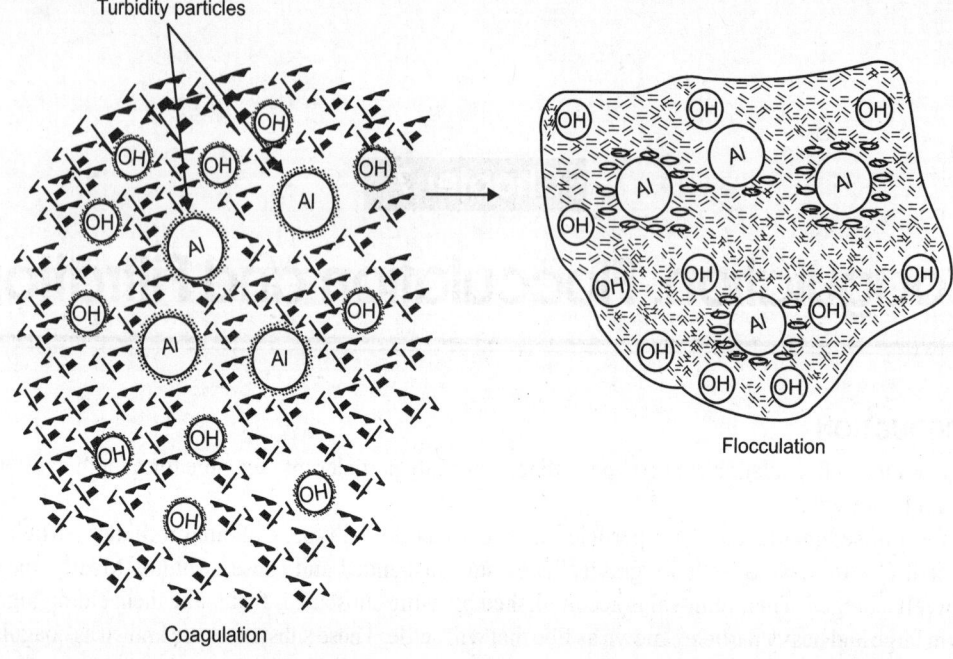

Fig. 7.1. Coagulation and flocculation process.

Chemical reactions:

$$Al_2(SO_4)_3 + 3Ca(OH)_2 \rightarrow 2Al(OH)_3\downarrow + 3CaSO_4$$

Alum   Lime   Floc

$$Fe_2(SO_4)_3 + 3Ca(OH)_2 \rightarrow 2Fe(OH)_3\downarrow + 3CaSO_4$$

Ferric sulphate   Floc

Inorganic salts: Inorganic salts of metals work by two mechanisms in water clarification. The positive charge of the metals serves to neutralise the negative charges on the turbidity particles. The metal salts also form insoluble metal hydroxides which are gelatinous and tend to agglomerate the neutralised particles. The most common coagulation reactions are as follows:

$$Al_2(SO_4)_3 + 3Ca(HCO_3)_2 = 2Al(OH)_3 + 3CaSO_4 + 6CO_2$$

$$Al_2(SO_4)_3 + 3Na_2CO_3 + 3H_2O = 2Al(OH)_3 + 3Na_2SO_4 + 3CO_2$$

$$Al_2(SO_4)_3 + 6NaOH = 2Al(OH)_3 + 3Na_2SO_4$$

$$Al_2SO_4)_3 \, (NH_4)_2SO_4 + 3Ca(HCO_3) = 2Al(OH)_3 + (NH_4)_2SO_4 + 3CaSO_4 + 6CO_2$$

$$Al_2(SO_4)_3 \, K_2SO_4 + 3Ca(HCO_3)_2 = 2Al(OH)_3 + K_2SO_4 + 3CaSO_4 + 6CO_2$$

$$Na_2Al_2O_4 + Ca(HCO_3)_2 + H_2O = 2Al(OH)_3 + CaCO_3 + Na_2CO_2$$

$$Fe(SO_4)_3 + 3Ca(OH)_2 = 2Fe(OH)_3 + 3CaSO_4$$

$$4Fe(OH)_2 + O_2 + 2H_2O = 4Fe(OH)_3$$

$$Fe_2(SO_4)_3 + 3Ca(HCO_3) = 2Fe(OH)_3 + 3CaSO_4 + 6CO_2$$

The effectiveness of inorganic coagulants is dependent upon water chemistry and in particular—pH and alkalinity. Their addition usually alters that chemistry. Table 7.1 illustrates the effect of the addition of 1 ppm of the various inorganic coagulants on alkalinity and solids concentration.

Table 7.1. Coagulant, acid and sulphate—1 ppm equivalents.

1 ppm formula or chemical	ppm Alkalinity reduction	ppm SO_4 increase	ppm Na_2SO_4 increase	ppm CO_2 increase	ppm Total solids increase
$Al_2(SO_4)_3 \cdot 18H_2O$	0.45	0.45	0.64	0.40	0.16
$Al_2(SO_4)_3 \cdot (NH_4)2SO_4 \cdot 24H_2O$	0.33	0.44	0.63	0.29	0.27
$Al_2(SO_4)_3 \cdot K_2SO_4 \cdot 24H_2O$	0.32	0.43	0.60	0.28	0.30
$FeSO_4 \cdot 7H_2O$	0.36	0.36	0.61	0.31	0.13
$FeSO_4 \cdot 7H_2O + (SCl)_2$	0.54	0.36	0.51	0.48	0.18
$Fe_2(SO_4)_3$	0.76	0.76	1.07	0.64	0.27
H_2SO_4 - 96%	1.00	1.00	1.42	0.88	0.36
H_2SO_4 - 93.2% (66°Be)	0.96	0.95	1.36	0.84	0.34
H_2SO_4 -77.7% (66° Be)	0.79	0.79	1.13	0.70	0.28
$NaSO_4$	–	0.64	0.95	–	1.00
$Na_2Al_2O_4$	Increase 0.54	–	–	Reduces 0.47	0.90

Factors Affecting Coagulation

Coagulant

Different sources of water need different coagulants, such as the following:

1. Filter alum: Filter alum, $Al_2 (SO_4)_3 \cdot 14 H_2O$ (aluminium sulphate), a powder, is one of the most commonly used coagulants. It is a good coagulant for hard waters with high alkalinity and pH 5.5–8.0. It is also available as liquid alum (Al_2O_3). Both provide Al^{+3} ions in the water. Normally, coagulation needs 1 mg/l of alum for every 5 NTU turbidity up to 30 NTU; and above that, 1 mg/l for every 10 NTU.
2. Activated alum: Activated alum is alum with about 9 per cent sodium silicate. It works as coagulant and coagulant aid.
3. Black alum: Black alum is alum containing activated carbon. It is applied for certain water with carbon adsorption requirements.
4. Ferric sulphate: Ferric sulphate [$Fe_2(SO_4)_3$] is the second most commonly used coagulant, which works for water with a pH range from 5 to 11.
5. Ferrous sulphate: Ferrous sulphate ($FeSO_4$) is useful for water with high pH (8.5–11).
6. Chlorinated copper: Chlorinated copper as, a mixture of ferric sulphate and ferric chloride ($FeCl_3$), has also been used for water with a pH range from 5 to 11.
7. Sodium aluminate: Sodium aluminate ($NaAlO_2$) is useful for hard water as it works both as a softener and a coagulant. Its solution is alkaline with pH 12.

Coagulant aids

These help the coagulation by creating better coagulation conditions, such as proper pH, alkalinity and particulate nuclei. Some of them act as secondary coagulants.

1. pH adjusting coagulant aids: pH adjusting coagulant aids include lime, sodium carbonate, sodium bicarbonate, sodium hydroxide, hydrochloric acid and sulphuric acid. The first four chemicals raise pH and alkalinity and the last two lower it.
2. Non-pH affecting coagulant aids: Non-pH affecting coagulant aids are substances that provide particulate matter as nuclei for coagulation (e.g. clay, sodium silicate and activated silica). They are also called weighting substances. These aids are useful when the turbidity is low and is hard to remove.
3. Coagulating aids acting as secondary coagulants: Coagulating aids acting as secondary coagulants are polymers, mostly cationic. They attract negatively charged turbidity particles. Depending on the quality of water, anionic or non-ionic polymers may work better for certain water.

pH

Effectiveness of a coagulant is generally pH dependent. Different water requires different coagulants based on its pH. Water with a colour will coagulate better at low pH (4.4–6) with alum.

Alkalinity

It is needed to provide anions, such as (OH⁻) for forming insoluble compounds to precipitate them out. It could be naturally present in the water or needed to be added as hydroxides, carbonates or bicarbonates, as coagulant aids. Generally, 1 part alum uses 0.5 parts alkalinity for proper coagulation.

Temperature

The higher the temperature, the faster the reaction and the more effective is the coagulation. Winter temperature will slow down the reaction rate, which can be helped by an extended detention time. Mostly, it is naturally provided due to lower water demand in winter.

Time

Proper mixing and detention-times are important.

Velocity

The higher velocity causes the shearing or breaking of floc particles and lower velocity will let them settle in the flocculation basins. Velocity around 1 ft/sec in the flocculation basins should be maintained.

Zeta potential

It is the charge at the boundary of the colloidal turbidity particle and the surrounding water. The higher the charge, the more is the repulsion between the turbidity particles, less the coagulation and *vice versa*. Higher zeta potential requires the higher coagulant dose. An effective coagulation is aimed at reducing zeta potential charge to almost 0.

Selection of a proper coagulant and a coagulant aid for a water supply is important. A jar test for water should be run to determine which coagulant and coagulant aid are economical and most effective.

FLOCCULATION

Flocculation is the clumping of microfloc particles to form large particles called floc. It is achieved by the gentle mixing of coagulated water, in tanks known as flocculation basins to allow further clumping of the coagulated matter and turbidity particles, to form large floc particles. Flocculation basins have slow mixing mechanical paddles, known as flocculators and baffles to provide adequate mixing and low velocity. These basins have a velocity about 1 ± 0.25 ft/sec and detention time of 15 to 45 minutes. The best floc is pinhead size and visible a few feet below the surface of coagulated water. Floc particles are heavy enough to settle to the bottom of basin by gravity. The flocculated water flows to the primary sedimentation basin for the next phase, the sedimentation.

The hardness removal is also achieved at this phase of treatment by using lime and soda ash. If chlorine dioxide is used for predisinfection, then chlorite removal can also be done here by using ferrous ions. For the proper chemical reactions, these chemicals should be applied in the following sequence: ferrous ions, alum, lime and then soda ash. The common coagulation and flocculation problems are shown in Table 7.2.

Table 7.2. Coagulation and flocculation problems and their solution.

Problems	Possible causes	Possible solutions
Poor floc formation	Inadequate coagulant dose	Run jar test; determine optimum dose and increase coagulant dose as required
	Improper detention time	Check required detention time by running jar test. Apply needed detention time, if possible, by adjusting flocculator's speed or changing flow rate
Flakey-feathery floc	Excess lime. Lime has a low solubility. Excess lime will precipitate as calcium hydroxide and form light floc	Run jar test; lower lime dose as required
	Inadequate coagulant dose. Coagulants form heavy floc	If excess lime dose is desirable, increase coagulant dose until floc quality is improved
Poor flocculation when optimum dose of coagulant is used	Improper mixing	Check rapid mix and mixer speed; adjust as needed
Poor floc formation under winter conditions with low water turbidity	Not enough turbidity for an effective flocculation	Try some weighting coagulant aid like clay or sodium silicate
	Improper detention time. Low temperature causes slower coagulation which needs longer detention time	Determine optimum detention time with jar test; apply it
Inadequate flocculation of yellowish water	Colour of water is due to decomposition of natural organic substances like leaves	Provide low pH and high dose of coagulant. Alum lowers pH by forming sulphuric acid in the water
Inadequate flocculation of summer water with low turbidity	Drought conditions. A lack of proper dilution factor and high concentration of minerals cause poor flocculation conditions	Run jar test by using alum and a weighting coagulant aid that will increase floc density and rate of coagulation by providing nuclei
Floc settles in coagulation basins	Excessive coagulant dose forms heavy floc	Run jar test to check coagulant dose and lower it as required

(Contd...)

Problems	Possible causes	Possible solutions
	Weighting coagulant aid dose is too high	Run jar test with and without coagulant aid to determine if coagulant aid is needed. Lower or stop feeding coagulant aid as required
	Velocity in basin is too low	Check velocity and flocculator's speed. Increase velocity as needed since too low velocity allows sedimentation of floc in basins

FILTRATION

Filtration is the mechanical removal of turbidity particles by passing the water through a porous medium, which is either a granular bed or a membrane. Filtration's purpose is to remove all the turbidity particles carried over from the sedimentation phase, thus producing a sparkling clear water with almost zero turbidity.

Thus, filtration is a fundamental unit operation that, separates suspended particle matter from water. Although industrial applications of this operation vary significantly, all filtration equipment operate by passing the solution or suspension through a porous membrane or medium, upon which the solid particles are retained on the medium's surface or within the pores of the medium, while the fluid, referred to as the filtrate, passes through.

In a very general sense, the operation is performed for one or both of the following reasons. It can be used for the recovery of valuable products (either the suspended solids or the fluid) or it may be applied to purify the liquid stream, thereby improving product quality or both. Examples of various processes that rely on filtration include adsorption, chromatography, operations involving the flow of suspensions through packed columns, ion exchange and various reactor engineering applications. In petroleum engineering, filtration principles are applied to the displacement of oil with gas (i.e. liquid-liquid separations), in the separation of water and miscible solvents (including solutions of surface-active agents) and in reservoir flow applications. In hydrology, interest is in the movement of trace pollutants in water systems, the purification of water for drinking and irrigation and to prevent saltwater encroachment into freshwater reservoirs. In soil physics, applications are in the movement of water, nutrients and pollutants into plants. In biophysics, the subject of flow through a porous media touches upon life processes such as the flow of fluids in the lungs and the kidney. Although there are numerous industry-specific applications of filtration, water treatment has historically and continues to be the largest general application of this unit operation.

The objective of this section is to provide an overview of filtration terminology and basic engineering principles, as well as calculation methods that describe the filtration process in a generalised way. The basis equations describing the generalised process of filtration have been around for nearly 100 years and with few refinements, continue to be applied to modern design practices.

FILTRATION DYNAMICS

When a suspension of solids passes through a porous media, the solid particles are collected on the feed side of the plate while the filtrate is forced through the media and carried away on the leeward side. A filter medium is, by nature, inhomogeneous, with pores non-uniform in size, irregular in geometry and unevenly distributed over the surface. Since flow through the medium takes place through the pores only, the micro-rate of liquid flow may result in large differences over the filter surface. This implies that the top layers of the generated filter cake are inhomogeneous and, furthermore, are established based on the structure and properties of the filter medium. Since the number of pore passages in the cake is large in comparison to the number in the filter medium, the cake's primary structure depends strongly on the structure of the initial layers. As a result, the cake and filter medium influence each other. Pores with passages extending all the way through the filter medium are capable of capturing solid particles that are smaller than the narrowest cross-section of the passage. This is generally attributed to the phenomenon of particle bridging or, in some cases, physical adsorption. Adsorption is the grouping together of molecules on the surface of a solid or liquid; such 'groupings' are the result of attractive forces between molecules. Activated carbons are highly porous; they contain mazes of interconnecting channels. An imbalance of molecular forces in the walls attracts many substances; these are physically held (adsorbed) by the carbon surfaces. After much use, the carbon may be regenerated and used again. Depending on the particular filtration technique, different filter media can be employed. Examples of common media are sand, diatomite, coal, cotton or wool fabrics, metallic wire cloth, porous plates of quartz, chamotte, sintered glass, metal powder and powdered ebonite. The average pore size and configuration (including tortuosity and connectivity) are established from the size and form of individual elements from which the medium is manufactured. On the average, pore sizes are greater for larger medium elements. In addition, pore configuration tends to be more uniform with more uniform medium elements, the fabrication method of the filter medium also affects average pore size and form. For example, pore characteristics are altered when fibrous media are first pressed together.

Pore characteristics also depend on the properties of fibers in woven fabrics, as well as on the exact methods of sintering glass and metal powders. Some filter media, such as cloths (especially fibrous layers), undergo considerable compression when subjected to typical pressures employed in industrial filtration operations. Other filter media, such as ceramic, sintered plates of glass and metal powders, are stable under the same operating conditions. In addition, pore characteristics are greatly influenced by the separation process occurring within the pore passages, as this leads to a decrease in effective pore size and consequently an increase in flow resistance. This results from particle penetration into the pores of the filter medium. The separation of solid particles from a liquid via filtration is a complicated process. For practical reasons filter medium openings are designed to be larger than the average size of the particles to be filtered. The filter medium chosen should be capable of retaining solids by adsorption. Furthermore, interparticle cohesive forces should be large enough to induce particle flocculation around the pore openings.

GRANULAR MEDIA FILTRATION

A granular media filter, generally, consists of a rectangular concrete structure with 4-feet-deep media formed of sand or a combination of sand, garnet, anthracite (crushed hard coal) and activated carbon (Fig. 7.2). The media are supported by a layer of gravel. Under the gravel is a drain system for the drainage of filter effluent, called filtrate. Mostly, a small amount of cationic polymer is applied to the filter influent for micro flocculation.

Polymer and turbidity particles form a very fine floc that accumulates on the top of the filter media and forms a straining mat (also called a surface cake) that removes the turbidity. Turbidity is removed by two mechanisms, straining and adsorption. Adsorption is acquiring the turbidity particles on the surface of micro floc. Most of the turbidity is removed in the top few inches of media (Fig. 7.3).

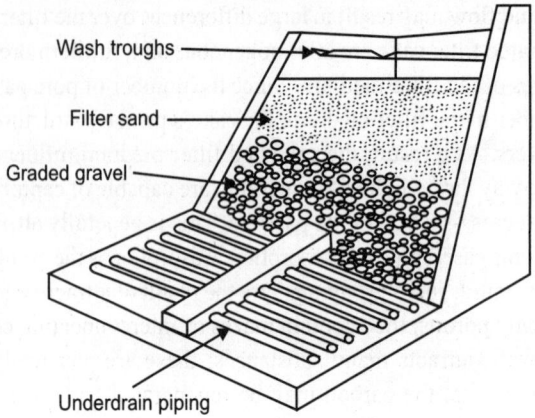

Fig. 7.2. Vertical section of sand filter.

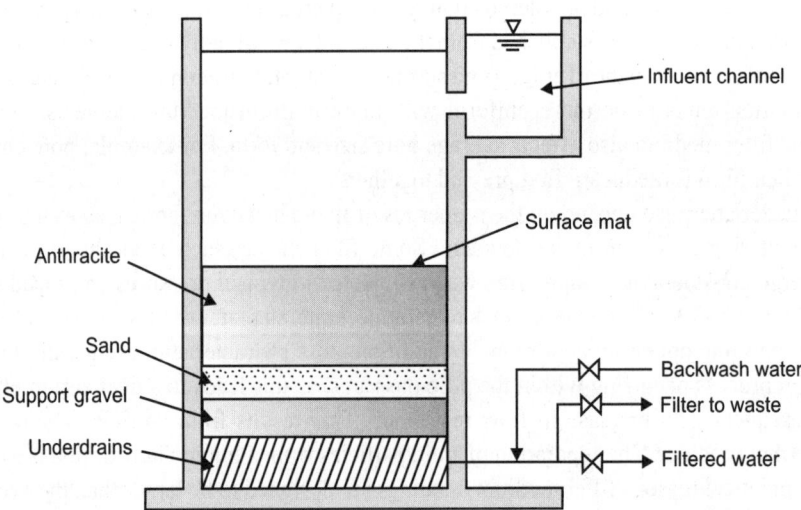

Fig. 7.3. Typical dual-media rapid filter.

There is a slightly high turbidity during the first 10 to 15 minutes of the filtration because the mat is not effectively formed. This is known as the ripening period, after which filtration is adequate. When there is too much build-up of the surface mat and filter interstices are plugged up, the rate of filtration decreases and turbidity starts going up. At this point, the filter needs backwashing.

Backwashing is the removal of filtered-out turbidity by reversing the flow through the filter (i.e., from the bottom upward). The time period from beginning filtration to the filter wash is called a filter run. The period from the start of filtration to the end of the backwashing is called a filter cycle. Turbidity of filter effluent and the resistance to flow, called head loss, are monitored continuously to determine the backwashing time and filter performance. Generally, a washed filter is taken out of service for at least 30 minutes for the proper settling of media before putting it back into operation.

A good filter operation removes more than 99 per cent of the feed water turbidity and produces a sparkling clear water with turbidity as low as 0.1 NTU or less.

Particle Size and Density

Particle size and density of a granular medium is expressed by three parameters: uniformity coefficient, effective size and specific gravity. The first two parameters are determined by sieving a sample of medium through a set of standard sieves with pore size as millimeters (mm). Two sieves are selected, one that allows 60 per cent of the media to pass through and retains 40 per cent and a second one that allows 10 per cent of the media to pass through and retains 90 per cent.

Uniformity coefficient is the ratio of the pore size of the first sieve to the second. Effective size is the pore size of the second sieve. If the pore size that allows 60 per cent of a medium to pass through is 0.75 mm and the pore size of the sieve that allows only 10 per cent to pass through is 0.45 mm, then, uniformity coefficient of this medium is 0.75 mm/0.45 mm = 1.66 and the effective size is 0.45 mm. Specific gravity is the ratio of the density of the medium to the density of water. It determines the vertical stratification of different media in the filter bed, with the lightest at the top and the heaviest at the bottom.

Types of Granular Filters Based on Media, Filtration Rate or Principle of Operation

The following list shows the types of filters:
1. Slow sand filters.
2. Rapid sand filters
3. High-rate sand filters.
4. Granular activated carbon multimedia filters
5. Pressure filters.

Each type of filter will be discussed next.

Slow sand filters

These filters were first used in 1829 to treat the London, England, water supply. A slow sand filter is a covered underground concrete structure with a 3 to 5-foot-deep sand bed and 6 to 18 inches of graded gravel, which has the largest size at the bottom and the smallest at the top (Fig 7.4). Effective size of sand particles is 0.25 to 0.35 mm, with the uniformity coefficient 2.5 to 3.5. Media are supported by the under drain system. The filter cover is at least 6 feet above the media. The filter is operated with 3 to 5-feet-deep water above the medium. Water flows slowly through the medium and leaves most of the turbidity particles in the top layer. Loading rate is 0.03 to 0.06 gallon per minute per square foot (gpm/ft^2) of the filter surface.

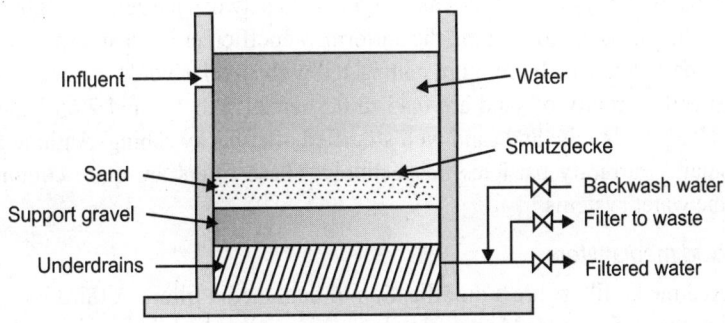

Fig. 7.4. Slow sand filter.

Turbidity particles form a surface mat that becomes sticky due to microbial activity. This mat is called *smutzdecke*, which is very effective to remove particles by straining, adsorption and microbial metabolism. After the filter run, which could be several days or even weeks, the filter is taken out of service and cleaned. For cleaning, the top layer of sand is scraped, washed and stored for replacement. The filter is cleaned several times by scraping the surface layer before replacing any sand. For an effective filtration, the minimum required depth of sand is 2 to 2.5 feet. There is no backwashing in these filters. These filters are effectively used for direct filtration of source water with very low (less than 1 NTU) turbidity such as pristine mountain streams or reservoirs.

Rapid sand filters

Unlike the slow sand filters, surface loading in these filters is 2 to 4 gpm/ft.^{2} and there is backwashing after the filter run. Sand depth, in these filters, is 2 to 3 feet. The particles have an effective size of 0.35 to 0.55 mm and uniformity coefficient of 1.6. Medium is supported on 18 inches of gravel, which is graded from 4 inches to pea size. The under drain system has a Leopold or Wheeler-type false bottom for an effective drainage of the filter effluent. To facilitate the uniform flow of the water, the Leopold system has blocks with small holes and the Wheeler system has conical rectangular cavities with balls. During filtration, there are about 30 inches of water above the medium. Free board, the distance between the surface of the medium and the lip of the backwash trough, is 24 to 27 inches to prevent any loss of medium during the backwashing. Filtration takes place in the top few inches of the medium. These filters are used to filter water with influent turbidity up to 5 NTU.

High-rate sand filters

Rapid sand filters can be modified to create high-rate sand filters. A coarser and lighter layer of anthracite is applied above the sand to allow the turbidity particles to penetrate deeper into the media. Due to deeper penetration of particulate matter, these filters allow a higher rate of filtration, longer filter runs and an effective and economical filtration. These filters are operated at 5 to 10 gpm/ft.^{2} loading. They have two or three media stratified according to their size, shape and specific gravity. The lightest and coarsest medium is at the top and the finest and heaviest medium is at the bottom. There are two types of high-rate filters: dual media and multimedia filters.

Dual media filters

Dual media filters have two media, which are anthracite and sand. Generally, the filter bed from top to bottom is formed of 18 to 30 inches of anthracite, 12 inches of sand and 12 inches of gravel. Anthracite, the crushed hard coal with angular particles and larger voids between particles, is lighter than sand. The effective size of anthracite is 0.6 to 0.7 mm, the uniformity coefficient is 1.6 and specific gravity is 1.55. Sand has rounded particles, which are more compacted with smaller voids. Effective size, uniformity coefficient and specific gravity of sand are 0.45 to 0.5 mm, 1.5 to 1.7 and 2.65, respectively. These specific gravities keep anthracite and sand well stratified after backwashing. Anthracite and sand trap the larger and smaller turbidity particles, respectively. These filters are quite common and popular among most of the water systems.

Triple media/mixed media filters

Triple media/mixed media filters are a modification of dual media filters. A third layer of the heaviest medium is applied under the sand. Mostly, this layer is garnet, which is heavier and finer than sand. Garnet has effective size of 0.2 to 0.3 mm and specific gravity of 4.2. From top to bottom, a typical triple media filter has 36 inches of anthracite, 18 inches of sand, 8 inches of garnet and 8 inches of gravel. Garnet removes the smallest turbidity particles. There is some mixing of the media at the interface of adjacent layers, which makes them mixed media filters.

Granulated activated carbon (GAC) multimedia filters

GAC filters have a layer of activated carbon on top of anthracite or sand. Activated carbon adsorbs various contaminants, such as tastes and odour-causing organics, THMs and synthetic organics. GAC is lighter than sand or anthracite and has an effective size of 0.55 to 0.65 mm with a uniform coefficient of 2.4. These filters have the problem of losing some carbon during the backwashing; therefore, backwashing is properly controlled to prevent the excessive loss of GAC. Commonly, backwashing causes 1 to 6 per cent GAC loss per year.

All granular media filters discussed to this point are gravity flow filters.

Pressure filters

In these filters, media are enclosed in a cylindrical steel tank and the water is forced under pressure through the filter. Media are either sand or diatomaceous earth.

Rapid sand pressure filter

Rapid sand pressure filter has an 18 to 24 inches-thick sand layer with gravel underneath. The filter rate is 2–5 gpm/ft^2. Being small, their use is limited to some industries and recirculation of swimming pool water.

Diatomaceous earth pressure filters

Diatomaceous earth pressure filters have diatomaceous earth medium. Diatomaceous earth is a light medium formed of commercially available diatom fossils with particle size of 5 to 50 micrometers (μm). As compared to several inches of sand, thickness of this medium is only 0.06 to 0.12 inches. Turbidity particles are retained on the surface and there is hardly any penetration of them into the medium. Generally, the filtration rate is 1 gpm/ft^2. These filters are used by the small water systems for low turbidity source water.

Filter Backwashing

There is no standard criterion for backwashing of a filter. Mostly, it is decided by the performance of the filter from effluent turbidly, head loss and filter run. For example, turbidity should not be more than 0.1 NTU, head loss should not be more than 6 feet (pressure as water height in feet) and filter run no longer than 24 hours. These are general guidelines, which vary from plant to plant.

Filter backwashing procedure

Following is a general step-by-step procedure for the manual filter wash:
1. Close the influent valve and let the water level drop to about 4 to 6 inches above the medium.
2. Close the effluent valve.
3. Gradually, start the surface wash system, which will loosen the surface mat of suspended material. Surface washing is done by revolving jets, by compressed air scrubbing or by mechanical rakes.
4. Open the backwash water valve gradually to prevent the media waste.
5. Open the waste-water drain valve. Wash until wash water is quite clean. Proper cleaning may take up to 10 minutes.

6. Stop the surface wash at least 2 minutes before closing the wash water valve.
7. Close the waste-water drain valve.
8. Let the media stratify properly.
9. To put the filter back in service, open the influent valve and then the effluent valve.

Modify this general procedure as required for a particular utility.

For uniformity of washing, a large number of plants have an automatic filter backwash system, which works fine; it needs to be monitored to make the necessary changes in its programming.

Proper backwashing requires about 50 per cent expansion of the fluidised (suspended) sand. The percentage of sand expansion is calculated by using a stick with small panes at different heights. The stick is placed on top of the media while washing the filter. The highest pane that gets some sand is the point to which sand is expanded. The percentage of expansion is the per cent of sand particles rising. It is the rise of sand divided by the depth of the sand media and then multiplied by 100. For example, if the rise is 15 inches and the depth of media is 30 inches, the expansion is (15 inch/30 inch) × 100 = 50 per cent.

Backwash volume should not be more than 2.5 per cent of the total water filtered.

Factors affecting granular media filtration

1. Turbidity: The less the turbidity in the filter influent, longer the filter run and better is the performance.
2. Media form: The coarser the media the less is the head loss, the longer is the run and *vice versa*.
3. Depth: The deeper the bed, the better is the filtration.
4. Backwashing: Proper backwashing is an important factor in the proper operation of a filter. Improper washing can cause the loss of media, mixing of media, formation of mud balls, cracks and craters. All these factors cause an inadequate filtration and a high-effluent turbidity.
5. Filtration rate: The higher the loading, the shorter the filter runs and less efficient is the filter.
6. Temperature: The higher the temperature, the better is the performance.
7. Water stability: In the lime softening plants, higher pH (above 9.3) and higher calcium carbonate content of water can cause deposition of calcium carbonate on the media particles. This build-up of calcium carbonate causes swelling of media and the formation of mud balls. Water needs to be stabilised by lowering the pH below 9.3. A controlled small amount of a polyphosphate, such as sodium hexametaphosphate, is applied as a sequestering agent to further correct this situation. Too much of a polyphosphate can cause excessive sloughing of calcium carbonate from the media particles, which causes higher turbidity and too little may not be enough for an adequate sequestering.
8. Polymer dose: A small dose (0.5–0.75 mg/l) of a polymer is helpful in forming a micro-floc mat to aid the filtration. A higher dose causes cracks in the filter mat and a lower dose does not form an effective microfloc.

MEMBRANE FILTRATION

This process is the passing of pre-treated water under pressure through a membrane to remove specific sized particles. A membrane is a very thin paperlike structure. Membranes can achieve the degree of treatment comparable to a conventional treatment plant. Membrane treatment is one of the best treatment technologies to meet the present and expected safety for drinking water act (SDWA) challenges. It is capable of removing most of the regulated contaminants.

Membrane Structure

Membranes are either hollow fine fiber (HFF) or spiral wound (SW) structures formed of cellulose acetate and synthetic materials, such as polypropylene or polyfuron.

Hollow fine fiber membrane

Hollow fine fiber membrane is a hollow fine hairlike tubular structure called fiber. The fiber is folded in a U form. A bundle of thousands of these U-like fibers is packed in a tubular pressure vessel, which increases the filtering surface tremendously. These pressure vessels, the tubes, make the hollow fiber membrane system very compact and efficient. Water flows through the membrane leaving concentrated water, called concentrate (retentate or waste) on the influent side and filtered water, called permeate on the other side of the membrane. Flow-through the membrane can run inside out or outside in. They are called cross flow and transverse flow, respectively. A cross-flow membrane has concentrate inside and permeate outside, while a transverse-flow membrane is just the opposite. Hollow fine fiber is a popular and commonly used form of membranes in water systems (Fig. 7.5).

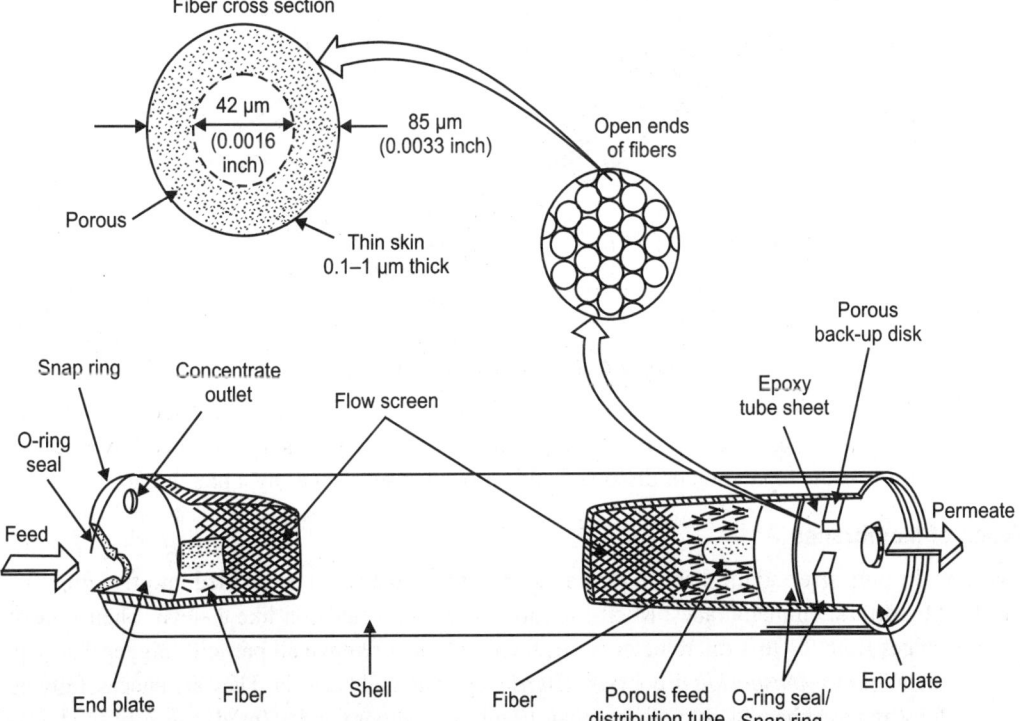

Fig. 7.5. Hollow fine fiber membrane.

Spiral wound membrane

Spiral wound membrane is a flat sheet rolled around a central tube, which collects permeate (Fig. 7.6). The sheet has an outside active membrane and a thick fibrous support or backing underneath. Usually, two membranes are placed back to back with a separator in between. They are rolled together to form a membrane element. These elements are housed in a cylindrical pressure vessel.

Mechanism of Particle Removal

There are three basic mechanisms: sieving, selective diffusion and charge repulsion:

1. Sieving: Each membrane has a uniform and specific pore size. All particles bigger than the pore size are sieved out/rejected. The smallest rejected particle is slightly bigger than the pore size. The smallest molecular weight that will be strained out is known as a cut-off molecular weight (COMW).

2. Selective diffusion: Passing only selected dissolved particles through the membrane is selective diffusion. For selective diffusion, a membrane needs to be semipermeable, which means it will allow only certain chemicals to diffuse through and all others will be rejected (e.g. reverse osmosis membranes).

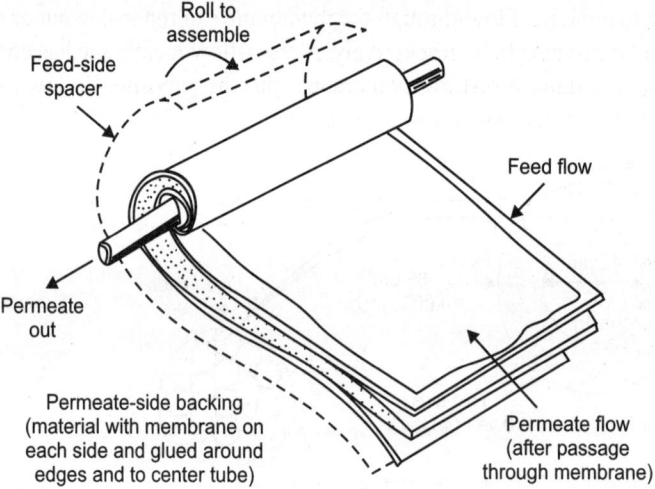

Fig. 7.6. Spiral-wound membrane.

3. Charge repulsion: Filtration uses a direct electric current; anions go to the anode and cations go to the cathode, e.g. electrodialysis membranes. Electrodialysis is dialysis aided by electrodes. Dialysis is the separation of dissolved and suspended substances by a membrane.

Types of Membranes

Based on the pore size and mechanism of functioning, membranes can be divided into five groups:

1. Microfiltration membranes: Microfiltration membranes function like a sieve. Their pore size ranges from 0.1 to 1 micrometer (µm); therefore, they remove all particles bigger than 1 µm, including *Cryptosporidium* oocysts, *Giardia* cysts and all bacteria. They are successfully used for water treatment plants with less than 12 million gallons per day (mgd) capacity and low raw water turbidity.

2. Ultrafiltration membranes: Ultrafiltration membranes are similar to the microfiltration membranes except the pore size is 0.003 to 0.1 µm to remove very small particles. They remove all particles bigger than this pore size, including viruses and THM formation precursors. They remove cysts and other pathogens by six logs, meaning 99.9999 per cent removal. They are more expensive due to the smaller pore size. The finer the pore size, the more effective the membrane and the more expensive it is.

3. Nanofiltration membranes: Nanofiltration membranes have nanometer (0.001 μm or 1 nm) pore size. They remove all the particles above nanometer size. Besides removing viruses, cysts and bacteria, they remove some dissolved substances.

4. Reverse osmosis membranes are semipermeable: They function as sieves and selective diffusion membranes due to osmosis, which allows some specific dissolved substances to pass through. Osmosis is the passage of water through a semipermeable membrane from the lower concentration of the dissolved substances to the higher concentration to equalise the concentration on both sides of the membrane. The force with which water flows through the membrane is called osmotic pressure. The greater the difference in concentration on two sides of the membrane, the higher the osmotic pressure and faster is the flow. In reverse osmosis, a pressure higher than the osmotic pressure is applied on the higher concentration side to force the water through the membrane in the reverse order.

These membranes will remove all the suspended particles larger than the pore size and only selective dissolved substances. These membranes remove substances like sodium, calcium, magnesium and other metal compounds. They are used to treat seawater that has total dissolved solids (TDS) in the range of 3.5 per cent (35000 mg/l) and other brackish water (TDS up to 3 per cent). Reverse osmosis has been used since 1960 for desalting brackish water. In the United States, the first municipal brackish water treatment plant was built in 1971 at Ocean Reef Club, Florida, to treat 0.6 mgd flow. At present, there are more than 100 reverse osmosis drinking water plants in the United States. It is a common process in Middle Eastern countries to treat salty water for drinking. Reverse osmosis is also used by the carwash industry to reduce the total dissolved solids in the municipal water supplies.

5. Electrodialysis membranes: Electrodialysis membranes use direct electric current to separate dissolved electrolytes from the water. Anions are collected at the anode and cations at the cathode, after passing through a resin membrane; Unlike all membrane systems, they are not pressure driven (Fig. 7.7).

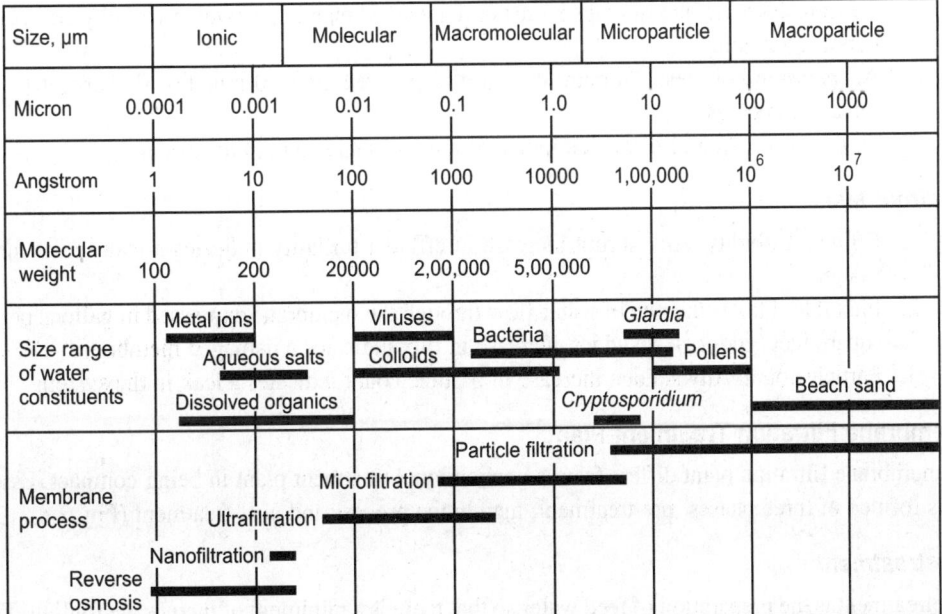

Fig. 7.7. Membranes and contaminant size.

Membrane Fouling

Membrane fouling is the clogging of the membrane by the filtered-out matter, which forms a layer called cake on the membrane surface. The degree of fouling depends on the quantity of particulate matter in the feed water and the level of its removal. The fouling is determined by applying indexes such as the silt density index (SDI) and the modified SDI. To determine SDI, two portions of 500 ml of the feed water are passed through a 0.45 μm pore size and 47 mm diameter membrane filter under 30 psig (pounds per square inch gauge pressure) of constant pressure. The time period is recorded (in minutes) for the initial and final 500 ml of filtration as t_i and t_f, respectively.

Final filtration is generally after 5, 10 or 15 minutes from the start of the initial filtration, depending on the quality of the feed water. This period is the third reading called the total time, t_t. SDI is calculated by the following formula:

$$SDI = [100\ (1 - t_i/t_f)]/t_t$$

where,

t_i = time to collect initial 500 ml.

t_f = time to collect final 500 ml.

t_t = total time, mostly 15 minutes as the maximum.

Membrane Integrity

Membrane integrity is the soundness of the membrane condition regarding leak damage. It is determined by various direct and indirect tests. There are important direct and indirect tests that are used for this purpose.

Direct tests

1. Bubble point: A 29.4 psi pressure is applied with a dilute surfactant (soap) solution. A 0.7 psi pressure decrease in 5 minutes indicates a damaged membrane, which will also be indicated by bubble formation.
2. Air pressure hold test: Pressure decrease is checked after 10 minutes. Perhaps, it is the most reliable direct test.
3. Sonic sensor: An online device indicates any unusual sound in the system.

Indirect tests

1. Effluent turbidity: Any abrupt increase in effluent turbidity indicates a leak in the membrane system.
2. Flux rate: Flux is the rate of water flow through the membrane expressed in gallons per day per square feet ($gpd/ft.^2$). A sudden increase in flux indicates a damaged membrane.
3. Particle count: Any sudden increase in particle count indicates a leak in the system.

Membrane Filtration Treatment Plant

A membrane filtration plant differs from a conventional treatment plant in being compact. Essentially, it is formed of three phases: pre-treatment, membrane process and post-treatment (Fig. 7.8).

Pretreatment

Pretreatment is the preparation of feed water so that there is a minimum of membrane fouling. The main pre-treatment methods are microstraining and conventional treatment up to final sedimentation.

1. Microstraining: Raw water is filtered with 5–13 μm pore size strainers without using any chemicals. This is a common practice for surface water with low turbidities.
2. Conventional treatment up to final sedimentation: Conventional treatment is used for turbid surface water with natural organic matter (NOM).

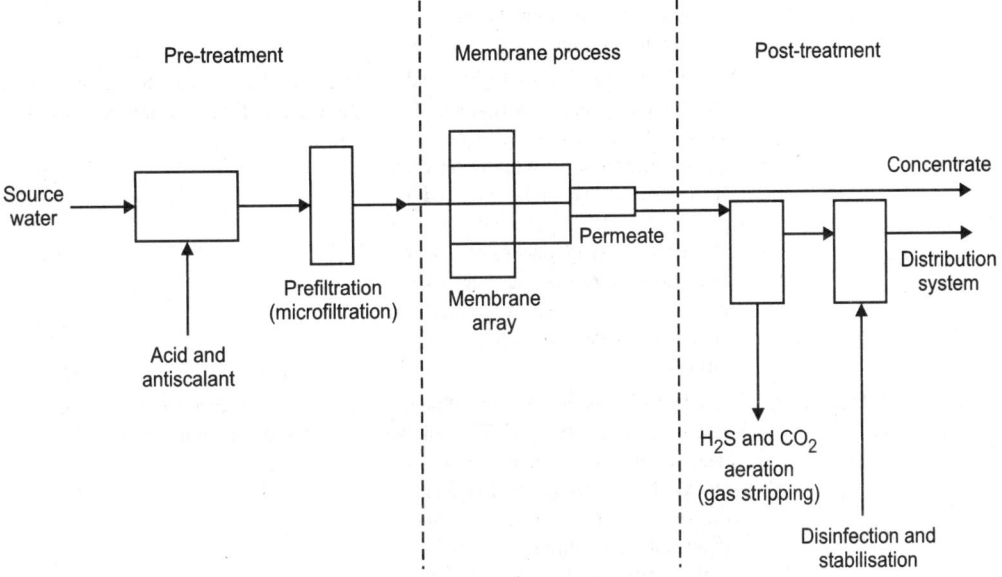

Fig. 7.8. A typical membrane filtration plant.

Membrane treatment

Membrane filtration uses an appropriate membrane system that takes into account source water quality and treatment requirements. Currently, there are three commonly used membrane systems: microfiltration, ultrafiltration and reverse osmosis. The first two are the most feasible alternatives to conventional water treatment to meet the requirements of the enhanced surface water treatment, the total coliform and the disinfectants and disinfection by-products rules. Here are some guidelines to select a membrane system for different source water:

1. Microfiltration: Microfiltration is used for surface water where conventional treatment requires pre-treatment, sedimentation and filtration.
2. Ultrafiltration: Ultrafiltration is used for surface water that needs the removal of very small particles such as viruses and dissolved organics.
3. Reverse osmosis: Reverse osmosis is suitable when water has a very high concentration of dissolved substances such as chlorides, nitrates and fluorides, in addition to other contaminants (e.g. salty water).

Posttreatment

In posttreatment, the effluent is treated to make sure that filtered water is safe and stable. Commonly, it needs aeration, pH adjustment and postdisinfection. To remove carbon dioxide and hydrogen sulphide, aeration is used. Some of the problems of filtration are shown in Table 7.3.

Table 7.3. Filtration problems and their solutions.

Problems	Possible causes	Possible solutions
Mud balls are in a sand filter	High level of $CaCO_3$ in water. Calcium carbonate forms aggregates in the surface mat sodium hexametaphosphate to the filter influent to keep calcium in solution	Lower the pH (below 9.3) of the influent water to convert $CaCO_3$ into soluble calcium bicarbonate. Also add 0.5–0.75 mg/l
	Abrupt high rate of washing. Mud balls are formed by faulty distribution of wash water caused by abrupt and quick opening of the wash water valve. It causes openings in the filter bed by pushing the media and letting some of the surface mat aggregates of $CaCO_3$ fall below the media's surface. They grow close to 1″ in size and can cause clogging of filter and high effluent turbidity	Start the filter washing by surface agitation and gradually increasing the wash water flow
Cracks are in surface mat of a sand filter	High dose of polymer to filter influent. Micro floc forms the gelatinous surface mat. When the mat becomes thick, it cracks at weak points due to water pressure. Cracks form water channels to let water pass through the media without effective filtration. Polymer feed for micro flocculation should be lowest effective dose—generally 0.25 to 0.75 mg/l.	Determine required polymer dose by jar test and decrease it as required
	Longer filter runs can cause a thicker mat which can crack at thinner places	Shorten the filter run adequately
There is jet action and sand boiling in a mixed media filter	Abrupt opening of the backwash water valve. Fast movement of backwash water in part of the filter bed can cause boiling of media particles resulting in media mixing and poor filtration	Open wash water valve slowly and partially for first few minutes of the washing
There is high turbidity in filter effluent of a conventional treatment plant when all other phases have no problems	High pH of the filter influent: pH above 9.4 causes fine calcium carbonate ($CaCO_3$) particles to go through filters to cause a high turbidity reading	Lower pH to 9.3 which will convert calcium carbonate to soluble calcium bicarbonate. Remember there is no insoluble calcium carbonate below pH 9.3.
	Overfeed of a polyphosphate: A dose of a polyphosphate such as sodium hexametaphosphate over 0.75 mg/l to the filter will cause excessive removal of calcium carbonate coating of the filter media which results in a high turbidity reading	Run a jar test; determine optimum dose of polyphosphate and apply. Generally, 0.5 mg/l dose is adequate

(Contd...)

Problems	Possible causes	Possible solutions
	Overfeed of a polymer. Overdose of a polymer to the filter influent will cause cracks and channels in the media for the turbidity particles to go through without filtration.	Check polymer dose to the filter influent. Usually, a dose above 0.5 mg/l can cause a high turbidity reading. Run jar test; determine optimum polymer dose and apply
	Air bubbles in the turbidity measuring cell. There is a higher concentration of dissolved gases in the water, especially in winter, which start coming out as gas bubbles at room temperature	Warm the sample to remove gases before taking the turbidity reading
	Scratched or smudged sample cell	Always use a scratch-free sample cell since scratches will also give a false high turbidity reading. Wipe the cell with soft tissue paper
Surface sweeps are being buried under the media	High calcium carbonate content. High calcium carbonate deposition on media particles will cause the swelling of media media with 5 per cent hydrochloric acid. The difference of the dried sample weight or volume before and after washing, times 100, is the per cent of swelling in weight and volume, respectively. It may require the acid washing of the whole media to reduce its volume. This will also improve the filter performance	Lower the pH below 9.3 and apply small amount of polyphosphate as previously discussed. The amount of swelling can be determined by washing a sample of
There is biofouling of a membrane	High bacterial count in feed water	Check bacterial count in pre-treated water and reduce it by proper pre-disinfection
	Membrane has been out of service too long. Some bacteria in and on a wet membrane will start to multiply if the membrane is out of service too long	Flush membrane with filtered water and sanitise it as recommended by the manufacturer
There is chemical scaling of the membrane	A high metallic (calcium, magnesium and iron) salt content. These metals will become concentrated in the feed water during filtration and precipitate pH of feed water is too high. This will cause more deposition of metallic salts	Use an antiscalant such as sodium hexa-metaphosphate or polyacrylate as recommended by the manufacturer

Lower the pH to dissolve the scaling substances |

Chapter 8

Ion Exchange and Carbon Adsorption

INTRODUCTION

Ion exchange and carbon adsorption are unrelated technologies and often have different objectives. They are, however, often times used in compliment to achieve high water quality attributes.

Ion exchange is a reversible chemical reaction wherein an ion (an atom or molecule that has lost or gained an electron and thus acquired an electrical charge) from solution is exchanged for a similarly charged ion attached to an immobile solid particle. These solid ion exchange particles are either naturally occurring inorganic zeolites or synthetically produced organic resins. The synthetic organic resins are the predominant type used today because their characteristics can be tailored to specific applications. An organic ion exchange resin is composed of high-molecular-weight polyelectrolytes that can exchange their mobile ions for ions of similar charge from the surrounding medium. Each resin has a distinct number of mobile ion sites that set the maximum quantity of exchanges per unit of resin. The industry application most familiar with ion exchange technology is metal plating. Most plating process water is used to cleanse the surface of the parts after each process bath. To maintain quality standards, the level of dissolved solids in the rinse water must be regulated. Freshwater added to the rinse tank accomplishes this purpose and the overflow water is treated to remove pollutants and then discharged. As the metal salts, acids and bases used in metal finishing are primarily inorganic compounds, they are ionised in water and could be removed by contact with ion exchange resins.

In a water deionisation process, the resins exchange hydrogen ions (H^+) for the positively charged ions (such as nickel, copper and sodium) and hydroxyl ions (OH^-) for negatively charged sulphates, chromates and chlorides.

Because the quantity of H^+ and OH ions is balanced, the result of the ion exchange treatment is relatively pure, neutral water. Ion exchange technology is applied in many other industry sectors, including the petroleum and chemical industries, as well as general waste-water treatment applications. The technology is most often compared to reverse osmosis, since both technologies are often aimed at similar objectives. In this regard, in addition to discussing ion exchange as a technology, we will also review some of the operational trade-offs and economics of the two processes in this chapter.

THEORY AND PRACTICE OF ION EXCHANGE

Water can contain varying concentrations of dissolved salts which dissociate to form charged particles called ions. These ions are the positively charged cations and negatively charged anions that permit the water or solution to conduct electrical currents and are therefore called electrolytes. Electrical conductivity is thus a measure of water, purity, with low conductivity corresponding to a state of high purity. The process of ion exchange is uniquely suited to the removal of ionic species from water supplies for several reasons. First, ionic impurities may be present in rather low concentrations. Second, modern ion-exchange resins have high capacities and can remove unwanted ions preferentially. Third, modern ion-exchange resins are stable and readily regenerated, thereby allowing their reuse. Other advantages ion exchange offers are: (i) the process and equipment are a proven technology. Designs are well developed into pre-engineered units that are rugged and reliable, with well-established, applications, (ii) fully manual to completely automatic, units are available, (iii) there are many models of ion-exchange systems on the market which keep costs competitive; (iv) temperature effects over a fairly wide range (from 0°C to 35°C) are negligible, and (v) the technology is excellent for both small and large installations, from home water softeners to large utility/industrial applications.

Ion exchange is a well-known method for softening or for demineralising water. Although softening could be useful in some instances, the most likely application for ion exchange in waste-water treatment is for demineralisation. Many ion-exchange materials are subject to fouling by organic matter. It is possible that treatment of secondary effluent for suspended-solids removal and possibly soluble organic removal will be required before carrying out ion exchange. Many natural materials and, more importantly, certain synthetic materials have the ability to exchange ions from an aqueous solution for ions in the material itself. Cation-exchange resins can, for example, replace cations in solution with hydrogen ions. Similarly, anion-exchange resins can either replace anions in solution with hydroxyl ions or absorb the acids produced from the cation-exchange treatment. A combination of these cation-exchange and anion-exchange treatments results in a high degree of demineralisation.

Since the exchange capacity of ion-exchange materials is limited, they eventually become exhausted and must be regenerated. The cation resin is regenerated with an acid; the anion resin is regenerated with a base. Important considerations in the economics of ion exchange are the type and amounts of chemicals needed for regeneration. Often, water to be demineralised is first passed through a cation-exchange material requiring a strong acid, usually sulphuric, for regeneration. The exchange material is referred to as strong acid resin. The amount of acid regenerant is somewhat more than the stoichiometric amount, possibly 100 per cent excess or more. If sulphuric acid is the regenerating acid, a waste brine is produced consisting of sulphates of the various actions in the water being treated. Because the partially treated water contains mineral acids, it is common to pass it next through an acid-absorbing resin or weak base resin. This resin can be regenerated with either a weak or strong base. The efficiency of regenerant use is quite high with these resins. If sodium hydroxide is the regenerating base, a waste brine is produced consisting of the sodium salts of the various anions in the water being treated. Certain anionic materials are not removed by the weak-base resin and must be further treated with strong-base resin if thorough demineralisation is desired. Regenerant usage by the strong-base resins is poorer than for the weak base resins. The reasons for applying this technology in the removal of mineral species should be quite apparent to those of you who work with applications involving heat exchange. Water problems in cooling, heating, steam generation and manufacturing are caused in large measure from the kinds and concentrations of dissolved solids, dissolved gases and suspended matter in the makeup water supplied. Table 8.1 lists the major objectionable ionic constituents present in many water supplies that can be removed by demineralisation. Prevention of scale and other deposits in cooling and boiling waters is best accomplished by removal of dissolved solids. Whereas in municipal water purification such removal is limited to the partial reduction of hardness and the removal of iron and manganese, in industrial water treatment it is often carried much further and may include the complete removal of hardness, the reduction or removal of alkalinity, the removal of silica or even the complete removal of all dissolved solids.

Table 8.1. Common ionic constituents contained in water.

Constituent of concern	Chemical designation	Resultant problems
Hardness	Calcium and magnesium salts in the forms of $CaCO_2$, Ca, Mg	This is the primary source of scaling in heat exchange equipment, boilers, pipelines/transfer lines, etc. Tends to form curds with soap and interferes with dyeing applications as well
Alkalinity	Bicarbonate (HCO_3), carbonate (CO_3) and hydrate (OH), expressed as $CaCO_3$	Causes foaming and carryover of solids with steam. Can cause embrittlement of boiler steel. Biocarbonate and carbonate generate CO_2 in steam, a source of corrosion
Free mineral acidity	H_2SO_4, HCl and other acids, expressed as $CaCO_3$	Causes rapid corrosion and deterioration of surfaces
Chloride	Cl^-	Interferes with silvering processes and increase TDS
Sulphates	$(SO_4)^=$	Results in the formation of calcium sulphate scale
Iron and manganese	Fe^{+2} (ferrous) Fe^{+3} (ferric) Mn^{+2}	Discolours water and results in the formation of deposits in water lines, boilers and other heat exchangers. Can interfere with dying, tanning, paper manufacture and various process works
Carbon dioxide	CO_2	Results in the corrosion of water lines, especially steam and condensate lines
Silica	SiO_2	Results in the formation of scale in boilers and cooling water systems can produce insoluble scale on turbine blades due to silica vapourisation in high pressure boilers (usually over 600 psi)

The two most frequently encountered water problems scale formation and corrosion are common to cooling, heating and steam-generating systems. Hardness (calcium and magnesium), alkalinity, sulphate and silica all form the main source of scaling in heat-exchange equipment, boilers and pipes. Scales or deposits formed in boilers and other exchange equipment act as insulation, preventing efficient heat transfer and causing boiler tube failures through overheating of the metal. Free mineral acids (sulphates and chlorides) cause rapid corrosion of boilers, heaters and other metal containers and piping. Alkalinity causes embrittlement of boiler steel and carbon dioxide and oxygen cause corrosion, primarily in steam and condensate lines.

Low quality steam can produce undesirable deposits of salts and alkali on the blades of steam turbines; much more difficult to remove are silica deposits which can form on turbine blades even when steam is satisfactory by ordinary standards. At steam pressures above 600 psi, silica from the boiler water actually dissolves in the gaseous steam and then reprecipitates on the turbine blades at their lower-pressure end.

In the operation of every cooling, heating and steam-generating plant, the water changes temperature. Higher temperatures, of course, increase both corrosion rates and scale-forming tendency. Evaporation in process steam boilers and in evaporative cooling equipment increases the dissolved-solids concentration of the water, compounding the problem.

In addition to the formation of scale or corrosion of metal within boilers, auxiliary equipment is also susceptible to similar damage. Attempts to prevent scale formation within a boiler can lead to makeup line deposits if the treatment chemicals are improperly chosen. Thus, the addition of normal phosphates to an unsoftened feed water can cause a dangerous condition by clogging the makeup line with precipitated calcium phosphate. Deposits in the form of calcium or magnesium stearate deposits, otherwise known as 'bathtub ring' can be readily seen and are caused by the combination of calcium or magnesium with negative ions of soap stearates.

Working of Ion Exchange

Ion exchangers are materials that can exchange one ion for another, hold it temporarily and then release it to a regenerant solution. In a typical demineraliser, this is accomplished in the following manner:

The influent water is passed through a hydrogen cation-exchange resin which converts the influent salt (e.g., sodium sulphate) to the corresponding acid (e.g. sulphuric acid) by exchanging an equivalent number of hydrogen (H^+) ions for the metallic cations (Ca^{+2}, Mg^{+2}, Na^+).

These acids are then removed by passing the effluent through an alkali regenerated anion-exchange resin which replaces the anions in solution (Cl^-, SO_4^-, NO_3^-) with an equivalent number of hydroxide ions. The hydrogen ions and hydroxide ions neutralise each other to form an equivalent amount of pure water. During regeneration, the reverse reaction takes place. The cation resin is regenerated with either sulphuric or hydrochloric acid and the anion resin is regenerated with sodium hydroxide. Figure 8.1 illustrates a basic scheme for ion exchange demineralisation.

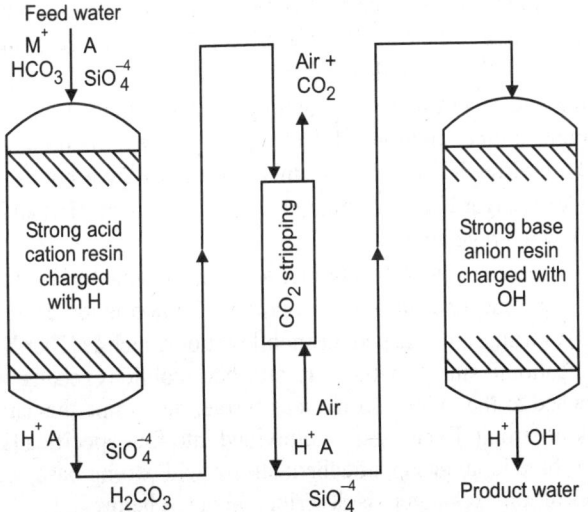

Fig. 8.1. Ion exchange demineralisation scheme.

There are various arrangements or equipment possible but in all cases, except in mixed-bed demineralisation, the water should first pass through a cation exchanger. In mixed demineralisation, the two exchange materials (that is, the cation-exchange resin and the anion-exchange resin) are placed in one shell instead of two separate shells. In operation, the two types of exchange materials are thoroughly mixed so that we have, in effect, a number of multiple demineralisers in series. Higher quality water is obtained from a mixed-bed unit than from a two-bed system (Fig. 8.2 for an example). Operation of cation and anion exchanges is shown in Fig. 8.3 (for fundamental processes) and Fig. 8.4 (operation modes for both cation/anion exchanges).

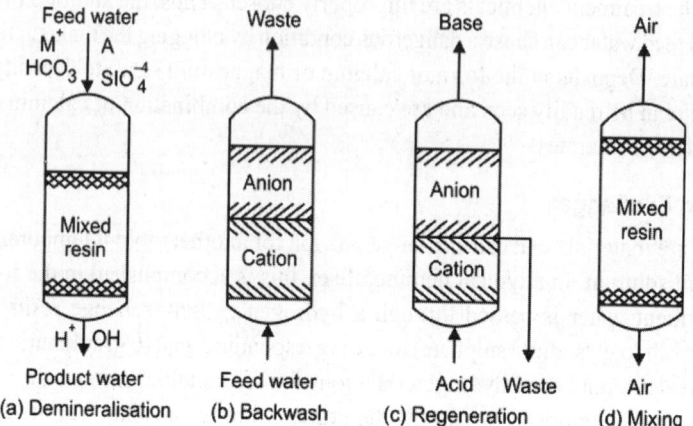

Fig. 8.2. Mixed resin demineralisation scheme.

To be suitable for industrial use, an ion-exchange resin must exhibit durable physical and chemical characteristics which are summarised by the following properties:

Functional groups

The molecular structure of the resin is such that it must contain a macroreticular tissue with acid or basic radicals. These radicals are the basis of classifying ion exchangers into two general groups: (i) cation exchangers, in which the molecule contains acid radicals of the HSO_3 or HCO type able to fix mineral or organic cations and exchange with the hydrogen ion H^+; and (ii) anion exchangers, containing basic radicals (for example, amine functions of the type NH_2) able to fix mineral or organic anions and exchange them with the hydroxyl ion OH^- coordinate to their dative bonds. The presence of these radicals enable a cation exchanger to be assimilated to an acid of form H-R and an anion exchanger to be a base of form OH-R when regenerated.

These radicals act as immobile ion-exchange sites to which are attached the mobile cations or anions. For example, a typical sulphonic acid cation exchanger has immobile ion-exchange sites consisting of the anionic radicals (SO) to which are attached the mobile cations, such as H^+ or Na^+. An anion exchanger similarly has immobile cationic sites to which are attached mobile (exchangeable) hydroxide anions (OR). The radicals attached to the molecular nucleus further determine the nature of the acid or base, whether it will be weak or strong. Exchangers are divided into four specific classifications depending on the kind of radical or functional group, attached; strong acid, strong base, weak acid or weak base. Each of these four types of ion exchangers is described in detail below.

Solubility

The ion-exchange substance must be insoluble under normal conditions of use. Most ion-exchange resins in current use are high molecular weight polyacids or polybases which are virtually insoluble in most aqueous and non-aqueous media. This is no longer true of some resins once a certain temperature has been reached. For example, some anion-exchange resins are limited to a maximum temperature of 105°F. Liquid ion-exchange resins exist also, yet we do not consider their applicability here and they also exhibit very limited solubility in aqueous solutions.

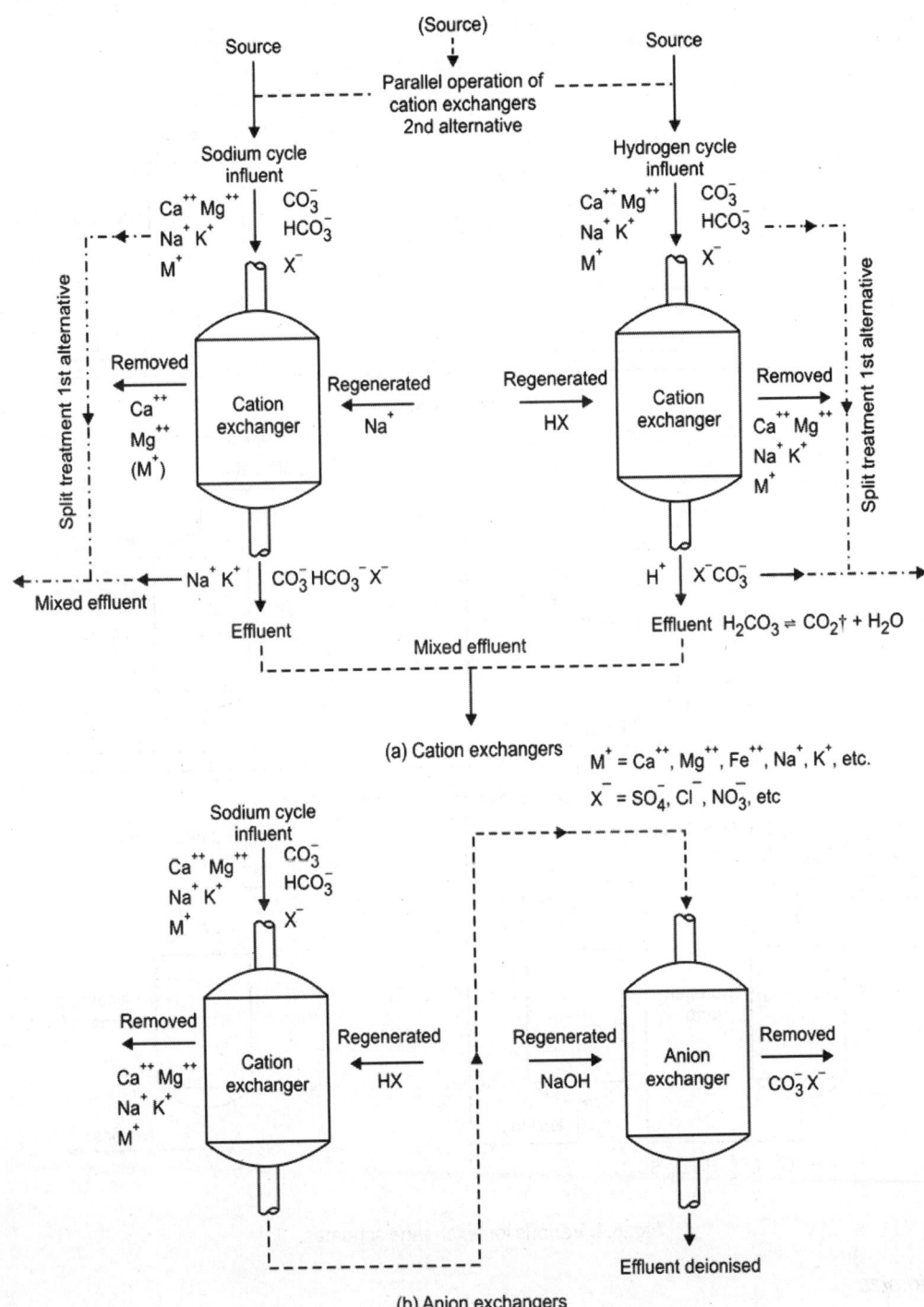

Fig. 8.3. Operational schemes of ion exchange.

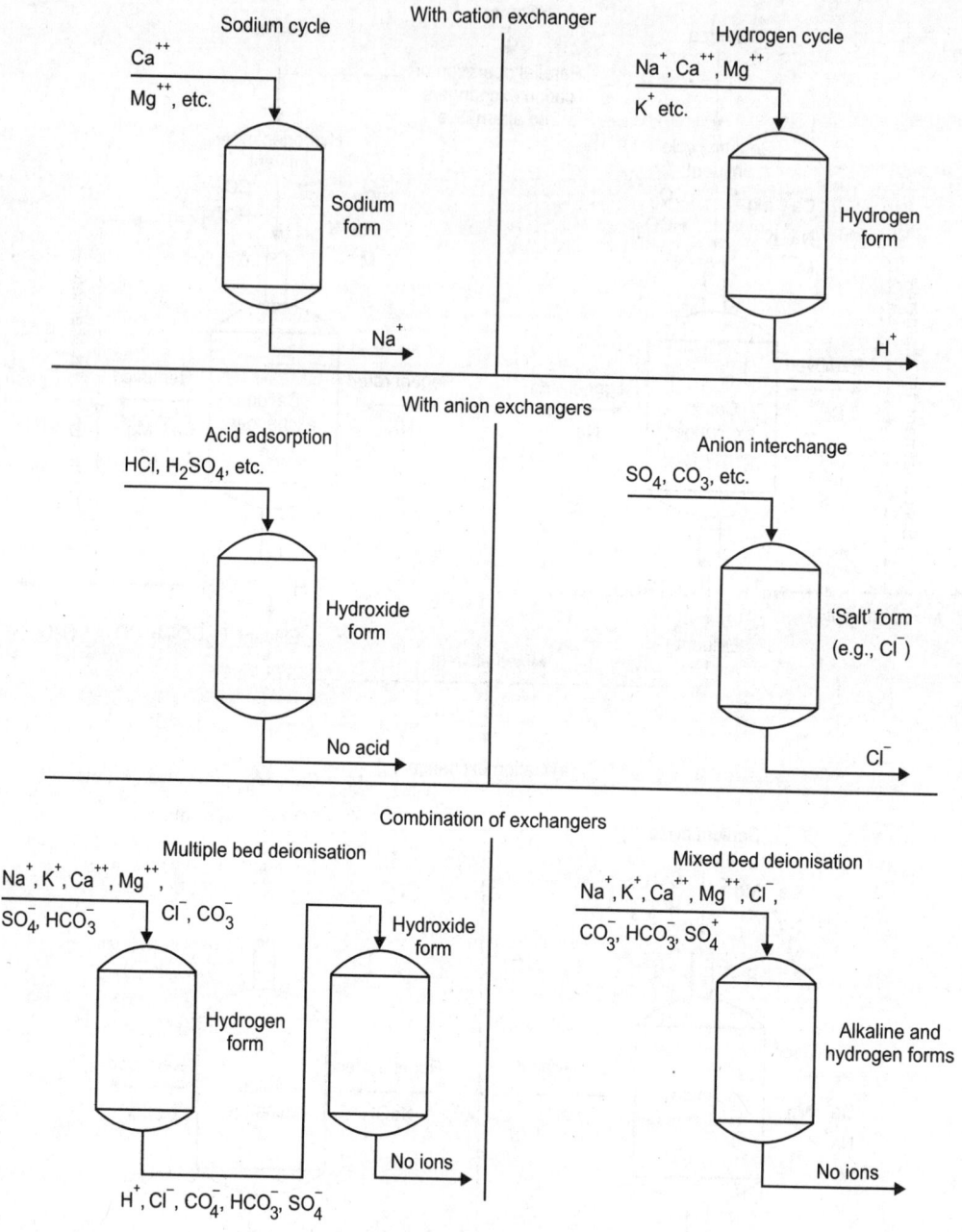

Fig. 8.4. Various ion exchange schemes.

Bead size

The resins must be in the form of spherical granules of maximum homogeneity and dimensions so that they do not pack too much, the void volume among their interstices is constant for a given type and the liquid head loss in percolation remains acceptable. Most ion-exchange resins occur as small beads or granules usually between 16 and 50 mesh in size.

Resistance to fracture

The ion or ionised complexes that the resins are required to fix are of varied dimensions and weights. The swelling and contraction of the resin bead that this causes must obviously not cause the grains to burst. Another important factor is the bead resistance to osmotic shock which will inevitably occur across its boundary surface, as there will be a salinity gradient of different magnitude during the cycle of the exchange material. The design of ion-exchange apparatus also must take into consideration the safe operation of the ion-exchange resin and avoid excessive stresses or mechanical abrasion in the bed, which could lead to breakage of the beads.

Types of Resins

As noted earlier, ion-exchange materials are grouped into four specific classifications depending on the functional group attached; strong-acid cation, strong-base anion, weak-acid cation or weak-base anion. In addition to these, we also have inert resins that do not have chemical properties.

Strongly acidic cation resins

Strongly acidic cation resins derive their exchange activity from sulphonic functional groups (HSO). The major cations in water are calcium, magnesium, sodium and potassium and they are exchanged for hydrogen in the strong acid cation exchanger when operated in the hydrogen cycle. The following stoichiometric equation represents the exhaustion phase and is written in the molecular form (as if the salts present were undissociated). It shows the cations in combination with the major anions, the bicarbonate, sulphate and chloride anions:

$$\begin{array}{ll} Ca\ 2HCO_2 & Ca \qquad\qquad\qquad 2H_2CO_3 \\ Mg \cdot SO_4 + 2RSO_3H \leftrightarrow 2RSO_3\ Mg + H_2SO_4 \\ Na\ 2Cl & Na \qquad\quad 2HCl \end{array}$$

Note that R represents the complex resin matrix. Because these equilibrium reactions are reversible, when the resin capacity has been exhausted it can be recovered through regeneration with a mineral acid. The strong-acid exchangers operate at any pH, split strong or weak salts, require excess strong-acid regenerant (typical regeneration efficiency varies from 25 per cent to 45 per cent in concurrent regeneration) and they permit low leakage. In addition, they have rapid exchange rates, are stable, exhibit swelling less than 7 per cent going from Na^+ to H^+ form and may last 20 years or more with little loss of capacity. These resins have found a wide range of application, being used on the sodium cycle for softening and on the hydrogen cycle for softening, dealkalisation and demineralisation.

Weakly acidic cation-exchange resins

Weakly acidic cation-exchange resins have carboxylic groups (COOH) as the exchange sites. When operated on the hydrogen cycle, the weakly acidic resins are capable of removing only those cations equivalent to the amount of alkalinity present in the water and most efficiently the hardness (calcium and magnesium) associated with alkalinity, according to these reactions:

$$\begin{array}{ll} Ca & Ca \\ Mg \cdot (HCO_3) + RCOOH \leftrightarrow 2RCOO\ Mg + H_2CO_4 \\ 2Na & 2Na \end{array}$$

These reactions are also reversible and permit acid regeneration to return the exhausted resin to the hydrogen form. The resin is highly efficient, for it is regenerated with 110 per cent of the stoichiometric amount of acid as compared to 200 to 300 per cent for strong-acid cation-exchange resins. It can be regenerated with the waste acid from a strong-acid cation exchanger and there is little waste problem during the regeneration cycle.

In order to prevent calcium sulphate precipitation when regenerated with H_2SO_4, it is usually regenerated stepwise with initial H_2SO_4 at 0.5 per cent. The resins are subject to reduced capacity from increasing flow rate, (above 2 gpm/ft) low temperatures and/or a hardness-alkalinity, ratio especially below 1.0.

Weakly acidic resins are used primarily for softening and dealkalisation, frequently in conjunction with a strongly acidic polishing resin. Systems which use both resins profit from the regeneration economy of the weakly acidic resin and produce treated water of quality comparable to that available with a strongly acidic resin.

Strongly basic anion-exchange resins

Strongly basic anion-exchange resins derive their functionality from the quaternary ammonium exchange sites. All the strongly basic resins used for demineralisation purposes belong to two main groups commonly known as type I and type II. The principal difference between the two resins, operationally; is that type I has a greater chemical stability and type II has a slightly greater regeneration efficiency and capacity. Physically, the two types differ by the species of quaternary ammonium exchange sites they exhibit. Type I sites have three methyl groups, while in type II, an ethanol group replaces one of the methyl groups. In the hydroxide form, the strongly basic anion will remove all the commonly encountered inorganic acids according to these reactions:

$$\left.\begin{array}{l} H_2SO_4 \\ 2HCl \\ H_2SiO_3 \\ H_2CO_3 \end{array}\right\} + 2ZOH \leftrightarrow \begin{array}{l} SO_4 \\ 2Cl \\ 2HSiO_3 + H_2O \\ 2HCO_3 \end{array}$$

Like the cation resin reactions, the anion-exchange reactions are also reversible and regeneration with a strong alkali, such as caustic soda, will return the resin to the hydroxide form. The strong-base exchangers operate at any pH, can split strong or weak salts, require excess high-grade NaOH for regeneration (with the typical efficiency varying from 18 to 33 per cent), are subject to organic fouling from such compounds when present in the raw water and to resin degradation due to oxidation and chemical breakdown. The strong-base anion resins suffer from capacity decrease and silica leakage increase at flow rates above 2 gpm/ft^3 of resin and cannot operate over 130°F to 150°F depending on resin type.

The normal maximum continuous operating temperature is 120°F and to minimise silica leakage, warm caustic (up to 120°F) should be used. Type I exchangers are for maximum silica removal. They are more difficult to regenerate and swell more (from Cl to OH form) than type II. The major case for selecting a type I resin is where high operating temperatures and/or very high silica levels are present in the influent water or superior resistance to oxidation or organics is required.

Type II exchangers remove silica (but less efficiently than type I) and other weak anions, regenerate more easily, are less subject to fouling, are freer from the odour of amine and are cheaper to operate than type I. Where free mineral acids are the main constituent to be removed and very high silica removal is not required, type II anion resin should be chosen.

Weakly basic anion resins

Weakly basic anion resins derive their functionality from primary (R-NH), secondary (R-NHR'), tertiary (R-N-R'2) and sometimes quaternary amine groups. The weakly basic resin readily absorbs such free mineral acids as hydrochloric and sulphuric and the reactions may be represented according to the following:

$$H_2SO_4 + 2ZOH = 2XZSO_4 + 2HO$$

Because the preceding reactions are also reversible, the weakly basic resins can be regenerated by applying caustic soda, soda ash or ammonia. The weak-base exchanger regenerates with a nearly stoichiometric amount of base (with the regeneration efficiency possibly exceeding 90 per cent) and can utilise waste caustic following strong-base anion-exchange resins. Weakly basic resins are used for high strong-acid waters (Cl, SO_4, NO_3) and low alkalinity, do not remove anions satisfactorily above pH 6, do not remove CO or silica, but have capacities about twice as great as for strong-base exchangers. Weak-base resins can be used to precede a strong-base anion resin to provide the maximum protection of the latter against organic fouling and to reduce regenerant costs.

The zeolite process should be compensated when: (i) the total hardness (TH) is greater than 400 ppm as $CaCO_3$, and (ii) the sodium salts (Na) are over 100 ppm as $CaCO_3$. Compensated hardness can be calculated from the following formula:

Compensated Hardness (ppm) = TH (ppm) × {9000/[9000 − Total Cations (ppm)]}

Express compensated hardness according to the following:

1. Next higher tenth of a grain up to 0.5 grains per gallon.
2. Next higher half of grain from 5.0 to 10.0 grains per gallon.
3. Next higher grain above 10.0 grains per gallon.

The salt consumption with a sodium cation exchange water softener ranges between 0.275 and 0.533 lbs of salt per 1000 grains of hardness, expressed as calcium carbonate, removed. This range is attributed to two factors: (i) the water composition, and (ii) the operating exchange value at which the exchange resin is to be worked. The lower salt consumption may be attained with waters that are not excessively hard nor high in sodium salts and where the exchange resin is not worked at its maximum capacity.

Inert resins

There also exists a type of resin with no functional groups attached. This resin offers no capacity to the system but increases regeneration efficiency in mixed-bed exchangers. These inert resins are of a density between cation and anion resins and when present in mixed-bed vessels help to separate cation and anion resins during backwash. Advantages of inert resins include: (i) classify cation and anion resins so that little or no mixing of cation or anion resin occurs before regeneration and a buffering mid-bed collection zone exists, (ii) improve regeneration efficiency, thereby reducing resin quantities needed, and (iii) protect against osmotic shock since the inert layer effectively prevents the exposure of cation resin to the caustic regenerant solution and the exposure of anion resin to the acid regenerant solution.

Ion Exchange Softening (Sodium Zeolite Softening)

This is one of the ion-exchange processes used in water purification. In this process, sodium ions from the solid phase are exchanged with the hardness ions from the aqueous phase. Consider a bed of ion-exchange resin having sodium as the exchangeable ion, with water containing calcium and magnesium hardness allowed to percolate through this bed. Let us denote the ion-exchange resinous material as RNa, where R stands for resin matrix and Na is its mobile exchange ion. The hard water will exchange Ca and Mg ions rapidly, so that water at the effluent will be almost completely softened. Calcium and magnesium salts will be converted into corresponding sodium salts.

The reaction will proceed toward the right-hand side to its completion until the bed gets completely exhausted or saturated with Ca and Mg ions. In order to reverse the equilibrium so that the reaction proceeds toward the left-hand side, the concentration of sodium ions has to be increased. This increase in sodium ions is accomplished by using a brine solution of sufficient strength so that the total sodium ions present in the brine are more than the total equivalent of Ca and Mg in the exhausted bed. This reverse reaction is carried out in order to bring the exhausted resin back to its sodium form. This process is known as regeneration. When the softener with the fresh resin in sodium form is put in service, the sodium ions in the surface layer of the bed are immediately exchanged with calcium and magnesium, thereby producing soft water with very little residual hardness in the effluent. As the process continues, the resin bed keeps exchanging its sodium ions with calcium and magnesium ions until the hardness concentration increases rapidly and the softening run is ended.

This softening process can be extended to a point where the hardness coming in and going out is the same. When this condition is reached, the bed is completely exhausted and does not have any further capacity to exchange ions. This capacity is called the total breakthrough capacity. In practice, the softening process is never extended to reach this stage as it is ended at some pre-determined effluent hardness, much lower than the influent hardness. This capacity is called the operating exchange capacity. After the resin bed has reached this capacity, the resin bed is regenerated with a brine solution.

The regeneration of the resin bed is never complete. Some traces of calcium and magnesium remain in the bed and are present in the lower-bed level. In the service run, sodium ions exchanged from the top layers of the bed form a very dilute regenerant solution which passes through the resin bed to the lower portion of the bed. This solution tends to leach some of the hardness ions not removed by previous regeneration. These hardness ions appear in the effluent water as leakage. Hardness leakage is also dependent on the raw water characteristics. If the Na/Ca ratio and calcium hardness are very high in the raw water, leakage of the hardness ions will be higher.

Resin Performance

Variances in resin performance and capacities can be expected from normal annual attrition rates of ion-exchange resins. Typical attrition losses that can be expected include: (i) strong cation resin: 3 per cent per year for three years or 10,00,000 gals/cu.ft, (ii) strong anion resin: 25 per cent per year for two years or 10,00,000 gals/cu.ft, and (iii) weak cation/anion: 10 per cent per year for two years or 7,50,000 gals/cu.ft. A steady falloff of resin-exchange capacity is a matter of concern to the operator and is due to several conditions.

Improper backwash

Blowoff of resin from the vessel during the backwash step can occur if too high a backwash flow rate is used. This flow rate is temperature dependent and must be regulated accordingly. Also, adequate time must be allotted for backwashing to insure a clean bed prior to chemical injection.

Channelling

Cleavage and furrowing of the resin bed can be caused by faulty operational procedures or a clogged bed or underdrain. This can mean that the solution being treated follows the path of least resistance, runs through these furrows and fails to contact active groups in other parts of the bed.

Incorrect chemical application

Resin capacities can suffer when the regenerant is applied in a concentration that is too high or too low. Another important parameter to be considered during chemical application is the location of the regenerant distributor. Excessive dilution of the regenerant chemical can occur in the vessel if the distributor is located too high above the resin bed. A recommended height is 3 inches above the bed level.

Mechanical strain

When broken beads and fines migrate to the top of the resin bed during service, mechanical strain is caused which results in channelling, increased pressure drop; or premature breakthrough. The combination of these resulting conditions leads to a drop in capacity.

Resin fouling

In addition to the physical causes of capacity losses there are a number of chemically caused problems that merit attention, specifically the several forms of resin fouling that may be found.

Organic fouling

Organic fouling occurs on anion resins when organics precipitate onto basic exchange sites. Regeneration efficiency is then lowered, thereby reducing the exchange capacity of the resin. Causes of organic fouling are fulvic, humic or tannic acids or degradation products of DVB (divinylbenzene) cross linkage material of cation resins. The DVB is degraded through oxidation and causes irreversible fouling of downstream anion resins.

Iron fouling

Iron fouling is caused by both forms of iron ions; the insoluble form will coat the resin bead surface and the soluble form can exchange and attach to exchange sites on the resin bead. These exchanged ions can be oxidised by subsequent cycles and precipitate ferric oxide within the bead interior.

Silica fouling

Silica fouling is the accumulation of insoluble silica on anion resins. It is caused by improper regeneration which allows the silicate (ionic form) to hydrolyse to soluble silicic acid which in turn polymerises to form colloidal silicic acid with the beads. Silica fouling occurs in weak-base anion resins when they are regenerated with silica-laden waste caustic from the strongbase anion resin unless intermediate partial dumping is done.

Microbiological fouling (MB)

Microbiological fouling (MB) becomes a potential problem when microbic growth is supported by organic compounds, ammonia, nitrates and so on which are concentrated on the resin. Signs of MB fouling are increased pressure drops, plugged distributor laterals and highly contaminated treated water.

Calcium sulphate fouling

Calcium sulphate fouling occurs when sulphuric acid is used to regenerate a cation exchanger after exhaustion by a water high in calcium. The precipitate of calcium sulphate (gypsum) that forms can cause calcium and sulphate leakage during subsequent service runs. Given a sufficient calcium input in the water to treat, calcium sulphate fouling is especially prevalent when the per cent solution of regenerant is greater than 5 per cent or the temperature is greater than 100°F or when the flow rate is less than 1 gpm/cu ft. Stepwise injection of sulphuric acid during regeneration can help prevent fouling.

Aluminium fouling

Aluminium fouling of resins can appear when aluminium floc from alum or other coagulants in pretreatment are encountered by the resin bead. This floc coats the resin bead and in the ionic form will be exchanged. However, these ions are not efficiently removed during regeneration so the available exchange sites continuously decrease in number.

Copper fouling

Copper fouling is found primarily in condensate polishing applications. Capacity loss is due to copper oxides coating the resin beads.

Oil fouling

Oil fouling does not cause chemical degradation but gives loss of capacity due to filming on the resin beads and the reduction of their active surface. Agglomeration of beads also occurs causing increased pressure drop, channelling and pre-mature breakthrough. The oil-fouling problem can be alleviated by the use of surfactants.

Operational Sequencing Considerations

The mode of operation for ion-exchange units can vary greatly from one system to the next, depending on the user's requirements. Service and regeneration cycles can be fully manual to totally automatic, with the method of regeneration being cocurrent, countercurrent or external. The exhaustion phase is called the service run. This is followed by the regeneration phase which is necessary to bring the bed back to initial conditions to cycle. The regeneration phase includes four steps: backwashing to clean the bed, introduction of the excess regenerant, a slow rinse or displacement step to push the regenerant slowly through the bed and finally a fast rinse to remove the excess regenerant from the resin and elute the unwanted ions to waste.

Service cycle

The service cycle is normally terminated by one or a combination of the following criteria:
1. High effluent conductivity.
2. Total gallons throughput.
3. High-pressure drop.
4. High silica.
5. High sodium.
6. Variations in pH.
7. Termination of the service cycle can be manually or automatically initiated.

Backwash cycle

Normally, the first step in the regeneration sequence is designed to reverse flow from the service cycle using sufficient volume and flow rate to develop proper bed expansion for the purpose of removing suspended material (crud) trapped in the ion-exchange bed during the service cycle. The backwash waste-water is collected by the raw water inlet distributor and diverted to waste via value sequencing. Backwash rate and internal design should avoid potential loss of whole bead resin during the backwash step. (Lower water temperature means more viscous force and more expansion.)

Regenerant introduction

This introduction of regenerant chemicals can be cocurrent or countercurrent depending on effluent requirements, operating cost and so on. Regenerant dosages (pounds per cubic foot), concentrations, flow rate and contact time are determined for each application. The regenerant distribution and collection system must provide uniform contact throughout the bed and should avoid regenerant hideout. Additional effluent purity is obtained with countercurrent systems, since the final resin contact in the service will be the most highly regenerated resin in the bed, creating a polishing effect.

Displacement slow rinse cycle

The final steps in the regeneration sequence are generally terminated on acceptable quality. Displacement, which precedes the rinse step, is generally an extension of the regenerant introduction step. The displacement step is designed to give final contact with the resin, removing the bulk of the spent regenerant from the resin bed.

Fast rinse cycle

The fast rinse step is essentially the service cycle except that the effluent is diverted to waste until quality is proven. This final rinse is always in the same direction as the service flow. Therefore, in countercurrent systems the displacement flow and rinse flows will be in opposite directions.

Sequence of Operation: Mixed Bed Units

In mixed-bed units, both the cation and the anion resins are mixed together thoroughly in the same vessel by compressed air. The cation and the anion resins being next to each other constitute an infinite number of cation and anion exchangers. The effluent quality obtainable from a well-designed and operated mixed-bed exchanger will readily produce demineralised water of conductivity less than 0.5 mmho and silica less than 10 ppb.

Service cycle

As far as the mode of operation is concerned, the service cycle of a mixed-bed unit is very similar to a conventional two-bed system, in that water flows into the top of the vessel, down through the bed and the purified effluent comes out the bottom. It is in the regeneration and the preparation of it that the mixed-bed differs from the two-bed equipment. The resins must be separated, regenerated separately and remixed for the next service cycle.

Backwash cycle

Prior to regeneration, the cation and the anion resins are separated by backwashing at a flow rate of 3.0 to 3.5 gpm/ft. The separation occurs because of the difference in the density of the two types of resin. The cation resin, being heavier, settles on the bottom, while the anion resin, being lighter, settles on top of the cation resin. After backwashing, the bed is allowed to settle down for 5 to 10 minutes and two clearly distinct layers are formed. After separation, the two resins are independently regenerated.

Regenerant introduction

The anion resin is regenerated with caustic flowing downward from the distributor placed just above the bed, while the cation resin is regenerated with either hydrochloric or sulphuric acid, usually flowing upward. The spent acid and caustic are collected in the interface collector, situated at the interface of the two resins. The regenerant injection can be carried out simultaneously as described or sequentially. In sequential regeneration, the cation-resin regeneration should precede the anion-resin regeneration to prevent the possibility of calcium carbonate and magnesium hydroxide precipitation, which may occur because of the anion-regeneration waste coming in contact with the exhausted cation resin. If this precipitation occurs, it can foul the resins at the interface. This becomes very critical when only the mixed-bed exchanger is installed to demineralise the incoming raw water.

In the case of sequential regeneration, during the caustic and acid injection period, a blocking flow of the demineralised water is provided in the opposite direction of the regenerant injection. This is required to prevent the caustic from entering the cation resin and acid from entering the anion resin. When regeneration is carried out simultaneously, acid and caustic injection flows act like blocking flows to each other and no additional blocking flow with water is needed. In a few sequential-type regeneration systems, acid is injected to flow downward through the central interface collector which now also acts as an acid distributor.

Rinsing and air mix cycles

After completion of the acid and caustic injection, both the cation and anion resins are rinsed slowly to remove the majority of the regenerant, without attempting to eliminate it completely. After the use of 7 to 10 gallons of slow rinse volume per cubic foot of each type of resin, the unit is drained to lower the water to a few inches above the resin bed. The resins are now remixed with an upflow of air. After remixing, the unit is filled completely with water flowing slowly from the top, to prevent anion-resin separation in the upper layers. The mixed-bed exchanger is then rinsed at fast flow rates. The conductivity of the effluent water may be very high for a few minutes and will then drop suddenly to the value usually observed in the service cycle. This phenomenon is characteristic of mixed beds and is due to the absorption of the remaining acid or caustic in different parts of the bed, by one or the other resin. This, no doubt, results in the loss of resin capacity, but this loss is negligible as compared to the length of the service cycle and the savings in the overall time required for regeneration.

Sequence of Operation: Softener Units

Following are the basic steps involved in a regeneration of a water softener:

Backwashing

After exhaustion, the bed is backwashed to effect a 50 per cent minimum bed expansion to release any trapped air from the air pockets, minimise the compactness of the bed, reclassify the resin particles and purge the bed of any suspended insoluble material. Backwashing is normally carried out at 5–6 gpm/ft. However, the backwash flow rates are directly proportional to the temperature of water.

Brine injection

After backwashing, a 5 per cent to 10 per cent brine solution is injected during a 30 minute period. The maximum exchange capacity of the resin is restored with 10 per cent strength of brine solution. The brine is injected through a separate distributor placed slightly above the resin bed.

Displacement or slow rinse

After brine injection, the salt solution remaining inside the vessel is displaced slowly, at the same rate as the brine injection rate. The slow rinsing should be continued for at least 15 minutes and the slow rinse volume should not be less than 10 gallons/cu ft of the resin. The actual duration of the slow rinse should be based on the greater of these two parameters.

Fast rinse

Rinsing is carried out to remove excessive brine from the resin. The rinsing operation is generally stopped when the effluent chloride concentration is less than 5–10 ppm in excess of the influent chloride concentration and the hardness is equal to or less than 1 ppm as CaCO.

Each arrangement will vary substantially in both operating and installed costs. Important factors for selection are:

1. Influent water analysis.
2. Flow rate.
3. Effluent quality.
4. Waste requirements
5. Operating cost.

Selectivity and General Considerations

As noted, ion exchange reactions are stoichiometric and reversible and in that way they are similar to other solution phase reactions. For example:

$$NiSO_4 + Ca(OH)_2 = Ni(OH)_2 + CaSO_4$$

In this reaction, the nickel ions of the nickel sulphate ($NiSO_4$) are exchanged for the calcium ions of the calcium hydroxide [$Ca(O)_2$] molecule. Similarly, a resin with hydrogen ions available for exchange will exchange those ions for nickel ions from solution. The reaction can be written as follows:

$$2(R\text{-}SO_3H) + NiSO_4 = (R\text{-}SO_3)2Ni + H_2SO_4$$

R indicates the organic portion of the resin and SO_3 is the immobile portion of the ion active group. Two resin sites are needed for nickel ions with a plus 2 valence (Ni^{+2}). Trivalent ferric ions would require three resin sites. As shown, the ion exchange reaction is reversible. The degree the reaction proceeds to the right will depend on the resins preference or selectivity, for nickel ions compared with its preference for hydrogen ions. The selectivity of a resin for a given ion is measured by the selectivity coefficient K, which in its simplest form for the reaction:

$$R^-A^+ + B^+ = R^-B^+ + A^+$$

is expressed as: K = (concentration of B^+ in resin/concentration of A^+ in resin) × (concentration of A^+ in solution/concentration of B^+ in solution).

The selectivity coefficient expresses the relative distribution of the ions when a resin in the A^+ form is placed in a solution containing B^+ ions. Table 8.2 gives the selectivity's of strong acid and strong base ion exchange resins for various ionic compounds. It should be pointed out that the selectivity coefficient is not constant but varies with changes in solution conditions. It does provide a means of determining what to expect when various ions are involved. As indicated in Table 8.2, strong acid resins have a preference for nickel over hydrogen. Despite this preference, the resin can be converted back to the hydrogen form by contact with a concentrated solution of sulphuric acid (H_2SO_4):

$$(R\text{--}SO_4)_2Ni + H_2SO_4 \rightarrow 2(R\text{--}SO_3H) + NiSO_4$$

As noted from above ammonia sodium hydrogen from above, but a little differently, this step is known as regeneration. In general terms, the higher the preference a resin exhibits for a particular ion, the greater the exchange efficiency in terms of resin capacity for removal of that ion from solution.

Greater preference for a particular ion, however, will result in increased consumption of chemicals for regeneration.

Table 8.2. Selectivity of ion exchange resins in order of decreasing preference.

Strong acid cation exchanger	Strong base anion exchanger
Barium	Iodide
Lead	Nitrate
Calcium	Bisulphate
Nickel	Chloride
Cadmium	Cyanide
Copper	Bicarbonate
Zinc	Hydroxide
Magnesium	Fluoride
Potassium	Sulphate
Ammonia	
Sodium	
Hydrogen	

Resins currently available exhibit a range of selectivity's and thus have broad application. As an example, for a strong acid resin, the relative preference for divalent calcium ions (Ca^{+2}) over divalent copper ions (Cu^{+2}) is approximately 1.5 to 1. For a heavy-metal-selective resin, the preference is reversed and favours copper by a ratio of 2.300 to 1.

Ion exchange resins are classified as cation exchangers, which have positively charged mobile ions available for exchange and anion exchangers, whose exchangeable ions are negatively charged. Both anion and cation resins are produced from the same basic organic polymers. They differ in the ionisable group attached to the hydrocarbon network. It is this functional group that determines the chemical behaviour of the resin. Resins can be broadly classified as strong or weak acid cation exchangers or strong or weak base anion exchangers.

Strong acid resins are so named because their chemical behaviour is similar to that of a strong acid. The resins are highly ionised in both the acid ($R-SO_3H$) and salt ($R-SO_3Na$) form. They can convert a metal salt to the corresponding acid by the reaction:

$$2(R-SO_3H) + NiCl_2 \rightarrow (R-SO_4), Ni + 2HCl$$

The hydrogen and sodium forms of strong acid resins are highly dissociated and the exchangeable Na^+ and H^+ are readily available for exchange over the entire pH range. Consequently, the exchange capacity of strong acid resins is independent of solution pH. These resins would be used in the hydrogen form for complete deionisation; they are used in the sodium form for water softening (calcium and magnesium removal). After exhaustion, the resin is convened back to the hydrogen form (regenerated) by contact with a strong acid solution or the resin can be convened to the sodium form with a sodium chloride solution. In the above, the hydrochloric acid (HCl) regeneration would result in a concentrated nickel chloride (NiCl) solution.

In a weak acid resin, the ionisable group is a carboxylic acid (COOH) as opposed to the sulphonic acid group (SO_3H) used in strong acid resins. These resins behave similarly to weak organic acids that are weakly dissociated. Weak acid resins exhibit a much higher affinity for hydrogen ions than do strong acid resins. This characteristic allows for regeneration to the hydrogen form with significantly less acid than is required for strong acid resins. Almost complete regeneration can be accomplished with stoichiometric amounts of acid. The degree of dissociation of a weak acid resin is strongly influenced by the solution pH. Consequently, resin capacity depends in part on solution pH. Figure 8.1 shows that a typical weak acid resin has limited capacity below a pH of 6.0, making it unsuitable for deionising acidic metal finishing waste-water.

Like strong acid resins, strong base resins are highly ionised and can be used over the entire pH range. These resins are used in the hydroxide (OH) form for water deionisation. They will react with anions in solution and can convert an acid solution to pure water:

$$R-NH_3OH + HCl \rightarrow R-NH_3Cl + HOH$$

Regeneration with concentrated sodium hydroxide (NaOH) converts the exhausted resin to the hydroxide form.

Weak base resins are like weak acid resins, in that the degree of ionisation is strongly influenced by pH. Consequently, weak base resins exhibit minimum exchange capacity above a pH of 7.0. These resins merely absorb strong acids: they cannot split salts.

In an ion exchange waste-water deionisation unit, the waste-water would pass first through a bed of strong acid resin. Replacement of the metal cations (Ni^{+2}, Cu^{+2}) with hydrogen ions would lower the solution pH. The anions (SO_4^{-2}, Cl^-) can then be removed with a weak base resin because the entering waste-water will normally be acidic and weak base resins sorb acids. Weak base resins are preferred over strong base resins because they require less regenerant chemical. A reaction between the resin in the free base form and HCl would proceed as follows:

$$R - NH_2 + HCl \rightarrow R - NH_3Cl$$

The weak base resin does not have a hydroxide ion form as does the strong base resin. Consequently, regeneration needs only to neutralise the absorbed acid: it need not provide hydroxide ions. Less expensive weakly 'basic reagents' such as ammonia (NH_3) or sodium carbonate can be employed. Chelating resins behave similarly to weak acid cation resins but exhibit a high degree of selectivity for heavy metal cations. Chelating resins are analogous to chelating compounds found in metal finishing waste-water; that is, they tend to form stable complexes with the heavy metals. In fact, the functional group used in these resins is an EDTAa compound. The resin structure in the sodium form is expressed as R-EDTA-Na. The high degree of selectivity for heavy metals permits separation of these ionic compounds from solutions containing high background levels of calcium, magnesium and sodium ions. A chelating resin exhibits greater selectivity for heavy metals in its sodium form than in its hydrogen form. Regeneration properties are similar to those of a weak acid resin; the chelating resin can be converted to the hydrogen form with slightly greater than stoichiometric doses of acid because of the fortunate tendency of the heavy metal complex to become less stable under low pH conditions. Potential applications of the chelating resin include polishing to lower the heavy metal concentration in the effluent from a hydroxide treatment process or directly removing toxic heavy metal cations from waste-waters containing a high concentration of non-toxic, multivalent cations. Table 8.3 shows the preference of a commercially available chelating resin for heavy metal cations over calcium ions. (The chelating resins exhibit a similar magnitude of selectivity for heavy metals over sodium or magnesium ions.) The selectivity coefficient defines the relative preference the resin exhibits for different ions. The preference for copper (shown in Table 8.3) is 2300 times that for calcium. Therefore, when a solution is treated that contains equal molar concentrations of copper and calcium ions, at equilibrium, the molar concentration of copper ions on the resin will be 2300 times the concentration of calcium ions; or, when solution is treated that contains a calcium ion molar concentration 2300 times that of the copper ion concentration, at equilibrium, the resin would hold an equal concentration of copper and calcium.

Table 8.3. Chelating cation resin selectivities for metal ions.

Metal	K_m/Ca[a]
Hg^{+2}	2800
Cu^{+2}	2300
Pb^{+2}	1200
Ni^{+2}	57
Zn^{+2}	17
Cd^{+2}	15
Co^{+2}	6.7
Fe^{+2}	4.7
Mn^{+2}	1.2
Ca^{+2}	1

[a] Selectivity coefficient for the metal over calcium ions at a pH of 4.

Equipment and Operation

Ion exchange processing can be accomplished by either a batch method or a column method. In the first method, the resin and solution are mixed in a batch tank, the exchange is allowed to come to equilibrium, then the resin is separated from solution. The degree to which the exchange takes place is limited by the preference the resin exhibits for the ion in solution. Consequently, the use of the resins exchange capacity will be limited unless the selectivity for the ion in solution is far greater than for the exchangeable ion attached to the resin. Because batch regeneration of the resin is chemically inefficient, batch processing by ion exchange has limited potential for application.

Passing a solution through a column containing a bed of exchange resin is analogous to treating the solution in an infinite series of batch tanks. Consider a series of tanks each containing 1 equivalent (eq) of resin in the X ion form. A volume of solution containing 1 eq of Y ions is charged into the first tank. Assuming the resin to have an equal preference for ions X and Y, when equilibrium is reached the solution phase will contain 0.5 eq of X and Y. Similarly, the resin phase will contain 0.5 eq of X and Y. This separation is the equivalent of that achieved in a batch process.

If the solution were removed from Tank 1 and added to Tank 2, which also contained 1 eq of resin in the X ion form, the solution and resin phase would both contain 0.25 eq of Y ion and 0.75 eq of X ion. Repeating the procedure in a third and fourth tank would reduce the solution content of Y ions to 0.125 and 0.0625 eq respectively. Despite an unfavourable resin preference, using a sufficient number of stages could reduce the concentration of Y ions in solution to any level desired. This analysis simplifies the column technique, but it does provide insights into the process dynamics. Separations are possible despite poor selectivity for the ion being removed. Most industrial applications of ion exchange use fixed-bed column systems, the basic component of which is the resin column (Fig. 8.5). The column design must:

1. Contain and support the ion exchange resin.
2. Uniformly distribute the service and regeneration flow through the resin bed.

3. Provide space to fluidise the resin during backwash.
4. Include the piping, valves and instruments needed to regulate flow of feed, regenerant and backwash solutions.

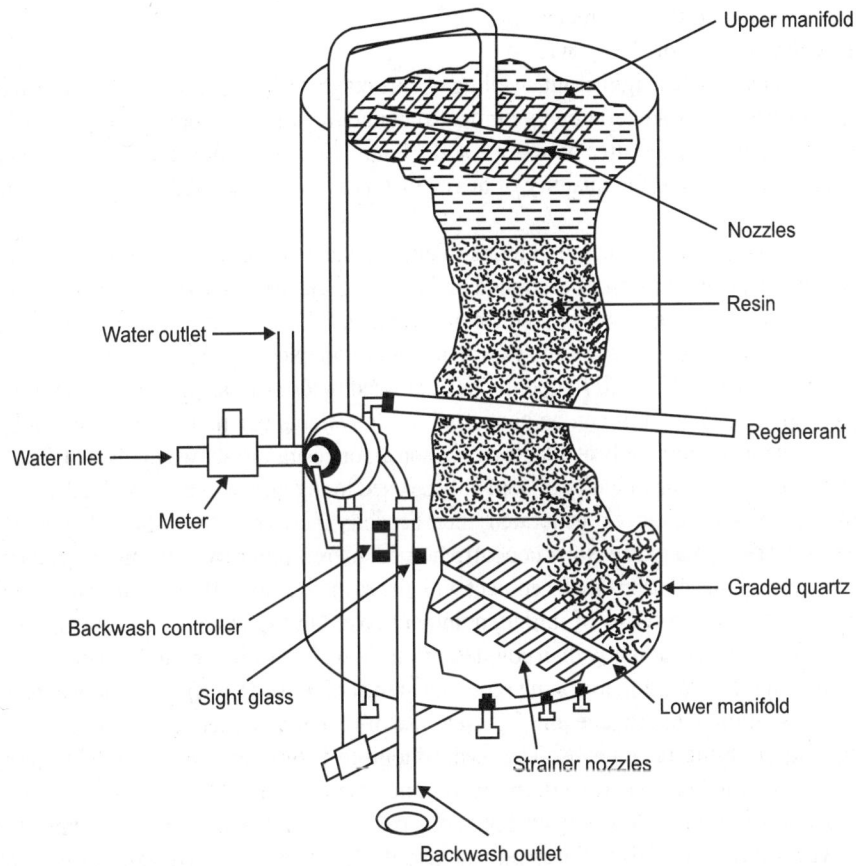

Fig. 8.5. Ion exchange unit.

After the feed solution is processed to the extent that the resin becomes exhausted and cannot accomplish any further ion exchange, the resin must be regenerated. In normal column operation, for a cation system being convened first to the hydrogen then to the sodium form, regeneration employs the following basic steps:

1. The column is backwashed to remove suspended solids collected by the bed during the service cycle and to eliminate channels that may have formed during this cycle. The back- wash flow fluidises the bed, releases trapped panicles and reorients the resin panicles according to size. During backwash the larger, denser panicles will accumulate at the base and the particle size will decrease moving up the column. This distribution yields a good hydraulic flow pattern and resistance to fouling by suspended solids.
2. The resin bed is brought in contact with the regenerant solution. In the case of the cation resin, acid elutes the collected ions and converts the bed to the hydrogen form. A slow water rinse then removes any residual acid.

3. The bed is brought in contact with a sodium hydroxide solution to convert the resin to the sodium form. Again, a slow water rinse is used to remove residual caustic. The slow rinse pushes the last of the regenerant through the column.
4. The resin bed is subjected to a fast rinse that removes the last traces of the regenerant solution and ensures good flow characteristics.
5. The column is returned to service.

For resins that experience significant swelling or shrinkage during regeneration, a second backwash should be performed after regeneration to eliminate channelling or resin compression. Regeneration of a fixed-bed column usually requires between 1 and 2 hours. Frequency depends on the volume of resin in the exchange columns and the quantity of heavy metals and other ionised compounds in the waste-water.

Resin capacity is usually expressed in terms of equivalents per liter (eq/l) of resin. An equivalent is the molecular weight in grams of the compound divided by its electrical charge or valence. For example, a resin with an exchange capacity of 1 eq/l could remove 37.5 g of divalent zinc (Zn^{+2}, molecular weight of 65) from solution. Much of the experience with ion exchange has been in the field of water softening: therefore, capacities will frequently be expressed in terms of kilograins of calcium carbonate per cubic foot of resin. This unit can be converted to equivalents per liter by multiplying by 0.0458. Typical capacities for commercially available cation and anion resins are shown in Fig. 8.4. The capacities are strongly influenced by the quantity of acid or base used to regenerate the resin. Weak acid and weak base systems are more efficiently regenerated; their capacity increases almost linearly with regenerant dose. Columns are designed to use either concurrent or countercurrent regeneration. In concurrent units, both feed and regenerant solutions make contact with the resin in a downflow mode. These units are the less expensive of the two in terms of initial equipment cost. On the other hand, concurrent flow uses regenerant chemicals less efficiently than countercurrent flow: it has higher leakage concentrations (the concentration of the feed solution ion being removed in the column effluent) and cannot achieve as high a product concentration in the regenerant. Efficient use of regenerant chemicals is primarily a concern with strong acid or strong base resins. The weakly ionised resins require only slightly greater than stoichiometric chemical doses for complete regeneration regardless of whether concurrent or countercurrent flow is used. With strong acid or strong base resin systems, improved chemical efficiency can be achieved by reusing a part of the spent regenerants. In strongly ionised resin systems, the degree of column regeneration is the major factor in determining the chemical efficiency of the regeneration process. To realise 42 per cent of the resin's theoretical exchange capacity requires 1.4 times the stoichiometric amount of reagent [2 lb HCl/ft^3 (32 g HCl/l)]. To increase the exchange capacity available to 60 per cent of theoretical increases consumption to 2.45 times the stoichiometric dose [5 lb HCl/ft^3 (80 g HCl/liter)]. The need for acid doses considerably higher than stoichiometric means that there is a significant concentration of acid in the spent regenerant. Further, as the acid dose is increased incrementally, the concentration of acid in the spent regenerant increases. By discarding only the first part of the spent regenerant and saving and reusing the rest, greater exchange capacity can be realised with equal levels of regenerant consumption. For example, if a regenerant dose of 5 lb HCl/ft^3 (80 g HCl/liter) were used in the resin system, the first 50 per cent of spent regenerant would contain only 29 per cent of the original acid concentration. The rest of the acid regenerant would contain 78 per cent of the original acid concentration. If this second part of the regenerant is reused in the next regeneration cycle before the resin bed makes contact with 5 lb/ft^3 (80 g/liter) of fresh HCl, the exchange capacity would increase to 67 per cent of theoretical capacity. The available capacity would then increase from

60 to 67 per cent at equal chemical doses. Figure 8.5 shows the improved reagent utilisation achieved by this manner of reuse over a range of regenerant doses. Regenerant reuse has disadvantages in that it is higher in initial cost for chemical storage and feed systems and regeneration procedure is more complicated. Still, where the chemical savings have provided justification, systems have been designed to reuse parts of the spent regenerant as many as five times before discarding them.

Cost Considerations and Comparisons to Reverse Osmosis (RO)

The technology that competes with ion exchange in waste-water application is reverse osmosis (RO), therefore it is appropriate to make some comparisons. Direct cost comparisons are not straightforward and requires comparison of some of the hidden cost parameters. Since there appear to be few detailed comparisons in the open literature, there exists the general impression that RO is more economical than ion exchange. Whereas this may be true in a number of applications, as a general rule this is not the case.

The following are factors that should be taken into consideration when making case-specific cost comparisons between the two technologies. First, ion exchange is generally run batch, whereas RO is essentially continuous. This implies a higher degree of operator attention for ion exchange. However, one must remember that RO membranes must be cleaned and this can be frequently depending upon the treatment application. Furthermore, although ion exchange systems are discontinuous, commercial systems are fully automated and hence operator attention can be brought to a minimum.

A second consideration is that RO tends to be sensitive to incoming suspended matter. Comprehensive and sometimes expensive pre-treatment technologies are generally needed with RO, whereas ion exchange is less sensitive to the suspended matter. Further, RO systems are sensitive to hardness, so that softening is usually required as a pre-treatment. As a rule, RO membranes cannot handle high silica waters.

RO systems are quite sensitive to certain temperature ranges. Most critical is the range from 25° to 15°C, where RO systems have been reported to loose up to 30 per cent performance. RO has increased salt passage if the temperature increases, whereas ion exchange is insensitive to temperature variations. With the new generation of high performance resins, ion exchange can be kept fairly small using short operating cycles and regeneration utilisation approaches stoichiometric theoretical values. This translates into lower running costs.

Between the two technologies, ion exchange can be thought of as more within the arena of a pollution prevention technology. We can state this because of the relative recoveries or yields. If we take a boiler application as an example, with ion exchange, the difference between 'net throughput' (the water you produce for the boiler) and 'gross throughput' (the water you consume is minimal. You simply need a few cubic meters for the dilution of regenerants and for rinse. Typically, for a medium TDS water, the wasted water is about 5 per cent or less. With TDS and older co-flow regeneration systems, it can reach or exceed 10 per cent. In comparison, with RO, only about 70 to 75 per cent of the water that is pumped into the system can be recovered. RO rejects large volumes of concentrate. Ion exchange removes all ions down to extremely low residuals. It does not remove non-ionic species, however (only partially). In contrast, RO removes all compounds based on their size. Very small ions or molecules, such as Na, Cl, CO_2 are only partially removed. Other ions (Ca, SO_4) are harmful to the membrane. Globally, RO is a partial demineralisation process, whereas complete demineralisation can be achieved with a simple ion exchange plant. To achieve the same salt residual as obtained with a simple ion exchange plant, a double-pass RO system is needed and is considerably more expensive. The company Rhom and Hass (go to www.rohmhaas.com for details) has done a cost comparison between several design cases for the two technologies. Although Rhom and Hass favours ion exchange because they are marketing their technology, the comparisons are well done and clearly support a cost advantage ion exchange. The following is one cost comparison summary. In this case, the plant specs are as follows:

1. Daily water production = 91 m^3/hr.
2. Operating basis = 300 days per year.
3. Treated water specifications: Conductivity = 1 µS/cm; Residula silica = 30 µg/liter as SiO_2.
4. Feed water salinity = 6.54 meq/liter (327 mg/liter as $CaCO_3$).

The comparison is made against a simple ion exchange plant versus a single pass RO system. The following is the overall cost estimate comparison, followed by a breakdown of the individual cost item details, including operating and energy costs. In reviewing these comparative costs, you see that there must definitely a significant cost savings in favour of ion exchange over RO and given the water conservation advantage or ion exchange over RO, clearly it is more within the arena of pollution prevention technologies. This is not necessarily the case for every situation and a detailed cost comparison for any investment needs to be made on a case-specific basis.

CARBON ADSORPTION IN WATER TREATMENT

Activated carbon is a crude form of graphite, the substance used for pencil leads. It differs from graphite by having a random imperfect structure which is highly porous over a broad range of pore sizes from visible cracks and crevices to molecular dimensions. The graphite structure gives the carbon it's very large surface area which allows the carbon to adsorb a wide range of compounds. Activated carbon can have a surface of greater than 1000 m^2/g. This means 5 grams of activated carbon can have the surface area of a football field.

Adsorption is the process by which liquid or gaseous molecules are concentrated on a solid surface, in this case activated carbon. This is different from absorption, where molecules are taken up by a liquid or gas. Activated carbon can made from many substances, containing a high carbon content such as coal, wood and coconut shells. The raw material has a very large influence on the characteristics and performance activated carbon.

The term activation refers to the development of the adsorption properties of carbon. Raw materials such as coal and charcoal do have some adsorption capacity, but this is greatly enhanced by the activation process. There are three main forms of activated carbon:

1. Granular activated carbon (GAC) — irregular shaped particles with sizes ranging from 0.2 to 5 mm. This type is used in both liquid and gas phase applications.
2. Powder activated carbon (PAC) — pulverised carbon with a size predominantly less than 0.18 mm (US Mesh 80). These are mainly used in liquid phase applications and for flue gas treatment.
3. Pelleted activated carbon — extruded and cylindrical shaped with diameters from 0.8 to 5 mm. These are mainly used for gas phase applications because of their low pressure drop, high mechanical strength and low dust content.

Activated carbon is also available in special forms such as a cloth and fibers. Activated Charcoal Cloth (ACC) represents a family of activated carbons in cloth form. These products are fundamentally unique in several important ways compared with the traditional forms of activated carbon and with other filtration media that incorporate small particles of activated carbon. ACC products are similar to the traditional activated carbon products in that they are 100 per cent activated carbon. This gives the products the same high capacity for adsorption of organic compounds and other odorous gases as the more traditional, pelletised, granular and powder forms of activated carbon. As with the traditional forms of activated carbons, ACC products can be impregnated with a range of chemicals to enhance the chemisorption capacity for selected gases. By being constructed of bundles of activated carbon filaments and fibers in a textile form, several important advantages are imparted to ACC. The diameter of these fibres is approximately 20 mm, so the kinetics for ACC products are similar to that of a very tine carbon particle. Gases and liquids can flow through the fabric and the accelerated adsorption kinetics mean that the ACC can retain the advantages of mass transfer zones associated with deeper filter beds. Faster adsorption rates mean smaller adsorption equipment and up to twenty times less carbon on line.

Adsorption is the process where molecules are concentrated on the surface of the activated carbon. Adsorption is caused by London Dispersion Forces, a type of Van der Waals Force which exists between molecules. The force act in a similar way to gravitational forces between planets. London Dispersion Forces are extremely short ranged and therefore sensitive to the distance between the carbon surface and the adsorbate molecule. They are also additive, meaning the adsorption force is the sum of all interactions between all the atoms. The short range and additive nature of these forces results in activated carbon having the strongest physical adsorption forces of any material known to mankind. All compounds are adsorbable to some extent. In practice, activated carbon is used for the adsorption of mainly organic compounds along with some larger molecular weight inorganic compounds such as iodine and mercury. In general, the adsorbability of a compound increases with: (i) increasing molecular weight, (ii) a higher number of functional groups such as double bonds or halogen compounds, and (iii) increasing polarisability of the molecule. This is related to electron clouds of the molecule (Fig. 8.6).

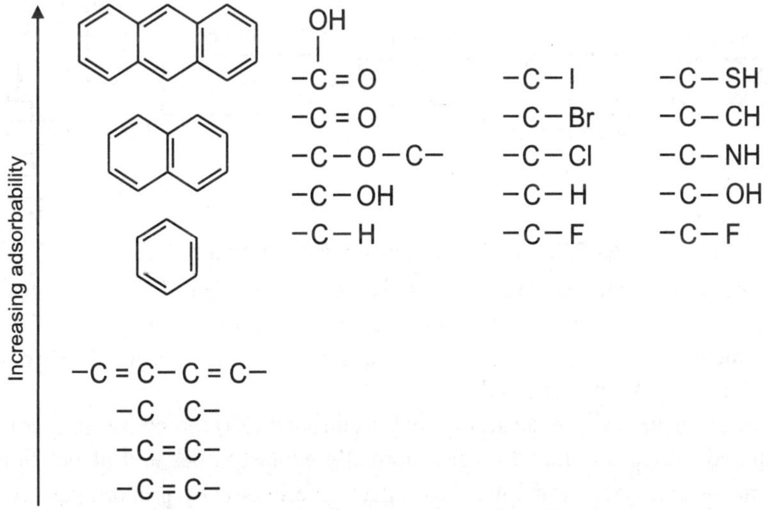

Fig. 8.6. Relative adsorptivity of organic materials.

The most common manufacturing process is high temperature steam activation though activated carbon can also be manufactured with chemicals. Along with the raw material, the activation process has a very large influence on the characteristics and performance of activated carbon. Figure 8.7 illustrates the production of granular activated carbons by steam activation of selected grades of pulverised and then reagglomerated bituminous coal.

Application

The way activated carbon is used is normally determined by the form of activated carbon. With granular and pelleted activated carbon, in most cases the carbon is installed in a fixed bed with the liquid or gas passing through the bed. The compounds to be removed are retained on the activated carbon. The carbon is used until exhaustion. It can then be: reactivated, normally off-site.

Functional groups

The molecular structure of the resin is such that it must contain a macroreticular tissue with acid or basic radicals. These radicals are the basis of classifying ion exchangers into two general groups: (i) cation exchangers, in which the molecule contains acid radicals of the HSO_3 or HCO type able to fix mineral or organic cations and exchange with the hydrogen ion H^+, and (ii) anion exchangers, containing basic radicals (for example, amine functions of the type NH_2) able to fix mineral or organic anions and exchange them with the hydroxyl ion OH^- coordinate to their dative bonds. The presence of these radicals enable a cation exchanger to be assimilated to an acid of form H-R and an anion exchanger to be a base of form OH-R when regenerated. *In-situ* regenerated. This is possible for most gas phase and some liquid phase applications.

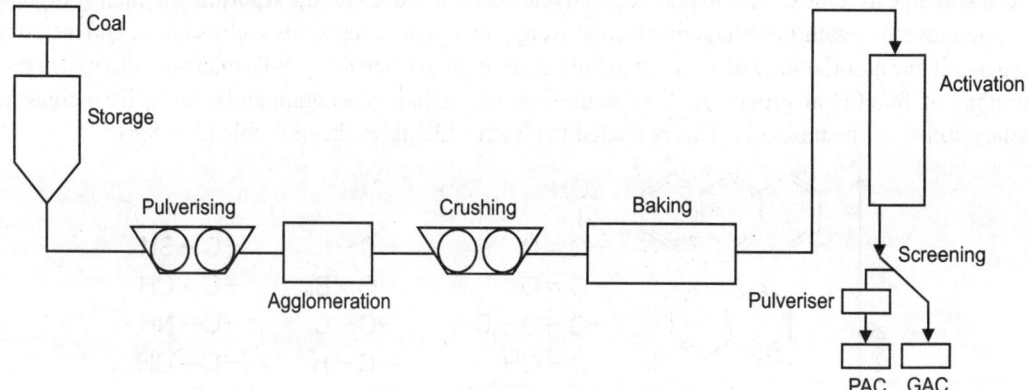

Fig. 8.7. Manufacture process for activated carbon.

It may also be replaced with new carbon and disposal of the exhausted carbon. Most adsorbers are pressure vessels constructed in carbon steel, stainless steel or plastic. Large systems for drinking water are often constructed in concrete. In some cases, a moving or pulsed bed adsorber is employed to optimise the use of the granular activated carbon.

The main factors in the design of an adsorption system are the: (i) carbon consumption — the amount of carbon required to treat the liquid or gas, normally expressed per unit of the fluid treated, and (ii) contact time — for a fixed flow rate, the contact time is directly proportional to the volume of carbon and is the main factor influencing the size of the adsorption system and capital cost.

With powder activated carbon, in most cases, the carbon, is dosed into the liquid, mixed and then removed by a filtration process. In some cases, two or more mixing steps are used to optimise the use of powder carbon. Powder activated carbon is used in a wide range of liquid phase applications and some specific gas phase applications such as Incinerator flue gas treatment and where it is bonded into filters such as fabrics for personnel protection.

Contaminants Activated Carbon

Activated carbon (AC) filtration is most effective in removing organic contaminants from water. Organic substances are composed of two basic elements, carbon and hydrogen. Because organic chemicals are often responsible for taste, odour and colour problems, AC filtration can generally be used to improve aesthetically objectional water. AC filtration will also remove chlorine. AC filtration is recognised by the Water Quality Association as an acceptable method to maintain certain drinking water contaminants within the limits of the EPA National Drinking Water Standards (Table 8.4).

Table 8.4. Water contaminants that can be reduced to acceptable standards by activated carbon filtration.

Primary drinking water standards contaminant	*MCL, mg/l
Inorganic Contaminants	
Organic arsenic complexes	0.05
Organic chromium complexes	0.05
Mercury (Hg^{+2}) inorganic	0.05
Organic mercury complexes	0.002
Organic contaminants	
Benzene	0.005
Endrin	0.0002
Lindane	0.004
Methoxychlor	0.1
1,2-dichloroethane	0.005
1,1-dichloroethylene	0.007
1,1,1-trichloroethane	0.200
Total trihalomethanes (TTHMs)	0.10
Toxaphene	0.005
Trichloroethylene	0.005
2,4-D	0.1
2,4,5-TP (Silvex)	0.01
Para-dichlorobenzene	0.075
Colour	15 colour units
Foaming agents (MBAS)	0.5 mg/l
Odour	3 threshold odour number

*Maximum contaminant level.

**Secondary maximum contaminant level.

AC filtration does remove some organic chemicals that can be harmful if present in quantities above the EPA Health Advisory Level (HAL). Included in this category are trihalomethanes (THM), pesticides, industrial solvents (halogenated hydrocarbons), polychlorinated biphenyls (PCBs) and polycyclic aromatic hydrocarbons (PAHs).

THMs are a by-product of the chlorination process that most public drinking water systems use for disinfection. Chloroform is the primary THM of concern. EPA does not allow public systems to have more than 100 parts per billion (ppb) of THMs in their treated water. Some municipal systems have had difficulty in meeting this standard.

The Safe Drinking Water Act mandates EPA to strictly regulate contaminants in community drinking water systems. As a result, organic chemical contamination of municipal drinking water is not likely to be a health problem. Contamination is more likely to go undetected and untreated in unregulated private water systems. AC filtration is a viable alternative to protect private drinking water systems from organic chemical contamination. Note that radon gas can also be removed from water by AC filtration, but actual removal rates of radon for different types of AC filtration equipment have not been established. This makes AC ideal for use both in industrial applications as well as in residential. Similar to other types of water treatment, AC filtration is effective for some contaminants and not effective for others. AC filtration does not remove microbes, sodium, nitrates, fluoride and hardness. Lead and other heavy metals are removed only by a very specific type of AC filter. Unless the manufacturer states that its product will remove heavy metals, one should assume that the AC filter is not effective in removing them.

AC works by attracting and holding certain chemicals as water passes through it. AC is a highly porous material; therefore, it has an extremely high surface area for contaminant adsorption. The equivalent surface area of 1 pound of AC ranges from 60 to 150 acres. AC is made of tiny clusters of carbon atoms stacked upon one another. The carbon source is a variety of materials, such as peanut shells or coal. The raw carbon source is slowly heated in the absence of air to produce a high carbon material. The carbon is activated by passing oxidising gases through the material at extremely high temperatures. The activation process produces the pores that result in such high adsorptive properties. The adsorption process depends on the following factors: (i) physical properties of the AC, such as pore size distribution and surface area, (ii) the chemical nature of the carbon source or the amount of oxygen and hydrogen associated with it, (iii) chemical composition and concentration of the contaminant, (iv) the temperature and pH of the water, and (v) the flow rate or time exposure of water to AC.

Make Note of Physical and Chemical Properties

Forces of physical attraction or adsorption of contaminants to the pore walls is the most important AC filtration process. The amount and distribution of pores play key roles in determining how well contaminants are filtered. The best filtration occurs when pores are barely large enough to admit the contaminant molecule. Because, contaminants come in all different sizes, they are attracted differently depending on pore size of the filter. In general AC filters are most effective in removing contaminants that have relatively large molecules (most organic chemicals). Type of raw carbon material and its method of activation will affect types of contaminants that are adsorbed. This is largely due to the influence that raw material and activation have on pore size and distribution.

Processes other than physical attraction also affect AC filtration. The filter surface may actually interact chemically with organic molecules. Also electrical forces between the AC surface and some contaminants may result in adsorption or ion exchange. Adsorption, then, is also affected by the chemical nature of the adsorbing surface. The chemical properties of the adsorbing surface are determined to a large extent by the activation process. AC materials formed from different activation processes will have chemical properties that make them more or less attractive to various contaminants. For example chloroform is adsorbed best by AC that has the least amount of oxygen associated with the pore surfaces. You can't possibly determine the chemical nature of an AC filter. However, this does point out the fact that different types of AC filters will have varying levels of effectiveness in treating different chemicals.

Contaminant Properties

As already noted, large organic molecules are most effectively adsorbed by AC. A general rule of thumb is that similar materials tend to associate. Organic molecules and activated carbon are similar materials; therefore there is a stronger tendency for most organic chemicals to associate with the activated carbon in the filter rather than staying dissolved in a dissimilar material like water. Generally, the least soluble organic molecules are most strongly adsorbed. Often the smaller organic molecules are held the tightest, because they fit into the smaller pores.

Concentration of organic contaminants can affect the adsorption process. A given AC filter may be more effective than another type of AC filter at low contaminant concentrations, but may be less effective than the other filter at high concentrations. This type of behaviour has been observed with chloroform removal. The filter manufacturer should be consulted to determine how the filter will perform for specific chemicals at different levels of contamination.

Conditions that Impact on Performance

Adsorption usually increases as pH and temperature decrease. Chemical reactions and forms of chemicals are closely related to pH and temperature. When pH and temperature are lowered many organic chemicals are in a more adsorbable form. The adsorption process is also influenced by the length of time that the AC is in contact with the contaminant in the water. Increasing contact time allows greater amounts of contaminant to be removed from the water. Contact is improved by increasing the amount of AC in the filter and reducing the flow rate of water through the filter.

AC filters can be a breeding ground for micro-organisms. The organic chemicals that are adsorbed to the AC are a source of food for various types of bacteria. Pathogenic bacteria are those that cause human diseases such as typhoid, cholera and dysentery. AC filtration should only be used on water that has been tested and found to be bacteria free or effectively treated for pathogenic bacteria. Other types of non-pathogenic bacteria that do not cause diseases have been regularly found in AC filters. There are times when high amounts of bacteria (non-pathogenic) are found in water filtered through an AC unit. Research shows little risk to healthy people that consume high amounts of non-pathogenic bacteria. We regularly take in millions of bacteria every day from other sources. However, there is some concern for certain segments of the population, such as the very young or old and people weakened by illness. Some types of non-pathogenic bacteria can cause illness in those whose natural defenses are weak. Flushing out bacteria that have built up in the filter can be accomplished by backflushing the AC filter prior to use. Water filtered after the initial flushing will have much lower levels of bacteria and ingestion of a high concentration of bacteria will have been avoided. Some compounds of silver have been used as disinfectants, especially in European operations. Silver has been added to certain AC filters as a solution to the bacteria problem. Unfortunately, product testing has not shown silver impregnated AC to be much more effective in controlling bacteria than normal AC filters. The areas that require definition when specifying and sizing a carbon adsorption system include:

Processing conditions:
1. Concentration of adsorbate.
2. Temperature of liquid stream.
3. pH of liquid stream.
4. Flow rates and operating frequency.
5. Pressure drop in system.

Characteristics of the adsorbate:
1. Relative molecular mass.
2. Solubility of the adsorbate.
3. Concentration relative to solubility limits.
4. Polarity of adsorbate.
5. Temperature of solution.

Selection of adsorbent for optimum efficiency:
1. Specific adsorption isotherm.
2. Selection of optimum activity level.
3. Cost sensitivity analysis.
4. Consideration of thermal reactivation.

We will now touch upon some of these factors. First, let's look at what we mean by system isotherm. Freundlich liquid phase isotherm studies can be used to establish the adsorptive capacity of activated carbon over a range of different concentrations. Under standard conditions, the adsorptive capacity of activated carbon increases as the concentration increases, until we reach a point of maximum saturation capacity. The Freundlich liquid phase isotherm can be used to determine the effect of solubility on the adsorptive capacity of activated carbon over a range of different concentrations. Phenol is highly soluble due to its polar nature whilst, in comparison, tetrachloroethylene (PCE) has a low solubility due to being non-polar. In the isotherms illustrated, the concentration of phenol is low relative to its solubility limit and consequently, the adsorptive capacity peaks at 18 per cent maximum. In comparison the concentration of tetrachloroethylene is relatively close to its solubility limit and accordingly, the adsorptive capacity is exceptionally good.

Initially, upon start-up of an adsorption unit, a slight increase in the effluent water's pH is observed. This is attributed to trace leaching of the soluble matter from the matrix of the activated carbon. Leaching can be controlled by application of efficient backwashing, which will readily remove any soluble materials. The system will then reach equilibrium with the pH of the feedwater.

Adsorption efficiency can be optimised by using finer particle size products which will improve the diffusion rate to the surface of the activated carbon. However, there is a tradeoff in using finer particles with pressure drop and hence energy use.

To best understand adsorptive solvent recovery we have to consider some fundamentals of adsorption and desorption. In a very general sense, adsorption is the term for the enrichment of gaseous or dissolved substances (the adsorbate) on the boundary surface of a solid (the adsorbent). On their surfaces adsorbents have what we call active centers where the binding forces between the individual atoms of the solid structure are not completely saturated. At these active centers an adsorption of foreign molecules takes place. The adsorption process generally is of an exothermal nature. With increasing temperature and decreasing adsorbate concentration the adsorption capacity decreases. For the design of adsorption processes it is important to know the adsorption capacity at constant temperature in relation to the adsorbate concentration.

Remember that this technology is versatile and is applied equally well to solvent recovery and pollution control applications in gas as well as liquid systems. Let's now focus attention on the applications in water treatment.

Applications

Applications of carbon adsorption go far beyond conventional water treatment applications which we will discuss in a general sense shortly. Table 8.5 provides a summary of the key applications of carbon adsorption systems for liquid phase applications.

The most common application of carbon adsorption in municipal water treatment is in the removal of taste and odour compounds. Figure 8.8 provides an example of a process flow diagram for a municipal water treatment plant. In this example water is pumped from the river into a flotation unit, which is used for the removal of suspended solids such as algae and particulate matter.

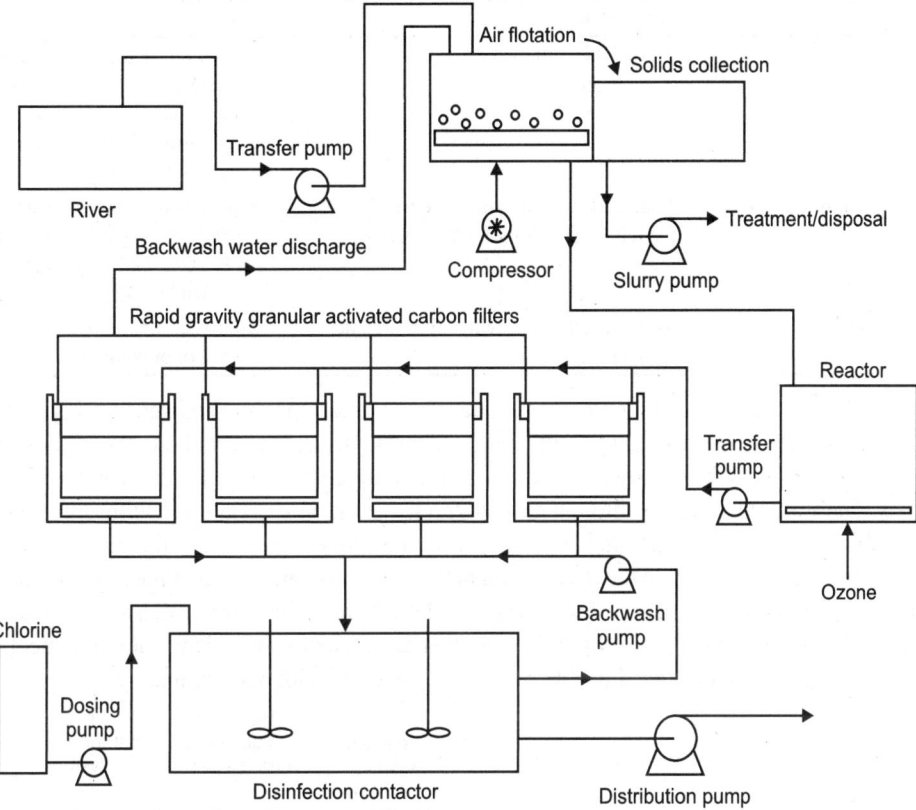

Fig. 8.8. Process flow sheet for municipal water plant.

Dissolved air is injected under pressure into the basin. This action creates micro bubbles which become attached to the suspended solids, causing them to float. This results in a layer of suspended solids on the surface of the water, which is removed using a mechanical skimming technique.

Table 8.5. Liquid phase applications of carbon adsorption.

Industry	Description	Use
Potable water treatment	Granular activated carbons (GAC) installed in rapid gravity filters	Removal of dissolved organic contaminants, control of taste and odour problems
Soft drinks	Potable water treatment, sterilisation with chlorine	Chlorine removal and adsorption of dissolved organic materials
Brewing	Potable water treatment	Removal of trihalomethanes (THM) and phenolics
Semi-conductors	Ultra-high purity water	Total organic carbon (TOC) reduction
Gold recovery	Operation of carbon in leach, carbon in pulp, and heap leach circuits	Recovery of gold from tailings dissolved in sodium cyanide
Petrochemical	Recycling of steam condensate for boiler feed water	Removal of oil and hydrocarbon contamination

(Contd...)

Industry	Description	Use
Groundwater	Industrial contamination of ground water reserves	Reduction of total organic halogens (TOX) and adsorbable organic halogens (AOX) including chloroform, tetrachloroethylene and trichloroethylene
Industrial waste-water	Process effluent treatment to meet environmental discharge standards	Reduction of total organic halogens (TOX), biological oxygen demand (BOD), and chemical oxygen demand (COD)
Swimming pools	Ozone injection for removal of organic contaminants	Removal of residual ozone and control of chloramine levels

The next step in the process involves the production of ozone bypassing high tension, high frequency electrical discharges through air in specially designed units. Ozone is injected into the water to provide bactericidal action and to break down the natural humic compounds that are the cause of taste and odour problems. The water then passes through a rapid gravity filtration system filled with activated carbon (GAC), which adsorbs the compounds resulting from the ozone treatment. Following adsorption, the water is disinfected for supply to the distribution network. Understand that treatment plants are unique, in many ways like oil refineries, i.e. design basis can be substantially different depending on the nature of the water being treated. Figure 8.9 provides another example of a municipal water treatment facility using PAC. Again the plant is used for the removal of taste and odour compounds.

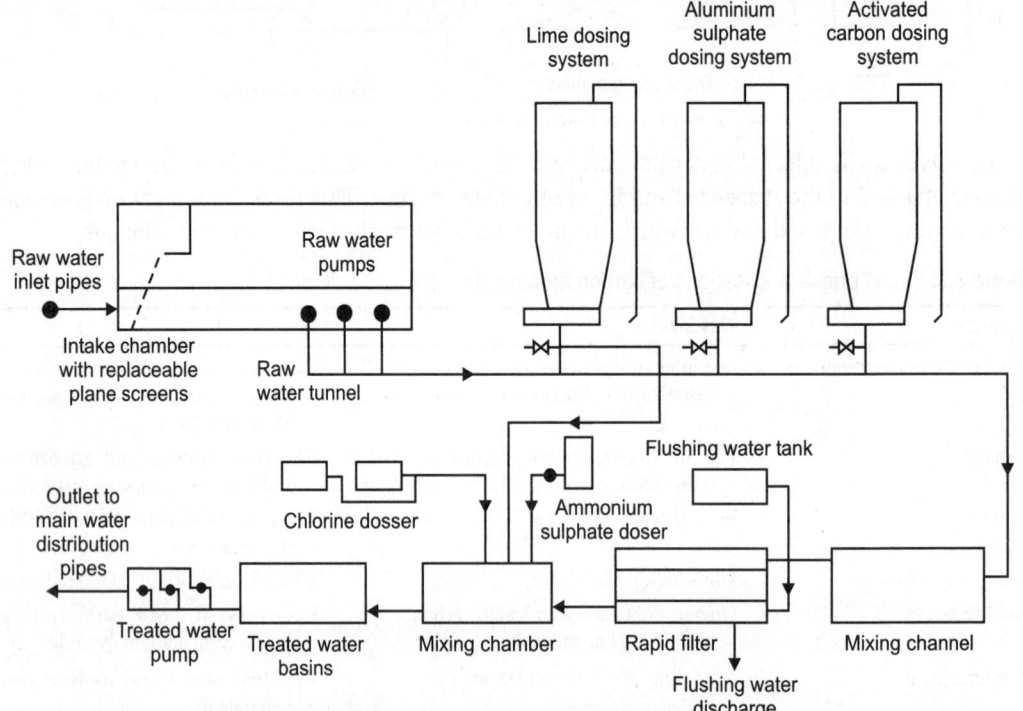

Fig. 8.9. Example of a municipal water treatment plant for taste and odour control.

There are regions where the treatment of water is intended for potable purposes is not necessary at all times during the year. The presence of taste, odour and naturally occurring toxins largely depends on the biological action in areas where lake or reservoir water supply is common. In these situations it is more cost effective to use intermittent dosing of activated carbon into the water during those times of the year where it is needed. The use of PAC is preferred in these cases, mainly because no costly fixed bed filtration equipment is required. The PAC can be dosed directly to existing flocculant tanks at a prescribed rate to achieve the level of pollutant removal required. Shown in Fig. 8.9, following the dosing of PAC the activated carbon is removed as part of the flocculation process or it can be filtered out by mechanical means. The final stage of water treatment is disinfection, whereupon the water is pumped to the distribution network.

Non-potable water treatment is also well within the economical applications of liquid phase adsorption systems. There, in fact, are so many unique examples of process water treatment throughout the chemical industry that we could go on for days discussing specify systems.

Figure 8.10 shows a process diagram for the removal of creosote and pesticides from the liquid phase in a timber treatment facility. A storage dosing tank is used for smoothing the flow from where the water is pumped into a chemical dosing system for pH adjustment. Then, ferric sulphate is added to form a precipitate with suspended solids, which is subsequently flocculated by the addition of polyelectrolyte.

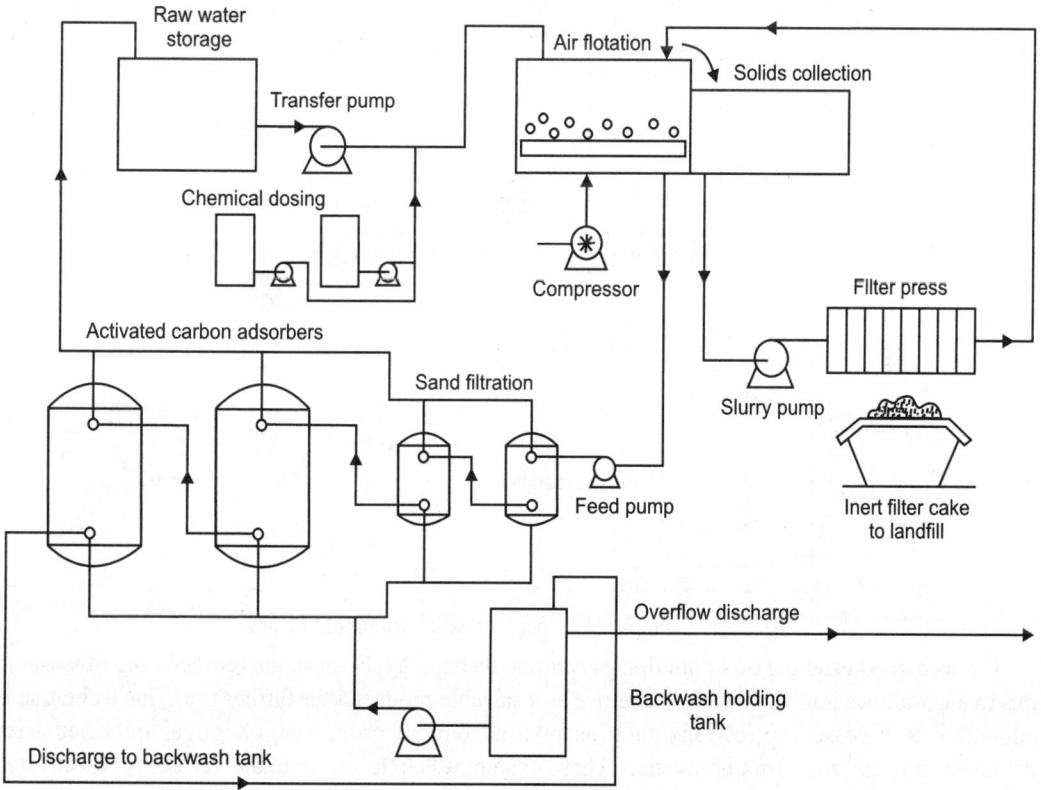

Fig. 8.10. Example of a process water treatment facility.

The water is then pumped through series operated sand filters, which provide the final stage of suspended solids removal and protect the granular activated carbon (GAC) filters from particulate contamination. Series operated GAC filters are then used to remove the dissolved creosote and pesticides from the water. To achieve compliance with specifications levels, water should be sampled and analysed after leaving the first GAC filter. The second GAC filter normally serves as a guard bed.

A final example of application and process layout is shown in Fig. 8.11. In this example the process relies on activated carbon to remove colour bodies from a recycled glucose intermediary prior to use in the production of confectionary. The glucose containing the colour taint must be mildly heated (to about 70°C), so that the normally solid product becomes less viscous and easier to pump. The syrup is pumped through a series of high efficiency filters (mechanical type) that remove entrained particulate matter and crystallised sugar formed during the heating process. Filtered syrup is then passed through columns containing GAC using a high residence time (a variation is simply the addition of PAC on an as-needed basis—this obviously has cost advantages for batch operated systems). During these stages the colour bodies are physically adsorbed by the activated carbon. When PAC is added to the process, the heated syrup is agitated. Following agitation, the syrup undergoes mechanical filtration to remove entrained PAC prior to the glucose being used to manufacture the confectionary.

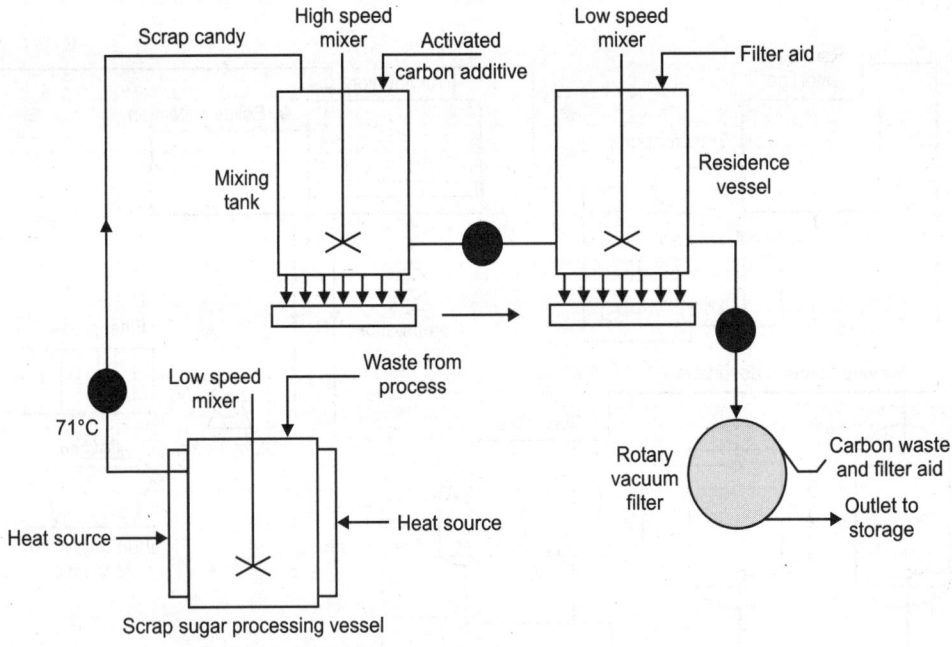

Fig. 8.11. Example of a decolourisation treatment facility.

This is a good example of a pollution prevention technology, because the reprocessing of waste in this manner allows it to be suitable for re-use as a saleable product after further use. This technique is adaptable in diverse applications such as pharmaceutical processes, chemical intermediaries manufacturing and soft drink production. These examples help to illustrate the versatility of activated carbon in standard water treatment applications. Another application which merits a distinct discussion is goundwater remediation.

Evaluating the Merits of Carbon Adsorption

Where activated carbon is a potential treatment technology, the first evaluation step is generally to run simple isotherms to determine feasibility. Isotherms are based on batch treatment where impurities reach equilibrium on available carbon surface. While such tests provide an indication of the maximum amount of impurity a GAC can adsorb, it cannot give definite scale up data for a GAC operation due to several factors:

1. In a GAC column, dynamic adsorption occurs along an adsorption wave front where the impurity concentration changes.
2. GAC rarely becomes totally exhausted in a column.
3. Ground GAC exhibits a significantly greater rate of adsorption than a normal GAC.
4. Effects of recycling with regeneration cannot be studied.

Because of these factors, pilot column tests should be conducted using the most promising carbons as indicated by the isotherms in order to give an accurate comparison of the carbons. In addition to this, by utilising reliable scaling-up calculations, pilot tests can be used for system sizing. Pilot column tests can be operated with columns arranged in parallel or in series. The choice of arrangement of the columns is dependent on the scope of the test; such as: (i) comparison of GAC grades and/or the effect of regeneration, and (ii) scale up to full plant design. In the first case, the columns should be arranged in parallel. In the second case, the columns arranged in series loaded with one GAC grade. The columns are mounted vertically and arranged as shown in Fig. 8.12 for the example of scale up tests and downflow operation. It is often good practice to operate columns in upflow as this reduces the opportunity for channelling.

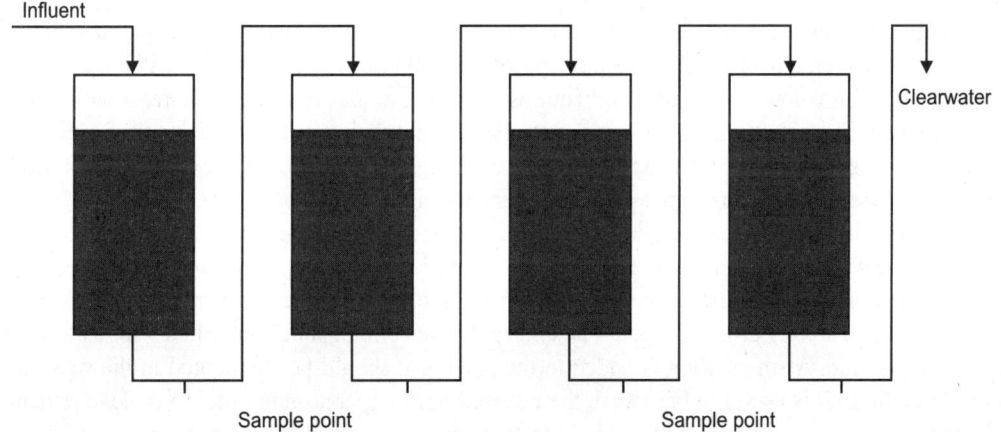

Fig. 8.12. Configuration of column scale-up tests.

It is also preferred where suspended solids create a high pressure drop or dissolved gases create bubbles in the carbon bed. For a downflow or percolation system, an influent line should be installed at the top of the column, with an effluent at the bottom. To prevent the column from draining during operation, the effluent line from the last column should extend from the bottom of the column to above the top of the column. This will keep the column filled with liquid at all times during operation and prevent siphoning from occurring. It is recommended that suspended solids be removed from the feed stream to a GAC column. If this is not possible in the scale-up design, then the effect of suspended solids should be included in the pilot run. In the upflow operation, most of the suspended solids work their way up through the GAC bed without a significant increase in pressure drop.

The carbon bed should be at least 60 cm deep with a 4 cm internal diameter. A smaller column is not recommended as the wall effect becomes significant. The carbon bed can be supported by glass wool, wire cloth, etc. Columns and fixtures can be constructed from glass, plastic, reinforced fiberglass or metal. Borosilicate glass is commonly used. It is essential that all columns used in the pilot system have at least the same internal diameter.

Contact flow rates, hourly space velocity (HSV) or the quantity of feed liquor, are expressed in the number of carbon bed volumes passing through the column per hour. A bed volume is the volume occupied by the carbon bed, including carbon volume and void volume. The recommended HSV range is between 0.1 and 3.0 depending upon the degree of purification required, the type and concentration of impurity, the nature of the process liquor and the pressure drop. Generally, high levels of purification, high impurity concentrations and/or high viscosities will require a lower HSV. As an example — a typical HSV, for decolourisation of starch based sweeteners is 0.25. The carbon will perform more efficiently, but with a tradeoff in the amount of liquid that can be processed through a column in a given period of time. On the other hand for the removal of traces of organics in drinking water and waste-water; a HSV range of 2 to 3 produces good results. As the flow rate and quantity of liquor are the most important controllable variables in developing design data, a feed pump suitable for accurate and continuous flow is required. Depending on the size of the pilot column system, the use of peristaltic, diaphragm, piston-type or centrifugal pumps are recommended. The feed pump should be used in combination with a volumetric or gravimetric flow control device. Before process liquor is delivered to the carbon, suspended matter should be removed, preferably by the same method planned for the plant system.

Adsorption to activated carbon is a function of diffusion rate. This means that the columns should be operated at a temperature at which the liquor approaches the viscosity of water but will not decompose or form too much colour. If the process requires operation at elevated temperature; a jacketed tube should be used. The column temperature is controlled by circulating water of the required temperature through the jacket. This requires a thermostatic bath enabling accurate temperature control. When loading the column, care should be taken to avoid entrapping air in the carbon bed.

Entrapped air can cause channelling during column operation, preventing complete contact of the process liquor with the carbon particles. In small columns, entrapped air can be avoided by pouring out the carbon in boiling water just before loading. Most of the excess water can be poured off, along with most of the fine carbon particles. About one quarter of the column should be filled with water before loading the carbon. As the carbon is added to the column it should be submerged in the water at all times. Sometimes it is useful to backwash the carbon prior to operation in order to remove remaining dust and entrapped air. It is recommended that a T-junction is used between dosing pump and the columns to allow any entrapped air to be released. Before starting the test, the complete system should be checked by running on water for several hours. After setting the appropriate flowrate, the liquor to be treated can be fed to the columns and this will displace the water. Several samples of the feed liquor should be taken over the duration of the test to highlight any drift in impurity concentration. Samples of effluent liquor after each column should be taken at regular time intervals or after a fixed number of processed bed volumes. If the accuracy of the analysis is affected by undissolved solids, all samples should be filtered prior to examination. For examination of purity levels of taken samples, standard purity test can be used.

When the effluent from the parallel columns or last column in series exceeds the purity requirement, the test should be stopped. It is not recommended to stop the liquor feed during the test. A procedure with intermittent daytime runs will give deviating test results in most cases. When you collect your data, tabulate the information on a spreadsheet. Effluent impurity concentrations for all columns should be plotted against elapsed time or processed bed volumes, generating 'Breakthrough curves'. The points at which the purity requirement is exceeded are defined as 'Breakthrough points'.

Comments on Both Technologies

Both technologies are extremely important to achieving high quality water characteristics and both are complex — each posing a different set of challenges in scaling up to commercial size operations. You will find that most equipment suppliers have the expertise to tailor their equipment and processes to specific applications, but that in many situations, pilot scale testing will be required.

Table 8.6 provides information of some representative organic chemicals that can be removed from water using activated carbon systems.

Table 8.6. Representative organic chemicals and typical retentivities on activated carbons.

Chemical	Formula	Molecular weight	Boiling point @760 mm Hg, °C	Average retentivity in % at 20°C and 760 mm Hg
Methane Series	C_nH_{2n+2}			
Methane	CH_4	16.04	−184	1
Ethane	C_2H_6	30.07	−86	1
Propane	C_3H_6	44.09	−12	5
Butane	C_4H_{10}	58.12	1	8
Pentane	C_5H_{12}	72.15	37	12
Hexane	C_6H_{14}	86.17	69	16
Heptane	C_7H_{16}	100.20	98.4	23
Octane	C_8H_{18}	114.23	125.5	25
Nonane	C_9H_{20}	128.25	150.0	25
Decane	$C_{10}H_{22}$	142.28	231.0	25
Acetylene series	C_nH_{2n-2}			
Acetylene	C_2H_2	26.04	−88.5	2
Propyne	C_3H_4	40.06	−23.0	5
Butyne	C_4H_6	54.09	27.0	8
Pentyne	C_5H_8	68.11	56.0	12
Hexyne	C_6H_{10}	82.14	71.5	16
Ethylene series	C_nH_{2n}			
Ethylene	C_2H_4	28.05	−103.9	3
Propylene	C_3H_4	42.08	−17.0	5
Butylene	C_4H_8	56.10	−5.0	8
Pentylene	C_5H_{10}	70.13	40.0	12
Hexylene	C_6H_{12}	84.16	64.0	−
Heptylene	C_7H_{14}	98.18	94.9	25
Octalene	C_8H_{16}	112.21	123.0	25

(Contd...)

Chemical	Formula	Molecular weight	Boiling point @760 mm Hg, °C	Average retentivity in % at 20°C and 760 mm Hg
Benzene series	C_nH_{2n-6}			
Benzene	C_6H_6	78.11	80.1	24
Toluene	C_7H_8	92.13	110.8	29
Xylene	C_8H_{10}	106.16	144.0	34
Isoprene	C_5H_8	68.11	34.0	15
Turpentine	$C_{10}H_{16}$	136.23	180.0	32
Naphthalene	$C_{10}H_8$	128.16	217.9	30
Phenol	C_6H_5OH	94.11	182.0	30
Methyl Alcohol	CH_3OH	32.04	64.7	15
Ethyl Alcohol	C_2H_5OH	46.07	78.5	21
Propyl Alcohol	C_3H_7OH	60.09	97.19	26
Butyl Alcohol	C_4H_9OH	74.12	117.71	30
Amyl Alcohol	$C_5H_{11}OH$	88.15	138.0	35
Cresol	C_7H_7OH	108.13	202.5	30
Methanol	$C_{10}H_{19}OH$	156.26	215	20
Formaldehyde	H_3CHO	30.03	−21.9	3
Acetaldehyde	CH_3CHO	44.05	21.0	7

Chapter 9

Water Sterilisation Technologies

INTRODUCTION

This chapter provides with a very basic overview of the principles and technologies associated with water purification or more specifically, sterilisation. In very simplistic terms, there are two general classes of technologies, namely those based on chemical methods and those based on non-chemical technologies. The major application is in purifying water for human consumption purposes.

WATERBORNE DISEASES

Untreated waters contain a number of harmful pollutants which give the water colour, taste and odour. These pollutants include viruses, bacteria, organic materials and soluble inorganic compounds and these must be removed or rendered harmless before the water can be used again. A breakdown of the documented outbreaks identifies acute gastroenteritis, hepatitis shigellosis, ciardiasis, chemical poisoning, typhoid fever and salmonellosis. Sources of contaminated water can be traced to semipublic water systems, municipal water systems and to individual water systems.

In cell culture, it has been shown that one virion can produce infection. In the human host, because of acquired resistance and a variety of other factors, the one virion/one infection possibility does not exist. Very little is known of the epidemiology of waterborne diseases. The current database is insufficient to determine the scope and intensity of the problem. The devastating effect of epidemics is sufficient to rank water-associated epidemics as a most important public health problem.

Viruses and bacteria may be eliminated by chemical methods or by irradiation and organic poisons may also be controlled. Inorganic matter must be removed by other means.

Viruses

Viruses are ultra-microscopic organisms. They are parasites; they need to infest a host in order to duplicate themselves. Viruses excreted with human and animal feces are called enteric viruses and more than 100 such organisms have been identified. As many as one million viruses can be found in one gram of excrement. The concentration in raw sewage varies over a wide range; as many as 4,63,500 infectious particles per liter of raw sewage have been detected. Viruses found in surface waters are introduced from three major sources. Viruses of human origin can be traced to untreated or inadequately treated domestic sewage. Runoffs from agricultural land, feedlots and forests introduce viruses from domestic and wild animals and birds. Plant viruses, insect viruses and other forms of life associated with the aquatic environment may also infect the waters.

Bacteria

In addition to viruses, bacteria (microscopic organisms that can reproduce without a host in the proper conditions) are also found in water. In general, damage to the human body from bacterial infection is due to the action of the toxins they produce. Bacteria found in water are derived from contact with air, soil, living and decaying plants and animals and animal excrements. Many of these bacteria are aerobic and anaerobic spore-forming organisms associated with varying densities of coliforms, fecal coliforms, fecal streptococci, staphylococci, chromogenic forms, fluorescent strains, nitrifying and denitrifying groups, iron and sulphur bacteria, proteus species and pathogenic bacteria. Many bacteria are of little sanitary significance and die rapidly in water. Fecal pollution adds a variety of intestinal pathogens. The most common genera found in water are salmonella, shigella, vibrio, mycobacterium, pasteurella and leptospira.

Contamination of Water

The circumstances under which water becomes contaminated are as varied as the ways water is taken internally. It is then conceivable that almost any virus could be transmitted through the water route. The increased use of water for recreational purposes increases the incidence of human contact with bodies of water and consequently, with waterborne viruses and bacteria. The major waterborne viruses among pathogens and the most likely candidates for water transmission, are the picornaviruses (from pico, meaning very small and RNA, referring to the presence of nucleic acid). The characteristics of picornaviruses are shown in Table 9.1.

Table 9.1. Picornavirus characteristics (very small RNA viruses).

Small spheres
RNA core, icosahedral form of cubic symmetry
Resistant to ether, chloroform and bile salts, indicating lack of essential lipids
Heat stabilised in presence of divalent cations (Molar $MgCl_2$)
Enteroviruses separated from rhinoviruses by acid lability of the latter viruses (inactivated at pH 3.0–5.0)

Among the picornaviruses are the enteroviruses (polioviruses, coxsackieviruses and echoviruses) and the rhinoviruses of human origin. Also included are enteroviruses from excrements of cattle, swine and other domesticated animals; and rhinoviruses of non-human origin, viruses of foot and mouth disease, teschen disease, encephalomyocarditis, mouse encephalomyelitis, avian encephalomyelitis and vesicular exantherm of pigs. Additionally, certain viruses can be transported by the water route because their vectors, water moulds and nematodes live in the soil and move with the movement of water. Plant pathogenic viruses also enter the water route and contribute to the problem, though this area has been given little attention in the past. Viruses associated with industrial abattoirs, meat packing, food processing, pharmaceutical and chemical operations are also a potential problem. All enteric viruses occur in sewage in considerable numbers and recent detection techniques make it possible to find these viruses in almost all streams that receive sewage effluents. Enteric viruses have been isolated from surface waters around the world. The contamination of surface water by enteric viruses appears to be ubiquitous.

Survival of Viruses

A variety of factors is responsible for the survival of viruses in water bodies. Some of the more significant ones are listed in Fig. 9.1. The survival of enteric viruses under laboratory conditions and in estuaries varies from a few hours to up to 200 days. Survival in winter is superior to that at summer temperatures. It is not known exactly what happens to these multitudes of viruses introduced in water bodies.

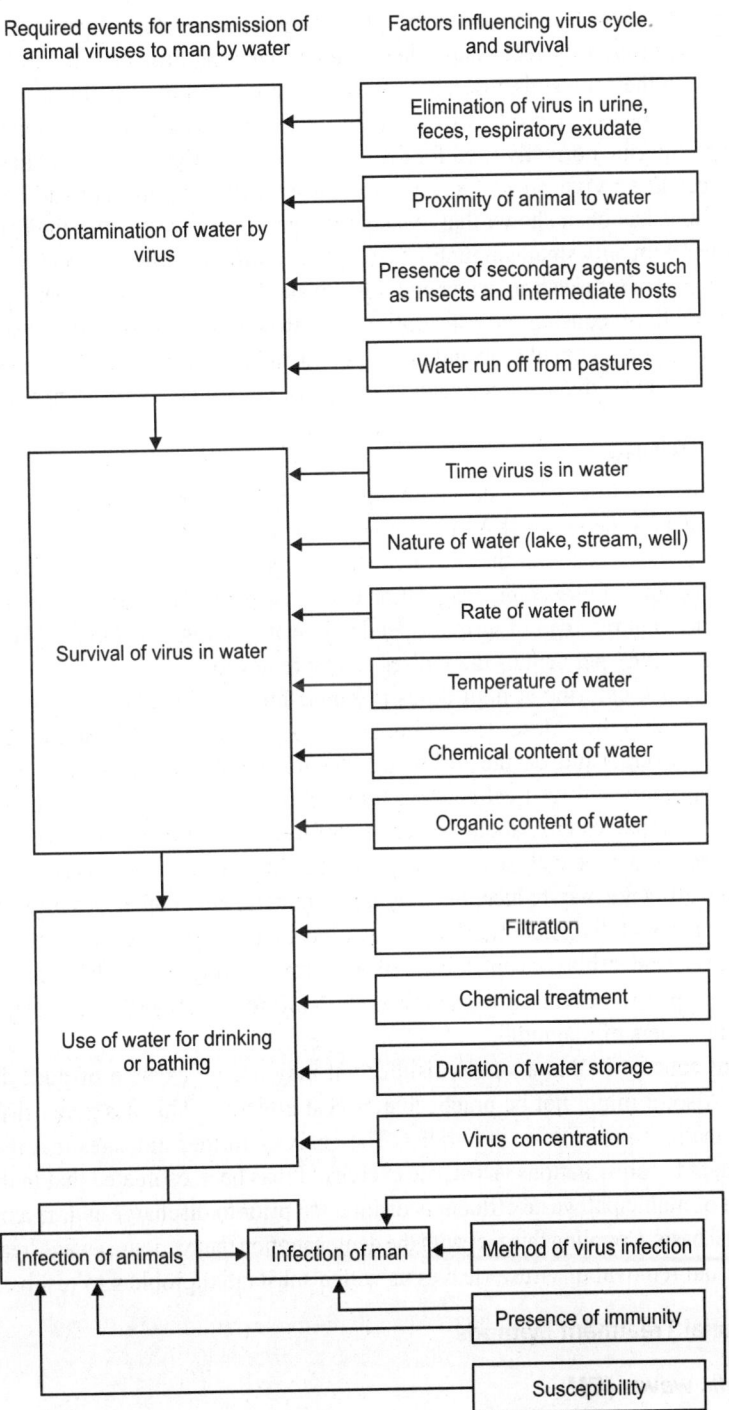

Fig. 9.1. Factors influencing virus survival in water.

The inability of rhinoviruses to withstand pH changes, temperature fluctuations and the lack of protective covering offered by feces and other organic materials probably makes the water route of minor importance in their transmission. These factors do not affect the enteroviruses, which are stable and persist in water for long periods. Coxsackieviruses, it has been found, are relatively resistant to concentrations of chlorine normally used for disinfection of bacteria in water. Studies have shown that enteric viruses easily survive present sewage treatment methods and may survive in waters for a considerable time. It has been shown that oysters incorporate poliovirus into their tissues even when grown in sea water with only small amounts of virus. The traditional processes and techniques currently in use for the removal of viruses from water and waste-water include methods effecting physical removal of the particles and those causing the inactivation or destruction of the organism. Among the first are sedimentation, adsorption, coagulation and precipitation and filtration. The second compasses high pH and chemical oxidation by disinfectants such as halogens. Let's take a look at these technologies.

TREATMENT OPTIONS

Primary treatment of municipal waste involving settling and retention removes very few viruses. Sedimentation effects some removal. Virus removal of up to 90 per cent (which is a minimal removal efficiency) has been observed after the activated sludge step. Further physical-chemical treatment can result in large reductions of virus titer, coagulation being one of the most effective treatments achieving as much as 99.9 per cent removal of virus suspended in water. If high pH (above 11) is maintained for long periods of time, 99.9 per cent of the viruses can be removed.

Of all the halogens, chlorine at high doses (40 mg/l for 10 minutes) is very effective, achieving 99.9 per cent reduction. Lower doses (for example, 8 mg/l) result in no decrease in virus.

As a result of several studies, the following conclusions regarding viruses in sewage warrant consideration: (i) primary sewage treatment has little effect on enteric viruses, (ii) secondary treatment with trickling filters removes only about 40 per cent of the enteroviruses, (iii) secondary treatment by activated sludge treatment effectively removes 90 per cent to 98 per cent of the viruses, and (iv) chlorination of treated sewage effluents may reduce, but may not eliminate, the number of viruses present.

The current concept of disinfection is that the treatment must destroy or inactivate viruses as well as bacillary pathogens. Under this concept, the use of coliform counting as an indicator of the effectiveness of disinfection is open to severe criticism given that coliform organisms are easier to destroy than viruses by several orders of magnitude.

An important concept is that a single disinfectant may not be capable of purifying water to the desired degree. Also, it might not be practicable or cost effective. This has given rise to a variety of treatment combinations in series or in parallel. The analysis further indicates that the search for the perfect disinfectant for all situations is a sterile exercise. It has been estimated that in the United States only 60 per cent of municipal waste effluent is disinfected prior to discharge and, in a number of cases, only on a seasonal basis. Coupling this fact with the demonstration that various sewage treatment processes achieve only partial removal of viruses leaves us with a substantial problem to resolve.

Non-conventional Treatment Methods

Electromagnetic waves (EM)

Electromagnetic radiation is the propagation of energy through space by means of electric and magnetic fields that vary in time. Electromagnetic radiation may be specified in terms of frequency, vacuum wavelength or photon energy. For water purification, EM waves up to the low end of the UV band will result in heating the water. (This includes infrared as well as most lasers.) In the visible range, some photochemical reactions such as dissociation and increased ionisation may take place. At the higher frequencies, it will be necessary to have thin layers of water because the radiation will be absorbed in a relatively short distance. It should be noted that the conductivity and dielectric constant of materials are, in general, frequency dependent. In case of the dielectric constant, it decreases as 1/wavelength. Hence, the electromagnetic absorption will vary with the frequency of the applied field. There may be some anomalies in the absorption spectra in the vicinity of frequencies that could excite molecules. At those frequencies, the absorption could be unusually large.

Ultraviolet radiation in the region between 0.2 μ to 0.3 μ has germicidal properties. The peak germicidal wavelength is around 0.26 μ. This short UV is attenuated in air and hence, the source must be very near the medium to be treated. The medium must be very thin as the UV will be attenuated in the medium as well.

X-rays and gamma rays are high-energy photons and will tend to ionise anything with which they collide. They could generate UV in air. At higher energies it is possible for the gamma rays to induce nuclear reactions by stripping protons or neutrons from nuclei. This could result in the production of isotopes and/or the production of new atoms.

Sound

A sound wave is an alteration in pressure, stress, particle displacement, particle velocity or a combination of these that is propagated in an elastic medium. Sound waves, therefore, require a medium for transmission; that is, they may not be transmitted in a vacuum. The sound spectrum covers all possible frequencies. The average human ear responds to frequencies between 16 Hz and 16 kHz. Frequencies above 20 kHz are called ultrasonic frequencies. Sound waves in the 50–200 kHz range are used for cleaning and degreasing. In water purification applications, ultrasonic waves have been used to effect disintegration by cavitation and mixing of organic materials. The waves themselves have no germicidal effect but, when used with other treatment methods, can provide the necessary mixing and agitation for effective purification.

Electron beams

The electron is the lightest stable elementary particle of matter known and carries a unit of negative charge. It is a constituent of all matter and can be found free in space. Under normal conditions, each chemical element has a nucleus consisting of a number of neutrons and protons, the latter equal in number to the atomic number of the element. Electrons are located in various orbits around the nucleus. The number of electrons is equal to the number of protons and the atom is electrically neutral when viewed from a distance. The number of electrons that can occupy each orbit is governed by quantum mechanical selection rules. The binding energy between an electron and its nucleus varies with the orbit number and in general the electrons with the shortest orbit are the most tightly bound. An electron can be made to jump from one orbit into another by giving it a quantum of energy. This energy quantum is fixed for any given transition and whether a transition will occur is again governed by selection rules. In other words, although an electron is given a quantum of energy sufficient to raise it to an adjacent higher state, it will not go up to that state if the transition is not permitted. In that case, it is theorised that if the electron absorbs the quantum, it will most probably go up to the excited state, remain there for a time allowed by the uncertainty principle, reradiate the quantum and return to its original state. If an electron is given a sufficiently large quantum of energy, it will completely leave the atom. The electron will carry off as kinetic energy the difference between the input quantum and the energy required to ionise. The remaining atom will now become a positively charged ion and the stripped electron will become a free electron. This electron may have sufficient energy when it leaves the atom (or it may acquire sufficient energy from an external field) to collide with another atom and strip it of an electron. This is the basis for electric discharge where free electrons are accelerated by an applied field and as they collide with neutral atoms, generate additional free electrons. This process avalanches as the electrons approach the positive electrode. At the same time, the positively charged ions are accelerated toward the negative electrode. In a vacuum, when a voltage is applied between two electrodes, electrons will move from the cathode to the anode. Of course, in a vacuum there will be no avalanching effects. Electrons are emitted from the cathode by a number of mechanisms:

1. Thermionic emission: Because of the non-zero temperature of the cathode, free electrons are continuously bouncing inside. Some of these have sufficient energy to overcome the work function of the material and can be found in the vicinity of the surface. The cathode may be heated to increase this emission. Also to enhance this effect, cathodes are usually made of or coated with, a low work-function material such as thorium.

2. Shottky emission: This is also a thermionic type of emission except that in this case, the applied electric field effectively decreases the work function of the material and more electrons can then escape.

3. High field emission: In this case, the electric field is high enough to narrow the work-function barrier and allow electrons to escape by tunneling through the barrier.

4. Photoemission: Electromagnetic radiation of energy can cause photoemission of electrons whose maximum energy is equal to or larger than the difference between the photon energy and the work function of the material.

5. Secondary emission: Electrons striking the surface of a cathode could cause the release of some electrons and hence, a net amplification in the number of electrons. This principle is used in the construction of photomultipliers where light photons strike a photoemitting cathode releasing photoelectrons. These electrons are subsequently amplified striking a number of electrodes (called dynodes) before they are finally collected by the anode.

Electromagnetism

In a high-gradient magnetic separator, the force on a magnetised particle depends on the intensity of the magnetising field and on the gradient of the field. When a particle is magnetised by an applied magnetic field, the particle develops an equal number of north and south poles. Hence, in a uniform field, a dipolar particle experiences a torque, but not a net tractive force. In order to develop a net tractive force, a field gradient is required; that is, the induced poles at the opposite ends of the particle must view different magnetic fields. In a simplified, one-dimensional case, the magnetomotive force on a particle is given by:

$$F_m = \mu(\delta H/\delta x) = MV(\delta H/\delta x) = \chi HV(\delta H/\delta x)$$

where, μ is the magnetic moment of the particle under field intensity, $H\delta H/\delta x$ is the field gradient. The magnetic moment μ is the product of the magnetisation of the particle and its volume ($\mu = MV$). And magnetisation is the product of the particle susceptibility, χ and the field intensity, H. In water purification, this magnetic force may be used to separate magnetisable particles.

Direct and alternating currents

Electrolytic treatment is achieved when two different metal strips are dipped in water and a direct current is applied from a rectifier. The higher the voltage the greater the force pushing electrons across the gap between the electrodes. If the water is pure, very few electrons cross the path between the electrodes. Impurities increase conductivity, hence decreasing the required voltage. Additionally, chemical reactions occur at both the cathode and the anode. The major reaction taking place at the cathode is the decomposition of water with the evolution of hydrogen gas. The anode reactions are oxidations by four major means: (i) oxidation of chloride to chlorine and hypochlorite, (ii) formation of highly oxidative species such as ozone and peroxides, (iii) direct oxidation by the anode, and (iv) electrolysis of water to produce oxygen gas.

Application

Electrolytic treatment

A great deal of interest was generated in the United States prior to 1930 in electrolytic treatment of waste-water, but all plans were abandoned because of high cost and doubtful efficiency. Such systems were based on the production of hypochlorite from existing or added chloride in the waste-water system. A great deal of effort has been made in re-evaluating such techniques.

Reduction in number of viable micro-organisms by adsorption onto the electrodes

Protein and micro-organism adsorption on electrodes with anodic potential has been documented. Micro-organism adsorption on passive electrodes (in the absence of current) has been observed with subsequent electrochemical oxidation. This does not appear to be a major route for inactivation.

Electrochemical oxidation of the micro-organism components at the anode

Oxidation of various viruses due to oxidation at the surface of the working electrode has been indicated, although the peak voltage used in many experiments would not be sufficient for the generation of molecular or gaseous oxygen.

Destruction of the micro-organisms by production of a biocidal chemical species

It has been shown that NaCl is not needed for effective operation in the destruction of micro-organisms. Biocidal species such as Cl, HO^-, O, ClO and HOCl occur but have very low diffusion coefficients. Hence, if this phenomenon occurs, the probability is that organisms are destroyed at the electrode surface rather than in the bulk solution.

Destruction by electric field effects

It has been observed that some organisms are killed in midstream without contact with the electrodes. The organisms were observed to oscillate in phase with the electric field. Hence, micro-organism kill can also be ascribed to changes caused by changing electromotive forces resulting from the impressed AC.

Electromagnetic separation

In the typical operation, a magnetised fine-particle seed (typically iron oxide) and a flocculent (typically aluminum sulphate) are added to the waste-water, prompting the formation of magnetic microflocs. The stream then flows through a canister packed with stainless steel wire and a magnetic field is applied. The stainless steel wool captures the flocs by magnetic forces.

OZONATION

Ozone has been used continuously for nearly 90 years in municipal water treatment and the disinfection of water supplies. This practice began in France, then extended to Germany, Holland, Switzerland and other European countries and in recent years to Canada and other developing countries. Ozone is a strong oxidising substance with bactericidal properties similar to those of chlorine. In test conditions it was shown that the destruction of bacteria was between 600 and 3000 times more rapid by ozone than by chlorine. Further, the bactericidal action of ozone is relatively unaffected by changes in pH while chlorine efficacy is strongly dependent on the pH of the water.

Ozone's high reactivity and instability as well as serious obstacles in producing concentrations in excess of 6 per cent preclude central production and distribution with its associated economies of scale.

In the electric discharge (or corona) method of generating ozone, an alternating current is imposed across a discharge gap with voltages between 5 and 25 kV and a portion of the oxygen is converted to ozone. A pair of large-area electrodes are separated by a dielectric (1–3 mm in thickness) and an air gap (approximately 3 mm). Although standard frequencies of 50 or 60 cycles are adequate, frequencies as high as 1000 cycles are also employed.

The mechanism for ozone generation is the excitation and acceleration of stray electrons within the high-voltage field. The alternating current causes the electron to be attracted first to one electrode and then to the other. As the electrons attain sufficient velocity, they become capable of splitting some oxygen molecules into free radical oxygen atoms. These atoms may then combine with O_2 molecules to form O_3.

Besides the disinfection of sewage effluent, ozone is used for sterilising industrial containers such as plastic bottles, where heat treatment is inappropriate. Breweries use ozone as an antiseptic in destroying pathogenic ferments without affecting the yeast. It is also used in swimming pools and aquariums. It is sometimes used in the purification and washing of shellfish and in controlling slimes in cooling towers. Ozone has also been shown to be quite effective in destroying a variety of refractory organic compounds.

ULTRAVIOLET RADIATION

It has been shown that:

1. Ultraviolet radiation around 254 mm renders bacteria incapable of reproduction by photochemically altering the DNA of the cells.
2. A fairly low dose of ultraviolet light can kill 99 per cent of the fecal coliform and fecal streptococcus.
3. Bacterial kill is independent of the intensity of the light but depends on the total dose.
4. Simultaneous treatment of water with UV and ozone results in higher micro-organism kill than independent treatment with both UV and ozone.
5. When ultrasonic treatment was applied before treating with the UV light, a higher bacteria kill was obtained.
6. The UV dose required to reduce the survival fraction of total coliform and fecal streptococcus to 102 (99 per cent removal) is approximately 4×10 ff Einsteins/ml.

Some limitations are associated with UV radiation for disinfection. These include: (i) the process performance is highly dependent on the efficacy of upstream devices that remove suspended solids, (ii) another key factor is that the UV lamps must be kept clean in order to maintain their peak radiation output and, (iii) A further drawback is associated with the fact that a thin layer of water (< 0.5 cm) must pass within 5 cm of the lamps.

One way of implementing the UV disinfection process at existing activated sludge plants involves suspending the UV lights (in the form of low-pressure mercury arc UV lamps with associated reflectors) above the secondary clarifiers. The effluent is exposed to the UV radiation as it rises over the wire in a thin film.

ELECTRON BEAM

The idea of using ionising radiation to disinfect water is not new. Ionising radiations can be produced by various radio-active sources (radioisotopes), by X-ray and particle emissions from accelerators and by high-energy electrons. The advances in reliable, relatively low-cost devices for producing high-energy electrons are more significant.

Unlike X-rays and gamma rays, electrons are rapidly attenuated. The maximum range of a 1 million-volt electron is about 4 m in air and about 5 cm in water. In transit in matter, an electron loses energy through collisions that ionise atoms and molecules along its path. Bacteria and viruses are destroyed by the secondary ionisation products produced by the primary traversing electron. The energetic electrons dissociate water into free radicals H^+ and OH^-. These may combine to form active molecules-hydrogen, peroxides and ozone. These highly active fragments and molecules attack living structures to promote their oxidation, reduction, dissociation and degradation. Studies have indicated that 4,00,000 rads would be adequate for sewage disinfection. At 100 ergs per gram rad, 4,00,000 rads would raise the temperature of the water or sludge by 1°C. At this dose, each cm^2 of moving sludge would receive about 12×10^{12} electrons, each electron producing some 30000 secondary ionisations.

BIOLOGY OF AQUATIC SYSTEMS

Before examining the various techniques for purifying water, an understanding of the key biological organisms is necessary. These key organisms include bacteria, algae, protozoa, crustaceans and fish. Bacteria and protozoa are the major groups of micro-organisms. There are a number of waterborne diseases of man caused by bacteria. Some of these organisms are used in evaluating the sanitary quality of water for drinking and recreational purposes.

Waterborne Diseases

There are a number of infectious, enteric (that is, intestinal) diseases of man which are transmitted through fecal wastes. Pathogens (disease-producing agents) include bacteria, viruses, protozoa and parasitic worms. Widespread diseases generally occur in regions where sanitary disposal of human feces is not practised. The most common waterborne bacterial diseases are typhoid fever (*Salmonella typhosa*), *Asiatic cholera* (vibrio comma) and bacillary dysentery (*Shigella dysenteriae*). The first of these is an acute infectious disease. Symptoms of typhoid fever are high fever and infection of the spleen, gastrointestinal tract and blood. For cholera, symptoms include diarrhoea, vomiting and severe dehydration. Dysentery produces diarrhoea, bloody stools and high fever. These diseases can cause death and are still prevalent in many underdeveloped nations. However, in this country, proper environmental control has virtually eliminated these problems. Waterborne outbreaks of infectious hepatitis have occurred. However, the main transmission mechanism is by person-to-person contact. The probability of outbreaks from municipally treated water supplies is low. Symptoms include loss of appetite, nausea, fatigue and pain. Also, a yellowish colour appears in the white of eyes and skin (yellow jaundice is an older term for the disease). It is generally not fatal except to individuals with weaker or older metabolisms. Amoebic dysentery is the most common enteric protozoal infection.

It is caused by Endamoeba histolytica and is transmitted by direct contact, food and through the water in tropical climates. It is not transmittable via water in temperate climates. Disinfection, as with all these diseases, is the safest means of prevention. Bilharziasis (or Schistosomiasis) is a parasitic disease generated by a small, flat worm that can infest the internal organs, such as the heart, lungs and liver and even the veins. Eggs of these worms existing in human abdominal organs can be transmitted to water via fecal discharges. Once in water, they hatch into miracida and enter into snails. They then develop into sporocysts that produce fork-tailed cercariae which eventually abandon their shells and attach onto humans. They bore through the skin, enter the bloodstream and eventually find their way to the internal organs to establish their homes. There is no immunisation for this disease. Many feel it is one of the world's worst health problems, particularly in agricultural regions of Africa and South America. Fortunately, this disease does not occur in the United States (the intermediate snail host just happens to be one of several specific species not found on the continental United States).

Bacteria consists of simple, colourless, one-celled plants that utilise soluble food. They are capable of self-reproduction without the aid of sunlight. As decomposers, they represent decaying organic matter in nature. They typically range in size from 0.5–5 and as such are only visible through a microscope. Individual bacteria cells take on various geometries. Typical configurations include spheres, rods or spirals. They may be single, in pairs, packets or chains.

Reproduction is by binary fission, meaning a cell divides into two new cells, each of which matures and divides again. Fission takes place every 1530 mill under ideal conditions. Ideal conditions mean that the growth environment has abundant food, oxygen and essential nutrients.

Bacteria are named according to a binomial system. The first word is the genus and the second is the species name. The most frequently referred to bacterium in the sanitary field is Escherichia *coli*. *E. coli* is a common coliform that can be used as an indicator of water's bacteriological quality. Under a microscope and magnified 1000 times, cells appear as individual short rods.

There are two major classifications of bacteria called heterotrophic and autotrophic. Heterotrophs, also called saprohytes, utilise organic substances both as a source of energy and carbon. Heterotrophs are further subclassified into three groups. Subclassifications are based on the bacteria's action toward free oxygen. Aerobes need free dissolved oxygen to decompose organics to derive energy for growth and reproduction. This can be described by:

$$\text{Organics} + \text{Oxygen} \rightarrow CO_2 + H_2O + \text{Energy}$$

The second subgroup are anaerobes, which oxidise organics in the absence of dissolved oxygen. This is accomplished by using the oxygen which is found in other compounds (such as nitrate and sulphate). Anaerobic behaviour can be described by the following reactions:

$$\text{Organics} + NO_3 \rightarrow CO_2 + N_2 + \text{Energy}$$

$$\text{Organics} + SO_4^- \rightarrow CO_2 + H_2S + \text{Energy}$$

Note also that:

$$\text{Organics} \rightarrow \text{Organic} + S + CO_2 + H_2O + \text{Energy}$$

and that the organic acids undergo further reaction:

$$\text{Organic acids} \rightarrow CH_4 + CO_2 + \text{Energy}$$

Facultative bacteria comprise the last group and use free dissolved oxygen when available. However, they can also survive in its absence (that is, they also gain energy from the anaerobic reaction). Heterotrophic bacteria decompose organics to obtain energy for the synthesis of new cells, respiration and motility. Some energy is lost in the process as heat. Autotrophic bacteria oxidise inorganic constituents for energy and utilise carbon dioxide as a source of carbon. The major bacteria types in this class are nitrifying, sulphur and iron bacteria. Nitrifying bacteria will oxidise ammonium nitrogen to nitrate. Sulphur bacteria perform a reaction given which causes crown corrosion in sewers. Water in sewers quite frequently turns septic and generates hydrogen sulphide gas by generating hydrogen sulphide. The H_2S generated absorbs in the condensation moisture on the sewer side walls and the crown of the pipe. Those sulphur bacteria able to survive at very low pH (pH < 1) oxidise weak H_2S acid to strong sulphuric acid. This oxidation reaction depletes the oxygen from the sewer air. Crown corrosion of concrete-lined systems can greatly reduce the structural integrity of piping and eventually cause walls to collapse. Iron bacteria oxidize soluble inorganic ferrous iron to insoluble ferric. Certain types of filamentous bacteria (Leptothrix and Crenothrix) deposit oxidised iron in the form of $Fe(OH)_3$ in their sheath. This produces yellow or reddish-coloured slimes. Water pipes are ideal environments for these type bacteria as they have an abundance of highly dissolved iron content to provide energy and bicarbonates to serve as a carbon source. As these micro-organisms mature and die, they decompose, generating obnoxious odours and foul tastes.

Fungi and Moulds

Fungi are microscopic non-photosynthetic plants which include in their classification yeast and moulds. Yeasts have a commercial value as they are used for fermentation operations in distilling and brewing. When anaerobic conditions exist, yeasts metabolise sugar, manufacturing alcohol from the synthesis of new cells. Alcohol is not manufactured under aerobic conditions and the yield of new yeast cells is greater. Filamentous forms of fungi are moulds. These best resemble higher orders of plant life, having branched or threadlike growths. They grow best in environments consisting of acid solutions with high sugar concentrations. Moulds are non-photosynthetic, multicellular, heterotrophic and aerobic. The growth of moulds can be suppressed by increasing the pH.

Algae, Protozoa and Multicellular Animals

Algae are microscopic photosynthetic plants. They are among the simplest plant forms, having neither roots, stems, nor leaves. Algae typically range from single-cell entities (which impart a green colour to surface waters) to branched forms that can be seen by the naked eye. The latter often appear as attached green slime on surface bodies of water. Diatoms refers to singlecelled algae which are housed in silica shells. The blue-green algae generally associated with water pollution are Anacystis, Anabaena and Aphanizomellon. Green algae are Oocystis and Pediastrum. Algae are autotrophic; that is, they use carbon dioxide or bicarbonates as sources of carbon. Inorganic nutrients of phosphate and nitrogen as ammonia or nitrate are also used. Some trace nutrients are also necessary (magnesium, boron, cobalt, calcium). The reaction or process by which algae propagate is known as photosynthesis. The products of photosynthesis are new plant growth and oxygen. The energy supplied to the reaction is derived from sunlight. Pigments biochemically convert solar energy into useful energy for plant reproduction and survival. In prolonged absence of sunlight, plant matter performs a dark reaction to exist. In this case, algae absorb oxygen and degrade stored food to produce yield energy for respiratory functions. The reaction rate for the dark reaction is much slower than photosynthesis. Macrophytes are aquatic photosynthetic plants (excluding algae). They often appear on surface bodies of water as floating, submerged and immersed aggregates. Floating plants are not anchored or rooted. In the animal kingdom, one of the simplest forms is the protozoan. Protozoa are single-celled aquatic animals that have relatively complex digestive systems. They use solid organic material as food and multiply by binary fission. They are aerobic organisms and digest bacteria and algae and consequently, play an essential role in the aquatic food chain. The smallest type are the flagellated protozoa which range in size from 10 μ to 50 μ. These have long hairlike strands which provide motility by a whiplike action. The amoeba is a member of the protozoa family. Rotifiers are simple, multicelled, aerobic animals. These metabolise solid food. Rotifiers are found in natural waters, stabilisation ponds and extended aeration basins in municipal treatment plants. Crustaceans are multicellular animals (about 2 mm in size). They are herbivores which ingest algae and are in turn eaten by fish.

Biological Growth Factors

Major factors affecting biological growth are temperature, nutrient availability, oxygen supply, pH, degree of sunlight and the presence of toxins. Bacteria are classified by their optimum temperature range for growth. For example, mesophilic bacteria grow best in a temperature range of 10°–40°C (optimum at 37°C). In general, the rate of biological activity almost doubles for every 10°–15°C rise in temperature within the range of 5°–35°C. Beyond 40°C, mesophilic activity drops dramatically and thermophilic growth is initiated (thermophilic bacteria have a range between 45°–75°C, with an optimum of about 55°C). Thermophilic bacteria are typically more sensitive to temperature variations.

Water Quality Test Methods

Determination of the bacteriological quality of water is not a straightforward analysis. The testing for a specific pathogenic bacteria can often lead to erroneous conclusions. Analyses for pathogenic bacteria are difficult to perform. In general, data are not quantitatively reproducible. As an example, if Salmonella was found to be absent from a water sample, this does not exclude the possible presence of Shigella, Vibrio or disease-producing viruses. The bacteriological quality of water is based on test procedures for non-pathogenic indicator organisms (principally the coliform group).

Coliform bacteria, typified by Escherichia coli and fecal streptococci (enterococci), reside in the intestinal tract of man. These are excreted in large numbers in the feces of humans and other warm-blooded animals. Typical concentrations average about 50,000,000 coliforms per gram. Untreated domestic waste-water generally contains more than 30,00,000 coliforms per 100 ml. Pathogenic bacteria and viruses causing enteric diseases originate from the same source (that is, fecal discharges of diseased persons). Consequently, water contaminated by fecal pollution is identified as being potentially dangerous by the presence of coliform bacteria.

Standards for drinking water specify that a water is safe provided that the test method does not reveal more than an average of one coliform organism per 100 ml. The number of pathogenic bacteria, such as Salmonella typhosa, in domestic waste-water is generally less than 1 per mil coliforms and the average density of enteric viruses has been measured as a virus-tocoliform ratio of 1:1,00,000. The die-off rate of pathogenic bacteria is greater than the death rate of coliforms outside of the intestinal tract of animals. Consequently, upon exposure to treatment, a reduction in the number of pathogens relative to coliforms will occur. Water quality based on a standard of less than one coliform per 100 ml is statistically safe for human consumption. That is, there is a high improbability of ingesting any pathogens. This is an Environmental Protection Agency (EPA) standard applicable only to processed water where treatment includes chlorination.

Coliform criteria for body-contact water use and recreational use have been established by most states. Upper limits of 200 fecal coliforms per 100 ml and 2000 total coliforms per 100 ml have been established. These values are only guidelines since there is no positive epidemiological evidence that bathing beaches with higher coliform counts are associated with transmission of enteric diseases. Some experts feel that these standards may be too conservative from a standpoint of realistic public health risk. Coliform standards applied to water used for swimming are linked to water-associated diseases of the skin and respiratory passages rather than enteric diseases. This, naturally, is entirely different than the purpose of the coliform standard for drinking water, which is related to enteric disease transmission. Here tighter restrictions are imperative, since a water distribution system has the potential of mass transmission of pathogens in epidemic proportions.

Water sample collection techniques differ depending on the source being tested. The minimum number of water samples collected from a distribution system which are examined each month for coliforms is a function of the population. For example, the minimum number required for populations of 1000 and 1,00,000 are 2 and 100 respectively. To ascertain compliance with the bacteriological requirements of drinking water standards, a certain number of positive tests must not be exceeded. When 10 ml standard portions are examined, not more than 10 per cent in any month should be positive (that is, the upper limit of coliform density is an average of one per 100 ml).

Coliforms are defined as all aerobic and facultative anaerobic, non-sporeforming species. Gram-stain negative rods ferment lactose and produce gas within 48 hours of incubation (at 35°C). The initial coliform analysis is the presumptive test which is based on gas production from lactose. In this test, 10 ml portions of water samples are transferred into prepared fermentation tubes using sterile pipettes. The tubes contain lactose or lauryl tryptose, broth and inverted vials. Inoculated tubes are placed in a warm-air incubator (at 35°C ± 0.5°C). Growth with the production of gas (the gas is identified by the presence of bubbles in the inverted vial) means a positive test. That is, it indicates that coliform bacteria may be present. A negative reaction, either no growth or growth without gas, excludes coliforms.

Such tests are employed to substantiate or refute the presence of coliforms in a positive presumptive test. In normal, potable water coliform testing, the test is confirmed using brilliant green bile broth. Occasionally, one may desire to run a completed test. This involves transferring a colony from an Endo (or EMB plate) to nutrient agar and into lactose broth. If gas is not produced in the lactose fermentation tube, the colony transferred did not contain coliforms and the test is negative. If gas is generated, a portion of growth on the nutrient agar is smeared onto a glass slide and prepared for observation under a microscope using the Gram-stain technique. If the bacteria are short rods, with no spores present and the Gramstain is negative, the coliform group is present and the test is completed. If the culture Gramstains positive (purple colour), the completed test is negative.

In examining surface water quality, an elevated-temperature coliform test is used to separate micro-organisms of the coliform group into those of fecal and non-fecal sources. This approach is applicable to studies of stream pollution, raw-water sources, waste-water treatment systems, bathing waters and general water quality monitoring. It is not recommended as a substitute for the coliform tests used in examination of potable waters.

The water analysis is incomplete unless the number of coliform bacteria present is determined as well. A multiple-tube fermentation technique can be used to enumerate positive presumptive, confirmed and fecal coliform tests. Results of the tests are expressed in terms of the most probable number (MPN). That is, the count is based on a statistical analysis of sets of tubes in a series of serial dilutions. MPN is related to a sample volume of 100 ml. Thus, an MPN of 10 means 10 coliforms per 100 ml of water.

For MPN determination, sterile pipettes calibrated in 0.1 ml increments are used. Other equipment includes sterile screw-top dilution bottles containing 99 ml of water and a rack containing six sets of five lactose broth fermentation tubes. A sterile pipette is used to transfer 1.0 ml portions of the sample into each of five fermentation tubes. This is followed by dispensing 0.1 ml into a second set of five. For the next higher dilution (the third), only 0.01 ml of sample water is required. This small quantity is very difficult to pipette accurately, so 1.0 ml of sample is placed in a dilution bottle containing 99 ml of sterile water and mixed. The 1.0 ml portions containing 0.01 ml of the surface water sample are then pipetted into the third set of five tubes. The fourth set receives 0.1 ml from this same dilution bottle. The process is then carried one more step by transferring 1.0 ml from the first dilution bottle into 99 ml of water in the second for another hundredfold dilution. Portions from this dilution bottle are pipetted into the fifth and sixth tube sets. After incubation (48-hour at 35°C), the tubes are examined for gas production and the number of positive reactions for each of the serial dilutions is recorded.

A final testing technique worth noting is the membrane filter method for coliform testing. This procedure involves passing a measured water sample through a membrane filter to remove the bacteria. The filter is then placed on a growth medium in a petri dish. The bacteria retained by the filter pad grow and establish a small colony. The number of coliforms present is established by counting the number of colonies and expressing this value in terms of number per 100 ml of water. This technique has been widely adopted for use in water quality monitoring studies, especially since it requires considerably less laboratory apparatus than the standard multiple-tubes technique. Also, this technique can be adapted to field studies.

Equipment needed to perform the membrane filter coliform test includes filtration units, filter membranes, absorbent pads, forceps and culture dishes. The common laboratory filtration unit consists of a funnel that fastens to a receptacle bearing a porous plate to support the filter membrane. The filterholding assembly can be constructed of glass, porcelain or stainless steel. It is sterilised by boiling, autoclaving or ultraviolet radiation. For filtration, the assembly is mounted on a side-arm filtering flask which is evacuated to draw the sample through the filter. For field use, a small hand-sized plunger pump or syringe is used to draw a sample of water through the small assembly holding the filter membrane.

Commercial filter membranes are normally 2 inch diameter disks with pore openings of 0.45 (± 0.02) R. This is small enough to retain microbial cells. Filters used in determining bacterial counts have a grid printed on the surface. To facilitate counting colonies, the filter membranes must be sterilised prior to use, either in a glass petri dish or wrapped in heavy paper. After sterilisation, the pads are placed in culture dishes to absorb the nutrient media on which the membrane filter is placed. During the testing, filters are handled on the outer edges with forceps that are also sterilised before use.

Glass or disposable plastic culture dishes are used. If glass petri dishes are employed, a humid environment must be maintained during incubation. This prevents losses of media by evaporation (the dishes have loose-fitting covers). Disposable plastic dishes have tight-fitting lids which minimise the problem of dehydration.

The size of the filtered sample is established by the anticipated bacterial density. An ideal quantity results in the growth of about 50 coliform colonies and not more than 200 colonies of all types. Often it may be difficult to anticipate the number of bacteria in a sample. Two or three volumes of the same sample must be tested. When the portion being filtered is less than 20 ml, a small amount of sterile dilution water is added to the funnel before filtration. This uniformly disperses the bacterial suspension over the entire surface of the filter. The filter-holding assembly is placed on a suction flask. A sterile filter is placed grid side up over the porous plate of the apparatus using sterile forceps. The funnel is then locked in place holding the membrane. Filtration is performed by passing the sample through the filter under partial vacuum. A culture dish is prepared by placing a sterile absorbent pad in the upper half of the dish and pipetting enough enrichment media on top to saturate the pad. M-Endo medium is used for the coliform group and M-FC for fecal coliforms. The filter is then removed from the filtration apparatus and placed directly on the pad in the dish. The cover is replaced and the culture is incubated (for 24-hour at 35°C). For fecal coliforms, incubation is performed by placing the culture dishes in watertight plastic bags and submerging them in a water bath at 44.5°C. Coliform density is calculated in terms of coliforms per 100 ml by multiplying the colonies counted by 100 and dividing this value by the milliliters of the sample filtered.

DISINFECTION BY CHLORINATION

Disinfection has received increased attention over the past several years from regulatory agencies through the establishment and enforcement of rigid bacteriological effluent standards. In upgrading existing waste-water treatment facilities, the need for improved disinfection as well as the elimination of odour problems are frequently encountered. Adequate and reliable disinfection is essential in ensuring that waste-water treatment plants are both environmentally safe and aesthetically acceptable to the public. Chlorine is the most widely used disinfectant in water and waste-water treatment. It is used to destroy pathogens, control nuisance micro-organisms and for oxidation. As an oxidant, chlorine is used in iron and manganese removal, for destruction of taste and odour compounds and in the elimination of ammonia nitrogen. It is, however, a highly toxic substance and recently concerns have been raised over handling practices and possible residual effects of chlorination.

Properties of Chlorine and Its Chemistry

Chlorine (Cl_2) is a greenish-yellow-coloured gas having a specific gravity of 2.48 as compared to air under standard conditions of temperature and pressure. It was discovered in 1774 from the chemical reaction of manganese dioxide ($MnNO_2$) and hydrochloric acid (HCl) by the Swedish chemist, Scheele, who believed it to be a compound containing oxygen. In 1810, it was named by Sir Humphrey Davy, who insisted it was an element (from the Greek work chloros, meaning greenish-yellow). In nature, it is found in the combined state only, usually with sodium as salt (NaCl), carnallite ($KMgCl_36H_2O$) and sylvite. Chlorine is a member of the halogen (salt-forming) group of elements and is derived from chlorides by the action of oxidising agents and, most frequently, by electrolysis. As a gas, it combines directly with nearly all elements. At 10°C, 1 volume of water dissolves about 3.10 volumes of chlorine; at 30°C, only 1.77 volumes of Cl_2 are dissolved in 1 volume of water.

In addition to being the most widely used disinfectant for water treatment, chlorine is extensively used in a variety of products, including paper products, dyestuffs, textiles petroleum products, pharmaceuticals, antiseptics, insecticides, foodstuffs, solvents, paints and other consumer products. Most chlorine produced is used in the manufacture of chlorinated compounds for sanitation, pulp bleaching, disinfectants and textile processing. It is also used in the manufacture of chlorates, chloroform and carbon tetrachloride and in the extraction of bromine.

As a liquid, chlorine is amber coloured and is 1.44 times heavier than water. In solid form, it exists as rhombic crystals. Various properties of chlorine are given in Table 9.2.

Chlorine gas is a highly toxic substance, capable of causing death or permanent injury due to prolonged exposures via inhalation. It is extremely irritating to the mucous membranes of the eyes and the respiratory tract. It will combine with moisture to liberate nascent oxygen to form hydrochloric acid. If both these substances are present in quantity, they can cause inflammation of the tissues with which they come in contact. Pulmonary edema may result if lung tissues are attacked. Chlorine gas has an odour detectable at a concentration as low as 3.55 ppm. Irritation of the throat occurs at 15 ppm. A concentration of 50 ppm is considered dangerous for even short exposures. At or above concentrations of 1000 ppm, exposure may be fatal. Chlorine can also cause fires or explosions upon contact with various materials. It emits highly toxic fumes when heated and reacts with water or steam to generate toxic and corrosive hydrogen chloride fumes.

Table 9.2. General properties of chlorine.

Symbol (as gas)		Cl_2
Atomic number		17
Atomic weight		35.453
Melting point (°C)		−101
Boiling point (°C)		−34.5
Liquid density (°C and 3.65 atm; g/l)		1.47
Vapour pressure (mmHg @ 20°C)		4800
Vapour density (@ STP: g/l)		2.49
Viscosity (micropoises) at		
Temperature	= 12.7°C	129.7
	= 20°	132.7
	= 50°	146.9
	= 100°	167.9
	= 150°	187.5
	= 200°	208.5

Chlorine is a strong oxidising agent and can be used to modify the chemical character of water. For example, it is used to control bacteria, algae and macroscopic biological-fouling organisms in condenser cooling towers. It is also used to alter the chemical character of some industrial process waters, such as the destruction of sulphur dioxide and ammonia, the reduction of iron and manganese and the reduction of colour (examples include bleaching operations in the pulp and paper industry and oxidation of organic constituents). In water chlorine hydrolyses to form hypochlorous acid (HOCl), as shown by the following reactions:

$$Cl_2 + H_2O = HOCl + H^+Cl^-$$

The hypochlorous acid undergoes further ionisation to form hypochlorite ions (OCl⁻):

$$HOCl = H^+ + OCl^-$$

Equilibrium concentrations of HOCl and OCl depend on the pH of the waste-water. Increasing the pH shifts the preceding equilibrium relationships to the right, causing the formation of higher concentrations of HOCl.

Chlorine may also be applied as calcium hypochlorite and sodium hypochlorite. Hypochlorites are salts of hypochlotous acid. Calcium hypochlorite [$Ca(OCl)_2$] represents the predominant dry form used in the United States. Calcium hypochlorite is commercially available in granular powdered or tablet forms. Either of these forms readily dissolves in water and contains approximately 70 per cent available chlorine. Sodium hypochlorite (NaOCl) is commercially available in liquid form at concentrations typically between 5 per cent to 15 per cent available chlorine. Hypochlorites react in water as follows:

$$NaOCl \quad \rightarrow \quad Na^+OCl^-$$
$$Ca(OCl)_2 \quad \rightarrow \quad Ca^{+2} + 2OCl^-$$
$$H^+ + OCL^- \quad \rightarrow \quad HOCl$$

The amount of HOCl plus OCl in waste-water is referred to as the free available chlorine. Chlorine is a very active oxidising agent and is, therefore, highly reactive with readily oxidised compounds such as ammonia. Chlorine readily reacts with ammonia in water to form chloramines.

$$HOCl + NH_3 \quad \rightarrow \quad H_2O + NH_2Cl \quad \text{(monochloramine)}$$
$$HOCl + NH_2Cl \quad \rightarrow \quad H_2O + NHCl_2 \quad \text{(dichloramine)}$$
$$HOCl + NHCl_2 \quad \rightarrow \quad H_2O + NCl_3 \quad \text{(trichloramine)}$$

The specific reaction products formed depend on the pH of the water, temperature, time and the initial chlorine-to-ammonia concentration ratio. In general, monochloramine and dichloramine are generated in the pH range of 4.5 to 8.5. Above pH 8.5, monochloramine usually exists alone. However, below pH 4.4, trichloramine is produced. When chlorine is mixed with water containing ammonia, the residuals developed produce a curve similar to the one shown in Fig. 9.2. The positive sloped line from the origin represents the concentration of chlorine applied or the residual chlorine if all of that applied appears as residual. The solid curve represents chlorine residuals corresponding to various dosages that remain after some specified contact time. The chlorine demand at a specified dosage is obtained from the vertical distance between the applied and residual curves. Chlorine demand represents the amount of chlorine reduced in chemical reactions (that is, it is the amount that is no longer available). For molar chlorine to ammonia-nitrogen ratios below 1, monochloramine and dichloramine are formed with their relative amounts dependent on pH and other factors. When higher dosages of chlorine are added, the chlorine-to-nitrogen ratio increases, resulting in an oxidation of the ammonia and a reduction of the chlorine. Three moles of chlorine react with two moles of ammonia, generating nitrogen gas and reducing chlorine to the chloride ion:

$$2\,NH_3 + 3Cl_2 \rightarrow N_2 + 6HCl$$

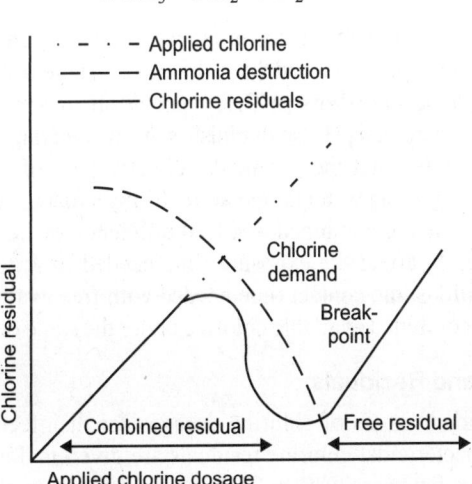

Fig. 9.2. Breakpoint chlorination curve.

Residuals of chloramine decline to a minimum value that is referred to as the breakpoint. When dosages exceed the breakpoint, free chloride residuals result. Breakpoint curves are unique for different water samples since the chlorine demand is a function of the concentration of ammonia, the presence of other reducing agents and the contact time between chlorine application and residual testing.

Germicidal Destruction

Chlorine's ability to destroy bacteria and various micro-organisms results from chemical interference in the functioning of the organism. Specifically, it is the chemical reaction between HOCl and the bacterial or viral cell structure which inactivates the required life processes. The high germicidal efficiency of HOCl is attributed to the ease by which it is able to penetrate cell walls. This penetration is comparable to that of water and is due both to its low molecular weight (that is, it's a small molecule) and its electrical neutrality. Organism fatalities result from a chemical reaction of HOCl with an enzyme system in the cell which is essential to the metabolic functioning of the organism.

The enzyme attacked is triosephosphate dehydrogenase, found in most cells and essential for digesting glucose. Other enzymes also undergo attack. However, triosephosphate dehydrogenase is particularly sensitive to oxidising agents. The OCl^- ion resulting from the dissociation is a relatively poor disinfectant because of its inability to diffuse through a micro-organism's cell walls. This is because of its negative charge. The sensitivity of bacteria to chlorination is well-known. However, the effect on protozoans and viruses has not been entirely delineated. Protozoal cysts and enteric viruses are more resistant to chlorine than are coliforms and other enteric bacteria.

Contact Time, pH and Temperature Effects

Hypochlorous acid and hypochlorite ion are known as free available chlorine. The chloramines are known as combined available chlorine and are slower than free chlorine in killing micro-organisms. For identical conditions of contact time, temperature and pH in the range of 6 to 8, it takes at least 25 times more combined available chlorine to produce the same germicidal efficiency. The difference in potency between chloramines and HOCl can be explained by the difference in their oxidation potentials, assuming the action of chloramine is of an electrochemical nature rather than one of diffusion, as seems to be the case for HOCl.

The effect of pH alone on chlorine efficiency is shown in Fig. 9.3. Chlorine exists predominantly as HOCl at low pH levels. Between pH of 6.0 and 8.5, a dramatic change from undissociated to completely dissociated hypochlorous acid occurs. Above pH 7.5, hypochlorite ions prevail; while above 9.5, chlorine exists almost entirely as OCl. Increased pH also diminishes the disinfecting efficiency of monochloramine.

It has also been demonstrated that the germicidal effectiveness of free and combined chlorine is markedly diminished with decreasing water temperature. In any situation in which the effects of lowered temperature and high pH value are combined, reduced efficiency of free chlorine and chloramines is marked. These factors directly affect the exposure time needed to achieve satisfactory disinfection. Under the most ideal conditions, the contact time needed with free available chlorine may only be on the order of a few minutes; combined available chlorine under the same conditions might require hours.

Chlorine Dosage Rates and Residuals

Table 9.3 gives recommended ranges of chlorine dosages for disinfection of various waste-waters. Recommended minimum bactericidal chlorine residuals are given in Table 9.4. Data in Table 9.4 are based on water temperatures between 20°C to 25°C after a 10 minute contact for free chlorine and a 60 minute contact for combined available chlorine.

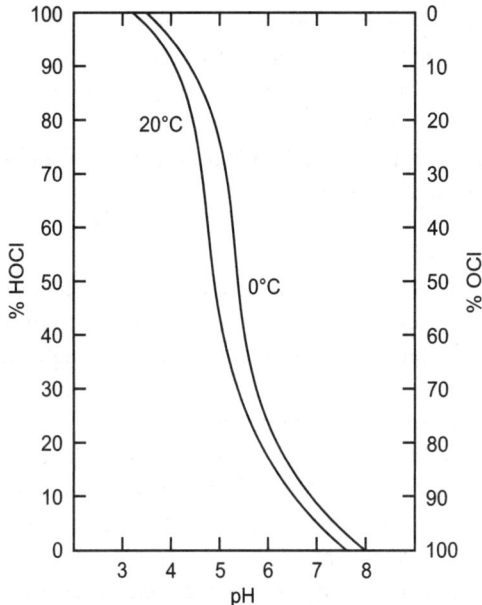

Fig. 9.3. Relative amounts of HOCl and OCl formed as a function of pH.

Table 9.3. Recommended chlorine dosage ranges.

Waste-water type	Chlorine dosage (mg/l)
Raw sewage	6–12
(Septic) raw sewage	12–25
Settled sewage	5–10
Chemical precipitation effluent	3–10
Trickling filter effluent	3–10
Activated sludge effluent	2–8
Sand filter effluent	1–5

Table 9.4. Minimum bactericidal chlorine residuals (mg/l).

pH value	Free available chlorine residual after 10 minutes contact	Combined available chlorine residual after 60 minutes contact
6.0	0.2	1.0
7.0	0.2	1.5
8.0	0.4	1.8
9.0	0.8	>3.0

The minimum residuals required for cyst destruction and inactivation of viruses are much greater. Although chlorine residuals in Table 9.4 are generally adequate, surface waters from polluted waterways are usually treated with much heavier chlorine dosages. Ordinary chlorination will destroy all strains of coli, aerogenes, pyocyaneae, typhsa and dysenteria.

In addition to these micro-organisms, three other types are readily destroyed: Enteric vegetative bacteria (Eberthella, Shigella, Salmonella and Vibrio species); Worms such as the block flukes (Schistosoma., species); Viruses (for example, the virus of infectious hepatitis). Each of these groups of organisms differs in its reaction with chlorine.

There is evidence that the comparative reaction of different organisms to one form of chlorine is not necessarily maintained relative to other forms.

Chlorination Systems

Water chlorination is carried out by using both free and combined residuals. The latter involves chlorine application to produce chloramine with natural or added ammonia. Anhydrous ammonia is used if insufficient natural ammonia is present in the waste-water. Although the combined residual is less effective than free chlorine as a disinfectant, its most common application is as a post-treatment following free residual chlorination to proved initial disinfection.

Free residual chlorination establishes a free residual through the destruction of naturally present ammonia. High dosages of chlorine applied during treatment may result in residuals that are esthetically objectionable or undesirable for industrial water use. Dechlorination is sometimes performed to reduce the chlorine residual by adding a reducing agent (called a dechlor). Sulphur dioxide is often used as the dechlor in municipal plants. Aeration by submerged or spray aerators also diminishes the residual chlorine concentration.

The chlorine used for disinfestation is available in three forms: liquified compressed gas, calcium hypochlorite or sodium hypochlorite and chlorine bleach solutions. Liquid chlorine is shipped in pressurised steel cylinders with sizes typically 100 and 500 lb; one-tonne containers are used in large installations. There are two types of chlorine dispensing system: direct feed and solution feed. The first involves metering dry chlorine gas and conducting it under pressure to the water. Solution-feed systems meter chlorine gas under vacuum and dissolve it in a small amount of water, forming a concentrated solution which is then applied to the water being treated. At 20°C, 1 volume of water dissolves 2.3 volumes of chlorine gas (about 7000 mg/l). At concentrations of total chlorine below 1000 mg/l, none of the gas exists in solutions as Cl_2; all of it is present as HOCl or dissociated ions calcium hypochlorite is a dry bleach which is available in granular and tablet forms. Calcium hypochlorite is relatively stable under normal conditions; however, it can undergo reactions with organic materials. It should be stored in an isolated area. Sodium hypochlorite is available in liquid form. It is marketed in carboys and rubber-lined drums for small quantities. Sodium hypochlorite solutions are highly corrosive, unstable and require storage at temperatures below 85°F. Sodium hypochlorite can either be delivered to the site in liquid form in 500–5000 gallon tank cars or trucks or manufactured on site. It is normally sold at a concentration of 12 per cent to 15 per cent by weight of available chlorine. It can be manufactured on site from salt or from sea water.

The main component in a chlorine gas feed is the variable orifice inserted in the feed line to control the rate of flow out of the cylinder. The orifice basically consists of a grooved plug sliding in a fitted ring. Feed rate is adjusted by varying the V-shaped opening. Since a chlorine cylinder pressure varies with temperature, the discharge through such a throttling valve does not remain constant without frequent adjustments of the valve setting. Also, conditions on the outlet side vary with pressure changes at the point of application. Therefore, a pressure-regulating valve is used between the cylinder and the orifice, with a vacuum-compensating valve on the discharge side. A safety pressure-relief valve is held closed by vacuum.

Chlorine feeders can be controlled either manually or automatically based on flow or chlorine residual or both. In manual mode, a continuous feed rate is established. This is satisfactory when chlorine demand and flow are relatively constant and where operators are available to make adjustments. Automatic proportional control equipment is used to adjust the feed rate to provide a constant pre-established dosage for all rates of flow. This is accomplished by metering the main flow and using a transmitter to signal a chlorine feeder. An analyser located downstream from the point of application is used to monitor the chlorinator. Combined automatic flow and residual control maintain a present chlorine residual in the water that is independent of the demand and flow variations. The feeder is designed to respond to signals from both the flow meter transmitter and the chlorine residual analyser. For hypochlorite solutions, positive-displacement diaphragm pumps (either mechanically or hydraulically actuated) are used. The hypochlorinator consists of a water-powered pump paced by a positive-displacement water meter. The meter register shaft rotates proportionately to the main line flow and controls a cam-operated pilot valve. This in turn regulates water now discharged of hypochlorite that is proportional to the main flow. Admitting main pressure behind the pumping diaphragm balances the water pressure in the pumping head. The advantage of this system is that the pump does not need electrical power. The hypochlorite dosage can be manually adjusted by changing the stroke length setting of the pump.

Chlorine Contact Tanks

The configuration of contact tanks can result in appreciable differences between actual and theoretical contact times. Contact times and germicidal efficiency depend on a number of parameters, the most important being the mixing characteristics of the basin. Proper designs must account for possible flow pattern elimination via short circuiting, acceptable dosage rates, optimum pH range and upstream removal of ammonia nitrogen.

Rapid dispersement of chlorine at the addition point increases chlorine contact and improves disinfection efficiency. Baffles can be designed to generate turbulence at the chlorine addition point and improve mixing. Baffled systems have the advantage of not requiring mechanical equipment. Mechanical mixing or air agitation can be used where plant hydraulics will not allow the use of baffles or where a portion of the existing basin can be converted to a mixing chamber and the remainder of the basin and/ or a long outfall sewer can be used to provide the needed contact time.

Toxic Effects of Chlorine

The toxicity of chlorine residuals to aquatic life has been well documented. Studies indicate that at chlorine concentration in excess of 0.01 mg/l, serious hazard to marine and estuarine life exists. This has led to the dechlorination of waste-waters before they are discharged into surface water bodies. In addition to being toxic to aquatic life, residuals of chlorine can produce halogenated organic compounds that are potentially toxic to man. Trihalomethanes (chloroform and bromoform), which are carcinogens, are produced by chlorination.

Chlorine Dioxide

Chlorine dioxide, discovered in 1811 by Davy, was prepared from the reaction of potassium chlorate with hydrochloric acid. Early experimentation showed that chlorine dioxides exhibited strong oxidising and, bleaching properties. In the 1930; the Mathieson, Alkali Works developed the first commercial process for preparing chlorine dioxide from sodium chlorate. By 1939, sodium chlorine was established as a commercial product for the generation of chlorine dioxide.

Chlorine dioxide uses expanded rapidly in the industrial sector. In 1944, chlorine dioxide was first applied for taste and odour control at a water treatment plant in Niagara Falls, New York. Other water plants recognised the uses and benefits of chlorine dioxide. In 1958, a national survey determined that 56 U.S., water utilities were using chlorine dioxided. The number of plants using chlorine dioxide has grown more slowly since that time.

At present, chlorine dioxide is primarily used as a bleaching chemical in the pulp and paper industry. It is also used in large amounts by the textile industry, as well as for the bleaching of flour, fats, oils and waxes. In treating drinking water, chlorine dioxide is used in this country for taste and odour control, decolourisation, disinfection, provision of residual disinfectant in water distribution systems and oxidation of iron, manganese; and organics. The principal use of chlorine dioxide in the United States is for the removal of taste and odour caused by phenolic compounds in raw water supplies.

Chlorine dioxide is a yellow-green gas and soluble in water at room temperature to about 2.9 g/l chlorine dioxide (at 30 mm mercury partial pressure) or more than 10 g/l in chilled water. The boiling point of liquid chlorine dioxide is 11°C; the melting point is –59°C. Chlorine dioxide gas has a specific gravity of 2.4. The oxidant is used in a water solution and is five times more soluble in water than chlorine gas. In addition, chlorine dioxide does not react with water in the same manner that chlorine does. Chlorine dioxide is volatile; consequently, it can be stripped easily from a water solution by aeration.

Chlorine dioxide has a disagreeable odour, similar to that of chlorine gas and is detectable at 17 ppm. It is distinctly irritating to the respiratory tract at a concentration of 45 ppm in air. Concentrations above 11 per cent can be mildly explosive in air. As a gas or liquid, it readily decomposes upon exposure to ultraviolet light. It is also sensitive to temperature and pressure, two reasons why chlorine dioxide is generally not shipped in bulk concentrated quantities. Chlorine dioxide has a much greater oxidative capacity than chlorine and is, therefore, a more effective oxidant in lower concentrations. Chlorine dioxide also maintains an active residual in potable water longer than chlorine does. It does not react with ammonia or with trihalomethane precursors when prepared with no free residual chlorine. Chlorine dioxide is prepared from feedstock chemicals by several methods. The specific method depends on the quantity needed and the safety limitations in handling the various feedstock chemicals. The most common processes are:

From sodium chlorite (NaClO$_2$)

1. Acid and sodium chlorite.
2. Gaseous chlorine and sodium chlorite.
3. Sodium hypochlorite, acid and sodium chlorite.

From sodium chlorate (NaClO$_3$)

1. The sulphur dioxide process.
2. The methanol process.

The first group of processes is more commonly used. The second group of processes is frequently used by industry where the quantities produced are much greater than in water utilities.

Oxidation of phenols with chlorine dioxide or chlorine produces chlorinated aromatic intermediates before ring rupture. Oxidation of phenols with either chlorine dioxide or ozone produces oxidised aromatic compounds as intermediates which undergo ring rupture upon treatment with more oxidant and/or longer reaction times. In many cases, the same non-chlorinated, ringruptured aliphatic products are produced using ozone or chlorine dioxide.

In oxidising organic materials, chlorine dioxide can revert back to the chlorite ion. In the presence of excess chlorine (or other strong oxidant), chlorite can be preoxidised to chlorine dioxide. Using large excesses of chlorine dioxide over the organic materials appears to favour oxidation reactions (without chlorination), but slight excesses appear to favour chlorination. When excess free chlorine is present with the chlorine dioxide, chlorinated organics usually are produced, but in lower yields, depending on the concentration of chlorine and its reactivity with the particular organic(s) involved. Treatment of organic compounds with pure chlorine dioxide containing no excess free chlorine produces oxidation products containing no chlorine in some cases, but products containing chlorine in others.

Under drinking water plant treatment conditions, humic materials and/or resorcinol do not produce trihalomethanes with chlorine dioxide. Also, saturated aliphatic compounds are not reactive with chlorine dioxide. Alcohols are oxidised to the corresponding acids.

The gaseous chlorine-sodium chlorite process for producing chlorine dioxide uses aqueous chlorine and aqueous sodium chlorite to produce a mixture of chlorine dioxide and chlorine (commonly as HOCl). Figure 9.4 shows such a system, consisting of a chlorine dioxide generator, a gas chlorinator, a storage reservoir for liquid sodium chlorite and a chemical metering pump. (Sodium chlorite solution can be prepared from commercially available dry chemical by adding it to water.) The recommended feed ratio of chlorine to sodium chlorite is 1:1 by weight. Additional chlorine can be injected into the reactor vessel without changing the overall production of chlorine dioxide.

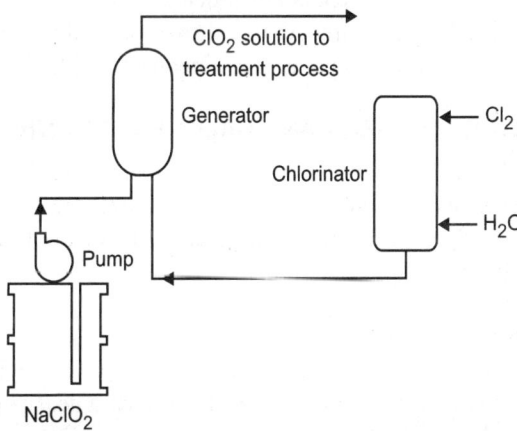

Fig. 9.4. Components of a gaseous sodium chlorite chlorine dioxide generation system.

A major disadvantage of this system is the limitation of the single-pass gas-chlorination phase. Unless increased pressure is used, this equipment is unable to achieve higher concentrations of chlorine as an aid to a more complete and controllable reaction with the chlorite ion. The French have developed a variation of this process using a multiple-pass enrichment loop on the chlorinator to achieve a much higher concentration of chlorine and thereby quickly attain the optimum pH for maximum conversion to chlorine dioxide. By using a multiple-pass recirculation system, the chlorine solution concentrates to a level of 5–6 g/l. At this concentration, the pH of the solution reduces to 3.0 and thereby provides the low pH level necessary for efficient chlorine dioxide production. A single pass results in a chlorine concentration in water of about 1 g/l, which produces a pH of 4 to 5. If sodium chlorite solution is added at, this pH, only about 60 per cent yield of chlorine dioxide is achieved. The remainder is unreacted chlorine (in solution) and chlorite ion. When upwards of 100 per cent yield of chlorine dioxide is achieved, there is virtually no free chlorite or free chlorine carrying over into the product water. The French system can be designed for variable-feed rates with automatic control by an analytical monitor. This has the advantages of eliminating the chlorine dioxide storage reservoir. Production can be varied by 20 equal increments. A 10 kg/hr. (530 lb/day) reactor can be varied in 0.5 kg/hr. (26.5 lb/day) steps over the range of 0–10 kg/hr. and this can be accomplished by automatic control with the monitor located in the main plant control panel.

Another approach to chlorine dioxide production is the acid-sodium chlorite system. The combination of acid and sodium chlorite produces an aqueous solution of chlorine dioxide without production of significant amounts of free chlorine. The acid-based process avoids the problem of differentiating between chlorine and chlorine dioxide for establishing an oxidant residual. This system uses liquid chemicals as the feedstock. Each tank has a level sensor to avoid overfilling. The tanks are installed below ground in concrete bunkers which are capable of withstanding an explosion. There are no floor drains in these bunkers. Any spillage must be pumped with corrosion-resistant pumps. Primary and backup sensors with alarms warn of any spillage. Because of the potential explosiveness, chemicals are diluted prior to the production of chlorine dioxide. The dilution is carried out on a batch basis controlled by level monitors. Proportionate quantities of softened dilution water along with the chemical reagents are pumped to mixing vessels by means of calibrated double-metering pumps. After the reactor is properly filled, an agitator within the container mixes the solution. Dilutions of 9 per cent HCl and 7.5 per cent sodium chlorite are produced in the chemical preparation process. The chlorine dioxide is subsequently manufactured on a batch basis. The final strength of the solution is about 20 per cent, 90 per cent to 95 per cent of this is chlorine dioxide and 4 per cent to 7 per cent is chlorine.

DISINFECTION WITH INTERHALOGENS AND HALOGEN MIXTURES

The interhalogen compounds are the bromine- and iodine-base materials. It is the larger, more positive halogen that is the reactive portion of the interhalogen molecule during the disinfection process. Although only used on a limited basis at present, there are members of this class that show great promise as environmentally safe disinfectants.

Properties of Bromine and Bromides

Bromine (from the Greek word bromos, meaning stench) has an atomic weight 79.909, atomic number 35, melting point –7.2°C and boiling point 58.78°C. As a gas it has a density of 7.59 g/l and as a liquid 3.12 g/l (20°C). It is a member of the halogen group of elements. Bromine is found mainly in the bromide form, widely distributed and in relatively small proportions. Extractable bromides occur in the ocean and salt lakes, brines or saline deposits left after these waters evaporated during earlier geological periods. The average bromide content of ocean water is 65 ppm by weight (about 308000 tonnes of bromine per cubic mile of sea water). The dead sea is one of the richest commercial sources of bromine in the world (containing nearly 0.04 per cent at the surface and up to 0.6 per cent at deeper levels). In the United States, major sources of bromine are the brine wells in Arkansas, Ohio and Michigan (bromide contents range from 0.2 per cent to 0.4 per cent).

Bromine is the only liquid nonmetallic element. It is a heavy, mobile, reddish-brown liquid that readily volatilises at room temperature to a red vapour having a strong pungent odour. Its disagreeable odour strongly resembles chlorine and has a very irritating effect on the eyes and throat. Bromine is readily soluble in water or carbon disulphide, forming a red solution. It is less active than chlorine but more so than iodine. Bromine unites readily with many elements and has a bleaching action.

The toxic action of bromine is similar to that of chlorine and can cause physiological damage to humans through inhalation and oral routes. It is an irritant to the mucous membranes of the eyes and upper respiratory tract. Severe exposures may result in pulmonary edema. Chronic exposure is similar to therapeutic ingestion of excessive bromides.

The most common inorganic bromides are sodium, potassium, ammonium, calcium and magnesium bromides. Methyl and ethyl bromides are among the most common organic bromides. The inorganic bromides produce a number of toxic effects in humans: depression, emaciation and in severe cases, psychoses and mental deterioration. Bromide rashes (called bromoderma) can occur especially on the facial area and resemble acne and furunculosis. This often occurs when bromide inhalation or administration is prolonged. Organic bromides such as methyl bromide and ethyl bromide are volatile liquids of relatively high toxicity. When any of the bromides are strongly heated, they emit highly toxic fumes.

Interhalogen Compounds and Their Properties

Interhalogen compounds are formed from two different halogens. These compounds resemble the halogens themselves in both their physical and chemical properties. Principal differences show up in their electronegativities. This is clearly shown by the polar compound ICl, which has a boiling point almost 40°C above that of bromine, although both have the same molecular weights. Interhalogens have bond energies that are lower than halogens and therefore in most cases they are more reactive. These properties impart special germicidal characteristics to these compounds. The principal germicidal compound of this group is bromine chloride. At equilibrium, BrCl is a fuming dark red liquid below 5°C. It exists as a solid only at relatively low temperatures. Liquid BrCl can be vapourised and metered as a vapour in equipment similar to that used for chlorine.

BrCl is prepared by the addition of equivalent amounts of chlorine to bromine until the solution has increased in weight by 44.3 per cent. The reaction is as follows:

$$Br_2 + Cl_2 \rightarrow 2BrCl$$

BrCl can be prepared by the reaction in the gas phase or in aqueous hydrochloric acid solution. In the laboratory, BrCl is prepared by oxidising bromide salt in a solution containing hydrochloric acid.

$$KBrO_3 + 2KBr + 6HCl \rightarrow 3BrCl + 3KCl + 3H_2O$$

BrCl exists in equilibrium with bromine and chlorine in both gas and liquid phases. Table 9.5 lists various physical properties of BrCl. Due to the polarity of BrCl, it shows greater solubility than bromine in polar solvents. In water, it has a solubility of 8.5 grams per 100 grams of water at 20°C (that is, 2.5 times the solubility of bromine; 11 times that of chlorine). Bromine chloride's solubility in water is increased greatly by adding chloride ions to form the complex chlorobromate ion, $BrCl_2$.

Table 9.5. Physical properties of BrCl.

Molecular weight	115.37
Melting point (°C)	−66
Boiling point (°C)	5
Density (g/cc), 20°C	2.34
Heat of fusion (cal/g)	17.6
Heat of vapourisation (cal/g)	53.2
Heat formation (kcal/mole)	0.233

(Contd...)

Heat capacity (cal./deg. mole, 298°K)	8.38
Entropy (cal./deg. mole, 298°K)	57.34
Dipole moment	0.56
Electrical conductivity (dm^{-1} cm^{-1})	–
Degree of dissociation (%, vapour 25°C)	21

Chemistry of Bromine Chloride

Various organic and inorganic species that act as reducing agents react with and destroy free halogen residuals during interaction with micro-organisms. Competitive reactions depend on the reactivity of the chemical species, temperature, contact time and pH. The quality of the effluent and the method of adding the disinfectant also help determine the specific reaction pathways. Bromine chloride is about 40 per cent dissociated into bromine and chlorine in most solvents. Because of its high reactivity and fast equilibrium, BrCl often generates products that result almost entirely from it. This is illustrated by the disinfectant products shown in Fig. 9.5. The major portion of the BrCl is eventually reduced to inorganic bromides and chlorides, with the exception of addition and substitution reactions with organic constituents.

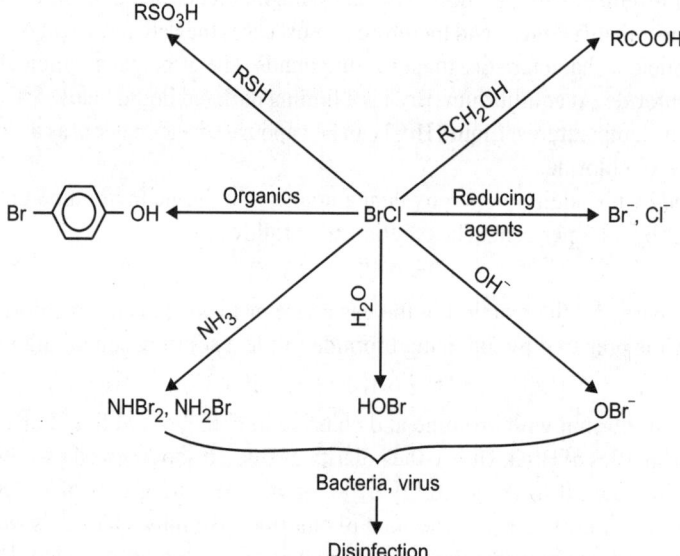

Fig. 9.5. Reactions in waste-water disinfection.

It should be noted that although BrCl is mainly a brominating agent that is competitive with bromine, its chemical reactivity makes its action similar to that of chlorine (that is, disinfection, oxidation and a bleaching agent). BrCl hydrolyses exclusively to hypobromous acid and if any hydrobromic acid (HBr) is formed by hydrolysis of the dissociated bromine, it quickly oxidises to hydrobromous acid via hypochlorous acid. Since hypohalous acid is a much more active disinfectant than the hypohalite ion, the effect of pH on ionisation becomes important. Hypobromous acid has a lower ionisation value than hypochlorous acid and this contributes to the higher disinfectant activity of BrCl compared with chlorine.

Bromine chloride also undergoes very specific reactions with ammonia and with organics. Monobromamine and dibromamine are the major products formed from reactions between BrCl and ammonia. These are unstable compounds in most conventional waste-water treatment plant effluents. In comparing the activities of bromamine versus chloramines, the effects of ammonia and high pH tend to improve the bromamine performance whereas the chloramine activity is reduced significantly. The reaction of ammonia with either BrCl or chlorine to form the halamine is very fast and generally goes to completion. As such, the presence of ammonia is essential to the disinfectant properties. Most sewage effluents typically have high ammonia concentrations in the range of 5–20 ppm. For such samples, the predominant bromine species (pH at 7 to 8) monobromamine and dibromamine are approximately equally distributed.

There are a large number of organics that undergo disinfection during the purification process. There are unfortunately a number of undesirable by-products and side reactions which occur with some of them. One is the reaction between chlorine and phenol, producing chlorophenols, which are suspected carcinogens. Chlorophenols have obnoxious tastes and are toxic to aquatic life even at very low concentrations. Brominated phenolic products which are formed in the chlorobromination of waste-water are generally more readily degraded and often less offensive than their chlorinated counterparts.

The major organic reactions of BrCl consist of electrophilic brominations of aromatic compounds. Many aromatic compounds do not react in aqueous solution unless the reaction involves activated aromatic compounds (an example being phenol). Bromine chloride undergoes free-radical reactions more readily than bromine.

Metal ions in their reduced state also undergo reactions with BrCl. Examples include iron and manganese.

$$Fe^{+2} + BrCl \rightarrow Fe^{+3} + Br^- + Cl^-$$

$$Mn + BrCl \rightarrow Mn^{+2} + Br^- + Cl^-$$

Waste-water occasionally contains hydrogen sulphide and nitrites. These contribute to higher halogen demands. Many of these reactions reduce halogens to halide salts.

Bromine chloride's reactivity with metals is not as great as that of bromine; however, it is comparable to chlorine. Dry BrCl is typically two orders of magnitude less reactive with metals than dry bromine. Most BrCl is less corrosive than bromine. Like chlorine, BrCl is stored and shipped in steel containers.

Disinfection with Bromine Chlorine

In chlorination, chlorine's reaction with ammonia forms chloramines, greatly reducing its bactericidal and virucidal effectiveness. The biocidal activity of monochloramine is only 0.02–0.01 times as great as that of free chlorine. Typical ammonia concentrations found in secondary sewage range from 5–20 ppm, which is about an order of magnitude greater than the amount needed to form monochloramine from normal chlorination dosages (which requires about 5–10 ppm). Therefore, monochloramine is the major active chlorine constituent in chlorinated sewage plant effluents. In contrast, BrCl ammonia reactions produce the major product bromamines. Bromamines have disinfectant characteristics which are significantly different than chloramines.

Toxicity of Aquatic Life

Bromamines are considerably less stable than chloramines in receiving waters. Bromamines tend to break down into relatively harmless constituents typically in under 60 minutes. Consequently, BrCl is less damaging to marine life than chlorine. Chloramines at concentrations below 0.1 ppm have resulted in fish kills. There are also indirect effects from chloramine contamination. For example, fish populations tend to avoid toxic regions, even at very low levels of concentrations. Consequently large areas of receiving waters can become unavailable to many species of fish and even cause blockage of upstream migrations during the spawning season. It should be noted that although chlorine efficiency is increased by nitrification, BrCl performance is not. Because of the high biocidal activity of bromamines, it is not necessary to utilise high concentrations and breakpoint conditions to achieve active halogen residuals, as is the case in chlorination. The breakpoint reaction with BrCl is achieved almost immediately in the presence of even slight excess amounts of bromine at pH levels of 7 to 8. There is, however, no need to reach the breakpoint to achieve good disinfectant properties with BrCl. In contrast, with chlorine it is necessary to add amounts in excess of the breakpoint to obtain sterilising characteristics.

Properties of Iodine

Iodine (from the Greek, *iodines*, meaning violet) has an atomic weight of 126.9044, atomic number 53, melting point 113.5°C and boiling point 184.35°C. As a gas, its density is 11.27 g/l and as a solid its specific gravity is 4.93 (20°C). This halogen was discovered by Courtois in 1811. It occurs sparingly in the form of iodides in sea water from which it is assimilated by seaweeds, in Chilean saltpeter and nitrate-bearing soil, in brines from ancient sea deposits and in brackish waters derived from oil and salt wells. Pure grades of iodine can be obtained from the reaction of potassium iodide with copper sulphate. Iodine is a grayish-black, lustrous solid that volatilises at ordinary temperatures to a blue-violet gas. It forms compounds with many elements. However, it is less active than many of the other halogens which displace it from iodides. Iodine dissolves readily in chloroform, carbon tetrachloride and carbon disulphide. It is only slightly soluble in water. Iodine is highly irritating to the skin, eyes and mucous membranes. Its effect on the human body is similar to that of bromine and chlorine. However, it is more irritating to the lungs.

Disinfection with Iodine Compounds

Two interhalogens having strong disinfecting properties are iodine monochloride (ICl) and iodine bromide (IBr). Iodine monochloride has found use as a topical antiseptic. It may be complexed with nonionic or anionic detergents to yield bactericides and fungicides that can be used in cleansing or sanitising formulations. These generally have a polymer structure which establishes its great stability, increased solubility and lower volatility. By reducing the free halogen concentration in solution, polymers reduce both the chemical and bactericidal activity. Complexes of ICl are useful disinfectants which compromise lower bactericidal activity with increased stability. Iodine monochloride is itself a highly reactive compound, reacting with many metals to produce metal chlorides. Under normal conditions it will not react with tantalum, chromium, molybdenum, zirconium, tungsten or platinum. With organic compounds, reactions cause iodination, chlorination, decomposition or the generation of halogen addition compounds. In water, ICl hydrolyses to hypoiodous and hydrochloric acids. In the absence of excess chloride ions, hypoiodous acid will disproportionate into iodic acid and iodine. Iodine bromide has a chemistry similar to ICl. Iodine bromide reacts with aromatic compounds to produce iodination in polar solvents and bromination in nonpolar solvents. It has complex chemical properties, as its solubility is increased more effectively by bromide than by chlorided ions. Primary hydrolysis takes place in the presence of hydrobromic acid. As a disinfectant, IBr is used in its complexed or stabilised forms. Unfortunately, it undergoes hydrolysis and dissociation reactions in aqueous solutions, both reactions being major limitations. Its disinfecting properties are similar to ICl and as in the case of ICl, germicidal activity should not be reduced by haloamine formation since bromamines are highly reactive and iodoamines are not generated. Upon application of prepared solutions to control micro-organisms, the complex releases IBr gradually. This process forms free iodine during the decomposition of IBr (the decomposition takes place as fast as the IBr is released).

Disinfection with Halogen Mixtures

Two approaches that have been investigated recently for disinfection are mixtures of bromine and chlorine and mixtures containing bromide or iodide salts. Some evidence exists that mixtures of bromine and chlorine have superior germicidal properties than either halogen alone. It is believed that the increased bacterial activity of these mixtures can be attributed to the attacks by bromine on sites other than those affected by chlorine. The oxidation of bromide or iodide salts can be used to prepare interhalogen compounds or the hypollalous acid in accordance with the following reaction:

$$HOCl + NaBr \rightarrow HOBr + HCl$$

It has been reported that the rate of bacterial sterilisation by chlorine in the presence of ammonia is accelerated with small amounts of bromides. As little as 0.25 ppm of bromamines can be significant under some conditions. However, if chloramines are produced prior to contact with bromide ions, the reaction and subsequent effect are reduced. Improved germicidal activity has also been shown for mixtures containing bromides and iodides with various chlorine releasing compounds. Bromide improves the disinfecting properties of dichloroisocyanuric acid and hypochlorite against several bacteria. Bromine-containing compounds are useful for their combined bleaching and disinfectant properties. There has been the concern that the use of interhalogen compounds in waste-water disinfection could produce unknown organic and inorganic halogen-containing substances. In the case of iodine, concern has been expressed over the physiological aspects in water supplies. Extensive studies have been reported on the role played by iodine and iodides in the thyroid glands of animals and man. Information on acute inhibition of hormone formation by excessive amounts of iodine is well known. Despite the fact that no strong evidence exists that iodine is harmful as a water disinfectant, only limited use has been attempted. Chronic bromide intoxication from continuous exposure to dosages above 3–5 gram is called bromism. Typical symptoms are skin rash, glandular excretions, gastrointestinal disturbances and neurological disturbances. Bromide can be absorbed from the intestinal tract and contaminate the body in a manner very similar to that for chloride. Brominated drinking water does not, however, significantly increase the amount of bromine admitted internally. The amount of additional bromine in chlorobrominated waters will not significantly increase human bromine concentrations nor result in bromism.

STERILISATION USING OZONE

Ozone (O_3) is a powerful oxidant and application to effluent treatment has developed slowly because of relatively high capital and energy costs compared to chlorine. Energy requirements for ozone are in the range of 10 to 13 kWh/lb generated from air, 4 kWh/lb from oxygen and 5.5 kWh/lb from oxygen-recycling systems. Operating costs for air systems are essentially the electric power costs; for oxygen systems the cost of oxygen must be added to the electrical cost.

Actual uses of ozone include odour control, industrial chemicals synthesis, industrial water and waste-water treatment and drinking water. Lesser applications appear in fields of combustion and propulsion, foods and pharmaceuticals, flue gas-sulphur removal and mineral and metal refining. Potential markets include pulp and paper bleaching, power plant cooling water and municipal waste-water treatment.

The odour control market is the largest and much of this market is in sewage treatment plants. Use of ozone for odour control is comparatively simple and efficient. The application is for preservation of environmental quality; in addition, alternative treatment schemes requiring either liquid chemical oxidants (like permanganate or hydrogen perioxide) or incineration can significantly increase capital and costs.

Ozone applications in the United States for drinking water are far fewer than in Europe. However, the potential market is large, if environmental or health needs ever conclude that an alternate disinfectant to chlorine should be required. Although energy costs of ozonation are higher than those for chlorination, they may be comparable to combined costs of chlorination dechlorination-reaeration, which is a more equivalent technique. One of ozone's greatest potential uses is for municipal waste-water disinfection.

Technical, economic and environmental advantages exist for ozone bleaching of pulp in the paper industry as an alternate to hypochlorite or chlorine bleaching which yields deleterious compounds to the environment.

Principles of Ozone Effluent Treatment

Ozone was first discovered by the Dutch philosopher Van Marun in 1785. In 1840, Schonbein reported and named ozone from the Greek word ozein, meaning to smell. The earliest use of ozone as a germicide occurred in France in 1886, when de Meritens demonstrated that diluted ozonised air could sterilise polluted water. In 1893, the first drinking water treatment plant to use ozone was constructed in Oudshorm, Holland. Other plants quickly followed at Wiesbaden and Paderborn in Germany. In 1906, a plant in Nice, France, was constructed using ozone for disinfection. Today, there are over 1000 drinking water treatment plants in Europe utilising ozone for one or more purposes. In the United States, the first ozonation plant was constructed in Whiting, Indiana, in 1941 for taste and odour control.

Over 100 years ago it had been demonstrated that ozone (O_3), the unstable triatomic allotrope of oxygen, could destroy moulds and bacteria and by 1892 several experimental ozone plants were in operation in Europe. In the 1920s, however, as a result of wartime research, during World War I, chlorine became readily available and inexpensive and began to displace ozone as a purifier in municipalities throughout the United States. Most ozone studies and developments were dropped at this time, leaving ozonation techniques, equipment and research at a primitive stage. Ozone technology stagnated and the development and acceptance of ozone for water and waste-water treatment was discontinued.

In addition to the popular use of chlorination as a waste-water disinfectant and the consequent technology lag in ozonation research, there was a third impediment to ozone commercialisation: the comparatively high cost of ozonation in relation to chlorination. Ozone's instability requires on-site generation for each application, rather than centralised generation and distribution. This results in higher capital requirements, aggravated by a comparatively large electrical energy requirement. Ozone's low solubility, in water and the generation of low concentrations, even under ideal conditions, also necessitates more elaborate and expensive contacting and recycling systems than chlorination.

In spite of such obstacles there is interest from time to time in the use of ozone, particularly for waste-water treatment. The technology for the destruction of organics and inorganics in water has not kept pace with the increasingly more sophisticated water pollution problems arising from greater loads, new products and new sources of pollutant entry into the environment and increased regulation. The growing trend toward water reuse and the fact that some highly toxic pollutants may be refractory to conventional treatment methods has spurred investigation into new treatments, including ozonation.

A significant impetus from time to time for developing new methods is dissatisfaction with chlorination. Chlorine affects taste and odour and produces chloramines and a wide variety of other potentially hazardous chlorinated compounds in waste-waters. It seriously threatens the environment with an estimated 1000 tonnes per year of chlorinated organic compounds discharged into U.S., waters (chloramines are not easily degradable and pose a hazard to the environment) and is questionable as a drinking water viricidal disinfectant. Ozone's development, on the other hand, could parallel a greater environmental awareness and a resulting demand for higher-quality effluents, as its potential for overcoming these problems is possible.

Properties of Ozone

Ozone is an unstable gas, having a boiling point of $-112°C$ at atmospheric conditions. Its molecular weight is 48. Ozone is partially soluble in water (approximately 20 times more soluble than oxygen) and has a characteristic penetrating odour which is readily detectable at concentrations as low as 0.01–0.05 ppm. Ozone is the most powerful oxidant currently available for use for waste-water treatment. Commercial generation equipment generates ozone at concentrations of 1 per cent to 3 per cent in air (that is, 2 per cent to 6 per cent in oxygen). Ozone is unstable in water, however, it is more stable in air, especially in cool, dry air.

As a strong oxidant, ozone reacts with a wide variety of organics. Ozone oxidizes phenol to oxalic and acetic acids. It oxidizes trihalomethane (THM) compounds to a limited extent within proper pH ranges and reduces their concentration by air stripping. Trihalomethanes are also oxidised by ozone in the presence of ultraviolet light. Oxidation by ozone does not result in the formation of THMs as does chlorination. A combination of ozone and ultraviolet light destroys DDT, malathion and other pesticides. However, high dosages and extended contact times that are not normally encountered in drinking water treatment are needed. Ozonised organic substances are usually more biodegradable and absorbable than the starting, unoxidised substances. When ozonation is employed as the final treatment step for potable water systems in water containing significant concentrations of dissolved organics, bacterial regrowth in the distribution system can occur. Consequently, ozonation is not typically used as the final treatment step but rather followed by granular activated carbon filtration and sometimes by the addition of a residual disinfectant. Humic materials are the precursors of THMs. Humic substances can be oxidised by ozonation. Under proper conditions significant reduction in THM formation can be realised when ozone is applied prior to a chlorination step. Because of ozone's instability, it is able to produce a series of almost instantaneous reactions when in contact with oxidisable compounds. One example follows:

$$O_3 + 2KI + H_2O \rightarrow I_2 + O_2 + 2KOH$$

In this reaction, iodine is liberated from a solution of potassium iodide. This reaction can be used to assess the amount of ozone in either air or water. For determination in air or oxygen, a measured volume of gas is drawn through a wash bottle containing potassium iodide solution. Upon lowering the pH with acid, titration is effected with sodium thiosulphate, using a starch solution as an indicator. There is a similar procedure for determining ozone in water.

A typical ozone treatment plant consists of three basic subsystems: feedgas preparation; ozone generation; and ozone/water contacting. Commercially, ozone is generated by producing a high-voltage corona, discharge in a purified oxygen-containing feedgas. The ozone is then contacted with the water or waste-water; the treated effluent is discharged and the feedgas is recycled or discharged.

Ozone's high reactivity and instability, as well as serious obstacles in producing concentrations in excess of 6 per cent, preclude central production and distribution with its associated economies of scale. The requirement for on-site generation and application of ozone must yield a cost-efficient, lowmaintenance operation in order to be useful. The feedgas employed in ozonation systems is either air, oxygen or oxygen-enhanced air. The particular selection of feedgas for each application is based on economics and depends on several factors: total quantity of ozone required; desired concentration of ozone in the feedgas; and fate (recycle or discharge) of the feedgas. For a given ozone generator with a specified power input and gas flow, two to three times as much ozone may be generated from oxygen as from air. The maximum concentration economically produced from air is about 2 per cent, while that generated from pure oxygen is approximately 6 per cent.

The use of higher concentrations of ozone provides two advantages: capital and operating costs per pound of ozone produced are substantially reduced and a greater concentration gradient for mass transfer of ozone is provided in the contacting step, yielding increased ozone-utilisation efficiency. These advantages, however, must be weighed against the increased cost of oxygen production. Air is generally employed in those applications requiring less than 50 pounds/day of low concentration ozone. If air is the feedgas, it must be dried and cooled to reduce accumulation of corrosive nitric acid and nitrogen oxides that occur as by-products when the dew point is above 40°C.

Ozone may be produced by electrical discharge in an oxygen-containing feedgas or by photochemical action using ultraviolet light. For large-scale applications, only the electric-discharge method is practical since the use of ultraviolet energy produces only low-volume, low-concentration ozone.

In the electric-discharge (or corona) method, an alternating current is imposed across a discharge gap with voltages, between 5 and 25 kV and a portion of the oxygen is converted to ozone. A pair of large-area electrodes is separated by a dielectric (1–3 mm in thickness) and an air gap (approximately 3 mm.) as shown in Fig. 9.6. Although standard frequencies of 50 or 60 cycles are adequate, frequencies as high as 1000 cycles are also employed.

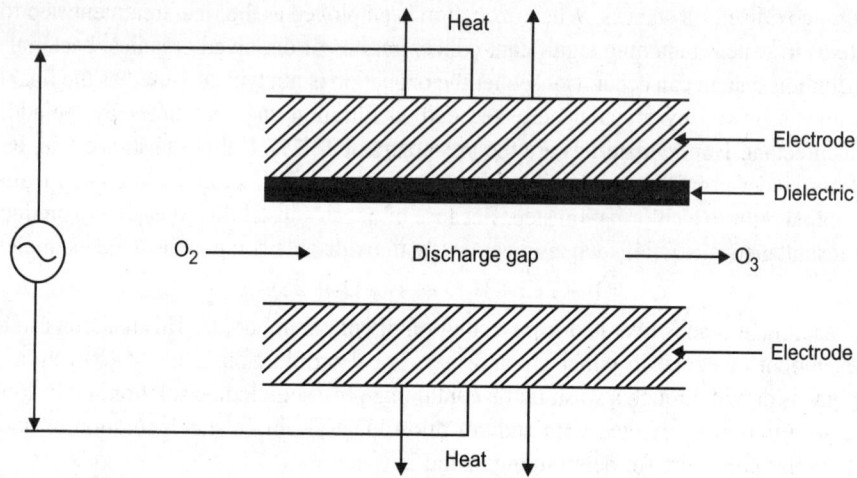

Fig. 9.6. Basic ozonator configuration.

Only about 10 per cent of the input energy is effectively used to produce the ozone. Inefficiencies arise primarily from heat production and to a lesser extent, from light and sound. Since ozone decomposition is highly temperature dependent, efficient heat-removal techniques are essential to the proper operation of the generator. The mechanism for ozone generation is the excitation and acceleration of stray electrons within the high-voltage field. The alternating current causes the electrons to be attracted first to one electrode and then the other. As the electrons attain sufficient velocity, they become capable of splitting some O_2 molecules into free radical oxygen atoms. These atoms may then combine with O_2 molecules to form O_3. Under optimum operating conditions (efficient heat removal and proper feedgas flow), the production of ozone in corona-discharge generators is represented by the following relationships, showing the factors to be considered in the design of these generators:

$$V \propto pg$$

$$(Y/A) \propto f \varepsilon V^2/d$$

where,

Y/A = ozone yield per unit area of electrode surface.
V = applied voltage.
p = gas pressure in the discharge gap.
g = discharge gap width.
f = frequency of applied voltage.
ε = dielectric constant.
d = thickness of the dielectric.

The following requirements will facilitate optimisation of the ozone yield:

1. The pressure/gap combination should be constructed so the voltage can be kept relatively low while maintaining reasonable operating pressures. Low voltage protects the dielectric and electrode surfaces. Operating pressures of 10–15 pounds per square inch gauge (psig) are applicable to many waste treatment uses.

2. For high-yield efficiency, a thin dielectric with a high-dielectric constant should be used. Glass is the most practical material. High-dielectric strength is required to minimise puncture, while minimal thickness maximises yield and facilitates heat removal.

3. For reduced maintenance problems and prolonged equipment life, high frequency alternating current should be used. High frequency is less damaging to dielectric surfaces than high voltage.

4. Heat removal should be as efficient as possible.

Basic configurations of ozone generators are shown in Figs. 9.6 and 9.7. The three designs are the Otto plate, the tube and the Lowther plate. The least efficient of these generators is the Otto plate, developed at the turn of the century. The tube and Lowther plate units include modern innovations in material and design. The Lowther plate generator is the most efficient configuration due in large measure to advantages in heat removal. In addition to ozone yield, the concentration of ozone is an important consideration. Ozone concentration from a generator is usually regulated by adjusting the flow rate of the feedgas and/or voltage across the electrodes.

Contactor design is important in order to maximise the ozone-transfer efficiency and to minimise the net cost for treatment. The three major obstacles to efficient ozone utilisation are ozone's relatively low solubility in water, the low concentrations and amounts of ozone produced from ozone generators and the instability of ozone. Several contacting devices are currently in use including positive-pressure injectors, diffusers and venturi units. Specific contact systems must be designed for each different application of ozone to waste-water. Further development in this area of gas-liquid contacting needs to be done despite its importance in waste treatment applications.

In order to define the appropriate contactor, the following should be specified:

1. The objective: disinfection biochemical oxygen demand (BOD) or chemical oxygen demand (COD) reduction to a particular level, trace refractory organics oxidation and so on.

2. Relative rates of competitive reactions: chemical oxidation, lysing bacteria, decomposition of ozone in aqueous solutions and so on.

3. Mass-transfer rate of ozone into solution.

4. Waste-water quality characteristics: total suspended solids organic loading and so on.
5. Operating pressure of the system.
6. Ozone concentration utilised.

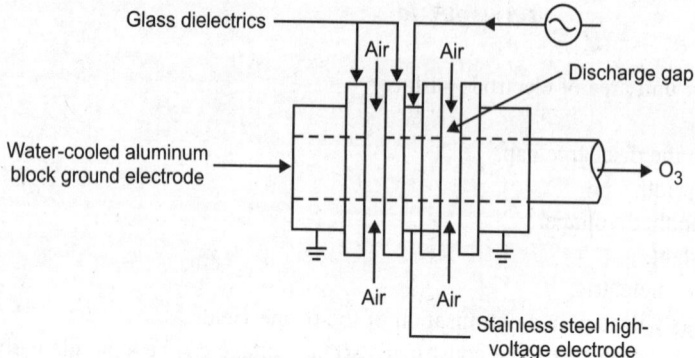

Otto plate-type generator unit

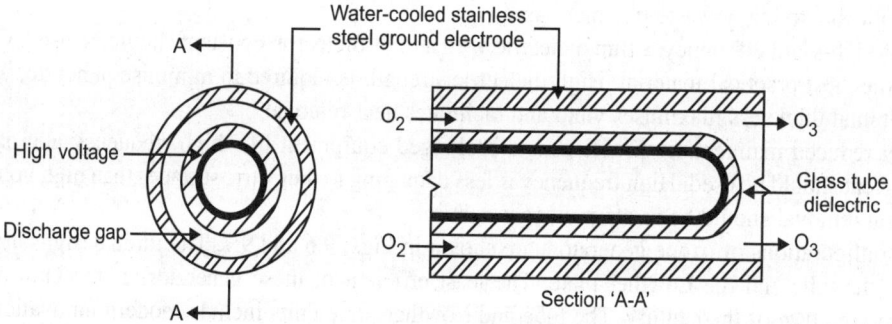

Tube-type generator unit

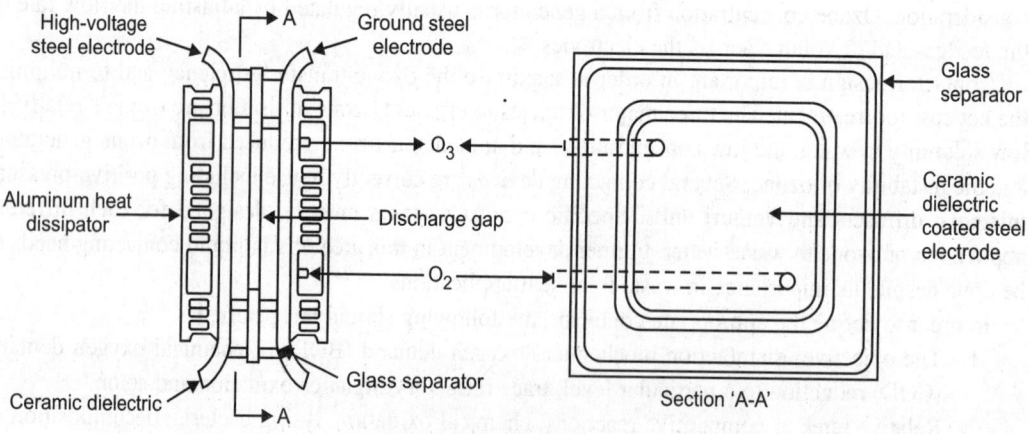

Lowther plate generator unit

Fig. 9.7. Types of ozone generators.

Other considerations for the contacting system itself include contactor type (for example, packed bed, sparged column); number and configuration of contactor stages; points of gas-liquid contact, whether the mix is cocurrent or countercurrent; and the construction materials used. It is clear that designing an ozonation system for even a relatively simple application requires a thorough understanding of many factors in order to employ sound engineering methods and optimisation techniques.

Applications

Market areas of interest to manufacturers of ozone systems, actual uses defined as those which have been in operation for some time and not including 'pilot' studies, arise in the following categories: odour control (sewage treatment and industrial), industrial chemicals synthesis, industrial water and waste-water treatment and drinking water disinfection.

Ozone has proven to be effective against viruses. France has adopted a standard for the use of ozone to inactivate viruses. When an ozone residual of 0.4 mg/l can be measured 4 minutes after the initial ozone demand has been met, viral inactivation is satisfied. This property plus ozone's freedom from residual formation are important considerations in the public health aspects of ozonation. When ozonation is combined with activated carbon filtration, a high degree of organic removal can be achieved. Concerning the toxicity of oxidation products of ozone and the removal of specific compounds via ozonation, available evidence does not indicate any major health hazards associated with the use of ozone in waste-water treatment.

Odour Control

The largest existing market for ozone systems is odour control. Much of this market is in sewage treatment plants. Industrial markets for ozone in odour control are smaller than for sewage treatment plants. Established applications include cooking odours at restaurants; pharmaceutical, fermentations; fish, meat and food processing; plastics and rubber processing; paint and varnish manufacture; and rendering plants. Nearly all of these industrial odour control applications use less than 100 lb/day of ozone and most use between 1–25 lb/day.

Industrial Chemical Synthesis

There has been only one major use for ozone today in the field of chemical synthesis: the ozonation of oleic acid to produce azelaic acid. Oleic acid is obtained from either tallow, a by-product of meat-packing plants or from tall oil, a by-product of making paper from wood. Oleic acid is dissolved in about half its weight of pelargonic acid and is ozonised continuously in a reactor with approximately 2 per cent ozone in oxygen; it is oxidised for several hours. The pelargonic and azelaic acids are recovered by vacuum distillation. The acids are then esterified to yield a plasticiser for vinyl compounds or for the production of lubricants. Azelaic acid is also a starting material in the production of a nylon type of polymer.

Industrial Water and Waste-water Treatment

The markets for ozone in industrial water and waste-water treatment are quite small. Industrial applications for ozone could grow. The use of ozone for treating photoprocessing solutions is a novel application that has been limited, but might grow. In this process, silver is recovered electrolytically; then the spent bleach baths of iron ferrocyanide complexes are ozonated. Iron cyanide complexes are stable to ozonation so that the ferrous iron is merely oxidised to ferric, which is its original form. Thus, the bleach is 'regenerated' and is ready for recycling and reuse by the photoprocessor.

Ideally, 20.2 pounds of the ferrocyanide can be converted to 11.7 pounds of ferricyanide by one pound of ozone. Indeed, ozone oxidation efficiency is nearly 100 per cent for ferrocyanide concentration above 1.0 g/l.

A typical ozone system consists of 100 g/hr, at a concentration of 1.0 per cent to 1.5 per cent in air fed to the bottom of bleach collection tanks through ceramic spargers (pore size of approximately 100 tonnes). The system contains air compression and drying equipment, automatic, control features and a flat-plate, air-cooled ozone generator. Regeneration of bleach wastes totalling about 10,000 gallons a year and recovery of other chemicals can also be cost effective.

Municipal Drinking Water

In the United States, Whiting, Indiana and Strasburg, Pennsylvania have used ozone in their drinking water treatment process. Other cities have run pilot studies. Ozone is used as a bleaching agent for miscellaneous items: petroleum, clays, wood products and chemical baths. It has been proposed as a bleaching agent for hair and as a disinfectant for oils and emulsions. Ozone is used to modify tryptophan and indigo plant juice. It is an important factor in colourfastness. The desulphurisation of flue gases by ozone has been considered an application where it promotes liquidphase oxidation. The operations are carried out with vanadium catalysts and the oxidation step is performed in gasfluidised beds. The desulphurising effect of ozone on light petroleum distillates has also been reported.

The use of ozone has been proposed in special ore-flotation processes. Two widely different applications involve hydraulic cement and the fabrication of coating on insulators.

The metallurgical applications include steel refining, electrochemical processes and gold recovery. The aggressive reactivity of ozone is evident in the corrosion of stainless steel and in chemical etching. Ozone has been examined as a potential source of high-energy oxidation and for combustion and propulsion applications.

Ozonation Equipment and Processes

Ozonation systems are comprised of four main parts, including a gas-preparation unit, an electrical power unit, an ozone generator and a contactor which includes an off-gas treatment stage. Ancillary equipment includes instruments and controls, safety equipment and equipment housing and structural supports. The four major components of the ozonation process are illustrated in Fig. 9.8.

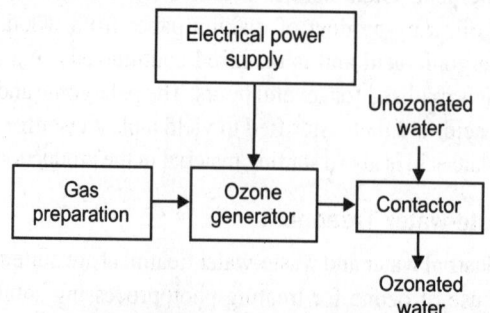

Fig. 9.8. Components of ozonation.

A high level of gas preparation (usually air) is needed before ozone generation. The air must be dried to retard the formation of nitric acid and to increase the efficiency of the generation. Moisture accelerates the decomposition of ozone. Nitric acid is formed when nitrogen combines with moisture in the corona discharge. Since nitric acid will chemically attack the equipment, introduction of moist air into the unit must be avoided. Selection of the air-preparation system depends on the type of contact system chosen. The gas-preparation system will, however, normally include refrigerant gas cooling and desiccant drying to a minimum dew-point of –40°C. A dew-point monitor or hygrometer is an essential part of any air preparation unit.

Conversion efficiencies can be greatly increased with the use of oxygen. However, the use of high-purity oxygen far ozone generation for disinfection is cost effective.

Electrical power supply units vary considerably among manufacturers. Power consumption and ozone-generation capacity are proportional to bath voltage and frequency. There are two methods to control the output of an ozone generator: vary voltage or vary frequency. Three common electrical power supply configurations are used in commercial equipment:

1. Low frequency (60 Hz), variable voltage.
2. Medium frequency (600 Hz), variable voltage.
3. Fixed voltage, variable frequency.

The most frequently used is the constant low-frequency, variable-voltage configuration. For larger systems, the 600 Hz fixed frequency is often employed as it provides double ozone production with no increase in ozone generator size.

The electrical (corona) discharge method is considered to be the only practical technique for generating ozone in plant-scale quantities. In principle, an ozone generator consists of a pair of electrodes separated by a gas space and a layer of glass insulator. An oxygen-rich gas is passed through the empty space and a highvoltage alternating current is applied. A corona discharge takes place across the gas space and ozone is generated when a portion of the oxygen is ionised and then becomes associated with nonionised oxygen molecules.

Typical horizontal tube-type ozone generator unit is preferred for larger systems. Water-cooled plate units are often used in smaller operations. However, these require considerably more floor space per unit of output than the tube-type units. The air-cooled Lowther plate type is a relatively new design. It has the potential for simplifying the use of ozone-generating equipment. However, it has had only limited operating experience in water treatment facilities.

After the ozone has been generated, it is mixed with the water stream being treated in a device called a contactor. The objective of this operation is to maximise the dissolution of ozone into the water at the lowest power expenditure. There is a variety of ozone contactor designs. Principal ones employed in waste-water treatment facilities include:

1. Multistage porous diffuser contactors, which involve a single application of an ozone-rich gas stream and application of fresh ozone gas to second and subsequent stages with off-gases recycled to the first stage.
2. Eductor-induced, ozone vacuum injector contactors, which include total or partial plant flow through the eductor; and subsequent stages with off-gases recycled to the first stage.
3. Turbine contactors, which involve positive or negative pressure to the turbine.
4. Packed-bed contactors, which include concurrent or countercurrent water/ozone-rich gas flow.

Two-level diffuser contactors, which involve application of ozone-rich gas to the lower chamber. Lower chamber off-gases are applied to the upper chamber. Offgas treatment from contactors is an important consideration. Methods employed for off-gas treatment include dilution, destruction via granular activated carbon, thermal or catalytic destruction and recycling.

Measurement and Control

Favourable operational economics and good management practices require high levels of control of the ozonation system. Depending on the specific process of ozone applications, plant size and design philosophy, the control system may be simple or complex. The trend in Europe is toward highly sophisticated and centralised control.

Several parameters should be measured to provide a fully operable ozonation system. There should be a means of providing full temperature and pressure profiles of the ozone generator feedgas from the initial pressurisation (by fan, blower or compressor) to the ozone generator inlet. Moisture content is also important. There should be a means of measuring the moisture content of the feedgas to the ozone generator. This procedure should be conducted with a continuously monitoring dew-point meter or hygrometer. Other parameters that require monitoring include:

1. Temperature, pressure, flow rate and ozone concentration of the ozone containing gas being discharged from all the ozone generators. This is the only effective method by which ozone dosage and the ozone production capacity of the ozone generator can be determined.
2. Power supplied to the ozone generators. The parameters measured include amperage, voltage, power and frequency, if this is a controllable variable.
3. Flow rate and temperature of the cooling water to all water-cooled ozone generators. Reliable cooling is important to maintain constant ozone production and to protect the dielectrics in the generation equipment.
4. There should be a means to monitor the several cycles of the desiccant drier, particularly the thermal-swing unit.

Analytical measurements of ozone concentrations must be made in the ozonised gas from the ozone generator, the contactor off-gases and the residual ozone level in the ozonised water. Methods of ozone measurement commonly used are the: simple 'sniff' test, Draeger-type detector tube, wet chemistry potassium iodide method, amperometric-type instruments, gas-phase chemiluminescence and ultraviolet radiation adsorption. The used of control systems based on these measurements varies considerably. The key to successful operation is an accurate and reliable residual ozone analyser. Continuous residual ozone monitoring equipment may be successfully applied to water that has already received a high level of treatment. However, a more cautious must be taken with the application of continuous residual ozone monitoring equipment for water that has only received chemical clarification because the ozone demand has not yet been satisfied and the residual, is not as stable. Ozone production must be closely controlled because excess ozone cannot be stored. Changes in process demand must be responded to rapidly. Ozone production is costly; underozonation may produce undesired effects and over-ozonation may require additional costs where off-gas destruction is used.

Operation and Maintenance

Ozonation equipment typically has low maintenance requirements. The air-preparation system requires frequent attention for air fluter cleaning/changing and for assuring that the desiccant is drying the air properly. However, both are usually simple operations. Two factors which impact ozone generator operation and maintenance are the effectiveness of the air-preparation system and the amount of time that the generator is required to operate at maximum capacity. Maintenance of the ozone generators is commonly scheduled once a year. However, many plants perform this maintenance every six months. Typically, one man-week is necessary to service an individual ozone generation unit of the horizontal tube type. Dielectric replacement due to failure as well as breakage during maintenance may be as low as 1 per cent to 2 per cent. An average tube life of ten years can be expected if a feed gas dew-point of –60' is maintained and if the ozone generator is not required to operate for prolonged periods at its rated capacity. Plate-type ozone generators use window glass, as dielectrics. However, the same attention to air preparation is taken as with the more expensive glass or ceramic tubes in order to avoid costly downtime. Operations and maintenance of the ozone contactor also requires attention. Turbines require electricity to power the drive motors, while porous diffusers require regular inspection and maintenance to insure a uniform distribution of ozone-rich gas in the contact chamber. It should be noted that serious safety problems exist with servicing some of these units. For example, even after purging the contact chambers with air, maintenance personnel entering the chambers should be equipped with a self-contained breathing apparatus, since the density of ozone is heavier than air and therefore is difficult to remove completely by air purging.

DISINFECTION/OXIDATION BY-PRODUCTS

Chemical disinfection became an integral part of municipal drinking water treatment over 100 years ago as a vital tool in achieving its principal objective: protection of public health. Oxidation, while not as vital to achieving public health objectives as disinfection, has also been accepted as an important part of drinking water treatment. Typically, oxidation is used to address aesthetic concerns such as colour, taste and odour, which impact consumer perception and acceptance of a water as fit for human consumption; Unfortunately, disinfection and oxidation produce a variety of by-products, which is the focus of this section. The discussion of disinfection and oxidation by-products is presented in five sections. Following an introduction to the subject, separate sections are devoted to the by-products formed by chlorine, ozone, chlorine dioxide and chloramines.

The use of chemicals for disinfection and oxidation is very common, occurring at nearly every drinking water treatment plant in the industrialised world. Familiarity with the by-products of the chemicals used for disinfection and oxidation is important, as these by-products can significantly impact consumer health. Some of these health effects have been identified in the recent past, and some of these health effects are not fully understood, making these by-products an ever-changing part of the drinking water treatment picture. An overview of oxidation by-products, including a historical perspective, known by-products, regulatory requirements, practical considerations, the chemistry of chlorine, ozone, chlorine dioxide, and chloramine by-product formation, means of controlling (i.e. reducing) their formation at a water treatment plant and means of removing these by-products, if possible, after they form in a water treatment plant are presented and discussed in this section.

Since the introduction of chlorine as a disinfectant in drinking water treatment at the turn of the twentieth century, chemical disinfection has been an integral part of municipal drinking water treatment. In addition to the use of chlorine as a microbial disinfectant, other benefits of chlorine—as well as other chemical disinfectants such as ozone and chlorine dioxide—were recognised, including the ability to eliminate colour and destroy many naturally occurring chemicals that cause objectionable taste and odour in the water. Consequently, water treatment plant operators commonly added as much disinfectant as necessary to achieve the desired aesthetic and microbial water quality.

In the early 1970s, researchers in the Netherlands and the United States were able to identify and quantify the formation of chloroform ($CHCl_3$) and other trihalomethanes (THMs) in drinking water and relate this formation to the use of chlorine during treatment. These early findings led to a large number of studies in the United States on the formation of these 'by-products' of chlorination. The studies included several monitoring surveys to assess the magnitude of the problem in drinking water treatment plants across the United States (i.e. occurrence studies), as well as studies to investigate how chlorination by-products were formed and what water quality and/or treatment conditions affected their formation.

It was also discovered that chloroform was not the only chemical formed as a result of the reaction of chlorine with natural organic matter (NOM) present in water and that in the presence of bromide ion (Br⁻) the reaction between chlorine, bromide, and NOM resulted in the formation of a mix of chlorinated and brominated chemical by-products. Using mass balance calculations it has been shown that the known chlorination by-products constitute between 30 and 60 per cent of the total organic halides (TOX) formed upon the reaction of chlorine with NOM and bromide. Therefore, much more work is needed to identify and characterise the remaining unknown by-products of chlorination.

The formation of disinfection by-products (DBPs) is not limited to chlorine disinfection. Haag and Hoigne reported the formation of bromate ion (BrO_3^-) when ozone was added to water containing bromide. Bromate was later identified as a suspected carcinogen. Ozone addition to natural water was also implicated in the formation of numerous organic by-products, such as aldehydes. Thus far, the presence of these ozone organic by-products in drinking water has not been determined to be a public health concern at the typical levels at which they are formed.

Another disinfectant/oxidant used in water treatment is combined chlorine. In the 1970s, the US EPA identified combined chlorine as an alternative to chlorine because it was believed not to form THMs. Later research showed that while monochloramine, the principle component of combined chlorine, is less reactive than free chlorine, it does react to form DBPs, although at much lower concentrations than are formed with free chlorine. The prominent category of DBPs formed due to chloramination of potable water are haloacetic acids (HAAs), mostly the dihalogenated species. The presence of bromide ion will both increase the production of HAAs and shift the speciation to the more brominated species, which are thought to present a greater carcinogen risk. In addition, Krasner and others reported the results of a 35 utility study in which they identified several new by-products, including cyanogen halides (e.g. CNCl), as by-products of chloramination.

However, no chloramine by-product was believed to be a significant public-health concern until Najm and Trussell reported that N-nitrosodimethylamine (NDMA) was a by-product of chloramine addition to drinking water and waste-water.

Chlorine dioxide forms by-products when added to water, such as chlorite (ClO_2^-) and chlorate (ClO_3^-) ions, both of which have been suspected to cause health effects. While the health effects of chlorate and chlorite continue to be a topic of debate, there was adequate concern that a limit on chlorite was adopted by the US EPA.

Known By-Products

It is now well established that all chemical disinfectants and oxidants currently used in water treatment form chemical by-products. Even among those known DBPs, the health effects of many are still uncertain. A listing of most of the known DBPs formed as a result of the use of chlorine, chloramines, ozone and chlorine dioxide during drinking water treatment are summarised in Table 9.6. While many of the DBPs listed in Table 9.6 have been detected in some treated water, they are typically present at very low concentrations.

Table 9.6. Known by-products of chlorine, combined chlorine (chloramines), ozone and chlorine dioxide application during drinking water treatment.

Class	By-product	Chemical Agent	Molecular formula
Trihalomethanes	Chloroform	Chlorine	$CHCl_3$
	Bromodichloromethane	Chlorine	$CHBrCl_2$
	Dibromochloromethane	Chlorine	$CHBr_2Cl$
	Bromoform	Chlorine, ozone	$CHBr_3$
	Dichloroiodomethane	Chlorine	$CHICl_2$
	Chlorodiiodomethane	Chlorine	CHI_2Cl
	Bromochloroiodomethane	Chlorine	$CHBrICl$
	Dibromoiodomethane	Chlorine	$CHBr_2I$
	Bromodiiodomethane	Chlorine	$CHBrI_2$
	Triiodomethane	Chlorine	CHI_3
Haloacetic acids	Monochloroacetic acid	Chlorine	$CH_2ClCOOH$
	Dichloroacetic acid	Chlorine	$CHCl_2COOH$
	Trichloroacetic acid	Chlorine	CCl_3COOH
	Bromochloroacetic acid	Chlorine	$CHBrClCOOH$
	Bromodichloroacetic acid	Chlorine	$CBrCl_2COOH$
	Dibromochloroacetic acid	Chlorine	$CBr_2ClCOOH$
	Monobromoacetic acid	Chlorine	$CH_2BrCOOH$
	Dibromoacetic acid	Chlorine	$CHBr_2COOH$
	Tribromoacetic acid	Chlorine	CBr_3COOH
Haloacetonitriles	Trichloroacetonitrile	Chlorine	$CCl_3C \equiv N$
	Dichloroacetonitrile	Chlorine	$CHCl_2C \equiv N$
	Bromochloroacetonitrile	Chlorine	$CHBrClC \equiv N$
	Dibromoacetonitrile	Chlorine	$CHBr_2C \equiv N$
Haloketones	1, 1-Dichloroacetone	Chlorine	$CHCl_2COCH_3$
	1, 1, 1-Trichloroacetone	Chlorine	CCl_3COCH_3
Aldehydes	Formaldehyde	Ozone, chlorine	$HCHO$
	Acetaldehyde	Ozone, chlorine	CH_3CHO
	Glyoxal	Ozone, chlorine	$OHCCHO$
	Methyl glyoxal	Ozone, chlorine	CH_3COCHO
Aldoketoacids	Glyoxylic acid	Ozone	$OHCCOOH$
	Pyruvic acid	Ozone	$CH_3COCOOH$
	Ketomalonic acid	Ozone	$HOOCCOCOOH$
Carboxylic acids	Formate	Ozone	$HCOO^-$
	Acetate	Ozone	CH_3COO^-
	Oxalate	Ozone	$OOCCOO^{2-}$
Oxyhalides	Chlorite	Chlorine dioxide	ClO_2^-
	Chlorate	Chlorine dioxide	ClO_3^-
	Bromate	Ozone	BrO_3^-

(Contd...)

Class	By-product	Chemical Agent	Molecular formula
Nitrosamines	N-Nitrosodimethylamine	Chloramines	$(CH_3)_2NNO$
Cyanogen halides	Cyanogen chloride	Chloramines	$ClCN$
	Cyanogen bromide	Chloramines	$BrCN$
Miscellaneous	Chloral hydrate	Chlorine	$CCl_3CH(OH)_2$
Trihalonitromethanes	Trichloronitromethane (Chloropicrin)	Chlorine	CCl_3NO_2
	Bromodichloronitromethane	Chlorine	$CBrCl_2NO_2$
	Dibromochloronitromethane	Chlorine	CBr_2ClNO_2
	Tribromonitromethane	Chlorine	CBr_3NO_2

Regulatory Requirements

After the discovery of THM formation, including chloroform from chlorine disinfection and the concerns about the health effects of chloroform [National Cancer Institute (NCI), 1976], the US EPA issued the THM Rule in 1979. The regulated THMs were chloroform ($CHCl_3$), bromodichloromethane ($CHBrCl_2$), dibromochloromethane ($CHBr_2Cl$) and bromoform ($CHBr_3$) and the THM Rule set a maximum contaminant limit (MCL) of 0.10 mg/l for the total sum of these four THMs (on a mass basis) in the distribution system. The THM Rule required water system monitoring at a minimum of four locations throughout the distribution system on a quarterly basis. Three monitoring sites were to be located at average hydraulic travel time through the system, while one site was to be located at the far reaches of the system representing maximum THM formation due to long travel time. The running annual average (RAA) of the average of the distribution system samples collected every quarter was not to exceed the MCL of 0.10 mg/l (100 µg/l). The rationale behind using a running annual average for compliance instead of an instantaneous THM concentration was based on the belief that the concern over THM consumption was related to chronic, not acute, adverse health effects. Therefore, the US EPA was attempting to reduce chronic exposure to THMs.

Until 1998, THMs were the only DBPs with a regulatory maximum limit in drinking water. In 1998, the US EPA issued Stage 1 of the Disinfectants/Disinfection By-products (D/DBP) Rule (US EPA, 1998). This rule reduced the limit on total THMs from 0.10 to 0.080 mg/l and added new limits on seven of the DBPs listed in Table 9.7. Only five of the total nine HAAs formed with chlorine and bromine were included in the D/DBP Rule because at the time, analysis of the remaining four HAAs was not feasible. Compliance with these new limits by water utilities began in January 2002. The D/DBP Rule also included a requirement to utilise chemical coagulation and clarification to maximise the removal of NOM through a conventional water treatment plant (i.e. enhanced coagulation). This requirement was based on the concept that minimising the concentration of NOM would reduce the amount of DBPs formed with subsequent chlorination.

In August 2003, the US EPA proposed Stage 2 of the D/DBP Rule (US EPA, 2003). Ultimately that new rule will maintain the limits for THM and HAAs at 0.100 and 0.080 mg/l, respectively, but will require that sampling sites be selected to represent the highest DBP levels in the system and that the compliance calculation be done via a locational running annual average (LRAA). The LRAA requires a running annual average in each sampling site, whereas the RAA currently in use requires a running annual average over all sampling sites in the entire system. Both the requirement for sampling at sites that represent the highest DBP levels in the system and the LRAA make the Stage 2 regulation more rigorous.

Table 9.7. Disinfection by-products regulated under stage 1 D/DBP rule.

By-product	By-product of	Regulatory limit, mg/l
Total THMs[a]	Chlorine	0.080
Five haloacetic acids (HAA5)[b]	Chlorine	0.060
Bromate (BrO$_3^-$)	Ozone	0.010
Chlorite (ClCO$_2^-$)	Chlorine dioxide	1.0

[a] Sum of four THMs: chloroform, bromodichloromethane, dibromochloromethane and bromoform.

[b] Sum of five HAAs: monochloroacetic acid, dichloroacetic acid, trichloroacetic acid, monobromoacetic acid and dibromoacetic acid.

The Stage 2 D/DBP Rule will be implemented in two steps. At first, systems will be required to conduct an initial distribution system evaluation (IDSE) to identify the locations in the system with the highest DBP concentrations. While this work is being done, systems will have to meet the existing limits using the traditional RAA calculation on the same sites used prior to the new regulation, but they will also be required to use the LRAA calculation on each of the existing sample sites. Compliance with the LRAA during this period will require that THM and HAAs levels at all these sites be below 0.120 and 0.100 mg/l, respectively. Once the IDSE is complete and the new sampling sites have been selected, utilities will be required to meet the more rigorous THM and HAAs limits of 0.100 and 0.080 mg/l.

There may be regulatory changes in the future for the ozonation by-product, bormate (BrO$_3^-$). From health effects data it appears that the 10^{-6} cancer risk level for bormate is 0.05 μg/l. The US EPA typically sets maximum contaminant levels (MCLs) for carcinogens at a level between their 10^{-6} and 10^{-4} cancer risk level meaning that the bromate MCL may be lowered from 10 μg/l (Stage 1 D/DBP Rule) to a level between 0.05 and 5 μg/l. (The current practical quantification limit for bromate is 10 μg/l.) Lowering the bromate MCL will make it exceedingly difficult to use ozone as a disinfectant when treating water containing measurable levels of bromide.

Practical Considerations

Minimising DBP formation needs to be balanced with the need to maximise disinfection. As DBPs are formed upon the reaction of various water constituents with disinfectants, one way of reducing DBP formation is to reduce the concentration of the disinfectant in the water and/or the time it is present in that water. Reducing the disinfectant concentration and/or contact time will directly result in reduced disinfection efficiency and thus reduced public protection against exposure to disease-causing micro-organisms. Therefore, any effort to reduce the formation of DBPs through minimising disinfectant concentration and/or contact time during water treatment must be balanced against the need to lower the microbial risk through adequate disinfection.

Primary Disinfection versus Residual Maintenance

As already discussed, disinfectants are used for two purposes: (i) primary disinfection, and (ii) residual maintenance (protecting the distribution system). The constraints for managing by-products are different for each one of these two purposes. Specifically, in the case of primary disinfection, it is sometimes possible to remove by-products after disinfection has been completed, as is the case with ozone or chlorine dioxide. In the case of residual maintenance, by-product removal is not practical because by-product formation takes place in the distribution system itself. Treatment for by-product removal would have to be provided at each individual residence. Fortunately, the strategy of removing the precursors to by-product formation works equally well in both cases.

CHLORINE BY-PRODUCTS

Chlorine is by far the most widely used disinfectant in drinking water treatment. It is also the disinfectant that forms the greatest variety of known by-products, most of which are listed in Table 9.6. Of the known chlorination by-products, the primary by-products of concern in drinking water treatment are THMs and HAAs. A cumulative distribution profile of average concentrations of total THMs (TTHM) and five HAAs (HAA5) in the distribution systems of approximately 360 US water treatment plants is shown in Fig. 9.9. As shown in Fig. 9.9, on average, 10 per cent of US distribution systems contained greater than 51 µg/l of HAA5 and greater than 73 µg/l of TTHM. Based on the data shown in Fig. 9.9, the majority of distribution systems in the United States contain HAAs and THMs, and slightly less than 10 per cent of distribution systems may not be meeting the Stage 1 D/DBP Rule.

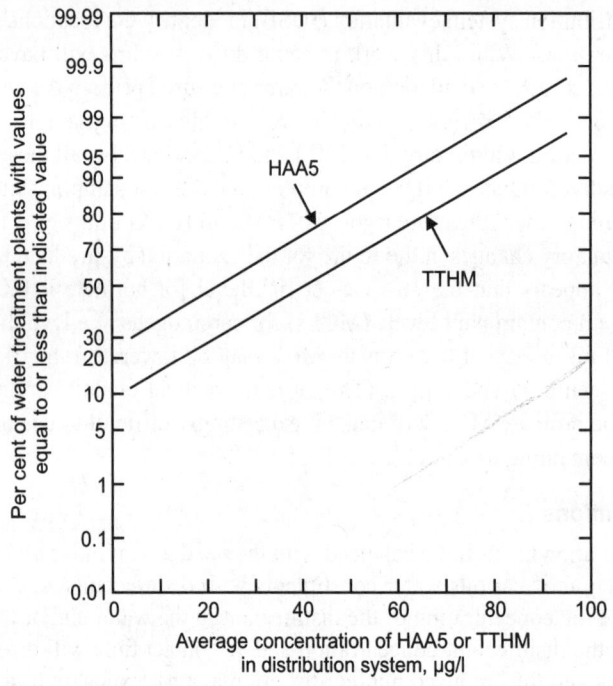

Fig. 9.9. Cumulative distribution profile of average TTHMs and HAA5 in approximately 360 US distribution systems monitored over a 9-month period beginning in summer 1997 and ending winter.

Chemistry of Formation

At an elementary level, chlorine reacts with NOM and bromide ions to form halogenated by-products, as shown in the following reaction:

$$\text{NOM with or without bromide} + \text{chlorine} \rightarrow \text{by-products} \qquad \text{... (9.1)}$$

According to Fuson and Bell the reaction between certain organic chemicals and chlorine to produce 'haloforms' has been known since 1822. However, it was only after the discovery of chloroform formation upon chlorination of drinking water in the early 1970s that the occurrence of haloform reactions in drinking water treatment was recognised. Many researchers have explored the mechanisms of THM and other by-product formation upon chlorination of natural water. Because NOM is complex and has unidentifiable structures in natural water, it has not been possible to identify and verify the specific reaction mechanisms between NOM and chlorine.

Formation of THMs and HAAs occurs at the same time when chlorine reacts with NOM. Ideal conditions for THM formation are different from the ideal conditions for HAA formation. It is generally believed that the reaction mechanism leading to the formation of THMs is base catalysed, meaning the reaction is catalysed by hydroxide ions (OH^-) present in the water and therefore proceeds faster at more alkaline pH values. However, the formation of HAAs is enhanced under acidic conditions. Therefore, pH will directly influence whether THM formation is favoured or HAA formation is favoured and regardless of the exact mechanisms involved in the formation of chlorination by-products, higher concentrations of chlorination by-products are formed with higher concentrations of organic precursor material, bromide ions (inorganic precursor) or chlorine, as shown in Eq. 9.1. The presence of bromide increases the amount of THMs that are formed when chlorine reacts with NOM. Bromide ions (Br^-) participate in the reaction between NOM and chlorine to form various by-products that have a mix of chlorine and bromine substitutions (e.g. bromodichloromethane, bromochloroacetic acid). The pathway of bromide THM formation is: (i) chlorine oxidation of bromide ions to form hypobromous acid (HOBr), (ii) HOBr dissociates depending on the pH to form hypobromite ion (OBr^-), and (iii) $HOBr/OBr^-$ and chlorine reacts with NOM resulting in the formation of a mixture of chlorinated and brominated by-products. The increase in THM formation when bromide is present occurs because bromide has a molecular weight of 79 g/mole compared to 35.5 g/mole for chloride.

Thus, an equal molar concentration of a highly brominated DBP (e.g. bromoform) compared to a purely chlorinated DBP (e.g. chloroform) will result in a significantly higher mass-based concentration of THMs.

The type of NOM present also influences the amount of DBP formation. Researchers have been able to fractionate NOM present in different water according to its molecular size and/or chemical characteristics. In natural water, NOM consists of humic (hydrophobic) and nonhumic (hydrophilic/polar) fractions. In most, but not all, natural water, hydrophobic NOM contributes more to THM and HAA formation than hydrophilic NOM on a mass/mass basis. Many conventional and advanced treatment processes preferentially remove hydrophobic NOM over hydrophilic NOM, so it may be important to characterise the source water NOM to assist in selecting the appropriate treatment process for the source water.

Since the discovery of chlorination by-products and the implementation of the THM Rule in 1979, a large amount of work has been conducted to identify and evaluate alternatives for reducing DBP formation (primarily THMs and HAAs). The following alternatives are available to water utilities for reducing chlorination by-products:

1. Use an alternate disinfectant/oxidant.
2. Reduce the free-chlorine contact time.
3. Reduce the concentration of NOM before chlorine addition.
4. Remove bromide before chlorine addition.
5. Change the pH of the water during chlorination.

Use of alternate disinfectants/oxidants

Chlorine is not the only oxidant/disinfectant available to treat water. Substitution of another oxidant/disinfectant that does not form the by-product of concern for chlorine may be a viable alternative. For example, neither ozone (O_3) nor chlorine dioxide (ClO_2)—both of which are strong oxidants and disinfectants—produces measurable THMs or HAAs when applied to natural water. If the source water contains a lot of humic acid that forms excessive amounts of chloroform, chlorine dioxide may be an appropriate alternative disinfectant to chlorine. The relative formation of chloroform with the addition of 10 mg/l of chlorine or chlorine dioxide to a solution containing 5 mg/l of humic acid is illustrated in Fig. 9.10. While 210 µg/l chloroform was formed with chlorination, the addition of up to 10 mg/l of chlorine dioxide formed no detectable levels of chloroform.

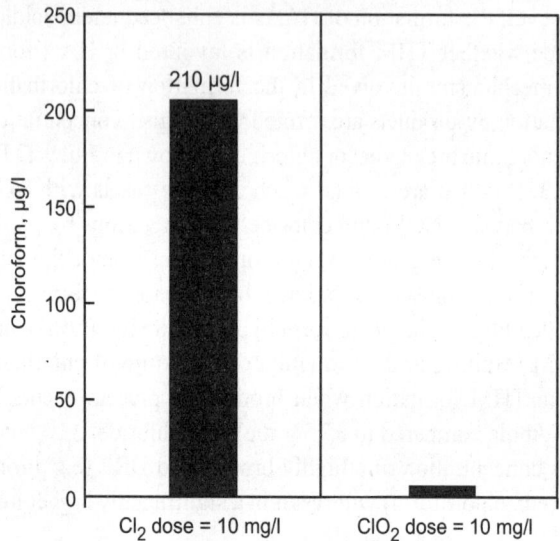

Fig. 9.10. Relative formation of chloroform with chlorine and chlorine dioxide addition to 5 mg/l humic acid at pH 7.0.

The use of alternative disinfectants should be thoroughly investigated. Other disinfectants like ozone and chlorine dioxide, while not forming THM and HAA by-products, do form other by-products, some of which are regulated and are of significant health concern. For instance, Trussell and Umphres showed that the addition of up to 10 mg/l ozone to a water containing 10 mg/l humic acid and 1 mg/l bromide ion formed less than 4 µg/l of THMs. However, Br^- originally present in the water could not be accounted for by summing the Br^- and HOBr present after ozonation. At the time it was not feasible to measure BrO_3^- at low levels. Since then it has been demonstrated that the addition of ozone to water containing bromide ions does result in the formation of bromate (BrO_3^-), also a significant health concern.

Potassium permanganate ($KMnO_4$) is another alternative to chlorine for oxidation at water treatment plants. While permanganate is a weak disinfectant, it has proven to be a good oxidant for the control of iron, manganese and sulphide. Iron is sometimes strongly complexed and hard to remove. Using potassium permanganate results in reduced formation of THM and HAA. Potassium permanganate addition to drinking water is not known to produce any regulated by-products, but $KMnO_4$ is a weak disinfectant, so additional chemical disinfection is typically necessary and may result in DBP formation.

Reduction of free-chlorine contact time

Minimising the free-chlorine contact time can minimise DBP formation. Reducing contact time may be achieved by placing the chlorine addition point near the end of the treatment train and adding ammonia to the treated water at the point where the chlorine disinfection requirements have been met. Ammonia reacts with free chlorine and forms combined chlorine, which is much less reactive with NOM. Experience from full-scale plants indicates that THMs and HAAs do continue to form under combined chlorine conditions, be it at a much lower rate compared to their formation under free-chlorine conditions.

Ammonia added at a mass ratio of less than 1:5 (between the ammonia-nitrogen dose and the residual chlorine concentration at the point of ammonia addition) will convert the free-chlorine residual to monochloramine according to the following reaction:

$$HOCl + NH_3 \rightleftarrows NH_2Cl + H_2O \qquad\qquad ... (9.2)$$

The conversion of free chlorine to combined chlorine is an excellent strategy for limiting the formation of many chlorination by-products, especially THMs and HAAs. The extent of THM and HAA reduction is a function of how long the free chlorine is allowed to react with the NOM to form the by-products before the ammonia is added to the water. Free-chlorine contact time with the water is necessary to achieve the target disinfection goals, unless an alternative disinfectant is used, such as ozone or chlorine dioxide. In that case, little to no free-chlorine contact time may be required before ammonia addition.

Reduction of concentration of natural organic matter (NOM) before chlorine addition

Based on Eq. 9.1, reducing the concentration of NOM will reduce the formation of chlorination by-products. The relationship between THM and HAA formation and NOM concentration is characterised in a study on Sacramento River water in USA. As shown in Fig. 9.11, the concentration of THMs and HAAs increase with increasing the NOM concentration when using chlorination. Therefore, reducing the total organic carbon (TOC) concentration in natural water results in a corresponding decrease in the formation of by-products during exposure to chlorine.

There are several methods for removing NOM during drinking water treatment, including:

1. Chemical coagulation and precipitation.
2. Adsorption on activated carbon.
3. Ion exchange.
4. Nanofiltration or reverse osmosis.

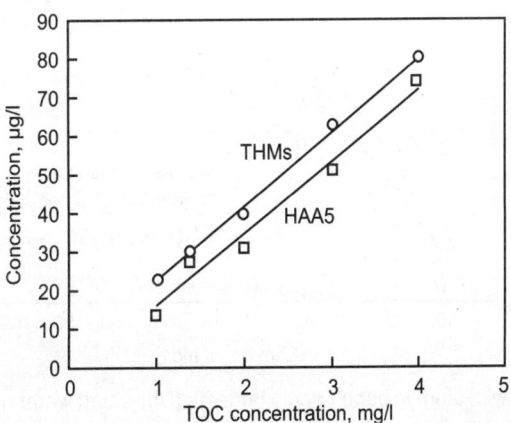

Fig. 9.11. Trihalomethane and HAA5 formation upon chlorination of Sacramento river water spiked with various doses of natural organic matter (Conditions: Temperature = 25°C, pH = 8.2; contact time = 3 hr., Br$^-$ < 0.010 mg/l. Chlorine dose was set to achieve a 3 hr., free chlorine residual of 0.5 to 1.5 mg/l).

Removal of NOM by chemical coagulation and precipitation

Chemical coagulation and precipitation is primarily achieved with the addition of inorganic coagulants, such as aluminium or ferric salts. These coagulants are mainly used for removing suspended material from water, but they can also adsorb a significant amount of NOM. When chemical precipitates are removed from the water through clarification and/or filtration, the adsorbed NOM is removed as well, reducing the concentration of NOM available to react with the chlorine added downstream for disinfection.

Alum is the most commonly used inorganic coagulant and has specific attributes that impact NOM removal. The minimum solubility of alum precipitate is around a pH of 6.3 at 25°C, resulting in an optimum NOM removal with alum at a pH ranging from 5.5 to 6.5, depending on the water temperature and total dissolved solids (TDS) concentration. Removal of NOM with alum can also occur at higher pH values, but higher alum doses are required to meet the same NOM removal that can be achieved at optimum pH. In instances of high-pH conditions at the point of coagulation, acid addition to lower the pH can help improve NOM removal.

The impact of pH on NOM removal is shown in Fig. 9.12. Three different scenarios of NOM removal with alum from a natural water sample were investigated in bench-scale jar tests. The three scenarios were: (i) without pre-addition of sulphuric acid, (ii) with pre-addition of 50 mg/l sulphuric acid, and (iii) with pre-addition of 100 mg/l sulphuric acid. Without acid addition to this water, an alum dose of about 90 mg/l is required to achieve 35 per cent reduction in the TOC concentration (resulting in a settled-water pH of about 7.0). With the addition of 50 mg/l sulphuric acid, the alum dose required to achieve the same TOC reduction is approximately 60 mg/l (with a settled-water pH of about 6.5), a 33 per cent reduction in coagulant usage. However, the residual aluminium ion must be monitored carefully so it does not exceed MCLs.

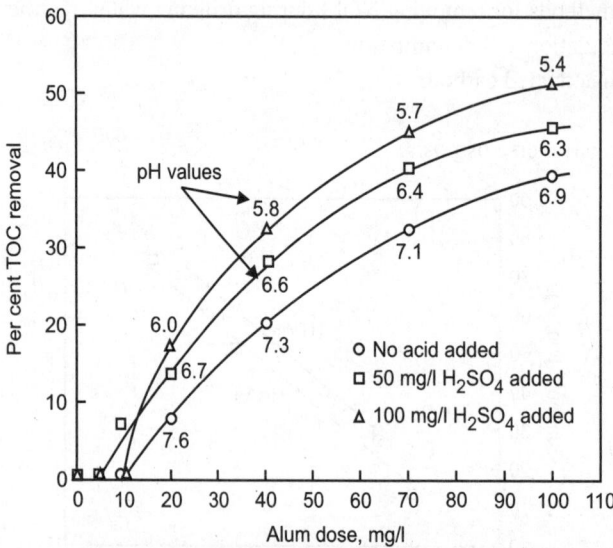

Fig. 9.12. Removal of NOM with alum in bench-scale jar tests (Untreated water quality: Temperature = 20°C, TOC = 9 mg/l, alkalinity = 160 mg/l as $CaCO_3$, turbidity = 3.8 NTU).

Using pH adjustment for NOM removal has a number of consequences that need to be considered before pH adjustment is adopted by a water treatment plant. These consequences include:

1. A lower alum dose, which will reduce the amount of sludge produced at the plant.

2. A lower settled-water pH, which will require a substantially higher dose of an alkaline chemical (such as lime or sodium hydroxide) to raise the pH of the finished water to acceptable levels (in the range of 8 to 8.5).
3. The high doses of acid and caustic will increase the TDS concentration in the finished water.
4. Costs associated with pH adjustment.

Removal of NOM by adsorption

A less common method of NOM removal is adsorption on activated carbon. A large number of studies have evaluated the removal of NOM with adsorption on activated carbon and have found that significant removal of NOM with activated carbon can be very costly. A significant fraction of the NOM is comprised of large molecular weight organic molecules that are poorly adsorbed on activated carbon resulting in the use of a large amount of activated carbon to remove a little NOM.

Removal of NOM by ion exchange

NOM in water is highly ionised, so it can be removed via ion exchange (IX), primarily using anion exchange resins (i.e. positively charged resins). Significant TOC reduction can be achieved when treating natural surface water with conventional anion exchange technology. The term 'conventional' refers to the typical packed-bed configuration of synthetic resin beads with approximate bead size ranging from 0.3 to 1.5 mm. When water is passed through a packed bed of IX beads, NOM is exchanged for chloride ions (Cl^-) or hydroxide ions (OH^-) present on the surface of the resin. The extent of NOM removal is a function of several water quality and resin-specific parameters. Depending on the values of these parameters, a TOC reduction of 50 per cent is typical with run lengths ranging from 500 to 5000 bed volumes (BVs) between regenerations. Upon exhaustion of the resin, it is regenerated with a mixture of a sodium chloride (NaCl) and sodium hydroxide (NaOH) solution. During regeneration, the NOM absorbed by the resin is replaced by chloride and hydroxide ions.

One limitation of the application of conventional IX technology for NOM removal is the slow process kinetics. One cause of the slow exchange process is the size of the bead. An alternative IX technology (MIEX) has been developed that uses a much smaller bead size (< 0.2 mm) added to the water at a treatment plant as a slurry, allowed to contact with the water for a short time and then settled to the bottom of the contactor, where it is collected, regenerated and reused. To increase the bead settling velocity in the contactor, the beads are magnetised which forces them to coalesce during settling to form larger masses, which then settle at a higher rate. The smaller bead size allows for faster removal kinetics, but due to its smaller size, it cannot be applied in a packed-bed configuration.

The advantage of IX technology for NOM removal is that it also removes bromide, which is the other THM or HAA precursor. While either type of IX technology can be designed and operated to achieve significant reduction in THM and HAA formation, disposal of the high-TDS regeneration brine remains the primary obstacle to the wider use of IX technology for NOM—or inorganics—removal in large-scale water treatment plants.

Removal of NOM by membranes

One other NOM removal option is high-pressure membrane filtration such as nanofiltration (NF) or reverse osmosis (RO) membranes. With pore sizes less than 1 nm, these membranes can reject more than 90 per cent of most NOM molecules. While NF membranes have been used successfully for groundwater treatment (mainly for the reduction of hardness caused by Ca^{2+} and Mg^{2+} ions), their use for surface water treatment is complicated by extensive pre-treatment requirements to prevent particulate fouling problems.

Removal of bromide before chlorine addition

In coastal regions, the intrusion of salt water and sea spray increase the concentration of bromide ions (Br⁻) in groundwater or fresh surface water bodies. Bromide is of importance as it is another precursors to many chlorination by-products. Preventing or reducing salt-water intrusion may reduce bromide concentrations and thus the levels of chlorinated by-products formed after chlorination. As discussed earlier, the presence of bromide along with NOM and chlorine increases the mass concentration of by-products formed. As shown in Fig. 9.13, the increase in bromide ion concentration from 0.12 to 0.44 mg/l increased the total THM levels formed in 3 hours of free-chlorine contact time from 80 to 142 µg/l, a 70 per cent increase. Increasing the bromide concentration resulted in a slight increase in the chlorine dose required to achieve the same 3 hours residual (see values in parentheses in Fig. 9.13). Therefore, the increase in THM formation was caused by both the higher bromide concentration and the higher chlorine dose added. The linear relationship between bromide concentration and TTHMs shown in Fig. 9.13 is not universal, as other work has shown a more logarithmic relationship between bromide concentration and THM formation in natural water.

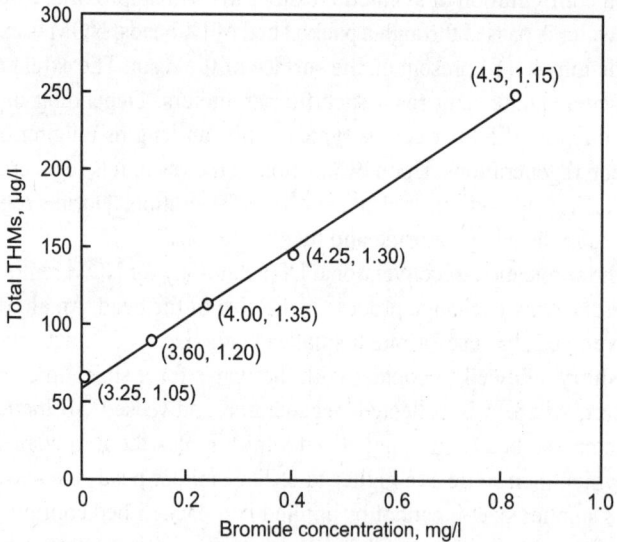

Fig. 9.13. Impact of bromide on THM formation (Conditions: Temperature = 25°C, pH = 8.2, contact time = 3 hr, TOC = 3.25 mg/l. Values in parenthesis represent the chlorine dose and the 3 hr chlorine residual for each data point).

Change of pH during chlorination

One possible measure for reducing the formation of specific chlorination by-products is changing the pH of the water where chlorine is added. Adjusting pH is especially applicable at treatment plants that utilise free chlorine to meet their disinfection requirements and then add ammonia to convert the free chlorine to combined chlorine before the water enters the distribution system. During the free-chlorine contact time, the pH of the water can either be increased or decreased to reduce the formation of a problem by-product. As noted earlier, increasing the pH of the water during chlorination increases the formation of THMs and reduces the formation of HAAs. Therefore, if a water treatment plant is easily meeting the regulatory requirements for THMs but is having a difficult time meeting the requirements for HAAs, then increasing the pH during the free-chlorine contact time can help reduce HAA formation. However, careful attention must be given to the formation of THMs as their levels will increase with increasing pH. Conversely, if compliance with the THM limit is more problematic than that with the HAA5 limit, then acid can be added to slightly reduce the pH of the water to reduce the formation of THMs during free-chlorine contact time while tolerating a slight increase in HAA formation. Increasing the water pH to reduce HAA formation impacts the efficiency of chlorine disinfection, which is a function of pH. Higher water pH values will require higher chlorine doses to meet the disinfection goals resulting in higher THM and HAA formation.

Removal of Chlorination By-products

While technologies exist to remove several chlorination by-products after they are formed, the application of these technologies at a water treatment plant is seldom practical compared to the other by-product reduction strategies discussed above. Chlorination by-products are organic chemicals, so they can be removed from water via several trace-chemical removal technologies. These include adsorption on activated carbon or for the more volatile by-products, removal with aeration and air stripping. Several researchers have evaluated the removal of chloroform with adsorption on GAC, but the adsorption capacity of GAC for chloroform is quite low requiring a high GAC replacement or regeneration frequency, possibly every few weeks. Similarly, air stripping is theoretically a viable approach for removing THMs from water because they are volatile chemicals. However, DBPs and THMs, in particular, continue to be formed after they are removed as long as there is a chlorine residual. For this reason, coupled with the fact that HAAs are not removed, air stripping is a non-viable alternative for DBP removal.

OZONE BY-PRODUCTS

Ozone is the second most frequently used primary drinking water disinfectant in the United States. It is also one of the most effective oxidants for the destruction of chemicals that cause colour, taste and odour in drinking water. When applied to some raw water, it has been shown to improve the downstream chemical coagulation and clarification, as well as media filtration. As the US water industry searched for alternatives to chlorination in the wake of the discovery of the formation of THMs upon chlorine addition to drinking water, many utilities saw ozone as an excellent, but expensive, alternative.

Ozone addition to natural water forms two types of by-products, as are listed in Table 9.6. Organic by-products that are formed from the breakdown of large-molecular-weight NOM molecules by the added ozone are the first type of by-product. The organic by-products are primarily aldehydes and ketoacids, which are not yet believed to have any significant adverse public health effects. The second type of ozone by-product is inorganic bromate (BrO_3^-). In the presence of bromide, ozonation can result in the formation of bromate, which is classified by the US EPA as a 'probable human carcinogen' with a 10^{-6} cancer risk level of 0.05 µg/l (US EPA, 1998). The following sections focus on the formation and control of these two types of ozonation by-products.

Chemistry of Formation

When added to water, ozone reacts with NOM and bromide to form various by-products. The kinetics of bromate formation with ozone are far more rapid (i.e. minutes) than those of THM formation with chlorine (i.e. hours to days). A simplified schematic of the primary reaction pathways of ozone with NOM and bromide is shown in Fig. 9.14. Reaction 1 in Fig. 9.14 represents the breakdown of NOM with ozone to form various organic by-products including aldehydes, ketoacids and others as listed in Table 9.6. While these by-products are believed to be benign at the levels formed during ozonation of drinking water, they are far more biodegradable than their 'parent' NOM molecules. The increase in the biodegradability of the organic material present in the water is of concern because it can promote bacterial growth in the distribution system. The total concentration of the organic by-products is gauged by measuring the biodegradable dissolved organic carbon (BDOC) concentration present in the water or the assimilable organic carbon (AOC) concentration, which is believed to represent the more readily biodegradable fraction of the BDOC.

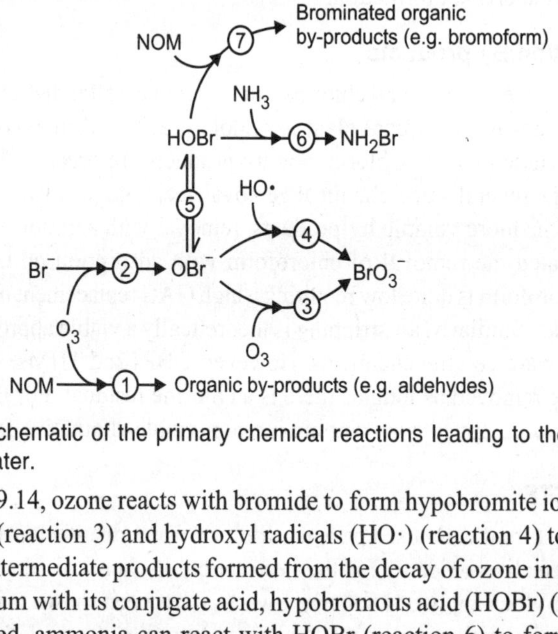

Fig. 9.14. A simplified schematic of the primary chemical reactions leading to the formation of ozonation by-products in natural water.

In reaction 2 in Fig. 9.14, ozone reacts with bromide to form hypobromite ions (OBr^-) which in turn reacts with both ozone (reaction 3) and hydroxyl radicals ($HO\cdot$) (reaction 4) to form bromate (BrO_3^-). Hydroxyl radicals are intermediate products formed from the decay of ozone in natural water. However, OBr^- is also in equilibrium with its conjugate acid, hypobromous acid (HOBr) (reaction 5). If present in the water being ozonated, ammonia can react with HOBr (reaction 6) to form bromamine (NH_2Br). HOBr can also react with NOM (reaction 7) to form brominated organic by-products, such as bromoform.

The reactions included in Fig. 9.14, especially those leading to bromate formation, are not the only reactions that take place between ozone, bromide, and NOM. The reactions leading to the formation of bromate are shown in Fig. 9.14 and include two primary steps: (i) the formation of OBr^-, and (ii) the subsequent formation of BrO_3^- via two parallel pathways, one where OBr^- reacts with molecular ozone (O_3) and one where OBr^- reacts with the hydroxyl radicals ($HO\cdot$). Based on the work by Gillogly it appears that the $HO\cdot$ pathway contributed to bromate formation far more than the molecular ozone pathway. The $HO\cdot$ pathway significance is supported by work that von Gunten and Hoigne reported, where the molecular ozone pathway contributed less than 30 per cent to the formation of bromate during ozonation, and work that Yates and Stenstrom performed, where the molecular ozone pathway generated less than 10 per cent of the bromate formed. In another study, Westerhoff confirmed that approximately 70 per cent of the bromate was formed through $HO\cdot$ mediated reactions.

The HOBr formed can react with NOM to form brominated organic by-products such as bromoform as shown in Fig. 9.14 (reaction 7). Reaction 7 is usually not significant for two reasons. First, bromide levels in natural water seldom exceed 0.5 mg/l. If 50 per cent of the bromide (0.5 mg/l) is converted to OBr^- and 50 per cent of the OBr^- is converted to $HOBr$, only 0.15 mg/l $HOBr$ will be formed (which is equivalent to about 0.11 mg/l as chlorine). Therefore, the amount of bromoform that would form is expected to be quite small; Trussell and Umphres showed that the addition of 10 mg/l ozone to a natural water containing 1 mg/l bromide formed less than 1 µg/l of bromoform.

The second reason for the unlikely formation of brominated organics is, while decreasing water pH favours the conversion of OBr^- to $HOBr$ (pK_a is 8.7 at 25°C) via the equilibrium reaction (reaction 5 in Fig. 9.14), it hinders the base-catalysed haloform reaction required to form bromoform. It has been shown that bromoform formation upon ozonation decreased from 37 µg/l at pH 6 to approximately 15 µg/l at pH 8.5 (bromide = 1 mg/l; DOC = 3.4 mg/l; ozone dose = 10.2 mg/l).

Formation Control

There is little that can be done to reduce the formation of the organic ozonation by-products other than using less ozone and/or removing NOM before adding ozone. Using less ozone is usually not realistic because the ozone dose is typically determined by other factors such as taste and odour control or disinfection requirements. Options for the removal of NOM are the same as those discussed earlier in this chapter for minimising the formation of chlorination by-products.

Utilising less ozone and/or reducing the concentration of bromide before ozonation will result in the formation of lower levels of bromate. Because using less ozone is usually not realistic as mentioned previously, the following two measures have been used to reduce the formation of bromate in some natural water: (i) pH depression; and (ii) ammonia addition.

pH depression

Lowering the pH of water during ozonation is the most reliable and proven method for reducing bromate formation upon ozonation of bromide-containing water. The rationale for this control strategy can be deduced from Fig. 9.14: The pK_a for the equilibrium reaction between $HOBr$ and OBr^- is 8.8 at 20°C. Therefore, at lower pH values, a greater portion of the OBr^- is converted to $HOBr$ (reaction 5 in Fig. 9.14), thus making it unavailable for the bromate formation reactions (reactions 3 and 4). The published literature includes many examples of the effect of pH on bromate formation, one of which is shown in Fig. 9.15. At an ozone dose necessary to achieve 0.5-log inactivation of Giardia cysts in the test water (3.4 mg/l), the bromate level formed at pH 8.0 was 2.6 times the level formed at pH 7.0 (13 µg/l compared to 5 µg/l). With an ozone dose necessary for 2-log inactivation of Giardia cysts (6.0 mg/l), the bormate level formed at pH 8.0 was 3.6 times greater than the bromate level formed at pH 7.0 (58 µg/l compared to only 16 µg/l). In other studies it has been shown that further reduction in bromate formation can be achieved as the pH is decreased to 6.5 or lower, although acid addition for bromate minimisation must be followed by caustic addition for corrosion control in the distribution system.

Ammonia addition

Because reaction 5 in Fig. 9.14 is an equilibrium reaction between $HOBr$ and OBr^-, a fraction of the oxidised bromide will exist as OBr^- even at low pH and still contribute to some bromate formation. To further minimise this fraction, ammonia can be added to serve as a 'sink' for $HOBr$ by transforming it to NH_2Br (reaction 6 in Fig. 9.14), thus creating a continuous driving force for the transformation of OBr^- to $HOBr$. The effect of ammonia addition on the formation of bromate in a natural water sample is shown in Fig. 9.16. Under ambient conditions (NH_3–N < 0.03 mg/l as N), adding 5 mg/l of ozone resulted in the formation of 26 μg/l bromate. Increasing the ammonia-nitrogen concentration to 0.7 mg/l (NH_3–N = 0.7 mg/l as N) decreased the bromate level formed to less than the detection limit of 5 μg/l, which is greater than 500 per cent reduction in bromate formation. Combining pH depression and ammonia: addition to some water may be used to reduce bromate formation by more than 90 per cent, as shown in Fig. 9.17.

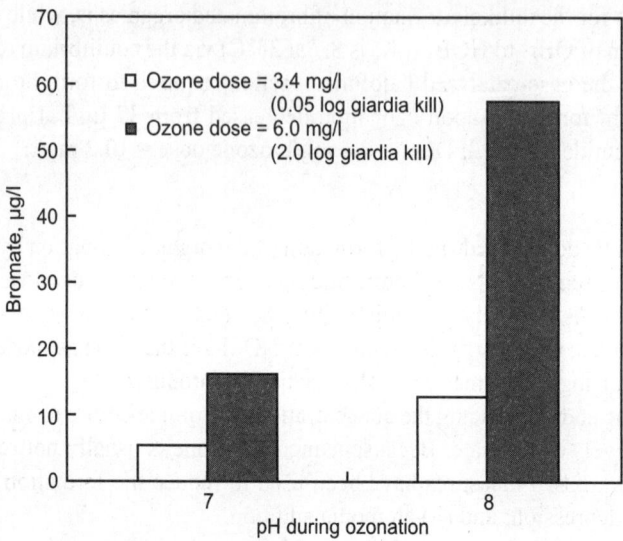

Fig. 9.15. Effect of pH on bromate formation during ozonation of natural water (TOC = 3.3 mg/l, Temperature = 8.5°C, Br^- = 0.5 mg/l, alkalinity at pH 8 = 80 mg/l as $CaCO_3$).

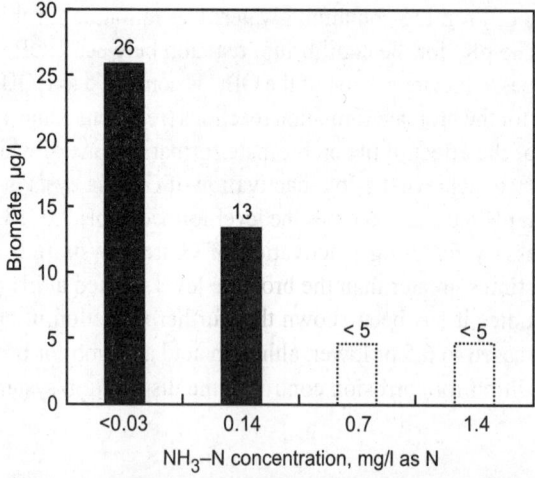

Fig. 9.16. Effect of ammonia on bromate formation during ozonation of natural water (TOC = 5 mg/l, ozone dose = 5 mg/l, pH = 7.0, temperature = 20°C, Br^- = 0.82 mg/l, alkalinity = 27 mg/l as $CaCO_3$).

Unfortunately, while pH depression has been demonstrated to reduce bromate formation, some studies do not show a significant impact of ammonia addition on bromate formation. The reason for the mixed ammonia results is not yet clear. For now, the impact of ammonia addition on bromate formation should be evaluated on a case-by-case basis.

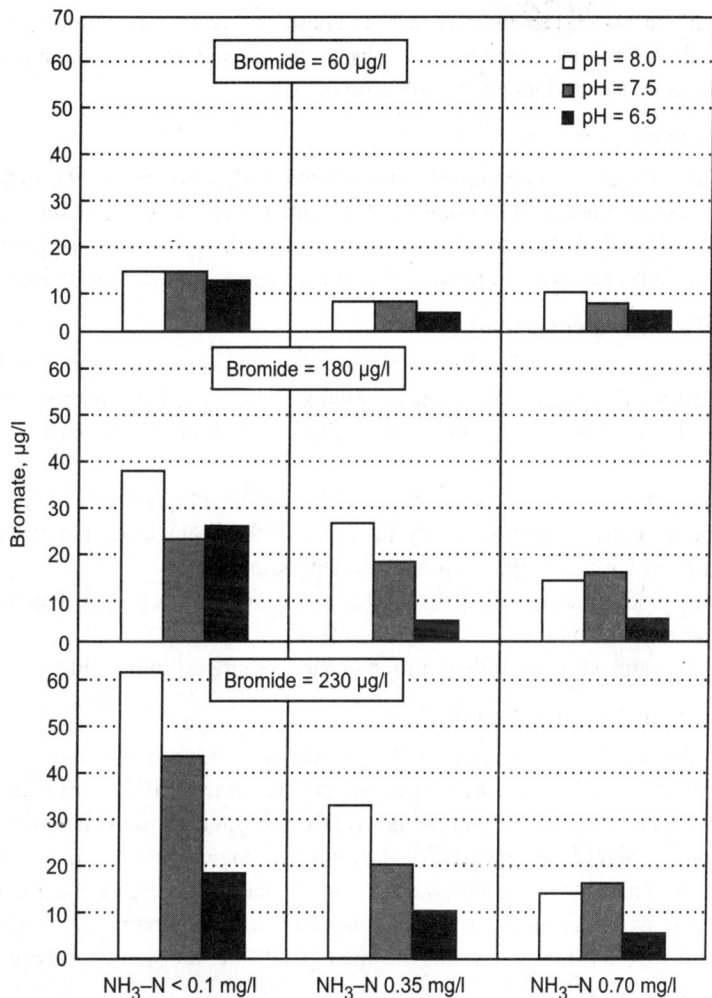

Fig. 9.17. Impact of pH, ammonia concentration and bromide concentration on the formation of bromate upon ozonation of Colorado River water. (Ozone dose = 2.25 ± 0.25 mg/l, temperature = 22°C, TOC = 3 mg/l).

Removal of Ozonation By-Products

Removal of the two main types of ozonation by-products, inorganic bromate and organic by-products, are discussed below:

Bromate

While the primary strategy for complying with a bromate standard is to minimise its formation in the first place, there are options for removing bromate from water after it is formed. The chemistry of bromate (BrO_3^-) is quite similar to that of nitrate (NO_3^-): both are monovalent anions, both are highly oxygenated, both are at the top of their respective oxidation scales and both anions are microbially reduced to benign forms under anoxic conditions. Therefore, just like nitrate, bromate can be removed from water through the following water treatment technologies: (i) ion exchange, (ii) membrane separation,

(iii) biological reduction, and (iv) chemical reduction. These four treatment technologies are not currently practical at full-scale for bromate removal, but a discussion of each of these treatment technologies is important for process understanding and possible future use.

Bromate removal by ion exchange

Anion exchange technologies can be used for bromate removal; however, similar to bromide, the low selectivity of most resins renders IX technology virtually impractical for bromate removal from drinking water. In addition, disposal of the spent high-concentration salt regeneration solution containing the eluted bromate is difficult because of the lack of viable handling and disposal options.

Bromate removal by membrane separation

Membrane separation processes using RO or NF membranes, can achieve greater than 90 per cent removal of a wide range of inorganic ions, including bromate. However, there are three primary drawbacks to the use of membranes for bromate removal, making membrane treatment not practical in this application.

The three primary drawbacks to bromate removal through membrane treatment are:
1. Membrane technology is more costly than any of the bromate formation control strategies discussed earlier in this chapter, such as pH depression.
2. Membranes produce a high-TDS residual stream that requires proper disposal (similar to IX brine disposal problems).
3. Special pre-treatment is typically required or the membranes foul rapidly.

Bromate removal by biological reduction

Bromate can be reduced biologically to bromide by denitrifying bacteria when dissolved oxygen (DO) concentrations are low (< 2.5 mg/l) and empty bed contact times (EBCT) are high (> 25 minutes). Similar to nitrification, bromate reduction is greatly inhibited by increased DO concentration. Unfortunately, this is a significant drawback to applying biological removal in a full-scale treatment plant because the DO concentration downstream of an ozonation process can be greater than 20 mg/l, especially when pure oxygen is used for ozone generation. The requirement for high EBCT is also a drawback as typical EBCT values in a water treatment plant filter are significantly lower than 25 min.

Bromate removal by chemical reduction

Chemical reduction of bromate to bromide can be achieved using reducing agents such as ferrous ions (Fe^{2+}) or the surface of activated carbon. Bromate can be reduced by Fe^{2+} under water treatment conditions according to the following reaction:

$$BrO_3^- + 6Fe^{2+} + 6H^+ \rightarrow Br^- + 6Fe^{3+} + 3H_2O \qquad ... (9.3)$$

Typical bromate reduction through the reaction shown in Eq. 9.3 is on the order of 40 to 80 per cent, depending upon the dose of Fe^{2+}. The unique aspect of using ferrous as a reducing agent is that it is oxidised to ferric (Fe^{3+}), which is a typical coagulant in water treatment. In plants that practice pre-ozonation, a ferrous salt can be added to the rapid mix chamber downstream of the pre-ozone contactor, which allows time for the reaction between ferrous ions and bromate to take place followed by the precipitation of the ferric coagulant formed. Even though the DO levels are high, because the rate of oxidation of Fe^{2+} with oxygen is quite slow, not all of the ferrous ions are oxidised. The residual iron passing through the treatment plant can cause aesthetic water quality problems, rendering this bromate removal option impractical.

Organic by-products

As noted earlier, ozonation of natural water forms various organic by-products such as aldehydes and ketoacids, all of which increase the concentration of the biodegradable organic matter (BOM) in the water. Due to the high biodegradation potential of these by-products, biological filtration downstream of ozonation has developed as the approach of choice for removing organic ozonation by-products. In typical dual-media filters (either anthracite-sand or GAC-sand), biological filtration will occur merely by allowing the water to pass through the filters without a disinfectant residual present.

Several factors affect the removal of BOM with biological filtration. These include: (i) BOM type and concentration; (ii) filter media type (i.e. GAC, anthracite, and/or sand), (iii) water-temperature, and (iv) and EBCT through the filter. Depending on these factors, steady-state biological performance will be reached within a maximum period of 1 to 2 months. Various studies have concluded that either GAC or anthracite, compared to sand, is necessary to achieve good attachment of the bacterial cells. Furthermore, research and full-scale operating experience have shown that anthracite performs as well as GAC in warm water for oxalate removal, as shown in Figs. 9.18 and 9.19. Data shown in Fig. 9.18 were gathered under relatively warm temperature conditions of 10 to 15°C and the anthracite-sand filter performed as well as the GAG-sand filter regardless of EBCT. Similar performance in warm conditions was reported by Price and Krasner. On the other hand, under cold temperature conditions as shown in Fig. 9.19, the GAC-sand biofilter was still capable of removing a fraction of the oxalate (albeit with high EBCT values), while no removal was achieved with the anthracite-sand biofilter. The operational data confirm that under relatively warm temperature conditions, anthracite-sand biofilters perform as well as GAC-sand biofilters.

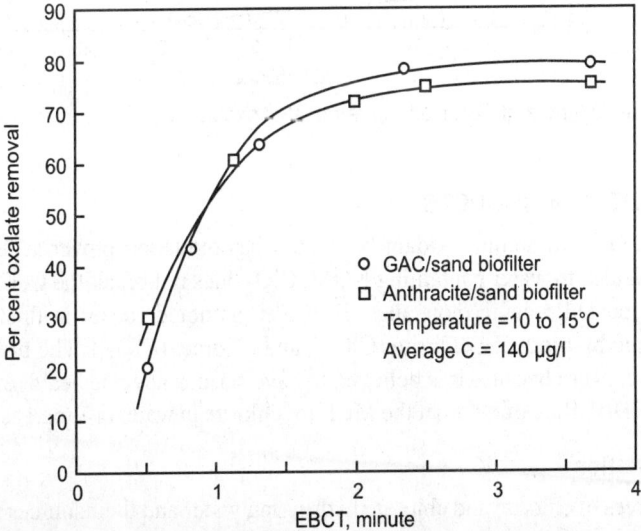

Fig. 9.18. Impact of media type and EBCT on the removal of oxalate with biofiltration under warm temperature conditions (Oxalate is a carboxylic acid formed at significant levels during the ozonation of drinking water).

Biologically active filters develop head loss similar to conventional filters, requiring backwashing on a regular basis. There are two types of concerns about backwashing of biological filters: (i) whether backwashing (especially with air scour) causes excessive detachment of the biomass off the filter media, and (ii) whether biofilters must be backwashed exclusively with non-chlorinated water to maintain their viability. Miltner and Ahmad evaluated these two issues and concluded that rigorous backwashing did not adversely impact biofiltration performance. While AOC removal was impaired immediately after backwashing with chlorinated water, its removal was back to normal levels within a few hours of filter run. Amritharajah concluded that air scour helped control the long-term buildup of head loss in a biofilter, which was confirmed by Teefy who noted that occasional backwashing with chlorinated water was necessary at a full-scale water treatment plant to prevent excessive biological growth, which otherwise results in shorter filter runs. The biofiltration performance achieved with filters backwashed with chlorinated water approximately every third wash.

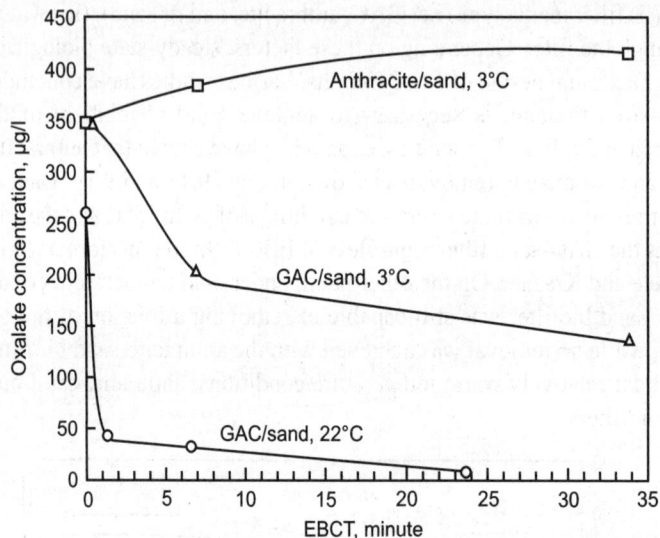

Fig. 9.19. Impact of media type and EBCT on the removal of oxalate with biofiltration under cold temperature conditions.

CHLORINE DIOXIDE BY-PRODUCTS

Chlorine dioxide (ClO_2) is a unique oxidant because it has oxidation power at least equal to that of chlorine, yet, when added to water containing NOM, ClO_2 does not break the C–C bond and therefore does not form halogenated organic molecules. However, as noted earlier in this chapter, ClO_2 does produce two inorganic by-products: chlorite (ClO_2^-) and chlorate (ClO_3^-). The presence of chlorite in drinking water is of concern because it is believed to have serious adverse health effects. In the United States, the Stage 1 D/DBP Rule of 1998 set the MCL for chlorite in water at 1 mg/l as shown in Table 9.7.

Chemistry of Formation

There are many sources of chlorite and chlorate in drinking water and their sources should be identified carefully before any mitigation measures are attempted. Chlorate can be present in the raw water entering a treatment plant due to agricultural use, as sodium chlorate has been used as a herbicide for almost a century. In addition, chlorate salts are also used commercially as oxidising agents in special industries, which can result in significant levels of chlorate in raw water. Chlorate is also a degradation product of chlorine in liquid hypochlorite solutions. Therefore, the use of liquid hypochlorite solutions at water treatment plants may result in significant degradation to chlorate and add significant amounts of chlorate into the water.

Chlorite and chlorate formation during chlorine dioxide generation

In plants using chlorine dioxide, there are two sources of chlorite and chlorate. The first is the process of chlorine dioxide generation itself. Typically, chlorine dioxide is generated by the controlled reaction between chlorine and sodium chlorite under very acidic conditions as follows:

$$2NaClO_2 + HOCl \rightarrow 2ClO_2 + H_2O + NaCl \qquad \qquad ... (9.4)$$

However, under excess chlorine conditions in the chlorine dioxide generator, chlorate can form according to the following simplified reaction:

$$NaClO_2 + HOCl \rightarrow NaClO_3 + HCl \qquad \qquad ... (9.5)$$

Chlorite and chlorate formation as by-products

If excessively high concentrations of sodium chlorite are used in the generator, residual chlorite may remain in the product solution. The residual chlorite will then be injected into the water stream with the chlorine dioxide. New developments in chlorine dioxide generation technologies have improved the generator efficiency and greatly minimised the formation of chlorite and chlorate.

The second source of chlorite and chlorate in plants using chlorine dioxide is the by-products of the decay of chlorine dioxide after it is applied to the water. The decay of chlorine dioxide produces chlorite and chlorate in two ways: The first is through the oxidation of various water constituents such as reduced iron, manganese or NOM. The reaction typically involves a one-electron transfer resulting in the formation of chlorite as follows:

$$ClO_2 + e^- \rightarrow ClO_2^- \qquad \qquad ... (9.6)$$

In addition, under high-temperature and/or high-pH conditions, chlorine dioxide also disproportionates to form chlorite and chlorate as follows:

$$2ClO_2 + 2OH^- \rightarrow ClO_2^- + ClO_3^- + H_2O \qquad \qquad ... (9.7)$$

Researchers have estimated that, in general, 50 to 70 per cent (by mass) of the chlorine dioxide applied during drinking water treatment is converted to chlorite. The formation of chlorite greatly limits the chlorine dioxide dose that can be applied during drinking water treatment knowing that the limit on chlorite is 1 mg/l, unless chlorite removal technologies are implemented downstream.

Formation Control

There is currently no measure that can be implemented at a water treatment plant to reliably reduce the formation of chlorite as a by-product of the decay of chlorine dioxide (Eq. 9.6), except through reducing the chlorine dioxide demand. Reducing ClO_2 demand is especially effective when chlorine dioxide is used to oxidise reduced substances such as iron, manganese and colour present in the source water. When chlorine dioxide is used for disinfection downstream of chemical precipitation, the chlorine dioxide demand can be reduced with improved NOM removal through coagulation or with improved iron and manganese pre-oxidation using other oxidants (e.g. permanganate). With a reduction in the chlorine dioxide demand, a lower chlorine dioxide dose will have to be added, thus reducing the amount of chlorite formed.

Removal of Chlorine Dioxide By-products

The uncontrolled production of chlorite is the primary obstacle facing widespread use of chlorine dioxide in drinking water treatment. Because implementing mitigation measures to reduce chlorite formation is not feasible, a significant amount of work has been conducted to evaluate options for the destruction of chlorite after it is formed. While many options exist for chlorite destruction, the following discussion focuses on the most feasible options for a full-scale water treatment plant: (i) reduction with ferrous ion, (ii) reduction with activated carbon, and (iii) oxidation with ozone.

Reduction with ferrous ion

Reduction with ferrous ion and activated carbon are based on the idea of reducing chlorite to chloride by the following half-reaction:

$$ClO_2^- + 4H^+ + 4e^- \rightarrow Cl^- + 2H_2O \qquad \qquad ... (9.8)$$

When ferrous (Fe^{2+}) is added to the water, it releases the two needed electrons to form ferric (Fe^{3+}) with the combined reaction being as such:

$$ClO_2^- + 4Fe^{2+} + 4H^+ \rightarrow Cl^- + 4Fe^{3+} + 2H_2O \qquad \qquad ... (9.9)$$

The mass ratio of ferrous ion to chlorite is 3.3:1 mg Fe^{2+}/mg ClO_2^- as shown in Eq. 9.9. The stoichiometry presented in Eq. 9.9 was validated by Latrou and Knocke who also showed that adding ferrous ions (as $FeSO_4 \cdot 7H_2O$) at a mass ratio of 3:1 (mg Fe^{2+}/mg ClO_2^-) resulted in virtually complete reduction of chlorite in less than 1 minute of reaction time (pH between 5 and 7; temperature between 5° and 25°C). These researchers also verified that the added ferrous ion was oxidised to ferric (Fe^{3+}) ion, which then enhanced downstream coagulation and flocculation by precipitating as $Fe(OH)_{3(s)}$. These findings were confirmed by Griese who, along with Latrou and Knocke showed that no chlorate was formed from the reaction between chlorite and ferrous ions. Subsequently, Hurst and Knocke verified this approach under alkaline conditions (pH between 8 and 10) and studied the effect of dissolved oxygen on the Fe^{2+} dose required. Based on these results, a mass ratio between 3.5:1 and 4:1 is more appropriate under high-O_2 conditions (> 5 mg/l) to satisfy the added demand for Fe^{2+}. Reduction with activated carbon is based on the fact that the surface of activated carbon is a good reducing agent and therefore, may be used to reduce chlorine to chloride. Chlorite is removed by GAC, but the removal efficiency increases rapidly over time and removal efficiency increases with increasing EBCT.

Oxidation with ozone

Oxidation with ozone is based on the chemistry of chlorite where chlorite is oxidised to chlorate ions with a strong oxidant such as ozone. Chlorine dioxide is a transient intermediate along the oxidation pathway from chlorite to chlorate. This is reasonable because the oxidation state of chloride in chlorine dioxide is (+4), which is between that in chlorite (+3) and that in chlorate (+5). Increasing the ozone dose oxidises the intermediate chlorine dioxide to chlorate. Water treatment plants that use chlorine dioxide as a raw-water pre-oxidant and utilise intermediate ozonation for disinfection can rely on the ozonation process to oxidise the chlorite by-product to chlorate. In plants that use chlorine dioxide followed by ozone, the chlorine dioxide dose can be increased in response to changing raw-water quality conditions without violating the chlorite standard in the finished water. This strategy is of limited value when chlorate is an issue.

CHLORAMINE BY-PRODUCTS

Converting free chlorine to chloramines via Eq. 9.2 has long been considered a cost-effective measure to reduce THM and HAA formation in chlorinated water. With tighter THM and HAA regulatory standards being implemented, an increasing number of water utilities are converting the secondary disinfectant in their distribution systems from free chlorine to combined chlorine. Chlorinated by-products also form with combined chlorine, however, the THM and HAA levels formed with combined chlorine are typically less than 20 per cent of those formed with free chlorine. The formation of THMs and HAAs with combined chlorine is likely attributed to the low concentration of free chlorine that is always in equilibrium with monochloramine and also to formation that takes place during the process of mixing.

Based on recent research, it has been found that the reaction of combined chlorine with organic matter present in some waters may result in the formation of NDMA as a by-product. The US EPA currently classifies NDMA as a 'probable human carcinogen' and has estimated its 10^{-6} cancer-risk level at 0.7 mg/l, which is well below the levels measured in chloraminated drinking water. At the present time an action limit of 10 mg/l has been set in the State of California, but no federal MCLs have yet been established.

Chemistry of Formation

The chemical mechanism leading to the formation of NDMA is not yet known, but it is a direct by-product of chloramine and not free chlorine as shown on Fig. 9.20. One proposed, but not confirmed, reaction mechanism is dimethylamine $[(CH_3)_2NH]$, which reacts with chloramine to form dimethylhydrazine $[(CH_3)_2N_2H_2]$, which in turn is oxidised to NDMA by chloramine. The research by Najm and Trussell found that the NDMA concentration increased with increasing chloramine dose, and the sequence of addition of chlorine and ammonia had no significant effect on the final levels of NDMA formed. Further unpublished research by MWH engineers on untreated water from the West Branch of the California State Aqueduct has shown that NDMA formation peaks at a pH of about 7 and at a Cl_2/NH_3 ratio of approximately 5:1 on a weight basis (Figs. 9.21a,b,c). Although significant NDMA formation occurs in the first few hours, it appears that the process continues at a slow pace for several days (Fig. 9.21d). As a result of this prolonged formation time, the highest NDMA levels in the system are likely to be formed where the travel time is the longest.

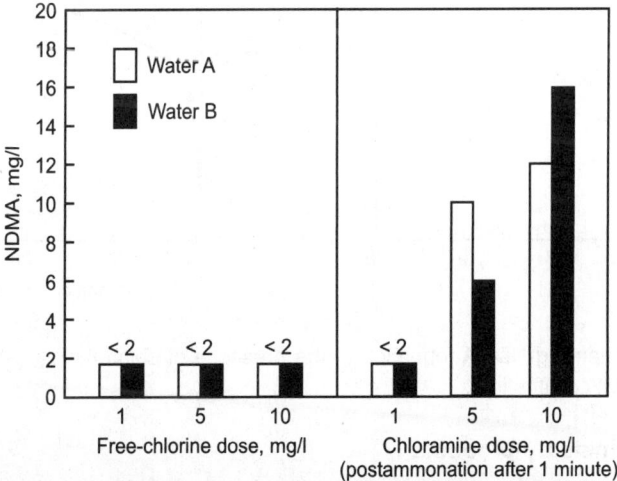

Fig. 9.20. NDMA formation after chlorine and chloramine addition to two test water (contact time = 24 hours).

In a survey of water systems conducted by the California Department of Health Services in 1999 it was found that 50 per cent of the system samples had no NDMA (MDL = 1 mg/l) and 5 per cent of the system samples had NDMA levels above 20 mg/l. Based on measurements made since that time, higher NDMA levels are generally observed downstream of ion exchange and following the use of cationic polymers produced from the Mannich reaction.

Formation Control

Alternatives for controlling the formation of NDMA can be gleaned from an examination of the circumstances that promote its formation. Generally, reducing the dose of chloramines, increasing the pH, and minimising the contact time in the distribution system are all practices that will reduce NDMA formation. It also seems likely that either minimising the use of polyDADMAC cationic polymers or selecting carefully among them will also reduce the levels formed. Care in recycling filter waste washwater has also been shown to be of significance.

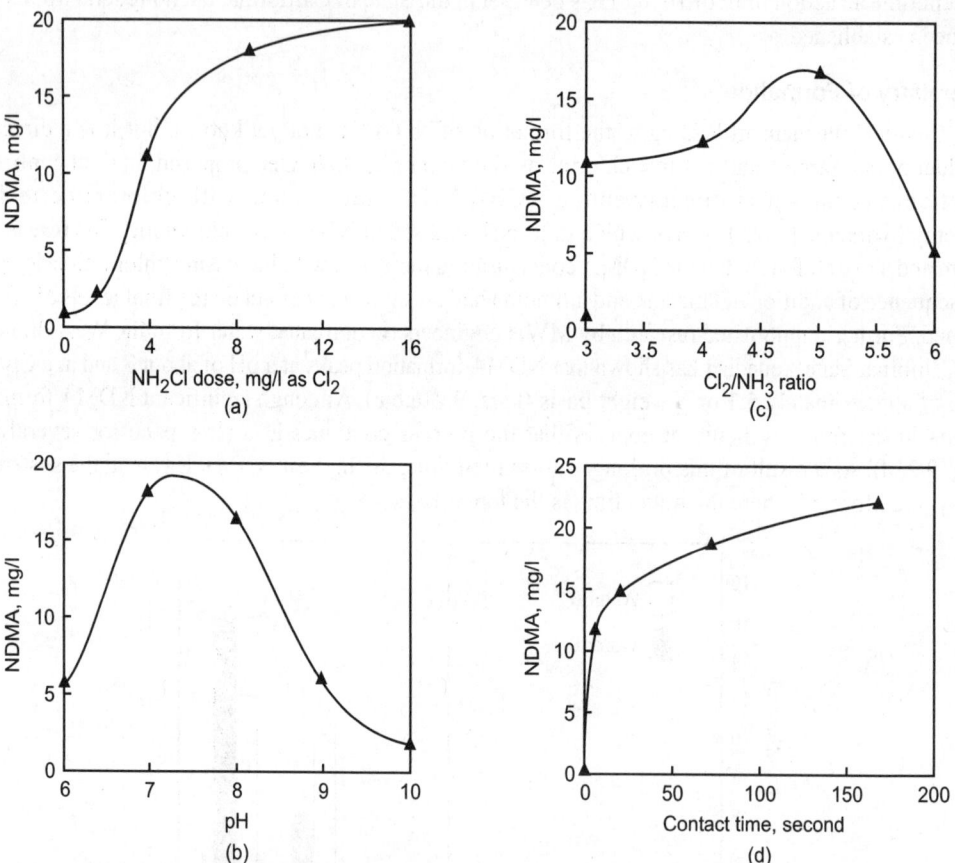

Fig. 9.21. Factors influencing NDMA formation in the presence of chloramines: (a) NH_2Cl dose, (b) pH, (c) Cl_2/NH_3 ratio and (d) contact time.

Removal of Chloramine By-products

The NDMA can be removed by reverse osmosis and photolysis. Although these alternatives may be practical in the treatment of contaminated groundwater or in the water 'exiting ion exchange processes', it is not a practical solution when the NDMA present results from the by-product reaction with chloramines used principally for residual maintenance rather than primary disinfection. As noted above, NDMA forms slowly in the distribution system, reaching a maximum concentration in the consumer's tap. As a result, any treatment for removal of NDMA would have to be applied at the consumer's tap. Preventing the formation of NDMA, as discussed above, may be the most economical approach.

SECTION III

Engineering Systems for Waste-water Treatment, Disposal and Analysis

Chapter 10

Basic Concepts of Waste-water Treatment and Disposal

INTRODUCTION

All industrial operations produce some waste-water which must be returned to the environment. Waste-waters can be classified as: (i) domestic waste-waters, (ii) process waste-waters, and (iii) cooling waste-waters. Domestic waste-waters are produced by plant workers, shower facilities and cafeterias. Process waste-waters result from spills, leaks and product washing. Cooling waste-waters are the result of various cooling processes and can be once-pass systems or multiple-recycle cooling systems. Once-pass cooling systems employ large volumes of cooling waters that are used once and returned to the environment. Multiple-recycle cooling systems have various types of cooling towers to return excess heat to the environment and require periodic blow down to prevent excess build-up of salts.

Domestic waste-waters are generally handled by the normal sanitary sewerage system to prevent the spread of pathogenic micro-organisms which might cause disease. Normally, process waste-waters do not pose the potential for pathogenic micro-organisms, but they do pose potential damage to the environment through either direct or indirect chemical reactions.

Some process wastes are readily biodegraded and create an immediate oxygen demand. Other process wastes are toxic and represent a direct health hazard to biological life. Cooling waste-waters are the least dangerous, but they can contain process waste-waters as a result of leaks in the cooling systems. Recycle cooling systems tend to concentrate both inorganic and organic contaminants to a point at which damage can be created.

WASTE-WATER CHARACTERISTICS

Waste-waters are usually classified as industrial waste-water or municipal waste-water. Industrial waste-water with characteristics compatible with municipal waste-water is often discharged to the municipal sewers. Many industrial waste-waters require pre-treatment to remove non-compatible substances prior to discharge into the municipal system. Characteristics of industrial waste-water vary greatly from industry to industry and consequently, treatment processes for industrial waste-water also vary, although many of the processes used to treat municipal waste-water are also used in industrial waste-water treatment.

Water collected in municipal waste-water systems, having been put to a wide variety of uses, contains a wide variety of contaminants. A list of contaminants commonly found in municipal waste-water along with their sources and their environmental consequences is given in Table 10.1

Table 10.1. Important waste-water contaminants.

Contaminant	Source	Environmental significance
Suspended solids	Domestic use, industrial wastes, erosion by infiltration/inflow	Cause sludge deposits and anaerobic conditions in aquatic environment
Biodegradable organics	Domestic and industrial waste	Cause biological degradation, which may use up oxygen in receiving water and result in undesirable conditions
Pathogens	Domestic waste	Transmit communicable diseases
Nutrients	Domestic and industrial waste	May cause eutrophication
Refractory organics	Industrial waste	May cause taste and odour problems, may be toxic or carcinogenic
Heavy metals	Industrial waste, mining, etc.	Are toxic, may interfere with effluent reuse
Dissolved inorganic solids	Increases above level in water supply by domestic and/or industrial use	May interfere with effluent reuse

Quantitatively, constituents of waste-water may vary significantly, depending upon the percentage and type of industrial waste present and the amount of dilution from infiltration/inflow into the collection system. Results of an analysis of a typical waste-water from a municipal collection system are given in Table 10.2.

Table 10.2. Typical analysis of municipal waste-water.

Constituent, mg/l	Concentration		
	Strong	Medium	Weak
Solids, total	1200	720	350
Dissolved, total	850	500	250
Fixed	525	300	145
Volatile	325	200	105
Suspended, total	350	220	100
Fixed	75	55	20
Volatile	275	165	80
Settleable solids, ml/l	20	10	5
Biochemical oxygen demand, 5-day, 20°C (BOD_5)	400	220	110
Total organic carbon (TOC)	290	160	80
Chemical oxygen demand (COD)	1000	500	250
Nitrogen (total as N)	85	40	20
Organic	35	15	8
Free ammonia	50	25	12
Nitrites	0	0	0
Nitrates	0	0	0
Phosphorus (total as P)	15	8	4
Organic	5	3	1
Inorganic	10	5	3

(Contd ...)

Constituent, mg/l	Concentration		
	Strong	*Medium*	*Weak*
Chlorides	100	50	30
Alkalinity (as CaCO$_3$)	200	100	50
Grease	150	100	50

The composition of waste-water from a given collection system may change slightly on a seasonal basis, reflecting different water uses. Additionally, daily fluctuations in quality are also observable and correlate well with flow conditions as noted in Fig. 10.1. Generally, smaller systems with more homogeneous uses produce greater fluctuations in waste-water composition.

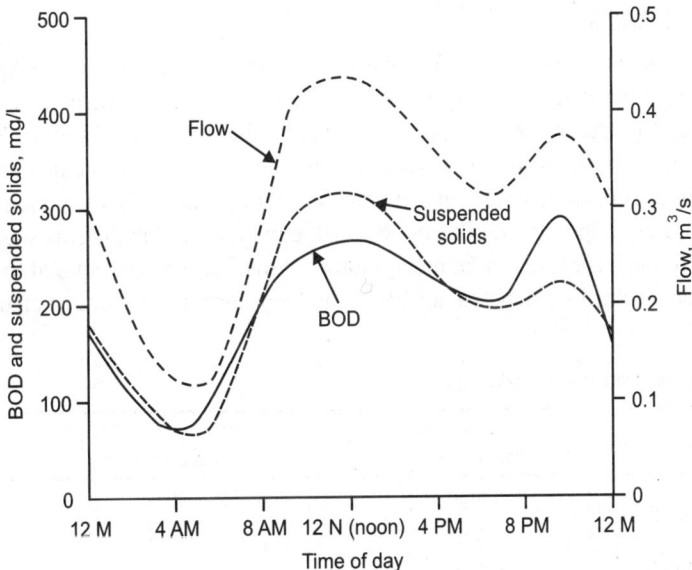

Fig. 10.1. Typical variation in flow, suspended solids and BOD$_5$ in municipal waste-water.

The most significant components of waste-water are usually suspended solids, biodegradable organics and pathogens. Suspended solids are primarily organic in nature and are composed of some of the more objectionable material in sewage. Body wastes, food waste, paper, rags and biological cells form the bulk of suspended solids in waste-water. Even inert materials such as soil particles become fouled by adsorbing organics to their surface. Removal of suspended solids is essential prior to discharge or reuse of waste-water.

Although suspended organic solids are biodegradable through hydrolysis, biodegradable material in waste-water is usually considered to be soluble organics. Soluble organics in domestic waste-water are composed chiefly of proteins (40 to 60 per cent), carbohydrates (25 to 50 per cent) and lipids (approximately 10 per cent). Proteins are chiefly amino acids, while carbohydrates are compounds such as sugars, starches and cellulose.

Lipids include fats, oil and grease. All of these materials contain carbon that can be converted to carbon dioxide biologically, thus exerting an oxygen demand. Proteins also contain nitrogen and thus a nitrogenous oxygen demand is also exerted. The biochemical oxygen demand test, is therefore, used to quantify biodegradable organics.

All forms of waterborne pathogens may be found in domestic waste-water. These include bacteria, viruses, protozoa and helminths. These organisms are discharged by persons who are infected with the disease. Although pathogens causing some of the more exotic diseases may rarely be present, it is a safe assumption that a sufficient number of pathogens are present in all untreated waste-water to represent a substantial health hazard. Fortunately, few of the pathogens survive waste-water treatment in a viable state. Traditional waste-water-treatment processes are designed to reduce suspended solids, biodegradable organics and pathogens to acceptable levels prior to disposal. Additional waste-water-treatment processes may be required to reduce levels of nutrients if the waste-water is to be discharged to a delicate ecosystem. Processes to remove refractory organics and heavy metals and to reduce the level of inorganic dissolved solids are required where waste-water reuse is anticipated.

EFFLUENT STANDARDS

The Water Pollution Control Act of 1972 mandated the Environmental Protection Agency to establish standards for waste-water discharges. Current standards require that municipal waste-water be given secondary treatment and that most effluents meet the conditions shown in Table 10.3 of the appendix. Secondary treatment of municipal waste-water is generally assumed to include settling, biological treatment and disinfection, along with sludge treatment and disposal. Thus, the principal components of municipal waste-water, suspended solids, biodegradable material and pathogens should be reduced to acceptable levels through secondary treatment. Industrial dischargers are required to treat their waste-water to the level obtainable by the 'best available technology' for waste-water treatment in that particular type of industry.

Table 10.3. Secondary treatment standards.

Characteristic of discharge	Unit of measurement	Average monthly concentration	Average weekly concentration
BOD$_5$	mg/l	30*†	45†
Suspended solids‡	mg/l	30*†	45†
Hydrogen-ion concentration	pH units	–	6.0–9.0§

* Or, in no case more than 15 per cent of influent value.

† Arithmetic mean.

‡ Treatment plants with stabilisation ponds and flows < 7570 m³/d (2 Mgal/d) are exempt.

§ Continuous, only enforced if caused by industrial waste-water or in-plant treatment.

The EPA regulations further define receiving streams as 'effluent-limited' and 'water-quality-limited'. An effluent-limited stream is a stream that will meet its in-stream standards if all discharges to that stream meet the secondary-treatment and 'best-available-technology' standards. Municipalities and industries discharging to effluent-limited streams are assigned discharge permits under the National Pollution Discharge Elimination System (NPDES); these permits reflect the secondary treatment and best-available-technology standards. A water-quality-limited stream would not meet the proposed in-stream standards, even if all discharges met secondary-treatment and best-available-technology levels.

TERMINOLOGY IN WASTE-WATER TREATMENT

The terminology used in waste-water treatment is often confusing to the uninitiated person. Terms such as unit operations, unit processes, reactors, systems and primary, secondary and tertiary treatment frequently appear in the literature and their usage is not always consistent. The meanings of these terms, as used in this text, are discussed in the following paragraphs. Methods used for treating municipal waste-waters are often referred to as either unit operations or unit processes. Generally, unit operations involve contaminant removal by physical forces, while unit processes involve biological and/or chemical reactions. The term reactor refers to the vessel or containment structure, along with all of its appurtenances, in which the unit operation or unit process takes place. Although unit operations and processes are natural phenomena, they may be initiated, enhanced or otherwise controlled by altering the environment in the reactor. Reactor design is a very important aspect of waste-water treatment and requires a thorough understanding of the unit processes and unit operations involved. A waste-water-treatment system is composed of a combination of unit operations and unit processes designed to reduce certain constituents of waste-water to an acceptable level. Many different combinations are possible. Although practically all waste-water-treatment systems are unique in some respects, a general grouping of unit operations and unit processes according to target contaminants has evolved over the years. Unit operations and processes commonly used in waste-water treatment are listed in Table 10.4 and are arranged according to conventional grouping. Actually, only a few waste-water-treatment methods fall completely into one category. Thus the usefulness of this classification system is somewhat compromised.

Table 10.4. Unit operations, unit processes and systems for waste-water treatment.

Contaminant	Unit operation, unit process or treatment system
Suspended solids	Sedimentation
	Screening and comminution
	Filtration variations
	Flotation
	Chemical-polymer addition
	Coagulation/sedimentation
	Land treatment systems
Biodegradable organics	Activated-sludge variations
	Fixed-film: trickling filters
	Fixed-film: rotating biological contactors
	Lagoon and oxidation pond variations
	Intermittent sand filtration
	Land treatment systems
	Physical-chemical systems
Pathogens	Chlorination
	Hypochlorination
	Ozonation
	Land treatment systems
Nutrients	
Nitrogen	Suspended-growth nitrification and denitrification variations
	Fixed-film nitrification and denitrification variations

(Contd ...)

Contaminant	Unit operation, unit process or treatment system
	Ammonia stripping
	Ion exchange
	Breakpoint chlorination
	Land treatment systems
Phosphorus	Metal-salt addition
	Lime coagulation/sedimentation
	Biological-chemical phosphorus removal
	Land treatment systems
Refractory organics	Carbon adsorption
	Tertiary ozonation
	Land treatment systems
Heavy metals	Chemical precipitation
	Ion exchange
	Land treatment systems
Dissolved inorganic solids	Ion exchange
	Reverse osmosis
	Electrodialysis

Municipal waste-water-treatment systems are often divided into primary, secondary and tertiary subsystems. The purpose of primary treatment is to remove solid materials from the incoming waste-water. Large debris may be removed by screens or may be reduced in size by grinding devices. Inorganic solids are removed in grit channels and much of the organic suspended solids is removed by sedimentation. A typical primary treatment system (Fig. 10.2) should remove approximately one-half of the suspended solids in the incoming waste-water. The BOD associated with these solids accounts for about 30 per cent of the influent BOD.

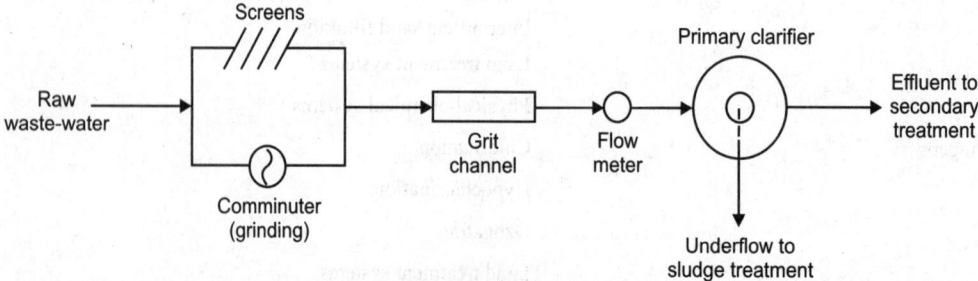

Fig. 10.2. Typical primary treatment system.

Secondary treatment usually consists of biological conversion of dissolved and colloidal organics into biomass that can subsequently be removed by sedimentation. Contact between micro-organisms and the organics is optimised by suspending the biomass in the waste-water or by passing the waste-water over a film of biomass attached to solid surfaces. The most common suspended biomass system is the activated-sludge process shown in Fig. 10.3(a). Recirculating a portion of the biomass maintains a large number of organisms in contact with the waste-water and speeds up the conversion process. The classical attached-biomass system is the trickling filter shown in Fig. 10.3(b). Stones or other solid media are used to increase the surface area for biofilm growth. Mature biofilms peel off the surface and are washed out to the settling basin with the liquid underflow. Part of the liquid effluent may be recycled through the system for additional treatment and to maintain optimal hydraulic flow rates.

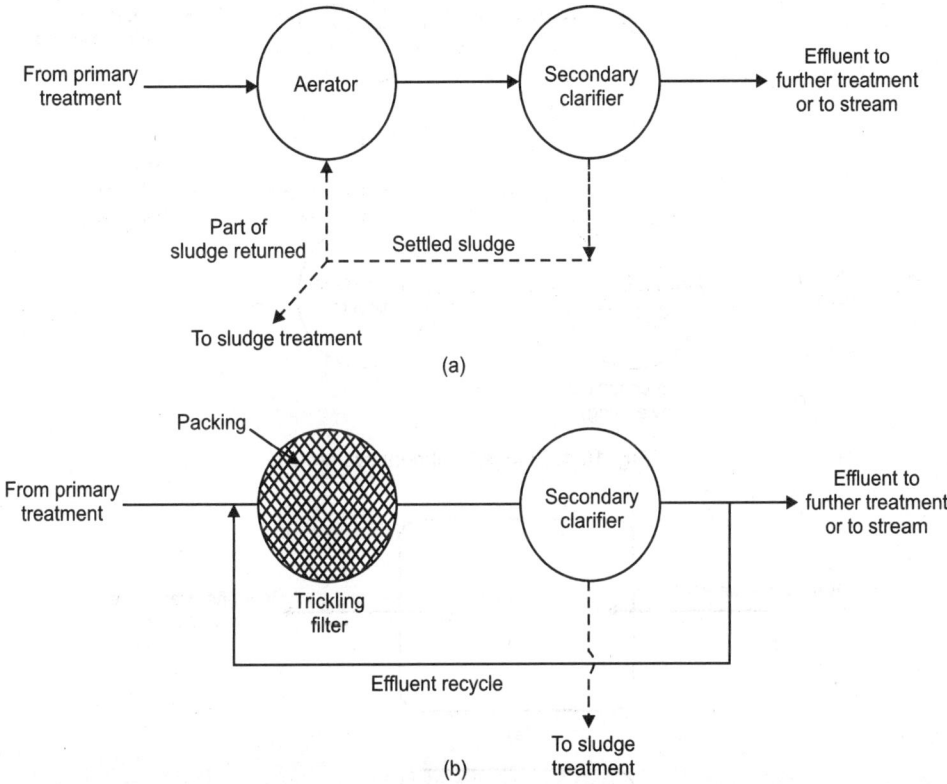

Fig. 10.3. Secondary treatment system: (a) activated sludge system; and (b) trickling filter system.

Secondary systems produce excess biomass that is biodegradable through endogenous catabolism and by other micro-organisms. Secondary sludges are usually combined with primary sludge for further treatment by anaerobic biological processes as shown in Fig. 10.4. The results are gaseous end products, principally methane (CH_4) and carbon dioxide (CO_2) and liquids and inert solids. The methane has significant heating value and may be used to meet part of the power requirements of the treatment plant. The liquids contain large concentrations of organic compounds and are recycled through the treatment plant. The solid residue has a high mineral content and may be used as a soil conditioner and fertiliser on agricultural lands. Other means of solids disposal may be by incineration or by landfilling. Sometimes primary and secondary treatment can be accomplished together, as shown in Fig. 10.5. The oxidation pond [Fig. 10.5(a)] most nearly approximates natural systems, with oxygen being supplied by algal photosynthesis and surface reaeration. This oxygen seldom penetrates to the bottom of the pond and the solids that settle are decomposed anaerobically. In the aerated lagoon system [Fig. 10.5(b)] oxygen is supplied by mechanical aeration and the entire depth of the pond is aerobic. Decomposition of the biomass occurs by aerobic endogenous catabolism. The small quantity of excess sludge that is produced is retained in the bottom sediments.

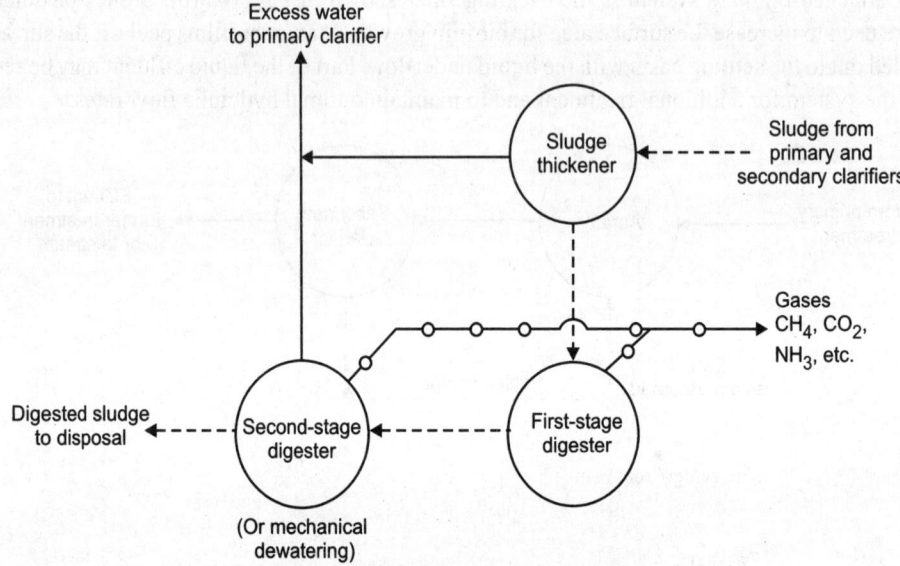

Fig. 10.4. Sludge treatment system.

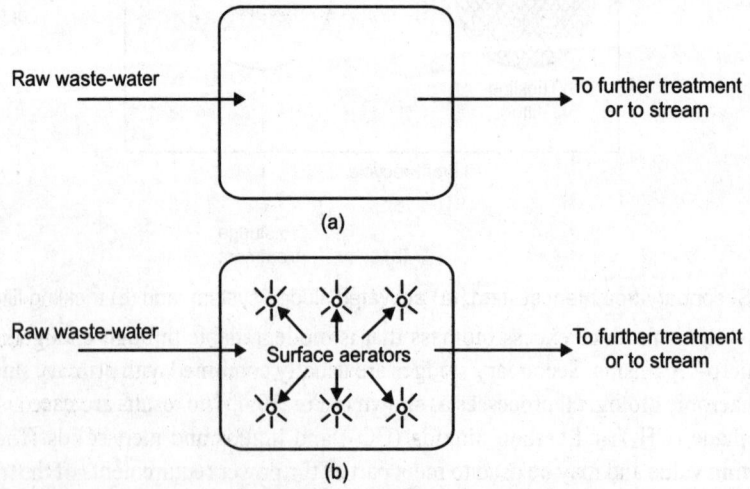

Fig. 10.5. Primary and secondary water treatment in combination: (a) oxidation pond and (b) aerated lagoon.

In most cases, secondary treatment of municipal waste-water is sufficient to meet effluent standards. In some instances, however, additional treatment may be required. Tertiary treatment most often involves further removal of suspended solids and/or the removal of nutrients. Solids removal may be accomplished by filtration and phosphorus and nitrogen compounds may be removed by combinations of physical, chemical and biological processes.

A careful inspection of Figs 10.2 through 10.5 leads to an interesting observation. The 'removal' processes in waste-water treatment are essentially concentrating or thickening, processes. Suspended solids are removed as sludges and dissolved solids are converted to suspended solids and subsequently become removable sludges. Hammer states that primary and secondary treatment, followed by sludge thickening, may concentrate organic material represented by 250 mg/l of suspended solids and 200 mg/l BOD in 375 litre of municipal waste-water (the average per capita contribution) to 2.0 litre of sludge containing 50,000 mg/l of solids. Most of the objectionable material initially in the waste-water is concentrated in the sludges and must be disposed of in a safe and environmentally acceptable manner. Vesilind notes that a majority of the expenses, effort and problems of waste-water treatment and disposal are associated with the sludges.

Design of waste-water-treatment systems is an important part of an environmental engineer's work. A thorough understanding of the unit operations and processes is necessary before the reactors can be designed.

Chapter 11

Primary and Secondary Treatment

INTRODUCTION

Waste-water contains a wide variety of solids of various shapes, sizes and densities. Effective removal of these solids may require a combination of unit operations such as screening, grinding and settling. Although no material is removed by the process, flow-measurement devices are essential for the operation of waste-water-treatment plants and are generally included in the primary system. Operations to eliminate large objects and grit, along with flow measurement, often referred to as preliminary treatment, are an integral part of primary treatment. Operations common to primary systems in most waste-water-treatment plants are described in the following section.

SCREENING

Water, when derived from the surface sources, may contain suspended matter which may range from floating debris such as sticks, branches, leaves, etc. to fine particles such as sand, silt, etc., causing turbidity. Screening devices are used to remove coarse solids from waste-water. Coarse solids consist of sticks, rags, boards and other large objects that often and inexplicably, find their way into waste-water collection systems. Because the primary purpose of screens is to protect pumps and other mechanical equipment and to prevent clogging of valves and other appurtenances in the waste-water plant, screening is normally the first operation performed on the incoming waste-water. Screens serve as a protective device for the remainder of the plant rather than as a treatment process.

Types of Screens

Screens are of two types: (i) coarse screens, and (ii) fine screens.

Coarse screens or bar screens

Coarse screens or bar screens are intended to intercept only grosser floating material. They are mostly in the form of bar grill. The bars are generally of 25 mm size and are spaced at 75 to 100 mm centres. Mostly, bars are kept inclined so that they can be cleaned easily with a rake. Trash racks are often included in dams and other intake structures. For the purposes of cleaning, they are placed on a slope of 3 to 6 vertical to 1 horizontal. Figure 11.1 shows three intake structures equipped with more or less self-cleaning screens.

In Fig. 11.1(a), the vertical screen is constructed of vertical wires or bars. The strained water enters the screen box, leaving behind leaves and other debris which drops below as sediment. In Fig. 11.1(b), straining is achieved by upward flow, leaving behind debris or waste. The inclined screen can be lifted into vertical position (shown by dotted lines), for cleaning. Figure 11.1(c) shows a self flushing inclined screen. In most of the cases, the common arrangement is to slide a pair of removable screens into vertical grooves in the walls and bottom of the inlet channel.

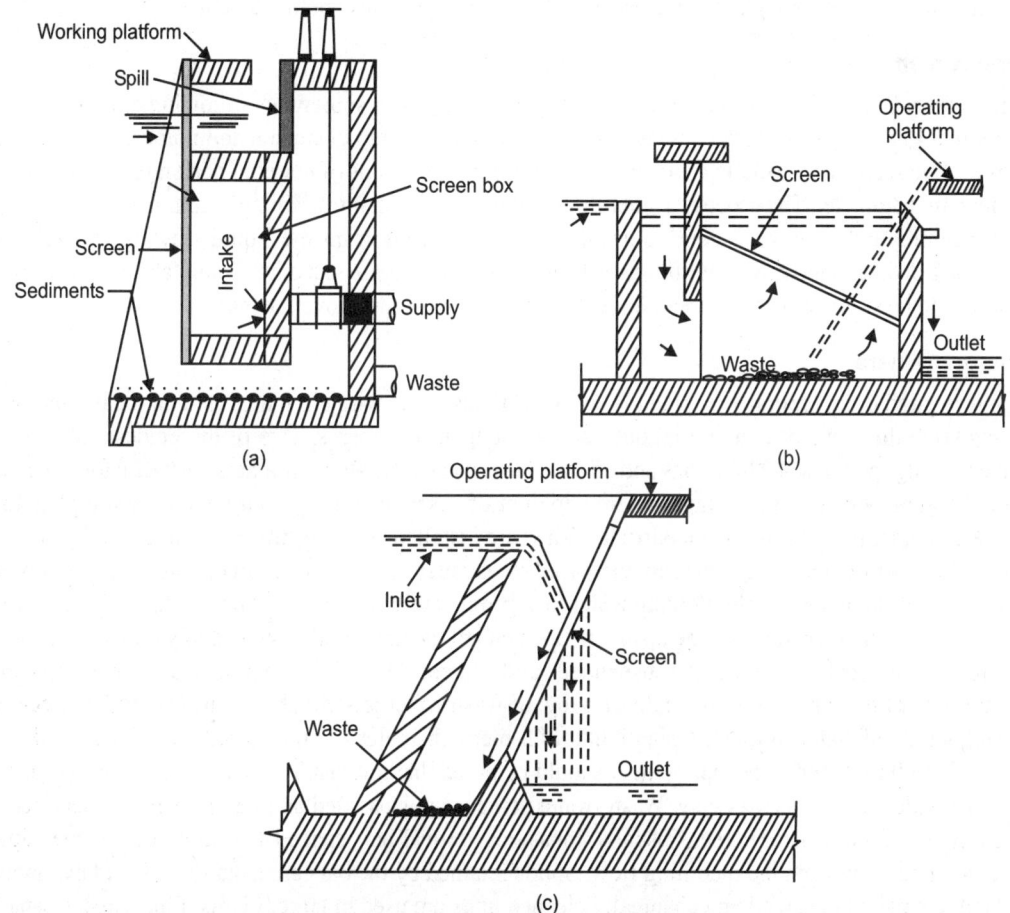

Fig. 11.1. Racks and screens of water intakes.

Head loss through racks and screens

The head loss through unobstructed screens depends upon the nature of their construction (open area, blocked area, shape of screen elements, etc.) as well as the approach velocity. Kirschmer has suggested the following empirical relationship and coefficients for racks with different shaped bears

$$h = \beta \, (w/b)^{4/3} \, h_v \cdot \sin \theta \qquad \qquad \dots (11.1)$$

where, h = head loss.

h_v = approach velocity head.

b = minimum width of clear opening between pairs of bars.

θ = angle of rack with horizontal.

β = a shape factor

= 2.42 for sharp-edged rectangular bars
= 1.83 for rectangular bars with semicircular upstream face.
= 1.79 for circular rods.

Some manufacturers recommend dropping the channel 150 to 300 mm across a bar screen. The maximum head loss through clogged racks and screens is generally below 80 cm.

Fine screens

Fine screens are used at surface water intakes sometimes alone, sometimes following a bar screen. In order that fine screens do not get clogged up, some device is setup to clean it continuously. Due to this reason, fine screens are usually arranged as endless bands or drums of material perforated with holes of about 6 mm diameter. The process is, therefore, known as 'automatic straining'. Automatic strainers are self-cleaning and they work continuously, eliminating solids from the liquid flowing through and disposing of these solids also continuously. In automatic strainers, a pan of straining fabric is submerged in water to be strained while the remainder of the fabric remains above the water.

Micro-strainers

A micro-strainer incorporates a specially woven stainless steel wire cloth mounted on the periphery of a revolving drum filled with continuous back washing arrangements. Two of the grades widely used have limiting apertures of 23 microns and 35 microns respectively. Micro-strainers are useful for screening stored waters which do not contain a large amount of suspended matter but which contain plankton, algae and other microscopic sized particles. They are installed to the upstream of rapid gravity or slow sand filters whose output may thereby increased by as much as 50 per cent. In operation, the revolving drum is kept submerged in the flowing water to approximately two-thirds of its depth. Raw water from lake or reservoir enters through the upstream open end of the drum and flows radially outwards through the microfabric, leaving behind the suspended solid content. The solids so strained are carried upwards on the side of the fabric, with the help of a row of wash water jets which may use about 1 per cent of total quantity of water strained. A single unit of 3-metre diameter × 3-metre wide drum may deal with about 50,000 to 80,000 litres per hour. It should be noted that microstraining cannot remove colour or finely divided matter such as clay. Waste-water screens are classified as fine or coarse, depending on their construction. Coarse screens usually consist of vertical bars spaced 1 or more centimetres apart and inclined away from the incoming flow. Solids retained by the bars are usually removed by manual raking in small plants, while mechanically cleaned units are used in larger plants. Fine screens usually consist of woven-wire cloth or perforated plates mounted on a rotating disk or drum partially submerged in the flow or on a travelling belt. Fine screens should be mechanically cleaned on a continuous basis. Typical screening devices are shown in Fig. 11.2.

Screening devices are contained in rectangular channels that receive the flow from the collection system. Manually cleaned devices should be readily accessible for cleaning and mechanically cleaned systems should be enclosed in suitable housing. Proper ventilation must be provided to prevent accumulation of explosive gases. A straight channel section should be provided a few meters ahead of the screen to ensure good distribution of flow across the screen. Hydraulically, flow velocity should not exceed 1.0 m/s (3.3 ft/s) in the channel, with 0.3 m/s (1 ft/s) considered good design. Head loss across the screen will depend on the degree of clogging. Clean bars and screens result in a head loss of less than 0.1 meter. Provisions should be made for a head loss of up to 0.3 meter for manually cleaned or for manually operated, mechanically cleaned screens.

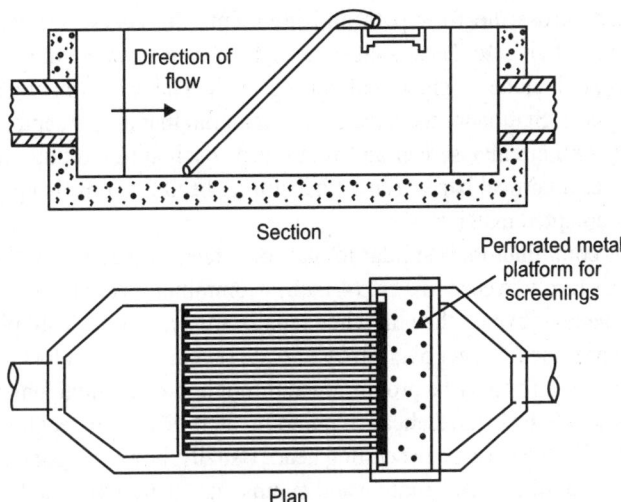

Fig. 11.2. Screening devices used in waste-water treatment: Manually cleaned bar rack.

The quantity of solids removed by screening depends primarily on screen-opening size. The quantity of screenings removed from a typical municipal waste-water as a function of the screen size is illustrated in Fig. 11.3. Screened solids are coated with organic material of a very objectionable nature and should be promptly disposed of to prevent a health hazard and/or nuisance condition. Disposal in a sanitary landfill, grinding and returning to the waste-water flow and incineration are the most common disposal practices.

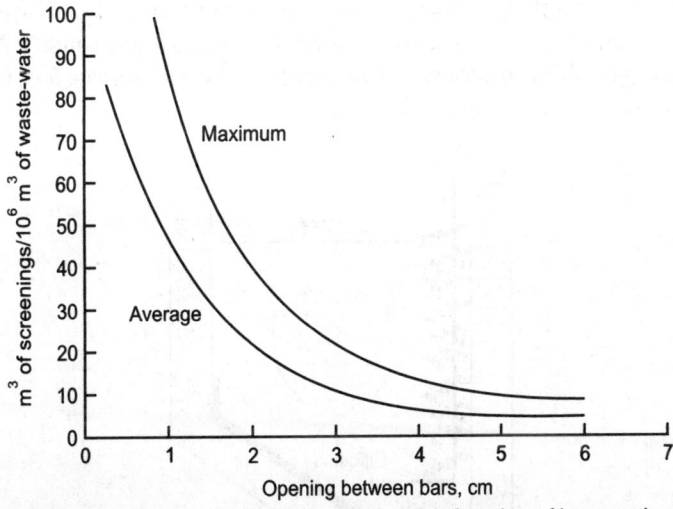

Fig. 11.3. Quantity of screening from municipal waste-water as a function of bar spacing using mechanically cleaned bar screens.

COMMINUTING

As mentioned above, screenings are sometimes shredded and returned to the waste-water flow. A hammermill device is most often used for this purpose. More often, a shredding device called a comminutor is located across the flow path and intercepts the coarse solids and shreds them to approximately 8 mm (¼ inch) in size. These solids remain in the waste-water. Many kinds of comminutors are available. Basic parts include a screen and cutting teeth. The screen may be a slotted drum that rotates in the vertical plane. Stationary teeth then shred material that is intercepted by the screen. Other types use a stationary semicircular screen and rotating or oscillating cutting teeth. Another device, called a barminutor, uses a vertical bar screen with a cutting head that travels up and down the rack of bars, shredding the intercepted material.

Channel design for comminutors is similar to that for screens. Since material does not accumulate on the device, head loss rarely exceeds 10 cm (4-inch). Comminutors are high-maintenance items and provisions should be made to bypass the unit when repairs are needed. In small plants, bypass through a bar screen is usually provided. Larger plants may operate several comminutors in parallel so that flow from one or more disabled units may be proportioned through the remaining units.

Shredding devices should be located ahead of pumping facilities at the treatment plant. Grit removal ahead of the shredder will save wear on the cutting head. Usually, however, grit chambers are located at or above ground level to facilitate grit handling and pumps may be necessary to lift the sewage to them. In this case, shredding is done ahead of the pumps and cutter wear must be tolerated.

GRIT REMOVAL

Municipal waste-water contains a wide assortment of inorganic solids such as pebbles, sand, silt, egg shells, glass and metal fragments. Operations to remove these inorganics will also remove some of the larger, heavier organics such as bone chips, seeds and coffee and tea grounds. Together, these compose the material known as grit in waste-water treatment systems. A schematic diagram of aerated grit chamber is shown in Fig. 11.4. Most of the substances in grit are abrasive in nature and will cause accelerated wear on pumps and sludge-handling equipment with which it comes in contact. Grit deposits in areas of low hydraulic shear in pipes, sumps and clarifiers may absorb grease and solidify. Additionally, these materials are not biodegradable and occupy valuable space in sludge digesters. It is, therefore, desirable to separate them from the organic suspended solids.

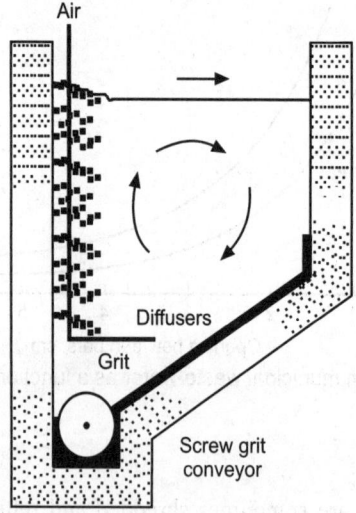

Fig. 11.4. Schematic diagram of an aerated grit chamber.

Because infiltration is a major source of inorganics, the quantity of grit varies with the type, age and condition of the pipe in the collection system. The type and quantity of industrial waste and the prevalence of domestic garbage grinders are also contributing factors. Quantities ranging from 4 to 200 $m^3/10^6\ m^3$ have been reported, with a typical value of around 15 $m^3/10^6\ m^3$ of waste-water. Grit removal facilities basically consist of an enlarged channel area where reduced flow velocities allow grit to settle out. Many configurations of grit tanks are available, with the most recent installations usually being channel-type or aerated rectangular basins. The deposited grit is removed by mechanical scrapers.

Hydraulically, grit chambers are designed to remove, by type-1 settling, discrete particles with diameters of 0.2 mm and specific gravity of 2.65. In channel-type, horizontal-flow grit chambers, it is important to maintain the horizontal velocity at approximately 0.3 m/s. A 25 per cent increase may result in washout of grit, while a 25 per cent reduction may result in retention of nontarget organics. Since a wide variation inflow rates may be encountered, the horizontal velocity must be artificially controlled. A proportioning weir on the effluent end of the tank [Fig. 11.5(a)] or a parabolic tank section [Fig. 11.5(b)] is often used to maintain steady flow at 0.3 m/s.

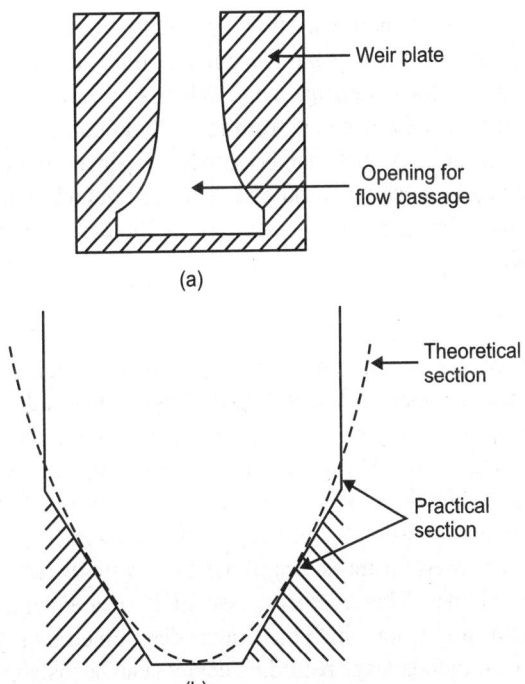

Fig. 11.5. Velocity control sections for horizontal grit channels: (a) proportioning weir and (b) parabolic channel section.

In larger treatment plants, the trend is toward aerated grit chambers. Turbulence created by the injection of compressed air keeps lighter organic material in suspension while the heavier grit falls to the bottom. Since roll velocity, rather than horizontal velocity, serves to separate the nontarget organics from the grit, artificial control of the horizontal velocity is not necessary. Adjustment of air quantities provides settling control. The design of aerated grit chambers is based on detention time at peak flow. Typical design parameters are shown in Table 11.1.

Table 11.1. Design parameter for aerated grit chambers.

Item	Value Range	Typical
Dimensions		
Depth, m	2–5	
Length, m	7.5–20	
Width, m	2.5–7.0	
Width-depth ratio	1:1–5:1	2:1
Detention time at peak flow, min.	2–5	3
Air supply,		
m^3/min. · m of length	0.15–0.45	0.3
Grit and scum quantities		
Grit, $m^3/10^3$ m^3	0.004–0.200	0.015

Aerated grit chambers may serve another useful purpose. If the sewage is anaerobic when it arrives at the plant, aeration serves to strip noxious gases from the liquid and to restore it immediately to an aerobic condition, which allows for better treatment. When an aerated grit chamber is used for this purpose, the aeration period is usually extended from 15 to 20 minutes.

Grit, particularly from channel-type grit chambers, may contain a sizable fraction of biodegradable organics that must be removed by washing or must be disposed of quickly to avoid nuisance problems. Grit containing organics must either be placed in a sanitary landfill or incinerated, along with screenings, to a sterile ash for disposal.

FLOW MEASUREMENT

Although the measurement of waste-water flows does not in itself result in removal of contaminants, it is an important adjunct to waste-water treatment. A knowledge of hydraulic loading rates is necessary for the operation of many of the reactors in a waste-water-treatment plant. Chemical additives, air volume, recirculation rates and many other operating parameters depend upon the hydraulic flow rate. Additionally, records of flows should be kept to establish trends in flow quantities for evaluation of infiltration/inflow quantities and to estimate future capacity needs.

The most common devices used for measuring flows in a waste-water-treatment plant are Parshal flumes and Palmer-Bowlus flumes. These devices, essentially open-channel venturi meters, have an established flow-head relationship from which the flow is determined by simply measuring the water elevation at a given point. Continuous-stage recording devices can be installed to provide flow records.

PRIMARY SEDIMENTATION

Primary sedimentation is a unit operation designed to concentrate and remove suspended organic solids from the waste-water. When primary treatment was considered sufficient as the total treatment, primary settling was the most important operation in the plant. Its design and operation were critical in reducing waste loads to receiving streams. With the current universal requirement for secondary treatment, primary sedimentation plays a lesser role. Indeed, many of the secondary waste-water-treatment unit processes are capable of handling the organic solids if good grit and scum removal are provided for in preliminary treatment.

Most of the suspended solids in waste-water are 'sticky' in nature and flocculate naturally. Primary settling operations proceed essentially as type-2 settling without the addition of chemical coagulants and mechanical mixing and flocculation operations. The organic material is slightly heavier than water and settles slowly, usually in the range of from 1.0 to 2.5 m/hr. Lighter materials, primarily oils and grease, float to the surface and must be skimmed off.

Primary sedimentation is accomplished in either long-rectangular tanks or circular tanks. Scum removal in rectangular tanks is accomplished by having the sludge scrapers penetrate through the surface as they return to the effluent end of the tank. Floating material is carried to a collection point some distance behind the effluent weirs where it is removed over a scum weir or by a transverse scum scraper. Circular tanks have a skimmer arm attached to the sludge-scraper drive mechanisms. The scum is wiped up an inclined apron and into a scum trough for removal. In both cases, a scum baffle should be located between scum removal facilities and the effluent weir. Separated scum is usually disposed of with screenings, unwashed grit, or digested sludge. Cross sectional diagrams of sedimentation tanks are shown in Figs 11.6 and 11.7.

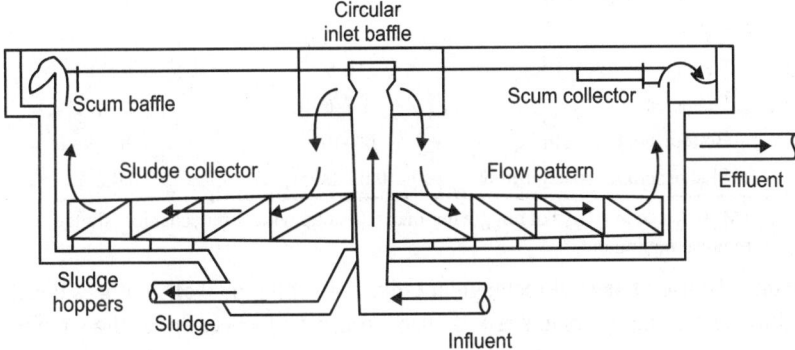

Fig. 11.6. Schematic diagram of a circular sedimentation tank.

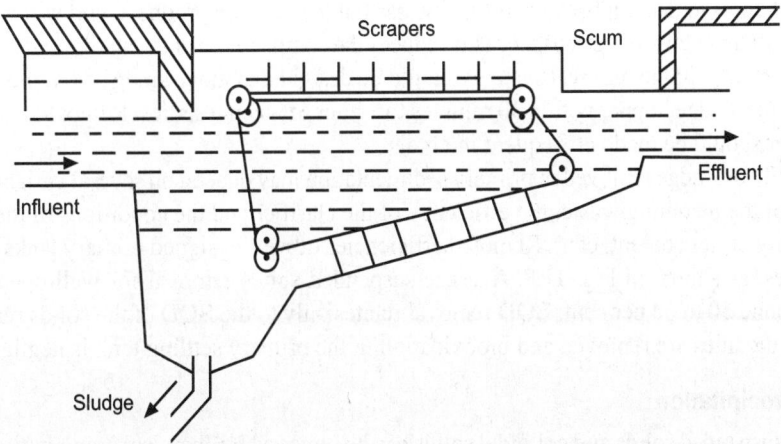

Fig. 11.7. Schematic diagram of a rectangular sedimentation tank.

Design criteria for primary sedimentation tanks are presented in Table 11.2.

Table 11.2. Design criteria for primary sedimentation tanks.

Parameter	Value	
	Range	Typical
Detention time, hour	1.5–2.5	2.0
Overflow rate, m³/m² · d		
Average flow	32–48	
Peak flow	80–120	100
Weir loading, m³/m · d	125–500	250
Dimensions, m		
Rectangular		
Depth	3–5	3.6
Length	15–90	25–40
Width*	3–24	6–10
Sludge scraper speed, m/min	0.6–1.2	1.0
Circular		
Depth	3–5	4.5
Diameter	3.6–60	12–45
Bottom slope, mm/m	60–160	80
Sludge scraper speed, r/min.	0.02–0.05	0.03

*Must divide into bays of not greater than 6.0 meter wide for mechanical sludge removal equipment.

In large plants, the use of several rectangular tanks with common walls reduces construction costs and space requirements. Smaller plants tend to use circular tanks because of the simplicity of sludge removal.

Sludge should be removed from the primary sedimentation tank before anaerobic conditions develop. If the sludge begins to decompose anaerobically, gas bubbles will be produced and will adhere to solid particles and lift them toward the surface. This reduces the compactness of the sludge and makes removal much less efficient. Sludge removal systems should be designed to move sludge from the farthest point in the tank to the sludge hopper within 30 minutes to 1 hour of when it settles. Removal from the hopper to the digester should be made at frequent intervals.

The quantity of sludge removed in primary sedimentation may depend on several variables, including the strength of the incoming waste, the efficiency of the clarifier and the conditions of the sludge (i.e., specific gravity, water content, etc.). Removal efficiencies of well-designed primary tanks depend upon overflow rates, as shown in Fig. 11.8. Average suspended-solids removal for well-operated systems should be around 50 to 60 per cent. BOD removal relates only to the BOD of the solids removed, since no dissolved organics are removed and biooxidation in the primary settling tank is negligible.

Chemical Precipitation

Lightweight suspended solids and colloidal solids can be removed by chemical precipitation and gravity sedimentation. In effect, the chemical precipitate is used to agglomerate the tiny particles into large particles that settle rapidly in normal sedimentation tanks. Aluminium sulphate, ferric chloride, ferrous sulphate, lime and polyelectrolytes have been used as coagulants. The choice of the coagulant depends upon the chemical characteristics of the particles being removed, the pH of the waste-waters and the cost and availability of the precipitants. While the precipitation reaction results in removal of the suspended solids, it increases the amount of sludge to be handled. The chemical sludge must be considered along with the characteristics of the original suspended solids in evaluating sludge-processing systems.

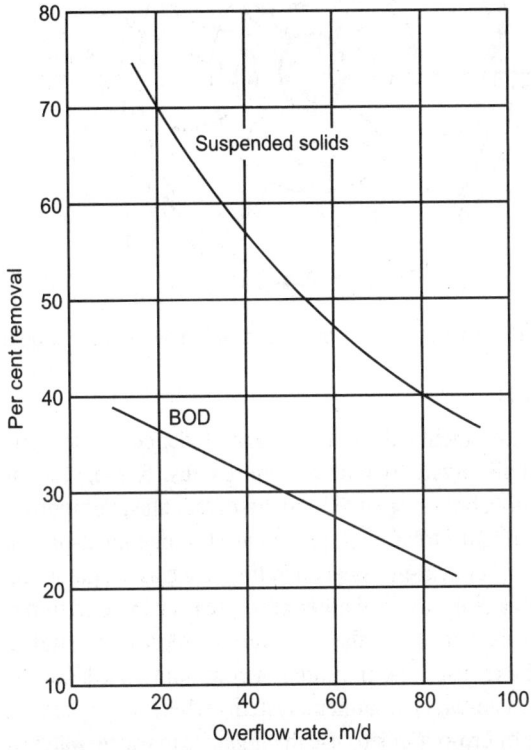

Fig. 11.8. Suspended solids and BOD removal as a function of overflow rate.

Normally, chemical precipitation requires a rapid mixing system and a flocculation system ahead of the sedimentation tank. However, in a circular sedimentation tank, the rapid-mixer and flocculation units are built into the tank. Schematic diagrams of chemical treatment systems are shown in Figs. 11.9 and 11.10.

Rapid mixers are designed to provide 30-s retention at average flow with sufficient turbulence to mix the chemicals with the incoming waste-waters. The flocculation units are designed for slow mixing at 20-min. retention. These units are designed to cause the particles to collide and increase in size without excessive shearing. Care must be taken to move the flocculated mixture to the sedimentation unit without disrupting the large floc particles.

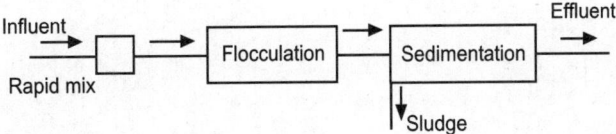

Fig. 11.9. Schematic diagram of a chemical precipitation system for rectangular sedimentation tanks.

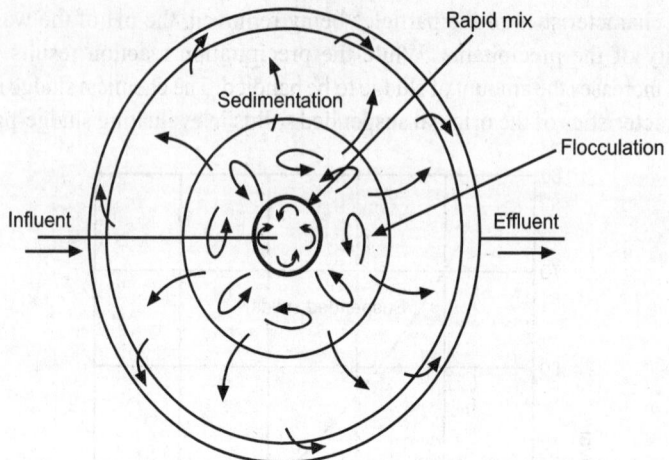

Fig. 11.10. Schematic diagram of a chemical precipitation system for circular sedimentation tanks.

SECONDARY TREATMENT

The effluent from primary treatment still contains 40 to 50 per cent of the original suspended solids and virtually all of the original dissolved organics and inorganics. To meet the minimum EPA standards for discharge, the organic fraction, both suspended and dissolved, must be significantly reduced. This organic removal, referred to as secondary treatment, may consist of chemical-physical processes or biological processes. Combinations of chemical-physical operations such as coagulation, microscreening, filtration, chemical oxidation, carbon adsorption and other processes can be used to remove the solids and reduce the BOD to acceptable levels. Currently, these operations represent a high-cost option with respect to both capital and operating expenses and thus are not commonly used. Biological processes are used in practically all municipal waste-water-treatment systems where secondary treatment is employed.

In biological treatment, micro-organisms use the organics in waste-water as a food supply and convert them into biological cells or biomass. Because waste-water contains a wide variety of organics, a wide variety of organisms or a mixed culture, is required for complete treatment. Each type of organism in the mixed sludge ponds function as settling basins with long retention times. The solids consolidate in the bottom while the supernatant is periodically removed from the top and recycled for retreatment. When the solids have accumulated to a preselected depth, the pond is taken out of service and allowed to dry out. The dried sludge is then removed for final disposal.

Chapter 12

Advanced Waste-water Treatment

INTRODUCTION

The quality of effluent provided by secondary treatment may not always be sufficient to meet discharge requirements. This is often the case when large quantities of effluent are discharged into small streams or when delicate ecosystems are encountered. In these instances, additional treatment to polish the effluent from secondary systems will be required or an alternative method of waste-water disposal must be found. Additional treatment, usually referred to as tertiary treatment, often involves the removal of nitrogen and phosphorus compounds, plant nutrients associated with eutrophication. Further treatment may be required to remove additional suspended solids, dissolved inorganic salts and refractory organics. Combinations of the above processes can be used to restore waste-water to potable quality, although at considerable expense. Referred to as reclamation, this complete treatment of waste-water can seldom be justified except in water-scarce areas where some form of reuse is mandated.

The term advanced treatment is frequently used to encompass any or all of the above treatment techniques and this term would seem to imply that advanced treatment follows conventional secondary treatment. This is not always the case, as some unit operations or unit processes in secondary or even primary treatment may be replaced by advanced-treatment systems. Advanced-treatment processes and operations are described in the following section of this chapter.

Because treatment systems are selected to meet discharge or reuse criteria with respect to specific parameters, the discussion is arranged according to treatment objectives.

NITROGEN AND PHOSPHOROUS CONTROL

Nitrogen compounds often move within the environment as they change form. Most of the problems caused by nitrogen compounds occur when they enter groundwater or surface water bodies. Nitrogen reaches fresh surface water through precipitation, dustfall, surface runoff, subsurface groundwater entry and the discharge of waste-water effluents. There are also blue-green algae and some bacteria which are able to fix nitrogen from nitrogen gas in the atmosphere.

Discharges of conventionally treated domestic and industrial waste-water effluents, high in nutrients, are the main sources of nitrogen pollution in the form of ammonia-nitrogen and sometimes organic-nitrogen. Nitrogen levels in industrial effluents vary.

Some industries posing significant nitrogen pollution problems include meat processing plants, milk processing plants, petroleum refineries, ice plants, fertiliser manufacturers, synthetic fibers facilities and ammonia scouring and cleansing operations.

The use of fertilisers, concentrated animal growth farms (feedlots) and stormwater runoff have also become sources of nitrogen pollution in agricultural areas. Subsurface drainage of agricultural lands carries excess nutrients to the receiving body of water (approximately 19 mg/l). Runoff from feedlots can contain up to 300 mg/l of ammonium due to the hydrolysis of urea and up to 600 mg/l of organic-nitrogen, compared to fertiliser alone, which can contribute anywhere from 1 to 9 mg/l total nitrogen. Stormwater runoff in urban areas also contributes nitrogen to surface waters. Combined sewer (storm and sanitary) overflows during rainstorms can add a significant occasional nitrogen load.

Nitrogen enters soil through precipitation, dustfall, application of waste-water or fertilisers, plant residues, composting or through fixation by bacteria directly from nitrogen gas in the atmosphere. Precipitation and dust fall contain nitrogen compounds due to the combustion of fossil fuels. The planting of legume crops (peas and beans) increases the bacterial fixation of nitrogen gas and can account for 25 per cent of the total nitrogen source in such areas.

Ninety per cent of nitrogen found in soil is in the organic form since it is derived from the decomposition residues of plants and animals. Most of the remaining 10 per cent of nitrogen is as an ammonium ion (NH_4^+) and tightly bound to soil particles. Septic tanks in heavily populated areas can contribute high nitrate levels to the soil at rates faster than can be assimilated. When nitrogen is supplied to soil at a faster rate than it can be assimilated by the soil's bacteria, it will be filtered down to groundwater pockets below. Marine environments receive nitrogen compounds mostly from land drainage. It is the nitrate form which is of concern in such environments.

Algae are the basic link in the conversion of inorganic compounds in water into organic matter. They perform this function through a mechanism known as photosynthesis in which inorganic compounds and carbon dioxide are converted to a carbohydrate energy source plus oxygen in the presence of light. The inorganic compounds required include those with hydrogen, nitrogen, phosphorous, sulphur, potassium, magnesium calcium and trace elements such as zinc, copper, iron and molybdenum. Since all of these chemicals except for phosphorous and nitrogen are usually present, phosphorous and nitrogen normally are considered to be the limiting nutrients.

Nitrogen reaching surface waters via the various pathways discussed serve to trigger and/or sustain the growth of algae. While some algal growth is desirable since by day photosynthesis removes carbon dioxide and produces oxygen, dense growths can create problems for the environment.

When the natural balance of nutrients is thrown off by an excess supply of nitrogen and other necessary compounds, algae will grow in large numbers creating a dense mat on the water surface known as an algal bloom or scum. Depending on the type of algae present, the water will take on a green, yellow, red, black or turbid appearance ruining the water for recreational purposes. The odours produced by the bacterial decomposition of algae further curb recreational uses.

Similarly, growth of other water plants and some diatoms (fresh and saltwater algae types which form silica shells) also is stimulated by an abundance of nutrients.

Algae growth in water supplies requires taste and odour removal through use of filtration followed by carbon adsorption. In addition, the water supply itself must be chemically treated to kill algae growths.

Algae produces oxygen by photosynthesis during the daylight; however, carbon dioxide is released at night during the algae's respiration process. While more oxygen is produced than respired, the presence of large quantities of algae can produce wide swings in the water's oxygen content, which can be harmful to other oxygen-dependent organisms living there.

Similar swings (diurnal) can be found in pH levels and alkalinity of the water as a result of CO_2 level changes due to algae photosynthesis and respiration.

During the night, respiration increases dissolved carbon dioxide levels, thus lowering the pH. Waters with high calcium content are able to buffer the pH effects of carbon dioxide reduction by the precipitation of calcium carbonate.

Dense algal growths limit photosynthesis to the top layer of the water. Depths of greater than three feet are shaded from the sunlight. The algae below die and are decomposed by bacteria, which in turn deplete the available oxygen.

In extreme cases, algae production leads to eutrophication, which is defined as a process of enrichment in which a water body is so fertilised with nutrients that aquatic productivity is greater than the decay rate. Eutrophication is more of a problem in relatively slow-moving waters such as surface streams, lakes and reservoirs where nutrients have time to build up.

Oxygen depletion in eutrophying water bodies has a snowball effect as other organisms die as a result, creating a further oxygen depletion problem by their own decay. Once eutrophication starts, it cannot be stopped.

Changes in organism distribution patterns occur in eutrophying water bodies. Predator-prey relationships change and 'trash' fish tend to flourish while desirable sport fish disappear, thus further altering the value of the water body.

Besides potentially destroying the aesthetic and recreational value of water bodies, there are some species of algae which can cause gastric disturbances in humans if the water is consumed or accidentally swallowed. Additionally, some people are allergic to planktonic algae. Certain blue-green algae and chlorella green algae species secrete substances toxic to fish, birds and pets. Toxic blue-green algae water blooms have been known to kill hundreds of birds in just a few hours. Nutrient imbalances in the marine environment result in the production of phytoplankton blooms and dense algal growths. Marine pests such as barnacles, mussels, tube worms, parasitic fungi and jellyfish swarms also result in nitrogen-rich environments.

Nitrogen compounds themselves are known to cause problems in both surface and groundwater. A survey conducted in Illinois indicated that the highest nitrate surface water levels occurred near agricultural areas. The nitrate form also occurs in groundwater where high concentrations can build up due to over-fertilisation or location of septic tanks near shallow wells. When water containing nitrates is used to prepare formulas for infants less than three months old, an occasional fatal blood disorder can occur called methemoglobinemia (or blue baby). Nitrate is transformed to nitrite in the infant's stomach, passes to the bloodstream and attaches itself to the haemoglobin molecule taking the place of oxygen with resulting suffocation. As a result of this danger, the US Public Health Service recommends a limit be placed on both nitrite and nitrate of 10 mg/l as nitrogen for drinking water. Nitrite (NO_2^-) also can be highly toxic to fish and other animals causing methemoglobinemia in them, too. Nitrites are very unstable ions which easily convert to nitrates however. The chloride ion has been shown to provide a protective effect against nitrite toxicity.

In general, nitrite is unimportant in waste-water or pollution. Nitrite indicates a past pollution which is or has been stabilised. Ammonia (NH_3) is a significant pollutant in raw water. It reacts with chlorine to form chloroamines which reduce chlorine disinfection properties. Ammonia can be toxic to fish at certain pHs. It exists as ammonia NH_3 at pHs above 7 and as an ammonium ion (NH_4^+) at pHs below 7. It is the ammonia molecule that is toxic to fish. A maximum concentration of 0.02 ppm has been set as a water quality standard for freshwater aquatic wildlife. Acute toxicity at a given pH will increase with corresponding increases in dissolved oxygen, carbon dioxide, temperature, or bicarbonate alkalinity levels within a range of 0.01 mg/l to 2.0 mg/l ammonia. At levels up to 25 mg/l, ammonia toxicity can affect all aquatic life. Diurnal pH fluctuations due to photosynthesis therefore can play a significant role in creating toxic conditions.

Nitrogen and algal growth can also interfere with industrial and water treatment operations. Nitrogen in the form of ammonia is corrosive to certain metals. Plankton and filamentous algae can clog sand filters in water treatment plants, cause foam when heated, corrode metal and create undesirable tastes, odours or oily substances which interfere with filter use. Algae growth is a problem in cooling towers as well. As with untreated carbonaceous matter, ammonia-nitrogen exerts an oxygen demand on receiving water bodies as it slowly oxidises to nitrite and then nitrate.

Nitrification of waste-waters has been a primary concern. BOD testing emphasises control of carbonaceous matter. Concern for control of nitrogen discharges in the environment has received attention and efforts have been directed toward determining design considerations for upgrading existing carbonaceous oxidation plants to accomplish nitrification as well and towards developing independent nitrification or nitrification-denitrification systems.

NITRIFICATION

Nitrification is the first of two stages in biological nitrogen removal in which ammonia-nitrogen is biologically oxidised to nitrate, a less objectable form which does not exert an oxygen demand on the receiving water. In the first step of nitrification, ammonium ions are oxidised to nitrite ions according to the following reaction in which 58 to 84 kcal/mole of ammonium is released:

$$NH_4 + 1.5O_2 \rightarrow H^+ + H_2O + NO_2$$

The bacteria responsible for this oxidation are usually nitrosomonas, although sometimes nitrosococcus can be involved. These bacteria are aerobic autotrophs. Autotrophs, unlike heterotrophs, which obtain their energy from the oxidation of carbonaceous (organic) matter, get their energy for growth from the oxidation of inorganic nitrogenous matter and use inorganic carbon rather than organic carbon, as heterotrophs do, for cell synthesis. Being aerobic, these autotrophs require the presence of oxygen to convert the nitrogen into a usable form.

In the second step of nitrification, the nitrite ion is further oxidised by nitrobacter bacteria to nitrate releasing only 15–21 kcal/mole of nitrite oxidised as follows:

$$NO_2^- + 5O_2 \rightarrow NO_3^-$$

Nitrobacter bacteria are aerobic autotrophs also. The energy freed by the nitrification reactions is used by the bacteria for growth. Since the nitrosomonas obtain more energy than the nitrobacter bacteria per mole of nitrogen oxidised, their mass in any nitrification system is greater. The nitrobacter bacteria require three times the substrate needed by nitrosomonas to get the same energy and therefore their population is 1/3 that of the nitrosomonas.

Nitrifiers grow at a much slower rate than heterotrophs. This growth rate difference can be measured by the BOD test. BOD_5, which represents biochemical oxygen uptake after five days, indicates the oxidation of carbon by heterotrophs. BOD_{20} similarly represents the final oxygen uptake after 20 days by bacteria through nitrification. While the nitrification reactions appear very simple, there are various intermediates and enzymes involved. The enzymes which control the rates of reactions conducted by the bacteria are sensitive to pH, temperature and substrate concentration. Conditions necessary for proper functioning of these enzymes are reflected in the overall cell preference. The enzymes are substrate-specific. Therefore, there can be many enzyme reactions involved in the bacteria cell synthesis. The enzymes must convert the nitrogen compounds into an amino acid form before they can be used directly by the bacteria. The growth of nitrosomonas is limited by the ammonia-nitrogen content of the waste-water, which in turn limits the nitrobacter's growth. When the food (substrate) supply is plentiful and other conditions are favourable, growth will increase unchecked. As the population begins to exceed the available food supply, the growth rate will decline. As the food supply becomes scarce, the bacteria begin to obtain their nutrition from the dead bacteria through lysis.

Growth rates also increase with rises in temperature. The growth of nitrobacter is more greatly influenced by temperature than that of nitrosomonas. Increases are exponential in nature, reaching a maximum at some optimum temperature and then quickly falling to zero once the optimum temperature is passed. The alkalinity and pH of a system are also important. The nitrification reaction releases carbon dioxide and free acid (H^+) during the oxidation of 1 mg ammonia to nitrate which destroys about 7 mg of alkalinity as calcium carbonate ($CaCO_3$). Depending on the alkalinity available, the reduction in $CaCO_3$ can have a depressing effect on the pH.

When the pH drops below 7, a considerable decrease in the nitrification rate will result. This is true for both acclimated and unacclimated systems, although the short-term effect on an acclimated system is less significant. It has been demonstrated that pH drops from 7.2 to 6.4 have no immediate adverse effects. Drops to 5.8, however, create significant reductions in the nitrification rate. An abrupt pH change from 5.8 back to 7.2 will cause an immediate rise in the nitrification rate. Therefore, pH has an inhibitory rather than toxic effect. Nitrifiers have been known to adapt to a pH range of 5.5 to 6.0. Many waste-waters do not have a sufficient alkalinity buffer and alkalinity maintenance becomes very important.

In order to obtain complete nitrification of a waste, 4.6 mg of oxygen is required for every mg of ammonia present. Generally, the nitrification rate will increase with an increase in dissolved oxygen of the system if other conditions are favourable. Studies have shown that rates are 10 per cent to 50 per cent lower at dissolved oxygen levels of 1 or 2 mg/l than at 4 to 7 mg/l.

Nitrification therefore is an important factor in stabilising the oxygen demand of the waste. Controlled biological treatment is necessary to obtain nitrification since the population of nitrifying organisms is minimal in surface waters.

Another advantage of nitrification is that a highly nitrified effluent is immune to petrification, thus helping to preserve the aesthetic quality of the receiving body of water.

Nitrification can be required when standards or limitations have been set on the receiving waters or effluent or where the reduction of the residual oxygen demand from ammonia is specifically required. The overall transformation of ammonia to nitrate will depend on how much organic nitrogen has been transformed to ammonia prior to the nitrification process. When the total nitrogenous content of the effluent must be reduced due to regulatory limitations and/or the growth of algae in the receiving water must be prevented or reduced, denitrification is required.

DENITRIFICATION

Denitrification is the second and final stage in the biological removal of nitrogen. With denitrification, nitrates are reduced to nitrogen gas. When methanol is used as a source of carbonaceous matter, the reaction for denitrification is:

$$5CH_3OH + 6H^+ + 6NO_3^- \rightarrow 5CO_2 + 3N_2 + 13H_2O$$

It is also possible for nitrites to be converted directly to nitrogen gas. The bacteria responsible for this transformation are heterotrophs, which derive their energy from organic chemicals through the reduction of nitrate or nitrite. These bacteria include *pseudomonas, achromobacter, bacillus* and *micrococcus*, which are facultative, meaning that they can survive with or without the presence of oxygen. The bacteria prefer oxidising organic matter with oxygen rather than by reducing nitrite or nitrate. Therefore, anaerobic (no oxygen) conditions must be maintained in a denitritication system.

With both nitrification and denitritication of wastes, nitrogen removals of 70–90 per cent can be obtained. While such removal will serve to reduce or prevent most algal growth and eutrophication, many blue-green algal blooms cannot be affected since these algae can fix nitrogen gas for their synthesis from the atmosphere.

BIOLOGICAL NITRIFICATION METHODS

Biological nitrification can be achieved by several means and through various add-on treatment and upgrading methods for new and existing treatment systems.

Domestic waste effluent has been one of the main contributors of nitrogenous compounds to our environment. Nitrogen in such wastes usually exists as organic-nitrogen or as free ammonia. The nitrate and nitrite concentrations are generally small in raw wastes in relation to the other forms. Typical values of nitrogen concentrations of raw domestic wastes are shown in Table 12.1.

Table 12.1. Typical values of nitrogen concentrations of raw domestic wastes.

Nitrogen Form		Waste effluents (m/l)	
	Strong	Moderate	Weak
Nitrogen as total nitrogen	85	40	20
Organic-nitrogen	35	15	8
Free ammonia	50	25	12
Nitrites	*	*	*
Nitrates	*	*	*

*negligible.

The most significant of the above compounds is ammonia since it can lower the dissolved oxygen of a receiving stream by nitrification. The ammonia content is derived from urea and to a lesser extent proteins. The organic-nitrogens are in the form of purines, pyrimidines, proteins urea and amino acids. Much of the organic-nitrogen is transformed to ammonia through hydrolysis before it reaches the wastewater treatment plant. Conversion of organic-nitrogen to ammonia continues to occur within a conventional treatment plant due to the actions of heterotrophic bacteria. The organic-nitrogen compounds usually are in a soluble form, while most of the ammonia is particulate. Primary sedimentation removes some of the particulate nitrogen forms, which are usually less than 20 per cent of the total nitrogen. Secondary sedimentation (clarification) removes another 10–20 per cent of the nitrogen. A domestic waste with predominantly the nitrate nitrogen form has been stabilised with respect to its oxygen demand and is considered to be an old waste.

After passing through conventional biological treatment, the secondary sanitary effluent will have a typical nitrogen content of 20–50 mg/l, indicating that nitrogen just passes through such systems. However, biological nitrification does not remove nitrogen any better than conventional treatment.

Biological nitrification can be achieved in conventional carbonaceous removal systems which have been modified to combine nitrification and by the addition of a separate tertiary system. Generally if the BODs/Total Kjeldahl Nitrogen (TKN) ratio is less than 3, a separate nitrification system must be added. If the BODs/TKN ratio is greater than 5, a combined system should work. There is no special recommendation for wastes with ratios between 3 and 5 at this time. There are varied opinions as to the merits of both methods. There are two basic concepts of biological treatment available. These are suspended growth, where the bacterial masses are suspended in a mixed liquor and separated via clarification and attached growth, whereby the bulk of the bacterial growth occurs on a plastic or stone media and solids separation is not necessary. The types of suspended growth systems available include various activated sludge setups. Attached growth systems include trickling filters, biodiscs and fluidised beds.

Activated Sludge Systems

Most of the studies done related to biological nitrification have been in the operation of activated sludge treatment systems, which is one of the major recognised effective methods. Important variables which have been studied extensively include the organic loading and sludge age, pH, dissolved oxygen, temperature and the presence of inhibitory substances which are discussed in this section.

The organic loading to a system is the single most important factor in nitrification. High organic loadings favour the growth of heterotrophic bacteria which then overrun the system. These bacteria have a faster rate of substrate oxidation than autotrophs. The heterotroph's faster growth rate is reflected in the oxygen uptake and sludge production as (bacterial) sludge is wasted faster than the nitrifiers can multiply. Therefore, with high organic (BODs) loadings, little nitrification will occur and at lower BODs loadings, approximately 10 mg/l, higher nitrification rates can occur. Increases in the organic loading of a waste can be compensated for by increasing the retention time of the activated sludge to prevent washouts of nitrifier populations before they can become established. Sufficient oxygen supplies also must be carefully monitored in such cases to supply both the carbonaceous and nitrogenous demands. These are important concerns in combined treatment systems. However, most of the carbonaceous oxygen demand already is removed when nitritication serves as tertiary treatment. A shock load of organics to a treatment plant would not have a significant impact on nitrification in a tertiary system.

Retention time is not important by itself, but may be used to moderate effects of changes in other parameters, such as organic loading, sludge age, dissolved oxygen and temperature. The time factor has a direct proportional relationship to the amount of nitrifiers which will be present. The average sludge retention time for a conventional activated sludge system is 3½ days. Six to ten days would be needed in a combined nitrification-carbon removal system to prevent washout of nitrifying populations.

Dissolved oxygen (DO) has a significant effect on nitrification. The stoichiometry of the nitrification reaction shows that four atoms of oxygen are needed to oxidise one molecule of ammonia to nitrate. This translates to a 50 per cent greater oxygen requirement for good nitritication of a typical domestic waste than is required for carbonaceous removal. Pilot-plant studies have shown that nitrification is possible at DOs of 1 mg/l and may not occur at all at DOs of 7 mg/l if other important factors are not favourable. Generally though, barring unfavourable conditions, higher DOs will increase the rate of nitrification.

Temperature affects bacterial metabolic activities, gas transfer rates (available DO) and settling characteristics of waste effluents. At temperatures above 40°C and below 5°C, nitrification rates are very slow. The optimum temperatures for nitrifying bacteria are 22°C and 30°C. Because the temperatures in summer and warm climates are within this optimum range, treatment plants can be operated at less favourable pHs and lower substrate levels that would be required during colder conditions to achieve the same degree of nitrification. In order to make up for the temperature difference, in winter, up to five times the summer detention time (capacity) may be required. Temperature deficiencies may also be made up by increasing the MLSS of the system and/or adjusting the pH.

Nitrification is most rapid when the pH is maintained at or slightly above neutral. Results have shown that the optimum pH for nitrification is 8.4. With all other conditions favourable, 90 per cent nitrification can be obtained at pHs of 7.8 to 8.9, but less than 50 per cent below 7.0 and above 9.8. Further reductions in nitrification rapidly occur below a pH of 6.0 and nitrification may cease entirely below pH 5.0. Nitrifiers have a low tolerance to the hydrogen ion concentration. Breakdowns of sludge flocs also have occurred when the pH drops below 7.0.

Inorganic loadings and to a lesser extent, the ammonia-nitrogen level of the waste, play a role in affecting nitritication rate. Studies have shown that at a given organic loading, increases in the influent ammonia concentration increase nitrate production levels. These increases are not proportional in nature. There is a point where the nitrate production will be limited.

The ammonia levels found in domestic waste-water are not sufficient to inhibit the rate of nitrification of such effluents. Figure 12.1 shows the rate of nitrification based on mixed liquor volatile suspended solids concentration, ammonia-nitrogen concentration, temperature and pH.

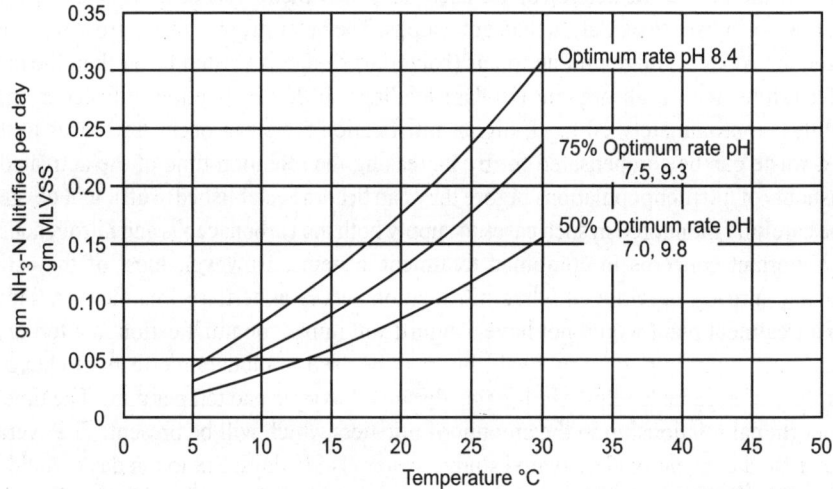

Fig. 12.1. Rate of nitrification versus temperature at various pH levels.

The compounds toxic to nitrifying bacteria are shown in Table 12.2.

Table 12.2. Compounds toxic to nitrifying bacteria.

Organics	Inorganics
Thiourea	ZN
Allyl-thiourea	OCN^{-1}
8-hydroxyquinoline	ClO_4^{-1}
Salicyladoxine	Cu
Histidine	Hg
Amino acids	Cr
Mercaptobenzthiazole	Ni
Perchloroethylene	Ag
Trichloroethylene	–
Abietec acid	–

Only inhibitory effects may be felt from heavy metals concentrations of 10–20 mg/l, provided that the pH is 7.5–8.0. Precipitated metals in the sludge can redissolve if the pH drops down resulting in a system upset. Industrial discharges which are unusually high in ammonia or nitrite can exert a temporary effect on the system also. To screen for a toxicity problem, batch oxygen uptake tests may be used and batch nitritication jar tests may then be run in order to determine the best pre-treatment.

Pre-treatment can afford some protection. Heavy metals may be removed by lime additions, carbon adsorption can be used for organics and two-stage systems are viable where the organic toxics are biodegradable. Perchloroethylene and trichloroethylene are not biodegradable and toxic to nitrifiers. For toxics that come and go, breakpoint chlorination may be used at the end of the system for added safety in ammonia removal.

In activated sludge systems, waste is biologically oxidised under aerobic conditions. Large and easily settleable solids are removed prior to entry of the waste effluent into a reactor. Air (oxygen) is supplied to the reactor by diffusion of mechanical aeration. After oxidation has occurred, the mass of bacteria which has grown is separated from the liquid in a settling tank or clarifier. Some of the solids are returned to the reactor while the rest are wasted. There are many variations of this method.

Combined Carbon and Nitrogen Removal Systems

The first nitrification processes developed were combined systems made by modifying extended aeration systems. Combining operations is advantageous in terms of cost for existing carbonaceous systems which can be upgraded to include nitrification. Combined carbon and nitrogen removal systems have a high proportion of influent organic loading relative to the ammonia-nitrogen concentration. As a result, the population of nitrifiers is small compared to heterotrophs. In addition, the conditions required for the carbon oxidising heterotrophs and nitrifying autotrophs are different and therefore operating parameters must be carefully controlled in combined systems. Combined systems should be based on the sludge growth rate or solids retention time. This generally means an additional oxygen supply, longer mean cell residence times (about 10 days) and operating temperatures of 21°C to 22°C.

Contact stabilisation systems are shown in Fig. 12.2, where sludge is re-aerated prior to being recycled with the influent, will not provide complete nitrification. Even though the solids can be retained for a longer time, there is an insufficient mass developed in the reactor.

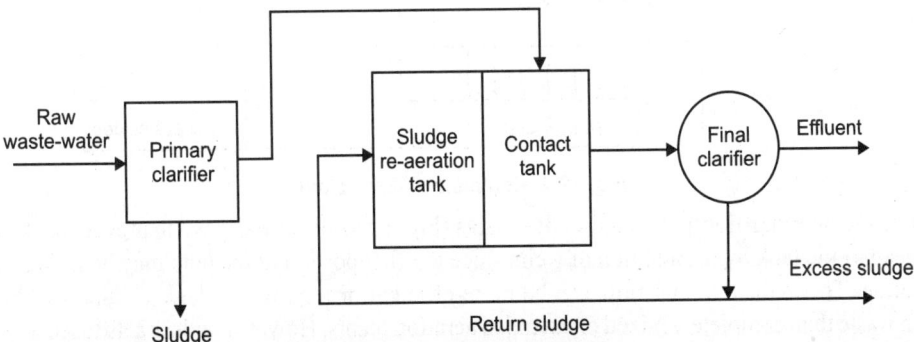

Fig. 12.2. Contact stabilisation plant.

The influent entering the last pass in step aeration systems (as shown in Fig. 12.3) may not have enough time for hydrolysation of the organic-nitrogen to ammonia to permit nitrification. This problem can be somewhat alleviated by setting up artificial sludge re-aeration zones in the first pass by not feeding influent to that section. However, back-mixing is not prohibited, neither is short-circuiting and it is possible for ammonia bleed-through to occur.

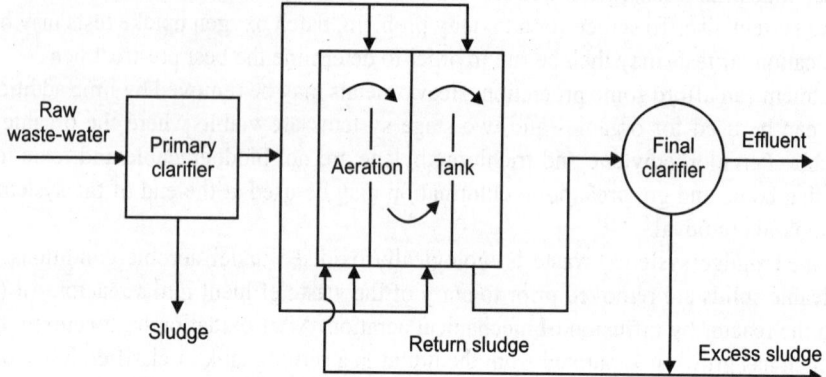

Fig. 12.3. Step aeration plant.

Extended aeration plants are usually operated at such long retention times that, except during cold temperatures below 5°C to 10°C, nitrification usually is obtained if the plant is operated properly. These systems are similar to completely mixed systems except that the hydraulic retention times are 24 to 48 hours rather than 2 to 8 hours.

Completely mixed systems (Fig. 12.4) can provide complete nitrification at typical domestic waste concentrations. In such systems, waste is distributed uniformly to all points within the aeration tank.

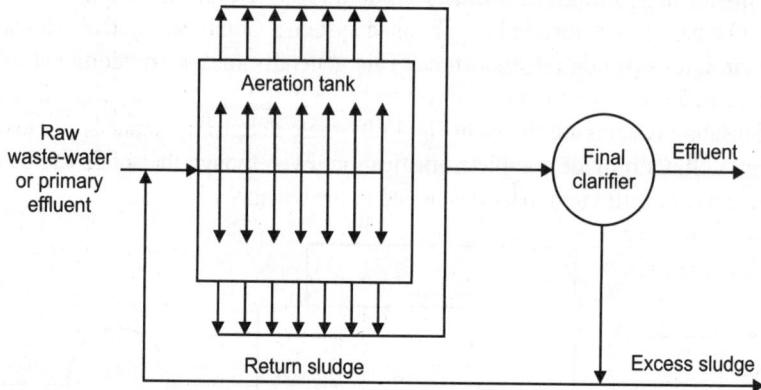

Fig. 12.4. Completely mixed plant.

Conventional (plug-flow) activated sludge plants (Fig. 12.5) can be designed to prevent back-mixing to the head of the tank by the addition of weirs since the first portion of the tank may be ineffective for nitrification. Theoretically, plug-flow can be more efficient or require less tank volume for the same strength waste than completely mixed or extended aeration plants. However, unless a diffused air system is installed, the carbonaceous oxygen demand can overpower the nitrifier's needs. If lime is used for flocculation before the plug-flow reactor, carbon dioxide should be added to avoid pH toxicity.

High-purity oxygen systems have been experimented with for nitrification. Since the cover prevents the escape of carbon dioxide, a buildup in the system occurs. pH levels of 6.0 are not uncommon. This has a depressing effect on the nitrification rate and even longer solids retention times are required. If pH is carefully maintained in the system, UNOX plants are no different in the degree of nitrification achievable than conventional aeration plants. The choice must be based on economic and social (odour) considerations.

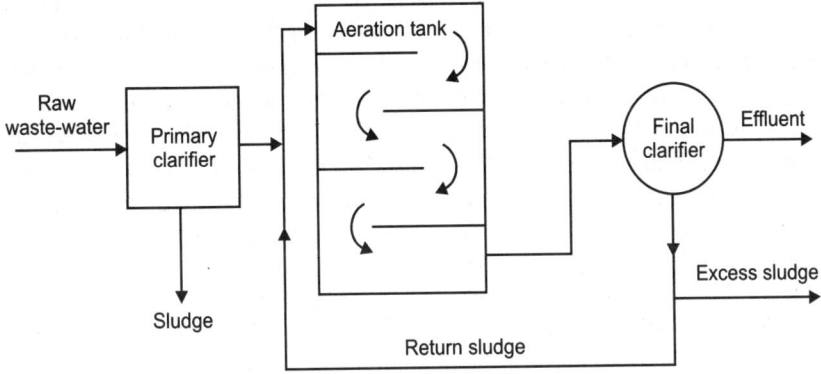

Fig. 12.5. Conventional activated sludge plant.

Two-Stage Carbon and Nitrogen Removal Systems

Physical separation of carbon and nitrogen removal functions can improve the control and efficiency of nitrification in certain cases. By reducing the BODs load in the first stage significantly to the influent ammonia concentration, more nitrifiers can be established. The value of separating the heterotrophic and autotrophic populations is realised in the reduced residence time required—6 days total versus 10 days in combined systems. Plug-flow systems are the favoured activated sludge method for obtaining nitrification. Lower effluent ammonia concentrations can be achieved than in completely mixed units.

Combined and Two-Stage Systems

Figure 12.6 is a typical two-stage activated sludge system.

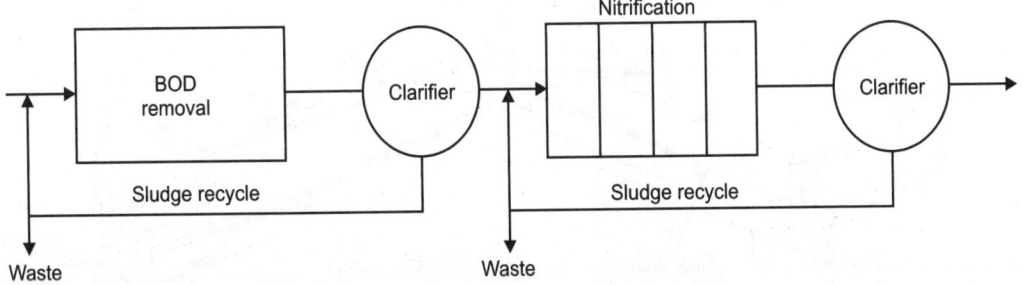

Fig. 12.6. Two-stage activated sludge system.

While it has been demonstrated that both types of systems can be operated to achieve complete nitrification, there are advantages and disadvantages to each type. Results of modifying activated sludge systems for nitrification have been inconsistent. Problems sometimes occur with rising sludge due to the long retention times required where denitrification begins to take place due to lack of oxygen. Nitrogen gas bubbles cause the sludge mass to rise.

Combined systems receiving a primary effluent with a weak waste (BOD_5) can be operated for nitrification satisfactorily with high MLSS down to temperatures of 10°C. However, two-stage systems are necessary in northern climates when waste-water temperatures often go below 18°C. It has been demonstrated that two-stage systems can handle seasonal load variations where combined systems cannot. Another advantage of combined rather than single-stage systems is the lower quantity of sludge (about half the amount), which must be handled and the normally better settling characteristics of the sludge produced. Two-stage systems tend to have more control problems since two systems are involved and the clarifier is the least stable component of a system. However, with careful monitoring, two-stage systems can be managed and a greater degree of control obtained over microbial processes.

Toxins may not be a problem to combined systems if the primary effluent is treated with a coagulant. Some people feel that toxic substances can be reduced by two-stage systems, but others feel that there is an advantage to sacrificing a carbonaceous system over a combined system. There are some substances, though, which may be toxic to nitrifiers yet biodegradable by heterotrophs; but, there are also indications that the toxic's advantages may not matter with domestic sewage. The first reactor in a two-stage system also may serve to reduce the possibility of organic surges to the nitrifying bacteria thus preventing overpowering carbonaceous bacterial growth. Combined systems require less land and capital expenditures than separate systems and have lower sludge disposal costs. However, the power costs for separate systems are less.

Trickling Filters

Trickling filter systems are the major type of attached growth system used to perform nitrification. Due to larger land requirements relative to activated sludge systems, trickling filters are often used for sanitary treatment in smaller cities of less than 10,000 people and less populated areas. Because of the relative stability of trickling filters compared to other biological systems, they are used for treating high-strength industrial wastes too. A diagram of a typical trickling filter is shown in Fig. 12.7.

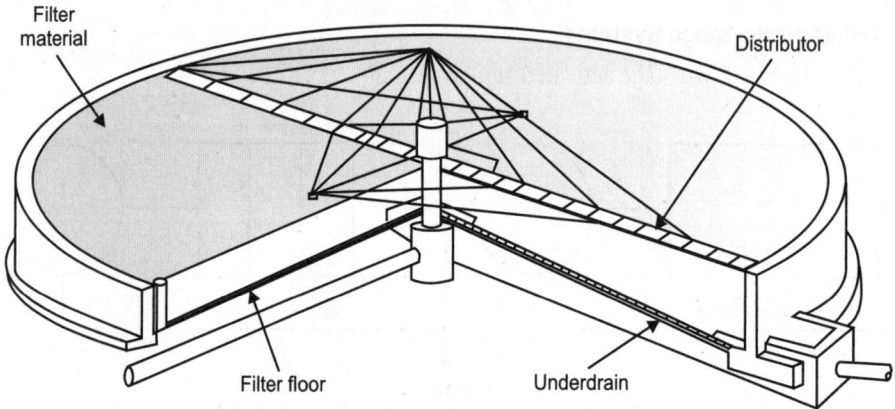

Fig. 12.7. Cutaway view of a trickling filter.

Trickling filters are usually circular in form. Waste effluent is distributed by rotary sprays over the media and is collected underneath by an underdrain system.

Media to which organisms attach can be made of rock, plastic or redwood. The liquid waste percolates down through the media and the substrate and inorganic and organic waste matter is assimilated by the organisms attached to the media. Aerobic degradation takes place on the outer portions of the biological film which develops on the media. As the mass of organisms becomes thicker, anaerobic conditions occur near the media surface. The surface organisms then die and are washed off periodically (sloughing).

Effluent peaks can be taken care of by designing the system with additional surface area and increasing recirculation rates during low-flow periods to keep the media from drying out. Clarification normally is not needed following trickling filters since the solids are maintained within the units.

Trickling filters are the homes of a varied assortment of organisms including: aerobic, anaerobic or facultative bacteria; fungi, during low pHs; algae; protozoans, worms, insect larvae and snails, which in turn feed on the bacteria. Due to the unstable characteristics of the slime, a kinetics theory for the biological activities has not yet been developed. Conclusions regarding nitrification in such systems are based on empirical results.

Nitrification and carbonaceous oxidation can occur simultaneously in trickling filters. It is better to use a media with a lower specific surface and higher voids, such as a maximum of 35 sq. ft./cu. ft., to prevent clogging in combined systems.

When nitrification is separate from carbonaceous oxidation, plugging is less of a problem for the nitrification system and application of a media with a high specific surface (up to 67 sq. ft./cu./ft.) is okay and reduces space requirements.

In two-stage systems where organic carbon and nitrogen activities occur separately, increases in nitrification have been proportional to increases in surface areas. The surface area requirements for two-stage systems increases greatly at temperatures of 7°C to 11°C than at 13°C to 19°C. Surface area requirements for nitrification also increase with the degree of ammonia-nitrogen reduction desired.

If ammonia removal must be below 2.5 mg/l, breakpoint chlorination may be added following the trickling filter, since the cost of removing such ammonia levels is much higher with trickling filters. The important variables in operating a trickling filter for either combined carbonaceous oxidation and nitrification or only nitrification as tertiary treatment include the organic loading, temperature, pH, dissolved oxygen and toxicants.

Organic loading

Organic loading has a significant effect on the ammonia content of the effluent. If it becomes too large, the media will be dominated by heterotrophs and significant nitrification will not occur. For combined systems, the organic loading must be reduced for cold weather operations, which increases the nitrification costs beyond those of adding a separate biological nitrification system or using a physical-chemical treatment. It has been demonstrated that nitrification efficiency levels of 75 per cent to 100 per cent can be achieved with BODs loadings of less than 10 lbs./1000 cu. ft./day. Efficiency diminishes at greater loadings.

Temperature

Greater nitrification can be achieved with higher temperatures. Temperatures within 15° to 30°C are preferable, with 30°C the optimum. Nitrification can be achieved down to 7°C, but as mentioned previously, cost factors make it impractical below 13°C. However, attached growth systems can compensate for cold temperatures better than suspended growth systems by thickening the slime.

pH

Nitrifying bacteria are limited by pH in attached growth systems as they are in suspended growth systems.

Dissolved oxygen

Oxygen mass-transfer limits in bacterial slimes may limit the nitrification reaction. In order to prevent oxygen from being the limiting factor, the dissolved oxygen supply must be 2.7 times the ammonia-nitrogen concentration. This can be achieved by increasing the recirculation rate to dilute the ammonia or by adding a high-purity oxygen to increase the oxygen transfer rate.

Toxicants

Nitrifying bacteria are subject to the same toxicant effects whether they live in an attached or suspended growth system. However, trickling filters can handle shock loads better than activated sludge.

Hydraulic loading

Hydraulic loading of the system can have a profound effect on the degree of nitrification attainable in a trickling filter. An increase from 10 MGAD to 30 MGAD has reduced nitrification from 72 per cent to 52 per cent.

Filter depth

Greater nitrification can be achieved at media depths of six feet.

Trickling Filters vs. Activated Sludge Systems

Most of the work on nitrification has been done with activated sludge systems; the theory of trickling filter operation is not as precise. However, biofilm models developed indicate that trickling filters can handle adverse loading and lower temperature conditions better than activated sludge systems. Trickling filters do not quickly show ammonia breakthrough with changes in loading rate.

Biodiscs

Biodiscs are attached growth systems which consist of a series of large-diameter plastic discs rotating on a horizontal shaft which runs across the top section of a trough-like reactor as shown in Fig. 12.8.

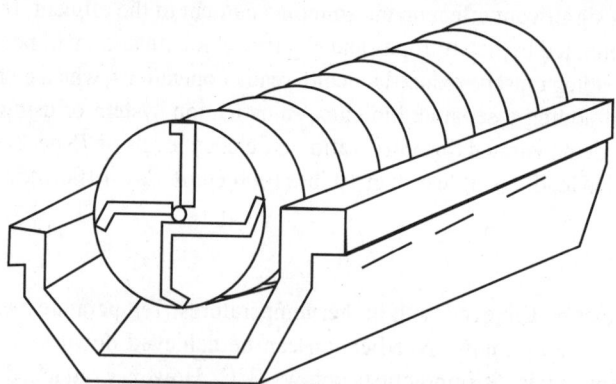

Fig. 12.8. Cutaway view of a biodisc unit.

Only 40 per cent of the discs are in the waste-water at any time. Biological films develop on the discs, which is the media. Such systems can be operated for carbonaceous oxidation or combined for nitrification-carbonaceous oxidation. In combined systems, organic oxidation occurs on the first discs and nitrification on the last. Nitrification does not begin until most of the BODs have been removed.

Problems have arisen in application due to diurnal variations or shock loads. These load changes create an increased organic loading and some or all of the nitrifying discs convert to heterotrophic colonies. Depending on the variations involved, this situation can be prevented by derating the disc loading or installing equalisation ahead of the system. This is the disadvantage which has limited acceptance of biodiscs for nitrification.

Temperature shows no effect on nitrification rates above 13°C. The discs are normally housed to reduce the effects of external temperatures, prevent algal growths and to keep out rain or hail which can shear growths off the discs.

Fluidised Beds

Work has been done regarding fluidised-bed systems. Their feasibility in large-scale applications has not as yet been accepted on a wide basis. However, they are discussed here since they have been shown to foster nitrification. Fluidised-bed systems are purported to combine the best features of activated sludge and trickling filter systems into one process. As in the trickling filters, the degradation organisms coat the media, which for fluidised beds is sand grains in suspension.

Fluidised beds can handle shock loads and toxics as can trickling filters, but there is minimal sloughing of growth. Secondary clarifiers are not needed. In fluidised beds, water is passed up through a bed of sand at a velocity high enough to impart motion to or fluidise the sand. An enormous surface area is obtained using sand—greater than 1000 sq. ft/cu. ft. of reactor.

Using pilot plants handling 80000 gpd domestic waste, 90 per cent nitrification and BODs have been obtained in 45 minutes with MLVSS concentrations between 8000 and 40,000 mg/l, in 5 per cent of the space required for a comparable conventional system. Oxygen depletion in areas of the reactor has been a significant problem in the combined nitrification organic removal application. In a study of the effectiveness of separate nitrification, 99 per cent nitrification was obtained with a raw waste containing 19.1 mg/l ammonia-nitrogen.

The nitrification performance of a fluidised bed decreased when the pH dropped below 6.0, even if sufficient alkalinity was available. Costs for operating and installing a fluidised-bed system are much less than costs for comparable conventional systems due to reduced space and retention time requirements.

ALTERNATIVE MEANS OF AMMONIA-NITROGEN CONTROL

An effluent ammonia-nitrogen problem or concern can be eliminated by several other less popular methods as described in this section.

Algal Ponds

These are shallow lagoons where intensive algal growths are cultivated under aerobic conditions. Algal ponds contain algae which reduce the nitrogen content of the effluent through photosynthesis as could occur in the receiving body of water if the effluent were discharged without such treatment. The algae is harvested from the pond along with the nitrogen which has been assimilated.

This form of nitrogen removal has its application in small cities with plenty of available land. It is a seasonal treatment method dependent on light and temperature. Ice and cold winter weather significantly reduce metabolic activities and ponds go anaerobic. During the spring which follows, hydrogen sulphide odours are released as ponds return to their aerobic states. As a result, algal ponds must be located as far as possible from existing and future residential communities. While construction and operating costs are low, the land costs and requirements can make ponds prohibitive.

Ion Exchange

Ion exchange involves the removal of ionic species (in this case, ammonium ions) from an aqueous phase. Nitrite, nitrate and organic-nitrogen cannot be removed by this method. For ammonium removal, effluent is passed through a column of clinoptilolite, a naturally occurring zeolite (resin) with a high selectivity for ammonium. Organics can foul the resin and must be removed prior to this treatment. When all available sites are taken up by the ammonia, breakthrough will occur and the resin must be regenerated. Ammonium removals between 90 to 97 per cent have been obtained by this method. There are some major disadvantages associated with ion exchange. These include the possibility of organic fouling and high regeneration costs. Additionally there is no ultimate disposal of the ammonium ion since it is contained in the waste brine from regeneration.

Land Disposal

Secondary effluent can be disposed of by spray irrigation providing less soil preparation and more crops for farmers, or as a soil conditioner for marginal or drastically disturbed land. Nitrogen removals between 30 per cent and 95 per cent can be obtained. Toxics management is important in land disposal due to the variability of the soil capacity to filter, buffer, absorb and chemically or biologically react with nitrogen. Land disposal can be a reliable method of nitrogen disposal if care is taken with application and utilisation. If the nitrogen loading rate is too high, plants and soil bacteria will not be able to assimilate all of it and an increase in groundwater nitrate will result. Other problems associated with land disposal include large land requirements, high management costs, climate dependence, potential health hazards from bacterial contamination in groundwater and accumulation of trace elements (toxicity).

Air Stripping

Air stripping has been used for ammonia removal from raw waste-water or digester supernatant. Ammonia-nitrogen is achieved by aeration of waste-water in a stripping tower. Effluent is pumped to the top of the tower and as it falls to the bottom, fans force air counter-current to the falling water. Ammonia is vapourised and discharged to the atmosphere. The ammonia must be in the molecular form of NH_3, not as an ammonium ion, NH_4^+. Up to 98 per cent removal can be obtained; however, residual levels of less than 5 ppm cannot be removed by this method. Nitrite, nitrate and organic-nitrogen levels are not affected by air stripping. Air temperature and effluent pH also affect the amount of ammonia which can be stripped. The pH must be raised to 10 or 11 with lime and the tower must be shut down during freezing weather. Problems in efficiency and scaling occur with cold weather operation. The ammonia removed from the effluent is discharged to the air. Rainfall washouts and nearby stormwater runoff can carry the ammonia to a receiving body of water. The net effect to the receiving water is not as bad though as it would be if the effluent were discharged directly.

Breakpoint Chlorination

In breakpoint chlorination or superchlorination, enough chlorine is added to oxidise ammonia-nitrogen to nitrogen gas. Approximately 10–20 mg/l of chlorine is needed to oxidise 1 mg/l of ammonia-nitrogen. With this method, ammonia-nitrogen levels can be brought down near zero. The effect of chlorine on organic-nitrogen is still uncertain. Nitrite and nitrate are not removed by this method and therefore breakpoint chlorination becomes a possible follow-up to incomplete biological nitrification where low or negligible levels of ammonia are required. Also, it should be noted that the chlorine is rarely added at the actual breakpoint.

Optimum breakpoint chlorination can be obtained at a pH of 10 and temperature of 30°C. Up to 90 per cent ammonia removal can be obtained in 4 to 60 hours. With chlorine gas, ammonia and to a lesser extent, organic-nitrogen, removal can be obtained. Results using chlorine dioxide are not very good for ammonia and are non-existent for organic-nitrogen.

The acidity produced by chlorination must be compensated for by lime or caustic soda addition, which increases the total dissolved solids of the effluent. Thus, breakpoint chlorination can be an expensive operation.

Biological nitrification can provide dependable ammonia removal in warm climates and warmer seasons of northern climates. If dependable nitrification is required in northern climates 365 days a year, it must be supplemented or replaced by physical-chemical treatment. Physical chemical treatments are not without drawbacks either. In many cases though, ammonia removal will be a seasonal requirement, lending itself to biological nitrification.

Nitrification may carry on to denitrification because nitrogen's nutrient effects are often harmful as well. The performance of denitrification systems are heavily depended upon the efficiency of nitrification.

Several different means of ammonia removal have been discussed. In selecting a method most suitable, the following factors should be considered:

1. Form and concentration of the influent nitrogen compounds.
2. Required effluent quality.
3. Other existing treatment processes.
4. Costs.
5. Degree of reliability required.
6. Flexibility of the system.

PHOSPHORUS

Phosphorus in waste-water may be present in three forms: orthophosphate, polyphosphate and organic phosphorus. Typically, the majority of phosphorus enters waste-waters from kitchen grinders, human wastes and inorganic phosphate compounds used in various household detergents. Phosphate control is of increasing concern because of its contribution to algae and aquatic growth and its interference with coagulation and lime-soda softening in concentrations as small as 0.2–0.4 mg/l.

Approximately 10 per cent of most phosphorus, which corresponds to the portion that is insoluble, is normally removed by primary settling. None of the phosphorus present in waste-water is gaseous at normal temperatures and pressures, so removal must be accomplished by precipitation. Chemicals used to form these insoluble precipitates are lime, alum, ferric chloride and sulphate. In some cases, polymers are added to lime and alum to enhance their flocculation characteristics.

Human wastes and kitchen wastes account for 30 per cent to 50 per cent of the phosphorus in domestic waste-water. Detergents containing phosphate builders account for the remaining 50 per cent to 70 per cent. Other phosphorus sources originate in industry where they are used to control corrosion and scaling. Also, discharges from potato processing plants, fertiliser wastes, animal feed lot wastes, dairy wastes, flour processing wastes and metal finishing wastes may contain high concentrations of phosphorus. Total phosphorus concentration in domestic raw waste-water (using 100 gallons per capita per day waste-water flow) is found to be about 10 mg/l. This figure can be used for rough design information when no phosphorus data are available.

The increased use of phosphates and the resulting discharge into receiving waters has been cited as being responsible for the stimulation of aquatic plant growth and for speeding up the eutrophication process in our lakes. As a result, more and more states have adopted effluent concentration limits ranging from 0.1 to 2.0 mg/l phosphorous with many limits established at 1.0 mg/l for 80 per cent to 95 per cent reduction. Therefore, there is a need for advanced waste-water treatment to remove phosphorus by chemical means since phosphorus removal obtainable by biological activity is limited. Ion exchange, reverse osmosis and recently, water hyacinths, contribute to phosphorus removal, but are more useful for the removal of nitrogen and dissolved inorganics.

Precipitation

During biological treatment, significant changes take place. As organic materials are decomposed, their phosphorus content is converted to orthophosphate. As a result, in a well-treated secondary effluent, a large fraction of the phosphorus is present as orthophosphate (PO_4), which is fortunate since it is the easiest form to precipitate. Materials found most practical for phosphorous precipitation are the ionic forms of aluminium, iron and calcium. Additions of polymers have also been used effectively in conjunction with lime [a form of calcium and alum (a form of aluminium)].

Aluminium Compounds—Chemical Reactions

The principle aluminium compound used in phosphorus precipitation is 'alum', a hydrated aluminium sulphate. Its reaction with PO_4 is as follows:

$$Al_2 (SO_4)_3 \cdot 14\ H_2O + 2\ PO_4 \rightarrow 2\ AlPO_4 + 3\ SO_4 + 14\ H_2O$$

As you can see, Al PO_4 precipitates out. From the previous relationship, it is determined that it takes 9.6 pounds of alum to remove one pound of phosphorus. But due to the solubility of Al PO_4, which is pH-dependent (optimum pH 5.5 to 6.5), bench, pilot and full-scale studies have shown that considerably higher than stoichiometric quantities of alum are necessary to meet phosphorus removal objectives as follows:

P reduction (%)	Alum: P weight ratio
75	13:1
85	16:1
95	22:1

The competing reaction, as follows, can partially account for the excess alum requirement:

$$Al_2 (SO)_4)_3 \cdot 14H_2O + 6HCO_3 \rightarrow 2Al(OH)_3 + 6CO_2 + 14H_2O + 3SO_4$$

Another form of aluminium which is used to precipitate PO_4 is sodium aluminate whose reaction with PO_4 is represented as follows:

$$Na_2O \cdot Al_2O_3 + 2PO_4 + 4H_2O \rightarrow 2AlPO_4 + 2NaOH + 6OH$$

In contrast to alum, the waste-water's pH is increased by the addition of sodium aluminate. Another possible source of aluminium is aluminium chloride, which is not as readily available as alum or sodium aluminate, but can be considered for phosphorus precipitation.

Dry Alum

Commercial dry alum, also known as 'filter alum', has the chemical formula $Al_2(SO_4)_3$. Alum is white to cream in colour and acidic in nature with a pH that varies between 3.0 and 3.5 in aqueous solutions having concentrations of 1 per cent to 10 per cent. Commercially available grades of dry alum include lump, ground, rice and powdered. Dry alum is not corrosive unless it absorbs moisture from the air as could be encountered in a humid atmosphere. Most municipal treatment plants use ground or rice alum because of their superior flow characteristics. Dry alum is generally stored in mild steel or concrete bins (with a 30-day supply), fed by conveyor into a dissolver and mixed to the proper dilution. Since alum in solution is corrosive, solution chambers should be constructed with a non-reactive material such as fiberglass or stainless steel. Solution flow from the dissolver to the point of application can be by gravity or pump.

Liquid Alum

Liquid alum is shipped in insulated tank cars or trucks. Liquid alum is heated in winter to prevent crystallisation, which can occur at $18°F$ with a Al_2O_3 strength of 8.3 per cent. Liquid alum is generally more economical than dry alum if the point of use is within 100 miles of the manufacturing plant.

Storing liquid alum is more difficult than dry alum since storage tanks, if outdoors, should be closed, heated and vented. Storage containers may be open if indoors. In any case, since liquid alum is corrosive, the storage tank must be constructed of an inert material. Feeding equipment can be similar to that of dry alum once it is mixed in the dissolver.

Dry Sodium Aluminate

Dry sodium aluminate, $Na_2Al_2O_4$, is non-corrosive with the pH of a 1 per cent solution being about 11.9. Requirements for $Na_2Al_2O_4$ are similar to those for dry alum. Precautionary measures to be taken are similar to those of strong alkalies. The main problem with sodium aluminate storage is it deteriorates upon exposure to air. Therefore, care must be taken to avoid the tearing of bags. Dry sodium aluminate is not available in bulk quantities. Dissolvers and feed equipment can be similar to those used for dry alum.

Liquid Sodium Aluminate

Liquid sodium aluminate is generally available in 30 gallon drums, tank trucks or tank cars. Liquid sodium aluminate is a strong alkali and should be handled with caution. Storage should be in shipping drums or in mild steel tanks. The storage containers should be heated in winter. Feeding equipment is similar to that used for alum.

Aluminium Chloride

Aluminium chloride, in most areas of the country, is not as readily available as the other compounds previously discussed. The majority of facilities using aluminium chloride are located near petrochemical refineries since these industries use three-fourths of the total production. Therefore, if a treatment facility is located near a refinery, a reliable economical source of aluminium chloride may be available. Solid aluminium chloride is off-white in colour and is derived from direct chlorination of scrap aluminium. This form is semi-pure anhydrous crystals. The most common form of liquid aluminium chloride contains 28 per cent aluminium chloride by weight. Shipping, storing, handling and feeding requirements of aluminium chloride are similar to those of alum.

Chemical Reaction of Iron Compounds

Both ferrous (Fe^{+2}) and ferric (Fe^{+3}) ions can be used in the precipitation of phosphate ions on a one-to-one mole ratio as follows:

$$Fe^{+3} \ PO_4^{-3} \rightarrow 4 \ FePO_4$$

The weight ratio of Fe^{+3} to P is 1.8:1 but in practice, just as in the case of aluminium, a larger amount of iron is required. The reaction with the ferrous ion is much more complicated and not fully understood. Ferrous sulphate; ferric sulphate, ferric chloride and ferrous chloride (pickle liquor) are the primary iron compounds used in phosphorus precipitation.

Ferric Chloride

Liquid ferric chloride is a staining, corrosive, dark brown oily liquid. The pH of a 1 per cent solution is 2. Ferric chloride solutions are normally shipped in 3000 to 4000-gallon truckload lots and in 4000 to 10,000-gallon carload lots. Shipping concentrations of ferric chloride vary since crystallisation temperatures increase as the concentration is increased. Storage tanks for ferrous chloride should have a free vent or vacuum relief valve and be made out of a non-corrodible material. Normally a 10-day to two-week supply should be kept on hand. Feeding equipment for ferric chloride can be similar to that used for alum except for the metering devices. Glass tube metering devices should not be used because of the tendency of ferric chloride to stain or deposit.

Ferrous Chloride (Waste Pickle Liquor)

$FeCl_2$ as a liquid is generally available in the form of waste pickle liquor from steel processing. Acidic in nature, $FeCl_2$ can vary from 1 per cent to 10 per cent in solution and usually averages about 1.5 per cent to 1.0 per cent in solution. Since ferrous chloride is not normally available on a continuous basis, storage and feeding equipment should be suitable for handling ferric chloride.

Ferric Sulphate

$Fe_2(SO_4) \cdot XH_2O$ is a dry partially hydrated product. Free acid of ferric sulphate is 2.5 per cent. Since ferric sulphate is actively corrosive in solution, it should be stored dry in shipping bags or in bulk in concrete or steel bins. Bin storage should be as tight as possible to avoid moisture absorption.

Feeding materials for transport of liquid ferrous chloride should be of a non-corrodible material, such as stainless steel, rubber, plastic, ceramic or lead. Dry feeding equipment is similar to that used for dry alum except the feeder should be of closed construction, thereby minimising water absorption.

Ferrous Sulphate

Copperas or ferrous sulphate $FeSO_4 \cdot 7H_2O$ is a by-product of pickling steel and is most common in dry form. When dissolved, ferrous sulphate is acidic. Ferrous sulphate is normally transported dry in bulk, bag or drum and is also available in bulk in a wet slate. Storage feeding and handling systems should be similar to those used for handling ferric sulphate. However, a general precaution should be taken against mixing ferrous sulphate with quicklime since mixing may produce high temperatures and the possibility of fire.

Lime Chemical Reactions

Calcium ions react with the phosphate ion to form hydroxyapatite as follows:

$$3HPO_4 + 5Ca + 4OH \rightarrow Ca_5(OH)(PO_4)_3 + 3H_2O$$

Lime dosage for phosphorus removal is generally not calculated since the lime dose is determined by other reactions that take place when the pH of the waste-water is raised.

Quicklime

CaO has a density of 55 to 75 lb/ft³ and is caustic (when in a slurry). A saturated lime solution has a pH of 12.4. Lime can be purchased bagged and in bulk. The CaO content of commercially available quicklime should not be used if it is less than 75 per cent because of excessive grit and difficulties in slaking.

Bulk lime should be stored in air-tight concrete bins having a 60° slope on the bin outlet. Lime feeding equipment is usually with a belt-type feeder emptying into a lime slaker. The slaker should be of the continuous type and should include one or more slaking compartments, a dilution compartment, a grit separation compartment and a continuous grit remover. A paste-type slaker should have a water-to-lime ratio of 2:1, an elevated temperature and a five-minute slaking time, whereas a detention-type slaker can operate with a water-to-lime ratio between 2.5:1 and 6:1 at moderate temperatures and a 10-minute slaking time.

The slaked lime is mixed in a holding tank and fed into the plant's system by a rotor-type feeder. A typical system is shown in Fig. 12.9.

Hydrated lime

Ca(OH)₂ is slaked lime and needs only enough water added to form milk or lime. The dust and slurry of hydrated lime is caustic in nature. Storage of hydrated lime is similar to that of quicklime except that bin agitation must be provided. Feeding equipment is usually gravimetric and dilution is not important, therefore, control of the amount of water used in the feeding operation is not considered necessary.

Polymers

Polymers are used in phosphorus control to enhance flocculation and settling. Characteristics of individual polymers vary widely and the manufacturer should be consulted for properties and availability. Polymers normally can be obtained in liquid or dry forms. Dry polymers are normally stored in bags and blended with water to obtain the recommended dilution for efficient action.

Liquid polymer systems differ from dry systems only in the equipment used to blend the polymer to the proper dilution. Liquid systems, in contrast to dry systems, require no ageing and simple dilution is the only requirement for feeding. Piping and accessories are normally stainless steel or plastic.

pH adjustment

Since the solubility of $AlPO_4$ and other phosphorus precipitating chemicals are pH-dependent, a discussion of two chemicals frequently used to raise the pH of waste-water is warranted.

Soda ash

Soda ash (Na_2CO_3) is available in two forms: light and dense soda ash. The pH of a 1 per cent solution of soda ash is 11.2. Soda ash is available in bulk, truck, box car and hopper car. Dense soda ash is generally used in municipal applications because of its superior handling characteristics. It has little dust, good flow characteristics and will not arch in the bin. It is relatively hard to dissolve, therefore, ample dissolver capacity must be provided.

Liquid caustic soda

Liquid caustic soda is shipped at two concentrations: 50 per cent and 73 per cent NaOH. The pH of a 1 per cent solution of caustic soda is 12.9. Shipment is normally by tank car or truck, which is transferred to storage and diluted as necessary for feeding. Liquid caustic soda crystallises at 53°F of when stored at 50 per cent strength. Therefore, storage tanks must be located indoors or provided with heating and suitable insulation if outdoors. If the NaOH is diluted to 20 per cent strength, its crystallisation temperature drops to –20°F. Caustic soda will tend to pick up iron when stored in steel vessels for extended periods, therefore, stainless steel, rubber, nickel alloys or plastics are normally used.

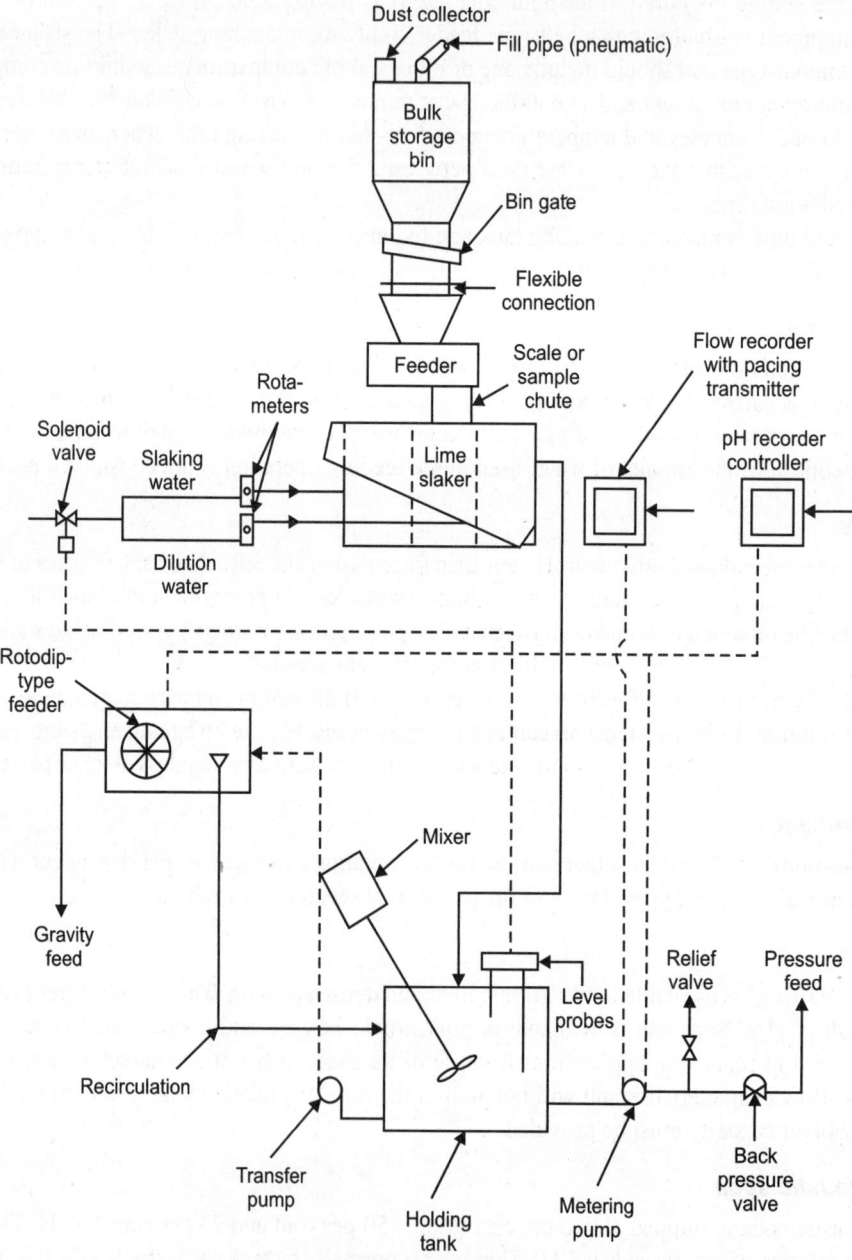

Fig. 12.9. Typical lime feed system.

Aquatics

The use of aquatic plants as a nutrient removal agent is not a new concept. The water hyacinth (*Eichlornia crassipes*) is the most researched plant in this area to date. The water hyacinth is a floating plant that covers vast areas of water surface and the plant interferes with navigation, causes flood control problems and restricts recreational activities like fishing, boating and water skiing.

Recently, 61 per cent removal of PO_4-P was accomplished by growing water hyacinths after a five-day detention time. However, most tests to date have shown that after 25 to 30 days of continuous operation, phosphate removal efficiency declined until only 5 per cent to 8 per cent removal efficiency was observed. Also, hyacinth removal efficiency is much less during the colder months. Research at Florida, found that the nutrient removal capacity of water hyacinths was directly related to pond surface area. In order to remove 44 per cent of the phosphorus, a one-million-gallon pond with 5.1 acres of water hyacinths was needed. For the small ponds used in the Gainesville test, influent valued ranged from 3.37 to 3.44 mg/l with effluent values of 1.82 to 1.86 mg/l. Effluent had a four-day detention time and the pond had a depth of one foot. The tests were also done with a deeper pond and it was found that the shallower pond had better PO_4-P removal efficiency. Nutrient uptake by the hyacinths was good during the area growth phase and vertical growth phase, but if the hyacinths' growth was not limited, lesser efficiencies in PO_4-P removal were noted when the pond became overgrown with hyacinths. This overgrowth could lead to anaerobic conditions in the pond. This experiment showed that nitrogen removal efficiency is better from water hyacinths than PO_4-P removal efficiency; however, some phosphorus removal is obtained.

SOLIDS REMOVAL

Removal of suspended solids and sometimes dissolved solids, may be necessary in advanced waste-water-treatment systems. The solids removal processes employed in advanced waste-water treatment are essentially the same as those used in the treatment of potable water, although application is made more difficult by the overall poorer quality of the waste-water.

Suspended Solids Removal

As an advanced treatment process, suspended-solids removal implies the removal of particles and flocs too small or too lightweight to be removed in gravity settling operations. These solids may be carried over from the secondary clarifier or from tertiary systems in which solids were precipitated.

Several methods are available for removing residual suspended solids from waste-water. Removal by centrifugation, air flotation, mechanical microscreening and granular-media filtration have all been used successfully. In current practice, granular-media filtration is the most commonly used process. Basically, the same principles that apply to filtration of particles from potable water apply to the removal of residual solids in waste-water. Differences in operational modes for application of these principles to waste-water filtration vs. potable water filtration may range from slight to drastic, however and the most commonly used waste-water filtration techniques are discussed below.

Sand filters have been used to polish effluents from septic tanks. Imhoff tanks and other anaerobic treatment units for decades. Because they are alternately dosed and allowed to dry, the term intermittent sand-filters has been applied to this type of unit. The process is essentially the slow sand filter. More recently, this type of filter has been applied to the effluent from oxidation ponds with considerable success. Effluent concentrations of less than 10 mg/l of BOD and suspended solids have been reported at filtering rates of 0.37 to 0.56 $m^3/m^2 \cdot d$. Filter runs in excess of 1 month are possible.

Use of intermittent sand filters in tandem with conventional secondary treatment has not been very successful. The nature of the solids from these processes results in rapid plugging at the sand surface, necessitating frequent cleaning and thus high maintenance costs. The use of intermittent filters for tertiary treatment is usually restricted to plants with small flows.

Granular-media filtration is usually the process of choice in larger secondary systems. Dual or multimedia beds prevent surface plugging problems and allow for longer filter runs. Loading rates depend on both the concentration and nature of solids in the waste-water. Filtering rates ranging from 12 to 30 m^3/m^2 day have been used with filter runs of up to 1 d.

Other recent innovations in filtration practices hold promise for advanced waste-water treatment. Moving bed filters have been developed which are continuously cleaned and the rate of cleaning can be adjusted to match the solids loading rate. Another modification called the pulsed-bed filter, uses compressed air to periodically break up the surface mat deposited on a thin bed of fine filter media. Only after a thick suspension of solids has accumulated on the bed, requiring frequent pulsing, is the filter backwashed. Both the moving bed and the pulsed-bed filters have the capability of filtering raw waste-water. A much higher percentage of solids can be removed by filtration than can be removed in primary settling. The filter effluent, containing lower levels of mostly dissolved organics, responds very well to conventional secondary treatment. The filtered solids can be thickened and treated by anaerobic digestion, with a resultant increase in overall methane production, a possible source of energy for use within the plant.

Dissolved Solids Removal

Both secondary treatment and nutrient removal decrease the dissolved-organic-solids content of waste-water. Neither process, however, completely removes all dissolved organic constituents and neither process removes significant amounts of inorganic dissolved solids. Further treatment will be required where substantial reductions in the total dissolved solids of waste-water must be made.

Ion exchange, microporous membrane filtration, adsorption and chemical oxidation can be used to decrease the dissolved solids content of water. These processes, were developed to prepare potable water from a poor-quality raw water. Their use can be adopted to advanced waste-water treatment if a high level of pre-treatment is provided. The removal of suspended solids is necessary prior to any of the processes. Removal of the dissolved organic material (by activated carbon adsorption) is necessary prior to microporous membrane filtration to prevent the larger organic molecules from plugging the micropores.

Advanced waste-water treatment for dissolved solids removal is complicated and expensive. Treatment of municipal waste-water by these processes can be justified only when reuse of the waste-water is anticipated.

Chapter 13

Root Zone Treatment Technology

INTRODUCTION

Rootzone treatment system (RZTS) are sealed filter beds consisting of a sand/gravel/soil system, occasionally with a cohesive element, planted with vegetation which can grow in wetlands. After removal of coarse and floating material the waste-water passes through the filter bed where biodegradation of the waste-water takes place.

Rootzone treatment systems (RZTS) use natural processes to effectively treat domestic and industrial effluents. This technology was developed during the seventies in Germany and since, then spread out all over the world. The process incorporates the self-regulating dynamics of a specially designed soil eco-system. RZTS are by now well known in temperate climates and are easy to operate on-site treatment facilities, which involve less installation, maintenance and operational costs than the conventional treatment methods. Also RZTS offer cost effective options for decentralisation of waste-water treatment.

The term Rootzone encompasses the interactions of various species of bacteria, fungi and other micro-organisms, the roots of wetland plants (helophytes), filter bed media, sun and, of course, water. The helophytes conduct oxygen through their stems into their root systems and create favourable conditions for the growth of aerobic micro-organisms. Since the process occurs in a deep filter bed, aerobic and anaerobic zones exist side by side. The waste-water enters the root zone horizontally or vertically and it passes through the system where the organic pollutants ' are decomposed biochemically by the micro-organisms present in the rhizosphere of the helophytes. The filter bed media are selected or mixed carefully to provide favourable conditions for both plants and bacterial growth and to ensure optimum hydraulic load.

The different types of Rootzone systems can be differentiated according to the following criteria:

Filter bed media   : Origin, composition, grain size distribution, hydraulic properties.
Flow direction   : Horizontal, vertical or hybrid forms.
Operation   : Continuous or intermittent feeding.
Planting   : Mono, or mixed cultures, single or multiple zone planting.
Position within the   As single or multilevel biological treatment stage (with or without
   treatment sequence :   pretreatment), as polishing stage after conventional treatment.

A classification scheme for waste-water treatment systems and the basic design of horizontal and vertical RZTS are shown in Figs 13.1 and 13.2.

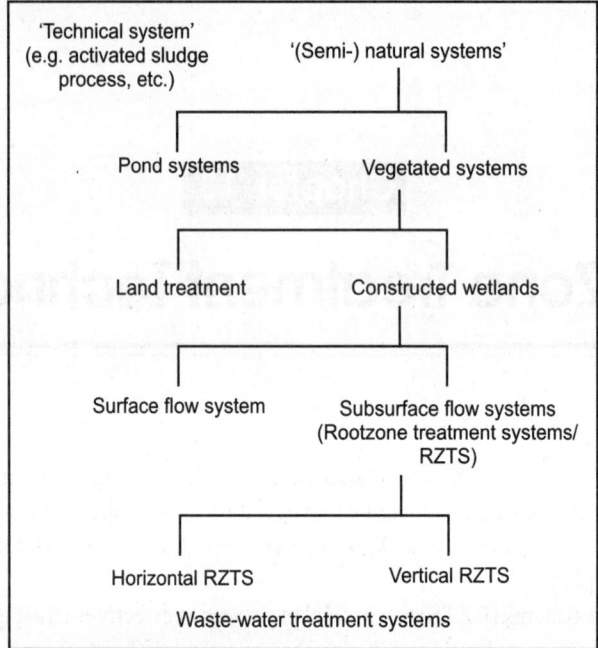

Fig.13.1. Classification scheme for waste-water treatment systems.

TREATMENT MECHANISM

The functional mechanisms in the soil matrix that are responsible for the mineralisation of biodegradable matter are characterised by complex physical, chemical and biological processes, which result from the combined effects of the filter bed material, wetland plants, micro-organisms and waste-water. The treatment processes are based essentially on the activity of micro-organisms present in the soil. Smaller the grain size of the filter material and consequently larger the internal surface of the filter bed higher would be the content of micro-organisms. Therefore the efficiency should be higher with finer bed material. This process, however is limited by the hydraulic properties of the filter bed; finer the bed material, lower the hydraulic load and higher the clogging tendency. The optimisation of the filter material in terms of hydraulic load and biodegradation intensity is therefore the most important factor in designing RZTS. The oxygen for microbial mineralisation of organic substances is supplied through the roots of the plants, atmospheric diffusion and in case of intermittent waste-water feeding through suction into the soil by the outflowing waste-water. The roots of the plants intensify the process of biodegradation also by creating an environment in the rhizosphere, which enhances the efficiency of micro-organisms and reduces the tendency of clogging of the pores of the bed material caused by an increase of biomass.

RZTS contain aerobic, anoxic and anaerobic zones. This, together with the effects of the rhizosphere causes the presence of a large number of different strains of micro-organisms and consequently a large variety of biochemical pathways are formed. This explains the high efficacy of biodegradation of substances that are difficult to treat. The filtration by percolation through the bed material is the reason for the very efficient reduction of pathogens, depending on the size of grain of the bed material and thickness of filter, thus making the treated effluent suitable for reuse.

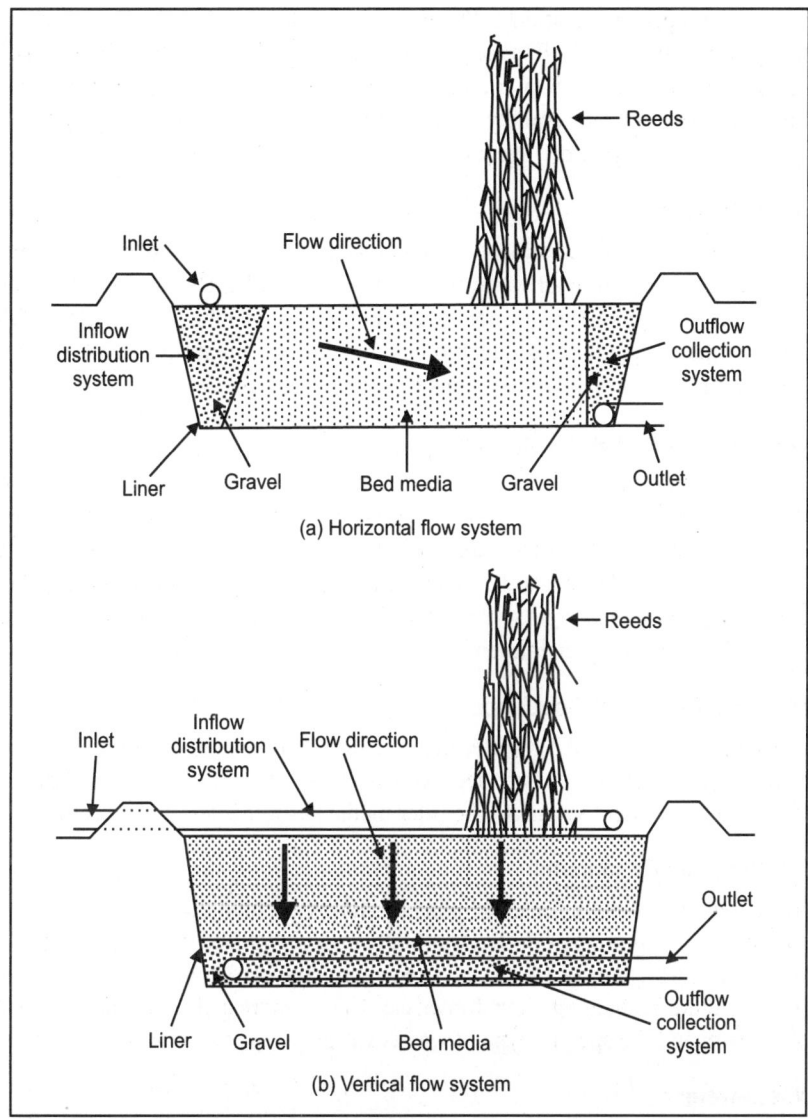

Fig. 13.2. Conceptual diagrams of Rootzone treatment systems

Conversion of nitrogen compounds (Nitrification/denitrification) occurs due to planned flow of waste-water through anaerobic and aerobic zones. Reduction of phosphorous depends on the availability of acceptors like iron compounds and the redox potential in the soil.

APPLICATIONS OF RZTS

Types of Waste-water

Domestic and industrial waste-water containing biodegradable matter is treatable with RZTS after sedimentation of particulate matter. Waste-water containing not easily biodegradable substances can be treated in RZTS by appropriate operation of the plant (e.g. prolonging the residence time of the waste-water or changing the conditions in the filter bed between aerobic and anaerobic environment). Stormwater can also be treated as well with RZTS after removal of settleable substances. RZTS have to be planned and designed in such a way that they comply with the environmental stipulations, standards and criteria, laid down by the concerned regulatory agencies.

Unit Size Range of RZTS

RZT Systems are very suitable for treatment of small quantities of waste-water (minimum 250 g BOD per day), since a homogeneous flow of waste-water through small RZTS can easily be maintained and complicated inflow distribution systems are not needed. For sewage treatment, up to 50 kg BOD per day is considered as a maximum quantity of effluent to be treated in one plant. In case of larger RZTS more complex waste-water inflow distribution systems have to be applied.

Decentralisation of Waste-water Treatment

The fact that the construction of small RZTS is easy makes it suitable for onsite treatment of domestic sewage avoiding long sewerage lines. Even treatment of the waste-water of a single house is possible. Thus small plots become usable for waste-water treatment, especially in situations where it is not possible to have a connection to a central waste-water plant such as hotels and tourist complexes, which are sited away from cities. In case of dispersed settlements, RZTS offer a feasible solution for decentralised sewage treatment after adequate pretreatment.

Position Within the Treatment Sequence

The potential design variations of RZTS offer the possibility to use them in the main treatment and in the polishing stage of the waste-water treatment sequence. Therefore RZTS can also be combined with high-tech and low-tech treatment systems. The filter media has to be designed accordingly.

Removal of Pathogenic Organisms

The high degree of removal of pathogens, including coliforms, protozoans and helminthes makes RZTS a feasible option for recycling of waste-water, especially in decentralised treatment systems. The treated effluent from properly designed RZTS can be used safely for gardening within domestic premises, toilet flushing, aquaculture, recharging groundwater, etc. The use of treated industrial waste-water depends on the type of industry and the specific type of waste-water.

Industrial Waste-water

For industrial waste-water the location and waste-water specific consideration have to be taken into account.

Limitations

RZTS are based on filtration mechanism, therefore, they are sensitive against clogging. Higher concentrations of settleable matter, oil and grease, especially mineral oil in the waste-water higher the tendency to surface clogging as well as internal clogging. Overloading of RZTS with organic matter can also cause clogging due to increased growth of bacteria within the filter matrix. These problems can be avoided by appropriate pretreatment of waste-water, design of the filter bed and proper operation of the system. RZTS are limited regarding the hydraulic loading, which depends on the design and quality of the filter material. This limitation has to be considered especially in case of irregular flow and in case of treatment of stormwater or polishing of already treated effluents. Heavy Metals from waste-water can accumulate in the RZTS depending on type of waste-water, bed material and species of plants. Heavy Metal containing waste-water should therefore be pretreated according to effluent limits given by the regulatory agencies. In very hot and dry climate the evaporation in RZTS can be so high that the waste-water is concentrated substantially. RZTS with smaller specific surface, for example vertical systems have to be used in cases where increase of TDS is undesirable. In case of waste-water containing toxic substances the same precautions are required as in other biological treatment systems (specific pretreatment).

PLANNING AND CONSTRUCTION OF RZTS

Site Selection

A distance of 5 to 20 m from the next residential building in case of domestic sewage is recommended, depending on the type of pretreatment, the size of the RZTS and the type of feeding of the plant beds. However in case of industrial effluents site-specific separation from residential areas is required. If properly designed and built, RZTS do not create any odour or nuisance in the vicinity. The location is to be selected in such a manner that drinking water sources are not impaired. The site must be safe from flooding.

The RZTS should be protected from unauthorised access. It must be possible to dispose of the treated effluent at the selected site as per standards set by the concerned regulatory agency. The following precautions shall be taken:

1. RZTS have to be marked clearly as waste-water treatment systems.
2. The site should be accessible for maintenance.
3. Natural slope should be used, to avoid the need for pumps.

The application of RZTS and the design depends also on the availability of suitable filling material for the filter beds. It is recommended to use locally available filter material to reduce the construction cost.

Pretreatment

RZTS require waste-water with low concentration of settleable and floating solids. Insufficient pretreatment can cause surface clogging in vertical systems and clogging in the infiltration area in horizontal systems. Sand settling devices, grease traps, gratings and sieves have to be used according to the characteristics of the raw waste-water. Industrial effluents have to be characterised fully before deciding upon adequacy and type of pretreatment.

In principle all proven pretreatment systems can be combined with RZTS. The following anaerobic pretreatment systems are especially suitable for small RZTS dealing with domestic sewage: (i) multi-compartment septic tank, (ii) imhoff tank, (iii) baffle reactor, and (iv) biofilm up-flow reactor.

In special cases the use of composting devices can be considered. If space is limited the use of trickling filters or other aerobic installations may become necessary in order to reduce the organic load of the waste-water. Industrial waste-water has to be assessed for the specific requirements of physico-chemical pretreatment. Specially designed Vertical RZTS can be used for pretreatment, and suitable measures should be taken for the disposal of primary sludge.

Filter Media

The filter media effective for the biological treatment must consist of sand/gravel, a carefully mixed soil or a comparable media. The design of the correct filter media according to the available material is the most important step in the design process. If the material is too coarse the waste-water will flow too fast; if it is too fine clogging and overflow will occur. Both cases cause poor treatment efficiency.

A remarkable reduction in permeability by deposition, sedimentation and bacterial growth has to be considered. High temperature as in tropical regions will influence this effect significantly.

Sand and gravel with rounded grains are ideal. Large and sharp edged particles can lead to damage of the liner. Media of relatively similar grain size, like river sand or sieved materials are best. The following filter bed parameter should be met:

Permeability (k_f)	:	$\approx 10^{-4} - 10^{-3}$ m/s for of domestic effluents,
		$\approx 10^{-6} - 10^{-3}$ m/s for industrial effluents.
Uniformity coefficient (U)	:	$d_{60}/d_{10} = \leq 5$; (ratio of grain sizes which contain 60 per cent and 10 per cent of the total weight)
Effective grain size (d_{10})		Should be ≥ 0.2 mm
Content of silt or bonded admixtures	:	Should be ≤ 5 per cent (if at all)

The grain size distribution is to be verified using soil analysis and a percolation test has to be performed before incorporation of the material in the beds.

The permeability of the media can be calculated by the following formula (after Hazen):

$k_{f[m/s]} = (d_{10})^2 : /100$; [$d_{10}$ in mm].

This gives a rough estimate; a safety factor of at least 5 should be chosen. In case of suitable locally available material of lower permeability the hydraulic and organic loading has to be adjusted accordingly. Rounded gravel/pebbles should be chosen, if available, for the infiltration area in horizontal RZTS and for the drainage area in both systems.

Depth of filter bed

The recommended depth of the filter media is:

For horizontal filters	:	50–100 cm
For vertical filters	:	60–120 cm

If a horizontal system is used, increasing the depth upto more than one meter is not useful, because of the limited root growth of the plants which is the only oxygenating factor. In vertical systems an increase in depth of more than 120 cm will enhance the treatment efficiency further, but this is normally limited due to increased cost of filter material.

Slope of Filter Bed

Both horizontal and vertical beds should have an even and flat surface to avoid the development of channels and pools and allow for evenly flooding in vertical systems. It also helps in horizontal systems to flood the surface at certain times to suppress weed growth.

Horizontal filter beds should have a defined infiltration area. This can be achieved by a reverse slope on the first part of the bed or a small earth wall transverse to the main flow direction at the end of the infiltration area. This helps also to keep sudden inflow peaks in the infiltration area. A slope on the base of the filter bed is normally not required. In special cases the bottom can have a slope up to 3 per cent.

Sealing of Filter Bed

RZTS have to be sealed with an impermeable layer at the bottom and the sides so that untreated or partly treated waste-water cannot infiltrate to the groundwater. Sealing is also required to recover the treated water for reuse and for compliance monitoring as per requirements of the regulatory agency. If the existing soil has a permeability coefficient $<10^{-8}$ m/s, no artificial sealing layer is necessary for sewage treatment applications. In this case a density test (after Procter) has to be performed. For industrial application, it is important to consider lining the system in all cases in order to prevent infiltration into the groundwater.

RZTS in soil with higher permeability require sealing of the bottom and sides. This can be achieved by:

1. Using concrete or plastic tank.
2. Providing plastic liner, UV resistant, if exposed to the sun, thickness ≥ 1 mm, root resistant, preferably from polyethylene or equivalent material. The liner has to be protected against damages caused by rocks of the existing soil and by sharp edged gravel of the drainage layer. Geotextiles may be used for prevention of such damages.
3. Providing clay sealing with a verified thickness of ≥ 30 cm. It has to be compacted properly.
4. Improvement of existing soil by admixture of bentonite or very fine clay (two layers of 20 cm each, mixed and compacted separately).

After finishing the sealing a leakage test should be carried out by filling the bed with water. If the loss is less than 2 mm overnight, the sealing is to be considered as satisfactory.

Dimensioning of the Filter Beds

For dimensioning RZTS for domestic waste-water the following inflow characteristics have to be considered:

1. BOD, settled, 27°C, 3 days, in grams/day.
2. Quantity of waste-water, in litres/day.

The BOD criteria (organic loading) ranges for dimensioning are:

For horizontal flow   10–30 g/BOD/m^2/day
For vertical flow   20–40 g/BOD/m^2/day

For industrial effluents specific recommendation according to type of industry are required. Future revisions of these guidelines will incorporate such recommendations. Parts near the lateral sides and the infiltration and drainage areas may not be included in the surface calculations.

The hydraulic load criteria range for dimensioning are:

For horizontal flow   40–100 L/m^2/day
For vertical flow   50–130 L/m^2/day

If the percolation is tested for the determination of the percolation crosssection and the bed geometry in case of Horizontal RZTS, a k_f - value reduced by a power of 10 should be applied. Hydraulic verification is indispensable. Hydraulic calculations are carried out according to the law of DARCY.

In fluid dynamics and hydrology, Darcy's law is a phenomenologically derived constitutive equation that describes the flow of a fluid through a porous medium.

Although Darcy's law (an expression of conservation of momentum) was originally determined experimentally by Henry Darcy, it has since been derived from the Navier-Stokes equations via homogenisation.

One application of Darcy's law is to water flow through an aquifer. Darcy's law along with the equation of conservation of mass are equivalent to the groundwater flow equation, one of the basic relationships of hydrogeology. Darcy's law is also used to describe oil, water, and gas flows through petroleum reservoirs.

Thus on applying Darcy's law:

$$Q = k_f \times i \times F$$

Q	[m³/s]	Flow
k_f	[m/s]	Permeability
i	[m/m]	Hydraulic gradient
F	[m²]	Effective cross section

The effective hydraulic load is highly dependent upon the characteristics of the bed media as well as the characteristics of the effluent. Therefore, these figures should be treated as indicative only. In case of combined systems or application of multi layer filter media, a verification of the applicability is to be provided. The climatic conditions of the location have to be taken into consideration for dimensioning RZTS. Considering the kinetics of enzymatic reactions, in tropical climate a higher efficiency of RZTS can be expected.

Construction Details

A freeboard of at least 20 cm (distance from bed surface to the upper edge of the lateral sealing) is to be provided. There should be free access to all operational points, like manholes, pumping stations, maintenance locations and sampling points. The access has to be constructed in a way, that crossing of the filter bed is avoided. RZTS should be designed in such a way that they are integrated into landscape as much as possible. Protective measures against the undesired water inflow are indispensable such as bunding all around.

Inlet and Outlet Constructions

Inlet structures must be so constructed that they distribute the incoming waste-water uniformly over the surface of the bed in case of Vertical RZTS, or across the infiltration cross-section in case of Horizontal RZTS, without leading to the formation of erosion furrows on the bed surface or to siltation or clogging of the filter media. Verification is to be carried out to prove the correct function of the inflow structures. The infiltration area has to be so calculated that overflow at normal operation is excluded. Hydraulic calculation of the infiltration section/area have to be done with a safety of one order of magnitude.

In horizontal filter beds the structures for distribution and collection of waste-water have to be designed and constructed in such a way that an even percolation of the whole bed matrix is achieved without short-circuiting. Typical inlet structures are based on gravity flow (weir constructions, leveled pipe outlets, dispersion systems through gravel layers) or on fluid dynamics (location and size of pipe orifices).

In vertical filter beds the distribution devices have to be designed and constructed for intermittent waste-water application and even feeding of the total bed surface. After each application the pipes of the inflow construction should run empty. This prevents bacterial growth and resulting clogging problems.

The outflow construction should have provision for adjusting the water level between the bottom of the bed and 10 cm above the surface of the bed. The construction of in- and outflow devices must allow for cleaning with mechanical or high pressure flushing tools. Outlet construction must allow for water sampling and examining as per the requirements of the concerned regulatory agencies.

Plantation

Selection of species

With the available experience the following list of species can be given:
1. *Phragmites australis* (reed).

2. *Phragmites karka* (reed).
3. *Arundo donax* (mediteranean reed).
4. *Typha latifolia* (cattail).
5. *Typha angustifolia* (cattail).
6. *Juncus (bulrush)*.
7. *Iris pseudacorus*.
8. *Schoenoplectus lacustris* (bulrush).

For horizontal RZTS in principle all helophytes can be used, which are deeprooted and oxygenate the rhizosphere through the roots. For vertical systems the plant selection is less critical, because the oxygen input is enhanced by the intermittent surface application.

Planting techniques

Planting of reeds can be done in the following way:

1. Reeds can be planted as rhizomes, seedlings or planted clumps.
2. Clumps can be planted during all seasons. ($2/m^2$)
3. Rhizomes grow best when planted in Pre-Monsoon. ($4-6/m^2$)
4. Seedlings should be planted in Pre-Monsoon ($3-5/m^2$)

Planting should be done from supporting boards to avoid compaction of the filter media. Initially the plants should be kept well watered, but not flooded. With well-developed shoots, the growth of weeds can be suppressed by periodical flooding. During the first growth period a sufficient supply of nutrients is required. If waste-water is used for initial watering, precautions like avoidance of stagnation have to be taken to inhibit the formation of H_2S within the filter bed.

OPERATION AND MAINTENANCE

Basic Considerations

Duration and degree of reduced treatment efficiency during the start-up period have to be stated in the technical report, which is part of the approval process. The operation of RZTS is expected to stabilise after 4–6 months after the plants are fully grown. A warranty for the planned function during the different seasons has to be provided by the supplier of the RZTS. RZTS require, servicing and maintenance. It is recommended to have a maintenance contract for functional control, system maintenance and care of the plants. Like other waste-water treatment plants RZTS need a comprehensive operation and maintenance manual for the operator, which covers all operational situations.

Operation Manual

The operation manual has to cover-among other things-the following subjects:

1. The type of waste-water application (continuous or intermittent flow) and the continuous or parallel operation of different filter beds.
2. Limits of organic and hydraulic loading.
3. Advice for the start-up phase.
4. Common problems and remedies.
5. Measures against unplanned overflow of the filter bed.
6. Measures against growth of weeds.

7. Measures to keep RZTS in stand-by condition for seasonal operation (e.g. hotels with seasonal operation).
8. Necessary measures for harvesting of plants, biomass removal, if required.
9. Operation during the different seasons of the year (monsoon, winter, summer).
10. Maintenance and removal of sludge or compost from pre-treatment devices.
11. Measures for renewal of clogged filter media.
12. Trouble shooting, correction of potential faults and nuisances.
13. Function control and maintenance of all technical parts of the plant, especially check of correct feeding and flow-through of waste-water, cleaning of feeding and drainage pipes, control of embankments and safety installations, check of surface and internal clogging, and control of vitality of the plants.
14. Instructions for self-control of the quality of final effluent according to the NOC.

Chapter 14

Membrane Technology

INTRODUCTION

A membrane is a layer of material which serves as a selective barrier between two phases and remains impermeable to specific particles, molecules, or substances when exposed to the action of driving force.

Membrane operation or membrane process is considered like a unit operation in chemical engineering.

In chemistry membrane operations use artificial membranes to separate mixtures. The membrane processes which are commonly used in water and waste-water are given below:

1. Pressure driven operations:
   (a) Microfiltration.
   (b) Nanofiltration.
   (c) Ultrafiltration.
   (d) Reverse osmosis.
   (e) Gas separation.
   (f) Pervaporation
2. Concentration driven operations:
   (a) Dialysis.
   (b) Osmosis.
3. Operations in electric potential gradient:
   (a) Electrodialysis.
   (b) Membrane electrolysis.
   (c) Electrophoresis
4. Operations in temperature gradient:
   (a) Membrane distillation.

PRESSURE DRIVEN OPERATIONS

The first commercially available membranes were designed as flat sheets rolled to form spiral wound membranes. These membranes could not tolerate solids and required high pressures to operate. The high operational cost of these membranes resulted in rare use and little municipal applications in the microfiltration mode. Spiral wound membranes are typically encountered in nanofiltration and reverse osmosis applications, and are commonly used for desalting brackish water and sea water for production of potable water.

Hollow fibre membranes were developed in the last decade as a means to address microfiltration needs while using low energy costs to operate. These membranes soon became an industry standard and a number of companies started manufacturing these high surface area membranes and applying them to the drinking water field (Fig. 14.1). Two types of pressure driven hollow fibre membranes are found:

1. Inside-out membranes, where the influent is fed inside the membrane's lumen and the clean water travels from the inside of the membrane to the outside.
2. Outside-in membranes where the influent is fed from the outside of the membrane and the clean water travels from the outside to the inside (lumen) of the membrane.

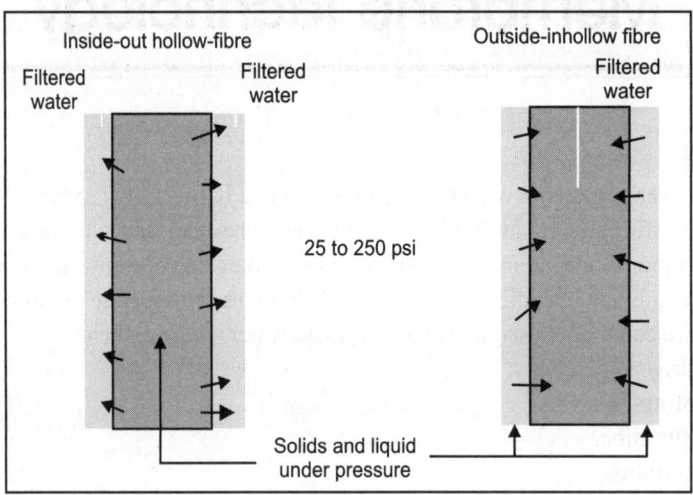

Fig. 14.1. Filtration modes—Hollow-fibre membranes.

All pressure driven, hollow-fibre membranes are installed within pressure vessels, necessary to apply the pressure for proper fluid transfer. Typical operational pressure for these membranes is 15 to 30 psi.

Microfiltration and Nanofiltration

Microfiltration is a filtration process which removes contaminants from a fluid (liquid and gas) by passage through a microporous membrane. A typical microfiltration membrane pore size range is 0.1 to 10 micrometres (μm). Microfiltration is fundamentally different from reverse osmosis and nanofiltration because those systems use pressure as a means of forcing water to go from low pressure to high pressure. Microfiltration can use a pressurised system but it does not need to include pressure.

Substances in the drinking water sources can be presented in the three main forms: suspended, colloidal and dissolved. The suspended matter which is commonly characterised by turbidity can be removed by most conventional treatment processes, the most familiar of which are chemically assisted coagulation followed by filtration or clarification and filtration. The coagulant dosage in this case is typically comparable to the turbidity level in the source.

The occurrence of *Cryptosporidium* and *Giardia* cysts and oocysts and other parasites in drinking water sources has uncovered a new field of application for membranes in the drinking water sector. The inefficiency of conventional filtration plants to filter and disinfect these pathogens from the potable water has forced engineers to look into new technologies. Membranes are the natural answer to solve their problem since these are absolute barriers to parasites which size exceed the membrane's pore size. Conventional treatment processes are also often not effective when colour and Total Organic Carbon (TOC) are present in higher levels in the feed water. As the colloidal and suspended portions of these components are relatively high, they are not readily removed by settling and gravity filtration.

Finally, high levels of iron and manganese in well waters have been difficult to treat with the conventional green sand approach and again, these have started to be good candidate plants for membrane technologies. Microfiltration and nanofiltration membranes are becoming more and more used in the drinking water field. For some applications, MF membranes are now seen as a proven technology. This includes *Cryptosporidium* and *Giardia* cysts and oocysts parasites removal and turbidity removal with microfiltration and colour and brackish water treatment with nanofiltration.

Advantages related to the use of membranes in drinking water treatment are: low energy requirements, absolute barrier effect to micro-organisms, lower chlorine requirement for disinfection, low chemical (if any) use, smaller footprint. The type of membrane used also influences some specific advantages. This chapter will present typical applications of both types of membranes in the drinking water field.

1. Removal of turbidity and parasites by direct microfiltration—disinfection MF.
2. Removal of Fe and Mn by combining oxidation with microfiltration.
3. Removal of colour and TOC by combining enhanced coagulation with microfiltration.
4. Removal of colour and TOC by nanofiltration.

Membrane filtration operates on the principle of particle separation based on a pore size and pore size distribution. Microfiltration membranes have pore sizes that vary from 0.075 micron to 3 microns. Depending on the membrane selected, it will allow to separate suspended solids over 0.45 microns, bacteria, cysts and many other parasites which diameter are larger than the larger pore size of the membrane. Nanofiltration membranes have pore sizes ranging from 0.005 microns to 0.001 microns and with such a small pore size are able to remove large molecular weight molecules such as humic acids and certain salts. This allows for production of a parasite and solids free water without the need of chemicals. Membranes are made of a number of materials, with ceramic, polymers and sintered metals being the most common types of membranes. Whereas ceramic and sintered metals are normally found in industrial applications, polymeric membranes are becoming a common tool for drinking water treatment and municipal applications.

Membranes require transmembrane pressure to drive the clean water through the membrane, leaving behind the concentrate containing the separated particles and solids. The transmembrane pressure required to operate membrane plants can be induced by pressure or by vacuum.

Similarly, there are a number of filtration pathways which are commonly found in membranes: Dead-end filtration, where the filtrate forms a cake as the filter gets plugged, cross flow filtration where the filtrate is moved away from the membrane, this avoiding rapid filter plugging and osmosis where the water is filtered through a semipermeable membrane. This section will focus on crossflow filtration membranes (Fig. 14.2).

Vacuum Driven Hollow Fibre Membrane—The ZeeWeed™ Membrane

The ZeeWeed™ based drinking water process is a revolutionary low energy membrane process that consists of outside-in hollow-fibre microfiltration modules immersed in raw feed-water. This microfilter has a 0.085 micron nominal and a 0.2 micron absolute pore size, ensuring that no particulate matter exceeding 0.2 microns will escape to the treated water stream.

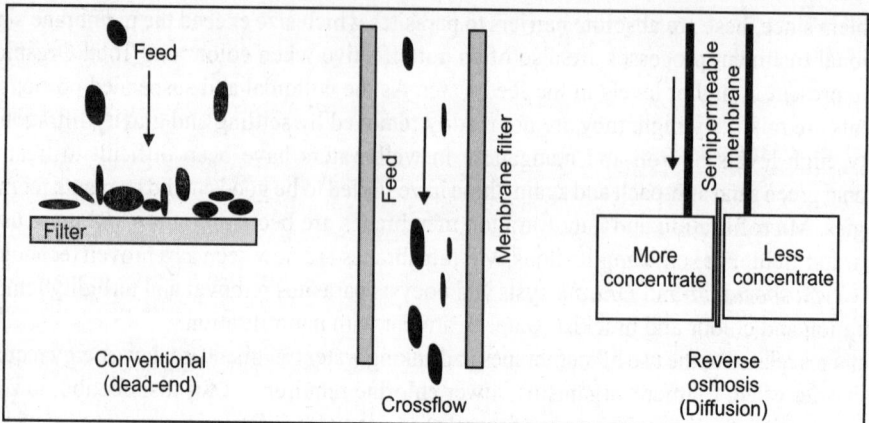

Fig. 14.2. Modes of filtration.

The membranes operate under a small suction created within the hollow fibres by a permeate pump. The treated water passes through the membrane, enters the hollow fibres and is pumped out to distribution by the permeate pumps. Air flow is introduced at the bottom of the membrane module to create a turbulence which scrubs and cleans the outside of the membrane fibres allowing them to function at a high flux rate. This air will also oxidise iron and other organic compounds, generating a better quality water than provided by microfiltration alone (Fig. 14.3).

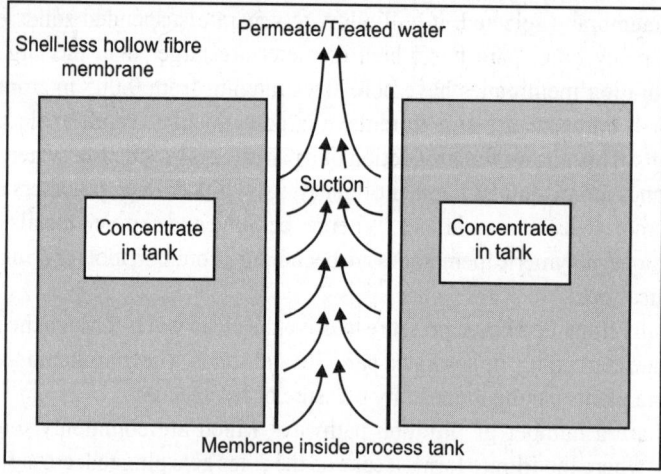

Fig. 14.3. Operational concept of an outside-in, Immersed, shell-less membrane.

Being an outside-in hollow fibre membrane, the plant does not need pretreatment, even if the feed water contains clays and fine particles. Therefore, in a single step, it replaces the coagulation, flocculation, clarification and sand filtration steps of conventional plants, but also eliminates the pretreatment required by spiral and inside-out membranes. A plant of this type operates with a process tank holding a set of immersed membranes. The water flows through the membranes and the permeate is pumped out. The air required to keep the membrane clean is generated by an air blower. The plant is easy to operate but also easy to assemble into small containerised plants which can be installed in small to large communities. The plant's process flow diagram is shown in Fig. 14.4.

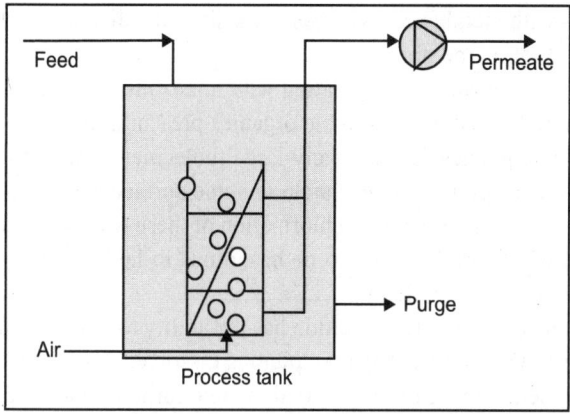

Fig. 14.4. Process flow diagram (PFD) of an immersed membrane microfilter.

Drinking water applications

Treatment with microfiltration membranes

Surface water treatment—disinfection by direct microfiltration: The use of a 0.2 microns microfiltration membrane in a drinking water filtration plant allows to address, in a single step, some of the most discussed current problems with current technologies:

1. The removal of *Giardia cysts, Cryptosporidium oocysts*, coliforms, and other parasites and suspended solids.
2. The reduction of viruses.
3. The reduction in use of disinfection chemicals.
4. The reduction of settling chemicals.
5. The reduction of sludge for disposal.

This type of treatment is achieved with any of the microfiltration membranes described above. Typical results obtained in drinking surface water treatment using microfiltration are presented in Table 14.1.

Table 14.1. Surface water treatment data—direct MF with an immersed membrane.

Feed water element	Treated water quality
Giardia and *Cryptosporidium*	Non-detectable >6 log removal
Coliforms	< 10 cfu/100 mL
Suspended solids	Non-detectable
Particle counts	< 3 particles/mL
Turbidity	< 0.1 NTU

Cryptosporidium and *Giardia* cysts are currently a big problem in unprotected surface water reservoirs. These cysts and oocysts are found as a result of contamination by human sewage but also by natural living organisms that defecate into the water. These parasites are just two amongst many make clear waters unhealthy to drink. It appears that everyday the WHO is finding more water parasites that are threatening human life.

The removal of cysts with membranes is an easy task since the diameter of these are larger than the diameter of most microfiltration membranes.

Surface water treatment—enhanced coagulation with microfiltration: Many surface drinking water supplies are highly coloured. The bulk of soluble organics present in natural water supplies consist of humic materials. These compounds are relatively large molecular weight polar organic compounds, which attribute the yellow to brown colour visible in some surface supplies. While these substances themselves do not cause any health concerns, chlorination of these waters can result in the formation of trihalomethanes (THM) which are believed to be hazardous to health, and which are coming under increasingly stringent government guidelines.

When combined to coagulation, microfiltration has the ability to remove colour and organic carbon from water sources. This is accomplished by precipitating dissolved organics into microflocs which can then be separated by the membrane. Colour and total organic carbon (TOC) are high in certain lake and river water supplies.

Microfiltration alone does not remove colour or TOC from the water. However, when combined with coagulation, these can be effectively removed, thus combining the absolute barrier advantage of MF with coagulation processes.

This unique process for colour, TOC and THM precursor removal has been developed using ZENON's immersed microfiltration membrane technology ZeeWeed®. The ability to build high solids levels in the process tank allows, by a combined mechanism of coagulation, co-precipitation and adsorption onto solids, to achieve high levels of TOC removal with lower dosages of coagulants. Two coagulants can be used: alum or iron chloride.

Depending on the water's chemistry, higher levels of removal can be obtained with higher dosages of coagulants and with adjusting the water's pH. Removals as high as 95 per cent colour removal and 85 per cent TOC removal are attainable with an optimised process. Process optimisation often requires pH adjustment which translates in the use of more chemicals and can be more difficult to operate in small plants. Vander-Venter has combined the used of immersed microfiltration with coagulant and powder activated carbon as a means to effectively remove natural organic matter (NOM) from surface water. Although this process is more difficult to operate, it significantly enhances the quality of the finished water with little chemical consumption.

Groundwater treatment by microfiltration: Well water often contain iron and manganese which need to be removed before human consumption. Many small communities rely on communal groundwater supplies, and require systems which assure removal of metals, turbidity, hydrogen sulphide and micro-organisms, while minimising chemical use and sludge production. Wells with high levels of iron and manganese are common in certain parts of the world, depending on the geological formation. Conventional technologies such as green sand and oxidation/settling are effective at low to medium concentrations. When well waters contain iron in excess of 5 mg/L and Manganese in excess of 1 mg/L, conventional technologies are no longer efficient due to filter blinding caused by the precipitated iron and iron bacteria films.

Furthermore, many wells under the influence of surface waters also contain micro-organisms, cysts and oocysts that need to be effectively removed for safe drinking water consumption. Deep wells also often contain H_2S and organics which also need to be removed, often resulting in a more complex treatment plant than required by these clear waters. The ZeeWeed membrane, due to its design features solves many of these problems without the addition of unnecessary steps (Table 14.2).

Table 14.2. Mechanisms for groundwater contaminant removal.

Contaminant removal	Removal mechanism.
Fe	Air oxidation
Mn	In-line oxidant admixing
Giardia, Cryptosporidium	Direct microfiltration
Turbidity, micro-organisms	Direct microfiltration
H_2S	Air scouring

The process flow diagram for the outside-in immersed membrane process for treatment of a complex groundwater is given in Fig. 14.5.

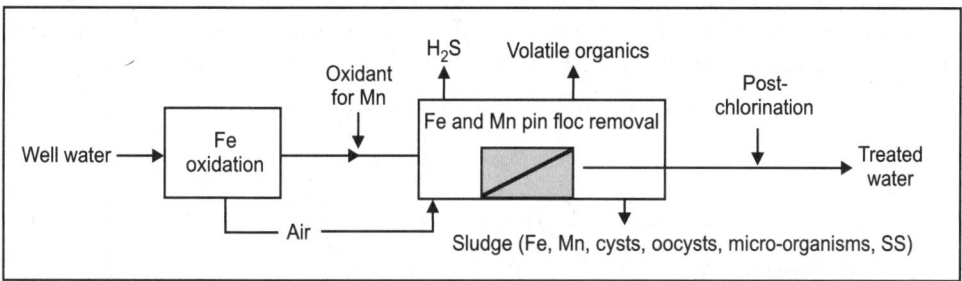

Fig. 14.5. Typical ZeeWeed treatment plant for a complex groundwater

Application of nanofiltration membranes for drinking water treatment

Nanofiltration separation incorporates a membrane with pore sizes and operating pressures, between the ultrafiltration and reverse osmosis membranes. They are typically operated at pressures in the range of 70 to 200 psi. Nanofiltration membranes prevent the passage of only a portion of the total dissolved solids (TDS) (primarily the divalent ions), and they remove most dissolved organic matter occurring in natural waters. Nanofiltration membranes are commonly used in the Municipal field for:

1. Desalting of Brackish waters.
2. Removal of organics and THM precursors from surface waters.

Nanofiltration membranes have a smaller pore size and can therefore remove organics as well as medium to large molecules from waters without the need of chemicals. The cost to pay for having smaller pores is the need of higher pressure to drive the clean water through the membrane this translates in higher energy needs.

Tighter porosity nanofiltration membranes also have the ability to remove a small percentage of salts from water and thus are used to desalt brackish waters. This is commonly seen in Florida, USA where the water's Total Dissolved Solids is too high for human consumption but low enough not to generate the high osmotic pressures requiring treatment by reverse osmosis. Desalting by nanofiltration is of little need in South America and will not be further discussed in this chapter.

Removal of colour and TOC by nanofiltration: Nanofiltration is commercially applied for treatment of coloured brackish waters, however systems are still at pilot or demonstration scale for applications using surface waters, which are typically variable in quality and turbidity. Nanofiltration membranes are commercially available, and improvements in membrane configuration and system design have recently taken place which will significantly improve it is cost-competitiveness.

Ultrafiltration

Ultrafiltration (UF) is a variety of membrane filtration in which hydrostatic pressure forces a liquid against a semipermeable membrane. Suspended solids and solutes of high molecular weight are retained, while water and low molecular weight solutes pass through the membrane. This separation process is used in industry and research for purifying and concentrating macromolecular (10^3–10^6 Da) solutions, especially protein solutions. Ultrafiltration is not fundamentally different from microfiltration, nanofiltration or gas separation, except in terms of the size of the molecules it retains. Ultrafiltration is applied in cross-flow or dead-end mode and separation in ultrafiltration undergoes concentration polarisation.

Ultrafiltration systems eliminate the need for clarifiers and multimedia filters for waste streams to meet critical discharge criteria or to be further processed by waste-water recovery systems for water recovery. Efficient ultrafiltration systems utilise membranes which can be submerged, back-flushable, air scoured, spiral wound UF/MF membrane that offers superior performance for the clarification of waste-water and process water. A simple diagram of ultrafiltration process is shown in Fig. 14.6.

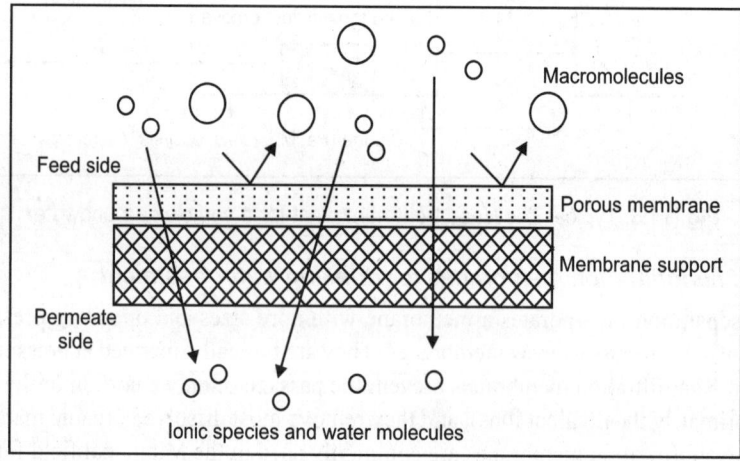

Fig. 14.6. Ultrafiltration process.

Membrane geometries

Spiral wound module: Consists of large consecutive layers of membrane and support material rolled up around a tube. Maximises surface area. Less expensive, however, more sensitive to pollution.

Tubular membrane: The feed solution flows through the membrane core and the permeate is collected in the tubular housing. Generally used for viscous or bad quality fluids. System is not very compact and has a high cost per m² installed

Hollow fibre membrane: The modules contain several small (0.6 to 2 mm diameter) tubes or fibres. The feed solution flows through the open cores of the fibres and the permeate is collected in the cartridge area surrounding the fibres. The filtration can be carried out either 'inside-out' or 'outside-in'.

Ultrafiltration module configurations

Pressurised system or pressure-vessel configuration: TMP (transmembrane pressure) is generated in the feed by a pump, while the permeate stays at atmospheric pressure. Pressure-vessels are generally standardised, allowing the design of membrane systems to proceed independently of the characteristics of specific membrane elements.

Immersed system. Membranes are suspended in basins containing the feed and open to the atmosphere. Pressure on the influent side is limited to the pressure provided by the feed column. TMP is generated by a pump that develops suction on the permeate side. Ultrafiltration, like other filtration methods can be run as a continuous or batch process

Applications

1. Dialysis and other blood treatments.
2. Concentration of milk before making cheese.
3. Fractionation of proteins.
4. Clarification of fruit juice.
5. Recovery of vaccines and antibiotics from fermentation broth.
6. Laboratory grade water purification.
7. Waste-water treatment.
8. Drinking water disinfection (including removal of viruses).
9. Removal of endocrines and pesticides combined with suspended activated carbon pretreatment.

Reverse Osmosis

Reverse osmosis (RO) is a filtration method that removes many types of large molecules and ions from solutions by applying pressure to the solution when it is on one side of a selective membrane. The result is that the solute is retained on the pressurised side of the membrane and the pure solvent is allowed to pass to the other side. To be 'selective', this membrane should not allow large molecules or ions through the pores (holes), but should allow smaller components of the solution (such as the solvent) to pass freely. Reverse osmosis is most commonly known for its use in drinking water purification from seawater, removing the salt and other substances from the water molecules. This is the reverse of the normal osmosis process, in which the solvent naturally moves from an area of low solute concentration, through a membrane, to an area of high solute concentration. The movement of a pure solvent to equalise solute concentrations on each side of a membrane generates a pressure and this is the 'osmotic pressure'. Applying an external pressure to reverse the natural flow of pure solvent, thus, is reverse osmosis. The process is similar to membrane filtration. However, there are key differences between reverse osmosis and filtration. The predominant removal mechanism in membrane filtration is straining, or size exclusion, so the process can theoretically achieve perfect exclusion of particles regardless of operational parameters such as influent pressure and concentration. Reverse osmosis, however, involves a diffusive mechanism so that separation efficiency is dependent on solute concentration, pressure, and water flux rate.

Process of reverse osmosis

Formally, reverse osmosis is the process of forcing a solvent from a region of high solute concentration through a semipermeable membrane to a region of low solute concentration by applying a pressure in excess of the osmotic pressure.

The membranes used for reverse osmosis have a dense barrier layer in the polymer matrix where most separation occurs. In most cases, the membrane is designed to allow only water to pass through this dense layer, while preventing the passage of solutes (such as salt ions). This process requires that a high pressure be exerted on the high concentration side of the membrane, usually 2–17 bar (30–250 psi) for fresh and brackish water, and 40–70 bar (600–1000 psi) for seawater, which has around 24 bar (350 psi) natural osmotic pressure that must be overcome. This process is best known for its use in desalination (removing the salt from sea water to get fresh water), but since the early 1970s it has also been used to purify fresh water for medical, industrial, and domestic applications. Osmosis describes how solvent moves between two solutions separated by a semipermeable membrane to reduce concentration differences between the solutions. When two solutions with different concentrations of a solute are mixed, the total amount of solutes in the two solutions will be equally distributed in the total amount of solvent from the two solutions. Instead of mixing the two solutions together, they can be put in two compartments where they are separated from each other by a semipermeable membrane. The semipermeable membrane does not allow the solutes to move from one compartment to the other, but allows the solvent to move. Since equilibrium cannot be achieved by the movement of solutes from the compartment with high solute concentration to the one with low solute concentration, it is instead achieved by the movement of the solvent from areas of low solute concentration to areas of high solute concentration. When the solvent moves away from low concentration areas, it causes these areas to become more concentrated. On the other side, when the solvent moves into areas of high concentration, solute concentration will decrease. This process is termed osmosis. The tendency for solvent to flow through the membrane can be expressed as 'osmotic pressure', since it is analogous to flow caused by a pressure differential. Osmosis is an example of diffusion. In reverse osmosis, in a similar setup as that in osmosis, pressure is applied to the compartment with high concentration. In this case, there are two forces influencing the movement of water: the pressure caused by the difference in solute concentration between the two compartments (the osmotic pressure) and the externally applied pressure.

Applications

Drinking water purification: Around the world, household drinking water purification systems, including a reverse osmosis step, are commonly used for improving water for drinking and cooking.

Water and waste-water purification: Rain water collected from storm drains is purified with reverse osmosis water processors and used for landscape irrigation and industrial cooling in Los Angeles and other cities, as a solution to the problem of water shortages. In industry, reverse osmosis removes minerals from boiler water at power plants. The water is boiled and condensed repeatedly. It must be as pure as possible so that it does not leave deposits on the machinery or cause corrosion. The deposits inside or outside the boiler tubes may result in under-performance of the boiler, bringing down its efficiency and resulting in poor steam production, hence poor power production at turbine. A simple reverse osmosis process is shown in Fig. 14.7.

Food industry: In addition to desalination, reverse osmosis is a more economical operation for concentrating food liquids (such as fruit juices) than conventional heat-treatment processes. Research has been done on concentration of orange juice and tomato juice. Its advantages include a lower operating cost and the ability to avoid heat-treatment processes, which makes it suitable for heat-sensitive substances like the protein and enzymes found in most food products.

Reverse osmosis is extensively used in the dairy industry for the production of whey protein powders and for the concentration of milk to reduce shipping costs. In whey applications, the whey (liquid remaining after cheese manufacture) is concentrated with RO from 6 per cent total solids to 10–20 per cent total solids before UF (ultrafiltration) processing. The UF retentate can then be used to make various whey powders, including whey protein isolate used in body-building formulations. Additionally, the UF permeate, which contains lactose, is concentrated by RO from 5 per cent total solids to 18–22 per cent total solids to reduce crystallisation and drying costs of the lactose powder.

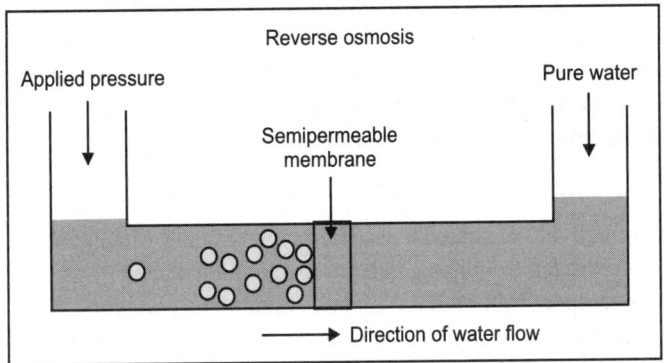

Fig. 14.7. Reverse osmosis process.

Hydrogen production: For small scale production of hydrogen, reverse osmosis is sometimes used to prevent formation of minerals on the surface of electrodes.

Desalination

Areas that have either no or limited surface water or groundwater may choose to desalinate seawater or brackish water to obtain drinking water. Reverse osmosis is the most common method of desalination, although 85 per cent of desalinated water is produced in multistage flash plants.

Large reverse osmosis and multistage flash desalination plants are used in the Middle East, especially Saudi Arabia. The energy requirements of the plants are large, but electricity can be produced relatively cheaply with the abundant oil reserves in the region. The desalination plants are often located adjacent to the power plants, which reduces energy losses in transmission and allows waste heat to be used in the desalination process of multistage flash plants, reducing the amount of energy needed to desalinate the water and providing cooling for the power plant. Sea water reverse osmosis (SWRO) is a reverse osmosis desalination membrane process that has been commercially used since the early 1970s.

Reverse osmosis — pros and cons

The semipermeable membrane used in reverse osmosis contains tiny pores through which water can flow. The small pores of this membrane are restrictive to such organic compounds as salt and other natural minerals, which generally have a larger molecular composition than water. These pores are also restrictive to bacteria and disease-causing pathogens. Thus, reverse osmosis is incredibly effective at desalinating water and providing mineral-free water for use in photo or print shops. It is also effective at providing pathogen-free water. In areas not receiving municipally treated water or at particular risk of waterborne diseases, reverse osmosis is an ideal process of contaminant removal.

The reverse osmosis process contains several downsides which make it an inefficient and ineffective means of purifying drinking water. The small pores in the membrane block particles of large molecular structure like salt, but more dangerous chemicals like pesticides, herbicides, and chlorine are molecularly smaller than water. These chemicals can freely pass through the porous membrane. For this reason, a carbon filter must be used as a complimentary measure to provide safe drinking water from the reverse osmosis process. Such chemicals are the major contaminants of drinking water after municipal treatment.

Another downside to reverse osmosis as a method of purifying drinking water is the removal of healthy, naturally occurring minerals in water. The membrane of a reverse osmosis system is impermeable to natural trace minerals. These minerals not only provide a good taste to water, but they also serve a vital function in the body's system. Water, when stripped of these trace minerals, can actually be unhealthy for the body. Reverse osmosis also wastes a large portion of the water that runs through its system. It generally wastes two to three gallons of water for every gallon of purified water it produces. Reverse osmosis is also an incredibly slow process when compared to other water treatment alternatives.

Gas Separation

Gas mixtures can be effectively separated by synthetic membranes. Membranes are employed in:
1. Separation of hydrogen from gases like nitrogen and methane.
2. Recovery of hydrogen from product streams of ammonia plants.
3. Recovery of hydrogen in oil refinery processes.
4. Separation of methane from biogas.
5. Enrichment of air by oxygen for medical or metallurgical purposes.
6. Enrichment of ullage by nitrogen in inerting systems designed to prevent fuel tank explosions.
7. Removal of water vapour from natural gas.
8. Removal of CO_2 from natural gas.
9. Removal of H_2S from natural gas.
10. Removal of volatile organic liquids (VOL) from air of exhaust streams.
11. Desiccation.

Usually nonporous polymeric membranes are used. There, vapours and gases are separated due to their different solubility and diffusivity in polymers. Polymers in glassy state, generally more effective for separation, predominantly differentiate in diffusivity. Small molecules of penetrants move among polymer chains according to the formation of local gaps by thermal motion of polymer segments. Free volume of the polymer, its distribution and local changes of distribution are of the utmost importance. Then diffusivity of a penetrant depends mainly on the size of its molecule.

Porous membranes can also be utilised for the gas separation. The pore diameter must be smaller than the mean free path of gas molecules. Under normal condition (100 kPa, 300 K) it is about 50 nm. Then the gas flux through the pore is proportional to molecules velocity, i.e. inversely proportional to square root of the molecule mass. It is known as Knudsen diffusion. Gas flux through a porous membrane is much higher than through nonporous one –3 to 5 orders of magnitude.

Separation efficiency is moderate—hydrogen passes four times faster than oxygen. Porous polymeric or ceramic membranes for ultrafiltration serve the purpose. Note, in case the pores are larger than the limit then viscous flow occurs, hence no separation. In special cases other materials can be utilised. Palladium membrane permits transport solely of hydrogen.

Pervaporation

As a separation technique, pervaporation occupies a special niche in the chemical industry—it is the only membrane process primarily used to purify chemicals. Currently, about one hundred pervaporation units are operating worldwide, most of them dehydrating solvents, such as ethanol and isopropanol. Now that pervaporation has been proven in these end-of-pipe applications, attention is turning to separations-closer to the chemical reaction step—more critical to production and promising much greater benefits. This shift in focus is accelerating the development of new, more robust membranes with better performance that can be used at higher temperatures. Over the next few years, pervaporation will be used increasingly to enhance reactor performance, either by purifying feeds or separating reaction products. Because pervaporation is a membrane process, these separations can be integrated with the reaction step, promising quantum improvements in reaction efficiencies, yields and process economics.

In pressure-driven membrane processes, such as ultra, micro and nanofiltration, the bulk component is purified by passing it through a porous membrane that holds back the minor component. The membrane acts rather like a filter or strainer. Reverse osmosis (RO) is similar, but uses nonporous membranes. In this case, the major component selectively permeates the membrane by preferential absorption, diffusion and desorption. The solute or minor component is held back. Pervaporation and vapour permeation processes are used in the reverse situation, i.e. when the membrane is preferentially permeated by the minor component. The bulk fluid is held back by the membrane. To get a pure product stream in this situation, almost all of the permeating component has to pass through the membrane, so pervaporation and vapour permeation resort to application of vacuum to the permeate side of the membrane. Very high pressure ratios can be achieved, so the minor component can be almost completely removed without excessive pressure difference across the membrane. Undue mechanical stresses on the membrane and equipment are avoided. Because substances that permeate nonporous membranes are reasonably volatile, application of vacuum always causes the permeate to be desorbed from the membrane in the vapour state. Hence, the term pervaporation is used if the feed to the membrane is liquid, since the contaminant appears to evaporate through the membrane. If the feed is vapour, or a gas/vapour mixture, the process is called vapour permeation. The best-performing industrial membranes permeate water in preference to other components, so filtration and RO processes are used typically to purify water. In contrast, pervaporation and vapour permeation are commonly used to remove water from organics.

Wide-ranging applications

Membranes are selective either by pore size (porous membranes) or because of their chemical affinity for the permeating component. By far, the majority of pervaporation membranes in commercial use are hydrophilic Most pervaporation membranes are therefore employed to dehydrate organics.

Although pervaporation and vapour permeation require significant driving forces to transport components through the membrane, the processes do not depend on particular vapour/liquid equilibria. Water is preferentially permeated from a stream irrespective of the other components present. In practice, pervaporation and vapour permeation are only competitive where distillation is difficult or costly.

Figure 14.8 shows where these processes are most usefully applied in dehydrating organics. The concentration and relative volatility of the organic are key variables.

Pervaporation is now coming of age as a separation process. Industrially, it has been introduced in end-of-pipe applications such as solvent recovery. As with other membrane processes, accumulated experience in both membrane fabrication, and design and operation of pervaporation units has matured the technology, such that it is now applied to separations that are integral to production. Although dehydrating the ethanol/water and isopropanol/water azeotropes account for the majority of all plants operating (roughly, four-fifths), most of the new plants being installed are in quite different applications.

Methanol and ethanol removal: Hydrophilic polymer membranes have also been developed that will permeate methanol, ethanol and, to some extent, isopropanol. (The lead photo shows a pervaporation plant for methanol treatment.) These compounds can be removed from less polar organics, although membrane selectivity is not as high as when permeating water.

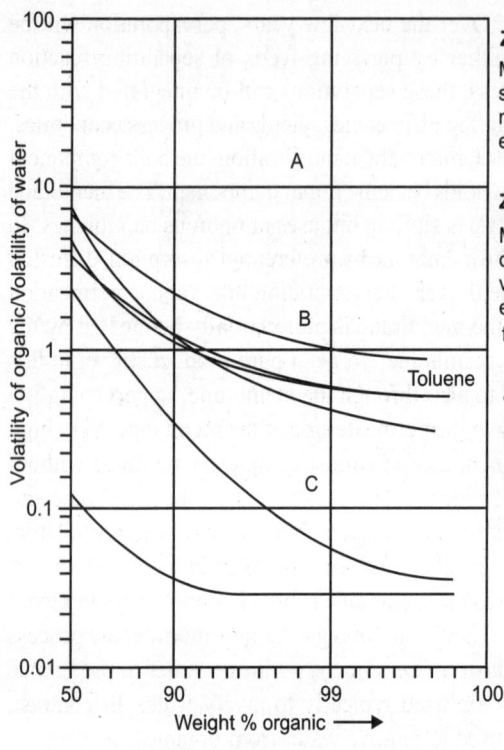

Zone A : Pervaporation can save energy

Mainly volatile organics contaminated with water. Can be separated by distillation, but energy requirements are high since major component evaporates. Pervaporation requires much less energy since only the contaminant water must be evaporated.

Zone B: Always pervaporate

Mainly organics contaminated with water; minimal volatility differences, difficult to distill. Classic use of pervaporation or vapor permeation with hydrophilic membranes for the most economical separation. The minor component (water) is removed from the organic without problems.

Zone C: Special situations only

Mainly high-boiling organics contaminated with water. Strip out the water, if necessary distill. Energy requirements are low since the minor component evaporates. Pervaporation is competitive only in special situations (e.g. foaming systems or where kinetic effects dominate).

Fig. 14.8. Pervaporation is mainly used to dehydrate organics.

Methanol forms azeotropes with many substances, particularly esters, and often cannot be recovered from spent solvents or reaction mixtures by simple distillation. Pervaporation provides a simple way to break these azeotropes. Used alone or in combination with distillation, such units provide an economical and reliable route to recover or remove methanol.

A separation scheme follows for methanol removal from a methanol rich methanol/ethyl acetate mixture. The mixture is distilled to the azeotrope, taking out pure methanol as bottom product. The overhead stream is passed directly to a vapour permeation unit that permeates a methanol-rich stream. This stream is condensed and passed back to the methanol column via the feed buffer.

Retentate from the vapour permeation unit, strongly depleted in methanol, can be fed directly to the ethyl acetate column. Pure ethyl acetate leaves this column as bottom product, while overhead azeotrope is sent to the vapour permeation unit.

Many solvent or ester/methanol mixtures can be separated using a similar scheme. If the feed is close to the azeotrope, then the methanol column can be dispensed with. If the capacity is small, the purification column for the second component may not be required, depending on the desired purity. One of the most promising uses of pervaporation technology is removing methanol from the products of transesterifications.

Continuous water removal from condensations—typical condensation reactions include:

$$\text{Esterification: } R'\text{-COOH} + HO\text{-}R = R'COOR + H_2O$$
$$\text{Acetalisation: } R'\text{-CHO} + 2HO\text{-}R = R'CH\text{-}(OR)_2 + H_2O$$
$$\text{Ketalisation: } R'R''CO + 2HO\text{-}R = R'R''C(OR)_2 + H_2O$$

Condensation reactions are normally equilibrium-limited so removal of coproduct water reduces production costs three ways:

1. Higher yield—lower reagent consumption.
2. Faster reaction—greater reactor throughput.
3. Purer products—less effort for product purification.

However, the optimum scheme for water removal from these reactions depends upon the relative volatilities of the reactants and products, and whether the units are operated batchwise or continuously. The following examples illustrate how pervaporation or vapour permeation can be used in particular situations.

Batch condensation of ethyl and propyl alcohols: Unless producing volatile products, the typical procedure for these reactions is to dissolve the acid in an excess of alcohol add catalyst, and then heat to drive off a water/alcohol mixture. This vapour is fed to a distillation column, the reaction water leaves as bottom product, and the alcohol/water azeotrope as top product. At the beginning of the reaction, the azeotropic mixture is fed back to the reaction and the reaction still proceeds at a reasonable rate. As the reaction progresses, the concentration of product in the excess alcohol approaches an equilibrium because of the water in the reactor; the reaction slows. At this point, recycle of wet alcohol is discontinued and fresh dry alcohol is added instead. Gradually coproduct water is distilled away, correspondingly more acid is reacted and a reasonably high yield is obtained, if sufficient time is available. The reaction product, excess alcohol and unused acid are then separated and a further batch is started. Normally, a batch is started with wet alcohol, generated at the end of the previous one.

There are three ways in which pervaporation or vapour permeation can be used to enhance such reactions:

Offline pervaporation (Fig. 14.9) is the simplest procedure and can be very effective. In this case, the alcohol/water azeotrope from the top of the column is collected in a tank during the latter part of the batch. The tank contents are continuously dehydrated by pervaporation and the resulting dry alcohol stored in a second tank.

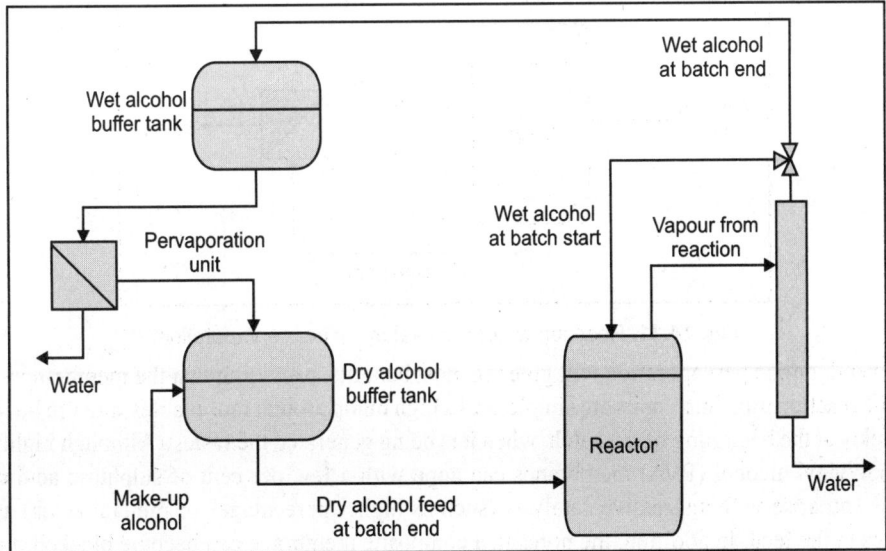

Fig. 14.9. Reaction water removal via offline pervaporation.

This material is fed to the process over the latter part of the batch. (Note: In Figs 14.9, 14.10 and 14.11, the horizontal line in the reactors represents a liquid line.)

Online vapour permeation (Fig. 14.10) is used to remove water directly from the product of the distillation column. Dry alcohol is continuously fed back to the reaction.

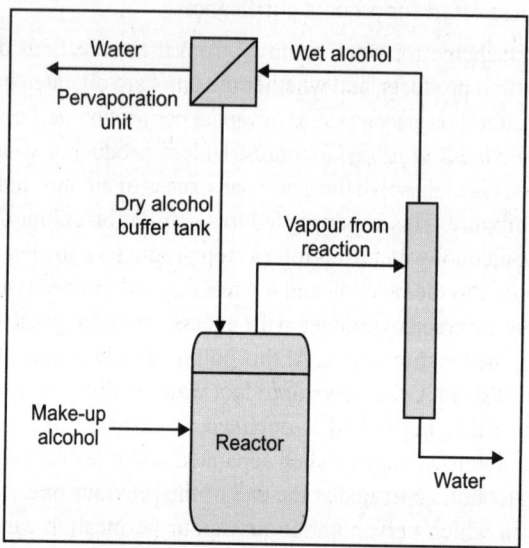

Fig. 14.10. Reaction water removal via online vapour permeation.

Online pervaporation (Fig. 14.11) replaces the distillation column altogether. The reaction mixture is continuously pumped through the pervaporation unit at a high rate and a drier stream is returned to the reactor.

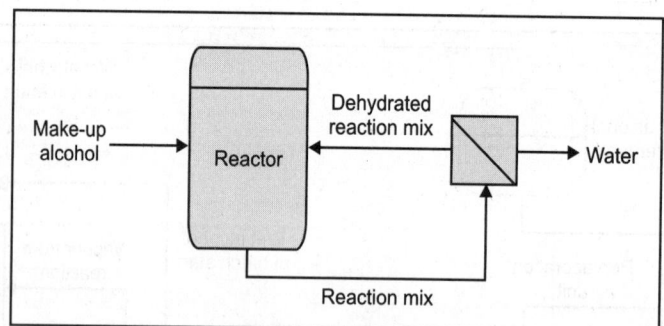

Fig. 14.11. Reaction water removal by online pervaporation.

In general, online pervaporation will give the most benefits, providing that the membrane is able to handle the reaction mix. Such units are simple, and a high pump around rate enables water to be removed very quickly at the beginning of the batch, when it is being generated the fastest. Although highly cross-linked polyvinyl alcohol (PVA) membranes can cope with a few per cent of sulphuric acid catalyst, problems can arise with aggressive catalysts (such as higher percentages of mineral acids) and with impurities in the feed. In addition, the pores in a composite membrane can become blocked if the acid or product permeate the membrane faster than they can evaporate into the permeate vapour.

Online vapour permeation avoids these problems, but the unit is constrained to operate at the reaction pressure (normally atmospheric), with the flowrate passing through the column. It is difficult to fully utilise such units throughout the cycle.

Offline pervaporation allows the membrane unit to be utilised at full capacity throughout the batch— in fact, it can also be used when the reactor is not operating. In addition, such a unit is easily coupled to other reactors. Such units enhance the economy of a batch processing operation, while providing a high degree of flexibility.

Continuous production of ethyl/propyl esters of low-volatility acids: The classical scheme for this is shown in Fig. 14.12. Acid is continuously fed into a reactor containing an excess of alcohol. A product mix is continuously drawn off from the reactor containing alcohol, water, ester and some unreacted acid. Three columns are then used to remove product ester and coproduct water. Unreacted alcohol is recycled back to the reaction.

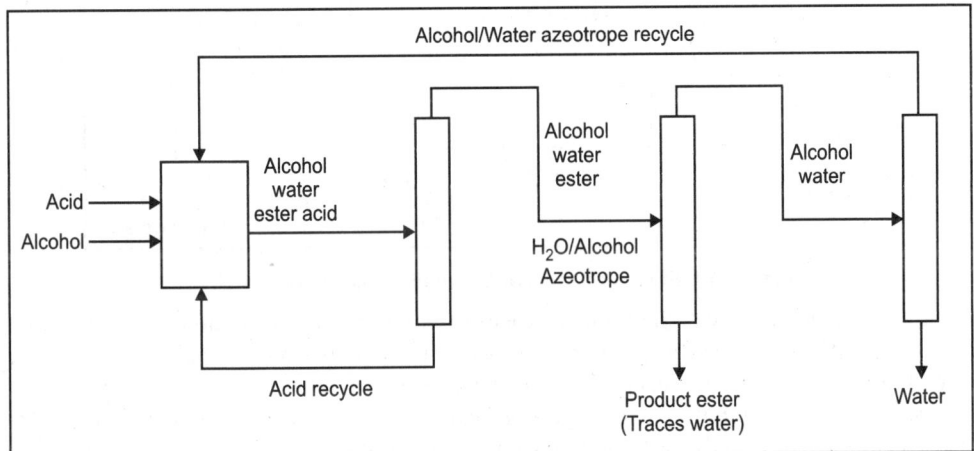

Fig. 14.12. Classical reaction/separation for esters.

In the first column, the least-volatile component, unreacted acid, is taken out at the bottom and recycled back to the reaction. Alcohol, water and ester pass overhead to the second column. Product ester is taken from the bottom of this column and water and alcohol are taken overhead. The third column is used to remove water, again taken out at the bottom. The overhead product from this column is the alcohol/water azeotrope, which is recycled to the reactor.

This scheme has two major drawbacks:

1. There is a high concentration of water in the reactor, requiring a large excess of alcohol to drive the reaction.
2. The recovered ester is contaminated with trace quantities of water, which hydrolyse the product, decreasing its purity, quality and usability in many situations.

Figure 14.13 shows a reaction scheme enhanced by continuously removing water directly from the reactor. In this case, the water is removed from the vapour phase. A vapour stream is sparged from the reactor and circulated through a vapour permeation membrane module, where water is selectively permeated through the membranes.

The membrane unit is sized such that all of the reaction water can be removed with the water/alcohol ratio just below the azeotropic composition.

The reaction mixture is similarly passed through the first column to remove excess acid. In the second column, ester is again taken out at the bottom. However, because relatively little water is in the feed to this column, the water is entrained out with the alcohol. The ester bottom product is contaminated with traces of alcohol instead of water. It will not hydrolyse, so premium product quality is assured.

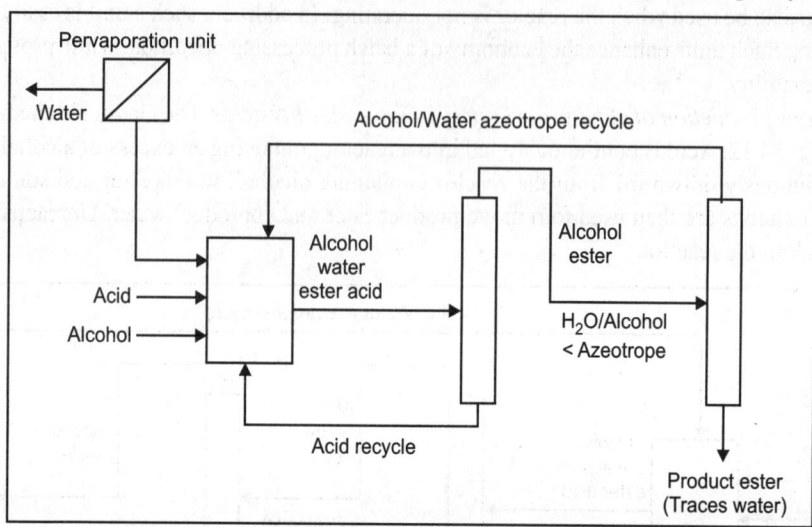

Fig. 14.13. Enhanced reaction/separation uses pervaporation.

Removing reaction water directly from the reactor improves reaction conditions — the reaction runs faster, lower residence times suffice, equipment costs are minimised and side reactions are reduced. No third column is required and product quality is better. A win on all counts.

Continuous esterification with a heterogeneous catalyst. Some condensation reactions are carried out using heterogeneous catalysts. Continuously removing water from such systems brings concrete benefits, particularly in yield. Figure 14.14 shows a scheme for removing water from a continuous esterification process that uses a heterogeneous catalyst. The process runs in four stages. Each stage includes a reactor where the components are brought close to equilibrium over the catalyst.

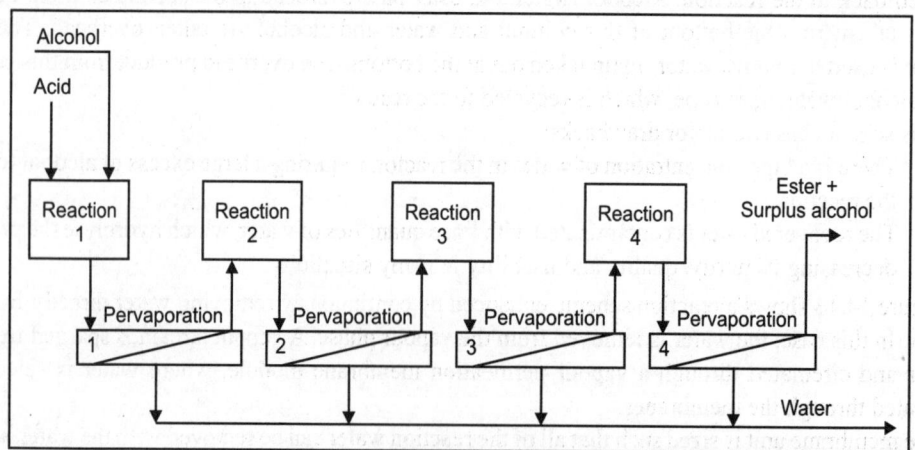

Fig. 14.14. Four-stage reaction/pervaporation cascade.

The mixture then flows through a pervaporation stage, where water generated in the reaction step is removed, shifting the reaction equilibrium. In the next reaction step, equilibrium is re-established and again the reaction water is removed. In four such stages, the reaction will be pushed far over to the right and consume nearly all of the feed supplied. Such a procedure not only maximises reagent usage but also minimises the separation work required to purify the product. A number of such plants are operating and more are under construction.

Transesterifications using methyl esters: Many esters are made using transesterification, because the milder conditions prevailing permit the reaction of components containing additional functional groups. In alcoholysis, a complex alcohol is reacted with a methyl ester, forming a complex ester and methanol. These reactions are all equilibrium-limited, so the reaction only proceeds if product or coproduct is removed.

Commonly, the reaction mix is distilled to remove the methanol. However, methanol forms azeotropes with many methyl esters, so driving out the methanol also removes a reactant.

Separation of these azeotropic mixtures by traditional means is difficult. However, polymer membranes with a low degree of cross-linking can be engineered that will preferentially permeate methanol. Figure 14.15 shows a vapour permeation unit installed to remove methanol from the top of a column, treating the boil-off from a transesterification reactor. The column is operated to condense overhead product close to the azeotropic point. Condensed liquid is refluxed through the column. Net overhead vapour is passed through the vapour permeation unit, which is sized to permeate methanol at the rate that it is generated in the reaction.

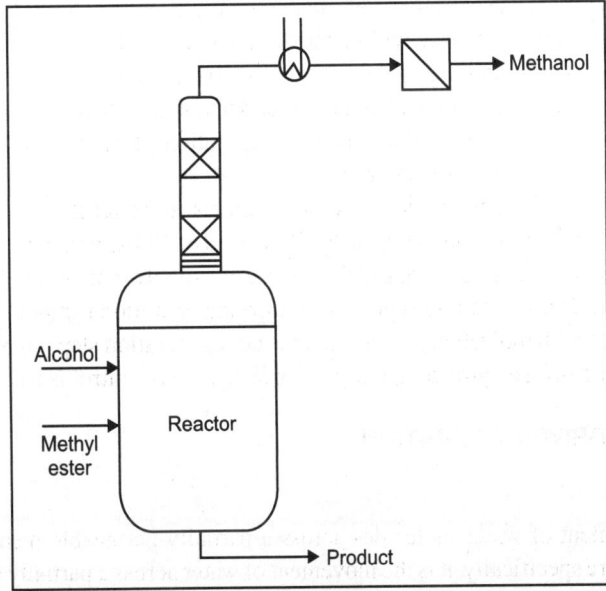

Fig. 14.15. Methanol removal from transesterification.

The recovered ester is recycled to the reaction. In the example shown, membrane area is saved by feeding ester only partially depleted in methanol back to the column. The remaining methanol is stripped from the stream as it passes down the column to the reactor.

Such enhancements directly impact the quantity of reagent required and so have a direct impact on the bottom line. Payback times for the separation equipment are typically short.

Continuous drying of reactor feed streams — replacing molecular sieves

In a typical solvent recovery application, dehydration is only required down to 0.5 per cent water or, in some cases, 0.1 per cent water. At this water content, the solvent will perform almost as well as a completely dry solvent. Conventional pervaporation plants have no problem reaching such end water contents.

However, some special situations arise when water should be completely removed from an organic stream. At most, only a few ppm are tolerated because whatever water is left in the organic will react in later processing steps and consume, for example, an expensive catalyst. Although molecular sieves have traditionally been used for these dehydration duties, pervaporation can also be applied and is now emerging as a strong competitor.

The perceived problem using pervaporation for ultra-drying is generating a sufficiently high vacuum on the permeate side of the membrane to remove all of the water from the retentate side. At 100°C, 10 ppm of water in ethanol generates a partial vapour pressure of only 0.01 mbar. Maintaining this level of vacuum on the permeate side of a pervaporation unit is hardly practical.

In practice, this problem can be avoided if tailored membranes are used. Composite polymer membranes are tailored by modifying the degree of cross-linking of the polymer in the thin separation layer. Highly cross-linked membranes exhibit high selectivity, but low flux; with weakly cross-linked membranes, the situation is reversed. During ultra-drying, a considerable amount of organic will copermeate with the water if the selectivity of the membrane is not too high. In fact, when removing water to the 10 ppm level, the permeate will be mostly organic—water will only be present at may be 500 ppm. In this situation, maintaining a total pressure of 10 mbar on the permeate side will result in a partial vapour pressure of water of 0.005 mbar, i.e. water will continue to permeate across the membrane and dehydration will proceed. Properly designed pervaporation plants with tailored membranes can easily reach ppm or even ppb levels of water.

Ceramic membranes allow extremely low water contents to be reached simply by operating the membrane at a higher temperature. Operating at 200°C instead of 100°C will increase the partial pressure of contained water by a factor of around 15, and drive correspondingly more water through the membrane.

Because pervaporation is a continuous process, the drying operation is much simpler and less error-prone than when drying with molecular sieves. There is no regeneration step, no nitrogen is required for regeneration, there is no off gas, product quality is consistent, and control is trivial.

CONCENTRATION DRIVEN OPERATIONS

Osmosis

Osmosis is the movement of water molecules across a partially-permeable membrane down a water potential gradient. More specifically, it is the movement of water across a partially permeable membrane from an area of high water potential (low solute concentration) to an area of low water potential (high solute concentration). It is a physical process in which a solvent moves, without input of energy, across a semipermeable membrane (permeable to the solvent, but not the solute) separating two solutions of different concentrations. Osmosis releases energy, and can be made to do work. Osmosis is a passive process, like diffusion.

Net movement of solvent is from the less-concentrated (hypotonic) to the more-concentrated (hypertonic) solution, which tends to reduce the difference in concentrations. This effect can be countered by increasing the pressure of the hypertonic solution, with respect to the hypotonic. The osmotic pressure is defined to be the pressure required to maintain an equilibrium, with no net movement of solvent. Osmotic pressure is a colligative property, meaning that the osmotic pressure depends on the molar concentration of the solute but not on its identity.

Osmosis is important in biological systems, as many biological membranes are semipermeable. In general, these membranes are impermeable to organic solutes with large molecules, such as polysaccharides, while permeable to water and small, uncharged solutes. Permeability may depend on solubility properties, charge, or chemistry, as well as solute size. Water molecules travel through the plasma cell wall, tonoplast (vacuole) or protoplast in two ways, either by diffusing across the phospholipid bilayer directly, or via aquaporins (small transmembrane proteins similar to those in facilitated diffusion and in creating ion channels). Osmosis provides the primary means by which water is transported into and out of cells. The turgor pressure of a cell is largely maintained by osmosis, across the cell membrane, between the cell interior and its relatively hypotonic environment.

Osmosis may occur when there is a partially permeable membrane, such as a cell membrane. When a cell is submerged in water, the water molecules pass through the cell membrane from an area of low solute concentration (outside the cell) to one of high solute concentration (inside the cell); this is called osmosis. The cell membrane is selectively permeable, so only necessary materials are let into the cell and wastes are left out. When the membrane has a volume of pure water on both sides, water molecules pass in and out in each direction at the exact same rate; there is no net flow of water through the membrane. In a solution, the concentration of water is dilluted (or lowered) by the presence of solute particles. If there is a solution on one side, and pure water on the other, there will be a higher concentration of water molecules on the pure water side of the membrane. Therefore, water molecules pass through the membrane from the pure water side toward the solution side more frequently than from the solution side going to the pure water side.

This will result in a net flow of water to the side with the solution. Assuming the membrane does not break, this net flow will slow and finally stop as the pressure on the solution side becomes such that the movement in each direction is equal: dynamic equilibrium. This could either be due to the water potential on both sides of the membrane being the same, or due to osmosis being inhibited by factors such as pressure potential or osmotic pressure.

Osmosis can also be explained using the notion of entropy, from statistical mechanics. Suppose a permeable membrane separates equal amounts of pure solvent and a solution. Since a solution possesses more entropy than pure solvent, the second law of thermodynamics states that solvent molecules will flow into the solution until the entropy of the combined system is maximised. Notice that, as this happens, the solvent loses entropy while the solution gains entropy. Equilibrium, hence maximum entropy, is achieved when the entropy gradient becomes zero, and dissolution takes place.

Pure water is more ordered than water in a solution; thus, from an entropic standpoint it takes some net energy to move a water molecule from a disordered solution and 'pack it in' with pure water. This is the same explanation as to why the disordered air does not spontaneously separate and order into oxygen and nitrogen, it would take energy for this to happen.

Additionally, particle size has no bearing on osmotic pressure, as this is the fundamental postulate of colligative properties.

Examples of osmosis

Osmotic pressure is the main cause of support in many plants. The osmotic entry of water raises the turgor pressure exerted against the cell wall, until it equals the osmotic pressure, creating a steady state. When a plant cell is placed in a hypertonic solution, the water in the cells moves to an area higher in solute concentration and the cell shrinks, and in doing so, becomes flaccid. This means the cell has become plasmolysed—the cell membrane has completely left the cell wall due to lack of water pressure on it; the opposite of turgid. Also, osmosis is responsible for the ability of plant roots to draw water from the soil. Since there are many fine roots, they have a large surface area, and water enters the roots by osmosis. Osmosis can also be seen when potato slices are added to a high concentration of salt solution. The water from inside the potato moves to the salt solution, causing the potato to shrink and to lose its 'turgor pressure'.

The more concentrated the salt solution, the bigger the difference in size and weight of the potato slice. In unusual environments, osmosis can be very harmful to organisms. For example, freshwater and saltwater aquarium fish placed in water of a different salinity than that to which they are adapted to will die quickly, and in the case of saltwater fish, dramatically. Another example of a harmful osmotic effect is the use of table salt to kill leeches and slugs.

Suppose an animal or a plant cell is placed in a solution of sugar or salt in water.

1. If the medium is hypotonic—a dilute solution, with a higher water concentration than the cell—the cell will gain water through osmosis.

2. If the medium is isotonic—a solution with exactly the same water concentration as the cell—there will be no net movement of water across the cell membrane.

3. If the medium is hypertonic—a concentrated solution, with a lower water concentration than the cell—the cell will lose water by osmosis.

Essentially, this means that if a cell is put in a solution which has a solute concentration higher than its own, then it will shrivel up, and if it is put in a solution with a lesser solute concentration than its own, the cell will expand and burst. Chemical gardens demonstrate the effect of osmosis in inorganic chemistry.

OPERATIONS IN ELECTRIC POTENTIAL GRADIENT

Electrodialysis

Electrodialysis (ED) is used to transport salt ions from one solution through ion-exchange membranes to another solution under the influence of an applied electric potential difference. This is done in a configuration called an electrodialysis cell. The cell consists of a feed (diluate) compartment and a concentrate (brine) compartment formed by an anion exchange membrane and a cation exchange membrane placed between two electrodes. In almost all practical electrodialysis processes, multiple electrodialysis cells are arranged into a configuration called an electrodialysis stack, with alternating anion and cation exchange membranes forming the multiple electrodialysis cells. Electrodialysis processes are different compared to distillation techniques and other membrane based processes (such as reverse osmosis) in that dissolved species are moved away from the feed stream rather than the reverse. Because the quantity of dissolved species in the feed stream is far less than that of the fluid, electrodialysis offers the practical advantage of much higher feed recovery in many applications.

In an electrodialysis stack, the diluate (D) feed stream, brine or concentrate (C) stream, and electrode (E) stream are allowed to flow through the appropriate cell compartments formed by the ion exchange membranes. Under the influence of an electrical potential difference, the negatively charged ions (e.g. chloride) in the diluate stream migrate toward the positively charged anode. These ions pass through the positively charged anion exchange membrane, but are prevented from further migration toward the anode by the negatively charged cation exchange membrane and therefore stay in the C stream, which becomes concentrated with the anions. The positively charged species (e.g. sodium) in the D stream migrate toward the negatively charged cathode and pass through the negatively charged cation exchange membrane. These cations also stay in the C stream, prevented from further migration toward the cathode by the positively charged anion exchange membrane. As a result of the anion and cation migration, electric current flows between the cathode and anode. Only an equal number of anion and cation charge equivalents are transferred from the D stream into the C stream and so the charge balance is maintained in each stream. The overall result of the electrodialysis process is an ion concentration increase in the concentrate stream with a depletion of ions in the diluate solution feed stream.

The E stream is the electrode stream that flows past each electrode in the stack. This stream may consist of the same composition as the feed stream (e.g. sodium chloride) or may be a separate solution containing a different species (e.g. sodium sulphate). Depending on the stack configuration, anions and cations from the electrode stream may be transported into the C stream, or anions and cations from the D stream may be transported into the E stream. In each case, this transport is necessary to carry current across the stack and maintain electrically neutral stack solutions.

Anode and cathode reactions

Reactions take place at each electrode. At the cathode,

$$2e^- + 2H_2O \rightarrow H_2(g) + 2OH^-$$

while at the anode,

$$H_2O \rightarrow 2H+ + \tfrac{1}{2}O_2\,(g) + 2e^-$$

or

$$2\,Cl^- \rightarrow Cl_2\,(g) + 2e^-$$

Small amounts of hydrogen gas are generated at the cathode and small amounts of either oxygen or chlorine gas (depending on composition of the E stream and end ion exchange membrane arrangement) at the anode. These gases are typically subsequently dissipated as the E stream effluent from each electrode compartment is combined to maintain a neutral pH and discharged or re-circulated to a separate E tank. However, some (e.g.) have proposed collection of hydrogen gas for use in energy production.

Applications: In application, electrodialysis systems can be operated as continuous production or batch production processes. In a continuous process, feed is passed through a sufficient number of stacks placed in series to produce the final desired product quality. In batch processes, the diluate and/ or concentrate streams are re-circulated through the electrodialysis systems until the final product or concentrate quality is achieved. Electrodialysis is usually applied to deionisation of aqueous solutions. However, desalting of sparingly conductive aqueous organic and organic solutions is also possible. Some applications of electrodialysis include:

1. Large scale brackish and seawater desalination and salt production.
2. Small and medium scale drinking water production (e.g. towns and villages, construction and military camps, nitrate reduction, hotels and hospitals).
3. Water reuse (e.g. industrial laundry waste-water, produced water from oil/gas production, cooling tower make-up and blowdown, metals industry fluids, wash-rack water).

4. Pre-demineralisation (e.g. boiler makeup and pretreatment, ultrapure water pretreatment, process water desalination, power generation, semiconductor, chemical manufacturing, food and beverage).
5. Food processing.
6. Agricultural water (e.g. water for greenhouses, hydroponics, irrigation, livestock).
7. Glycol desalting [e.g. antifreeze/engine-coolants, capacitor electrolyte fluids, oil and gas dehydration, conditioning and processing solutions, industrial heat transfer fluids, secondary coolants from heating, venting, and air conditioning (HVAC)].
8. Glycerine purification.

The major application of electrodialysis has historically been the desalination of brackish water or seawater as an alternative to RO for potable water production and seawater concentration for salt production (primarily in Japan). In normal potable water production without the requirement of high recoveries, RO is generally believed to be more cost-effective when total dissolved solids (TDS) are 3000 parts per million (ppm) or greater, while electrodialysis is more cost-effective for TDS feed concentrations less than 3000 ppm or when high recoveries of the feed are required.

Another important application for electrodialysis is the production of pure water and ultrapure water by electrodeionisation (EDI). In EDI, the purifying compartments and sometimes the concentrating compartments of the electrodialysis stack are filled with ion exchange resin. When fed with low TDS feed (e.g. feed purified by RO), the product can reach very high purity levels (e.g. 18 Megohms). The ion exchange resins act to retain the ions, allowing these to be transported across the ion exchange membranes. The main usage of EDI systems such as those supplied by Ionpure or SnowPure are in electronics, pharmaceutical, power generation, and cooling tower applications.

Membrane Electrolysis

Membrane electrolysis (ME) is a membrane process driven by an electrolytic potential. It is used for the removal of metallic impurities from plating, anodizing, etching, stripping, and other metal finishing solutions. This technology uses an ion exchange membrane and an electrical potential applied across the membrane. The membranes are ionpermeable and selective, permitting ions of a given electrical charge to pass through. Cation exchange membranes allow only cations, such as copper or aluminium to pass through, while anion exchange membranes allow only anions, such as sulphates or chromates to pass through.

Features

1. Maintains process solutions for consistent performance and quality.
2. Maintains solutions for constant production rates.
3. Accurate, high-quality etching and chrome conversion coating results.
4. Reduces reject rate (reduces reprocessing).
5. Reduces waste-water treatment and waste disposal.
6. Reduces fresh process solution usage.

Applications

Typical applications include:

1. Purification and regeneration of chromium plating baths.
2. Recycling and maintenance of chrome conversion coating solutions.
3. Reactivation and metal removal from deoxidising solutions.

Figure 14.16 shows etching solution performance.

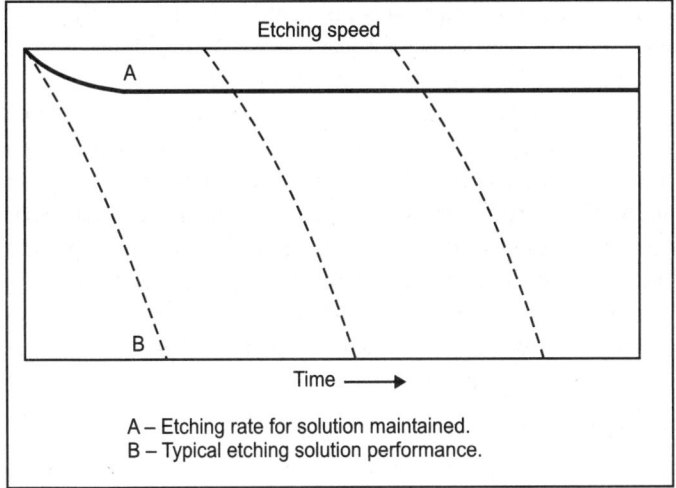

Fig. 14.16. Etching solution performance.

Electrophoresis

Electrophoresis is the motion of dispersed particles relative to a fluid under the influence of a spatially uniform electric field. This electrokinetic phenomenon was observed for the first time in 1807 by Reuss, who noticed that the application of a constant electric field caused clay particles dispersed in water to migrate. It is ultimately caused by the presence of a charged interface between the particle surface and the surrounding fluid. The dispersed particles have an electric surface charge, on which an external electric field exerts an electrostatic Coulomb force. According to the double layer theory, all surface charges in fluids are screened by a diffuse layer of ions, which has the same absolute charge but opposite sign with respect to that of the surface charge. The electric field also exerts a force on the ions in the diffuse layer which has direction opposite to that acting on the surface charge. This latter force is not actually applied to the particle, but to the ions in the diffuse layer located at some distance from the particle surface, and part of it is transferred all the way to the particle surface through viscous stress. This part of the force is also called electrophoretic retardation force.

OPERATIONS IN TEMPERATURE GRADIENT

Membrane Distillation

Membrane Distillation (MD) is a water separation/purification technique, where water is transported between hot and a cool streams separated by a hydrophobic membrane. The exchange of water vapour relies on a small temperature difference between the two streams. It is an alternative to other conventional methods such as reverse osmosis (RO) or simple distillation. Working under mild conditions, MD offers a number of advantages over RO and simple distillation.

Any waste or low quality heat source can be utilised to heat up the hot stream, and the carrying out of the operation is also quite simple due to the fact that the process takes places at normal pressure.

Application of membrane distillation in environmental protection

The driving force for membrane processes may have quite a different character. Very often it is a pressure difference that affects the mass transport through a membrane. In other cases it is a concentration gradient or an electrical potential gradient. Membrane distillation is a thermally driven process. Although thermally driven processes have been known for many years, the membrane distillation process is still considered a new, promising membrane operation. This process has been studied since the 1960's. Development in membrane manufacturing in the 1980's allowed us to obtain commercial membranes with desired properties. Improvements in module design and better understanding of phenomena occurring in a layer adjacent to a membrane also contributed to renewed interest in MD. In comparison with other separation operations, MD has very important advantages: practically a complete rejection of dissolved, nonvolatile species, lower operating pressure than pressure-driven membrane processes, reduced vapour space (practically to membrane thickness) compared to conventional distillation. Low operating temperature (considerably below boiling point) of a feed enables the utilisation of waste heat as a preferable energy source. The possibility of utilising of alternative energy sources such as solar, wave or geothermal energy is particularly attractive.

Principle of membrane distillation (MD)

Membrane distillation (MD) is a process in which a microporous, hydrophobic membrane separates aqueous solutions at different temperatures and compositions. The temperature difference existing across the membrane results in a vapour pressure difference. Thus, vapour molecules will be transported from the high vapour pressure side to the low vapour pressure side through the pores of the membrane. According to terminology for membrane distillation, the name membrane distillation should be applied for membrane operation having the following characteristics:

1. The membrane should be porous.
2. The membrane should not be wetted by process liquids.
3. No capillary condensation should take place inside the pores of the membranes.
4. Only vapour should be transported through the pores of the membrane.
5. The membrane must not alter vapour equilibrium of the different components in the process liquids.
6. For each component the driving force of the membrane operation is a partial pressure gradient in the vapour phase.

The principle of direct contact membrane distillation is presented in Fig 14.17. The process essentially involves the following steps:

1. Evaporation of volatile compounds of a feed at the warm feed/membrane interface.
2. Transfer of vapour through the membrane pores.
3. Condensation of the permeate at the membrane/cold distillate interface.

Figure 14.18 illustrates different MD configurations commonly used to obtain the required driving force. In all solutions the membrane is directly exposed to the warm solution, but the method of permeate condensation is different. In direct contact MD (Fig. 14.18a) the cold distillate is in direct contact with the membrane and vapour transported through the membrane condenses directly in a stream of cold distillate. In the gas-gap MD system (Fig. 14.18b), the permeate is condensed on a cooling surface. In this case, the total length of vapour diffusion is the sum of membrane thickness and air gap. The condensed distillate does not have to be in contact with the membrane. In a low pressure MD system (Fig. 14.18c), the pressure is applied on the distillate side and the condensation of the permeate takes place outside the module. In the last MD system, a sweeping gas is applied and permeate condensation occurs outside the module (Fig. 14.18d).

If the term 'membrane distillation' is used without any specification, this term applies to the direct contact membrane distillation. The separation mechanism is based on the vapour/liquid equilibrium.

This means that the component with the highest partial pressure will exhibit the highest permeation rate. MD is a highly selective operation for nonvolatile species, such as ions, colloids, macromolecules, which are unable to evaporate and diffuse across the membrane. The solutes are completely rejected and the permeate is then pure water. When volatile species are present in the feed they will be also transported through the membrane. According to the vapour/liquid equilibrium, permeate composition depends on the composition and temperature of a feed.

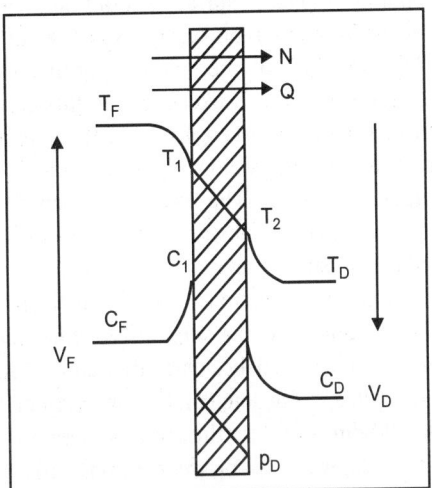

Fig. 14.17. The principle of direct contact membrane distillation.

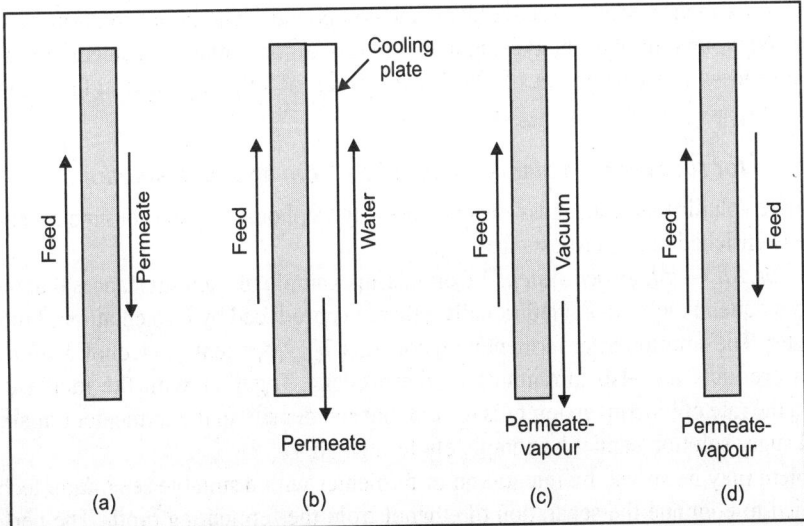

Fig. 14.18. Membrane distillation configuration: (a) direct contact MD, (b) air-gap MD, (c) low pressure MD, and (d) sweeping gas MD.

Membrane characteristics: The membranes used in MD rather act as a physical support for a liquid/vapour interface, but their choice for the process is very important. The membranes have to meet several requirements simultaneously. The presence of only the vapour phase in the membrane pores is a necessary condition for MD. Therefore, hydrophobicity of the membrane (i.e. its non-wettability which prevents the bulk liquid transport across the membrane) plays an essential role in this process. The MD membranes are prepared from hydrophobic polymers such as polytetrafluoroethylene (PTFE), polypropylene (PP), or poly(vinylidene fluoride) (PVDF), which have a low surface energy. Moreover, these polymers exhibit excellent chemical resistance and good thermal stability.

A contact angle measured on a hydrophobic surface should be greater than 90°. For example, the parameter measured on PTFE or PVDF membrane surface was 108° or 107°, respectively.

Membranes prepared from hydrophobic polymers do not permit for the flow of liquid through the membrane until a critical penetration pressure is exceeded. In this case the liquid may penetrate the membrane pores and non selective flow is observed. Liquid entry pressure of water (LEPW) is a very important parameter which determines the magnitude of pressure which should not be exceeded. LEPW can be calculated using the Laplace equation. The presence of surfactants or organic solvents may significantly reduce the liquid surface tension, causing membrane wetting. The wetted membrane must be completely dried and cleaned before subsequent use.

Concentration of nonvolatile acids: Investigations have shown the possibility of MD application for concentration of diluted nonvolatile acids such as sulphuric or phosphoric. The temperature and the concentration polarisation phenomena decreased the achievable acid concentration due to a high viscosity of concentrated solutions. The raw phosphoric acid can be concentrated to about 32 per cent P_2O_5. The permeate flux then decreased to 100 dm^3/dm^2d (at the feed temperature equal to 333 K).

Phosphogypsum from apatite Kola, waste by-product from the manufacture of phosphoric acid by the wet-process route, can be a source of rare earth elements. Experimental results show that the membrane distillation process can be applied for concentration of sulphuric acid solution obtained after apatite phosphogypsum extraction used to recover lanthane compounds. The initial solution was 16 per cent H_2SO_4, saturated by apatite phosphogypsum and contains, among other substances, 1.6 g La_2O_3. The concentration process was carried out till the concentration of sulphuric acid in the solution reached about 40 per cent.

MD application for recovery of volatile compounds from aqueous solutions

Substances more volatile than water pass easier across a hydrophobic microporous membrane; therefore, the permeate is enriched in these substances.

Fermentation/MD—integrated system: Ethanol is an example of such substances that preferentially vapourise from aqueous solutions. Traditionally, ethanol is produced by fermentation of biomass in the batch fermenter. The solution after fermentation contains 7–19 per cent of alcohol. Unfortunately, the fermentation products are also inhibitants of the process. Together with the increase of ethanol concentration the rate of bioconversion falls to zero, the cell density in the fermenter remains low and a concentrated sugar solution cannot be completely fermented.

This problem may be solved by integration of fermenter with a suitable separation technique. MD may by applied to continue the separation of ethanol from the fermenting broth. The performance of fermentation in the membrane bioreactor allows for a considerable acceleration of its course and increases its efficiency through the selective removal of fermentation products.

HCL recovery from industrial effluents: Before electroplating, the metal surface has to be clean. Different acids such as HCl, HNO_3 or H_2SO_4 are used for these purposes as a pickling liquor to remove surface oxides. The spent pickling liquors contain residual acid and suitable salts. Their composition depends on metal objects. Frequently they contain very harmful heavy metals. The components must be removed, recovered or recycled for environmental reasons.

A neutralisation procedure is used traditionally. Our experiments with membrane distillation show that it is possible to recover hydochloric acid used for pickling. Moreover, salt can be separated from the spent solution after its concentration to the supersaturated state.

During the MD process both water vapour and gaseous HCl are transported through pores of the membrane from the warm feed to cold distillate, whereas salt was retained in the feed. The vapour was condensed directly in a cold distillate and gaseous HCl was then dissolved. Vapour composition is mainly affected by a hydrochloric acid concentration in the feed and its temperature. At low acid concentration in the feed the permeate was pure water. The rise of acid concentration in the feed above 19 per cent caused a substantial increase in HCl molar flux through the membrane.

The presence of salt in the feed containing HCl changes the results of MD due to a change of vapour composition. Accumulated salt in the feed decrease HCl solubility (desalting out effect), thus the molar HCl flux was then higher than in the case of hydrochloric acid solution without salt. The retention coefficient of the salt was 99.8 per cent, thus permeate was pure hydrochloric acid with a concentration which can be significantly higher than in the feed. The results of the experiments show that MD may be a promising method of HCl recovery from industrial effluents (Fig. 14.19).

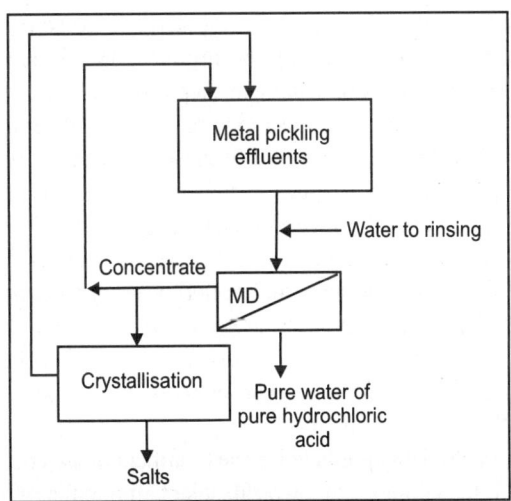

Fig. 14.19. Schematic diagram for HCl recovery from metal pickling effluents using MD.

A part of a concentrated solution may be recycled to favour the HCl desorption. The useful products will be pure water, pure hydrochloric acid, and metal salts after crystallisation from the supersaturated feed. The acid could be used for eatching of the fresh elements.

Chapter 15

Treating the Sludge

INTRODUCTION

The similarity between a water treatment facility and any manufacturing operation, whether it be a rubber producing plant or an auto-making facility or an iron and steel plant, are the reliance upon combinations of unit processes and unit operations that work in harmony to produce a high quality product. But that is where the similarity ends. A normal manufacturing operation aims not only to produce a high quality product, but efficient businesses strive to do so by eliminating or minimising their wasteful by-products—simply because those by-products have little to no market value and add cost to production. If our cost of production is higher, then profit margins are lower. That in fact is the basis for pollution prevention and waste minimisation practices of modern-day industry.

A water treatment facility differs in this regard because the primary objective is to produce high quality water by removing or destroying as much of the contaminants as possible. We cannot produce high quality water without generating the wasteful by-product, sludge, very often in large quantities. Water treatment plants are simply pollution control technologies, whether they are applied to industrial applications or municipal. That does not mean that pollution prevention practices are not appropriate for water treatment plants—they most certainly are and can minimise solid waste generation. But understand that we cannot eliminate the wasteful by-product of sludge as one might try and do if we had an manufacturing facility and we identified another technology to make our product and eliminate a wastestream generated by the older technology.

Sludge or solid waste is unavoidably produced in the treatment of water containing suspended solids. There are, however different technologies that we can select among that will indeed concentrate these solids and thereby reduce the volumes that we ultimately must dispose of. In addition, some sludge can be stabilised and treated, which can impart a low, but none-the-less marketable value to this waste. These technologies and practices do indeed constitute pollution prevention and waste minimisation programmes within water treatment plant operations and they can have a very significant and positive impact on the overall costs of the operation. This brings us to a collection of technologies that focus on: (i) sludge concentration, (ii) sludge stabilisation, and (iii) sludge handling and disposal. Some technologies fall into the category of pollution prevention, while others are within the normal arena of solid waste management and disposal.

Note that pollution prevention or P2 technologies, as in other industry sectors, are not necessarily the preferred choices. Specific technology selection quite often depends on localised conditions. By this, we mean the properties of the sludge, the volumes handled and the comparative costs between technologies and or practices. In a very general sense, pollution prevention technologies are only appropriate when they are financially attractive for an operation. Like any other engineering project, the investment into a technology that falls within the pollution prevention arena must have financial attractiveness. An alternative way of stating this is that there are indeed situations where more conventional methods resulting in large volumes of sludge are more cost effective than a leading-edge technology that minimises or reduces sludge volumes. The financial attractiveness of an investment needs to be assessed on a case by case basis. This chapter focuses on sludge processing and post-processing technologies. Where appropriate, we will point out which technologies may be considered P2.

When we think of sludge, what automatically comes to mind is sewerage. Water carriage systems of sewerage provide a simple and economical means for removing offensive and potentially dangerous wastes from household and industry. The solution and suspension of solids in the transporting of water produces sewage. Thus, the role of solids and sludge removal at Sewage Treatment Plants is apparent. Sludge removal is complicated by the fact that some of the waste matters go into solution while others are colloidal or become finely divided in their flow through the sewage system. Ordinarily, less than half of such waste remains in suspension in a size or condition that can be separated by being strained out, skimmed off or settled out. The remainder must then be precipitated out by chemical means, filtered mechanically or be subjected to biological treatment whereby they are either removed from the water or changed in character as to be rendered innocuous. Sewage contains mineral and organic matter in suspension (coarse and fine suspended matter), in colloidal state (very finely dispersed matter) and in solution. Living organisms, notably bacteria and protozoa, find sewage to be an abundant source of food and their lives' activities result in the decomposition of sewage.

Sewage becomes offensive due to its own instability together with the objectionable concentration of suspended materials. In addition, the potential presence of disease producing organisms makes sewage dangerous. Removal or stabilisation of sewage matters may be accomplished in treatment works by a number of different methods or by a suitable combination of these methods.

While sewage sludge is rich in nutrients and organic matter, offering the potential for applications as a biosolid (discussed below) or it has a heating value making it suitable for incineration, many industrial sludge are often unsuitable for reuse. A more common practice with industrial sludge is to try and identify a reclaim value, i.e. if the sludge can be concentrated sufficiently there may be portion of this waste which is reclaimable or may enter into a recycling market.

STABILISATION AND CONDITIONING

Pre-Stage Basics

Before sludge undergoes treatment such as dewatering or thickening, it must be stored and pretreated. Sludge storage is an important, integral part of every waste-water sludge treatment and disposal system. Sludge storage provides many benefits including equalisation of sludge flow to downstream processes, allowing sludge accumulation during times of non-operation of sludge-processing facilities and allowing a uniform feed rate that enhances thickening, conditioning and dewatering operations.

Sludge is stored within waste-water treatment process tankage, sludge treatment process systems, or separately in specially designed tanks. Sludge can be stored on a short-term or a long-term basis. Small treatment plants, where storage time may vary from several to 24 hours, may store sludge in waste-water clarification basins or sludge-thickening tanks. Larger plants often use aerobic digester, facultative lagoons and other processes with long detention times to store sludge. The pretreatment of sludge is often necessary before dewatering or thickening can take place. It includes degritting and grinding. Sludge degritting involves the installation of grit removal and processing facilities at the head works where raw waste-water first enters the treatment plant. As a result, there is reduced wear on influent pumping systems and primary sludge pumping, piping and thickening systems. Sludge grinding involves shearing of large sludge solids into smaller particles. This method is used to prevent problems with operation of downstream processes. Inline grinders reduce cleaning and maintenance down time of equipment. The grinders can shear sludge solids to 6 to 13 mm, depending on design requirements.

Sludge-pumping systems play an important part in waste-water treatment plants, particularly those operations experiencing average flows of greater than 1 million gallons per day (mgd). There are different types of pumps within this process. Typical advantages of kinetic pumps for sludge transport include lower purchase cost, lower maintenance cost due to wear, less space used and availability of both dry-well and submersible pumps. Advantages of positive displacement pumps include improved process control and pumping capability at high pressure and low flow.

Sludge cake storage (where a cake is the dewatered solid part of sludge) provides similar benefits for downstream disposal alternatives, like composting and incineration, to sludge storage which is used for thickening and dewatering. Storage of sludge cakes increases operational reliability, evens outflow fluctuations and allows accumulation when downstream operations are not in service. Bins or hoppers are used to store sludge cakes. These can be made of any size form several cubic meters to 380 cubic meters capacity. Existing sludge dewatering operations can produce cakes that are 15 to 40 per cent solids. These cakes range in consistency from pudding to damp cardboard. Since they will not flow by gravity in a pipe or channel, sludge cakes must be transported by one of the following methods: mechanical conveyors such as flat or troughed belt, corrugated belt or Archimedes screw; gravity drop from dewatering equipment into storage hoppers directly below; and pumping by positive displacement pumps.

Before any of the sludge can proceed to dewatering or thickening processes, it must be conditioned. Sludge conditioning involves chemical or thermal treatment to improve the efficiency of the downstream processes. Chemical conditioning involves use of inorganic chemicals or organic polyelectrolytes or both. The most commonly used inorganic chemicals are ferric chloride and lime. Organic polymers are used for both sludge-thickening and dewatering processes. Their advantage over inorganics is that polymers do not greatly increase the amount of sludge production: 1 kg of inorganic chemicals added will produce 1 kg of extra sludge. The disadvantage of polymers is their relatively high cost. There are several important factors that affect conditioning of sludge. They include: sludge characteristics, sludge handling and sludge coagulation and flocculation. The fundamental purpose of sludge conditioning is to cause the aggregation of fine solids by coagulation and inorganic chemicals, flocculation with organic polymers or both. A critical design parameter in conditioning is dosage. Selection of the right dosage of a chemical conditioner is critical for good performance. The dosage affects the solids content of sludge cakes as well as solids capture rate and solids disposal cost. Dosage is determined from pilot studies, bench tests and on-line tests. In the following sections we will cover the basics of sludge stabilisation and then conditioning. Our objective is to gain a working knowledge of these operations and to build our vocabulary.

Chemical Stabilisation

Chemical stabilisation is a process whereby the sludge matrix is treated with chemicals in different ways to stabilise the sludge solids. Two common methods employed are lime stabilisation and the use of chlorine.

The lime stabilisation process can be used to treat raw primary, waste activated, septage and anaerobically digested sludge. The process involves mixing a large enough quantity of lime with the sludge to increase the pH of the mixture to 12 or more. This normally reduces bacterial hazards and odour to a negligible value, improves vacuum filter performance and provides satisfactory means of stabilising the sludge prior to ultimate disposal.

Stabilisation by chlorine addition has been developed and is marketed under the registered trade name 'Purifax'. The chemical conditioning of sludge with chlorine varies greatly from the more traditional methods of biological digestion or heat conditioning. First, the reaction is almost instantaneous. Second, there is very little volatile solids reduction in the sludge. There is some breakdown of organic material and formation of carbon dioxide and nitrogen; however, most of the conditioning is by the substitution or addition of chlorine to the organic compound to form new compounds that are biologically inert.

Stabilisation via Aerobic Digestion

Aerobic digestion is an extension of the activated sludge aeration process whereby waste primary and secondary sludge are continually aerated for long periods of time. In aerobic digestion the micro-organisms extend into the endogenous respiration phase. This is a phase where materials previously stored by the cell are oxidised, with a reduction in the biologically degradable organic matter. This organic matter, from the sludge cells is oxidised to carbon dioxide; water and ammonia. The ammonia is further converted to nitrates as the digestion process proceeds. Eventually, the oxygen uptake rate levels off and the sludge matter is reduced to inorganic matter and relatively stable volatile solids.

The primary advantage of aerobic digestion is that it produces a biologically stable end product suitable for subsequent treatment in a variety of processes. Volatile solids reductions similar to anaerobic digestion are possible. Some parameters affecting the aerobic digestion process are:

1. The rate of sludge oxidation.
2. Sludge temperature.
3. System oxygen requirements.
4. Sludge loading rate.
5. Sludge age.
6. Sludge solids characteristics.

Aerobic digestion has been applied mostly to various forms of activated sludge treatment, usually 'total oxidation' or contact stabilisation plants. However, aerobic digestion is suitable for many types of municipal and industrial waste-water sludge, including trickling filter humus as well as waste activated sludge. Any design for an aerobic digestion system should include an estimate of the quantity of sludge to be produced, the oxygen requirements, the unit detention time, the efficiency desired and the solids loading rate. Aerobic digestion tanks are normally not covered or heated, therefore, they are much cheaper to construct than covered, insulated and heated anaerobic digestion tanks. In fact, an aerobic digestion tank can be considered to be a large open aeration tank. Similar to conventional aeration tanks, the aerobic digesters may be designed for spiral roll or cross roll aeration using diffused air equipment. The system should have sufficient flexibility to allow sludge thickening by providing supernatant decanting facilities. The advantages most often claimed for aerobic digestion are:

1. A humus-like, biologically stable end product is produced.
2. The stable end product has no odours; therefore, simple land disposal, such as lagoons, is feasible.
3. Capital costs for an aerobic system are low, when compared with anaerobic digestion and other schemes.

4. Aerobically digested sludge usually has good dewatering characteristics. When applied to sand drying beds, it drains well and redries quickly if rained upon.

5. The volatile solids reduction can be equal to those achieved by anaerobic digestion.

Supernatant liquors from aerobic digestion have a lower BOD than those from anaerobic digestion. Most tests indicated that BOD would be less than 100 ppm. This advantage is important because the efficiency of many treatment plants is reduced as a result of recycling high BOD supernatant liquors. There are fewer operational problems, with aerobic digestion than with the more complex anaerobic form because the system is more stable. As a result, less skillful and costly labour can be used to operate the facility. In comparison with anaerobic digestion, more of the sludge basic fertiliser values are recovered.

The major disadvantage associated with aerobic digestion is high power costs. This factor is responsible for the high operating costs in comparison with anaerobic digestion. At small waste treatment the power costs may not be significant but they certainly would be at large plants. Aerobically digested sludge does not always settle well in subsequent thickening processes. This situation leads to a thickening tank decant having a high solids concentration. Some sludge do not dewater easily by vacuum filtration after being digested aerobically. Two other minor disadvantages are the lack of methane gas production and the variable solids reduction efficiency with varying temperature changes.

In a typical plant operation the pollutants dissolved in the waste-water or that would not settle in the primary clarifiers flow on in the waste-water to the Secondary treatment process. Secondary treatment further reduces organic matter (BOD_5) through the addition of oxygen to the waste-water which provides an aerobic environment for micro-organisms to biologically break down this remaining organic matter. This process increases the per cent removals of BOD and TSS to a minimum of 85 per cent. A secondary treatment facility can be comprised of Oxygenation Tanks, Pure Oxygen Generating Plant, Liquid Oxygen Storage Tanks, Secondary Clarifiers, Return Sludge Pumping Station and Splitter Box, Sludge Thickeners and Pumping Station, Sludge Dewatering Building Addition and modifications to the existing Service Water Pumping Station. The Pure Oxygen Generation System often incorporates a pressure swing adsorption (PSA) system oxygen generating system A PSA system will provide a certain amount (as tonnes per day) of pure oxygen to the oxygenation system. As backup to the oxygen generating system, spare oxygen storage tanks containing liquid oxygen can be included in the design. Figure 15.1 illustrates what an aeration reactor looks like.

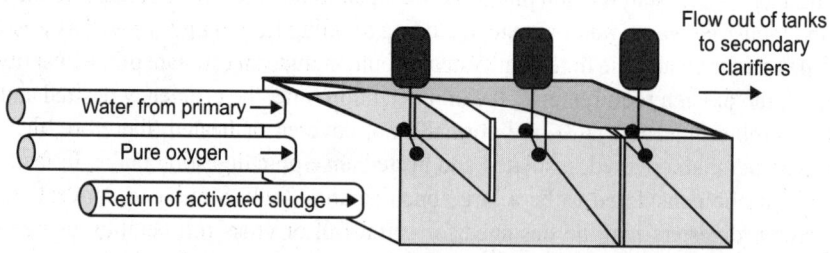

Fig. 15.1. Example of aeration reactors.

The oxygenation system is comprised of several covered oxygenation tanks, mechanical mixing system and pressure-controlled oxygen feed and oxygen purity-controlled venting system. The primary effluent enters the head end of the tanks where it mixes with return activated sludge which consists of micro-organisms 'activated' by the organic matter and oxygen. This combination of primary effluent and return sludge forms a mixture known as 'Mixed Liquor'. This mixed liquor is continuously and thoroughly mixed by the mechanical mixer in each tank. The oxygen gas produced in the PSA system is introduced into the first stage of each tank and then remains to contact with the mixed liquor throughout the oxygenation system. Secondary clarifiers are used in this process.

Once the mixed liquor goes through the complete oxygenation process, it flows to four secondary clarifiers where the biological solids produced during the oxygenation process are allowed to settle and be pumped back to the head of the system. These settled solids being pumped, called return activated sludge, mix with the primary effluent to become mixed liquor. Since the population of micro-organisms is growing some micro-organisms in the return activated sludge are removed from the system. This solids waste stream is called waste activated sludge (WAS) and flows to the secondary gravity thickener for solids processing. The cleaned waste-water flows over the weir of the secondary clarifier and on to the disinfection (chlorination)-process. The activated sludge process describes is an aerobic, suspended growth, biological treatment method. It employs the metabolic reactions of micro-organisms to produce a high quality effluent by oxidation and conversion of organics to carbon dioxide, water and biosolids (sludge). Basically the system speeds up nature and supplies oxygen so the aquatic environment will not have to. High concentrations of micro-organisms (compared to a natural aquatic environment) in the activated sludge use the pollutants in the primary treated waste-water as food and remove the dissolved and non-settleable pollutants from the waste-water. These pollutants are incorporated into the micro-organisms bodies and will then settle in the secondary clarifiers. Oxygen needs to be supplied for the micro-organisms to survive and consume the pollutants.

Stabilisation via Anaerobic Digestion

The purpose of digestion is to attain both of the objectives of sludge treatment—a reduction in volume and the decomposition of highly putrescible organic matter to relatively stable or inert organic and inorganic compounds. Additionally, anaerobic sludge digestion produces a valuable by-product in the form of methane gas (the primary constituent of natural gas, which we can burn for heat or convert to electricity). Sludge digestion is carried out in the absence of free oxygen by anaerobic organisms. It is, therefore, anaerobic decomposition. The solid matter in raw sludge is about 70 per cent organic and 30 per cent inorganic or mineral. Much of the water in waste-water sludge is 'bound' water which will not separate from the sludge solids. The facultative and anaerobic organisms break down the complex molecular structure of these solids setting free the 'bound' water and obtaining oxygen and food for their growth.

Anaerobic digestion involves many complex biochemical reactions and depends on many interrelated physical and chemical factors. For purposes of simplification, the anaerobic degradation of domestic sludge occurs in two steps. In the first step, acid forming bacteria attack the soluble or dissolved solids, such as the sugars. From these reactions organic acids, at times up to several thousand ppm and gases, such as carbon dioxide and hydrogen sulphide are formed. This is known as the stage of acid fermentation and proceeds rapidly. It is followed by a period of acid digestion in which the organic acids and nitrogenous compounds are attacked and liquefied at a much slower rate.

In the second stage of digestion, known as the period of intensive digestion, stabilisation and gasification, the more resistant nitrogenous materials, such as the proteins, amino-acids and others are attacked. The pH value must be maintained from 6.8 to 7.4. Large volumes of gases with a 65 or higher percentage of methane are produced. The organisms which convert organic acids to methane and carbon dioxide gases are called methane formers. The solids remaining are relatively stable or only slowly putrescible, can be disposed of without creating objectionable conditions and have value in agriculture. The whole process of sludge digestion may be likened to a factory production line where one group of workers takes the raw material and conditions it for a second group with different 'skills' who convert the material to the end products.

In a healthy, well operating digester, both of the above stages are taking place continuously and at the same time. Fresh waste-water solids are being added at frequent intervals with the stabilised solids being removed for further treatment or disposal at less frequent intervals. The supernatant digester liquor, the product of liquefaction and mechanical separation is removed frequently to make room for the added fresh solids and the gas is, of course, being removed continuously.

While all stages of digestion may be proceeding in a tank at the same time with the acids produced in the first stage being neutralised by the ammonia produced in subsequent stages, best and quickest results are obtained when the overall pH of 6.8 to 7.4 predominates. The first stage of acid formation should be evident only in starting up digestion units. Once good alkaline digestions is established, the acid stage is not apparent unless the normal digestion becomes upset by overloading, poisonous chemicals or for other reasons. It is critical to the overall process to maintain balanced populations of acid formers and methane formers. The methane formers are more sensitive to environmental conditions and slower growing than the acid forming group of bacteria and control the overall reactions.

The progress of digestion can be measured by the destruction of organic matter (volatile solids) by the volume and composition of gases produced, by the pH, volatile acids and alkalinity concentration. It is recommended that no on parameter or test be used to predict problems or control digesters. Several of the following parameters must be considered together.

The reduction of organic matter as measured by the volatile solids indicates the completeness of digestion. Raw sludge usually contains from 60 to 70 per cent volatile solids while a well digested sludge may have as little as 50 per cent. This would represent a volatile solids reduction of about 50 per cent. Volatile solids reduction should be measured weekly and trended. Downward trends in volatile solids reduction might mean:

1. Temperature too low and/or poor temperature control.
2. Digester is overloaded.
3. Ineffective mixing of digester contents.
4. Grit and/or scum accumulations are excessive.
5. Low volatile solids in raw sludge feed.

A well digested sludge should be black in colour, have a not unpleasant tarry odour and when collected in a glass cylinder, should appear granular in structure and show definite channels caused by water rising to the top as the solids settle to the bottom.

For domestic waste-water in a normally operating digestion tank, gas production should be in the vicinity of 12 cu.ft. of gas per day per lbs of volatile matter destroyed. This would indicate that for a 50 per cent reduction of volatile matter, a gas yield of six cu.ft. per lbs of volatile matter added should be attained. The quantity of gases produced should be relatively constant if the feed rate is constant.

Sharp decreases in total gas production may indicate toxicity in the digester. The gas is usually about 70 per cent methane, about 30 per cent carbon dioxide and inert gases such as nitrogen. An increasing percentage of carbon dioxide may be an indication that the digestion process is not proceeding properly. In plants with primary and secondary digester, raw sludge is pumped to the primary digester displacing partially digested sludge. The major portion of the digestion with the greatest gas yield is in the primary digester.

Volatile acids (mainly acetic acid) are generated by the acid forming bacteria as a result of the initial breakdown of the sludge solids. The volatile acids concentration indicates digestion progress and is probably the best warning sign of trouble. In a well operating digester, the volatile acids concentration should be measured weekly and remain fairly constant. Sudden increases in volatile acids means digester trouble. During periods of digester imbalance, volatile acids should be measured daily.

Bicarbonate alkalinity indicates the buffering capacity of the sludge, the ability to keep the pH constant and the ability to neutralise acids. Normally, the bicarbonate alkalinity varies between 1500 and 6000 mg/l (as calcium carbonate). The ratio between the volatile acids and the bicarbonate alkalinity concentrations is an excellent process indicator. Normally if the ratio of volatile acids concentration (mg/liter) to bicarbonate alkalinity (mg/liter) < 0.25, the digester is operating properly. A rising volatile acids to bicarbonate alkalinity ratio means possible trouble. Sometimes either decreasing the sludge feed to digester or resting the digester will correct the problem.

Since digestion is accomplished by living organisms, it is desirable to provide an environment in which they are most active and carry on their work in the shortest time. The environmental factors involved are moisture, temperature, availability of proper food supply, mixing and seeding, alkalinity and pH. To these might be added the absence of chemicals toxic to the organisms. Moisture is always adequate in waste-water sludge.

It has been found that sludge digestion proceeds in almost any range of temperature likely to be encountered, but the time taken to complete digestion varies greatly with the temperature. Also rapid changes in temperature are detrimental. Digester temperature should not vary more than ± 2°F per day. Pumping excessive quantities of thin sludge can cause significant decreases in digester temperature. Thin, dilute sludge with a high moisture content also waste digester space and reduce solids retention time. The methane forming organisms are extremely sensitive to changes in temperature. At a temperature of 55°F, about 90 per cent of the desired digestion is completed in about 55 days. As the temperature increases, the time decreases, so that at 75°F the time is cut to 35 days, at 85°F to 26 days and at 95°F to 24 days. The theoretical time for sludge digestion at 95°F is one half that at 60°F. Of course the figures are average, not exact figures for all sludge of varying, composition. These digestion times may be materially reduced in digesters provided with efficient mixing of thickened sludge.

The proper amount of food must be provided for the digester organisms. This is in the form of volatile sludge solids from the various waste-water treatment units. The total volume of raw sludge pumped to the digester, the rate at which it is pumped and the degree to which it is made available to all of the different groups of organisms are vital factors in efficient digester operation. If too much sludge is added to a digester, the first or acid stage, predominates to such an extent that the environment becomes unfavourable for the organisms responsible for the second stage of digestion, the balance of the whole digestion process is upset and the digester is said to be overloaded. If this is due to unbalanced plant design whereby the digester capacity is too small in relation to the sludge producing units, the only solution is to provide additional digester capacity.

There are, however, other factors which, can upset the balance of the digestion process and which are under the control of the operator. In heated digesters, failure to maintain uniform temperatures in the digester within the proper range will upset the digestion process. Adding fresh solids in large volumes at widely separated intervals or removing too much digested sludge, at one time will result in temporary overloading. Avoid shock loadings of solids. In unheated digestion tanks similar conditions are to be expected seasonally and during winter months digester organisms are almost dormant, so that with the advent of warm weather there is in the digester an excessive accumulation of almost raw sludge solids.

This, together with the normally slower digestion in unheated tanks, necessitates storage capacity twice that needed in heated digesters.

The organisms in a digester are most efficient when food is furnished them in small volumes at frequent intervals. Fresh sludge solids should therefore be pumped to the digester as often as practical, at least twice a day for the smallest plants and more frequently where facilities and operator's attention are available. This, of course, fits in with the proper schedule of removing sludge from settling units before it becomes septic.

In starting a digester unit, quickest results can be obtained by putting in it at the start some digested sludge if this is obtainable from another digester or a nearby plant. In this way all stages of digestion can be started almost simultaneously instead of by successive stages. This seeding supplies an adequate number of organisms of the methane forming type to consume the end products of the first stage and in this way the unit will 'ripen' in the shortest time. After normal operation has been established, seeding of the fresh solids as added to the digester by mixing them with the digesting sludge greatly improves the rate of digestion.

Role of Mixing

Mixing plays an important role in digester operation. Without well-mixed systems, the processes cannot have acceptable levels of efficiency. There are a number of methods or combination of methods whereby proper mixing is attained. These include:

1. Stirring by rotating paddles and scum breaker arms.
2. Forced circulation of sludge and/or supernatant by pumps or by draft tubes with impeller.
3. Discharge of compressed sludge gas from diffusers at the bottom of the digestion tank.

Mixing may be either intermittent or continuous, but however effected it provides all working organisms their proper food requirements and helps maintain uniform temperature. Intermittent mixing allows separation and removal of supernatant from a single stage digester. With continuous mixing the digestion proceeds at a higher rate throughout the entire tank, thus reducing the tank capacity needed. Such continuous mixing requires a second digester or storage tank into which digesting sludge may be moved to make room for fresh sludge in the first digester and to make possible separation and removal of supernatant in the secondary digester.

Sludge Conditioning Using Chemicals

Sludge conditioning is a process whereby sludge solids are treated with chemicals or various other means to prepare the sludge for dewatering processes. Chemical conditioning (sludge conditioning) prepares the sludge for better and more economical treatment with vacuum filters or centrifuges. Many chemicals have been used such as sulphuric, acid, alum, chlorinated copperas, ferrous sulphate and ferric chloride with or without lime and others. The local cost of the various chemicals is usually the determining factor. In recent years the price of ferric, chloride has been reduced to a point where it is the one most commonly used. The addition of the chemical to the sludge lowers or raises its pH value to a point where small particles coagulate into larger ones and the water in the sludge solids is given up most readily. There is no one pH value best for all sludge. Different sludge such as primary, various secondary and digested sludge and different sludge of the same type have different optimum pH values which must be determined for each sludge by trial and error. Tanks for dissolving acid salts, such as ferric chloride, are lined with rubber or other acid-proof material. Intimate mixing of sludge and coagulant is essential for proper conditioning.

Feeders are also necessary for applying the chemicals needed for proper chemical conditioning. The most frequently encountered conditioning practice is the use of ferric chloride either alone or in combination with lime. The use of polymers is rapidly gaining widespread acceptance. Although ferric chloride and lime are normally used in combination, it is not unusual for them to be applied individually. Lime alone is a fairly popular conditioner for raw primary sludge and ferric chloride alone has been used for conditioning activated sludges. Lime treatment to a pH of 10.4 or above has the added advantage of providing a significant degree (over 99 per cent) of disinfection of the sludge. Organic polymer coagulants and coagulant aids have been developed in the past 20 years and are rapidly gaining acceptance for sludge conditioning. These polymers are of three basic types:

1. Anionic (negative charge)—serve as coagulants aids to inorganic aluminium and iron coagulants by increasing the rate of flocculation, size and toughness of particles.
2. Cationic (positive charge)—serve as primary coagulants alone or in combination with inorganic coagulants such as aluminium sulphate.
3. Nonionic (equal amounts of positively and negatively charged groups in monomers)— serve as coagulant aids in a manner similar to that of both anionic and cationic polymers.

The popularity of polymers is primarily due to their ease in handling, small storage space requirements and their effectiveness. All of the inorganic coagulants are difficult to handle and their corrosive nature can cause maintenance problems in the storing, handling and feeding systems in addition to the safety hazards inherent in their handling.

Sludge Conditioning by Thermal Methods

There are two basic processes for thermal treatment of sludge. One, wet air oxidation, is the flameless oxidation of sludge at temperatures of 450° to 550°F and pressures of about 1200 psig. The other type, heat treatment, is similar but carried out at temperatures of 350° to 400°F and pressures of 150 to 300 psig. Wet air oxidation (WAO) reduces the sludge to an ash and heat treatment improves the dewaterability of the sludge. The lower temperature and pressure heat treatment is more widely used than the oxidation process.

When the organic sludge is heated, heat causes water to escape from the sludge. Thermal treatment systems release water that is bound within the cell structure of the sludge and thereby improves the dewatering and thickening characteristics of the sludge. The oxidation process further reduces the sludge to ash by wet incineration (oxidation). Sludge is ground to a controlled particle size and pumped to a pressure of about 300 psi. Compressed air is added to the sludge (wet air oxidation only), the mixture is brought to a temperature of about 350°F by heat exchange with treated sludge and direct steam injection and then is processed (cooked) in the reactor at the desired temperature and pressure. The hot treated sludge is cooled by heat exchange with the incoming sludge. The treated sludge is settled from the supernatant before the dewatering step. Gases released at the separation step are passed through a catalytic after-burner at 650 to 705°F or deodourised by other means. In some cases these gases have been returned through the diffused air system in the aeration basins for deodourisation.

An advantage of thermal treatment is that a more readily dewaterable sludge is produced than with chemical conditioning. Dewatered sludge solids of 30 to 40 per cent (as opposed to 15 to 20 per cent with chemical conditioning) have been achieved with heat treated sludge at relatively high loading rates on the dewatering equipment (2 to 3 times the rates with chemical conditioning). The process also provides effective disinfection of the sludge. Unfortunately, the heat treatment process ruptures the cell walls of biological organisms, releasing not only the water but some bound organic material. This returns to solution some organic material previously converted to particulate form and creates other fine particulate matter. The breakdown of the biological cells as a result of heat treatment converts these previously particulate cells back to water and fine solids. This aids the dewatering process, but creates a separate problem of treating this highly polluted liquid from the cells. Treatment of this water or liquor requires careful consideration in design of the plant because the organic content of the liquor can be extremely high.

The WAO process also aims to reduce sludge volume. The other thermal sludge conditioning method is best-known as sludge pasteurisation and deserves more than just a brief overview.

Sludge Pasteurisation Process

This process is really sludge disinfection. Its aim is the destruction or inactivation of pathogenic organisms in the sludge. Destruction is defined as the physical disruption or disintegration of a pathogenic organism, while inactivation is defined as the removal of a pathogen's ability to infect.

In the United States procedure to reduce the number of pathogenic organisms are a requirement before sale of sludge or sludge-containing products to the public as a soil conditioner or before recycling sludge to croplands. Since the final use of disposal of sludge may differ greatly with respect to health concerns and since a great number of treatment options effecting various degrees of pathogen reduction are available, the system chosen for the reduction of pathogens should be tailored to the specific application. Thermal conditioning of sludge in a closed, pressurised system destroys pathogenic organisms and permits dewatering. The product generally has a good heating value or can be used for landfilling or fertiliser base. In this process, sewage sludge is ground and pumped through a heat exchanger and sent with air to reactor where it is heated to a temperature of 350–400°F. The processed sludge and air are returned through the heat exchanger to recover heat. The conditioned slurry is then discharged to a gravity thickener where the vapours are separated and the solids are concentrated (thickened). The treatment process renders the sludge easily dewaterable without the addition of chemicals.

After thickening, a variety of sludge handling and disposal options are available. For example, the thickened sludge can be applied directly to land. If liquid disposal is not applicable to a specific project, the thickened sludge can be dewatered by centrifugation, vacuum filtration or filter pressing. The dewatered residue can then be landfill or incinerated. These options are discussed further on.

Thermal sludge conditioning and its effects on the chemical and physical structures of waste-water sludge can be best understood from analyses of typical sewage sludges. Waste-water sludge is a complex mixture of waste solids forming a gelatinous mass that is nearly impossible to dewater without further treatment. The organic fraction of the sludge consists of lipids, proteins and carbohydrates, all bound by physical-chemical forces in a predominantly water-gel-like structure. When the sludge is heated under pressure to temperatures above 350°F (176.5°C), the gel-like structure of sludge is destroyed, liberating the bound water. Dewatering by filtration without chemical conditioning is then a simple matter.

There are several characteristics of thermally conditioned sludge which have an important effect on the cost of plants utilising thermoconditioning. Various factors, such as thickening properties, dewatering properties, heavy metal distribution, heating value, volatile solids solubilisation and others, all have a major impact on the evaluation of various process alternatives and ultimate disposal. Thickening and dewatering properties vary depending on the type of sludge. In general, vacuum filtration rates vary from 2–15 lbs/ft^2/hr., with cake moistures ranging from 50–70 per cent. Lower values (2–4 lbs/ft^2/hr.) are observed for high proportions of waste-activated sludge; the higher values (up to 15 lbs/ft^2/hr.) are observed for sludge which are predominantly primary sludge.

Similar results have been obtained for filter pressing. Mixtures of primary and waste-activated sludge of relatively the same proportions which have been thermally conditioned and dewatered at rates from 2540 lbs dry solids per ft/hr., with cake moistures ranging from 50–60 per cent. Heating values of thermally conditioned sludge cake are typically about 12500 Btu/lbs. No marked differences in heating values have been found for different types of sludge. Sludge conditioned with ferric chloride and lime have been found to have heating values in the range of 9000–10000 Btu/lbs. The lower values, experienced for chemically conditioned sludge could be due to:

1. Selective solubilisation of materials of lower heating value in the thermal conditioning process.
2. Enclothermic reactions with the conditioning materials.
3. Operational differences in the analytical methods used for determining the heating value of the volatile content.

Solubilisation of a fraction of the influent-suspended solids can occur as a result of thermal conditioning. In low-pressure, wet-air oxidation, some of the organics present are oxidised as well. Solubilisation of the volatile suspended solids produces a supernatant or filtrate of relatively high organic strength.

Ash solubilisation and volatile suspended-solids oxidation also decrease the solids loads to downstream solids-handling units.

There are several advantages that thermoconditioning has over chemical conditioning. These include the sterility of the end product and a residue that can be readily thickened. Bacteria are numerous in the human digestive tract; humans excrete up to 10^{13} coliform and 10^{16} of other bacteria in their feces every day. The most important of the pathogenic bacteria are listed in Table 15.1, together with the diseases they cause which may be present in municipal waste-water treatment sludges. Table 15.2 lists potential parasites in waste-water sludge.

Table 15.1. Pathogenic human bacteria potentially in waste-water sludge.

Species	Disease
Arizona hinshawii	Arizona infection
Bacillus cereus	B. cereus gastroenteritis; food poisoning
Vibrio cholerae	Cholera
Clostridium perfringens	C. perfringens gastroenteritis; food poisoning
Clostridium tetani	Tetanus
Escherichia coli	Enteropathogenic E. coli infection; acute diarrhoea
Leptospira sp	Leptospirosis; Swineherd's disease
Mycobacterium tuberculosis	Tuberculosis
Salmonella paratyphi, A, B, C	Paratyphoid fever
Salmonella sendai	Paratyphoid fever
Salmonella sp (over 1500 serotypes)	Salmonellosis; acute diarrhoea
Salmonell typhi	Typhoid fever
Shigella sp.	Shigellosis; bacillary dysentery; acute diarrhoea
Yersinia enterocolitica	Yersinia gastroenteritis
Yersinia pseudotuberculosis	Mesenteric lymphadenopathy

Table 15.2. Pathogenic human and animal parasites found in waste-water sludge.

Species	Disease
Protozoa	
Acanthamoeba sp	Amoebic meningoencephalitis
Balantidium coli	Balanticliasis, Balantidial dysentery
Dientamoeba fragilis	Dientamoeba infection
Entamoeba histolytica	Amoebiasis; amoebic dysentery
Giardia lamblia	Giardiasis
Isospora bella	Coccidiosis
Naegleria fowleri	Amoebic meningoencephalitis
Toxoplasma gordii	Toxoplasmosis
Nematodes	
Ancyclostoma dirodenale	Ancylostomiasis; hookworm disease
Ancyclostoma sp	Cutaneous larva migrans
Ascaris lumbricoides	Ascariasis; roundworm disease; *Ascaris pneumonia*
Enterobius vermicularis	Oxyuriasis; pinworm disease
Necator americanus	Necatoriasis; hookworm disease
Strongyloides stercoralis	Strongyloidiasis; hookworm disease
Toxocara canis	Dog roundworm disease, visceral larva migrans
Toxocara cati	Cat roundworm disease; visceral larva migrans
Trichusis trichiura	Trichuriasis; whipworm disease
Helminths	
Diphyllobothrium latum	Fish tapeworm disease
Echinococcus granulosis	Hydatid disease
Echinococcus multilocularis	Aleveolar hydatid disease
Hymenolepsis diminuta	Rat tapeworm disease
Tymenolepsis nana	Dwarf tapeworm disease
Taenia saginata	Taeniasis; beef tapeworm disease
Taenia solium	Cysticercosis; pork tapeworm disease

Sludge stabilisation processes are ideally intended to reduce putrescibility, decrease mass and improve treatment characteristics such as dewaterability. Many stabilisation processes also accomplish substantial reductions in pathogen concentration. Sludge digestion is one of the major methods for sludge stabilisation. Well-operated digesters can substantially reduce virus and bacteria levels but are less effective against parasitic cysts. The requirement for pasteurisation is that all sludge be held above a predetermined temperature for a minimum time period. Heat transfer can be accomplished by steam injection or with external or internal heat exchangers. Steam injection is preferred because heat transfer through the sludge slurry is slow and not dependable. Incomplete mixing will either increase heating time, reduce process effectiveness or both. Overheating or extra detention times are not desirable, however, because trace metal mobilisation may be increased, odour problems will be exacerbated and unneeded energy will be expended. Batch processing is preferable to avoid reinoculations if short circuiting occurs.

In components include a steam boiler, a pre-heater, a sludge heater, a high-temperature holding tank, blow-off tanks and storage basins for the untreated and treated sludge.

Pasteurisation is employed extensively in Western Europe. As examples, in Germany and Switzerland it is required before application of sludge to farmlands during the spring-summer growing season. Based on Western European experience, heat pasteurisation is a proven technology, requiring skills such as boiler operation and understanding of high-temperature and pressure processes. Pasteurisation can be applied to either untreated or digested sludge with little pre-treatment. Digester gas, available in many plants, is an ideal fuel and is usually produced in sufficient amounts to disinfect locally produced sludge. The disadvantages of this process include significant odour problems and the need for large sludge storage facilities following the process. The storage facilities are not only a problem because of space requirements, but they offer the opportunity for bacterial pathogens to regrow if the sludge becomes reinoculated.

A pasteurisation process should be designed to provide uniform minimum temperature of at least 70°C for at least 30 minutes (note—some literature sources argue for higher minimum temperatures). Batch processing is necessary in order to prevent short-circuiting and recontamination of the sludge, especially by bacteria. In-line mixing of steam and sludge is normally, practiced to ensure uniform heat transfer among the sludge mass.

This practice also eliminates the need to mix the sludge while it is held at the pasteurisation temperature. Figure 15.2 provides us a glimpse of what important process components are in a sludge pasteurisation process.

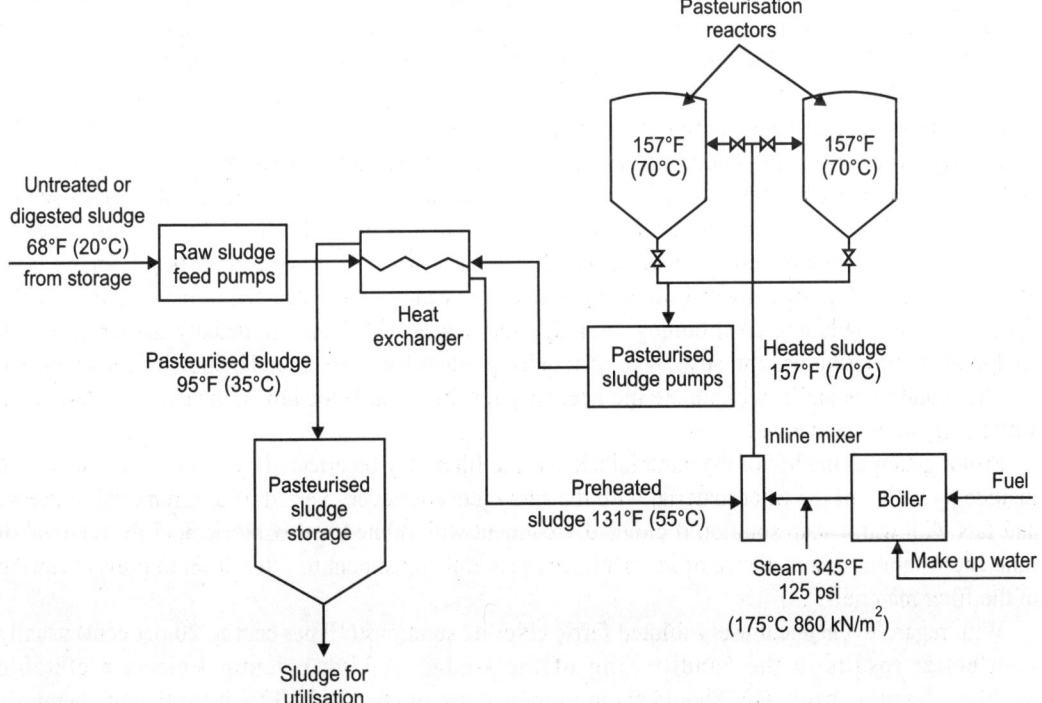

Fig. 15.2. Sludge pasteurisation process with heat recovery.

Blending

Stabilisation can be aided by the technique of blending sludge. Blending is a process where two or more types of sludge are 'blended' together to facilitate a higher sludge solids concentration and more homogenous mixture of sludge prior to dewatering. Blending operations tends to decrease the chemical demand for conditioning and dewatering sludge. The blending operation usually takes place in sludge holding tanks normally where primary sludge is mixed with waste activated sludge. The amounts of the sludge to be blended can only be found by experimentation, with the final results being seen at the dewatering operations.

SLUDGE DEWATERING OPERATIONS

Another term for dewatering the sludge is sludge thickening. The objective is to concentrate the sludge and quite frankly—make it as dry as economically possible for post processing and disposal purposes. There are both mechanical and thermal techniques for achieving this. Among the mechanical processes used to dewater sludge are belt filter presses and drum filters (vacuum technologies), pressure filter presses and centrifugation.

Vacuum Filtration

The vacuum filter for dewatering sludge is a drum over which is laid the filtering medium consisting of a cloth of cotton, wool, nylon, dynel, fiber glass or plastic, or a stainless steel mesh or a double layer of stainless steel coil springs. The drum with horizontal axis is set in a tank with about one quarter of the drum submerged in conditioned sludge. Valves and piping are so arranged that, as a portion of the drum rotates slowly in the sludge, a vacuum is applied on the inner side of the filter medium, drawing out water from the sludge and holding the sludge against it. The application of the vacuum is continued as the drum rotates out of the sludge and into the atmosphere. This pulls water away from the sludge, leaving a moist mat or cake on the outer surface. This mat is scraped, blown or lifted away from the drum just before it enters the sludge tank again. The common measure of performance of vacuum filters is the rate in pounds per hour of dry solids filtered per square foot of filter surface. For various sludge this rate may vary from a low of 2.5 for activated sludge to a high of 6 to 11 for the best digested primary sludge. The moisture content in the sludge cake also varies with the type of sludge from 80 to 84 per cent, for raw activated sludge to 60 to 68 per cent for well digested primary sludge. While operating costs, including conditioning of sludge for vacuum filtration, are usually higher than with sludge beds, filtration has the advantage of requiring much less area, is independent of seasons and weather conditions and can eliminate the necessity for digestion since raw sludge can be dewatered sufficiently to be incinerated.

Prolongation of the life of the material used as the filter may be effected by proper care. Such care includes washing of the filter material with the spray jets after every period of use, removal of grease and fats with warm soap solution if clogged, treatment with diluted hydrochloric acid for removal of lime encrustations, maintenance of scraper bade in careful adjustment to filter drum to prevent tearing of the filter material.

With regard to chemical use—diluted ferric chloride solutions (10 per cent to 20 per cent) usually give better results in the conditioning of the sludge. A high calcium lime is preferable or sludge filtration work. One should avoid excessive use of chemicals. The quantities of chemicals used for conditioning can be frequently reduced by careful control of the mixing and flocculation equipment. The maintenance of a uniform vacuum is necessary for satisfactory operation. Loss or fluctuations in vacuum usually indicate a break in the filter material, poorly conditioned sludge or uneven distribution of the sludge solids in the filter pan.

Rotary drum precoat filter

This machine is used to polish solutions having traces of contaminating insolubles, so it is not a dewatering machine *per se*, but its use is often integrated into the process. To polish the solution the drum deck is pre-coated with a medium of a known permeability and particle size that retains the fines and produces a clear filtrate. The following materials are used to form the pre-coat bed: Diatomaceous Earth (or Diatomite) consisting of silicaceous skeletal remains of tiny aquatic unicellular plants; Perlite consisting of glassy crushed and heat-expanded rock from volcanic origin; Cellulose consisting of fibrous light weight and ash less paper like medium; Special ground wood is becoming popular in recent years since it is combustible and reduces the high cost of disposal. There are nowadays manufacturers that grind, wash and classify special timer to permeabilities which can suit a wide range of applications. These materials when related to pre-coating are wrongly called filter-aids since they do not aid filtration but serve as a filter medium in an analogy to the filter cloth on a conventional drum filter. The Pre-coat Filter is similar in appearance to a conventional drum filter but its construction is very different. The scraper blade on conventional drum filters is stationary and serves mainly to deflect the cake while it is back-blown at the point of discharge. The scraper on a pre-coat filter, which is also called 'Doctor Blade', moves slowly towards the drum and shaves-off the blinding layer of the contaminants together with a thin layer of the pre-coating material. This movement exposes continuously a fresh layer of the pre-coat surface so that when the drum submerges into the tank it is ready to polish the solution. The blade movement mechanism is equipped with a precision drive having an adjustable advance rate of 1 to 10 mm/hr. The selected rate is determined by the penetration of fines into the pre-coat bed which, in turn, depends on the permeability of the filter aid. Once the entire pre-coat is consumed the blade retracts at a fast rate so that the filter is ready for a new pre-coating cycle. The cake discharges on conventional drum filters by blow-back hence a section of the main valve's bridge setting is allocated for this purpose. On pre-coat filters the entire drum deck is subjected to vacuum therefore there are two design options:

1. A conventional valve that is piped, including its blow-back section, to be open to vacuum during polishing. When the pre-coat is consumed its blow-back section is turned on to remove the remaining pre-coat heel over the doctor blade.
2. A valveless configuration in which there is no bridge setting and the sealing between the rotating drum and the stationary outlet is by circumferential 'o' rings rather than by a face seal used on conventional valves.

Pressure Filtration

Pressure filtration is a process similar to vacuum filtration where sludge solids are separated from the liquid. Leaf filters probably are the most common type of unit. Like vacuum filtration, a porous media is used in leaf filters to separate solids from the liquid. The solids are captured in the media pores; they build up on the media surface; and they reinforce the media in its solid-liquid separation action. Sludge pumps provide the energy to force the water through the media. Lime, aluminium chloride, aluminium chlorohydrate and ferric salts have been commonly used to condition sludge prior to pressing. The successful use of ash pre-coating is also prevalent. Minimum chemical costs are supposed to be the major advantage of press filters over vacuum filters. Leaf filters represent an attempt to dewater sludge in a small space quickly. But, when compared to other dewatering methods, they have major disadvantages, including: (i) batch operation, and (ii) high operation and maintenance costs. Some other types of pressure filters include hydraulic and screw presses, which while effective in dewatering sludges, have a major disadvantage of usually requiring a thickened sludge feed. Sludge cakes as high as 75 per cent solids using pressure filtration have been reported.

Centrifuge Dewatering

Centrifuges are machines that separate solids from the liquid through sedimentation and centrifugal force. In a typical unit sludge is fed through a stationary feed tube along the centerline of the bowl through a hub of the screw conveyor. The screw conveyor is mounted inside the rotating conical bowl. It rotates at a slightly lower speed than the bowl. Sludge leaves the end of the feed tube, is accelerated, passes through the ports in the conveyor shaft and is distributed to the periphery of the bowl. Solids settle through the liquid pool, are compacted by centrifugal force against the walls of the bowl and are conveyed by the screw conveyor to the drying or beach area of the bowl.

The beach area is an inclined section of the bowl where further dewatering occurs before the solids are discharged. Separated liquid is discharged continuously over adjustable weirs at the opposite end of the bowl. The important process variables are: (i) feed rate, (ii) sludge solids characteristics, (iii) feed consistency, (iv) temperature, and (v) chemical additives. Machine variables are: (i) bowl design, (ii) bowl speed, (iii) pool volume, and (iv) conveyor speed. Two factors usually determine the success of failure of centrifugation—cake dryness and solids recovery. The effect of the various parameters on these two factors are listed below:

To increase cake dryness	To increase solids recovery
Increase bowl speed	Increase bowl speed
Decrease pool volume	Increase pool volume
Decrease conveyor speed	Decrease conveyor speed
Increase feed rate	Decrease feed rate
Decrease feed consistency	Increase temperature
Increase temperature	Use flocculents
Do not use flocculents	Increase feed consistency

Centrifugation has some inherent advantages over vacuum filtration and other processes used to dewater sludge. It is simple, compact, totally enclosed, flexible, can be used without chemical aids and the costs are moderate. Industry particularly has accepted centrifuges in part due to their low capital cost, simplicity of operation and effectiveness with difficult-to-dewater sludges. The most effective centrifuges to dewater waste sludges are horizontal, cylindrical—conical, solid bowl machines. Basket centrifuges dewater sludges effectively but liquid clarification is poor. Disc-type machines do a good job of clarification but their dewatering capabilities leave much to be desired. Centrifuges are being installed in more and more waste-water treatment plants for the following reasons: (i) the capital cost is low in comparison with other mechanical equipment, (ii) the operating and maintenance costs are moderate, (iii) the unit is totally enclosed so odours are minimised, (iv) the unit is simple and will fit in a small space, (v) chemical conditioning of the sludge is often not required, (vi) the unit is flexible in that it can handle a wide variety of solids and function as a thickening as well as a dewatering device, (vii) little supervision is required, and (viii) the centrifuge can dewater some industrial sludges that cannot be handled by vacuum filters.

The poor quality of the centrate is a major problem with centrifuges. The fine solids in centrate recycled to the head of the treatment plant sometimes resist settling and as a result, their concentrations in the treatment system gradually build up. The centrate from raw sludge dewatering can also cause odour problems when recycled. Flocculents can be used to increase solids captures, often to any degree desired; as well as to materially increase the capacity (solids loading) of the centrifuges. However, the use of chemicals nullifies the major advantage claimed for centrifuges—moderate operating costs. As noted, three basic-types of centrifuges are disc-nozzle, basket and solid bowl. The latter two types have been used extensively for both dewatering and thickening. The disk-nozzle centrifuge is seldom used for dewatering sludge, but is used more for sludge thickening in the industrial sector. Because the solid bowl design has undergone major improvements throughout the history of its use, this method is used more than any other to dewater sludge. Because of recent improvements in solid bowl centrifuge design, solid concentrations can reach 35 per cent. The solid bowl conveyor centrifuge operates with a continuous feed and discharge rates. It has a solid-walled imperforated bowl, with a horizontal axis of rotation. These centrifuges are enclosed, so they have a limited odour potential compared with other dewatering methods. The laydown area, access area and centrifuge required space for a large machine (200 m to 700 gpm of sludge feed) is approximately 400 square feet. Compared to other mechanical dewatering machines, this space is significantly smaller. An example of a continuous horizontal solid-bowl centrifuge is illustrated in Fig. 15.3. It consists of a cylindrical rotor with a truncated cone-shaped end and an internal screw conveyor rotating together. The screw conveyor often rotates at a rate of 1 to 2 rpm below the rotor's rate of rotation. The suspension enters the bowl axially through the feed tube to a feed accelerated zone, then passes through a feed port in the conveyor hub into the pond. The suspension is subjected to centrifugal force and thrown against the bowl wall where the solids are separated. The clarified suspension moves toward the broad part of the bowl to be discharged through a port. The solid particles being scraped by the screw conveyor are carried in the opposite direction (to the small end of the bowl) across discharge ports through which they are ejected continuously by centrifugal force. As in any sedimentation centrifuge, the separation takes place in two stages: settling (Fig. 15.3, in the right part of the bowl) and thickening or pressing out of the sediment (left-hand side of the bowl).

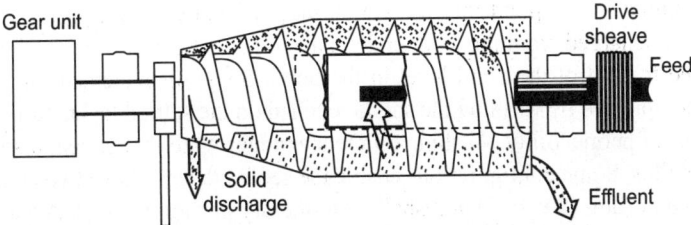

Fig. 15.3. Continuous solid-bowl centrifuge.

Because the radius of the solid discharge port is usually less than the radius of the liquid overflow at the broader end of the bowl, part of the settled solids is submerged in the pond.

The remainder, closer to the center, is inside the free liquid interface, where they can drain before being discharged. The total length of the 'settling' and 'pressing out' zones depends on the dimensions of the rotor. Their relative length can be varied by changing the pond level through suitable adjustment of the liquid discharge radius. When the pond depth is lowered, the length of the pressing out zone increases with some sacrifice in the clarification effectiveness. The critical point in the transport of solids to the bowl wall is their transition across the free liquid interface, where the buoyancy effect of the continuous phase is lost. At this point, soft amorphous solids tend to flow back into the pond instead of discharging. This tendency can be overcome by raising the pond level so that its radius is equal to or less than, that of the solids discharge port. In reality, there are no dry settled solids. The solids form a dam, which prevents the liquid from overflowing. The transfer of solids becomes possible because of the difference between the rotational speed of the screw conveyor and that of the bowl shell. The flights of the screw move through the settled solids and cause the solids to advance. To achieve this motion, it is necessary to have a high circumferential coefficient of friction on the solid particles with respect to the bowl shell and a low coefficient axially with respect to the bowl shell and across the conveyor flights. These criteria may be achieved by constructing the shell with conical grooves or ribs and by polishing the conveyor flights. The conveyor or differential speed is normally in the range of 0.8 per cent to 5 per cent of the bowl's rotational speed.

The required differential is achieved by a two-stage planetary gear box. The gear box housing carrying two ring gears is fixed to and rotates with the bowl shell. The first stage pinion is located on a shaft that projects outward from the housing. This arrangement provides a signal that is proportional to the torque imposed by the conveyor. If the shaft is held rotational (for example, by a torque overload release device or a shear pin), the relative conveyor speed is equivalent to the bowl rotative speed divided by the gear box ratio. Variable differential speeds can be obtained by driving the pinion shaft with an auxiliary power supply or by allowing it to slip forward against a controlled breaking action. Both arrangements are employed when processing soft solids or when maximum retention times are needed on the pressing out zone. The solids handling capacity of this type centrifuge is established by the diameter of the bowl, the conveyor's pitch and its differential speed. Feed ports should be located as far from the effluent discharge as possible to maximise the effective clarifying length. Note that the feed must be introduced into the pond to minimise disturbance and resuspension of the previously sedimented solids. As a general rule, the preferred feed location is near the intercept of the conical and cylindrical portions of the bowl shell. The angle of the sedimentation section with respect to the axis of rotation is typically in the range of 3° to 15°. A shallow angle provides a longer sedimentation area with a sacrifice in the effective length for clarification. In some designs, a portion of the conveyor flights in the sedimentation area is shrouded (as with a cone) to prevent intermixing of the sedimented solids with the free supernatant liquid in the pond through which they normally would pass. In other designs, the clarified liquid is discharged from the front end via a centrifugal pump or an adjustable skimmer that sometimes is used to control the pond level in the bowl. Some displacement of the adhering virgin liquor can be accomplished by washing the solids retained on the settled layer, particularly if the solids have a high degree of permeability. Washing efficiency ranges up to 90 per cent displacement of virgin liquor on coarse solids. Some configurations enable the settled layer to have two angles; comparatively steep in the wetted portion (10°–15°) and shallow in the dry portion (3°–5°). A wash is applied at the intersection of these angles, which, in effect, forms a constantly replenished zone of pure liquid through which the solids are conveyed. The longer section of a dry shallow layer provides more time for drainage of the washed solids. In either washing system, the wash liquid that is not carried out with the solids fraction returns to the pond and eventually discharges along with the effluent virgin liquor.

Estimating capacities of tubular and solid-bowl centrifuges

When a rotating centrifuge is filled with suspension, the internal surface of liquid acquires a cylindrical geometry of radius R_1, as shown in Fig. 15.3. The free surface is normal at any point to the resultant force acting on a liquid particle. If the liquid is lighter than the solid particles, the liquid moves toward the axis of rotation while the solids flow toward the bowl walls. The flow of the continuous liquid phase is effectively axial. A simplified model of centrifuge operation is that of a cylinder of fluid rotating about its axis. The flow forms a layer bound outwardy by a cylinder, R_2 and inwardly by a free cylindrical surface, R_1 (Fig. 15.4). This surface is, at any point, normal to a resulting force (centrifugal and gravity) acting on the solid particle in the liquid. The gravity force is, in general, negligible compared to the centrifugal force and the surface of liquid is perpendicular to the direction of centrifugal force.

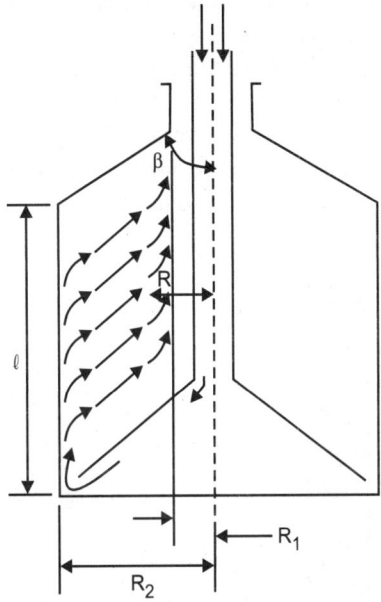

Fig. 15.4. Centrifuge operation.

If the particle's density is lower than that of the liquid, the path of the liquid will be centripetal, as illustrated in Fig. 15.5.

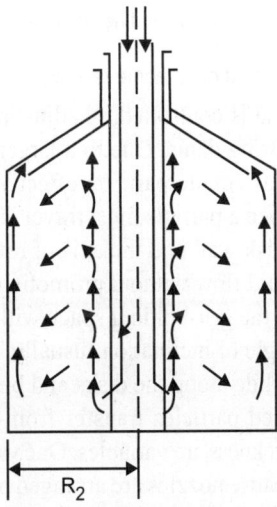

Fig. 15.5. Operation when particle density is less than liquid density.

Disk-bowl centrifuges

Disk-bowl centrifuges are used widely for separating emulsions, clarifying fine suspensions and separating immiscible liquid mixtures. Although these machines are generally not applied to waste-water applications and are more usually found in food processing, they can find niche applications in water treatment. More sophisticated designs can separate immiscible liquid mixtures of different specific gravities while simultaneously removing solids. Figure 15.6 illustrates the physical separation of two liquid components within a stack of disks. The light liquid phase build up in the inner section and the heavy phase concentrates in the outer section. The dividing line between the two is referred to as the 'separating zone'. For the most efficient separation this is located along the line of the rising channels, which are a series of holes in each disk, arranged so that the holes provide vertical channels through the entire disk set. These channels also provide access for the liquid mixture into the spaces between the disks. Centrifugal force causes the two liquids to separate and the solids move outward to the sediment-holding space.

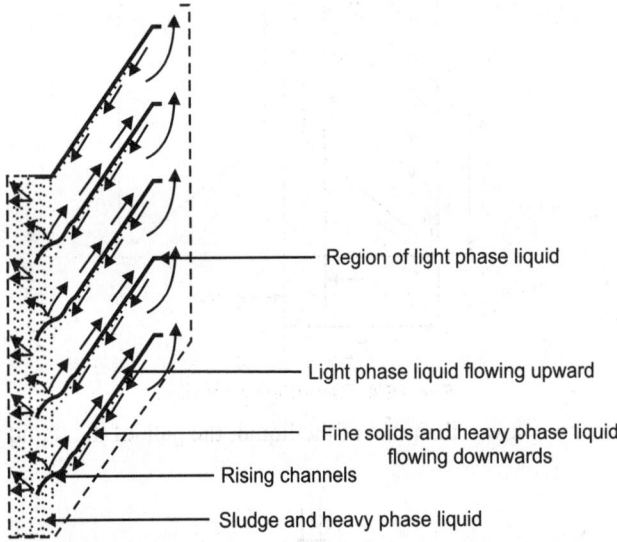

Region of light phase liquid

Light phase liquid flowing upward

Fine solids and heavy phase liquid flowing downwards

Rising channels

Sludge and heavy phase liquid

Fig. 15.6. Separation is achieved by use of stack discs.

The position of the separating zone is controlled by adjusting the back pressure of the discharged liquids or by means of exchangeable ring dams. Due to a larger diameter, the disk bowl operates at a lower rotational speed than its tubular counterpart. Its effectiveness depends on the shorter path of particle settling. The maximum distance a particle must travel is the thickness of the spacer divided by the cosine of the angle between the disk wall and the axis of rotation. Spacing between disks must be wide enough to accommodate the liquid flow without promoting turbulence and large enough to allow sedimented solids to slide outward to the grit-holding space without interfering with the flow of liquid in the opposite direction. The disk angle of inclination (usually in the range of 35° to 50°) generally is small to permit the solid particles to slide along the disks and be directed to the solids-holding volume located outside of the stack. Dispersed particles transfer from one layer to the other; therefore, the concentration in the layers and their thickness are variables. One variation of the disk-type bowl centrifuge is the nozzle centrifuge, so named because nozzles are arranged on the periphery or on the bottom of the bowl in a circle that is smaller in diameter than the bowl peripheral diameter.

Hydroclones

We can think of these machines as low-energy centrifuges. Hydroclones are employed for the separation of solid particles from medium to low-viscosity liquids. Like their cyclone counterparts used in gas cleaning applications, hydroclones are simple in design and the degree of separation can be altered by either varying loading conditions or changing geometric proportions. Unlike other types of solid-liquid separating equipment, they are better suited for classifying than for clarifying because high shearing stresses in a hydroclone promote the suspension of particles which oppose flocculation. However, by properly specifying dimensions and operating conditions, they can be used as thickeners in such a manner that the underflow contains mostly solid particles, while the clear overflow constitutes the largest portion of the liquid.

The fluid vortices and flow patterns characteristic of gas-cyclone operations are equally descriptive of liquid-hydroclones. However, the density differences between particles and liquids are significantly smaller than for gas-solid systems. For example, the density of water is approximately 800 times greater than that of air. This means that high fluid-spinning velocities cannot be employed in hydroclones as excessive pressure drop becomes a limitation. Obviously, the efficiency of hydroclones is low in comparison to gas cyclones. The design features of a hydroclone are illustrated in Fig. 15.7.

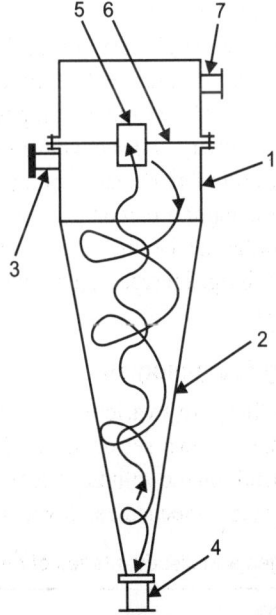

Fig. 15.7. Features of a hydroclone.

It consists of an upper short cylindrical section (1) and an elongated conical bottom (2). The suspension is introduced into the cylindrical section (1) through the nozzle (3) tangentially, whence the fluid acquires an intensive rotary motion. The larger particles, under the action of centrifugal force, move toward the walls of the apparatus and concentrate on the outer layers of the rotating flow. Then they move spirally downward along the walls to the nozzle (4), through which the thickened slurry is evacuated. The largest portion of liquid containing small particles (clear liquid) moves in the internal spiral flow upward along the axis of the hydroclone. The cleared liquid is discharged though the nozzle (5) and fixed at the partition (6) and the nozzle (7). The actual flow pattern is more complicated than described because of radial and closed circulating flows. Because of peripheral flow velocities, the liquid column formed at the hydroclone axis has a pressure that is below atmospheric. The liquid bulk flow limits the upward flow of small particles from the internal side and has a significant influence on the separating effect. Hydroclones are applied successfully for classification, clarification and thickening of suspensions containing particles from 5 to 150 μm in size. The smaller the hydroclone diameter, the greater the centrifugal forces developed and consequently, the smaller the size particles that can be separated.

Thickeners

Thickening is practiced in order to remove as much water as possible before final dewatering of the sludge. It is usually accomplished by floating the solids to the top of the liquid (floatation) or by allowing the solids to settle to the bottom (gravity thickening). Other method of thickening are by centrifuge, gravity belt and rotary drum thickening, as already described. These processes offer a low-cost means of reducing the volumetric loading of sludge to subsequent steps. In the flotation thickening process air is injected into the sludge under pressure. The resulting air bubbles attach themselves to sludge solids particles and float them to the surface of an open tank. The sludge forms a layer at the top of the tank which is removed by a skimming mechanism. This process increases the solids concentration of activated sludge from 0.5–1 per cent to 3–6 per cent.

Gravity thickening has been widely used on primary sludge for many years because of its simplicity and inexpensiveness. In gravity thickening, sludge is concentrated by the gravity- induced settling and compaction of sludge solids. It is essentially a sedimentation process. Sludge flows into a tank that is similar to the circular clarifiers used in primary and secondary sedimentation.

The solids in the sludge settle to the bottom where a scraping mechanism removes them to a hopper. The type of sludge being thickened has a major effect on performance. The best results can be achieved with primary sludge. Purely primary sludge can be thickened from 1–3 per cent to 10 per cent solids. As the proportion of activated (secondary) sludge increases, the thickness of settled solids decreases. There are various designs for sludge thickeners.

Comparing Mechanical Dewatering Technologies

As we see from the above descriptions there are a variety of technologies from which to select from for sludge dewatering operations. Each has its own set of advantages, disadvantages and limitations in operating ranges. Selection greatly depends on the volumes and nature of the sludge. Table 15.3 provides a relative comparison between, the principle mechanical dewatering techniques.

Table 15.3. Comparison of the advantages and disadvantages of mechanical thickening technologies.

Technology or method	Advantages	Disadvantages
Gravity	Simple	Potential for obnoxious and harmful odours
	Low operating and maintenance costs	Thickened sludge concentration limited for WAS
	Low operator attention and moderate training requirements	High space requirements for WAS
	Minimal power consumption	

(Contd ...)

Technology or Method	Advantages	Disadvantages
Dissolved Air Flotation	Effective for WAS (waste activated sludge)	Relatively high power consumption
	Can work without conditioning chemicals	Thickening solids concentration limited
	Relatively simple equipment components	Potential for obnoxious and harmful odours
		High space requirements
Centrifugation	Low space requirements	Best suited for continuous operations
	Effective for WAS	Sophisticated maintenance requirements
	Minimum housekeeping and odour problems	Relatively high power consumption
	Highly thickened concentrations available	Relatively high capital cost
Rotating Drum Filter	Low space requirements	Can be polymer dependent
Low capital cost	Sensitive to polymer type	
	Relatively low power consumption	Housekeeping requirements high
	High solids capture achievable	Potential for obnoxious and harmful odours
		Moderate operator attention and training requirements
Gravity belt thickener	Low space requirements	Housekeeping requirements high
	Relatively low power consumption	Can be polymer dependent
	Relatively low capital cost	Moderate operator attention and training requirements
	Can achieve high thickened concentrations and solids capture with minimum power	Potential for obnoxious and harmful odours

Drying Beds

This is one of two common methods of dewatering based upon thermal energy. Drying beds are generally used for dewatering of well digested sludges. Attempts to air dry raw sludge usually result in odour problems. Sludge drying beds consist of perforated or open joint drainage pipe laid within a gravel base. The gravel is covered with a layer of sand. Partitions around and between the drying beds are generally open to the weather but may be covered with ventilated greenhouse type enclosures where it is necessary to dewater sludge in wet climates.

The drying of sludge on sand beds is accomplished by allowing water to drain from the sludge mass through the supporting sand to the drainage piping and natural evaporation to the air. As the sludge dries, cracks develop in the surface allowing evaporation to occur from the lower layers which accelerates the drying process.

There are many design variations used for sludge drying beds, including the layout of the drainage piping, thickness and type of materials in the gravel and sand layers and construction materials used for the partitions. The major variation is whether or not the beds are covered. Any covering structure must be well ventilated. In the past, some beds were constructed with flat concrete bottoms for drainage without pipes, but this construction has not been very satisfactory. Asphalt concrete (blacktop) has been used in some drying beds.

The only sidestream is the drainage water. This water is normally returned to the raw sewage flow to the plant or to the plant headworks. The drainage water is not normally treated prior to return to the plant. Experience is the best guide in determining the depth of sludge to be applied, however, typical application depth is 8 to 12 inches. The condition and moisture content of the sludge, the sand bed area available and the need to draw sludge from digesters are factors to consider. It is not advisable to apply fresh sludge on top of dried sludge in a bed.

The best time to remove dried sludge from drying beds depends on a number of factors, such as subsequent treatment by grinding or shredding, the availability of drying bed area for application of current sludge production, labour availability and, of course, the desired moisture content of the dried sludge. Sludge can be removed by shovel or forks at a moisture content of 60 per cent, but if it is allowed to dry to 40 per cent moisture, it will weigh only half as much and is still easy to handle. If the sludge gets too dry (10 to 20 per cent moisture) it will be dusty and will be difficult to remove because it will crumble as it is removed. Many operators of smaller treatment plants use wheelbarrows to haul sludge from drying beds. Planks are often laid on the bed for a runway so that the wheelbarrow tyre does not sink into the sand. Wheelbarrows can be kept close to the worker so that the shovelling distance is not great. Most plants use pick-up trucks or dump trucks to transport the sludge from the drying bed. Dump trucks have the advantage of quick unloading.

Where trucks are used, it is best to install concrete treadways in the sludge drying bed wide enough to carry the dual wheels since the drying bed can be damaged if the trucks are driven directly on the sand. The treadways should be installed so that good access is provided to all parts of the beds. If permanent treadways have not been installed, heavy planks may be placed on the sand. Large plants will normally utilise mechanical equipment for handling the dried sludge. Some communities have encouraged public usage of the dried sludge. In some cases users are allowed to remove the sludge from the beds, but this may not be satisfactory in many cases. Local regulations should be reviewed before attempting to establish a public utilisation programme.

Sludge Lagoons

This is a technique that relies both on the settling characteristics of sludge and solar evaporation. The considerable labour involved in sludge drying bed operation may be avoided by the use of sludge lagoons. These lagoons are nothing but excavated areas in which digested sludges allowed to drain and dry over a period of months or even a year or more. They are usually dug out by bulldozers or other dirt-moving equipment; with the excavated material used for building up the sides to confine the sludge. Depths may range from two to six feet. Areas vary and although drainage is desirable, it is not usually provided. Digested sludge is drawn as frequently as needed, with successive drawings on top of the previous ones until the lagoon is filled. A second lagoon may then be operated while the filled one is drying.

After the sludge has dried enough to be moved, a bulldozer or a tractor with an end-loader, may be used scoop out the sludge. In some locations it may be pushed from the lagoon by dozers into low ground for fill. Lagoons may be used for regular drying of sludge, re-used after emptying or allowed to fill and dry, then levelled and developed into lawn.

They can also be used as emergency storage when the sludge beds are full or when the digester must be emptied for repair. In the latter case it should be treated with some odour control chemicals, such as hydrated or chlorinated lime. The size of the lagoon depends upon the use to which it will be put. Lagoons may take the place of sludge beds or provide a place for emergency drawings of sludge, but they may be unsightly and even unwanted on a small plant site. However, they are becoming more popular because they are inexpensive to build and operate.

Although lagoons are simple to construct and operate, there can be problems associated with sizing them. These problems largely arise from uncertainty in estimating the solar evaporative capacity.

In semi-arid regions evaporation ponds are a conventional means of disposing of waste-water without contamination of ground or surface waters. Evaporation ponds as defined herein will refer to lined retention facilities. Successful use of evaporation for waste-water disposal requires that evaporation equal or exceed the total water input to the system, including precipitation. The net evaporation may be defined as the difference between the evaporation and precipitation during any time period. Evaporation rates are to a great extent dependent upon the characteristics of the water body. Evaporation from small shallow ponds is usually considered to be quite different than that of large lakes mainly due to differences in the rates of heating and cooling of the water bodies because of size and depth differences. Additionally, in semi-arid regions, hot dry air moving from a land surface over a water body will result in higher evaporation rates for smaller water bodies.

The evaporation rate of a solution will decrease as the solids and chemical composition increase. Depending upon its origin, evaporation pond influent may contain contaminates of various amounts and composition. Decreases in evaporation rates compared to fresh water rates can seriously increase the failure potential of ponds designed on freshwater evaporation criteria. Designers of settling ponds and lagoons that rely on evaporation need to know the probability level of their designs being exceeded. Confidence limits for published evaporation normals have not been given, nor have analyses been made of the effects of uncertainty in the estimated normals or of the temporal variation of net evaporation. Definition of the spatial and temporal distribution of parameters such as evaporation and precipitation is difficult in mountainous regions.

A concern is that the application of many of the empirical equations, based on climatological data, for estimating evaporation have not been thoroughly tested for high altitude conditions. In particular, the ability of these equations for defining the variability of evaporation basically is unknown. Historically, pan data is the most common means for defining free water evaporation. However, the density of evaporation pan stations is much less than that of weather stations.

Many methods exist for either measuring or estimating evaporative losses from free water surfaces. Evaporation pans provide one of the simplest, inexpensive and most widely used methods of estimating evaporative losses. Long-term pan records are available, providing a potential source of data for developing probabilities of net evaporation. The use of pan data involves the application of a coefficient to measured pan readings to estimate evaporation from a larger water body. Among the most useful methods for estimating evaporation from free water surfaces are the methods which use climatological data. Many of these equations exist, most being based directly upon the a method which was originally intended for open water surfaces, but is now commonly applied to estimates of vegetative, water use.

Of concern is that there is very little information often available concerning the effects of common waste waters on evaporation rates. As noted, the evaporation rate of a solution will decrease as the solids and chemical concentrations increase. However, the overall effects on evaporation rates of dissolved constituents as well as colour changes and other factors of waste-water are largely unknown.

Evaporation from surface ponds are usually based upon estimates of annual net evaporation. Calculation of annual evaporation rates requires estimates during periods when the surface may be frozen. Most studies related to cold weather evaporation have been concerned with snow rather than ice. In general, the evaporation from a snow pack is usually much less than the amount of melting that occurs. Considering the large percentage of the annual evaporation which occurs during the warmer months and the overall uncertainties involved in estimates of evaporation from water surfaces, the amount of evaporation from frozen ponds during winter can reasonably be neglected in calculating annual evaporation.

A more important consideration is the evaporation which occurs during winter from ponds which may remain unfrozen due to the introduction of warm waste-water. In these cases, water temperature will influence the evaporation rates. However, the low value of the saturation vapour pressure of the air above any water body will limit evaporation. For lined ponds, evaporation will be confined mainly to the water surface area. Evaporation from the soil and vegetation on the banks surrounding the pond should be minimal. However, for ponds which have appreciable seepage to the surrounding area, evaporation from this area will be dependent upon the type and amount of vegetation, as well as the moisture content of the upper soil layers.

If water losses from the surrounding area are a major component of the total evaporative losses of the pond, then soil moisture conditions will be expected to be high. Under non-limiting soil moisture conditions vegetative moisture losses are often defined as 'potential' losses. Evaporative losses in this case would not be expected to differ greatly from free water evaporation. The literature recommends in fact that lake evaporation be used as a measure of potential evapotranspiration. Thus, for high soil moisture conditions, evaporation rates calculated for the water surface should be applicable to the surrounding area. The influence upon evaporation of vegetative growth within a pond is uncertain. The literature is inconclusive as to whether vegetation will increase or decrease evaporation compared to an open surface. It appear that the effect may be somewhat dependent upon the size of the water body. Literate studies indicate vegetation will decrease evaporation for extensive surfaces with the effect being less for smaller surface areas.

It is very possible, however, that the introduction of vegetation upon the surface of a water body of more limited extent may increase its evaporative water loss, but only while the vegetation remains in a healthy, robust condition. Thus, the effect of the presence of vegetation appears to range from being a water conservation mechanism to that of increasing evaporation. In either case, the potential effects appear to be quite large with reported ratios of vegetative covered to open water evaporation under extreme conditions ranging from 0.38 to 4.5. In most instances, this ratio would be expected to be much closer to unity.

VOLUME REDUCTION

As title implies, we will now focus our attention to those technologies aimed at reducing the volume of the final form of the sludge. Dewatering or thickening technologies can only bring us so far in concentrating the form of the waste. Ultimately, we must find ways of either disposing of this waste or in using it. We will discuss applications later on. Of immediate concern is how we can reduce the volume of so-called 'dry' sludge, at solids contents ranging anywhere from 30 to 60 per cent, even further.

Incineration

In all types of incinerators, the gases from combustion must be brought to and kept at a temperature of 1250°F to 1400°F until they are completely burned. This is essential to prevent odour nuisance from stack discharge. It is also necessary to maintain effective removal of dust, fly-ash and soot from the stack discharge. This may be done by a settling chamber, by a centrifugal separator or by a Cottrell electrical precipitator. The selection depends on the degree of removal efficiency required for the plant location. All types of sludge, primary, secondary, raw or digested sludge, may be dried and burned. Raw primary sludge with about 70 per cent volatile solids contains about 7800 Btu per pound of dry solids and when combustion is once started will burn without supplementary fuel, in fact an excess of heat is usually available. Digested sludge may or may not require supplementary fuel, depending on the moisture content of the cake and per cent volatile solids or degree of digestion. Raw activated sludge generally requires supplementary fuel for drying and burning. In all cases, supplementary fuel is necessary to start operation and until combustion of the solids has been established.

Incineration of sludge has gained popularity throughout the world, especially at large plants. It has the advantages of economy, freedom of odour, independence of weather and the great reduction in the volume and weight of end product to be disposed of. There is a minimum size of sewage treatment plant below which incineration is not economical. There must be enough sludge to necessitate reasonable use of costly equipment. One of the difficulties in operating an incinerator is variations in tonnage and moisture of sludge handled.

There are two major incinerator technologies used in this process. They are: (i) the multiple hearth incinerator, and (ii) the fluidised bed incinerator. An incinerator is usually part of a sludge treatment system which includes sludge thickening, macerations, dewatering (such as vacuum filter, centrifuge, or filter press), an incinerator feed system, air pollution control devices, ash handling facilities and the related automatic controls. The operation of the incinerator cannot be isolated from these other system components. Of particular importance is the operation of the thickening and dewatering processes because the moisture content of the sludge is the primary variable affecting the incinerator fuel consumption.

Incineration may be thought of as the complete destruction of materials by heat to their inert constituents. This material that is being destroyed is the waste product (i.e., the sludge). Sewer sludge as sludge cake normally contains from 55 to 85 per cent moisture. It cannot burn until the moisture content has been reduced to no more than 30 per cent. The purpose of incineration is to reduce the sludge cake to its minimum volume, as sterile ash. There are three objectives incineration must accomplish:

1. Dry the sludge cake.
2. Destroy the volatile content by burning.
3. Produce a sterile residue or ash.

There are four basic types of incinerators used in waste-water treatment plants. They are the multiple hearth incinerator, the fluid bed incinerator, the electric furnace and the cyclonic furnace. Each system has its own distinct method of incineration and while one may be more cost efficient, another may have more of an environmental impact.

The multiple hearth incinerator is the most prevalent incinerator technology for the disposal of sewage sludge due to its low ash discharge. Sludge cake enters the furnace at the top. The interior of the furnace is composed of a series of circular refractory hearths, which are stacked one on top of the other. There are typically five to nine hearths in a furnace. A vertical shaft, positioned in the center of the furnace has rabble arms with teeth attached to them in order to move the sludge through the mechanism.

Each arm is above a layer of hearth. Teeth on each hearth agitate the sludge, exposing new surfaces of the sludge to the gas flow within the furnace. As sludge falls from one hearth to another, it again has new surfaces exposed to the hot gas. At the top of the incinerator there is an exit for flue gas, an end product of sludge incineration. At the bottom of the furnace there is an exit for the ashes.

In fluid bed incinerator air is introduced at the fluidising air inlet at pressure of 3.5 to 5 psig. The air passes through openings in the grid supporting the sand and creates fluidisation of the sand bed. Sludge cake is introduced into the bed. The fluidising airflow must be carefully controlled to prevent the sludge from floating on top of the bed. Fluidisation provides maximum contact of air with sludge surface for optimum burning. The drying process is practically instantaneous. Moisture flashes into steam upon entering the hot bed. Some advantages of this system are that the sand bed acts as a heat sink so that after shutdown there is minimal heat loss. With this heat containment, the system will allow startup after a weekend shutdown with need for only one or two hours of heating. The sand bed should be at least 1200°F when operating (Fig. 15.8). The electric furnace is basically a conveyor belt system passing trough a long rectangular refractory lined chamber. Heat is provided by electric infrared heating elements within the furnace. Cooling air prevents local hot spots in the immediate vicinity of the heaters and is used as secondary combustion air within the furnace. The conveyer belt is made of continuous woven wire mesh chosen of steel alloy that will withstand the 1300 to 1500°F temperatures. The sludge on the belt is immediately levelled to one inch. The belt speed is designed to provide burnout of the sludge without agitation.

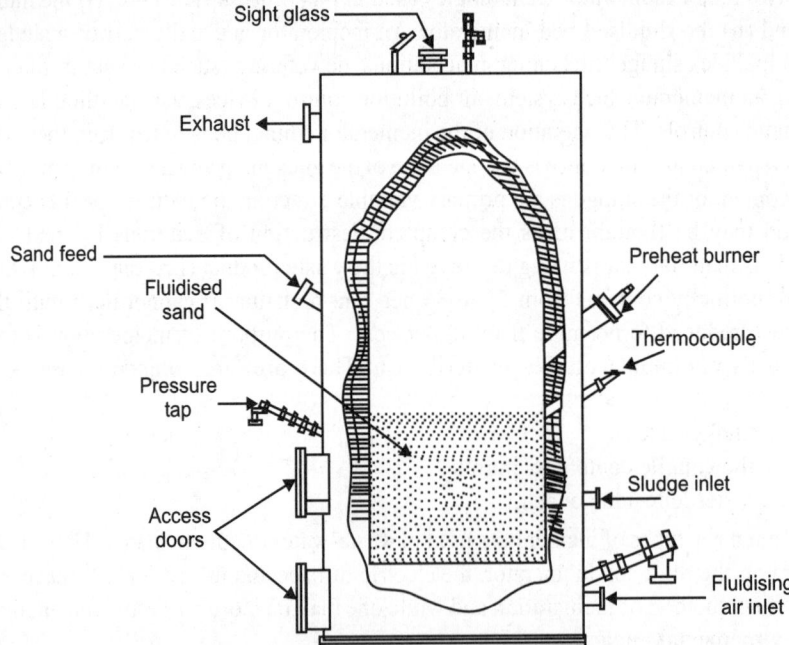

Fig. 15.8. Cross-section of a fluid bed incinerator.

The cyclonic furnace is a single hearth unit where the hearth moves and the rabble teeth are stationary. Sludge is moved towards the center of the hearth where it's discharged as ash. The furnace is a refractory lined cylindrical shell with a domed top. The air, heated with the immediate introduction of supplemental fuel creates a violent swirling pattern which provides good mixing of air and sludge feed. The air, which later turns into flue gas, swirls up vertically in cyclonic flow through the discharge flue in the center of the doomed roof. One advantage of these furnaces is that they are relatively small and can be placed in operation, at operating temperature within an hour.

A good question for us to ask at this stage is what does a sludge treatment plant do with the ash that is discharged out of the furnace? As ash falls into a wet sump, turbulence is created by the entrance of water. This turbulence is necessary so that the ash doesn't collect and cake up.

This water containing the ash is pumped into a holding pond or lagoon, with a residence time of at least 6 hours. During this time, 95 per cent of the ash will have settled to the bottom and the overflow is taken back to the treatment plant. There has to be a minimum of two lagoons with one being used to hold the ash-water discharge and the other for drying. When dry, the ash is hauled to a landfill or used for concrete. Mixing one part of ash to four parts cement will produce a slow-setting concrete with no loss in strength.

A serious environmental impact that incineration has is on the air. An incinerator's smoke discharge or flue gas should be colourless. Flue gas is an emission mainly made up of nitrogen, carbon dioxide and oxygen. There are traces of chloride and sulphides in the gas and if these levels become too high, they could cause the possibility of corrosion. With respect to the colour of the discharge again, if there is a significant amount of particulate matter in the emission, it will be detected by colour. The stream can range from a black to white appearance and will have a pale yellow to dark brown trail. The discharge should also have no discernable odour and there should be no detectable noise due to incinerator operation at the property line. Unfortunately coloured emissions and odour problems do occur and treatment plants take the proper actions to correct it. Air pollution controls are critical factors that add significant costs onto these technologies.

When dealing with incinerators, fuel is generally the most expensive part of the process from an operational standpoint. There should be a ratio calculated before hand that represents the amount of fuel used for the amount of sludge inputted. If there is a significant change to the amount of fuel consumed, it could mean that there is a problem in the fuel supply system, air flow to the incinerator or that an extensive furnace cleaning is in order. Minimal cost of operation and equipment maintenance is another economic parameter for sludge incineration. Preventive maintenance is the single most important factor in reduction of operating costs.

Essentially costs can be related to one basic parameter, namely—the lower the moisture content is in the sludge, the less expensive the incinerator will be to operate. Also incinerators are bought based on what moisture level of sludge they are going to be effective with. Some incinerators can burn out sludge with 20 per cent moisture levels and some cannot.

Comments on Incineration

Today, waste-water solids are more complex and include sludge from secondary and advanced waste treatment processes. These sludge are more difficult to dewater and thereby increase fuel requirements for combustion. Due to environmental concerns with air quality and the energy crisis, the use of high-temperature processes for combustion of municipal solids is being scrutinised. More efficient solids dewatering processes and advances in combustion technology have renewed an interest in the use of high-temperature processes for specific applications.

High-temperature processes should be considered where available land is scarce, stringent requirements, for land disposal exist, destruction of toxic materials is required or the potential exists for recovery of energy, either with waste-water solids alone or combined with municipal refuse. High-temperature processes have potential advantages over other methods which include:

1. Maximum volume reduction: Reduces volume and weight of wet sludge cake by approximately 95 per cent, thereby reducing disposal requirements.
2. Detoxification: Destroys or reduces toxics that may otherwise create adverse environmental impacts.
3. Energy recovery: Potentially recovers energy through the combustion of waste products, thereby reducing the overall expenditure of energy.

Disadvantages of high-temperature processes include:

1. Cost: Both capital and operation and maintenance costs, including costs for supplemental fuel, are generally higher than for other disposal alternatives.
2. Operating problems: High-temperature operations create high maintenance requirements and can reduce equipment reliability.
3. Staffings: Highly skilled and experienced operators are required for high-temperature processes. Municipal salaries and operator status may have to be raised in many locations to attract the proper personnel.
4. Environmental impacts: Discharges to atmosphere (particulates and other toxic or noxious emissions), surface waters (scrubbing water) and land (furnace residues) may require extensive treatment to assure protection of the environment.

Combustion is the rapid exothermic oxidation of combustible elements in fuel. Incineration is complete combustion. Classical pyrolysis is the destructive distillation, reduction or thermal cracking and condensation of organic matter under heat and/or pressure in the absence of oxygen. Partial pyrolysis or starved-air combustion, is incomplete combustion and occurs when insufficient oxygen is provided to satisfy the combustion requirements. Combustion of waste-water solids, a two-step process, involves drying followed by burning. A value commonly used in sludge incineration calculations is 10000 Btu per pound of combustibles. It is important to clearly understand the meaning of combustibles. For combustion processes, solid fuels are analysed for volatile solids and total combustibles. The difference between the two measurements is the fixed carbon. Volatile solids are determined by heating the fuel in the absence of air. Total combustibles are determined by ignition at 1336°F (725°C). The difference in weight loss is the fixed carbon. In the volatile-solids determination used in sanitary engineering, sludge is heated in the presence of air at 1021°F (550°C). This measurement is higher than the volatile-solids measurement for fuels and includes the fixed carbon.

Numerically, it is nearly the same as the combustible measurement. If volatile, solids are used in the sense of the fuels engineer, it will be followed parenthetically by the designation fuels usage. If the term volatile solids or volatiles is used without designation, it will indicate sanitary engineering usage and will be used synonymously with combustibles. The amount of heat released from a given sludge is a function of the amounts and types of combustible elements present.

The primary combustible elements in sludge and in most available supplemental fuels are fixed carbon, hydrogen and sulphur. Because free sulphur is rarely present in sewage sludge to any significant extent and because sulphur is being limited in fuels, the contributions of sulphur to the combustion reaction can be neglected in calculations without compromising accuracy. Similarly, the oxidation of metals contributes little to the heat balance and can be ignored. Solids with a high fraction of combustible material (for example, grease and scum) have high fuel values. Those which contain a large fraction of inert materials (for example, grit or chemical precipitates) have low fuel values. Chemical precipitates may also exert appreciable heat demands when undergoing high-temperature decomposition. This further reduces their effective fuel value.

Table 15.4 provides a summary of typical chemical reactions that take place during combustion, along with heating values of the reactions.

Table 15.4. Chemical reactions occurring during combustion.

Reaction	High heat value of reaction
$C + O_2 \rightarrow CO_2$	–14100 Btu/lbs of C
$C + \frac{1}{2}O_2 \rightarrow CO$	–4000 Btu/lbs of C
$CO + \frac{1}{2}O_2 \rightarrow CO_2$	–4400 Btu/lbs of CO
$H_2 + \frac{1}{2}O_2 \rightarrow H_2O$	-61100 Btu/lbs of H_2
$CH_4 + 2O_2 \rightarrow CO_2 + 2H_2O$	–23900 Btu/lbs of CH_4
$2H_2S + 3O_2 \rightarrow 2SO_2 + 2H_2O$	–7100 Btu/lbs of H_2S
$C + H_2O$ (gas) $\rightarrow CO + H_2$	–4700 Btu/lbs of C
Sludge combustibles $\rightarrow CO_2 + H_2O$	–10000 Btu/lbs of combustibles

The following are experimental methods from which sludge heating value may be estimated or computed:

1. Ultimate analysis—an analysis to determine the amounts of basic feed constituents. These constituents are moisture, oxygen, carbon, hydrogen, sulphur, nitrogen and ash. In addition, it is typical to determine chloride and other elements that may contribute to air emissions or ash-disposal problems. Once the ultimate analysis has been completed, Dulong's formula can be used to estimate the heating value of the sludge. Dulong's formula is:

$$Btu/lbs = 14544C + 62208(H_2 – O_2/8) + 4050S \qquad ... (15.1)$$

where, C, H_2, O_2 and S represent the weight fraction of each element determined by ultimate analysis. This formula does not take into account endothermic chemical reactions that occur with chemically conditioned or physical-chemical sludge. The ultimate analysis is used principally for developing the material balance, from which a heat balance can be made.

2. Proximate analysis—a relatively low-cost analysis in which moisture content, volatile combustible matter, fixed carbon and ash are determined. The fuel value of the sludge is calculated as the weighted average of the fuel values of its individual components.

3. Calorimetry—this is a direct method in which heating value is determined experimentally with a bomb calorimeter. Approximately 1 gram of material is burned in a sealed, submerged container. The heat of combustion is determined by noting the temperature rise of the water bath. Several samples must be taken and then composited to obtain a representative 1 gram sample. Several tests should be run and the results must be interpreted by an experienced analyst. New bomb calorimeters can use samples up to 25 grams and this type of unit should be used where possible.

The preceding tests give approximate fuel values for sludge and allow the designer to proceed with calculations which simulate operations of an incinerator. If a unique sludge will be processed or unusual operating conditions will be used, pilot testing is advised. Many manufacturers have test furnaces especially suited for pilot testing.

Wet Air Oxidation

When the organic sludge is heated, heat causes water to escape from the sludge. Thermal treatment systems release water that is bound within the cell structure of the sludge and thereby improves the dewatering and thickening characteristics of the sludge. The oxidation process further reduces the sludge to ash by wet incineration (oxidation). Sludge is ground to a controlled particle size and pumped to a pressure of about 300 psi. Compressed air is added to the sludge (wet air oxidation only), the mixture is brought to a temperature of about 350°F by heat exchange with treated sludge and direct steam injection and then is processed (cooked) in the reactor at the desired temperature and pressure. The hot treated sludge is cooled by heat exchange with the incoming sludge. The treated sludge is settled from the supernatant before the dewatering step. Gases released at the separation step are passed through a catalytic after-burner at 650° to 705°F or deodourised by other means. In some cases these gases have been returned through the diffused air system in the aeration basins for deodourisation. The same basic processes are used, for wet, air oxidation of sludge by operating at higher temperatures (450° to 640°F) and higher pressures (1200 to 1600 psig). The wet air oxidation (WAO) process is based on the fact that any substance capable of burning can be oxidised in the presence of water at temperature between 250° and 700°F. Wet air oxidation does not require preliminary dewatering or drying as required by conventional air combustion processes. However, the oxidised ash must be separated from the water by vacuum filtration, centrifugation or some other solids separation technique.

Wet-air oxidation (also called liquid-phase thermal oxidation) is not a new technology; it has been around for over forty years and has already demonstrated its great potential in waste-water treatment facilities. Despite this, there are some very important issues that remain to be addressed before a wet oxidation process can be scaled-up: the kinetics of oxidation of many important hazardous compounds are as yet unavailable, to mention only one among them. However, the kinetic models that predict solely the disappearance rate of mother compounds usually reported in the open literature are not enough; what is needed is a model capable of predicting complete conversion of all organic species present in a waste-water. Such models have to rely on the use of lumped parameters such as total organic carbon (TOC), chemical oxygen demand (COD) and biochemical oxygen demand (BOD). To point out a reaction engineering problem associated with the designing of a well established subcritical wet oxidation reactor, one can assume the TOC reduction to be linearly dependent on both reactant, organic compounds and oxygen.

For kinetic analysis and design purposes, the species originally present in a waste-water or produced during the course of oxidation are conveniently divided at least into three lumps: (i) original compounds and relatively unstable intermediates; (ii) high molecular mass organic acids; and (iii) low molecular mass organic acids. As it has been shown for catalytic cracking, the lumped oxidation kinetics in many cases also obey the power-law form, with the exponent for a continuous-stirred tank reactor (CSTR) being lower than that for a plug-flow reactor (PFR) or a batch reactor (BR). It has been demonstrated that the kinetic behaviour of a reactive mixture of organics in a batch system is governed by the most refractory lump, i.e. the lump of low molecular mass acids, while this is not the case with CSTR. (Consequently, the lumped kinetics developed from BR data cannot be used for predicting TOC conversions in CSTR.)

The Catalytic Wet Air Oxidation (CWAO) process is capable of converting all organic contaminants ultimately to carbon dioxide and water and can also remove oxidisable inorganic components such as cyanides and ammonia. The process uses air as the oxidant, which is mixed with the effluent and passed over a catalyst at elevated temperatures and pressures. If complete COD removal is not required, the air rate, temperature and pressure can be reduced, therefore reducing the operating cost. CWAO is particularly cost-effective for effluents that are highly concentrated (chemical oxygen demands of 10,000 to over 1,00,000 mg/litre) or which contain components that are not readily biodegradable or are toxic to biological treatment systems. CWAO process plants also offer the advantage that they can be highly automated for unattended operation, have relatively small plant footprints and are able to deal with variable effluent flow rates and compositions. The process is not cost-effective compared with other advanced oxidation processes or biological processes for lightly contaminated effluents (COD less than about 5000 mg/liter). The CWAO process is a development of the wet air oxidation (WAO) process. Organic and some inorganic contaminants are oxidised in the liquid phase by contacting the liquid with high pressure air at temperatures which are typically between 120°C and 310°C.

In the CWAO process the liquid phase and high pressure air are passed co-currently over a stationary bed catalyst. The operating pressure is maintained well above the saturation pressure is maintained well above the saturation pressure of water at the reaction temperatures (usually about 15–60 bar) so that the reaction takes place in the liquid phase. This enables the oxidation processes to proceed at lower temperatures than those required for incineration. Residence times are from 30 minutes to 90 minutes and the chemical oxygen demand removal may typically be about 75 to 99 per cent. The effect of the catalyst is to provide a higher degree of COD removal than is obtained by WAO at comparable conditions (over 99 per cent removal can be achieved) or to reduce the residence time. Organic compounds may be converted to carbon dioxide and water at the higher temperatures; nitrogen and sulphur heteroatoms are converted to molecular nitrogen and sulphates. The process becomes autogenic at COD levels of about 10000 mg/l, at which the system will require external energy only at start-up. A simplified process diagram of the wet air oxidation process is shown in Fig. 15.9. Typical wet oxidation applications have a feed flow rate of 1 to 45 ml/hr (5 to 100 gpm) per train, with a chemical oxygen demand (COD) between 10000 and 1,00,000 mg/liter. Wet air oxidation can involve any or all of the following reactions:

$$\text{Organics} + O_2 \rightarrow CO_2 + H_2O + RCOOH$$
$$\text{Sulphur species} + O_2 \rightarrow SO_4^{-2}$$
$$\text{Organic Cl} + O_2 \rightarrow Cl^{-1} + CO_2 + RCOOH$$
$$\text{Organic N} + O_2 \rightarrow NH_4^{+1} + CO_2 + RCOOH$$
$$\text{Phosphorus} + O_2 \rightarrow PO_4^{-3}$$

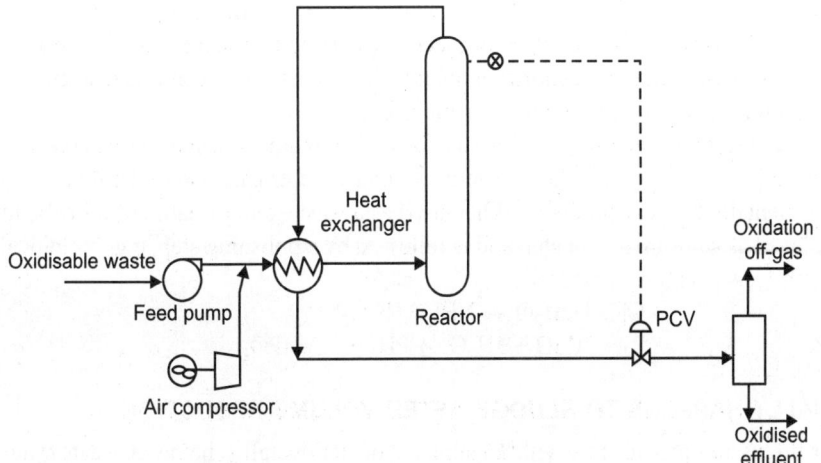

Fig. 15.9. Process scheme for wet air oxidation (WAO).

Note that RCOOH denotes short chain organic acids such as acetic acid which make up the major fraction of residual oxidation intermediates in a typical wet oxidation effluent. Properties of wet oxidation liquid effluent include: negligible NO_x and SO_2, negligible particulate matter and some VOCs, depending on the waste. Wet oxidation is a mature technology with a long history of development and commercialisation. Wet oxidation is applicable to numerous types of waste and is used commercially for the treatment of high strength industrial waste-water, ethylene and refinery spent caustic sludge. There are two other processes that we should mention that are used in conjunction with the WAO process. The first of these is thermal sludge conditioning/low pressure oxidation (LPO). Thermal sludge conditioning is used for the conditioning of biological sludge for dewatering. Thermal conditioning is accomplished using temperatures of 175° to 200°C (350° to 400°F). The low temperature allows for low operating pressures. Thermal conditioning is most commonly used for municipal waste-water treatment sludge. It has also been applied to industrial sludge processing. The technology is applicable to any organic sludge which is difficult to dewater or that contains pathogenic components. The LPO process heats sludge to a point where the biosolids break apart, releasing much of the water trapped within the cell structures, allowing filter presses, vacuum filters, belt presses and other dewatering technologies to perform their jobs more effectively. This process along with dewatering achieves a 90 to 95 per cent sludge volume reduction, while at the same time destroys any pathogens in the sludge. The second process used in conjunction with WAO is wet air regeneration. This is a liquid phase reaction in water using dissolved oxygen to oxidise sorbed contaminants and biosolids in a spent carbon slurry, while simultaneously regenerating the powdered activated carbon. The regeneration is conducted at moderate temperatures of 400° to 500°F and at pressures from 700 to 1000 psig. The process converts organic contaminants to CO_2, water and biodegradable short chain organic acids; sorbed inorganic constituents such as heavy metals can be converted to stable, non-leaching forms that can be separated from the regenerated, carbon. The technology can be more cost and energy efficient than incineration and the regeneration is accomplished in a slurry without NO_x, SO_x or particulate air emissions.

Applications of Hydrolysis

In addition to the WAO process, hydrolysis, a technology similar to wet oxidation, can be applied for the treatment of waste-waters when oxygen is not a necessary reactant. In hydrolysis, certain constituents of waste-waters and sludges can react directly with water at elevated temperatures and pressures to yield a treated effluent which is detoxified or meets the desired treatment objective. Waste-waters and sludges which contain cyanide, phosphorus or other hydrolysable constituents, can potentially be treated by hydrolysis without the addition of an oxidising agent.

Cyanide can react with water to yield formate ion and ammonia. Phosphorus can react with water to yield phosphate ion. A variety of other waste-water constituents can also be treated by hydrolysis to yield environmentally friendly products. When used as a waste-water treatment process, hydrolysis is usually employed as a pre-treatment step and is followed by a polishing step, e.g. biological treatment. Common hydrolysis reactions include:

$$CN^- + 2H_2O \rightarrow NH_3 + HCO_2^-$$
$$P_4 + 3CaO + 3H_2O \rightarrow PH_3 + 3CaHPO_2$$

WHAT FINALLY HAPPENS TO SLUDGE AFTER VOLUME REDUCTION

The sidebar discussion provides us with a summary of the overall scheme of waste-water treatment covered over the last several hundred pages. At the end of the day, what we are left with is ultimate sludge. If we choose incineration, we still have a solid waste left to deal with ash. If we choose another route to sludge volume reduction, we still have a solid waste residue to deal with. There is no ultimate destruction of sludge, only ultimate sludge that we are left with. The final engineering solution we need to devise is how to ultimately handle this waste. It simply boils down to whether we select a so-called pollution prevention related technology or a final disposal option for the solids waste. In the remaining sections we will explore the options available to us.

An Overview of the Options

The solids that result from waste-water treatment may contain concentrated levels of contaminants that were originally contained in the waste-water. A great deal of concern must be directed to the proper disposal of these solids to protect environmental consideration. Failure to do this may result in a mere shifting of the original pollutants in the waste stream to the final disposal site where they may again become free to contaminate the environment and possibly place the public at risk. A more reasonable approach to ultimate solids disposal is to view the sludge as a resource that can be recycled or reused that concept embodies the spirit of pollution prevention (Fig. 15.10).

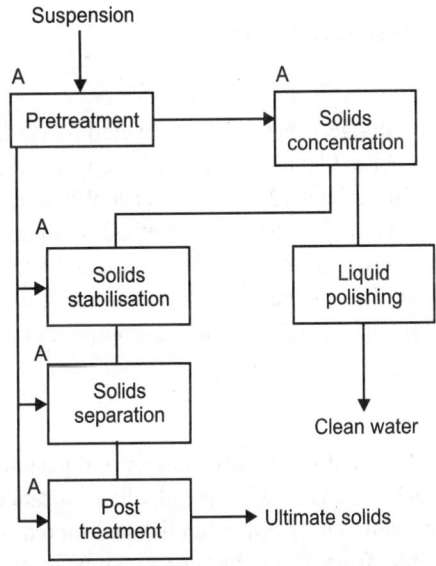

A—Addition of additives

Fig. 15.10. Summarising solid-liquid separation technology.

As already noted, all the sludge produced at a treatment plant (whether it be sewage or industrial in origin) must be disposed of ultimately. Treatment processes such as have been described may reduce its volume or so change its character as to facilitate its disposal, but still leave a residue which in most cases must be removed from the plant site. Like the liquid effluent from the treatment plant, there are two broad methods for the disposal of sludge:

1. Disposal in water.
2. Disposal on land.

This applies regardless of whether or not the sludge is treated to facilitate or permit the selected method of disposal. Before discussing the specific options in detail, let's first take a look at the big picture for each option. Disposal in water is one option to consider. This is an economical but not common method because it is contingent on the availability of bodies of water adequate to permit it. At some sea-coast cities, sludge either raw or digested is pumped to barges and carried to sea (the context of these discussions is strictly sewage) to be dumped in deep water far enough offshore to provide huge dilution factors and prevent any ill effects along shore. In the past few years there has been an increased problem of pollutional loads, well above safe standards, affecting the south-shore beaches on Long Island, facilitating the closing of the beaches to the public.

Some of these pollutional loads have been attributed to sludge deposits coming to shore form offshore sludge barging operations. Where barged to sea, the value of some treatment such as thickening or digestion, depends on the relative cost of the treatment and savings in cost by barging smaller volumes, or the value of gas produced by digestion. Overall, this is an environmentally unfriendly option and the bottom line is that it is no different that straight landfill and in fact can be more environmentally damaging. It is plain and simple an end-of-pipe solution. Under land disposal the following methods may be included:

1. Burial.
2. Fill.
3. Application as fertiliser or soil conditioner.

Burial

Burial is used principally for raw sludge, where, unless covered by earth, serious odour nuisances are created. The sludge is run into trenches two to three feet wide and about two feet deep. The raw sludge in the trenches should be covered by at least 12 inches of earth. Where large areas of land are available, burial of raw sludge is probably the most economical method of sludge disposal as it eliminates the costs of all sludge treatment processes. It is, however, rarely used and even then as a temporary makeshift because of the land area required. The sludge in the trenches may remain moist and malodorous for years so that an area once used cannot be reused for the same purpose or for any other purpose for a long period of time.

Fill

The option of using sludge for fill is confined almost entirely to digested sludge which can be exposed to the atmosphere without creating serious or widespread odour nuisances. The sludge should be well digested without any appreciable amount of raw or undigested mixed with it. Either wet or partially dewatered sludge, such as obtained from drying beds or vacuum filters can be used to fill low areas. Where wet sludge is used the area becomes a sludge lagoon. When used as a method of disposal, the lagoon area is used only until filled and then abandoned. When used as a method of treatment, the sludge after some drying, is removed for final disposal and the lagoon reused. Lagoons used for disposal are usually fairly deep. Sludge is added in successive layers until the lagoon is completely filled. Final disposal of digested sludge by lagoons is economical as it eliminates all dewatering treatments. It is applicable, however, only where low waste areas are available on the plant site or within reasonable piping distance. They are frequently used to supplement inadequate drying bed facilities. Dewatered digested sludge from drying beds and vacuum filters can be disposed of by filling low areas at the plant site or hauled to similar areas elsewhere without creating nuisances. The ash from incinerators is usually disposed of by using it for fill. Where fill area is available close to the incinerator, the ash can be made into a slurry with water when removed from the ash hopper and pumped to the point of disposal. If the fill area is remote, the ash should be sufficiently wet to suppress the dust and transported by truck or railroad cars to the point of disposal. It should be clear that the above options for sludge still are temporary solutions and they still have environmental trade-offs. In the end, they too represent environmentally unfriendly solutions and are end-of-pipe disposal technologies that add costs to treatment.

Well what about pollution prevention type technologies? The two we will explore in some detail are Soil Conditioning or Fertiliser and Composting.

Sewage sludge contains many elements essential to plant life, such as nitrogen, phosphorous, potassium and in addition, at least traces of minor nutrients which are considered more or less indispensable for plant growth, such as boron, calcium, copper, iron, magnesium, manganese, sulphur and zinc. In fact, sometimes these trace elements are found in concentrations, perhaps from industrial wastes, which may be detrimental. The sludge humus, besides furnishing plant food, benefits the soil by increasing the water holding, capacity and improving the tilth, thus making possible the working of heavy soils into satisfactory seed beds. It also reduces soil erosion. Soils vary in their requirements for fertiliser, but it appears that the elements essential for plant growth may be divided into two groups: those which come from the air and water freely and those which are found in the soil or have to be added at certain intervals. In the first group are hydrogen, oxygen and carbon in the second group are nitrogen, phosphorous and potassium and several miscellaneous elements usually found in sufficient quantities in the average soil, such as calcium, magnesium, sulphur, iron, manganese and others. The major fertilising elements are nitrogen, phosphorous and potassium and the amount of each required depends on the soil, climatic conditions and crop. Nitrogen is required by all plants, particularly where leaf development is required. Thus, it is of great value in fertilising grass, radishes, lettuce, spinach and celery. It stimulates growth of leaf and stem. Phosphorous is essential in many phases of plant growth. It hastens ripening, encourages root growth and increases resistance to disease. Potassium is an important factor in vigorous growth. It develops the woody parts of stems and pulps of fruits. It increases resistance to disease, but delays ripening and is needed in the formation of chlorophyll. Dried or dewatered sewage sludge makes an excellent soil conditioner and a good, though incomplete fertiliser, unless fortified with nitrogen, phosphorous and potassium. Heat-dried, raw activated sludge is the best sludge product, both chemically and hygienically, although some odour may result from its use. Heat-dried, digested sludge contains much less nitrogen and is more valuable for its soil conditioning and building qualities than for its fertiliser content. For some crops it is deleterious. It is practically odourless when well digested. Sludge cake from vacuum filters, because of its pasty nature, cannot be readily spread on land as a fertiliser or soil conditioner. It must be further air-dried. At some plants the sludge cake is stockpiled on the plant site over winter. Freezing, thawing and air drying result in a material which breaks up readily. Digested sludge has been said to be somewhat comparable to farm manure in its content of fertiliser constituents, their relative availability and the physical nature of the material.

Before sludge digestion was so widely adopted, the application of raw sludge to fields was sometimes detrimental because the grease content was difficult for the soil to absorb and caused it to become impervious. In digested sludge, however, fat has been reduced and become so finely divided that it does not adversely affect the porosity of the soil. The continued use of digested sludge tends to lower the pH value of soil and it is recommended that either lime or ground limestone be applied occasionally. In some tests it has been found that activated sludge used as an organic carrier for added inorganic forms of nitrogen, has given better results for crops with a short growing season than activated sludge alone. The inorganic nitrogen is quickly available while that from the organic portion is available more slowly and lasts over a period of time.

There is a potential hazard of transmission of parasitic infections with air-dried digestion sludge as a result of handling the sludge or from sludge contaminated vegetables eaten raw. Spreading of digested sludge in the fall and allowing it to freeze in cold climates in the winter is believed helpful in killing these organisms. Heat-dried sludge is considered safe for use under all conditions because of the destructive action of heat upon bacteria.

Let's talk about composting. A good compost could contain up to 2 per cent nitrogen, about 1 per cent phosphoric acid and many trace elements. Its most valuable features, however, are not its nutrient content, but its moisture retaining and humus forming properties. Many types of micro-organisms are involved in converting the complex organic compounds such as carbohydrates and proteins into simpler materials, but the bacteria, actinomycetes and fungi, predominate. These organisms function in a composting environment that is optimised by copying the natural decomposition process of nature where, with an adequate air supply, the organic solids are biochemically degraded to stable humus and minerals. Compost is generally considered as a material to be used in conjunction with fertiliser, rather than as a replacement for fertiliser unless it is fortified with additional chemical nutrients. Compost benefits the soil by replenishing the humus, improving the soil structure and providing useful nutrients and minerals. It is particularly useful on old, depleted soils and soils that are drought-sensitive. In horticulture applications, compost has been useful on heavy soils as well as sandy and peat soil. It has been commonly applied, to parks and gardens because it increases the soil water absorbing capacity and improves the soil structure.

All composting processes attempt to create a suitable environment for thermophilic facultative aerobic micro-organisms. If the environmental conditions for biological decomposition are appropriate, a wide variety of organic waste can be composed. The most important criteria for successful composting are: (i) complete mixing of organic solids, (ii) nearly uniform particle size, (iii) adequate aeration, (iv) proper moisture content, (v) proper temperature and pH, and (vi) proper, carbon-nitrogen ratio in the raw solids. The smaller the particles, the more rapidly they will decompose; size is controlled by grinding. Air is necessary for aerobic organisms to function in a fast, odour-free manner. Aeration is enhanced by blending wastes to form a porous solids structure in the composting materials. Some composting systems use blowers while others aerate by frequent turning of compost placed in windows and bins. The solids to be composted must not, of course, contain high concentrations of materials toxic to the decomposing micro-organisms. A proper moisture content is the most important composting criteria. Micro-organisms need moisture to function but too much moisture can cause the process to become anaerobic and develop the characteristic odour and slow decomposition rate associated with anaerobic processes.

Composting mixtures, should have a pH near 7 (neutral) for optimum efficiency. The temperatures vary a great deal but those in the thermophilic range (greater than 110°C) produce a more rapid rate of decomposition than those in the lower mesophilic range. Higher temperatures also cause a more efficient destruction of pathogenic organisms and weed seeds. An essential requirement of the composting process is control of the ratio of carbon to nitrogen in the raw materials. Micro-organisms need both carbon and nitrogen, but they must be available in the proper amounts of decomposition or will be prolonged. The time required to complete composting varies, depending on the climate, materials composted, the degree of mechanisation, whether the process is enclosed and the desired moisture content of the final product. Composting detention times from a couple of weeks to several months have been reported.

Many types of wet solids have been successfully used in composting operations. These include sewage sludge, cannery solids, pharmaceutical sludge and meat packing wastes. Sewage sludge has been frequently used as an additive when composting dry refuse and garbage. It enhances the composting operation because: (i) it serves as a seeding material to encourage biological action, (ii) it helps to control the moisture content in the composting mixture, (iii) it enhances the value of the compost by contributing nitrogen and other nutrients, and (iv) it can be used to control the important carbon/nitrogen ratio. Normally, blending sewage sludge with other compost raw materials required prior dewatering of the sludge. If the dewatering step is omitted, the moisture content of the mixture is too high and odours develop. Reducing sludge moisture from 90 to 70 per cent by vacuum filtration or centrifugation allows good aerobic composting with garbage at a blended moisture content of 53 per cent. In favourable climates, the composting of digested sludge with sawdust, straw and wood shavings has been successful.

Pollution Prevention Options

The balance of our discussions focus on the pollution prevention technologies for sludge management and use. As we have seen, sewage sludge has many characteristics that are good for soils and plants, if applied properly. Research has shown that the organic matter in sludge can improve the physical properties of soil. Reused sludge is also considered biosolids, which is a slightly more attractive name. Used as a soil additive, sludge improves the bulking density, aggregation, porosity and water retention of the soil. When added properly, sludge enhances soil quality and makes it better for vegetation. Vegetation also benefit from the nitrogen, phosphorus and potassium in sludge. When applied to soils at recommended volumes and rates, sludge can supply most of the nitrogen and phosphorus needed for good plant growth, as well as magnesium and many other essential trace elements like zinc, copper and nickel.

There are alternative systems to the marketability of biosolids from waste-water treatment plants. In fact, there are more than a dozen systems encompassing Class A pathogen-reduction technologies, but among these the most promising and widely used are alkaline stabilisation, thermal drying and composting. We only briefly mentioned alkaline stabilisation, but in reality this is a variation of sludge pasteurisation. The basic process uses elevated pH and temperature to produce a stabilised, disinfected production. The two alkaline stabilisation systems most common in are a lime pasteurisation system and a cement kiln dust pasteurisation system. The lime pasteurisation product has a wet-cake consistency, while the kiln dust pasteurisation has a moist solid like consistency. Both products can be transported to agricultural areas for ultimate use.

In contrast, composting processes utilise a mixture of solids and yard waste under controlled environmental conditions to produce a disinfected, humus-like product. Three common composting systems are a horizontal agitated reactor, a horizontal nonagitated reactor and an aerated static pile system (nonproprietary). Compost can be marketed as a soil conditioner in competition such products as peat, soil and mulch. Although a large potential market exits, significant effort is required to penetrate this market.

The lime stabilisation system has advantages of low capital costs, process reliability, flexibility and operability. The main disadvantage attributed to this system are questionable product marketability because of the uncertain availability of suitable agricultural land in some parts of the country where the product could be locally marketed. The steam drying alternative has the advantages of small facility land requirements, good public acceptance and favourable product marketability. The disadvantages of this system included relatively high capital costs, reduced expansion flexibility and complex operational requirements. These advantages and disadvantages apply to all of the thermal drying alternatives.

Land application is the largest beneficial use for sewage sludge. Since municipal sludges are a by-product of the foods, we eat, they contain important nutrients such as nitrogen, phosphorus and potassium. Proper land application provides a way to recycle these nutrients and return them to the soil safely. Sludge can also be processed into heat dried pellets that are marketed as fertilisers and soil conditioners. The pelletisation process also reduces disease causing organisms. Golf courses, parks, cemeteries, nurseries and municipal landscaping projects provide markets for such pelletised sludge products.

Composting is another way to recycle nutrients and organic matter in sludge. The benefits from using sludge composts include increased water and nutrient holding capacity and increase aeration and drainage of soils. Composted sludge's also provide the soil with low levels of plant nutrients. More and more countries are turning to composting as a method to beneficially manage sludge's.

There are concerns that land application of sludge will result in an increase of pathogenic bacteria, viruses, parasites, chemicals and metals in drinking water reservoirs, aquifers and the food chain. This raises additional concerns of cumulative effects of metals in cropped soils. Research shows that if metals such as zinc, copper, lead, nickel, mercury and cadmium are allowed to build up in soils due to many applications of sludges over the years, they could be released at levels harmful to crops, animals and humans. While some of these metals are necessary micronutrients, at higher levels they may be harmful to plants, particularly those grown on acid soils (soils with a low pH). Cadmium, a suspected carcinogen and mercury cause even greater concern because of their toxic effects on animals and humans. Likewise, synthetic organic compounds such as dioxins and PCBs, if present, cause concern about ecological and human health impacts. The degree of risk depends directly on the initial sludge quality, the way the sludges are processed and how the amended soil is managed during and after land application. Current enviromental regulations requires sludge treatment processes to reduce pathogens prior to land application. Further EPA mandate specific limits for metals contained in sludge. Since metal concentrations depend mainly on the type and amount of industrial waste that flows into the waste-water treatment system, strongly enforced pre-treatment and source control programmes could effectively reduce the metals content of sludge.

Providing proper employee training and applying the best management practices will yield the best sludge use programme. The fate of sludge components is also influenced by factors such as climate (rainfall and temperature), soil management (irrigation, drainage, liming, fertilisation and addition of amendments) and composition of the sludge. In the past, the success of land application has been hurt by the mismanagement of important factors such as soil pH. For example, the uptake of many metals, such as cadmium, is related to soil pH. If pH drops below a certain level, heavy metals will be released, increasing the chances of leaching and plant uptake. In, addition, nutrient contamination of surface waters through non-point source pollution needs to be carefully monitored. While not a concern for human health and the environment, odours associated with poorly managed sludge application can be a serious concern to those living near application sites. Prompt incorporation of sludges and sludge products into the soil and avoidance of stockpiling can help to prevent odour problems. It is essential for sludge management programmes to have knowledgeable, staff available to teach people how to apply and monitor the sludges and the treated area correctly.

In general, researchers agree that the effects of organic compounds, certain pesticides and metals are not dangerous when managed properly at regulated levels. However, they caution that additional study of organic compounds and long-term fate of materials is needed before unlimited application of sludge can occur safely on all lands.

Sludge landfill can be defined as the planned disposal of waste-water solids including sludge, grit and ash at a designated site where it is buried and monitored. The sludge is delivered to the landfill by trucks that pick up the sludge from the waste-water treatment plants. There are several different types of landfilling, these are all listed below under disposal methods, but the most frequent method used is dewatering then burial. This method is done by the plant dewatering the sludge then trucks pick up the sludge which is approximately 80 per cent moisture and 20 per cent solids. The trucks then dump the sludge into the landfill, where tractors bury the sludge using one of two special burial techniques. These techniques utilise space most efficiently and develop a grade for drainage of precipitation.

Many municipalities and state regulatory agencies do not want sludge to be landfilled. Most states require special permission to do so. Landfills must be monitored regularly with monitoring wells and a few other environmental safety measures. The municipality are the state determine where and how the sludge will be disposed of. Once they are designated to be a part of land use, the sludge is either landfilled or if it is usable or the right grade, which is usually grade A, the sludge is used for composting. The essential difference between land application and landfill is that land application leads to treatment or assimilation, while landfill leads to containment and only for an unspecified time.

A landfill has two major drawbacks, these drawbacks are leachate and the gases of decomposition. Siting and design of landfill operations to avoid disturbing water quality should be based on geological and hydrological considerations. The disposal options we have available to us are:

1. Dumped in sand and gravel within open pits previously dug by bulldozer, pits then filled to control odour and other problems.
2. Dumped at a site and levelled.
3. Dumped on top of fill and mixed with refuse during compaction.
4. Dumped into pit.
5. Dewatered by the treatment plant, moved to landfill, dumped and immediately buried.
6. Only air-dried digested sludge accepted.
7. City landfill disposal of sludge unregulated.

The most important factor of a landfill is to build it properly so that the environment is not disturbed in any fashion. There are several components to the design of a environmentally friendly landfill. These components are that the landfill should be placed on a compacted low permeable medium, preferably a clay layer. This layer is then covered by a impermeable membrane which is then covered by a granular substance to act as a secondary drainage system. Layers upon layers are built up, while each layer is separated by a granular membrane. This is done over and over again until the entire landfill is full. Then they cap off the landfill to prevent excess amounts of surface water from entering. The design of the landfill layers and the mound are:

1. An above grade containment mound, sloped to support the weight of the waste and cover.
2. A liner system across the base to retard entry of water and subsequent percolation of leachate.
3. A leachate collection and removal system that is drained freely by gravity, with drainage above ground.
4. A cover system consisting of a layer with gas collection equipment, a composite liner, a drainage liner and a permanent vegetative cover.
5. A monitoring system.

To determine the cost analysis of landfilling sludge you must evaluate the steps preceding it. After the sewage treatment plant has treated the sludge they send it to a dewatering site. This site reduces the sludge to 20 per cent solid and 80 per cent water. The actual cost of operating a dewatering facility depends upon size and technology. The two contaminants of environmental concern from refuse disposal are gas and leachate. The leachate is generated because of the water that penetrates the landfill and the gas is due to the decomposing of the organic matter. Gas production from the organic matter begins before it is actually landfilled, the principal gases that are generated from the decomposing matter is carbon dioxide and methane. Carbon dioxide is important in the surrounding areas water quality, because it is soluble in water, unlike the other gases that can be produced within the landfill which are insoluble. When carbon dioxide is dissolved in water it lowers the pH, which creates a corrosive environment. It also creates an increase in water hardness.

Usually the effects of carbon dioxide are at a maximum during the first few months of decomposition and could continue on for a few years. As time goes on carbon dioxide values decrease and pose a lesser problem as the years go by. Leachate production within a landfill depends on the amount of water that enters the landfill. Leachate results when the amount of water entering exceeds the amount of water that can be retained by the waste. This is a major reason why site investigation and soil characteristics are so very essential in landfill design. The primary causes of excess water intrusion are due to a raise in groundwater elevation. Another consideration that should be evaluated is the topography and the climate of the area, because these two factors can cause a dramatic impact on the landfill if they are not assessed properly. The best approach to leachate management is to prevent or limit its production from the beginning. This is why proper design and elaborate research of an area are so very essential to a landfill and its operation.

Biosolids Regulations

The EPA has developed comprehensive biosolids use and disposal regulations, which are organised in five parts. These parts are general provisions, land application, surface disposal, pathogens and vector attraction reduction and incineration. Parts of the regulations which address standards for land application, surface disposal and incineration practices consist of general requirements, pollutant limits, operational requirements, management practices, frequency of monitoring, record keeping and reporting requirements for biosolids processing facilities to abide by. Regulatory considerations play a key role in determining how to efficiently use sludge. The EPA has adopted a sludge management policy intended to encourage the beneficial use of sludge while protecting public health and the environment. The revisions follow many years of sludge used in field studies that analysed the effect of toxic elements in land-applied sludge and sludge composts. The regulations and technical support documents are interpreted by some scientists to reinforce the safety of using sludge on both agricultural and non-agricultural lands while ensuring the protection of soils, water quality, the food chain and human health. However, scientific uncertainties remain particularly with respect to the long-term and ecological safety of sludge application. New York State has adopted an integrated waste management policy that involves a hierarchy of solid waste management methods intended to reduce dependency on landfills for waste disposal. The hierarchy is incorporated into New York's Environmental Conservation Law in order of preference: Reduction, recycle/reuse, incineration and landfilling. This policy is the cornerstone of the state's solid waste management programme. Several components of the programme assist local governments in managing their wastes safely and efficiently. These regulations contain specific guidelines for land application and other sludge management options that must be considered by a municipality or purveyor during its planning process. These regulations are currently under revision.

Land applied biosolids must meet quality requirements. Land applications includes all forms of applying bulk or bagged biosolids to land for beneficial uses at argonomic rates (rates designed to provide the amount of nitrogen needed by the vegetation while minimising the amount that passes below the root zone). These include application to agricultural land; pasture and range land; non-agricultural land such as forests; public contract sites such as parks and golf courses; disturbed land such as mine spoils, construction sites and gravel pits; and home lawns and gardens. Selling or giving away biosolids products is addressed under land application of domestic septage (liquid or solid material is addressed under land application of domestic septage (liquid or solids material removed from septic tank). The person who prepares biosolids for land application or applies biosolids to the land must obtain and provide the necessary information needed to comply with the rule. For example, the person who prepares bulk biosolids that are land applied must provide the person who applies it to land with all information necessary to comply with the rule, including the total nitrogen concentration of the biosolids. The rule establishes two levels of biosolids quality with respect to heavy metal concentration—pollutant ceiling concentrations and pollutant concentrations ('high quality' biosolids); two levels of quality with respect to pathogen densities—class A class B; and two types of approaches for meeting vector attraction reduction-biosolids processing or the use of physical barriers (Vector attraction reduction reduces the potential for spreading infectious disease agents by vectors, that is, flies, rodents and birds).

To qualify for land application, biosolids must meet the pollutant ceiling concentrations, class B requirements for pathogens and vector attraction reduction requirements. Bulk biosolids applied to lawns and home gardens must meet the pollutant concentration limits, class A pathogen reduction requirement and vector attraction reduction using biosolids processing. Bulk biosolids applied to agricultural and non-agricultural land must meet at a minimum the pollutant ceiling concentrations and cumulative pollutant loading, at least class B pathogen reduction requirements and one of the vector attraction reduction requirements.

Management practices that apply to land applied biosolids (other than 'exceptional quality' biosolids products) include:

1. No application to flooded, frozen or snow covered ground.
2. No application at rates above argonomic rates (reclamation projects may be excepted).
3. No application if threatened endangered species are adversely affected.
4. Labelling of biosolids that are sold or given away.
5. A required 10 meter buffer from US waters.

If the biosolids are of 'exceptional quality'—that is, they meet the pollutant concentration limits, class A pathogen reduction requirements and a vector attraction processing option—they are usually exempt. However, when biosolids meeting class B pathogen reduction requirements are applied to the land, additional site restrictions are required. Table 15.5 provides a summary of the land application pollution limits for biosolids as they currently stand.

Table 15.5. Land application pollutant limits[a].

Pollutant	Ceiling concentration limits,[b] mg/kg	Cummulative pollutant loading rates, kg/ha	High quality pollutant concentration limits,[c] mag/kg	Annual pollutant loading rates, kg/ha/yr.
Arsenic	75	41	41	2.0
Cadmium	85	39	39	1.9
Chromium	3000	3000	1200	150
Copper	4300	1500	1500	75
Lead	840	300	300	15

(Contd ...)

Pollutant	Ceiling concentration limits,[b] mg/kg	Cummulative pollutant loading rates, kg/ha	High quality pollutant concentration limits,[c] mag/kg	Annual pollutant loading rates, kg/ha/yr.
Mercury	57	17	17	0.85
Molybdenum	75	18	18	0.90
Nickel	420	420	420	21
Selenium	100	0	36	5.0
Zinc	7500	2800	2800	140

[a]EPA part 503 standards. All weights are on a dry-weight basis.

[b]Absolute values.

[c]Monthly averages.

Chapter 16

Analysis of Water

INTRODUCTION

Water is vital for life. Not only do we need water to drink, to grow food and to wash, but it is also important for many of the pleasant recreational aspects of life.

Some of the important uses of water are in: (i) domestic water supply, (ii) industrial water supply, (iii) effluent and waste disposal, (iv) fishing, (v) irrigation, (vi) navigation, (vii) power production, (viii) recreation, e.g. sailing and swimming.

Each different use has its own requirements over the composition and purity of the water and each body of water to be used will need to be analysed on a regular basis to confirm its suitability. The types of analysis could vary from simple field testing for a single analyte to laboratory-based, multi-component instrumental analysis. Water is found naturally in many different forms. In the liquid state it is found in rivers, lakes and groundwater (water held in rock formations) and also as sea water and rain. As a solid, it is found as ice and snow. Water in the vapour state is found in the atmosphere. You will certainly be familiar with the fact that sea water contains large quantities of dissolved material in the form of inorganic salts but it may come as a surprise that nowhere in the environment can you consider water to be chemically pure. Even the purest snow contains components other than water.

The composition of water continuously changes as it travels in the environment. Sampling at a large number of locations is therefore necessary to monitor these changes. Careful choices of locations, sampling time, and sample storage procedures are necessary for reliable monitoring. The quality of water can be assessed by using measurements relating to the overall effects of groups of compounds or ions (water quality parameters), as well as by analysis of the major individual components. Methods for both types of determination have been described. These include both volumetric and instrumental methods.

Some of the constituents found in natural river water are: (i) ions derived from commonly occurring inorganic salts, e.g. sodium, calcium, chloride and sulphate ions, (ii) smaller quantities of ions (e.g. transition metal ions) derived from less common inorganic salts perhaps derived from leaching of mineral deposits, (iii) insoluble solid material, either from decaying plant material or inorganic particles from sediment and rock weathering, (iv) soluble or colloidal compounds derived from the decomposition of plant material, and (v) dissolved gases which includes oxygen, which is so vital in supporting aquatic life. Dissolved gases occur through contact with the atmosphere and through respiration and photosynthesis. A fast flowing turbulent river will usually be saturated in atmospheric gases. Respiration of aquatic animals releases energy from foodstuffs, consuming oxygen and producing carbon dioxide, as follows:

$$C_6H_{12}O_6 + 6O_2 \rightarrow 6CO_2 + 6H_2O + energy \qquad \text{... (16.1)}$$
$$\text{glucose}$$

Photosynthesis by plants reverses this process, producing organic compounds and oxygen from carbon dioxide by using sunlight as an energy source:

$$6CO_2 + 6H_2O + hv \rightarrow C_6H_{12}O_6 + 6O_2 \qquad \qquad \text{... (16.2)}$$

Oxygen levels in water are depleted by slow oxidation of organic and in some cases, inorganic material. The presence of large quantities of oxidisable organic material (e.g. from sewage effluents) is often the most serious form of pollution in watercourses.

Ions commonly found in the mg l^{-1} concentration range are shown in Table 16.1. Others (e.g. fluoride ions) may occur depending on the mineral deposits in the locality.

Table 16.1. Ions found in mg l^{-1} concentrations in natural waters.

Concentration range (mg l^{-1})	Cations	Anions
0–100	Ca^{2+}, Na^+	Cl^-, SO_4^{2-}, HCO_3^-
0–25	Mg^{2+}, K^+	NO_3^-
0–1	Fe^{2+}, Mn^{2+}, Zn^{2+}	PO_4^{3-}
0–0.1	Other metal ions	NO_2

Figure 16.1, shows typical comparative analyses for rain water, river water and sea water. You will find similar ions in all three, with the only difference being the concentration range. Sea water contains the common ions at the g l^{-1} level, whereas for river and rain water the values are at the mg l^{-1} level. All are easily measurable with modern instrumentation.

The situation would be a little different if we tabulated the less common species. The range of ions (particularly metal ions) would be limited in river water by the chemical composition of the rocks over which it was flowing.

On the other hand, sea water contains trace quantities of virtually every element, with the highest concentrations being found close to the surface and in coastal areas. This is a very complicated analytical matrix indeed.

CONCENTRATION RATIOS OF THE RAIN, RIVER AND SEA WATER

Natural Processes which will Affect Constituents

A detailed comparison of the concentration of ions from a large number of rivers, compared with the concentrations in sea water (which appears depleted in a number of elements, including calcium), is one of the methods of studying the complexities of marine chemistry.

Water authorities often feel it necessary to analyse a river at many locations along its course. This is because the composition of water is never static. It changes by interaction with the atmosphere and crust and by chemical and biological processes occurring within the water. This does not even include the possibility of extra material being added in the form of pollution. Let us consider a river flowing from its source to the sea. Even at its source, water will contain dissolved salts from the passage of water through the earth to form the river.

Weathering of rocks

This will produce an increase in inorganic salt content. The composition may also be affected by interaction with material on the river bed. Clays, often found on river beds, are natural ion exchangers.

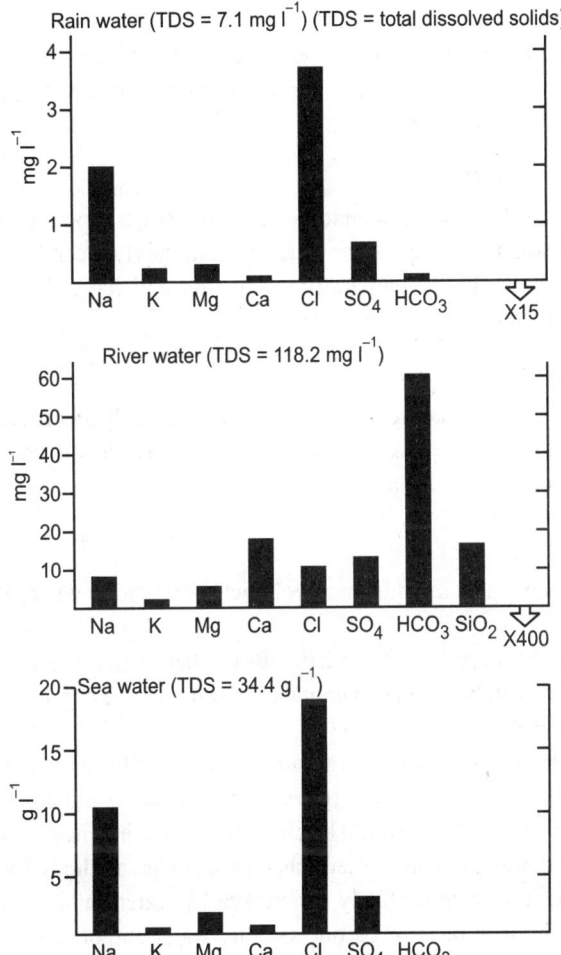

Fig. 16.1. Typical comparative analyses for rain water, river water and sea water; note the different scales for each bar graph.

Sedimentation of suspended material

As the river progresses downstream it will generally become less turbulent and so less capable of supporting suspended material.

Effect of aquatic life

Consumption and production of oxygen and carbon dioxide is well-known phenomena. Living plants will also absorb nutrients (including nitrate and phosphate) necessary for growth. The death and decay of organisms will release ions and also produce suspended material. This will slowly decompose into simpler chemical compounds. If the process proceeded to completion in the presence of oxygen, the final products would be carbon dioxide and water. At the same time, the oxygen concentration would fall. If the oxygen concentration was already low, then the final products would include ammonia and methane. Dense beds of vegetation can also very effectively filter out suspended solids.

Aeration

The generation of oxygen by plants is not the only method by which the gas enters water. There is continuous transfer of gases between the atmosphere and water. The oxygen can replenish the oxygen removed by oxidation of organic material.

Volatilisation and evaporation

Low-relative-molecular-mass organic compounds tend to have a high vapour pressure and will be readily lost from water. A significant percentage of the water itself in the river can be lost through evaporation (the rate depending on the ambient temperature) and this will have the effect of increasing the concentration of all dissolved material in the river.

Additional water volumes

Any water entering from tributaries or directly from overland flow will alter the analytical concentrations and may bring new constituents to the river. Similar considerations allow to understand the composition of waters in other areas in the environment.

GROUNDWATER

Groundwater is sub-surface water in soils and geological formations where the ground has become saturated with water.

The groundwater could be more concentrated in salts leached from mineral deposits. During passage through the earth, the water will have been in contact with degradable organic material. This can lower the oxygen content of the water.

Even if you disregard the introduction of new compounds by pollution any environmental water will contain a large number of components. In fact, if you start considering components which may be found at trace levels (less than mg l^{-1}) the task would be almost impossible as new components are constantly being identified in natural waters. This present chapter includes methods for the analysis of major components of water which may be routinely undertaken by water authorities. Even so, it would be unusual for all of the methods to be used on one sample. Water authorities or others undertaking the analyses will in general have a reasonable idea of what species to expect in the water. Unless there is a specific reason for more complete analysis, the analytical scheme will usually be restricted to components which are likely to cause environmental problems or exceed prescribed limits. Remember that consideration of the analytical process has to start with sampling and sample storage.

SAMPLING

Let us now consider developing a sampling programme for a river. The sample or samples (often only 250 or 500 ml each) must be representative of the whole body of water requiring analysis. The sample must also be kept in such a manner that the concentration of the species to be analysed is unchanged during transportation and storage.

1. Before starting, decide on what analyses are required. The analytical techniques to be used will affect the sample size taken, the type of sample bottle and also the method of storage. It will be too late to alter these by the time you get back to the laboratory. You should also confirm that laboratory time is available for analysis of the samples. Sample preservation times should be kept to a minimum (hours to days, depending on the analysis).

2. Decide on a sampling programme. We have already discussed how the composition of natural water is always changing. Sometimes the variation in composition may be periodic:

    Seasonal—the concentration is affected by natural growth processes.

    Weekly—a pollutant may only be emitted from a factory during the working week.

    Daily—the concentration of some components may be changed due to biological processes needing the presence of sunlight.

    Regular variations in concentrations are: (i) dissolved oxygen; and (ii) nitrate.

    Oxygen is produced by photosynthesis in daytime but is consumed by respiration or by oxidation of organic material continuously. There will be a continuous but slow replenishment from the atmosphere. A drop in oxygen concentration during the night would be expected.

    Variation of nitrate would be more complex. This is a nutrient which is necessary for growth and so if there were no additional inputs it would decrease in the spring growing season and increase in winter; however, if a farmer put an excessive amount of nitrate-containing fertiliser on a neighbouring field, there would be a sudden increase in any river into which the field drained.

3. Decide on the total number of samples you are taking, remembering that each location should be sampled in duplicate. Although it is good practice to start by taking as many samples as you feel necessary for complete monitoring, you do also have to take into consideration the time required for the analyses. It is very common to severely underestimate the time involved in the laboratory analyses.

    A further consideration if there is to be any statistical treatment of results is that there are sufficient samples for the treatment to be significant.

4. Decide on the location of the sampling and the sampling apparatus. If you are to take samples regularly from one location, the first consideration must be ease of access. Remember that the weather may not always be perfect. Surface water sampling requires little sophistication in sampling apparatus (often directly into a sample bottle or a bucket) but the surface may not be the best location for sampling. It may not provide the most representative sample. There also is the possibility of contamination by surface pollutants. Surface contamination can be largely overcome by inserting the bottle upside down in the water and inverting to fill it from just below the surface. Ideally, however, the river should be sampled further underneath the surface, in its main flow and at similar depths for each sample. A simple sub-surface sampler would be a weighed, stoppered bottle on an attachment line. The stopper is removed at the required depth by a cable. More complex designs, such as the Van Dorn sampler shown in Fig. 16.2, are open cylinders with valves at each end and produce less disturbance to the river on sampling. The sampler is sealed by using a weight (messenger) dropped down the attachment line to activate the valve mechanism.

    If you are monitoring the effect of a discharge into a river, samples should be taken far enough downstream for the discharge to be completely mixed. Samples taken further upstream would be unrepresentative as the analysis would depend on how much the discharge had mixed with the river.

5. Decide on the sample volume to be taken to the laboratory and the sample storage containers. The latter are usually made of glass or polyethylene. However, these materials (and those in the container top) are not as inert as you might think. Polyethylene containers may leach organic compounds into the sample, while glass bottles can leach inorganic species (sodium, silica

and other components of the glass). How much you fill the container is also important. If you are analysing volatile material or dissolved gases, the container must always be full. For other components, it is beneficial not to fill the container completely as the contents can then be more easily mixed before analysis. Try attempting to mix the contents of a completely full container. At this stage, it would also be worthwhile to check the equipment used at all stages of the sampling procedure to ensure that nothing will introduce contamination—the sampler itself, funnels and any tubing used. If you are sampling from a motorised boat without care, the boat itself could introduce contamination into the water and may disturb the sediment.

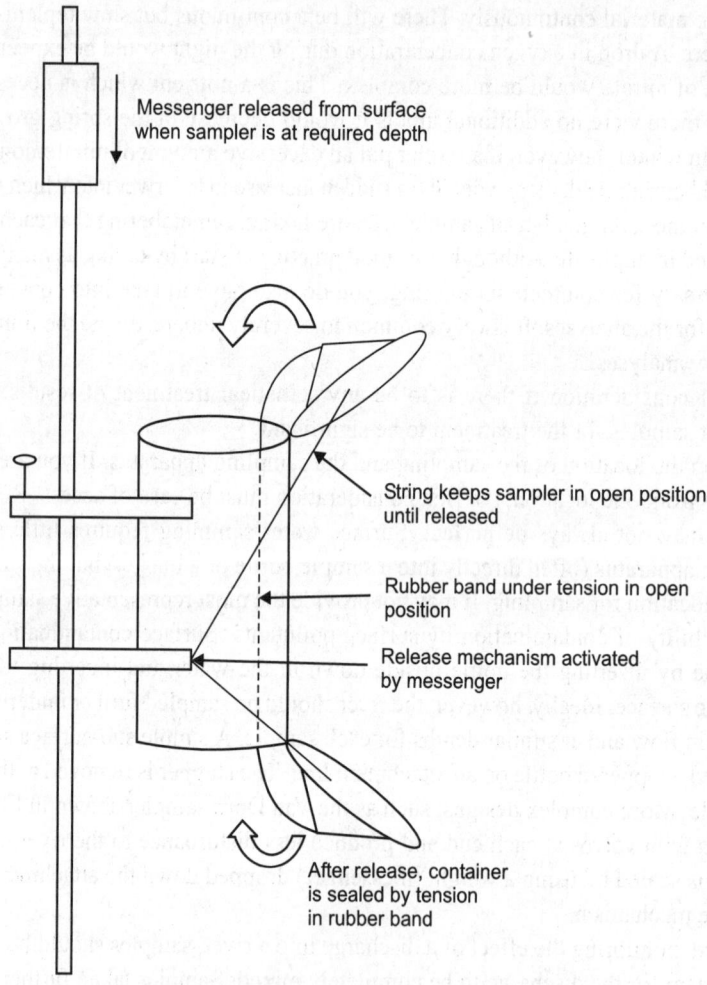

Messenger released from surface when sampler is at required depth

String keeps sampler in open position until released

Rubber band under tension in open position

Release mechanism activated by messenger

After release, container is sealed by tension in rubber band

Fig. 16.2. Schematic of a Van Dorn water sampler.

6. Decide on the method of storage of samples. Standard methods are available for most components to minimise analyte loss. The method varies according to the physical and chemical properties of the species.

For example:

Nitrate – store at 4°C to lower biological degradation.

Pesticides – store in the dark to avoid photochemical decomposition.

Metal ions – acidify the sample to prevent adsorption of metal ions on to the sides of the container.

Phenols – add sodium hydroxide to lower the volatility.

For some analyses (e.g. biochemical oxygen demand), no preservation is possible and the analysis should be performed as soon as possible after sampling, keeping the sample cool during transportation to the laboratory.

You should note that this may mean different sample storage conditions and storage containers for each analysis.

After all of these considerations, one can start sampling.

MEASUREMENT OF WATER QUALITY

This section discusses techniques which are usually intended to provide a measurement relating to the overall effect of groups of compounds or ions rather than to measure concentrations of individual components. These were originally conceived as simple and convenient methods to assess waste water but are now in widespread use to monitor long-term changes in environmental waters. You will find that many of the techniques involve titrations or use spectrometry.

Suspended Solids

We can all visualise streams so full of suspended material that the water is opaque, and where no visible life could possibly exist. This represents an extreme case of high solids loading. Any natural water will contain some suspended solids, but often the material is of such a small particle size that it cannot be easily seen. It is only when you look at two samples of water, one of which you consider 'clean' and the other which has been filtered to sub-micron level, that you can see the difference. The filtered water glistens, while the 'clean' water suddenly looks distinctly dirty. Even if the particles are chemically inert, their physical properties could cause problems.

Physical problems

Physical problems may be caused by suspended solids: (i) they cut down light transmission through the water and so lower the rate of photosynthesis in plants, and (ii) in less turbulent parts of the river some of the solids may sediment out, thus smothering life on the river bed.

Analysis of suspended solids is by using typically, a glass fibre filter disc with a 1.6 μm pore size would be used with a Hartley filter funnel (Fig. 16.3). The paper is clamped inside the funnel to prevent any part of the sample escaping around the side of the filter.

Dissolved Oxygen and Oxygen Demand

All animal life in a river is dependent on the presence of dissolved oxygen. More subtle requirements for a healthy river also include the presence of oxygen for the whole ecosystem. We have already seen how the presence of organic matter can remove oxygen from water by oxidation. Although the process can be written down as a simple chemical reaction, it is in fact, a microbiological process, known as aerobic decay. This converts the major elements present in plant matter (C, H, N, S) into CO_2, H_2O, NO_3^- and SO_4^{2-}, respectively.

It will perhaps come as a surprise that even if no oxygen is present in the water, organic material will still be broken down. Instead of the material being oxidised, it is reduced. The process is once again microbiological and is known as anaerobic decay. In this case, the final products are CH_4, NH_3 and H_2S.

Consideration of the products of anaerobic decay (in particular, their toxicity, smell and flammability) show that this condition should be avoided at all costs in environmental waters.

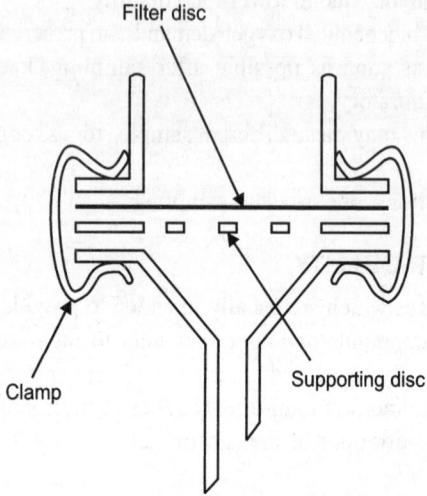

Fig 16.3. Schematic of a Hartley funnel.

The solubility of oxygen in water is low. Saturated water at 25°C and 1 atm pressure contains 8.54 mg l^{-1} oxygen. The sensitivity of fish to low oxygen is very species-dependent. Salmon can only survive under almost saturated conditions, trout to about 1.5 mg l^{-1}, while carp and tench are more resistant, surviving down to about 0.3 mg l^{-1} oxygen. It is easy to deplete the oxygen content if any material is present which would react rapidly with the oxygen. Such material could be organic, as already discussed, but could also be inorganic. Iron in the form of Fe^{2+} can deplete oxygen by oxidation to Fe^{3+}. Natural replenishment by oxygen from the atmosphere can be very slow.

Two distinct analyses which could be useful if monitoring environment waters for oxygen:

1. A direct measurement of the oxygen concentration in the sample. This would give an indication of the health of the river at a particular location and at the time of sampling. It would be of less use for assessing the overall health of a river as the oxygen level can vary dramatically with location and with time.

2. A measurement of the amount of material which, given time, could deplete the oxygen level in the river. This is known as the oxygen demand, and gives an indication of the possibility of oxygen depletion which will occur if the oxygen is not replenished. Such a measurement would be much more suitable for determining the overall health of the river since the oxygen demand is unlikely to change suddenly.

The analytical techniques used for dissolved oxygen measurement can also be used to measure oxygen demand and so these will be discussed first.

Dissolved oxygen

The determination of oxygen can be either by titration (Winkler method) or by use of an electrode sensitive to dissolved oxygen. The results are either expressed as a simple concentration (mg l^{-1}) or as a percentage of full saturation. The concentration of oxygen in saturated water is dependent on the temperature, pressure and salinity of the water and would need either to be established from published tables or determined experimentally. The first problem to overcome is transport of the sample to the laboratory. Without modification to the sample, this would cause sufficient agitation to the water to saturate the sample with oxygen from the air, regardless of its original content.

In the Winkler method, the oxygen is 'fixed' immediately after sampling by reaction with Mn^{2+}, added as manganese (II) sulphate, together with an alkaline iodide/azide mixture:

$$Mn^{2+} + 2OH^- + \tfrac{1}{2}O_2 \rightarrow MnO_2(s) + H_2O \qquad \text{.... (16.3)}$$

The iodide is necessary for the analytical procedure in the laboratory and the azide is present to prevent interference from any nitrite ions which can oxidise the manganese (II) ion. The sample completely fills the bottle to ensure no further oxygen is introduced. After transport to the laboratory, the sample is acidified with sulphuric or phosphoric acid.

This produces the following reaction:

$$MnO_2 + 2I^- + 4H^+ \rightarrow Mn^{2+} + I_2 + 2H_2O \qquad \text{... (16.4)}$$

The released iodine can then be titrated with sodium thiosulphate using a starch indicator:

$$I_2 + 2S_2O_3^{2-} \rightarrow S_4O_6^{2-} + 2I^- \qquad \text{... (16.5)}$$

The electrode method is used for field measurements of dissolved oxygen and can also be employed in the laboratory for determination of the Biochemical oxygen demand (see below). Several types of systems are available for this purpose, including the Mackereth cell shown in Fig. 16.4. In the latter, the current generated by the cell is proportional to the rate of diffusion of oxygen through the membrane, which is in turn proportional to the concentration of the oxygen in the sample. The reactions involved are as follows:

at the cathode $\qquad \tfrac{1}{2}O_2 + H_2O + 2e \rightarrow 2OH^- \qquad \text{... (16.6)}$

at the anode $\qquad Pb + 2OH^- \rightarrow PbO + H_2O + 2e \qquad \text{... (16.7)}$

Instruments usually read oxygen directly with a scale from 0–100 per cent saturation and are calibrated by setting 100 per cent with fully aerated water and 0 per cent with water with no oxygen content (sodium sulphite is added to the water). This calibration must be made each time that the electrode is used.

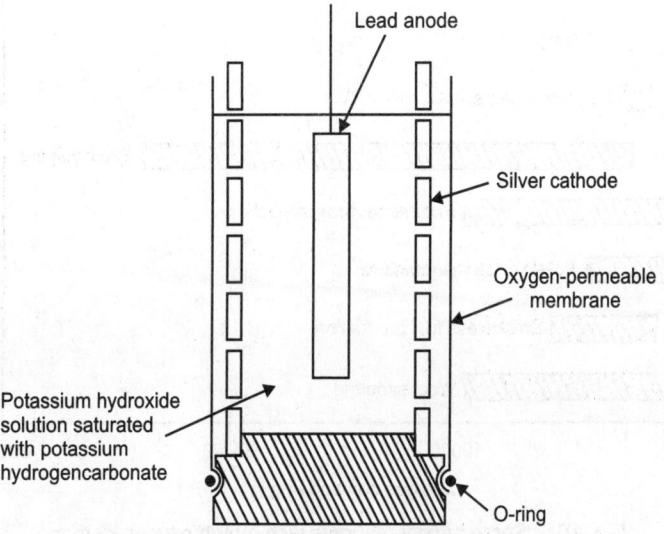

Fig. 16.4. Schematic of a Mackereth cell.

Oxygen demand

This can be measured by a number of methods. We will compare these after each of them has been described.

Biochemical oxygen demand

The method used to measure the biochemical oxygen demand (BOD) attempts to replicate the oxidation conditions found in the environment. In this, the dissolved oxygen level of a fully aerated water sample is first determined by either of the methods previously described. The measurement is repeated on a sample after it has been left for five days in the dark in a completely filled container and under standard conditions designed to be ideal to promote microbiological activity (20°C, after adjustment of the pH to between 6.5 and 8.5, with the possible addition of salts containing magnesium, calcium, iron (III) and phosphate as nutrients). Care has to be taken with contaminated or treated waters that no compounds are present which would lower the microbial activity, e.g. chlorine. The latter can be removed by the addition of sodium bisulphite.

If the sample is expected to have a high oxygen demand, a dilution should be made with well-aerated water (whose oxygen content is known). Ideally, 30 per cent or more of the oxygen should remain at the end of the analysis. The diluent should include nutrient salts, with distilled water alone not being satisfactory. If, for any reason, the sample is thought to be sterile, a seed sample of sewage may be added.

If there is no dilution of the sample, then we can write the following:

$$\text{BOD} = (\text{initial oxygen concentration-final oxygen concentration})\text{mg l}^{-1} \qquad \text{... (16.8)}$$

Typical BOD values for unpolluted water are of the order of a few mg l^{-1}. Many seemingly innocuous effluents have a very high oxygen demand, as shown in Fig. 16.5. If you remember that the saturated oxygen level in water is of the order of 8 mg l^{-1}, then you will be able to see how the introduction of a small quantity of high-strength effluent can deplete the oxygen in many times its own volume of water.

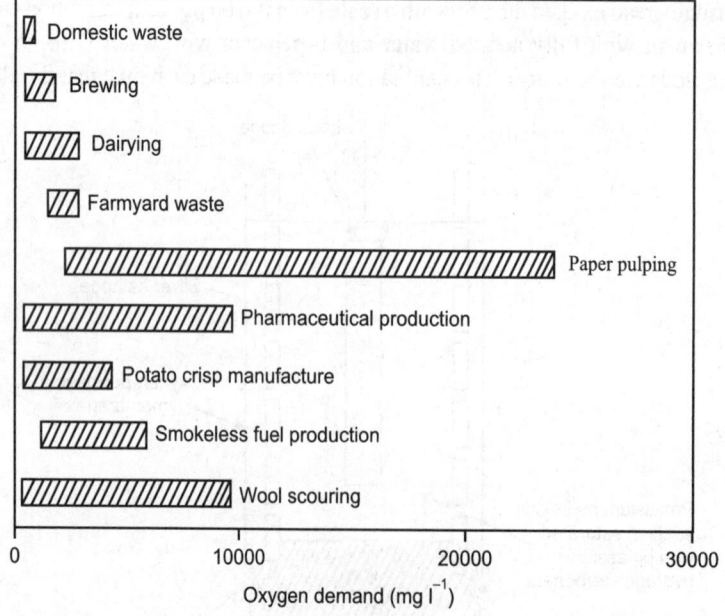

Fig. 16.5. Some typical effluents with a high oxygen demand.

Chemical oxygen demand

The term 'Chemical oxygen demand' (COD) relates to a family of techniques which involve reacting the sample with excess oxidising agent. After a fixed period, the concentration of unreacted oxidising agent is determined either by spectrometry or titration. The quantity of oxidising agent used can be calculated and the oxygen equivalent then determined. Such methods include the following:

Measurement of the two-hour dichromate value

Here, the sample is refluxed with excess potassium dichromate in concentrated sulphuric acid for 2 hours:

$$Cr_2O_7^{2-} + 14H^+ + 6e \rightarrow 2Cr^3 + 7H_2O \qquad \qquad ... (16.9)$$

Silver sulphate may be included to catalyse the oxidation processes of alcohols and low-molecular-weight acids.

Chloride ions give a positive interference by the reaction shown in the following equation. The interference is reduced by the addition of mercury (II) sulphate, with a chloro complex being formed:

$$Cr_2O_7^{2-} + 6Cl^- + 14H^+ \rightarrow 2Cr^{3+} + 3Cl_2 + 7H_2O \qquad \qquad ... (16.10)$$

If the excess dichromate is determined by titration, then iron (II) ammonium sulphate can be used:

$$6Fe^{2+} + Cr_2O_7^{2-} + 14H^+ \rightarrow 6Fe^{3+} + 2Cr^{3+} + 7H_2O \qquad \qquad ... (16.11)$$

In high-throughput laboratories, commercially available kits may be used which determine the unused dichromate by measuring the absorbance at 620 nm. Semi-micro systems are available which allow the test to be performed by using 2 ml aliquots of sample, heating a sealed tube with premixed reagents. With such systems, 25 analyses can be performed simultaneously.

Permanganate tests

In these methods, excess potassium permanganate is added under specified conditions, which can range from three minutes on a steam bath to four hours at room temperature. The unreacted permanganate can be determined by any of a number of techniques including the liberation of iodine, followed by titration of the latter with thiosulphate, as follows:

$$2MnO_4^- + 16H^+ + 10I^- \rightarrow 2Mn^{2+} + 8H_2O + 5I_2 \qquad \qquad ... (16.12)$$

$$I_2 + 2S_2O_3^{2-} \rightarrow S_4O_6^{2-} + 2I^- \qquad \qquad ... (16.13)$$

The confusing number of variations of this method leads to limitation in its use since it is very difficult to obtain comparative inter-laboratory data. The 'three-minute' variant of the test does, however, provide a rapid method of testing a specific water for its oxidising ability.

A comparison of the BOD and COD tests is given in Table 16.2.

Table 16.2. Comparison of various BOD and COD tests.

BOD	COD[a]
Five-day analysis time	Rapid analysis
Closely related to natural processes	Less relationship to natural processes
Difficult to reproduce, both within laboratories and between laboratories	Good reproducibility
Care has to be taken with polluted water	Can analyse heavily polluted water

[a]All of these tests will be affected by the presence of inorganic reducing or oxidising agents, with the former giving positive results, and the latter (possibly) leading to negative results.

Relationship of oxygen demand to specific concentrations

If a single organic compound was present in the water and the oxidation reactions proceeded to completion, the above methods would give an accurate measurement of its concentration. The determination of known amounts of a single compound can be used in the laboratory to test experimental procedures. Potassium hydrogenphthalate is often used, which is oxidised according to the following equation:

$$C_8H_5O_4K + 15/2O_2 \rightarrow 8CO_2 + 2H_2O + K^+ + OH^- \qquad \text{... (16.14)}$$

Total Organic Carbon

None of the oxygen demand methods give a precise estimation of the total organic carbon (TOC) loading of the water. A number of techniques are available which can achieve this. All involve the oxidation of the organic matter to carbon dioxide, after prior acidification to remove interference from carbonates. The methods used include the following:

1. Injection of a small quantity of water into a gas stream passing through a heated tube to carry out the oxidation. Measurement using this technique is possible to the mg l^{-1} level.
2. Wet oxidation by using potassium peroxydisulphate at room or elevated temperatures. This method is about 100 times more sensitive than the heated tube oxidation approach.

The carbon dioxide can then be measured either by absorption in solution and measurement of its conductivity, by reduction to methane and analysis of this gas by flame ionisation detection or by direct measurement by infrared spectrometry.

Attempts are often made to replace BOD and other oxygen demand measurements with TOC. To understand this, the following advantages should be noted:

1. It is a rapid technique.
2. It would be expected to give highly reproducible results.
3. It can be easily automated, either for laboratory analysis or for on-line monitoring of effluents.

pH, Acidity and Alkalinity

The pH is related to the number of hydrogen ions, in solution by the following relationship:

$$pH = -\log_{10}a(H^+) \qquad \text{... (16.15)}$$

where, $a(H^+)$ is the hydrogen ion activity.

At the low concentrations of hydrogen ions and low ionic strengths which are typical of unpolluted environmental samples, the hydrogen ion activity is approximately equivalent to the hydrogen ion concentration. Some typical pH values found with environmental water samples are shown in Fig. 16.6.

The unpolluted rain water is slightly acidic is due to the presence of dissolved carbon dioxide, as follows:

$$H_2O + CO_2 \text{ (gas)} \rightleftharpoons H_2O \cdot CO_2\text{(solution)} \rightleftharpoons H^+ + HCO_3^- \rightleftharpoons 2H^+ + CO_3^{2-} \qquad \text{... (16.16)}$$

Hardness in water is due to the presence of polyvalent metal ions, e.g. calcium and magnesium, arising from dissolution of minerals. For instance, the dissolution of limestone involves the equilibria shown in Eqs 16.17 and 16.18 below. From these equilibria, you should be able to see that the water will then be slightly alkaline:

$$CaCO_3 \rightleftharpoons Ca^{2+} + CO_3^{2-} \qquad \text{... (16.17)}$$

$$CO_3^{2-} + H_2O \rightleftharpoons HCO_3^- + OH^- \qquad \text{... (16.18)}$$

The biological effect of a change in pH can most easily be seen by the sensitivity of freshwater species to acid conditions. Population of salmon start to decrease below pH 6.5, perch below pH 6.0, and eels below pH 5.5, with little life possible below pH 5.0. The eradication of life can result from a change of little more than 1 pH unit. Chemical effects are also observed. Since a decrease in pH increases the solubility of metals. The use of lead piping for domestic water supplies becomes of greater concern as the water becomes more acidic. The weathering of minerals, such as limestone or dolomite, by water becomes more rapid with a decrease in pH.

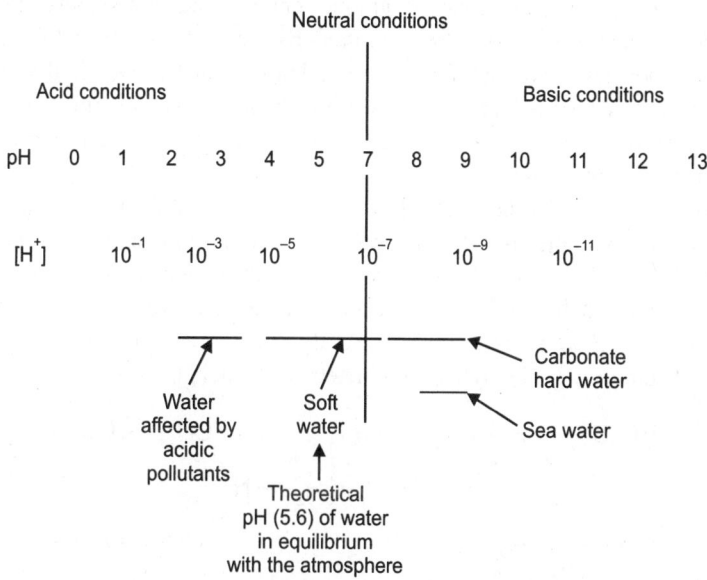

Fig. 16.6. Typical pH ranges for environmental water.

A typical procedure for the measurement of pH involves calibration with two buffer solutions spanning the expected pH of the sample, followed by measurement of the sample.

The procedures for 'Alkalinity' and 'Acidity' measure, by titration, the quantity of acid or base needed, respectively, to change the pH of a sample to 4.5, corresponding to the methyl orange end-point. From a chemical point of view, this gives a measurement of the buffer capacity (resistance to change in pH) of the water. This resistance to change could be caused, for instance, by the presence of carbonate or hydrogencarbonate ions, as shown by Eqs 16.19 and 16.20 below. Indeed, the units of alkalinity and acidity are expressed as mg l^{-1} $CaCO_3$, regardless of the true species producing the effect:

$$CO_3^{2-} + H^+ \rightleftharpoons HCO_3^- \qquad \text{... (16.19)}$$

$$HCO_3^- + H^+ \rightleftharpoons H_2O \cdot CO_2 \, l \, H_2O + CO_2 \qquad \text{... (16.20)}$$

A high buffer capacity is a useful feature if an acidic or basic pollutant is being added to water, as this will lessen the pH change of the receiving water.

Water Hardness

The term water hardness will be very familiar to those who live in an area where there are high concentrations of calcium and magnesium in water supply. The effects noticed include the following:

1. Deposition of a white solid whenever the water is heated. This is commonly seen as the 'furring-up' of kettles. This may also load to blockage of hot water pipes and a decrease in the efficiency of industrial heat exchangers.

2. The formation of scum whenever soap or washing powder is added to water. Sometimes, coloured spots are produced on clothes. No detergent action can occur until all of the hardness has been removed from the water.

The effects are generally produced by the presence of polyvalent metal ions in the water from the weathering of minerals. Usually, this is almost entirely due to calcium and magnesium ions, although others such as aluminium, iron, manganese and zinc ions may make a small contribution. It is the transition metal ions which produce the staining often observed. The minerals producing the hardness are often based on carbonates [limestone ($CaCO_3$) and dolomite ($CaCO_3 \cdot MgCO_3$)] or sulphates [gypsum ($CaSO_4$)]. It is only the hardness derived from carbonates which gives rise to solid deposition ('carbonate' hardness or 'temporary' hardness). Hardness which does not produce this effect is known as 'non-carbonate' or 'permanent' hardness.

However, a small degree of hardness does have some beneficial effects. For example, the alkalinity lowers the solubility of toxic metals, while the buffering action of the carbonate hardness lessens the effect of acidic pollutants. This buffer effect increases with the concentration of hardness in the water. In addition, there is evidence that hard water is beneficial to health, particularly in the reduction of heart disease and it certainly is more pleasant to drink. Analysis is normally performed by complexometric titration using the disodium salt of ethylenediaminetetraacetic acid (Fig. 16.7).

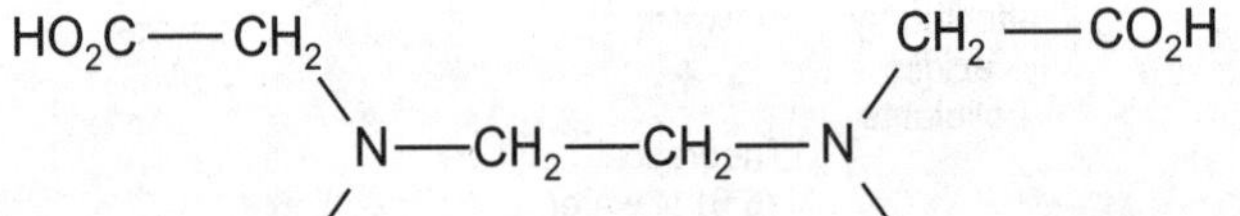

Fig. 16.7. Structure of ethylenediaminetetraacetic acid.

This forms a 1:1 complex with divalent metal ions, according to the following:

$$M^{2+} + H_2EDTA^{2-} \rightleftharpoons M(EDTA)^{2-} + 2H^+ \qquad \qquad ... (16.21)$$

where, H_2EDTA^{2-} is the di-anion derived from the acid.

To determine both calcium and magnesium by titration, the pH has to be buffered at pH 10. The end-point is detected by using an indicator such as Erichrome Black T. A calcium-only value can be found by titrating at a higher pH. Under these conditions, the magnesium would precipitate as $Mg(OH)_2$.

The titration estimates the total divalent metal as a molar concentration. Many non-chemists are unfamiliar with molar concentrations and so the quantity is often re-expressed in more familiar terms. However, it would be impossible to convert the value into the more familiar weight concentrations (mg l^{-1}) without knowing the precise individual concentrations of calcium, magnesium and the other ions. Even then, you would not be able to quote a single figure for the total hardness—just a table of individual concentrations. In order to overcome this, the total hardness is expressed in mg l^{-1} units as if it were all calcium carbonate, even if it is due to calcium sulphate, magnesium carbonate or any other polyvalent metal salt.

Electrical Conductivity

In order to know the total inorganic salt content in a sample. A simple method would be to evaporate the sample to dryness and then weigh the resulting solid. Large volumes of sample would, however, need to be evaporated, thus making the technique less attractive than at first thought. It would be much more convenient if an electrode could simply be placed in the sample to make the measurement. The closest method to this ideal situation is the use of a conductivity cell for dissolved ions, as illustrated in Fig. 16.8.

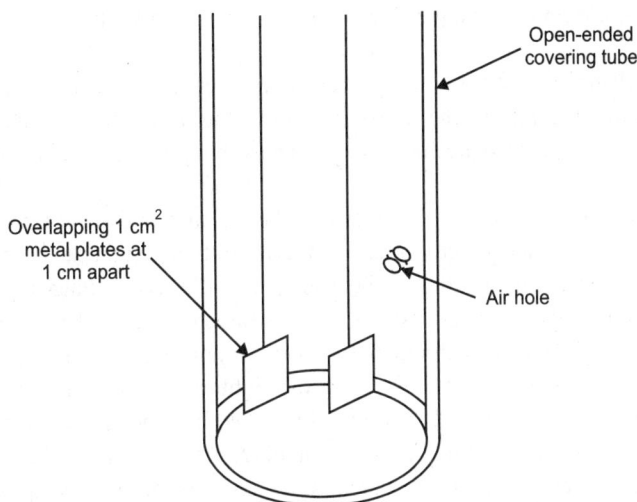

Fig. 16.8. Schematic of a typical conductivity cell.

Using this method, a low-voltage alternating current is applied across the electrodes. The resistance of the liquid between the electrodes is measured, and is converted to conductivity according to the following formula:

$$K = \frac{L}{AR} \qquad \text{... (16.22)}$$

where, K is the conductivity, L the distance between the electrodes (cm), A the surface area of the electrodes (cm^2) and R the resistance [ohm = siemens (S)$^{-1}$] (note that the siemen is the SI unit of electric conductance).

The units of conductivity applicable to environmental samples are μS cm^{-1}, with a typical value of 200 μS cm^{-1} being found for a soft water with a significant ionic salt content. The cell is calibrated by using solutions of known conductivity. Conductivity is highly temperature-dependent and so care has to be taken that calibration solutions and the unknown sample are at the same temperature. A standard temperature of 25°C is often used. The relationship between conductivity and total salt content is not simple. All ions having the same charge have approximately the same conductivity, but unfortunately most environmental waters contain ions with different charges in varying concentrations. If a series of waters of roughly similar composition is known, an approximate conversion can be made. For many waters in USA and European countries, the following equation is valid:

$$\text{Total salt concentration} = A \times \text{conductivity (mg l}^{-1}) \qquad \text{... (16.23)}$$

where, A is a constant in the range 0.55 to 0.80.

TECHNIQUES FOR THE ANALYSIS OF COMMON IONS

This section discusses the application of techniques to quantify ions present in the mg l^{-1} range. Many of the instrumental methods for ions within the mg l^{-1} concentration range need little sample preparation. The instrumental method then becomes just one part of a more complex analytical procedure.

Ultraviolet and Visible Spectrometry

This spectroscopic technique is based on Beer-Lambert law.

At sufficiently low concentrations, the Beer-Lambert law is followed:

$$A = \varepsilon c l \qquad \qquad ...(16.24)$$

where, A is the absorbance of radiation at a particular wavelength [$= \log(I_0/I)$], I_0 the intensity of the incident radiation, I the intensity of the transmitted radiation, ε the proportionality constant (molar absorptivity [$1 \text{ mol}^{-1} \text{ cm}^{-1}$)], c the concentration of the absorbing species (mol l^{-1}) and 1 the pathlength of the light-beam (cm).

The Beer-Lambert law is fundamental to many of the techniques that we will be discussing in the following sections. The instruments used to measure the absorption of light can range from sophisticated laboratory instruments which can operate over the whole ultraviolet/visible range to portable calorimeters employing natural visible light, which are used as field instruments. This makes absorption spectrometry one of the most useful and versatile techniques for an environmental analyst.

After all, none of the common ions in water absorb light in the visible region of the spectrum. In addition, the only ions commonly found in water which absorb in the ultraviolet range above 200 nm are nitrate and nitrite. The main use of the technique involves the analysis of light-absorbing derivatives of these ions. This can be carried out for almost all of the common anions (except sulphate), as well as ammonia. We can summarise as follows:

Analysis by direct absorption of: nitrate.

Analysis after formation of derivative: (i) chloride, (ii) fluoride, (iii) nitrate, (iv) nitrite, and (v) phosphate.

As an example of such an approach, the procedure for phosphate involves the addition of a mixed reagent (sulphuric acid and ammonium molybdate, ascorbic acid and antimony potassium tartrate) to a known volume of sample, making up to the working volume, shaking and leaving for 10 minutes. A blue-coloured phospho-molybdenum complex is produced and its absorbance is measured at 725 nm.

Quantification

Ultraviolet/visible spectrometry is the first technique where, at low concentrations, there is a simple linear correlation between the instrument response and the concentration of the unknown.

This technique for a quantitative analysis can make up a series of standard solutions of known concentration of the unknown and from this construct a calibration curve. The concentration should be within the range over which the Beer-Lambert law applies and thus a straight-line graph will be produced. Above this range, the calibration will no longer be linear and the solutions should be diluted. The 'best-fit' calibration line can readily be calculated by using the method of least-squares found on standard PC spreadsheets or even the most basic scientific calculators. The absorbance of the unknown can then be measured and from this the concentration calculated.

This procedure is known as calibration by external standards. We will find instances in the following sections where the sample matrix can affect the response of the instrument and so 'external standards' may not be the best method to employ.

High-throughput laboratories

You will find a great diversity of instrumentation based on these chemistries for high-throughput laboratory analysis. A number of instruments are based on continuous flow, with a schematic of a typical system being shown in Fig. 16.9. Instead of prior mixing of the reagents for each analysis, streams of each reagent (segmented by air bubbles to diminish premature mixing effects) in narrow-bore tubes are mixed by combining the flows at a T-junction or within a (mixing) cell. A sample is introduced from an automatic sampler as a continuous flow into the reaction stream. The combined flow is then led into a spectrophotometer and the absorption measured. The flows of all of the reagents and samples are controlled from a multi-channel peristaltic pump (Fig. 16.10).

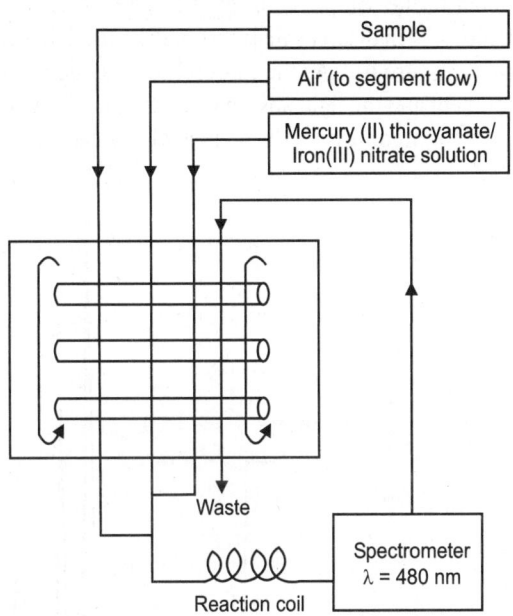

Fig. 16.9. Schematic of a typical continuous-flow system used for the analysis of chloride ions.

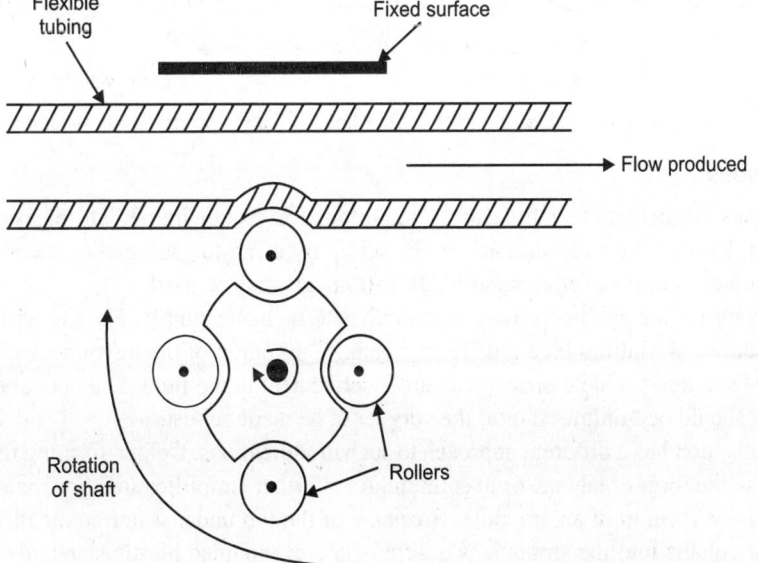

Fig. 16.10. Schematic of the operation of a peristaltic pump.

Other instruments are based on flow-injection techniques where individual aliquots of sample are injected into a continuous flow of water. Colour-forming reagents are then added also via a continuous-flow system. The mixing of the reagents and samples is dependent on the length and diameter of the tubing. After time being allowed for the formation of the colour, the absorbance of the solution at a specific wavelength is then measured. The response of the instrument is in the form of a peak, with the peak height being proportional to the sample concentration (Fig. 16.11). If there is a requirement to analyse more than one of the ions, then discrete analysers may be used. In such systems, the samples are introduced into vials on a rotating carousel. As the carousel rotates, reagents are added and mixed, time is allowed for colour development and the light absorbance at a specific wavelength is then determined. If the instrument is suitably configured, several different analyses can be performed simultaneously on samples by one instrument.

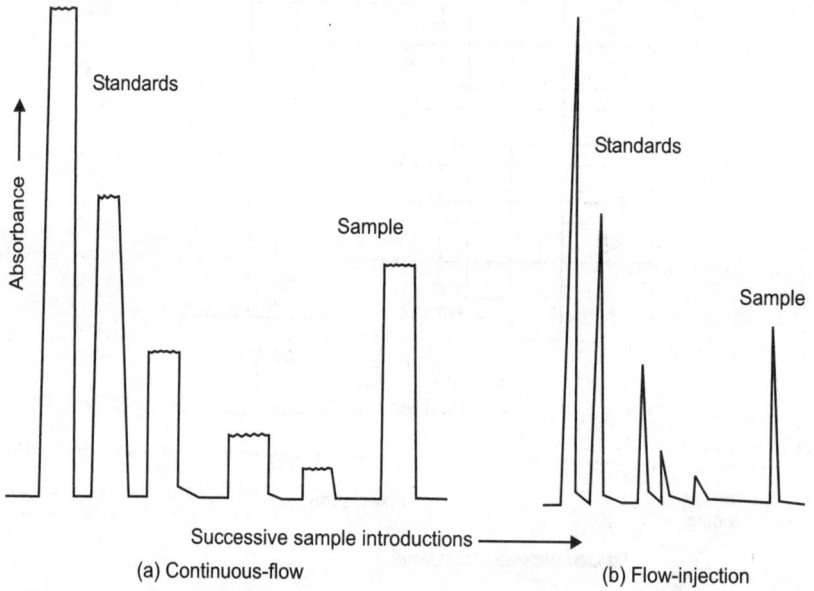

Fig. 16.11. Typical outputs obtained from (a) continuous-flow; and (b) flow-injection analysers.

Field techniques

Field techniques are becoming increasingly important for giving immediate measurement of ion concentrations. Un-manned field stations can be set up by using the automatic procedures described above. Alternatively, portable (often hand-held) instruments may be used.

The procedure for the colour-forming reaction has to be made simple. No one wishes to perform complicated analytical routines on a muddy riverbank. Calibration of the instrument should avoid the use of standard solutions, which, once again, are inconvenient in the field. The optical components of the instrument should be minimised or at the very least, be made robust.

Each manufacturer has a different approach to such modifications. Colour-forming reagents may be pre-measured in the form of tablets, or in solution. As a further simplification, one manufacturer seals the reagents under vacuum in an ampoule. Breakage of the top under water automatically draws the correct sample volume into the ampoule. Coloured glass or moulded plastic standards are often used rather than solutions.

These can be in the form of a disc. One manufacturer's design contains glasses of different optical density (Fig. 16.12). The disc is rotated through the light beam until the colour of the standard glass matches that of the unknown. Alternatively, a moulded plastic cube may be used which has a stepped side to provide a number of possible pathlengths (and hence absorbances). The most simple procedure for quantification is by visual comparison of the colour of the standards and the unknown using available sunlight.

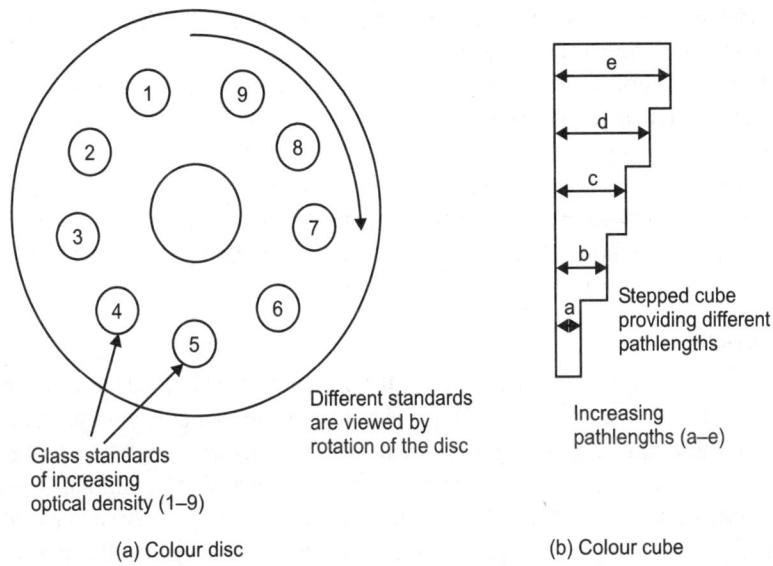

(a) Colour disc   (b) Colour cube

Fig. 16.12. Typical designs of colour standards: (a) colour disc, (b) colour cube.

Alternatively, portable spectrometers are available, often housed in briefcases, along with titration equipment and pH and conductivity electrodes—these are known as 'water quality' test kits.

A note on units

The most obvious way of expressing the concentration of ions is as the mass of the ion per unit volume (mg l^{-1}), but sometimes you will find other units, most notably as the concentration of the major element within the ion. This alternative method is most common for the nitrogen-containing ions. Nitrate, nitrite and ammonium are often expressed as mg l^{-1} of NO_3^-, NO_2^- and NH_4^+, respectively, but can all be expressed as mg l^{-1} of nitrogen (mg l^{-1} N). It then becomes easy to compare the relative concentrations of species without having to use molarities. If all of the ammonia in a water sample which contains a concentration of 2 mg l^{-1} (expressed as nitrogen) is totally converted into nitrate, then the water will contain a nitrate concentration of 2 mg l^{-1} (also expressed as nitrogen). This is easier for a non-specialist to understand than by saying that 3.09 mg l^{-1} NH_4^+ will produce 8.86 mg l^{-1} NO_3^-. Difficulties can arise because the two systems are sometimes used in parallel.

Emission Spectrometry (Flame Photometry)

Emission spectrometry relies on the principle that, for some metals at low concentrations, the intensity of light emitted from an electronically excited atom (usually produced by introduction of the sample into aflame) is proportional to the concentration of the excited species. Simple and inexpensive instrumentation is available, often known as 'Flame Photometers'.

Flame photometry seems almost ideally suited to the analysis of environmental water samples.

1. Although the use of flame photometry is limited to a few alkali metal and alkaline-earth ions, this includes sodium, potassium and calcium, three of the four major cations present in water.
2. The linear concentration ranges [0–10 mg l^{-1} (for sodium and potassium) and 0–50 mg l^{-1} (for calcium)] are within that expected for environmental water samples. Little sample preparation is needed.
3. The instrument is simple to use and the only laboratory requirements are a gas supply (natural gas is adequate) and a source of vacuum. This can be easily installed in temporary laboratories for analysis close to the sampling site.

It is a pity that flame photometers cannot be used to analyse the fourth common ion, i.e. magnesium, as all of the routine analytical requirements for metal ions could then be satisfied by this simple method. Analysis for magnesium is usually carried out by using atomic spectrometry.

The major disadvantage of flame photometry is the variation of the response of the instrument with time (i.e. drift). Great care has to be taken to ensure that calibration of the instrument and the analytical measurements are performed quickly after each other. It is also good practice to repeat the calibration after the analysis to check that no variation has occurred.

Ion Chromatography

The methods we have looked at so far have been for the analysis of individual ions, but sometimes a complete analysis of all of the ions in the sample is needed. Chromatographic separation of the ions is an obvious approach. Liquid chromatography would seem particularly useful since the species to be analysed are already in solution. From your reading elsewhere, you will be familiar with the principles of high performance liquid chromatography (HPLC) and how its application over the last three decades has expanded to include virtually all soluble ions and compounds. The major application in environmental analysis has been for inorganic anions. Several variations of the liquid chromatographic technique have been developed which normally use specialised 'ion chromatographs'. The most sensitive systems are often those which use a technique known as ion suppression (Fig. 16.13).

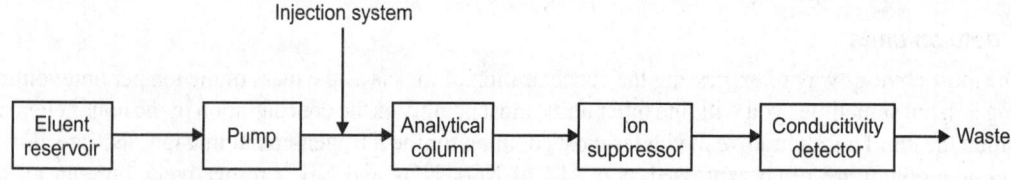

Fig. 16.13. Major components of a suppressed ion chromatographic system.

The separation of the anions is achieved by using an ion-exchange column (length 10–25 cm, 3–4.6 mm i.d.), usually based on poly(styrene–divinylbenzene) or another organic polymer, with an eluent typically containing sodium hydroxide or a sodium carbonate/hydrogencarbonate buffer. Detection of the analyte ions is achieved by monitoring the increase in conductivity of the eluent produced by the ions as they pass through the detector. In order to maximise detection sensitivity, prior to passing to the detector, all buffer ions have to be removed from the eluent as these would contribute to the background conductivity. The sodium ions in solution are replaced by hydrogen ions. The hydroxide ions react to form water (Eq. 16.25). Carbonate and hydrogencarbonate react to form carbon dioxide Eqs 16.26 and 16.27, which has little conductivity in solution.

Hydroxide eluents: $$OH^- + H^+ \rightarrow H_2O$$ ... (16.25)

Carbonate/hydrogencarbonate eluents:

$$HCO_3^- + H^+ \rightleftharpoons H_2O + CO_2 \qquad\qquad ... (16.26)$$

$$CO_3^{2-} + 2H^+ \rightleftharpoons H_2O + CO_2 \qquad\qquad ... (16.27)$$

The suppressor has to provide, uninterruptedly, precisely the correct number of protons for the neutralisation. There are a number of methods used to achieve this. All of these are based on the ion-exchange process.

One manufacturer uses a continuous suppression system, as shown in Fig. 16.14. The eluent passes between cation-exchange membranes, through the detector cell, and is finally recycled on the outside of the membranes. The H$^+$ ions necessary to replace the Na$^+$ ions in the fresh eluent are generated by electrolysis of the recycled eluent, with the H$^+$ ions being generated at the cathode.

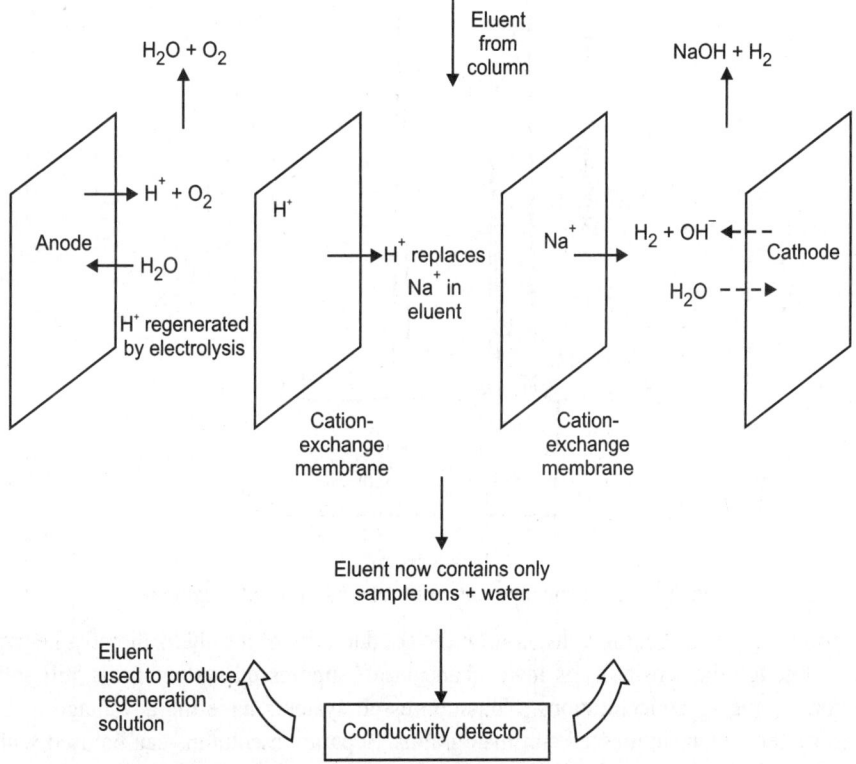

Fig. 16.14. Schematic of a continuous eluent suppression system.

Other manufacturers use cation-exchange columns which need periodic regeneration. This can be achieved without interruption of the analytical operation of the chromatograph. In one system, there is a carousel of regeneration columns with one column being regenerated while another one is in use. Another system regenerates the column while the following sample is being loaded. Disposable regeneration columns may also be used.

The response plotted in the chromatogram is conductivity. As the latter is directly proportional to the concentration of the ion, quantification can be simply carried out by comparison of the peak area of the unknown with that of a standard of similar concentration, i.e. by external standards.

For ions of interest in environmental water found at mg l^{-1} concentrations and with a suppressed system the sample would need to be diluted before injection. This, along with filtration, is often the only sample preparation necessary and common ions in water can be determined within the space of a few minutes (Fig. 16.15).

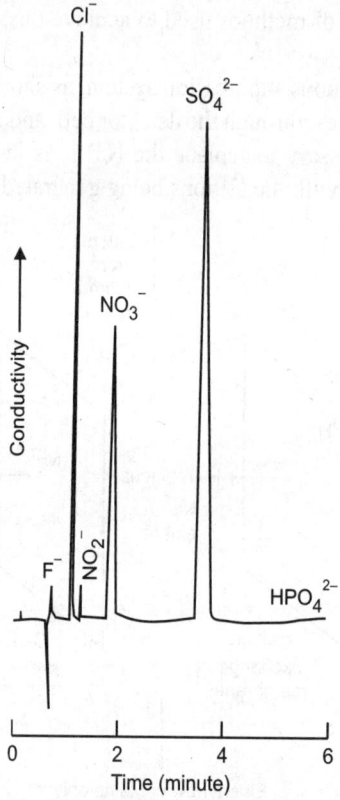

Fig. 16.15. Typical chromatogram of a natural water sample.

Non-suppressed ion chromatographs monitor the conductivity of the eluent directly, i.e. without the suppressor. Although the sensitivity is lower than that of suppressed systems, it is still sufficient to determine ions at mg l^{-1} concentrations. Non-suppressed systems have the advantage of being less complex instruments than suppressed chromatographs. Separation columns can be used with a wider range of eluents. The instruments resemble conventional liquid chromatographs (the components being simply pump, analytical column and detector) but often they contain no metal components in contact with the eluent and with the pumps designed to operate at lower pressures than is necessary for conventional HPLC. This reflects the lower back pressures found with ion chromatography when compared with reversed phase liquid chromatography.

Although most analysis nowadays would use specialised 'ion chromatographs', conventional high performance liquid chromatography may still find some application. Methods developed for conventional HPLC can use either a reversed phase column and ion-pair techniques or an ion-exchange column. Ultraviolet absorbance and conductivity detectors are used. When a conductivity detector is employed, the system becomes similar to the specialised chromatographic set-up without ion suppression which has already been described. The sensitivity is lower in comparison with the specialised system, although it is still sufficiently high to analyse common anions at mg l^{-1} concentrations.

Although the most common use of chromatography is for anions, similar methods have been developed for specialised 'ion chromatographs' for the separation of the common cations (Na^+, K^+, NH_4^+, Ca^{2+}, Mg^{2+} etc.) in a single isocratic run. A typical eluent would be methanesulphonic acid. This would allow ion suppression similar to that used for anions, although this may not be necessary at typical natural water concentrations.

Examples of the use of other techniques

We have now discussed the most widely used methods for analysis of the common ions. There are, however, a few frequently used techniques which have not yet been covered. We will look at the analysis of three species — ammonia, fluoride and sulphate — to exemplify these techniques.

Ammonia

Ammonia is the only alkaline gas commonly found in environmental water. If extracted from the sample, the ammonia can be determined by a simple acid-base titration. Magnesium oxide is added to the sample to make it slightly alkaline. The ammonia is then present predominantly in the form of NH_3, rather than the less volatile NH_4^+. Ammonia is then distilled off (Fig. 16.16) and absorbed into boric acid solution. The boric acid sharpens the end-point of the subsequent titration with standard acid.

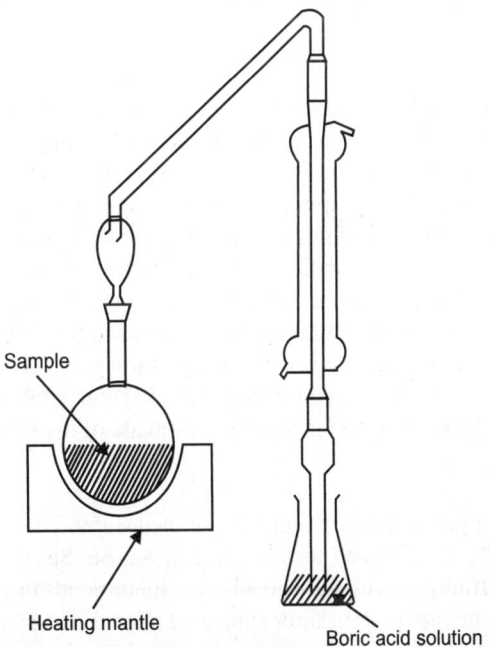

Fig. 16.16. Schematic of the apparatus used for ammonia determination.

For rapid screening of samples, it is possible to use an ion-selective electrode (i.e. an electrode whose potential, measured with respect to a reference, is proportional to the log of the activity of one particular ion). A combination pH electrode is simply an ion-selective electrode responsive to hydrogen ions and a reference electrode housed in a single body. Ion-selective electrodes are available for most common ions and gases which dissolve as ionic species, although they do have some limitations. Many are prone to interference from other species and thus have poor precision. Even the pH electrode has taken many years of development to produce reliable responses. All of them respond to ionic activity rather than concentration, and so it is essential to add a large excess of an ionic salt to both the standard solutions and the unknown in order that the ionic strength of each solution is identical.

Ammonia electrodes are of the gas-sensing type. The ammonia diffuses through a permeable membrane and causes a pH change in a small volume of internal solution, which is sensed by a 'glass' electrode. Prior to measurement, concentrated sodium hydroxide solution is added to the samples and standards. This serves to increase the pH to above 11 to ensure that the ammonia is in the unprotonated form, and also to provide a constant ionic strength. The ammonia electrodes respond only to gaseous alkaline gases. For most environmental applications (except in the analysis of heavily polluted water), there will then be little possibility of interference. Calibration is by external standards.

Fluoride

A second electrode which has found widespread use for water analysis is that which detects fluoride ions. This is a solid-state electrode where the electrical potential is generated by migration of the ion through a doped lanthanum fluoride crystal. This once again gives extremely high specificity to the analyte ion, with the only pre-treatment necessary being the addition of buffer solution to maintain constant pH and ionic strength. Alternative techniques for fluoride determination are spectrometry and ion chromatography.

Sulphate

There is no direct calorimetric method available for sulphate and ion-selective electrodes for the ion are not very reliable; in fact, the only direct instrumental method is by using ion chromatography. Virtually every other method is based on precipitation of an insoluble sulphate. Barium or 2-aminoperimidinium salts are used for the precipitation. The precipitate formed may then be weighed for a direct determination of the sulphate. This represents one of the few remaining important applications of gravimetric analysis.

Other methods using insoluble salt precipitation are indirect, estimating the excess cation after precipitation of the sulphate. Excess barium may be determined by titration or by atomic absorption spectrometry Excess 2-aminoperimidinium ions may be estimated by visible spectrometry.

None of these methods would appear ideal for a high-throughput laboratory. For most samples, sulphate would be the only major sulphur-containing species. Total sulphur in solution, as determined by an elemental sulphur analyser, will then give a good estimate of the sulphate concentration.

Trace Pollutants

Several trace elements (few ppm or less) are found in polluted water. The most dangerous among them are the heavy metals, e.g. Pb, Cd, Hg and metalloids, e.g. As, Se, Sb, etc. As mentioned earlier, the heavy metals have a great affinity for sulphur and attack sulphur bonds in enzymes, thus immobilising the latter. Other vulnerable sites are protein carboxylic acid (COOH) and amino ($-NH_2$) groups. Heavy metals bind to cell membrane, affecting transport processes through the cell wall. They also tend to precipitate phosphate biocompounds or catalyse their decomposition.

The trace elements in natural waters and waste-waters are summarised in Table 16.3. The sources of heavy metals in surface waters are shown in Fig. 16.17. Street dust containing heavy metals, e.g. Pb, represent an important source of metal input to surface waters. Metals are contributed by industrial effluents as domestic sewage. All these sources may be routed by way of sewage treatment works which reduce significantly the amount of metal discharged. In developing countries, however, such sewage treatment works hardly exist or function.

Table 16.3. Toxic trace elements in natural water and waste-water.

Elements	Sources	Effects and significance
Arsenic	Mining by-product, pesticides, chemical waste	Toxic, possibility carcinogenic
Cadmium	Industrial discharge, mining waste, metal plating, water pipes	Replaces zinc biochemically, causes high blood pressure, kidney damage, destruction of testicular tissue and red blood cells, toxicity to aquatic biota
Beryllium	Coal, nuclear power and space industries	Acute and chronic toxicity, possibly carcinogenic
Boron	Coal, detergent formulations, industrial wastes	Toxic to some plants
Chromium	Metal plating, cooling tower water, additive (chromate) normally found as Cr (VI) in polluted water	Essential trace elements; possibly carcinogenic as Cr (VI)
Copper	Metal plating, industrial and domestic waste, mining, mineral leaching	Essential trace elements, not very toxic to animals, toxic to plants and algae at moderate levels
Fluorine (Fluoride ion)	Natural geological sources, industrial waste, water additive	Prevent tooth decay at about 1 mg/l, causes mottled teeth and bone damage at about 5 mg/l
Lead	Industry, mining, plumbing, coal, gasoline	Toxic (anaemia, kidney disease, nervous disorder), wild-life destroyed
Manganese	Mining industrial waste, acid mine drainage, microbial action on manganese minerals at low pE	Relatively non-toxic to animals, toxic to plants at higher levels, stains materials (both—room fixtures and clothing)
Mercury	Industrial waste, mining, pesticides, coal	Highly toxic
Molybdenum	Industrial waste, natural sources	Possibly toxic to animals, essential for plants
Selenium	Natural geological sources, sulphur, coal	Essential at low levels but toxic at higher levels
Zinc	Industrial waste, metal plating, plumbing	Essential in many metalloenzymes, toxic to plants at higher levels

Another major cause of concern is the presence in water of a number of non-bioaccumulative organic compounds with adverse toxicological properties. For many years, there has been much concern over compounds suspected of being carcinogens.

A typical example would be chloroform which can be produced in trace quantities during the disinfection of water by chlorination and which is thought to be harmful at $\mu g \, l^{-1}$ concentrations. Of more recent concern is the large number of compound types considered to be endocrine disruptors. These compounds can range from pesticides, through components of common plastics, to active ingredients in the contraceptive pill.

In the early days of instrumental analysis the concentrations would have been beyond the capabilities of the available instrumentation and techniques but developments since then have made such analyses routine. This is partly due to the development of more sensitive instrumentation, but also through the development of suitable pre-treatment processes. This is required to remove potential interferences and, for many techniques, to increase the analyte concentration to within the instrument sensitivity.

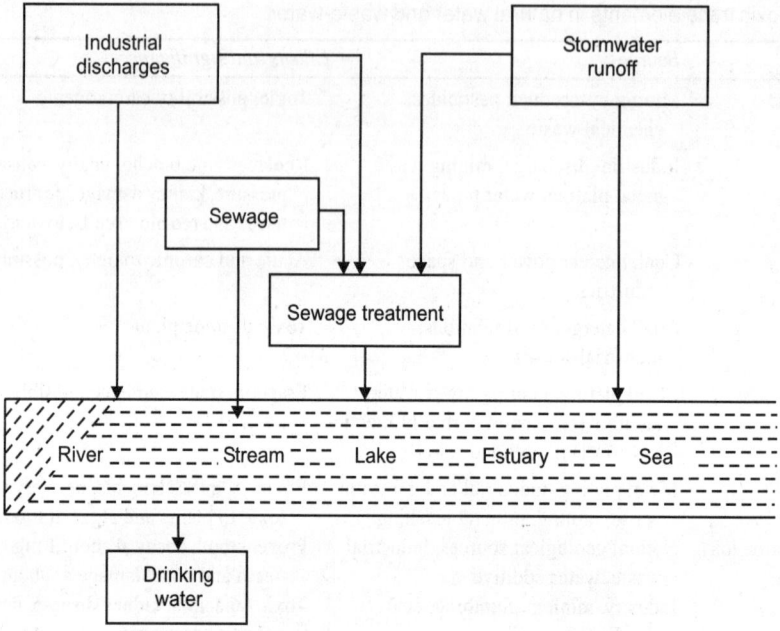

Fig. 16.17. Sources of heavy metals in surface waters.

Organic Trace Pollutants

The range of organic compounds which may be found in environmental waters includes the following:
1. Naturally occurring compounds from decaying organic material.
2. Pollutants discharged or escaping into the environment.
3. Degradation and inter-reaction products of the pollutants.
4. Substances introduced during sewage treatment.

Analysis

Typical analyses could include:
1. Analysis of individual compounds or groups of compounds of environmental concern.
2. Total analysis of all organic components above the limit of detection. This is an enormous task and at the lower end of the concentration range there will almost invariably be unidentified components.
3. Field screening for specific pollutants prior to laboratory analysis.
4. Qualitative identification of trade products in spillages or discharges.

The properties of compounds causing widespread environmental problems include toxicity, slow biodegradation and the ability to bioaccumulate within organisms. The types of organic compound may be:
1. Pesticides, particularly those containing chlorine.
2. Chlorinated solvents.
3. Polychlorinated biphenyls.

4. Dioxins.
5. Endocrine disruptors.

For more localised pollution problems, we could extend our list of concern to include virtually every organic compound currently in use or production, together with their reaction and degradation products.

For the purpose of grouping into suitable analytical techniques, organic pollutants are often classified as being either 'volatile' (e.g. chloroform) and semi-volatile (e.g. most pesticides). The two groups may have different extraction and clean-up methods.

Analysis of complex mixtures of organics would normally involve the chromatographic separation of the components. As most organic compounds have significant volatilities even at room temperature, gas chromatography would be expected to be a useful technique.

The alternative of high performance liquid chromatography is used only where there are advantages over established gas chromatographic methods, although the number of applications of this technique is increasing.

A major area where non-chromatographic methods are used is in the determination of groups of compounds such as phenols and also of classes of detergents, where the total concentration of the group of substances is required rather than the concentration of individual compounds.

Groups of organic campounds can be analysed by ultraviolet/visible absorption spectrometry which appears ideal. Absorptions are broad and the molar absorptivities often vary little between compounds within groups. A single absorption measurement could be used to determine the total concentration of the group. Although there may be suitable volumetric techniques for individual groups of compounds they would not be sufficiently sensitive for concentrations in the μg l^{-1} range.

The current desire for field screening has lead to novel approaches which may not have widespread use in other areas of chemical analysis. These include the use of immunoassays. After dealing with sample storage and extraction, this section will then look at gas chromatographic methods and later discuss the other techniques.

GUIDELINES FOR STORAGE OF SAMPLES AND THEIR SUBSEQUENT ANALYSIS

Considerations necessary for organic trace pollutants contain the following list:

1. The volatility of organic compounds: Even high-relative-molecular-mass compounds (e.g. pesticides) have a significant vapour pressure at room temperature. Storage containers should be completely filled and kept at sub-ambient temperatures; 4°C is often specified in analytical procedures. The latter is the temperature of a normal domestic refrigerator.
2. Microbial degradation: Storage at 4°C will lower microbial activity; storage below 0°C (i.e. deep freeze) will lower this still further.
3. Photolytic decomposition: Many potential analytes (e.g. organochlorine pesticides) are photosensitive in dilute aqueous solution. Therefore, the samples should be stored in the dark.
4. Contamination from the container: Glass bottles should be used as bottles made of organic polymers will leach, potentially interfering monomers and additives into the sample.
5. Loss of analyte on to container walls: Low-solubility organic compounds can be adsorbed on to the container walls. This problem cannot be fully overcome. The best method of minimising the effect is to proceed with the analysis as quickly as possible. Many procedures specify a maximum storage time.

The sample volumes which are required depend on the concentration of the analyte. Although the chromatographic techniques used involve the injection of just a few microliters of solution, and spectrophotometric analysis a few milliliters, the solutions may first have been extracted from several liters of sample.

The precautions necessary to avoid either contamination or loss of material at these low concentrations, during the subsequent analysis, are often not appreciated.

Precautions

A few typical precautions should indicate the caution that is necessary:

1. The analysis should be performed in a laboratory as free as possible from the analyte. Remember that many of these trace contaminants are solvents frequently found in analytical laboratories.
2. Any stock solvents should be safeguarded, minimising exposure to the atmosphere and avoiding sample withdrawal with potentially contaminated pipettes or syringes.
3. Samples and working standards should be placed well away from more concentrated solutions or stock solvents.
4. As traces of pesticides are commonly found in laboratory solvents, pesticide free grade solvents should be used for these analyses.
5. Glassware should be scrupulously cleaned or new, if at all possible.

Such is the problem of contamination that the practical lower limits of detection can often be limited by the background concentrations of the analyte (or of interfering components) in the reagents or laboratory atmosphere.

Extraction techniques for chromatographic analysis

Extraction of the compound of interest from the aqueous sample into an organic solvent is commonplace before any chromatographic analysis. The major reasons for this are as follows:

1. To separate unwanted components present in large excess.
2. To separate minor components which have overlapping peaks with the components of interest.
3. To concentrate the components of interest.

For some samples, the extraction may be the only pre-treatment necessary before injection into the chromatograph while for more complex samples it may be just one stage of a multi-stage process. Most of the techniques described may be integrated with the chromatographic stage and subsequent data handling. There is no one method of choice. The best method will be dependent on the following:

1. The chemical and physical properties of the compounds being determined and potential interferences.
2. The choice of gas or liquid chromatography as the separative technique.
3. Whether solvent-free methods are preferred. Such methods remove the concern of possible contamination of the laboratory and its atmosphere (health effects and cross-contamination of other samples), contamination of aqueous waste and the cost of disposal of the waste solvent.
4. The number of samples to be analysed. If you have a large number of samples and are working in a well-equipped laboratory, the techniques which are fully integrated with the chromatograph may be preferable. In a smaller laboratory, dedicated instruments may not be justifiable and simpler methods may be preferred.
5. Whether you would wish to perform the field extractions.

The extraction methods are common in many areas of chemical analysis. Try thinking of methods you have already come across before studying the following sections. These methods are summarised in Fig. 16.18.

Solvent extraction

In this method, the water sample is shaken with an immiscible organic solvent in which the components are soluble. Hexane and petroleum ether are the most common extraction solvents, although oxygenated and chlorinated solvents are sometimes used. The organic layer is separated and after drying, is injected into the chromatograph. The extractions can be made selective towards acidic and basic components by altering the pH of the aqueous layer. If the sample is acidified, the basic components are less likely to be extracted, for example:

$$RNH_2 + HCl \qquad\qquad \rightarrow \qquad\qquad RNH_3^+ \ Cl^- \qquad\qquad ... (16.28)$$

amine, soluble in non-polar solvents $\qquad$ amine hydrochloride, less soluble in non-polar solvents

Similarly, if the sample is made basic, acidic components are less likely to be extracted, for example:

$$RCO_2H + NaOH \qquad\qquad \rightarrow \qquad\qquad RCO_2^- \ Na^+ \qquad\qquad ... (16.29)$$

carboxylic acid soluble in non-polar solvents $\qquad$ Carboxylate salt, less soluble in non-polar solvents

(a) $\qquad$ (b) $\qquad$ (c) $\qquad$ (d)

(e) $\qquad$ (f) $\qquad$ (g)

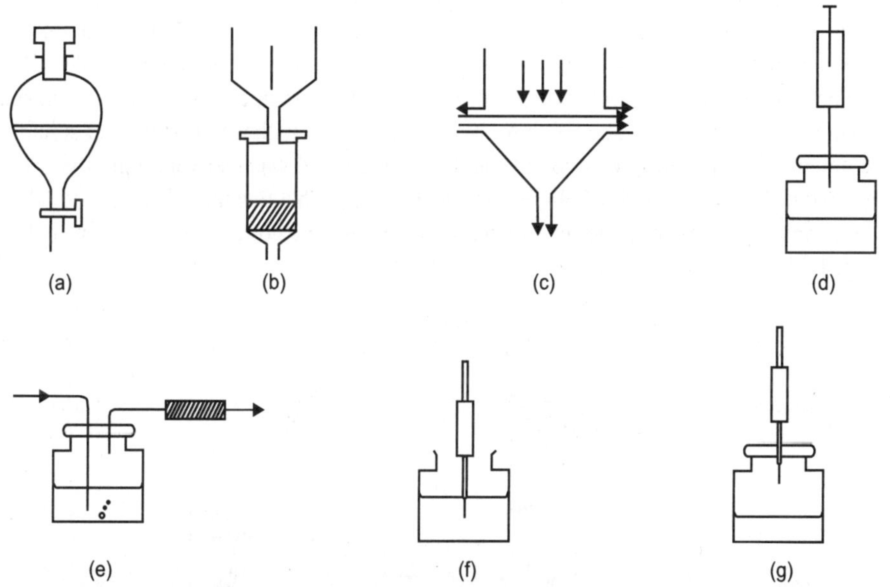

Fig. 16.18. Summary of extraction methods: (a) solvent extraction, (b) solid-phase extraction — cartridge, (c) solid-phase extraction — disc, (d) head-space analysis, (e) purge and trap, (f) solid-phase microextraction — direct, (g) solid-phase microextraction — head-space.

When making the choice of extraction solvent, the response of the chromatographic detector should always be considered. Hexane or petroleum ether will appear as the predominant peak in the subsequent chromatogram if a gas chromatograph with flame ionisation detection is used. The least interference will be caused if the solvent peak appears before the analyte peaks but there is still a potential problem with peaks resulting from trace impurities in the solvent. Because of this, even analytical-grade solvents may have to be redistilled prior to use.

If a selective chromatographic detector is being used, it is possible to use an extraction solvent for which the detector has low sensitivity, e.g. hexane or petroleum ether for electron capture detection with gas chromatography. Aromatic solvents should be avoided if liquid chromatography with ultraviolet detection is to be used.

If an unsuitable solvent cannot be avoided (e.g. if a chlorinated or oxygenated solvent is required with subsequent electron capture detection) and the analyte has low volatility, it is possible to evaporate the extract to dryness and redissolve the residue in a compatible solvent. It is, however, better to avoid this if at all possible. Liquid-liquid extraction has long been seen as the standard extraction method but in more recently developed procedures it is being replaced by newer techniques such as solid-phase extraction (SPE). These are more rapid and can more easily be automated.

Solid-phase extraction

The use of this technique has rapidly increased over the past few years and when developing new procedures may often be the first-choice method. A short disposable column containing 100–500 mg of adsorbent material is used here. The column packing is usually a reversed-phase material similar to that used in high performance liquid chromatography columns. The use of an ODS packing material (which contains octadecylsilane groups chemically bonded on to a silica support) is common. Other materials are available, including ion exchangers and adsorbents such as 'Florisil'. Before use, conditioning of the column is usually necessary—this is carried out by passing a small volume of methanol through the column. Preconditioned columns are, however, commercially available. The water sample is then passed through the column by applying mild suction or pressure. The organic components of the sample are retained on the packing material. The column can then be washed with water or another suitable solvent to remove potentially interfering compounds, and air dried if the wash solvent is immiscible with the following solvent. The compounds of interest are then eluted with a few milliliters of a suitable organic solvent. As sample volumes could be several hundred milliliters, a concentration factor of about 100 is routine (Fig. 16.19).

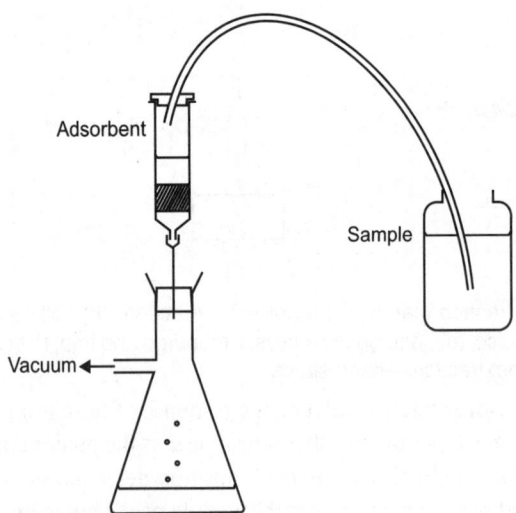

Fig. 16.19. Solid-phase extraction with large sample volumes.

A wide range of solvents can be used. The best extractants are often those where the polarity of the solvent matches that of the extractant, e.g. hexane could be used for non-polar organochlorine pesticides. The subsequent stage of the analytical procedure could also influence the choice of solvent. Methanol or acetonitrile are often used if liquid chromatography is to be used as the separation method. The procedure for solid-phase extraction is summarised in Table 16.4.

Table 16.4. Steps in a solid-phase extraction process.

Condition column with methanol

Load sample

Wash column with water

Pass air through the column to remove as much water as possible (an option if the elution solvent is immiscible with water)

Elute with a suitable organic solvent

Manifolds are available which allow processing of a number of samples simultaneously. In addition, SPE set-ups can be directly coupled to HPLC systems.

Advantages of solid-phase extraction

Advantages of solid-phase extraction over liquid-liquid extraction are:

1. It is a very rapid process and can easily be automated.
2. High concentration factors can easily be achieved.
3. Solvent consumption can be much lower.

There are, however, instances where liquid-liquid extraction may still be the method of choice. These are usually when there are solids present or there is a high loading of organic material in the sample (e.g. humic acid) which could block or overload the column.

A further development is the use of extraction discs where the adsorbent material is held within the fiber structure of a polytetrafluoroethylene (PTFE) filter disc. After pre-washing the disc with a portion of the final eluting solvent and conditioning with methanol, the extraction procedure is simply to pass the sample, by suction, through the filter. The extracted components are then eluted by using a suitable solvent. The advantages of discs over columns include a higher sample throughput (several hundred milliliters of sample may need to be passed through the filter if a high concentration factor is needed) and the lower likelihood of the filter clogging with particles. Some standard methods now include liquid-liquid extraction and solid-phase extraction as alternative procedures.

Head-space analysis

In this technique, the water sample is placed in a container with a septum seal in the lid and an air space above the sample. The most simple procedure is then, after allowing for the air to equilibrate with the water, to inject an air sample (containing volatile organic components) into the gas chromatograph. This technique overcomes problems found in liquid-liquid extraction resulting from solvent interference. The sensitivity towards a particular component will, however, be dependent on its volatility, favouring low-molecular-mass, neutral components. The overall sensitivity of the technique may be increased by heating the sample. Be aware, however, that you are also increasing the vapour pressure of the water and care should be taken to check the water compatibility of the chromatographic column.

Purge and trap techniques

These techniques extract the volatile organic content from the sample by using a purge gas stream. In many instruments, the organics are collected in a short tube of adsorbent material such as activated charcoal or a porous polymer (e.g. 'Tenax'). After the collection period, the tube is flash-heated to release the organics into the gas chromatograph. Other instruments collect the volatile components into a secondary liquid nitrogen cold trap. Rapid heating of this trap then releases the organics into the chromatograph.

Solid-phase microextraction

This technique could be seen as using both the principles of solid-phase extraction and head-space sampling. A fiber which is originally contained within a syringe needle (Fig. 16.20) is exposed either to the stirred sample or to the head-space above the sample. The fiber typically consists of fused silica with a coating of polydimethylsiloxane, or alternatively polyacrylate, with the phase being chosen according to the compound being determined. The dissolved components partition between the sample and fiber. After equilibration is complete (2–15 minutes for liquid samples), the fiber is withdrawn into the syringe needle for storage prior to analysis.

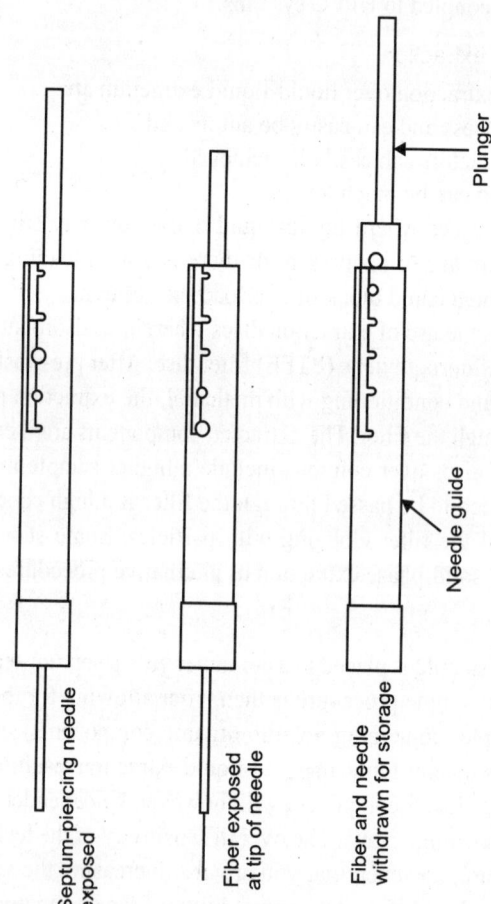

Fig. 16.20. Schematic of the solid-phase microextraction process.

The method cannot only be used for volatile organics but also for semi-volatile pollutants, such as chlorinated pesticides. Different fibers are used. A smaller depth coating (7 μm) is suitable for the semi-volatile compounds and a thicker (100 μm) coating for volatiles. Fibers are also available for the extraction of polar organic compounds (e.g. phenols) which are often very difficult to extract by other techniques. In the case of phenols, sample modification (e.g. lowering the pH and adding sodium chloride) increases the extraction efficiency.

Subsequent analysis can be carried out by either gas or liquid chromatography. With gas chromatography (GC), the fiber is directly introduced into the GC injector inside the syringe and re-exposed once the needle has pierced the injection septum. Most GC systems can be used without modification. Desorption of the organics takes place into the carrier gas, although this can take 20–30 seconds. In order to overcome this problem, a technique known as *cryo-focusing* is used. The sample is condensed on to the top of the column held at a low temperature, typically 40°C. Rapid heating of the column then releases the sample. If the subsequent chromatographic method is high performance liquid chromatography (HPLC), then the compounds can be desorbed by immersion of the fiber into a suitable solvent. The solution is then injected into the chromatograph. Injection systems are also available which permit the introduction of the fiber directly into the mobile phase, where the latter flows along the length of the fiber on to the head of the analytical column.

The fibers can be re-used as many as 50–100 times. The advantages of the technique include its simplicity and low cost of apparatus. As the complete extract is introduced into the chromatograph, this can lead to 100–700× lower detection limits than liquid-liquid extraction. No solvent is injected and short narrowbore columns can be used with gas chromatography. These columns would become flooded with solvent if used after liquid-liquid extraction. The fibers can cope with high levels of contamination and so they can be used for dirty samples such as waste water. One disadvantage of the method is that it is an equilibration technique. Extraction of each compound will be different and so calibration is necessary for each of these. In addition, changes in composition of the water samples could alter the extraction equilibria and hence the extraction efficiency.

Gas chromatography

Chromatographic separation of a mixture occurs by the differential partition of the components between a stationary phase and a mobile phase. In gas-liquid chromatography, the mobile phase is a gas and the stationary phase is a liquid adsorbed on, or chemically bonded to a solid. The main components of a gas chromatograph are shown in Fig. 16.21.

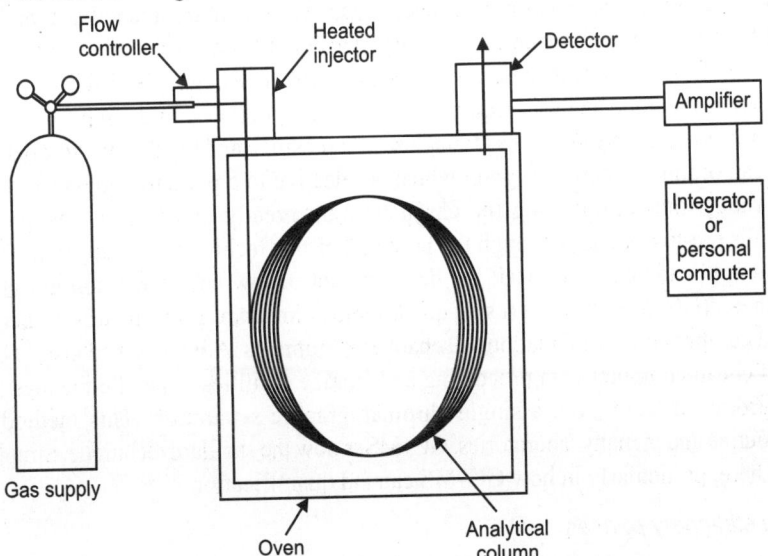

Fig. 16.21. Major components of a gas chromatograph.

Gas chromatography has the advantage over other chromatographic techniques of combining high separation efficiencies with the availability of highly specific and sensitive detectors. A high proportion of the separations required can be performed by using just a few stationary phases. The wide range of phases which are available does, however, permit the development of columns for specific problem separations. We will discuss the columns and detectors used for water analysis first of all, and then present some examples of complete analytical procedures (including sample pre-treatment).

Detectors

The most common detectors used for environmental trace analysis are listed in Table 16.5.

Table 16.5. Common detectors used in gas chromatography.

Detector	Typical application
Flame ionisation detector	Sensitive universal detector for organic compounds
Electron capture detector	Highly sensitive, specific detector responding to atoms with a high electron affinity, e.g. chlorine. Typical analytes are chlorinated pesticides and chlorinated solvents
Hall electrolytic conductivity detector	Highly sensitive, specific detector for halogens, nitrogen and sulphur. Typical analytes are pesticides and trihalomethanes
Thermionic detector	Element-specific detector for compounds containing nitrogen and phosphorus. Typical analytes include pesticides
Flame photometric detector	Element-specific detector for compounds containing sulphur and phosphorus. Typical analytes include pesticides
Photo-ionisation detector	Specific to compounds with aromatic rings or double bonds. Typical analytes include industrial solvents
Mass spectrometric detector	Highly specific, sensitive detector for all organic compounds. Can also be used for peak identification

The electron capture detector has held a special place within environmental analysis since many of the compounds of concern contain chlorine atoms. Had it not been for the development of this highly sensitive and specific detector (for some compounds, 10–100 times more sensitive than flame ionisation detection) much of the trace analysis required for these compounds would not have been possible.

More recently, mass spectrometry has found use as a sensitive and highly selective detection method. The detector can produce a chromatogram which is selective to a particular mass (or more accurately mass/charge ratio), thus simplifying the chromatogram greatly and to some extent lessening the requirements for pre-treatment. Although the potential of this technique was apparent for many years, its widespread application had to await the development of low-cost bench-top gas chromatograph/mass spectrometer (GC–MS) systems (using quadrupole or ion-trap spectrometers) rather than the more expensive and cumbersome combination of separate instruments. Advances had also to be made in the availability of cheap computer data processing and storage facilities to handle the massive amount of information produced from even a single chromatographic separation. This method is becoming increasingly routine and in many laboratories, GC–MS is now the standard technique. Simple applications are described here, particularly in how GC–MS can aid quantification.

Columns and stationary phases

The range of columns available is extensive. In the choice of column for a particular application, not only does the chromatographer have to consider the most appropriate stationary phase but also the column dimensions. The latter not only affects the separation efficiency, but must also be considered for compatibility with the detector being used, the method of sample introduction, and the sample type.

The column types available can be divided into the following, listed in order of decreasing separation efficiency:

1. Narrow-bore capillary columns: Typical dimensions: length, 30–60 m; i.d., 0.2 mm; flow, 0.4 ml min^{-1} He.
2. Wide-bore capillary columns: Typical dimensions: length, 15–30 m; i.d., 0.53 mm; flow, 2.5 ml min^{-1} He.
3. Packed columns: Typical dimensions: length, 2 m; i.d., 2 mm; flow, 20 ml min^{-1} He.

Most recent analytical methods for water analysis use the first two types of column, but you may occasionally find packed columns in long-established methods or for less-demanding applications.

Narrow-bore columns offer the greatest detection sensitivity and are used for analyses close to the limits of detection. The low carrier gas-flow rate is well suited for applications where mass spectrometric detection is used. However, direct sample injection on to the column by using a syringe is not possible as the column would become overloaded. A splitting device is necessary for the introduction of the sample.

Wide-bore columns have a larger sample capacity and direct syringe injection is possible. The sample may also be introduced from a sample concentration system such as a 'purge-and-trap' device. The greater sample capacity may be required if a low-sensitivity detector is being used. Wide-bore columns are also less affected by contamination from non-volatile components in the sample and so find a use with highly contaminated samples, such as waste-water.

Many of the organic compounds of environmental interest are of high relative molecular mass and have low volatilities. High oven temperatures are necessary for these and consequently silicone polymers are often the favoured stationary phases. Poly(ethylene glycol) columns are also popular. As with other uses of gas chromatography, the best separation efficiencies are achieved when the stationary phase has a similar polarity to the components of the analyte. Fuel oils are separated on non-polar columns (e.g. dimethylsilicone), pesticides and chlorinated solvents are often separated on medium-polarity columns (e.g. diphenyl/dimethyl silicone), whereas 2,3,7,8-tetrachlorodibenzo-p-dioxin can be separated from its isomers by using highly polar columns (c.g. cyanopropyl silicone).

The stationary phase may be adsorbed or chemically bonded on to the column walls of capillary and wide-bore columns, or on to a support material in packed columns. For analyses close to the limit of detection and at high oven temperatures, column bleeding may become a significant factor. The use of low-loaded columns (0.1–0.25 μm film thickness) or chemically bonded phases may reduce this effect. A higher loading of columns (1–5 μm film thickness) is possible at lower temperatures for the analysis of volatile compounds. Thicker films have higher sample capacities for highly concentrated components, but there is a corresponding decrease in column efficiency when compared to thinner films.

Injection methods

If you are using a narrow-bore capillary, then a device is needed to reduce the microliter volumes injected by syringe to the nanoliter volumes which the column can accept without overloading. A number of techniques are available, including those summarised below. The first two methods use a split/spilt-less injector (an injection system which can be used in either of the two modes), whereas the third requires a modified form which has simple temperature programming.

With split injection, the sample from the syringe is introduced into a vapourising chamber which is maintained at a high temperature and has a lateral throughflow of gas. Only a small fraction of the sample enters the column, with the rest escaping to the atmosphere through an outlet valve.

With split-less injection, the full sample is vapourised before introduction into the column, which is held at a temperature below the boiling point of the solvent. This concentrates ('focuses') the sample in a small section of the capillary so that when the temperature programme is begun the solvent elutes as a narrow band without interfering with the analyte peaks. Apart from venting the gases at the end of the transfer on to the column, the whole sample is transferred. This makes the technique more sensitive than the split method.

The final technique, i.e. large-volume injection, is useful for trace analysis, as a concentration stage is included. Up to 250 μl of sample is slowly injected on to a cold short column which may be packed by capillaries or packing material. Most or all of the solvent is slowly vapourised (20–30 seconds) before a more rapid heating to transfer the concentrated sample on to the column. The latter is held at a low enough temperature to focus the sample in the capillary. The chromatographic separation starts on commencement of the temperature programme.

With a suitable choice of analytical column, simple extraction may be a sufficient pre-treatment for the direct injection of the extract into the chromatograph. For instance, the UK HMSO method for halomethanes simply uses the extraction of the compounds into petroleum ether and injection of the extract directly into the chromatograph. The chromatograms of such extractions may be complex (particularly if flame ionisation detection is being used), with a single extraction stage usually having insufficient specificity to simplify the chromatograms greatly. Indeed, a simple extraction with injection of the extract into the chromatogram is often used as a survey method to identify organic compounds in water. Alternative extraction methods would be suitable for halomethanes. Halomethanes are examples of volatile organics. Head-space analysis, purge and trap or solid-phase microextraction would be suitable.

For the analysis of individual components (often semi-volatiles) expected to be found at low concentrations (e.g. pesticides), further pre-treatment may be necessary.

The major stages in any pre-treatment scheme are: (i) extraction; (ii) clean-up to remove interfering components; and (iii) concentration of extract.

Until the successful development of solid-phase extraction, solvent extraction had been the most often used technique for stage one. The low volatility of many of the compounds of interest in this category renders the alternative vapour-phase extraction methods difficult. The clean-up stages will invariably be chromatographic, often using column chromatography. This may involve more than one separation stage. To illustrate the method, it is easiest to study one analysis in detail. For this, an analytical scheme for the commercial pesticide, DDT has been taken from the European Standard Method, EN ISO 6468 (1996) with necessary ammendments.

Analysis of DDT

This was the first synthetic insecticide to come into widespread use. It was introduced after the Second World War, and although now controlled or banned in many areas of the world (particularly in the West), it is now a universal contaminant. In common with most commercial products, the insecticide is not a single chemical compound—the major active component (p,p'-DDT) only consisting of 70–80 per cent of the total content. One of the minor components, p,p'-DDD (similar in structure to p,p'-DDT, but with a –CHCl$_2$ side-chain rather than –CCl$_3$) is, in fact, more toxic to insects than p,p'-DDT.

When considering environmental samples, a number of decomposition and metabolic products will also be present. In fact, for many samples, the highest-concentration component is not p,p¢-DDT but its primary metabolic product, DDE.

The following chromatographic peaks are expected in DDT analysis:

1. Components of technical DDT

    p,p′-DDT (70–80 per cent)

    o,p′-DDT (15–20 per cent)

    p,p′ -DDD (1–4 per cent)

2. Decomposition products

    p,p′-DDE (aerobic decomposition)

    p,p′-DDD (anaerobic decomposition)

Thus, we have a multi-component mixture even without the presence of any other compounds expected in the water sample. Interfering components in a typical sample could include other pesticides and polychlorinated biphenyls (PCBs). These often have similar extraction properties to the DDT components.

A typical pre-treatment would be as follows:

1. Extraction of the organic components into hexane, with a l l sample being extracted into three aliquots (30 + 20 + 20 ml) of solvent.

2. Drying the combined extracts by using a column containing 5 gram sodium sulphate. The chromatographic columns in the subsequent stages of the procedure are deactivated by the presence of water and so drying the extract at the earliest possible stage is essential.

3. Further concentration of the extract to a 1 ml volume. This could be by a number of methods, including using a Kuderna-Danish evaporator or a rotary evaporator. The sample is then placed on to the top of the first chromatographic column described below.

4. Clean-up of the extract by column chromatography.

    (a) Alumina-alumina/silver nitrate column: This column contains a bottom layer of alumina/ silver nitrate, a layer of alumina and a top layer of sodium sulphate. Alumina is a polar column material and will retain polar components in the extract. The silver nitrate helps retain compounds containing unsaturated carbon-carbon bonds. Non-polar material, including the DDT components, is eluted by using 30 ml of hexane. The extract is next reduced in volume to l ml and then a 100 μl portion is added to the top of the second column.

    (b) Silica gel column: This is a less polar column than the first and can be used to separate potential non-polar interferences from the sample. First 10 ml of hexane are passed through the column, eluting the PCBs, with the DDT components being retained on the column, followed by 8 ml of a 90 per cent hexane/10 per cent toluene solvent mixture. DDT is then eluted with a more polar solvent mixture (12 ml of 10 per cent diethyl ether in hexane).

The eluates are then re-concentrated to 1 ml or less before injection into the chromatograph. Try calculating the overall concentration factor of the pre-treatment process. You should come up with the factor of 100× if the final solution volume is 1 ml. A typical chromatogram is shown in Fig. 16.22. The detection limit for each component is approximately 10 ng l^{-1}.

The above procedure is just one method of pre-treatment. Other chromatographic methods may be used, such as preparative-scale thin layer chromatography (TLC) or solid-phase extraction (SPE). Each of these methods will still, however, be made up of the same individual stages of extraction, concentration and removal of selected interfering components. Clean-up of the extract simplifies the subsequent chromatogram.

This is simply protection of the column and detector from contamination. Without clean-up, the column lifetime will be shortened and the detector sensitivity lowered. Cleaning detectors to restore the sensitivity can be very time-consuming.

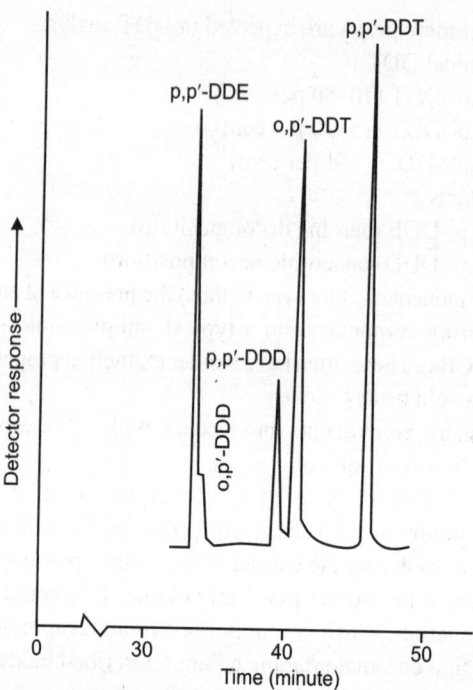

Fig. 16.22. Chromatographic separation of DDT components using a 25 m × 0.32 mm i.d. methylsilicone capillary column with a temperature gradient to 220°C.

Fingerprinting oil spills

If a film of oil is discovered on water, the first question likely to be asked is:

'What is it? Petrol? Fuel oil? Paraffin?'

The next question might be:

'Where did it come from?'

The commercial products mentioned above are complex mixtures of organic compounds. The precise composition of the mixture can vary from sample to sample, and so even if a complete quantitative analysis of every component in the mixture were undertaken, there would still be much difficulty in interpretation of the data.

A simpler procedure is to produce a chromatogram under standard conditions (column packing, flow rate, column temperature, etc.) and to compare the trace either with a library of reference materials, or preferably a sample of the material suspected to have been discharged. Capillary columns are necessary for resolution of individual components. Often, the correspondence of retention times and the overall envelope shape of the chromatogram will be sufficient to characterise the effluent. Hydrocarbon fuels give chromatograms with regularly spaced peaks (consecutive members of homologous series of compounds within the fuel). Lubricating oils have fewer resolved peaks. Natural product (vegetable) oils have simpler chromatograms with few individual peaks. Further information can be obtained if individual components in the material can be identified. Thus, the presence of simple polyaromatic species, such as anthracene, will identify coke-oven fractions.

Sample preparation from water containing low concentrations of hydrocarbons is simply to extract the material with a suitable volatile solvent (e.g. diethyl ether). After washing and drying, the extract is concentrated by using a dry nitrogen stream. With heavily polluted water, the organic material is separated by extractive distillation with toluene. The oil can then be recovered by fractional distillation from the toluene.

Complications can occur in interpretation of the chromatogram with material which has not been sampled immediately after discharge. The compositions of oil spills change with time (see Fig. 16.23). Volatile components of the oil evaporate, with, in general, lower-molecular-mass components disappearing first. The oil will also slowly be biodegraded, with the rate of degradation of a particular component being dependent on its chemical structure. A straight-chain hydrocarbon, for example, will be degraded more quickly than its branched-chain isomer. The determination of trace components and their relative concentrations has been found useful for assisting identification. Polynuclear aromatic hydrocarbons and their alkylated derivatives have been used as reference compounds in the finger-printing of oil spills. Although GC apparently seems a very simple method of identification of spillages, it is, in fact a skilled task needing a great deal of experience.

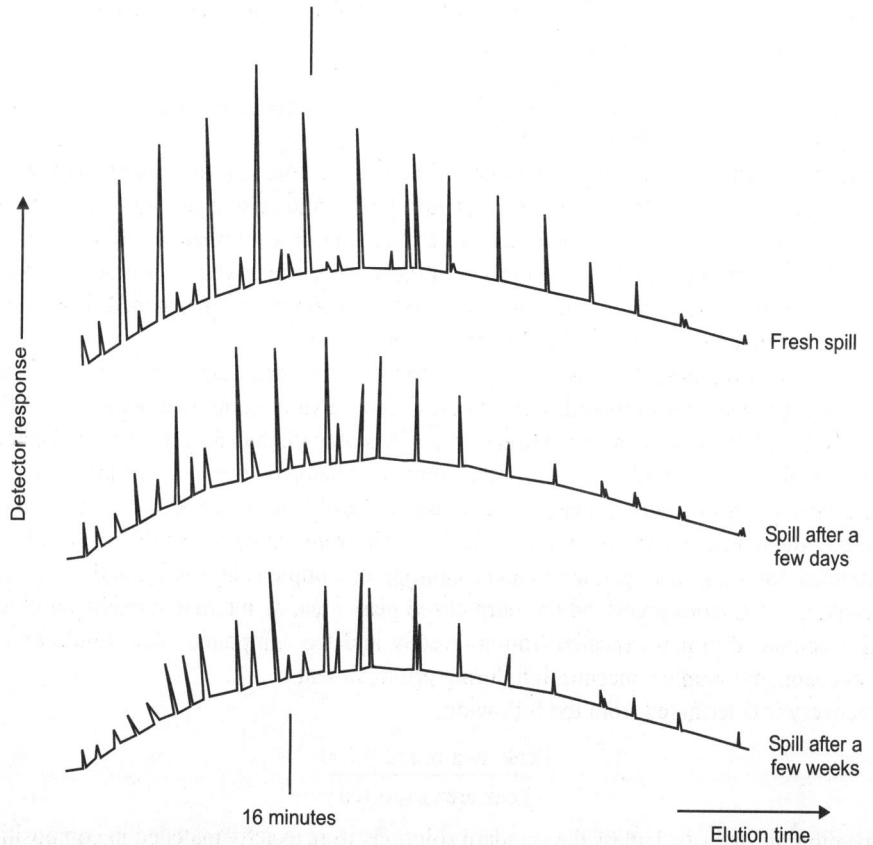

Fig. 16.23. Typical envelope shape in a chromatogram of an oil spill and its change with age of spill.

Quantification

When you look at a number of gas chromatography standard methods, it almost seems that each has a different procedure to determine concentrations. Most are variations of one or more of the methods described below. We will start with the most simple method, discuss the problems with this approach, and then consider some ways of overcoming the problems.

External standards

The simplest and most obvious method is to compare the peak area of the compound in the unknown solution with the areas of a series of solutions which have been used to form a calibration curve (or, if the calibration has been shown to be linear, with a single solution of known concentration close in value to the unknown), i.e. calibration is by external standards as discussed earlier.

Internal standards

The above problem can be overcome by adding an internal standard. This is a compound which will produce a chromatographic peak close to, but resolved from, the unknown species. An accurately known amount of the standard is added to a fixed volume of the unknown solution and to each of the external calibration solutions. Any variation in injection volume would show up in a change in peak area of the internal standard. There are a number of methods by which a correction may be applied.

The normalisation procedure plots the following:

$$\frac{\text{Peak area}}{\text{Internal standard peak area}} \text{Versus concentration}$$

Quantification if there is sample pre-treatment: There is the potential for loss of analyte during the pre-treatment. The procedures described so far will not take this into account and any loss of the unknown will result in a low analytical value. One method to overcome this problem would be to use external standards and to submit them to the same clean-up procedure as the unknown. Losses during clean-up would be assumed to be the same for the standards and the unknown. Internal standards would again be added prior to chromatography to overcome injection problems.

During method development and as a quality control step, the percentage recovery would need to be determined. The procedure for this is described below. Such a value should be as close to 100 per cent as possible, although for some complex extractions, values of more than 60 per cent may be acceptable.

Percentage recovery: A known amount of the compound being determined is added to a blank. The latter could be a synthetic solution made up as close as possible to the expected sample composition (not including the unknown) or it could be a field sample from which the analyte has been extracted (a 'preextracted' sample). After extraction and clean-up, the sample is injected into the chromatograph and the peak area (or more precisely, the normalised peak area, as internal standards are necessary) produced is compared to that expected from a directly injected compound. You should remember to take into account any sample concentration during pre-treatment.

The recovery is determined from the following:

$$\frac{\text{Peak area found} \times 100}{\text{Peak area expected}}$$

Isotope dilution analysis: Unless the standard solutions were exactly matched in composition to the unknown, there would still be the possibility of errors due to the sample matrix changing the recovery efficiency. In order to overcome this, you would need to have the standards and the unknown in the same matrix, i.e., the standard solutions added to the sample. This is possible if mass spectrometric detection is used. The standard is an identical compound to the unknown except that one or more atoms have been isotopically substituted this is often a deuterated compound. The peaks corresponding to the unknown and the standard will have the same retention times, but can be distinguished by detection at the two mass/charge ratios corresponding to the unsubstituted compound and the isotopically substituted standard. This is illustrated in Fig. 16.24.

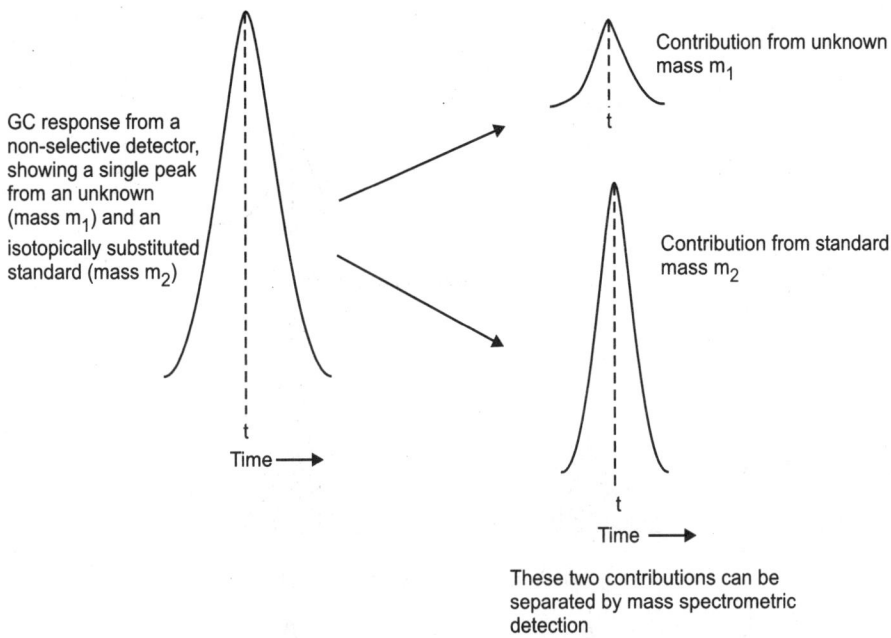

GC response from a non-selective detector, showing a single peak from an unknown (mass m_1) and an isotopically substituted standard (mass m_2)

Time ⟶

Contribution from unknown mass m_1

t

Contribution from standard mass m_2

t

Time ⟶

These two contributions can be separated by mass spectrometric detection

Fig. 16.24. Illustration of the principle of isotope dilution analysis.

In practice, allowances have to be made in the calculation for the compound and the substituted compound not being isotopically pure, i.e. detection at any one mass would include contributions from both the sample and the 'spike' (see Fig. 16.25). The isotopic abundances of the unspiked sample and the spike need to be precisely known from the mass spectra of the individual components [scans (a) and (b) in Fig. 16.25]. The concentration in the original sample can then be calculated from equations derived from the ratios given in the following equation:

Intensity ratio (mass m_1/mass m_2) of spiked sample

$$= \frac{\text{Number of unknown molecules at mass } m_1 + \text{number of spiked molecules at mass } m_1}{\text{Number of unknown molecules at mass } m_2 + \text{number of spiked molecules at mass } m_2} \quad \text{... (16.30)}$$

where, the number of unknown (or spiked) molecules at mass m_x (x = 1 or 2) is given by:

$$\frac{\text{Total mass of unknown (or spike)} \times \text{fractional molecular abundance at mass } m_x}{\text{Molar mass of compound (or spike)}} \quad \text{... (16.31)}$$

You should note that although isotope dilution analysis overcomes many of the problems found in other methods there should ideally be one isotopic standard for each compound being determined. However, this may increase the cost of the analysis quite considerably.

Liquid chromatography

For several years, liquid chromatography (LC) had the role of separation of classes of compounds which was difficult to achieve by using gas chromatography. Its application is now widening, largely due to the advent of bench-top LC-MS systems and the increasing use of solid-phase extraction which is particularly suited to interfacing with LC equipment. The lower separation efficiency of LC in comparison to that of GC is largely offset by the high selectivity of mass spectrometric detection.

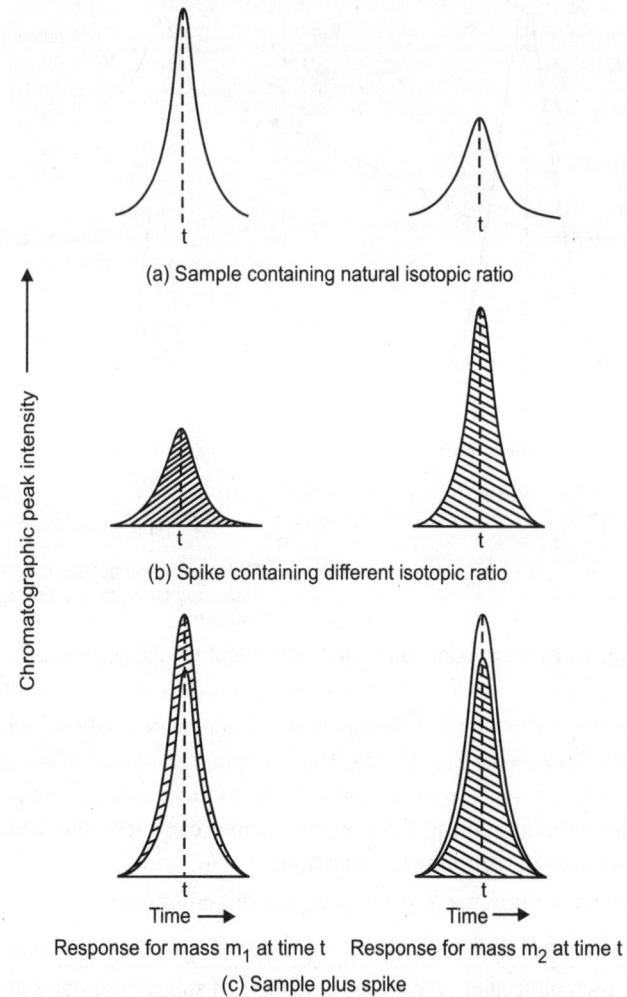

Fig. 16.25. Quantification by isotope dilution analysis in practice.

Longer established methods which use LC concern groups of compounds which can be determined using specific detectors or by derivatisation.

Conductivity detection has found widespread use for inorganic ions. Low molecular-mass carboxylic acids (e.g. formic and acetic acids) have very similar physical properties to the inorganic acids and ion chromatography provides a convenient alternative to gas chromatography for these acids.

One group for which fluorescence detection has high sensitivity are polynuclear aromatic hydrocarbons (PAHs). Some examples of these are shown in Fig. 16.26. They are highly carcinogenic compounds which are produced in trace quantities whenever fossil fuels are burnt. Typical water extracts could include up to 70 PAHs with a total concentration of around 1 μg l^{-1}. In order to monitor these low concentrations, sample preconcentration is needed. Solid-phase extraction, using an octadecylsilane (ODS) column, or a combination of ODS and amino-type columns, has been used. Sensitivity can be maximised if the detector is capable of changing the excitation and detection wavelengths throughout the chromatographic run, since each component has different optimum settings. The range of wavelengths used is 270–300 nm for excitation and 330–500 nm for detection.

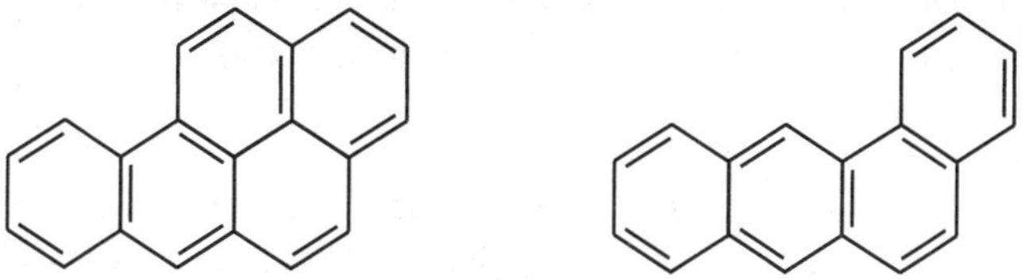

| Benzo[a]pyrene | Benz[a]anthracene |

Fig. 16.26. Some typical PAHs found in environmental samples.

Fluorescent derivatives can be made from non-fluorescent or weakly fluorescent compounds. Phenols and *N*-methylcarbamate pesticides (Fig. 16.27) are often analysed in this way. The procedure for *N*-methylcarbamates uses postcolumn derivatisation. The HPLC eluent is hydrolysed with sodium hydroxide at 95°C, thus producing methylamine. The latter is then reacted with *o*-phthalaldehyde and 2-mercaptoethanol to produce the fluorescent derivative. The fluorescent excitation wavelength is 230 nm and detection is > 418 nm, giving a limit of detection of approximately 1 μl g^{-1} per component for a 400 μl sample, injected without preconcentration.

Type	Example	Structure
N-Methylcarbamate	Carbofuran	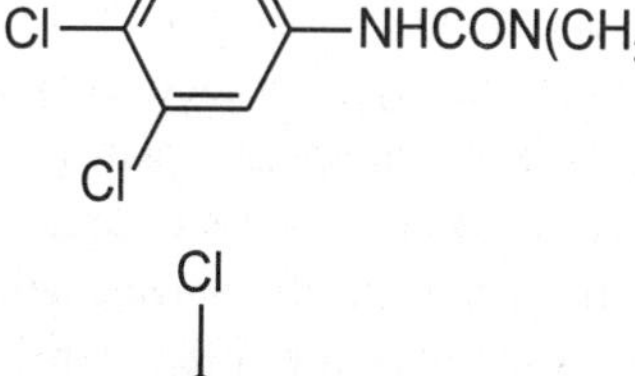
Urea	Diuron	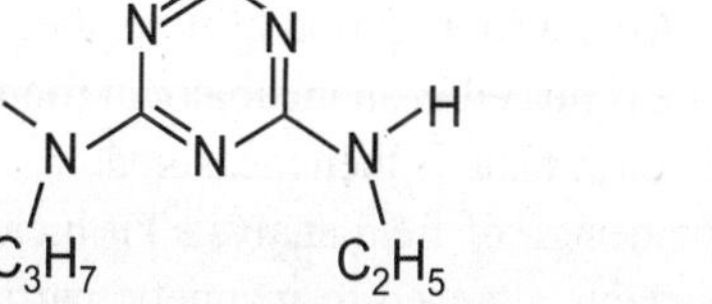
Triazine	Atrazine	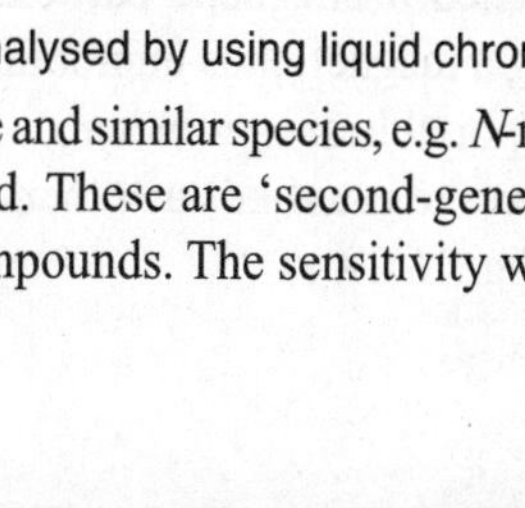

Fig. 16.27. Some examples of pesticides that can be analysed by using liquid chromatography.

HPLC with ultraviolet detection is sometimes used for these and similar species, e.g. *N*-methylcarbamate, urea and triazine pesticides can be analysed by this method. These are 'second-generation' pesticides which have been developed to replace organic halogen compounds. The sensitivity with UV detection is lower than that achieved by fluorescence measurements and preconcentration (solvent extraction or solid-phase extraction) has to be used prior to injection. This form of detection is also less specific than fluorescence and there is a greater possibility of chromatographic interference from other components in the sample. As with the case of phenols, the development of liquid chromatographic methods often stems from the difficulties encountered with analyses using gas chromatographic techniques. In many cases, this may be attributed to the polarities of the molecules (e.g. phenols and N-methylcarbamates) or their thermal labilities (e.g. N-methylcarbamates and phenylureas).

Immunoassay

The techniques described so far involve the use of complex laboratory equipment and often long pre-treatment stages. Ideally, an analyst would like to achieve the required sensitivity and specificity with simpler equipment and without any pre-treatment being required. Field analysis would also be desirable.

Part of the solution to this problem could be the use of immunoassay but as a separate test has to be designed for each analyte, it will never be the complete answer. Field kits (necessary apparatus, reagents and calibration standards for a specific number of analyses) are available in the μg/l range with an analysis time of 10–15 minutes. Laboratory kits are available for individual compounds in the ng l^{-1} range and are typically capable of handling 40 samples in a period of two hours. The methods commercially available include the analysis of individual pesticides (e.g. atrazine, carbofuran and paraquat). BTEX compounds [benzene-toluene-ethylbenzene-xylene(s)], total petroleum hydrocarbon (TPH), PCBs and PAHs, with the list continually expanding. Several of these methods are now approved by the US environmental protection agency (EPA). The use of these kits can be very simple, requiring little background knowledge. More thorough knowledge is necessary to understand the potential applications and limitations of the immunoassay process and the almost bewildering number of variations of the basic technique. Chemical and biological principles will both need to be understood and the techniques used sometimes seem to be more at home in a life sciences rather than a pure chemical laboratory. First of all, let us look at a simple method. Later, we will look at the background principles behind the method in an attempt to understand its particular merits.

Methodology

Most field and laboratory kits use a technique known as competitive ELISA (enzyme-linked immunosorbent assay). For laboratory analyses, reactions take place in the wells of a microtitreplate (Fig. 16.28). These are plastic plates which contain typically 40, 48 or 96 wells for the simultaneous analysis of the samples and standards. An automatic scanner (microtitreplate reader) measures their light absorbance at specific wavelengths. This apparatus is commonplace in biomedical laboratories. The wells of the plate are filled with 100 μl sample or standards in duplicate. The reagents are then added. After a short period of time, the plate is then washed with water, further reagents are added and the plate is placed in an incubator at room temperature for a period of up to one hour. The absorption of light in each plate is then measured.

One design of field analysis kit includes individual pre-coated tubes, while another manufacturer has reagents attached to magnetic particles. The latter can be separated from the reagent and wash solutions by using a magnet, thus immobilising the particles on the walls of the tube. Light absorbance is measured by a portable spectrometer.

The response curve is unlike any others you are likely to have come across, with a typical example being shown in Fig. 16.29.

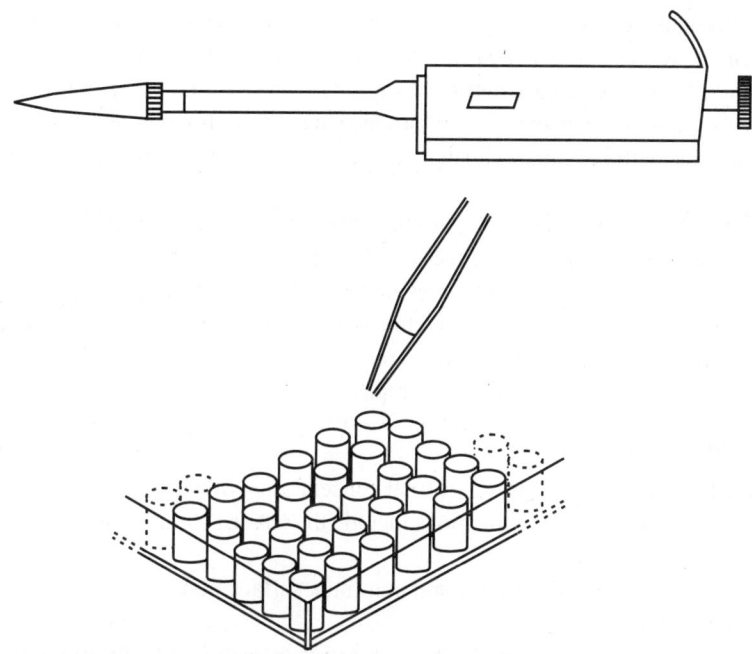

Fig. 16.28. A micropipette and microtitreplate used in immunoassay.

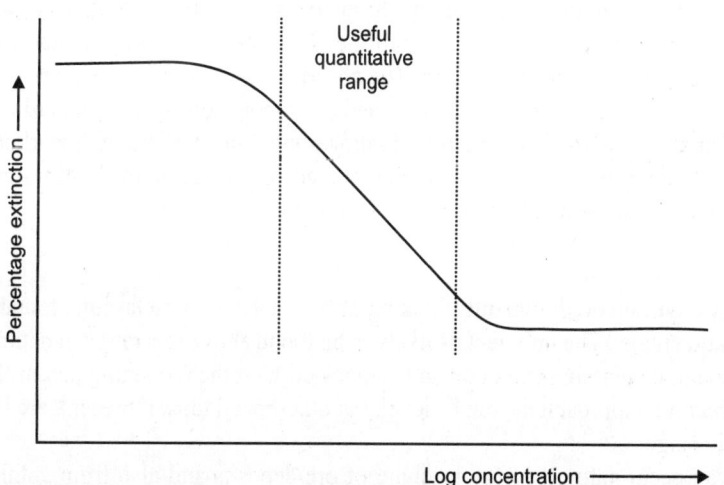

Fig. 16.29. A typical response curve found in immunoassay.

Development of tests and implications for analyses

An essential stage in the development of the kits is the production and isolation of the antibodies. The initial immune response can only be produced by molecules with an M_r greater than about 10000, much larger than most pollutant molecules. Derivatives (conjugates) of the initial molecule (hapten) must first be produced by covalently bonding the latter to a carrier protein. Antibodies are generated by injection of the conjugate into a laboratory animal. After a few weeks; the antibody can be harvested from samples of the blood serum of the animal. Sufficient antibodies will be produced for several thousand kits. Monoclonal antibodies and genetically engineered antibodies are now becoming more common. These are single chemical reagents of a defined composition with constant specificity characteristics and can be mass produced. As no two animals will produce identical antibodies, even 'identical' tests from different manufacturers will have to be considered as different analytical techniques and will need to be assessed separately.

The antibodies recognise molecules according to their molecular shape and bind at specific sites in the molecule. During the development process, tests have to be conducted to ensure the recognition sites ensure specificity for the compound being analysed. There is the possibility of 'cross-reactivity' with other compounds with a similar shape and functional groups. Cross-reactivity can, in fact, be used to advantage in some kits, e.g. the triazine pesticide test kit, which are designed to respond to groups of chemicals rather than to individual groups of compounds.

Spectrometric methods

Often, a technique is required to measure the total concentration of a group of compounds, rather than individual concentrations. Such determinations include the analysis of the following:

1. Total phenols.
2. Surfactants (total, anionic, cationic and non-ionic surfactants).
3. Total hydrocarbon(s).

Visible spectrometry is often used for phenols and surfactant analysis after the formation of derivatives. Chromatographic methods have in the past not been used as they give too much information!

The simplicity of the method can be seen by the analysis of anionic surfactants using a 'Methylene-Blue' method. Under basic conditions, a salt is formed between the methylene blue and the surfactant and this salt can be extracted into chloroform. The absorbance of the extract is measured in the visible region (at 652 nm) and the concentration determined by comparison with a standard calibration curve.

Infrared spectrometry is used for the total hydrocarbon content. The hydrocarbons are extracted from the acidified water by using a non-hydrocarbon solvent (e.g. carbon tetrachloride) and the absorption measured at 2920 cm^{-1}, corresponding to the C–H stretching frequency.

Metal Ions

In this section, we will be predominantly looking at the analysis of metal ions found in the $\mu g \ l^{-1}$ to $mg \ l^{-1}$ concentration range. The only metals likely to be found above this range in natural waters are the four ions (i.e., sodium, potassium, calcium and magnesium). Of the remaining metals, iron, manganese and zinc can sometimes approach the $mg \ l^{-1}$ level, but other metal ions, if present, are likely to be at the lower end of this range.

Metal ions can occur naturally from leaching of ore deposits and also from anthropogenic (man-made) sources. Such sources include metal refining, industrial effluents and solid waste disposal. Much solid waste, including power station fly-ash, sewage sludge and harbour dredgings, contains significant concentrations of metal ions (upto 1000 mg kg^{-1} total metal) which can leach into solution if in contact with water.

This area of analysis is currently dominated by techniques which can be grouped together under the general title of Atomic Spectrometry. The main individual techniques are as follows:

1. Flame atomic absorption spectrometry (Flame AAS).
2. Graphite furnace atomic absorption spectrometry (GFAAS).

3. Inductively coupled plasma-optical emission spectrometry (ICP-OES).
4. Inductively coupled plasma-mass spectrometry (ICP-MS).

These will be discussed, along with some other methods, in the following sections, showing the relative merits of each technique and their potential applications.

Storage of samples for metal ion analysis

1. Polyethylene bottles are less likely to contaminate the sample with metal ions than glass bottles. The only exception to the use of polyethylene bottles is for mercury analysis when glass bottles should be used. Mercury ions readily react with many organic materials.
2. The sample should be acidified to minimise precipitation of metal ions. A typical procedure is the addition of 2 ml of 5 mol l^{-1} hydrochloric acid per liter of sample.
3. Scrupulous cleaning of bottles is important. This usually includes an acid washing stage to ensure complete removal of trace metals. In the case of aluminium, the concern over contamination extends to glassware used in the subsequent analysis. You are often advised to pre-leach glassware with dilute nitric acid and to reserve glassware solely for aluminium determinations. Such a procedure would, in fact, be good practice for all metal analyses.

Pretreatment

Most routine analyses require the total metal content of the sample, regardless of its chemical nature. Pretreatment can include evaporation to dryness and redissolution in acid, partial evaporation with acid or digestion with acid at an elevated temperature for several hours. This is to dissolve suspended material and ensure that the metal is present as the free ion.

The more modern techniques we will be discussing (GFAAS, ICP-OES and ICP-MS) are sufficiently sensitive and interference-free for the majority of samples to require no further pre-treatment.

Most of the other analytical techniques require an extraction/concentration step for trace analyses. This may be a separate solvent extraction stage, as with flame AAS and some visible spectrometric methods, or may be a concentration stage in the analytical technique itself (Ion Chromatography and Anodic Stripping Voltammetry). Such a step can also serve to remove potentially interfering ions which may be present in far greater concentrations than the analyte.

The most common method proceeds with the formation of a neutral complex with an organic ion and extraction of this into an organic solvent (simple metal salts or ionic complexes would not extract). Upto a twenty times increase in concentration is possible in a single stage. The complexing agent used depends on the subsequent analytical procedure, and this will be discussed in the relevant sections. Other extraction/concentration methods include the use of chelating or ion-exchange columns. The metal ions are first held on the column, either by complex formation with the column packing material (chelating column) or by ion exchange. The ions are then eluted as a concentrated extract with an appropriate solvent, often an aqueous buffer.

Atomic spectrometry

We will start with a discussion of the technique you are probably most familiar with—flame atomic absorption spectrometry (Flame AAS)—and then show how the other atomic spectrometric techniques overcome problems found in its use for trace metal analysis.

Flame atomic absorption spectrometry

A schematic of a flame atomic absorption spectrometer is shown in Fig. 16.30 below.

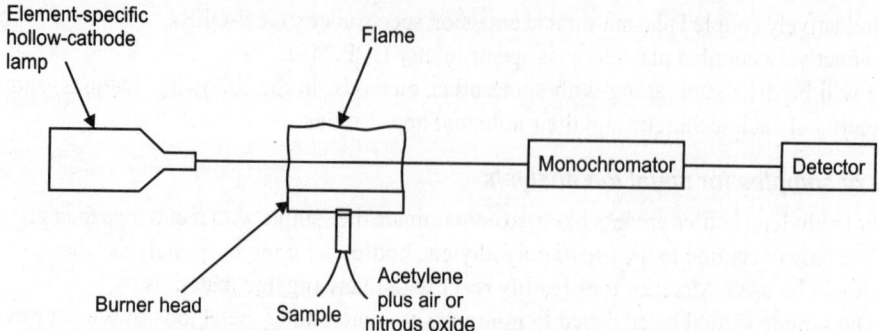

Fig. 16.30. Schematic of a flame atomic absorption spectrometer.

In this technique, a light beam of the correct wavelength to be specific to a particular metal is directed through a flame. The flame atomises the sample, producing atoms in their ground (lowest) electronic energy state. These are capable of absorbing radiation from the lamp.

Although the equipment appears completely different from other forms of absorption spectrometry, the law by which the absorption of light is related to concentration is similar to that we have used already for the absorption of ultraviolet and visible radiation. The concentration range over which the law applies for flame atomic absorption spectrometry is usually 0–5 mg l^{-1}. Over the last three decades, atomic absorption spectrometry has dominated routine analysis of metal ions in aqueous samples at mg l^{-1} and higher concentrations.

Atomic absorption is indeed a sensitive technique and if it is used for the more common ions already discussed above, the water samples would have to be diluted before analysis. Magnesium is analysed by flame atomic absorption, often after sample dilution. If the technique is used for sodium or potassium analysis, lower-sensitivity absorption lines, rather than the highest-sensitivity lines, would be used in addition to diluting the sample. Atomic emission (flame photometry) is, however, the preferred technique for these ions.

In high-throughput laboratories, low-concentration samples (< 1 mg l^{-1}) would normally be determined by the techniques described later in this section, particularly ICP-OES and ICP-MS. If these are not available, flame AAS can be used with sample preconcentration. This may simply involve partial evaporation of the acidified sample for zinc, iron and manganese analyses. Solvent extraction has been routinely used for other metals. Since atomic absorption analysis is relatively free from interference from other trace metal ions (i.e. the presence of other materials usually has little effect on the accuracy of the analysis), the extraction need not be highly specific to any one particular metal. In fact, it may be beneficial to be able to use a single complexing agent for several metals since the extraction stage is the most time-consuming part of the analytical procedure. Ammonium pyrrolidinedithiocarbamate (APDC) is often used as it forms stable complexes with most transition metals, if the pH is correctly adjusted. As an example, the optimum pH for lead extraction is 2.3. After extraction of the analyte in an organic phase, the organic phase is aspirated directly into the flame. The increase in the sensitivity is above that which is expected from the simple concentration factor. This is due to the increased aspiration rate resulting from the lower viscosity of the organic solvent in comparison to water.

A number of disadvantages of solvent extraction/flame atomic absorption, are:

1. It is very time-consuming.
2. The sensitivity may still be insufficient for low-concentration metal ions.
3. The risk of sample contamination is considerably increased.

To overcome these problems, other atomic spectrometric techniques have been applied to trace metal analysis.

Flameless atomic absorption

By replacement of the flame by other methods of atomising the sample, the sensitivity can be increased sufficiently to remove the need for sample preconcentration. For most metals, this would mean the use of graphite furnace atomisation (also known by the more general term 'electrothermal atomisation'), as shown in Fig. 16.31, but, as we will see later, this is not the only method possible.

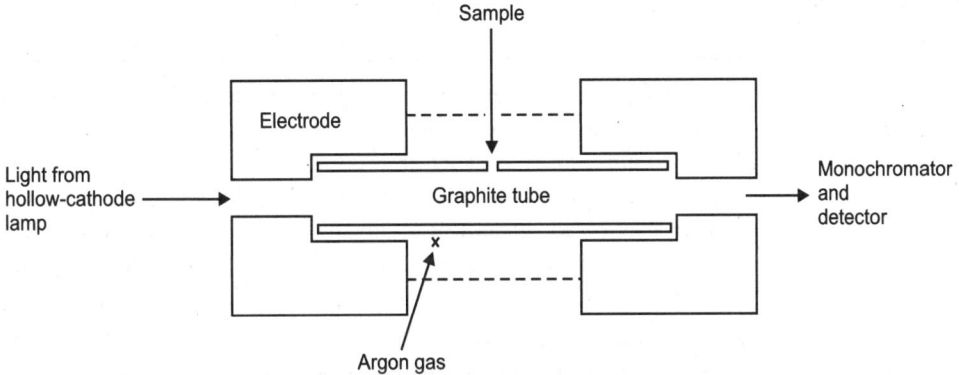

Fig. 16.31. Schematic of a graphite furnace.

Graphite furnace AAS involves injecting a sample (up to 25 µl) into a small graphite tube (2–3 cm × 5–10 mm) which is heated in pre-programmed stages, as follows:

1. Drying.
2. Decomposition.
3. Atomisation.

The absorbance of a light beam shone through the cell is measured during the atomisation stage. The optimum temperatures and duration of each stage are metal-dependent, with a complete programme taking 2–3 minutes.

A comparison of flame and graphite furnace atomic absorption spectroscopies is presented in Table 16.6. As you can see from this table, the chief advantage of flameless AAS arises from removing the necessity of preconcentration of the sample. An extraction stage may still sometimes be necessary for complex samples in order to reduce potential interferences, as in the case of sea water analysis. One major source of error is background interference, which results from light scattering by solid particles within the beam. The scattering is highly dependent on wavelength, as follows:

$$\text{Scattering} \propto \frac{1}{\lambda^4}$$

$$... \ (16.32)$$

where, λ is the wavelength of radiation.

The analytical wavelengths used for lead and cadmium are towards the far end of the available ultraviolet range and so analyses for these elements are highly susceptible to interference. Automatic background corrections should always be used for these elements. An analytical wavelength of 283.3 nm is also often preferred for lead, rather than the more sensitive 217 nm wavelength, as this lessens the effect of light scattering.

Table 16.6. A comparison of the advantages of flame and flameless (graphite furnace) atomic absorption spectroscopies.

Advantages of solvent extraction/flame AAS	Advantages of graphite furnace AAS
Simple technique	Increased sensitivity ($\mu g\ l^{-1}$ concentrations)
The solvent extraction stage can be used to remove potential interferences	Decreased overall analytical time as the solvent-extraction stage is not usually necessary
More readily available equipment	Smaller samples required
Shorter instrument time	Unattended operation is possible
Lower instrument cost	Reduced risk of sample contamination

Background effects can also occur from other sources according to the analyte being considered, including the presence of thermally stable molecular ions. The absorbance can be highly structured (e.g. narrow absorption bands within a much broader absorption). There are a number of methods available for background correction, as shown in Table 16.7. Due to the different principles on which the corrections are based, there may be advantages or disadvantages for each type according to the application. Revalidation may be necessary if a different technique is used to that specified in a standard method.

Table 16.7. Common background correction methods.

Method	Feature
Deuterium lamp (continuum method)	A second absorbance measurement over a slightly larger wavelength range than the atomic absorption which gives the background reading
Zeeman	An intense magnetic field splits the absorption into magnetic components at slightly different wavelengths. Absorbance measurements with and without the magnetic field can be processed to correct for the background
Smith-Hieftje	A pulse increase in the lamp current removes the atomic absorption, leaving only the background

Other flameless atomisation techniques can be used for specific elements. Inorganic mercury salts can be chemically reduced by using tin (II) chloride or sodium borohydride. The elemental mercury produced is then swept by a stream of nitrogen or air into a gas cuvette for absorption measurement in a modified spectrometer. Tin, lead and a number of metalloids (As, Se, etc.) can be reduced by sodium borohydride to volatile hydrides which are swept from the sample by a gas stream. Mild heating breaks down the hydrides to produce the elements in their ground states suitable for absorbance measurements.

Quantification

The major advantage of atomic absorption over other techniques is often stated as its lack of interference, particularly between metals. All that would appear necessary for quantification would be to use external standards to produce a calibration graph.

There are, however, a number of factors which will affect the accuracy of the analysis. One of these is chemical, typically where refractory salts are formed between the metal and an anion. These interferences are well-known and usually described in instruction manuals accompanying the spectrometer. A typical concern for environmental samples is shown by the effect of phosphate on calcium, which decreases the absorption due to the formation of insoluble and refractory calcium phosphate. Similar problems can occur in the presence of sulphate and silicate ions. These problems can be overcome by adding a small quantity of release agent to each solution. A 10 per cent lanthanum solution is often used. The lanthanum preferentially reacts with the phosphate. Alternatively, an EDTA solution can be used. In this case, the EDTA complexes with the calcium.

For more complex analytes, other factors may affect the accuracy. These include physical effects where the viscosity or surface tension of the solution is altered. Such properties will affect the aspiration of the solution into the flame and hence the measured absorbance.

The method of standard addition is often used to overcome this problem. A calibration curve is produced from a series of sample solutions which have been increased in concentration by adding known amounts of the metals ion being determined. This, of course, will increase the measured absorbance. The easiest way to achieve this for trace work where you do not wish to dilute the sample is to add small volumes of higher-concentration standards so that the change in overall volume is negligible. The amount of metal added needs to be chosen so that the increase in absorbance is of the same order as that of the original sample. If you add, by chance, an amount of metal ion which will double its concentration, then the absorbance will double. It is perhaps a little less obvious to see how you calculate the unknown concentration from a series of additions. You do this by plotting a graph of the concentration increase against the absorbance. A linear plot should be produced, but, of course, the line does not pass through the origin. There will still be absorption from the metal ions in the untreated sample (i.e. the y-axis intercept). The concentration of the sample is found from the x-axis intercept, with the latter being the negative value of the sample concentration.

The development of the inductively coupled plasma (ICP) techniques for water analysis can be seen as an attempt to overcome these problems. At the same time, they maintain the advantages of graphite furnace AAS of being sufficiently sensitive not to require a pre-concentration stage and also in not using flammable or explosive gases. This permits unattended, 24 hours, operation. In both methods, the sample is atomised in a plasma flame at 6000–10000 K (Fig. 16.32). This is generated by a flowing stream of argon which is ionised by an applied radiofrequency (RF) field.

Inductively coupled plasma-optical emission spectrometry (ICP-OES)

With this technique, the emission spectrum is monitored. Simultaneous ICPOES can determine 60 or more elements at once by monitoring at pre-set wavelengths. This includes halogens and some other non-metals and metalloids, as well as metals. Sequential spectrometers, which are more common for water analysis, restrict themselves to a smaller number of elements, determined by the requirements of the analysis, measured in succession by rapid changes in the detection wavelength. The total analysis time is still fast, typically 5 seconds per element. A further advantage of ICP-OES is its wide dynamic range (approximately 10^5), which means that trace metals can be measured simultaneously with higher-concentration species.

In common with other emission techniques, there is the problem of spectral overlap from different elements, as an element will produce many more lines in its emission spectrum than in its corresponding absorption spectrum. The choice of the analytical wavelength is based on freedom from interference as well as sensitivity. For routine water analysis this problem has largely been overcome, with sensitive and interference-free lines being well documented. Quantitative analysis can be performed by using external standards after first confirming that the chosen wavelength is free from interference. For quality control, monitoring can be at two wavelengths, which of course should produce identical results if there is no interference at either of the wavelengths.

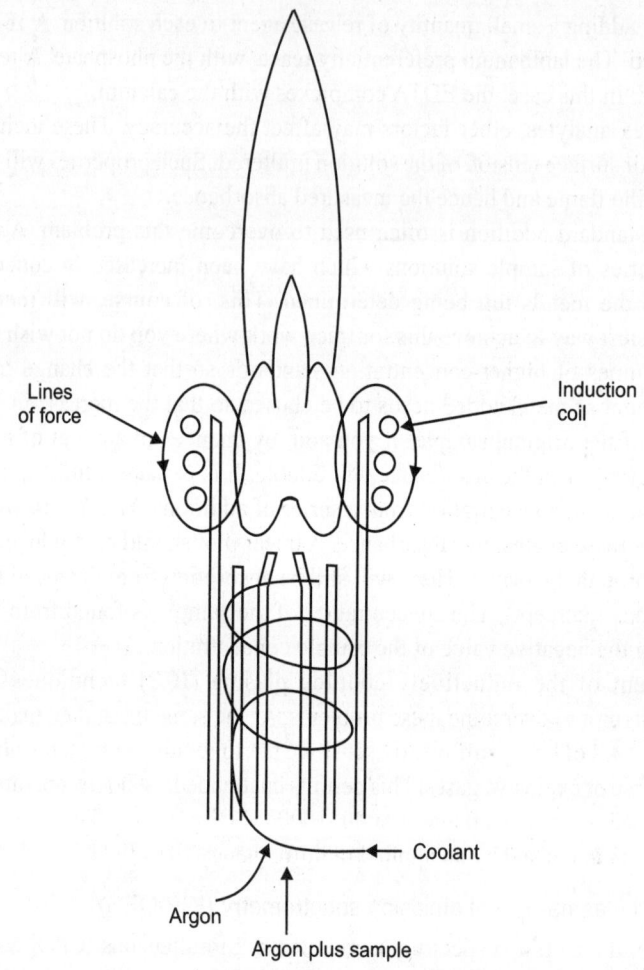

Lines of force

Induction coil

Coolant

Argon

Argon plus sample

Fig. 16.32. Schematic of a plasma flame unit used in ICP techniques for atomisation of samples.

For many years, the sensitivity of ICP-OES lay between those of the flame and furnace AAS techniques for most elements, thus making the technique useful for most, but not all, of the major components of water. Considerable effort was made to improve the sensitivity by changes in the spectrometer design. The major improvement lay in the relative position of the detector with respect to the plasma (Fig. 16.33). Originally, this was at right angles to the plasma, so giving a short pathlength through the flame. The sensitivity is increased by 8–10× (i.e. to ca. furnace AAS sensitivity) by moving the detector to an axial position.

Problems which needed to be overcome included the effect of the plasma tail on the optics. One method of overcoming this is by diverting the plasma tail away from the optics by a radial flow of gas. Organic solutions cannot be used with axial flow detection. The linear range of the instrument is unchanged, but is moved to lower concentrations. As a consequence, both types of instrument are still in use today. Some instruments are capable of operating in either mode, with the axial configuration being reserved for applications needing higher sensitivities.

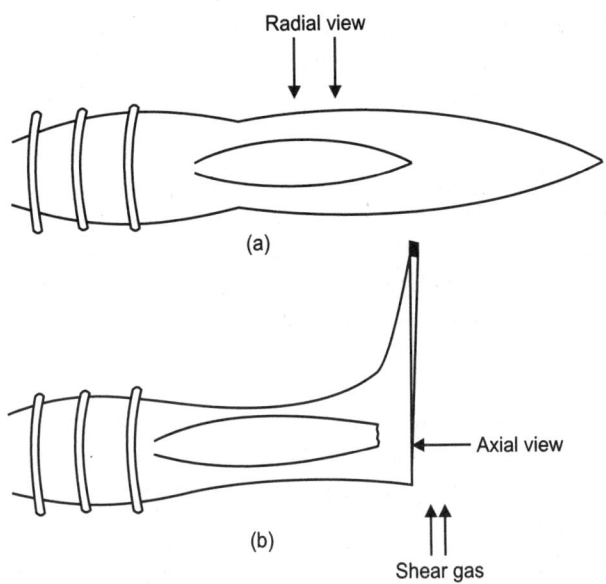

Radial view

(a)

Axial view

(b)

Shear gas

Fig. 16.33. Relative positions of the detector in ICP-OES: (a) radial; (b) axial.

Inductively coupled plasma-mass spectrometry (ICP-MS)

A more recent development is to use the inductively coupled plasma as an ion source for a mass spectrometer. For routine applications, this is usually a quadrupole spectrometer as is commonly found in a GC-MS system. The mass spectra of inorganic mixtures are simple in comparison to the more familiar organic compound spectra and fewer interferences occur in metal analysis. Although the technique is not strictly simultaneous — the ions being determined sequentially — determination of some 20 elements is possible within a period of 4 seconds. The sensitivity is slightly lower than that of graphite furnace AAS but is still sufficient to determine trace metal ions at below 1 μg l^{-1} in aqueous samples. The linearity range is 6–8 orders of magnitude, according to the particular application.

Since each metal is determined according to its mass/charge ratio, there would seem to be, at first sight, little chance of interference in the technique. High concentrations of salts can cause deposition of solids in the instrument. Although there are few instances where there are problems with isotopes with the same mass/charge ratios, molecular ions can be formed, particularly with refractory elements. For example, $^{44}Ca^{16}O^+$ has the same mass/charge ratio as $^{60}Ni^+$, while $^{40}Ar^{35}Cl^+$ could interfere with $^{75}As^+$. Most of these problems can be eliminated by simple pre-treatment of the sample to remove the potential interference before introduction into the ICP-MS system. Flow-injection techniques may be used to automate the process and may also be used to alleviate the problem of high salt concentrations if simple dilution is not possible. Avoidance of the use of hydrochloric acid for acidification will lessen problems with chlorine interference.

The use of mass spectrometric detection makes possible the use of the isotope dilution techniques discussed earlier in this chapter. The method would have to be applied to each isotope in turn and is time-consuming if large numbers of samples and elements have to be determined. As a consequence, for routine samples external standards are sometimes used after quality control checks to confirm the lack of any interferences. The standards may be matrix-matched to minimise problems caused, for instance, by viscosity differences. Standard addition is also used.

Each development in atomic spectrometry has brought with it a significant increase in instrument capital cost. The cost of AA instrumentation is generally according to the following:

Flame AAS < Furnace AAS < ICP-OES < ICP-MS

In addition, ICP techniques have significantly higher running costs due to consumption of the argon necessary to generate the plasma. The advantages of ICP techniques are, however, so great that ICP-OES has for several years been the preferred technique for the high-concentration metal analysis in major water analysis laboratories, with ICP-MS being used for lower-concentration metals. Atomic absorption methods find a role in smaller laboratories where the sample throughput is insufficient to justify the additional capital and running costs of ICP techniques.

Visible spectrometry

Until the widespread use of atomic spectrometric techniques, visible spectrometry was the most commonly used technique for metal ion analysis. Standard methods were developed for all commonly found metal ions. These methods use colour-forming complexing agents. Selectivity in the analysis is achieved in two different ways, as follows:

1. Solvent extraction is sometimes used. Chromium is analysed as the diphenylcarbaside complex after extraction into a trioctylamine/chloroform mixture. This gives a limit of detection of 5 μg l^{-1} in the original sample. The complexing agent dithizone can be used for 17 metals. The selectivity is achieved by precise control of pH and the use of masking agents.

2. Alternatively, a colour-forming complexing agent can be used which is sufficiently sensitive and selective for use in the aqueous sample without extraction being necessary. Some examples are shown below in Table 16.8.

Table 16.8. Some examples of colour-forming complexing agents.

Metal	Reagent	Limit of detection $(\mu g\ l^{-1})$
Iron (II)	2,4,6-Tripyridyl-1,3,4-triazine	60
Manganese	Formaldoxime	5
Aluminium	Pyrocatechol violet	13

A number of these techniques have been adapted for use with portable colorimeters (e.g. iron, manganese, chromium and copper) and it is perhaps in this area that such techniques have the most widespread current usage.

It is useful to consider why such a well-established technique as visible spectrometry could become largely superseded by atomic methods:

1. Atomic methods are more rapid.

2. Although visible spectrometric pre-treatment is generally simple when analysing relatively unpolluted water samples (rivers and lakes), they may become complex and time-consuming with more complicated samples such as sewage effluents.

3. Visible spectrometry is often affected by interference from other elements. This can be illustrated by the determination of iron using 2,4,6-tripyridyl-1,3,5-triazine. The concentration effects observed on a true value of 1000 mg l^{-1} iron are shown in Table 16.9.

Table 16.9. Concentration effects on the measured ion content $(1000 \text{ mg } l^{-1})$ of an iron sample.

Additional ion	Effect
100 mg l^{-1} sulphate	-0.020 mg l^{-1}
2 mg l^{-1} cadmium	$+0.009$ mg l^{-1}
10 mg l^{-1} lead	-0.026 mg l^{-1}

Nonetheless, visible spectrometry remains a frequently used technique and would be the method of choice when atomic methods are unavailable.

Anodic stripping voltammetry

A number of electrochemical methods are sufficiently sensitive to determine the low levels of metal ions typically found in the environmental water samples without any separate preconcentration. Anodic stripping voltammetry (ASY) has found particular use in environmental analysis, where at least 19 metals can be analysed in this way.

The apparatus used consists of an electrolytic cell containing a working electrode (a mercury drop, or a thin film of mercury deposited on a glassy carbon electrode), a reference electrode and a counter electrode. A three-electrode system is used so that current and applied potential can be measured independently. This attempts to compensate for the change in potential drop due to the resistance of the test solution during the analysis. The latter would affect the measurement in a two-electrode system.

The sample is placed in the cell along with a supporting electrolyte (e.g. 0.1 mol l^{-1} acetate buffer at pH 4.5). Nitrogen or argon is bubbled through the solution to remove dissolved oxygen, which would otherwise interfere in the analysis.

The working electrode is held at a small negative potential with respect to the reference while the solution is stirred. Reduction of the metal ions to the free metal occurs at the working electrode. Under controlled conditions of deposition time and stirring rate, the quantity of metal deposited on the electrode is proportional to its original concentration:

$$M^{2+} + 2e \rightarrow M \qquad \qquad \text{... (16.33)}$$

After a predetermined time, the potential of the electrode is slowly changed in the positive direction. At specific potentials, depending on the metal and supporting electrolyte, each metal is oxidised and returned back into solution as follows:

$$M \rightarrow M^{2+} + 2e \qquad \qquad \text{... (16.34)}$$

This process is monitored by plotting the current change between the working and counter electrodes against the potential (Fig. 16.34). The height of the peak in the curve is proportional to the concentration of the metal.

The supporting electrolytes necessary for individual metals are tabulated in standard texts. Electrolytes are often acidic, as their potentials become little affected by minor changes in the sample composition. Complexing agents (e.g. acetate) are sometimes included to stabilise particular oxidation states or to move the stripping potential of the metal away from potential interferences. Four metals of major environmental concern, i.e. copper, lead, cadmium and zinc, can, however, be analysed in a single scan by using the acetate buffer mentioned earlier. Quantification is by standard addition. As an electrochemical method, the technique determines only free metal ions in solution, plus, to some extent, loosely associated complexes.

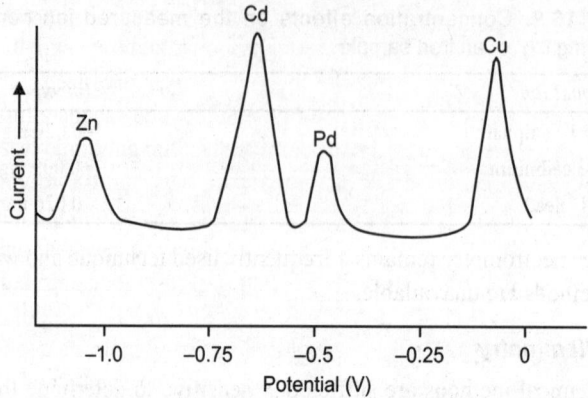

Fig. 16.34. A typical anodic stripping voltammogram.

However, anodic stripping voltammetry does have at least one disadvantage. The laboratory method is slow, with stripping times varying between 30 seconds 30 minutes and during such periods the apparatus is devoted to a single sample. Compare this with atomic spectrometry, where the instrumental time is only a few seconds per sample. The time taken for analysis has, however, been overcome with commercially available field instruments. These utilise disposable electrodes and microprocessor control which automatically takes the solution through the scanning cycle. Sample pre-treatment is by addition of salts in tablet form. The conditioning breaks down any complexed forms of the metal and so the concentration output is of the total dissolved metal. A complete lead and copper analysis at the μg 1^{-1} level can be performed in 3 minutes.

If the total metal content is required with laboratory apparatus, sample pre-treatment is also necessary. This may range from simple acidification to UV irradiation in order to destroy any potential complexing agents. By performing the analysis with and without pre-treatment, a measure of the free and complexed metal ions can be made, which would not be possible by using atomic spectrometry. This makes anodic stripping voltammetry a useful research tool, but because of its relative slowness in the laboratory, limited in its application for routine analysis.

Liquid chromatography

As environmental waters invariably contain a large number of metal ions and often at similar concentrations, you might think that liquid chromatography would be a frequently used analytical technique. Your argument might be that it should be possible to determine all of the metal ions present by using a single sample injection into the chromatograph. Atomic methods still, however, dominate metal analysis. Liquid chromatography only finds use in areas where atomic spectrometry is not ideal.

1. Complex matrices: Extraction techniques are often necessary when complex samples are analysed by AAS in order to remove interfering components. This extends the time taken to perform an analysis considerably.

2. Analysis of mixtures of uncommon elements: AAS determines individual elements by using a different lamp for each one. Additional and perhaps unsuspected, elements will not be detected. With a correct choice of column and eluent, these would be seen as additional peaks in a liquid chromatographic analysis. The need to change lamps for each element may also mean that AAS is a slower technique than chromatography for complex mixtures. It is more difficult to obtain the hollow-cathode lamps necessary for AAS for the more uncommon elements.

3. Quantification of different chemical forms of the ion: In certain instances, ion chromatography can separate and quantify the chemical forms. Atomic absorption spectrometry is unable to distinguish the different species that may be present.

If we extend our comparison to ICP techniques, then many of the perceived advantages may still hold. Interferences in complex samples may still be found. Sequential plasma emission spectrometers detect a limited number of elements and unsuspected elements may still be missed. Different chemical forms are not distinguished. ICP techniques are, however, more rapid for multi-element analyses.

Chromatographic methods using both dedicated ion chromatographs and conventional HPLC have been developed. The most sensitive method for transition metals in complex mixtures using a dedicated chromatograph is known as Chelation ion chromatography. This method involves the use of two preconcentration columns (Fig. 16.35) and spectrometric detection after mixing with a derivatisation agent, i.e. 4-(2-pyridylazo)resorcinol. The detection limits are 0.2–1 μg l^{-1} with a 20 ml sample volume.

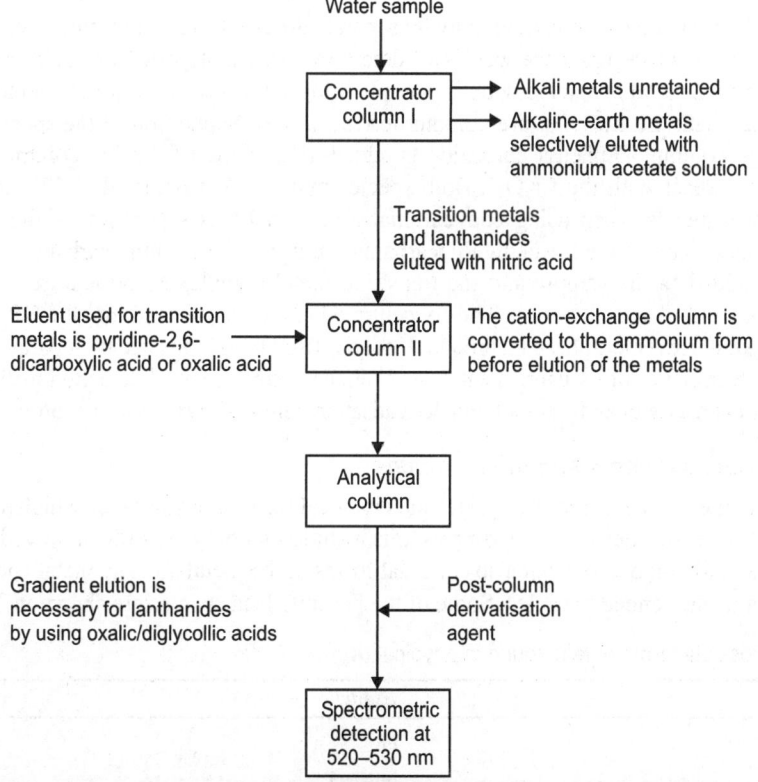

Fig. 16.35. Schematic of a chelation ion chromatography system.

After acid digestion to ensure that the metal is present as the free ion, the sample is added on to the first column, eluted on to the second, and then on to the analytical column. Although we can say that ion chromatography removes the need for a separate preconcentration stage, you can see that preconcentration still occurs within the instrument as an identifiable analytical step. Separation of all common transition metals takes less than 14 minutes, with a typical chromatogram being shown in Fig. 16.36. Furthermore, separation of the lanthanide elements requires less than 12 minutes to carry out.

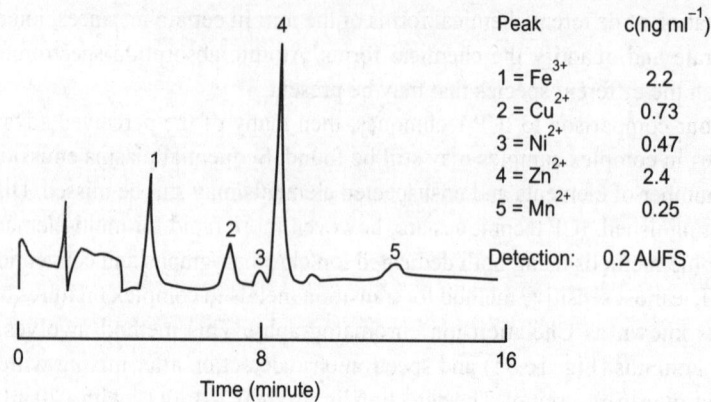

Fig. 16.36. A typical chelation ion chromatogram, obtained for a sample of 32 gram of sea water.

A common application of chromatography for separate different chemical forms of the same element is for Cr^{3+} and $Cr_2O_7^{2-}$. Once again, the species are determined by visible spectrometry after derivatisation. Pyridine dicarboxylic acid is reacted with the sample prior to the analytical separation to form a stable anionic complex with the Cr^{3+} (i.e. pre-column derivatisation). Separation of the species employs an anion-exchange column. Diphenyl carbazide is added after elution from the column (post-column derivatisation) to react with the $Cr_2O_7^{2-}$. Both species may be then detected at 520 nm. A common method for trace metals when using conventional HPLC involves separation of the thiocarbamate complexes. Excess thiocarbamate (diethylthiocarbamate and pyrrolidine dithiocarbamate salts have both been used) is added to the sample and the transition metal complexes formed are concentrated by liquid-liquid or solid-phase extraction. The concentrated extract is injected into the HPLC system and separated by using a reversed-phase technique. Underivatised metal ions can be separated by reversed-phase ion-pair techniques or by using a cation-exchange column. Several detection methods have been used, including conductivity and post-column derivatisation with 4-(2 pyridylazo)resorcinol.

Metal speciation: a comparison of techniques

Speciation is defined as the different physical and chemical forms of a substance which may exist in the environment. When considering water samples, this includes not only the truly dissolved metal ions (as free metal ions or as complexes), but also colloidal forms of the metal and any metal contained within, or adsorbed on to, suspended particles. Some of the possible lead species are shown in Table 16.10.

Table 16.10. Possible forms of lead found in a typical river.

Species	Example	Physical form
Free metal	Pb^{2+}	Solution
Ion-pair	$PbHCO_3^+$	Solution
Complexes with organic pollutants	Pb^{2+}/EDTA	Solution
Complexes with natural acids	Pb^{2+}/fulvic acid	Suspension
Ion absorbed on to colloids	Pb^{2+}/Fe(OH)$_3$	Colloidal
Metal within decomposing organic material	Pb in organic solids	Solid
Ionic solids	Pb^{2+} held within clays	Solid
	$PbCO_3$	Solid

The great diversity of species may be found. These include not only well defined ions and compounds, but also loosely bonded complexes and adsorbed species. The free metal ion often only comprises a small percentage of the total content. The interconversion between species is slow and for many purposes they can be considered as being distinct chemical forms.

For a number of metals, there may also be concern over the different organic derivatives in the environment. An example would be tributyl tin which has been used in anti-fouling paint formulations for ships hulls and its dibutyl and monobutyl degradation products. Other metals with important organic derivatives include lead and mercury.

The transport of each species in the environment will be different and they will also have different toxicological properties. As an example, let us consider the behaviour of metals within a stream and in the associated sediment. Any decaying vegetation will increase the metal loading in the stream water. Since the organic acids produced as part of the decay process will form soluble coordination complexes with the metals. The toxicity of the stream water, however, may not be increased as much as you might expect. As a very general rule, metal complexes have lower toxicities than their corresponding free metal ions. If the metal has more than one stable oxidation state in water, there may also even be differences in behaviour between different oxidation states. For example, chromium in the form of $Cr_2O_7^{2-}$ has a greater toxicity than Cr^{3+}. It would appear that the $Cr_2O_7^{2-}$ ion can enter cells via routes which permit entry of the similarly sized SO_4^{2-} ion. Such a route would not be possible for the positively charged Cr^{3+} ion.

Each of the analytical techniques described in this chapter will 'respond' in a different manner to the species in solution — this is summarised in Table 16.11. If one of these techniques is included as part of a more lengthy analytical procedure, the pre-treatment stages may also alter the species being analysed. Any filtration, for instance, will remove particulate matter.

Table 16.11. Response of various analytical techniques to different metal species.

Technique	Response
Atomic spectrometry	All the metal species in the sample, i.e., the total metal is determined
Visible absorption spectrometry	Free metal ions, plus ions released from complexes by the colour-forming reagent
Anodic stripping voltammetry	Free metal ions analysed, plus any ions released from complexes during analysis. The total is often referred to as the 'total ASV-labile content'
Liquid chromatrography	Non-labile species can sometimes be determined separately
Gas chromatrography	Organic derivatives can be determined separately

Speciation may be investigated by taking advantage of the different responses of the analytical techniques, and the effect of pre-treatment. The most common method is to perform several ASV analyses with different pre-treatment stages. A simple two-step procedure would be to perform the analysis on samples with and without ultraviolet irradiation, thus giving a value for the free metal (or more precisely, the total ASV-labile content) and total metal content, respectively.

The complete chemical characterisation of a sample would be exceedingly complex and time-consuming. When you remember that the total metal concentration may not be greater than a few $\mu g\, l^{-1}$, you will realise that you may also be reaching the detection limits of the available techniques. This aspect of environmental trace metal analysis is currently of great interest and improved techniques are continually being reported in the literature.

SECTION IV

Recycling and Reuse of Waste-water from Various Industries

Chapter 17

Chemical and Allied Industries

INTRODUCTION

In any country, economic development and environment protection should go hand in hand. There cannot be economic development (i.e. the improvement of the standard of living of the common man) without increasing the production of goods and services. Every such activity of production has its own associated problem in the balance of ecology and environment in general. Chemical industries, in particular, are being looked at with awe and suspicion in this respect. It should be the responsibility of the chemical manufacturers to use only such technologies which contribute the least to the upsetting or downgradation of the ecology and environment. Efforts are to be made in improving the technologies to better the performance.

The environmental impact of an existing process can be reduced by using clean technologies, which will, in turn, mean higher costs. Manufacturers wanting to determine the environmental implication of their activities need to look beyond the factory fences and consider the impact of obtaining the raw materials and subsequent use of their products.

PETROLEUM REFINERY AND PETROCHEMICALS

The term petroleum is used to describe the mixture of hydrocarbons in oil, including the gases above the liquid in oil wells and the gases and solids which are dissolved in the liquid. Petroleum was formed in remote periods of geological time from the remains of living organisms. It is, therefore, a fossil fuel.

Waste Generation in Petroleum Refinery

For convenience of discussion, wastes related to the petroleum industry have been grouped according to the following activities:

1. Crude oil producing and handling, including the removal of crude oil from the ground to tankage.
2. Refining.
3. Transportation and marketing, including the movement of crude oil from oil field tankage to the refinery, the movement of refined products to distribution terminals and sales stations and the operation of sales stations.

Condensate waters

Condensate waters, as referred to herein, originate from distillate separators, running tanks and barometric condensers. It has been reported that condensate waters from distillate separators may contain one or more of such compounds as organic and inorganic sulphides, normal or acid sulphites and sulphates, sulphonic acids and their salts, mercaptans, amines, amides, quinalines and pyridines, naphthenic acids, phenols, etc. They may also contain chemicals used for corrosion prevention such as ammonia, caustic soda, calcium hydroxide, etc. Not all these substances will be found in a specific waste-water at the same time. Waste of this type may also contain suspended matter such as coke, iron sulphide, silica, metallic oxides, soaps, emulsions, sulphonic, naphthenic acids and insoluble mercaptides, etc.

It has been possible in some instances to reduce the quantity of condensate wastes through the use of 'dirty water' cooling tower systems for barometric condensers and jet vacuum pumps. In other cases, surface condensers have been substituted for barometric condensers and provided on jet vacuum pumps to control the quantity and strength of this type of waste. Reduction in the quantity of condensate wastes not only lessens the waste disposal problem but also may reduce the loss of valuable water-soluble material.

In case where the water-soluble constituent forms a constant boiling mixture with water, it is sometimes possible to reduce losses by azeotropic distillation. In one case, a phase separation on cooling made it possible to reduce solvent losses from condensate waters by the use of two azeotropic distillations, one for the solvent rich layer and one for the water rich layer.

The installation of a stainless steel mat (called demister) in an asphalt stripper at a propane deasphalting unit has prevented the carry over of asphalt in the tower overhead. As a result, spray condenser water which condenses the stripping steam, can be recycled to a cooling tower. Approximately 500 gpm of asphalt contaminated water has been eliminated from the sewer in this manner.

Most of the hydrogen sulphide in these wastes and a high percentage of the ammonia can be successfully removed by steam stripping.

Acid wastes

Sulphuric acid is used extensively in the petroleum industry both as a treating agent and a catalyst. Other acids and acid salts also used as catalysts include hydrofluoric acid, phosphoric acid, aluminum chloride and zinc chloride. Acid bearing wastes originate from the acid treatment of gasoline, white oils, lubricating oils and waxes; from the handling of acid sludges and the recovery or manufacture of acid; from the alkylation of motor fuel stocks; from the use of acidic catalysts; and from special chemical manufacturing. The wastes occur as rinse waters, scrubber discharges, spent catalyst, sludges, condensate waters and miscellaneous discharges resulting from sampling procedures, leaks and spills and shutdowns. Acid wastes from pump gland leakage and sampling may be collected locally and returned to the process. Substantial reductions in loss of acid to the plant sewers have been attained by this procedure.

Petroleum products are treated with sulphuric acid to improve colour, odour and stability and to reduce gum content and remove sulphur. Acid requirements and concentrations vary widely with the characteristics of the hydrocarbon and the objective in treating. In general, acid requirements vary from 0 to 60 pounds per barrel using acid concentrations of 85 to 99.9 per cent.

The acid sludges produced by the sulphuric acid treatment of oils vary considerably depending on the amount of acid used and the characteristics of the treated stock. The consistency of the sludge may vary from that of a low viscosity fluid to that of a highly viscous material that will solidify on standing. The specific gravity of representative acid sludges may vary from about 1.2 to 1.8 and the sulphuric acid content from about 25 to 80 per cent.

The characteristics and quantity of sludge produced afford utilisation opportunities to most refineries. The methods of utilisation are varied and include — burning immediately as fuel, treatment to produce by-products such as oils, tars, asphalts, resins, fatty acids, sulphonic acids, ammonium sulphate, metallic sulphate, coke and other materials, and regeneration for acid recovery. Although some refineries use acid sludges as fuel alone, after fluxing with fuel oil, after neutralisation with alkaline sludges or alkali or after some combination of these procedures, the probable need of special facilities to minimise corrosion, the high cost of handling and the atmospheric pollution problem due to sulphur dioxide limit the application of this method. The sludge formed from the acid treatment of light oils is sometimes neutralised with alkaline wastes and discharged to the plant sewers. This practice is not recommended if other alternates are available. Utilisation of acid sludges to recover sulphuric acid offers the best means of handling the disposal problem at most refineries. It is possible to manufacture 50 per cent or more of a complete refinery's acid requirements from acid sludge. The processes for the recovery of sulphuric acid are basically either hydrolysis or thermal decomposition processes. Sulphur is recovered by a modification of the thermal decomposition process.

Upon hydrolysis, sulphuric acid sludges resolve into two phases—a weak sulphuric acid layer and a tarry acid oil layer. The latter is usually burned as a fuel. About 40 to 65 per cent of the acid used can be recovered by concentrating the weak acid layer. This material is black in colour but contains a relatively small amount of carbonaceous matter. The addition of water in the hydrolysis operation is controlled so that acid concentration of the water layer is not less than 40 per cent. A number of different type concentrators have been used for this material. Concentration is generally carried out by direct contacting of the acid with hot gases at atmospheric pressure or by indirect heat transfer which utilises steam or a similar heat source in a unit which may be operated under vacuum. In some cases, the weak acid produced by hydrolysis may be used as a source of acid in waste treatment operations such as breaking emulsions, neutralising alkaline wastes, etc. or it may be used to dilute stronger acid to the desired strength instead of using water. In the thermal decomposition process, the sludge is decomposed to sulphur dioxide, water and coke in a continuously operating kiln at temperatures about 500°F. The sulphur dioxide is converted to sulphuric acid by one of the conventional catalytical contact processes or into free elemental sulphur. Experience has been that from 85 to 90 per cent of the sulphur in the sludge can be recovered in this manner. The sludge conversion-contact acid plant method can produce white acids of 98 to 99.9 per cent concentration or higher. Black acids of the same strength can be produced by feeding weak acids obtained by hydrolysis to the towers in the place of clean water. This type of recovery plant has the advantage that clean acid can be produced, if desired and that auxiliary equipment can be installed for the conversion of hydrogen sulphide to sulphuric acid and facilities can be provided for burning sulphur for acid manufacture. Unfortunately, this type of recovery plant is not in general economically applicable to smaller refineries for production under about 25 tons per day.

The sludge conversion-contact acid plant has its own waste disposal problems. The waste-water streams may contain sulphur dioxide, sulphuric acid, coke and oil-in-water emulsions. In one refinery, a sulphur dioxide bearing waste from a contact cooler in the acid recovery plant is used in the treatment of a petroleum sulphonate waste. Sulphur dioxide may also be stripped from waste-water and utilised. The coke is usually recovered by sedimentation, dried and burned as fuel. Emulsions may be broken using alum or other materials and the oil recovered as fuel.

Sulphuric acid and hydrofluoric acid waste may originate from the alkylation of feed stocks for the manufacture of aviation gasoline. The sulphuric acid alkylation process produces about 50 barrels of approximately 90 per cent acid per 1000 barrels of product. In view of threatened shortages of sulphuric acid, many refineries are re-using and regenerating spent acid from their alkylation operations. In some cases the spent acid is returned to the acid supplier. Applications in the refinery for re-use include acid treatment of naphthas, furnace oil, lubricating oil and waxes.

Waste caustics

Waste caustics as referred to herein originate from the caustic washing of light oils to remove mercaptans, hydrogen sulphide and other acidic materials that may occur naturally in crude oil or any of its fractions or may be produced by a variety of processing methods. The quantity of waste caustic produced will vary greatly depending on the characteristics of the crude and the methods of processing. The constituents of waste caustics responsible for their pollution characteristics include mercaptans, thiophenol, thiocresols, phenols, cresols, disulphides, alkylsulphides, the sodium salts of any one of a number of saturated mono-acids, naphthenic acids or sulphonic acids and other materials.

The satisfactory disposal of waste caustic is a major problem at most refineries. The problem is lessened in some cases by process replacement or improvement to achieve more efficient utilisation of available treating capacity. Regenerative systems have been used to eliminate the discharge of spent caustic from mercaptan and hydrogen sulphide removal operations. Likewise, the 'down grading' use of spent solutions from one process as a treating agent in another has proven advantageous. For example, discarded solutions, after having been made essentially lead free, may be used as prewashes for other treating operations such as scrubbing of hydrogen sulphide from distillates and gases.

Spent caustics from the treating of motor stock from catalytical cracking processes, commonly called carbolates, have been sold as such for the recovery of phenolic constituents. In some cases, this waste is first acidified to release 'acid oils' which are separated from the waste layer and sold for their phenolic content. Acid oils from carbolate solutions are approximately 90 per cent phenolic and amount to about 50 per cent of the volume of the carbolate. The low acid oil content and the heterogenous nature of the constituents make spent caustic from treating a poor source of recoverable by-products. Studies at one refinery indicate that the acid oil content of this material is 3 to 10 per cent by volume and is 60 per cent phenolic and 40 per cent naphthenic. Acid oils may be released or 'sprung' from spent caustic solutions by neutralisation with mineral acids and acid gases. Frequently waste acidic material such as spent acids and acid sludges from other refinery operations may be used for this purpose. The sulphides are converted to hydrogen sulphide which may be burned or recovered and the acid oils are burned as fuel. The water layer from this reaction is usually drawn to the sewer, preferably at a low rate since it still contains a relatively high concentration of oxygen consuming materials. Flue gases have been used in some cases for the treatment of waste caustics. The caustic solutions are neutralised by the flue gases at temperatures of 160° to 180°F to about pH 5 to release the acid oils, unoxidised hydrogen sulphide and mercaptans and certain weak organic acids. During this process the easily oxidised mercaptides and sulphides are partially oxidised. The acid oils separated from the solution collected are used for fuel or a source of special chemicals. The hydrogen sulphide and odourous materials released to the gas stream must be conducted to burning facilities for atmospheric pollution control. The remaining water solution contains a mixture of carbonates, bicarbonates, sulphates, sulphites, thiosulphates, phenols, etc. which may require further oxidative treatment prior to discharge to natural waters.

Alkaline water

Alkaline water, as differentiated from alkaline condensate water and waste caustics, may originate from the washings of neutralised acid treated oils, the washings of caustic treated oils, the dehydration of treated light oils, the aqueous tank bottoms of stored caustic treated and washed gasolines, vessel and tower washings at times of shutdowns and miscellaneous sources.

The alkaline water referred to above, originating from continuous treating processes, contribute to the general pollution load but do not create a major problem. The intermittent discharge of aqueous tank bottoms and the washing of towers and vessels at times of shutdown can cause pollution peaks requiring special attention.

At one refinery where biological treatment is to be provided for the plant effluent, intermittent flows of alkaline water, such as washings from batch treaters and shutdown drainage is discharged to holding tanks from which they can be drawn to the sewer at a controlled rate.

Cooling water

Cooling water makes up nearly the entire volume of waste-water from petroleum refining operations. Since these wastes may become oil-contaminated due to equipment failure. Thus it is necessary to provide oil separation facilities as insurance against accidental pollution. Consequently, uncontaminated cooling water is generally turned into a common oil carrying sewerage system. However, when the standards of waste-water quality necessitate costly 'effluent polishing', the separate collection and disposal of cooling waters which are subject to infrequent pollution has proved economical at some refineries. Likewise, the size of effluent treatment facilities has been substantially reduced by decreasing the quantity of waste-water through the use of recirculating cooling systems. At one refinery requiring effluent treatment, the quantity of waste-water has been reduced from 24 mgd to 8 mgd by the installation of cooling towers.

The use of dirty water cooling tower systems and the elimination of barometric condensers and jet vacuum pumps, emulsions also serve to keep sizes of treatment facilities at a minimum. Where water is scarce, refineries have used sewage treatment plant effluents as a source of water supply and in other cases have used their own treated effluents.

Waste-water from Refinery

Waste-waters in a refinery originate from the following processes:
1. Storage and transportation.
2. Fractionation by pressure and vacuum distillation.
3. Reforming.
4. Cracking: catalytic and thermal.
5. Hydro-desulphurisation.
6. Solvent treating processes for lube oil manufacture.
7. Utilities function—steam boiler, cooling water, etc.

The water use and waste-water generation in the oil refineries in the country are noted to be greatly influenced by the type of cooling system, once through or recirculation adopted. It may be noted that in refineries having once through cooling system due to an enormous amount of water used, because of easy availability of sea water, a huge amount of valuable hydrocarbon is lost along with the waste-waters. These refineries have only primary oil separation facilities and taking undue advantage of the large quantity of waste-water, discharge the further recoverable oil along with the effluent. This will result in low concentration of oil in the effluent but the quantum oil of discharged is indeed high. Quantities of pollutants from any process section is a function of type or source of process feed stock and the type of processes involved, e.g. crude oils from different sources, will probably produce different quantities of pollutants, thus cracking processes will produce more phenols. Pollutants in the waste-water depends on the amounts of steam, process water and cooling water used in a refinery. Aqueous wastes will be more if a refinery uses more stripping steam and once-through cooling water than the refinery which uses air cooling.

The approximate quantum of water used by different unit operation in refinery is:

1. 30 l/kg of oil for processing and cooling in once-through basis.
2. 1 l/kg of oil by recycling and to increase the efficiency of the processes.
3. 0.3 l/kg of oil, when air cooling is used in place of water cooling.

Correlation of plant effluent analyses should be on the basis of kg of pollutant per unit quantity of plant feedstock, e.g. a particular cracking plant charged with a particular feed stock will produce a fixed amount of water soluble phenols per thousand barrels of feed. The amount of stripping steam and the amount of cooling water discharge will determine the phenol concentration in the cracking plant aqueous effluent.

Solvents and effluents

Refineries generally use solvents like toluene, methyl ethyl ketone, benzene for removing waxes and furfural for the removal of aromatics during the manufacture of lubricating oil. During the operation with these solvents starting from receiving in the refineries to final operation there is every possibility for the loss of these through spillage, solvent extraction and maintaining the proper ratio. The solvents lost by these processes ultimately accumulates in the effluent water system through surface drains with washout waters. Removal of these solvents contamination from the effluent water is a typical specialised job.

Solvent contaminated water streams from the units in the refineries gets diluted in the total influent water which undergoes the various treatment. The physical treatment removes free oils and suspended solids and chemical treatment removes emulsified oils, sulphides, etc. while the biological treatment removes the dissolved degradable hydrocarbons. So there is every possibility that the solvents remain unremoved in the final effluent water unless special treatment procedures are followed. On the contrary the chemicals like furfural are harmful for the biological treatment through trickling filter and activated sludge because the micro-organisms responsible for biodegradation gets damaged through furfural.

Waste-water Treatment

Removal of pollutants from oil refinery can be classified in three types of treatments:

Physical treatment

Waste-water may contain coarse, suspended and floating solids, grease, etc. These need to be removed before waste-water is subjected either to chemical or biological treatment. Common unit operations of physical treatments are bar screens, grinders, grit chambers, grease traps, flocculation, sedimentation, flotation, chemical precipitation, sludge pumping. This treatment basically removes inert material which may hinder the subsequent treatments.

Chemical methods

After removal of grit and floating matter, suspended and dissolved organic matter are removed. Important unit operations and processes involved in the chemical treatment of waste-water are: Chemical coagulation, flocculation and sedimentation.

Biological methods

Biological treatment unit is primary meant for removal of pollutants like phenol, residual sulphide and BOD and also the non-recoverable oil present in the secondary effluent. Bacterial seeding and fertilisation of the oily waste with appropriate bacterial species will accelerate biological degradation provided that dissolved oxygen and sufficient time are available.

Petrochemical and Allied Products

It is difficult to make any general statements on the petrochemical industry owing to the fact that product mix, raw materials and production technology vary from one installation to another. The inputs to the petrochemical industry are almost all petroleum-based products. The primary inputs are ethylene, propylene, butadiene, benzene, toluene, xylenes, ammonia and methanol. These are produced from the following raw materials: petroleum distillates, propane, ethane, natural gas, coal, shale oil and biomass.

Water use

The variation in inputs, processes and products corresponds to a wide range of water uses both in quantity and quality and even more so in a wide range of waste streams. The variety means that such indicators as specific water use depend strongly on the various processes and water-treatment technologies, so that an overall analysis is difficult. However, a general perspective of the alternative technologies and the resulting water use and waste streams will be provided. Possible recycling and reuse of waste-water and effluent treatment practices will also be considered. The three basic uses for water in the petrochemical industry are: cooling water, which accounts for 80 per cent of intake water; steam generation, which uses 5 per cent of intake water; and process water, which uses 15 per cent of intake water.

Cooling water

A large amount of heat is generated in the petrochemical industry. This heat is removed from the processes through contact or non-contact cooling with water. The exact amount of water needed is a function of the particular process employed. In closed (non-contact) systems, the quality of the water is an important factor for the maintenance of the cooling system. In contact cooling the water quality is very important to prevent contamination of the processes outputs.

Steam generation

There are three major uses of steam generated within a petrochemical plant:
1. Non-contact process heating.
2. Power for a variety of purposes.
3. As a diluent, stripping medium or source of vacuum using steam jet injectors.

Process water

Although process water accounts for only 15 per cent of total water use, it is vital to the production of petrochemical products. As more of the cooling systems become closed cycles, the percentage of process water used will increase. Petrochemical processes have been categorised by the USEPA into four groups on the basis of water use within the process.

Sources and characteristics of waste-water

In the section on water use it was seen that water in the petrochemical industry is used in a variety of ways; the resulting raw waste-water streams are just as varied. The major sources of waste-water in the petrochemical industry are:
1. Raw materials themselves.
2. Products remaining in the solution after separation.

3. By-products produced during reactions.
4. Spills, washdowns and vessel clean-outs.
5. Cooling tower and boiler blow-down, steam condensate and water-treatment processes.
6. Storm water run-off.
7. Sanitary systems.

Wastes from these sources have a variety of characteristics that include high COD concentrations, high total dissolved solids, high COD/BOD ratios and contamination by heavy metals or compounds inhibiting biological treatment. Except for spills, storm-water run-off and sanitary systems the sources listed are found in the four process groups listed in the section on process water.

Waste-water treatment practices

The three main components of petrochemical wastes are biological substances, other chemical substances that are only slightly biodegradable or not at all, and suspended matter.

Water pollution control in petrochemical industries

Huge volume of water is required for various purposes in manufacture of petrochemicals. Water is used for cooling, process operations, fire fighting, drinking, housekeeping, steam raising, etc. Whole water after use reappears as effluents which requires treatment and disposal.

Treatment of phenolic waste-water

Increasingly stringent regulation for the quality of drinking water has resulted in an enhanced interest in the decontamination of water, waste-waters and polluted industrial effluents for phenolics. Phenolic compounds contain hydroxyl groups attached to aromatic nucleus. They may contain other functional groups such as alkyl, methoxy, halo, carbonyl groups. At very low concentrations, they impart taste and odours in drinking water and may taint the flavour of fishes grown in contaminated waters. Aquatic life is adversely affected by phenolics at ppm levels. These effects have led to the specifications of acceptable limits for phenolics by Bureau of Indian Standards as following: Public water supply, 0.005 mg/l, industrial effluents to be discharged to surface waters, 1.0 mg/l; and industrial/trade effluents to be discharged to public sewers, 5.0 mg/l.

Sources of phenolic waste-water

Generally the level of phenolics in domestic waste-waters are low. Many industries produce phenolic waste-waters. These include petrochemical industries petroleum refineries, coal cooking and coal gasification, resin manufacturing, dye synthesis, wood preserving plants, pulp and paper mills and aircraft manufacture.

Treatment methodologies

Commonly used methods for the treatment of phenol bearing waste-waters include solvent extraction, physical adsorption, chemical oxidation and aerobic biological processes. Alternate biological treatment methods such as anaerobic and anoxic processes and enzymatic approach for the removal of phenolic are receiving wide attention only recently. These methods are discussed briefly with their merits and demerits.

Solvent extraction

Solvent extraction procedure is practiced for the recovery of phenols from concentrated industrial waste-waters. The criteria for choosing the appropriate solvent for extraction are: (i) low water solubility,

(ii) no emulsion formation with water, and (iii) must be easily regenerable. Among many solvents which include octane, mixtures of benzene with butylacetate and isopropyl, isobutyl acetate and isopropyl ether two are extensively used for dephenolising waste-waters. Solvent extraction process can typically reduce phenol concentrations from as high as 17,000 to 920 mg/l. Dephenolised waste-water will have to be treated by other suitable methods for further removal of phenols. Adsorption method can be used for the removal of phenols from contaminated drinking water sources as well as for waste-waters that contain moderately low phenol concentrations. It is a separation process in which phenols are transferred from aqueous phase to the surface of adsorbent. Extensively used adsorbent is activated carbon. Commercially available activated carbons are derived from natural materials such as coconut shell, wood charcoal, which are carbonised at controlled conditions so as to have precise surface properties and high surface area. These materials are costly and need to be regenerated and reactivated for further repeated use. Generally used-procedure for the regeneration of activated carbon is burning at 600°C under controlled conditions of moisture and oxygen, which is highly energy intensive. About 5–10 per cent material is lost during regeneration. Other source materials, such as straw, used rubber tyres, fertiliser waste slurry have been tried in an attempt to produce low cost, disposable activated carbons. These materials have exhibited adsorptive properties similar to commercially available activated carbons. Activated carbon derived from fertiliser waste slurry has been successfully used for the removal of phenolics from synthetic as well as actual waste-waters from oil refineries.

Chemical oxidation

Many oxidants like ozone, H_2O_2 and permanganate have been used for the chemical oxidation of phenols. It has been reported that ozonisation can result in structural modifications of pollutants, which make them amenable for biodegradation. Requirement of around 7.5 kg of ozone per kg of phenol present is reported for the oil refineries waste-waters, if effluent quality of 0.2 mg/l is to be achieved. Reaction between phenol and hydrogen peroxide is slow under non-catalysed conditions. However in the presence of Fe(II) ions H_2O_2 oxidises phenol to hydroquinone and catechol, which are further oxidised to quinones, carboxylic acid and finally to CO_2. Fenton's Reagent [H_2O_2 = Fe (II) Catalyst, pH 2–4] is reported to be the cheapest oxidation system as compared to ozone, chlorine dioxide and potassium permanganate. If the waste-water contains phosphates, catalytic action is reduced and the oxidation of phenols cannot be achieved. Phenolic effluents from paint and, pharmaceutical manufacturing have been treated with Fenton's Reagent.

Enzymatic treatment

Enzymatic approach for the treatment of phenolic waste-waters is now attracting wide attention. Use of enzymes which include peroxidases, tyrosinase and laccases are now being explored to convert soluble phenolics to insoluble polyphenolic precipitates, which can then be removed by filtration.

Biological methods

Biological treatment processes are preferred over abiotic methods as they lead to the complete mineralisation of phenol to CO_2 and other inorganic components. Physical and enzymatic methods convert phenol from one form/phase to another, which still requires further treatment. Biological methods can be classified as: (i) aerobic, (ii) anoxic, and (iii) anaerobic depending on the environment present in the treatment system. Generally biological processes are used after dephenolising the concentrated phenolic waste-waters by solvent extraction. However, there are also a few reports on the application of anaerobic and anoxic biological treatment processes for treating concentrated waste-water.

LEATHER AND TANNERY

Tanning is the chemical process that converts animal hides and skins into leather. The term hide is used for the skin of large animals (e.g. cows or horses), while skin is used for that of small animals (e.g. sheep). The hide is composed of three layers: epidermis, dermis and subcutaneous.

Liquid Wastes

In the development of the tanning industry, water plays a vital role as the industry consumes large quantities of water. Approximately 30–40 litres of water is used for processing one kg of raw hide/skin into finishing leather. Most of the Indian tanneries which are located near the river banks or natural water bodies draw surface water. Groundwater from their own open wells/tube wells existing within their premises is also used by some tanneries. Most of the traditional tanneries do not have overhead water tanks for proper distribution system. Water is being pumped directly to the process and in a few tanneries, it is stored in open cement lined pits and ground level tanks. In general, the quantity of water usage and nature of waste-water discharge varies from process to process and tannery to tannery and from time to time. Most of the discharges are intermittent. The average water usage and waste-water discharge per kg of hide/skin for different process are as follows: (i) raw to E.I. 25–30 l/kg of raw weight, (ii) raw to wet blue 25–30 l/kg of raw weight, (iii) raw to finish 30–40 l/kg of raw weight, (iv) E.I. to finish 40–50 l/kg of E.I. weight, and (v) wet blue to finish 20–25 l/kg of wet blue weight.

Most of the tanneries neither have proper drainage system for collection of the waste-water nor any effluent treatment system. The waste-water is discharged from various sectional operations intermittently and it takes its own course to the nearby low-lying area neighbouring land, pond, street, roadside, etc.

Characteristics of waste-water

Characteristics of the effluents vary from tannery to tannery and in any one tannery with respect to time. The waste-water from beamhouse process, viz. soaking, liming, deliming, etc. are highly alkaline, containing decomposing organic matter, hair, lime, sulphide and organic nitrogen with high BOD and COD. The waste-water from tanyard process, viz. pickling, chrome tanning are acidic and coloured. Vegetable tan waste-water contain high organic matter. The chrome tanning wastes contain high amounts of chromium mostly in the trivalent form. The details of the tanning operations, water and other chemicals used, general constituents in the waste-water are furnished in Table 17.1.

Table 17.1. Details of tanning process, water usage, chemicals used and general constituents of waste-water.

Important operations in tanning process	Mode of operation waste-water discharge in M³/ton of skin/hide processed	Approx. qty. of water used	Important chemicals	General constituents of waste-water
Soaking	Pits/paddles	9.0–12.0	Wetting, emulsifying agents and bactericidal agent	Olive green in colour, obnoxious smell, contains soluble proteins, suspended matter and high amount of chlorides
Liming	Pits/paddles	2.5–4.0	Lime and sodium sulphide	Highly alkaline, contains high amount of sulphides, ammoniacal nitrogen, suspended solids, hair, pulp and dissolved solids

(Contd ...)

Important operations in tanning process	Mode of operation waste-water discharge in M³/ton of skin/hide processed	Approx. qty. of water used	Important chemicals	General constituents of waste-water
Deliming	Paddles/pits/ drums	2.5–4.0	Ammonium salts, enzymatic bates	Alkaline, contains high amount of organic matter and ammoniacal nitrogen
Vegetable tanning	Pits/drums	1.0–2.0	Vegetable tanning material	Highly coloured, acidic and has a characteristic offensive odour
Pickling and chrome tanning	Drums	2.0–3.0	Common salt, acid, basic chrome salt	Coloured, acidic, contain high amount of trivalent chromium, TDS and chlorides
Dyeing and fat liquoring	Drums	1.0–1.5	Dyes and fatty oils	Coloured, acidic, dyes and oil emulsions
Composite waste-water including washing (raw to finish process)	–	30.0–40.0	–	Alkaline, coloured contains soluble proteins, chromium, high TDS, chlorides, sulphides, suspended solids, etc.

Characteristics of the composite waste-water

The characteristics of the composite waste-water is governed by the following factors:
1. Intermittent discharge of waste-water from different sectional operations.
2. Wide variation in the volume and quality of waste-water from section to section.
3. Partial operations in one tannery and balance operation in another tannery.

It would be difficult to arrive at a realistic characteristic range of the composite effluents to be discharged by various tanning units. However, from the analysis of the waste-water samples collected from various tanneries located in Kolkata and Tamil Nadu region, the general characteristics range of the composite waste-water from raw to finishing process is given in Table 17.2. The wide variation of BOD, COD, chromium, sulphide and other parametres exhibited in raw to finish composite waste-water is due to the variation in the process, changes in the type, quantity and quality of chemicals used for the process, fluctuations in the volume of water used for process and washings.

Impact of Liquid Wastes on Environment

Tannery effluents contain vegetable tannins and non-tannins which exert oxygen demand. They also contain high amounts of protein, especially when a hair pulping unhairing system is used. These proteins are biologically degradable and exert high BOD. The discharge of untreated waste-waters in water courses may affect the physical, chemical and biological characteristics of the water and deplete dissolved oxygen from the water bodies. The high oxygen demand of tannery wastes is due to proteins, fatty matter and tannins. High pH, excessive alkalinity, suspended matter, sulphides are injurious to fish and other aquatic life in streams. Sulphides present in tannery waste-waters can cause unpleasant odour problems, react with iron and other metals causing black precipitate, render the water unfit for industrial uses and affect fish and other aquatic life. The sulphide toxicity to fish increases as the pH value is lowered. Nitrogen and phosphorus from tannery effluents encourage uncontrolled growth of algae and other aquatic plants in water bodies. High amounts of chloride present in tannery waste-waters can make the receiving water less suitable for drinking, industrial and agricultural purposes.

Table 17.2. Characteristics of composite waste-water from 'raw hide to chrome tanning finished leather'.

Parametre	Concentration range
pH	7.50–8.50
Alkalinity as $CaCO_3$	1100–2000
BOD 5 days @ 20°C	1200–2500
COD	3000–6000
Chlorides as Cl^-	4500–6500
Total solids	17000–25000
Dissolved solids	14000–20500
Suspended solids	3000–4500
Sulphides as S^{2-}	20–40
Total chromium	80–250

Note: All values except pH are expressed in mg/l.

Suspended solids both inorganic and organic present in tannery waste-waters may settle in a stream and affect fisheries by covering the bottom of the stream thereby destroying bottom fauna necessary for fish as food or reduce the spawning ground of fisheries. In general the toxic effects of chromium salts particularly hexavalent chromium salts towards aquatic life varies with species, temperature, valence of chromium and the complex synergistic and antagonistic effects due to other factors such as hardness. The effluent from vegetable tanning is coloured and contain some amounts of non-biodegradable matter. When these waste-waters are discharged into streams it is reported that the colour attributed persisted for a long time. The discharge of untreated tannery wastes into water course may also increase the turbidity of water thereby reducing light penetration and impairing photosynthetic activity of aquatic plants. Groundwater has been found to be affected where waste-water from tanneries are ponded or lagooned spread out on land or discharged into dry river beds. The groundwater is reported to be rendered unfit for drinking and irrigation where the tanneries are concentrated together. When the tannery waste gains access to cultivable lands or when the lands are irrigated with such waste, fertility of the soil is reported to have been affected. It may change the characteristics of the soil and interfere with the intake of water by plants. When tannery waste-waters are applied to land, the soil productivity decreases.

Control of Water Pollution

Tannery effluents if disposed off without any treatment either on land or in inland surface waters, may create severe problems leading to damage of the environment. Tannin, trivalent chromium, proteinous matter, sulphides and high BOD/COD of the waste-water call for proper treatment. Tannery wastes in general, can be treated to get the desired end results in the following main stages:

1. In-plant measures.
2. Primary treatment.
3. Chemical treatment.
4. Secondary treatment

In-plant control measures

In-plant measures include reduction of water consumption in the tannery, process modifications to reduce pollutional load of the waste being generated, segregation of soak liquor and lastly, recovery of the by-products and reuse of various process liquors.

Reduction of water usage in a tannery

To reduce the volume of the effluent, the water usage in tanneries can be considerably reduced by:
1. Better housekeeping.
2. Alteration of processes and low float systems to use less water.
3. Separation of cleaner fractions of the waste for direct reuse without treatment.
4. Recycle after complete or partial treatment.

Process modifications to reduce pollutional load

The quantities of pollutants which are inevitable in tannery waste-water are as follows:

Hide salt	150 g/kg
Hair protein	40 g/kg
Hide protein	25 g/kg
Hide fat and carbohydrate	15 g/kg
Dirt and manure	5 g/kg
Organic solids from cleaning and premises	3 g/kg

Recovery and utilisation of by-products

The various wastes generated by the tannery have great potential for their reuse. Tannery unhairing effluent and acid are the two waste liquids individually difficult to dispose off or to utilise, but if combined together, useful products can be recovered from this combination. These precipitated products have high contents of essential amino acids such as cysteine, lysine, valine, leucine, etc.

Primary treatment

Most of the tanneries do not have any treatment facilities. In tanneries where treatment of effluents is carried out, it is limited to the mixing of the various effluents, followed by sedimentation. Even those operations are not carried out satisfactorily. Screening for removal of coarser impurities, hair and fleshing followed by settling for atleast 4 hours in a continuous flow settling tank form the essential primary treatment of tannery waste-waters.

Secondary biological treatment

Low cost technology and conventional treatment systems (both aerobic and anaerobic or in combination) can be employed for the secondary biological treatment of equalised and settled tannery wastes. Tannery effluent can also be satisfactorily treated in admixture with sewage in a sewage treatment plant provided the proportion of tannery effluents is not high.

Anaerobic systems

Some of the tanneries are using anaerobic lagoons with success. Experiments on anaerobic lagoon with the settled waste-water gave percentage reduction varied from 42.4 per cent at a BOD loading of 0.41 kg/m^3/day to 85.5 per cent at a BOD loading of 0.14 kg/m^3/day. It was found that ten days detention time would be sufficient to bring BOD reduction of about 85 per cent. If the final effluent has to be discharged on land, it would be sufficient to treat the settled waste in anaerobic lagoons for 10 days.

Anaerobic lagoon treatment could reduce BOD by 88.5 per cent but colour removal could not be obtained. Anaerobic treatment of vegetable tannery waste-water is practically feasible at shorter time using fixed film reactors. The success in treatment is strongly dependent on the high concentration of active biomass retained within the reactor and on provisions of sufficient contact time between active biomass and waste-water. The increased sludge retention enhances the rate of conversion of organic matter to methane and carbon dioxide. The reactors, having different methods of sludge retention investigated for the treatment of vegetable tannery waste-water corroborate the above facts.

Combined process

For complete treatment to meet the standards of discharge into river, anaerobic treatment followed by aerobic process is necessary. The treatment of raw tannery waste-water having BOD concentrations of 2000–4000 mg/l is not economically feasible to treat by aerobic method due to oxygen transfer limitations. The effluent produced after anaerobic treatment of tannery waste-water has COD and BOD concentrations permissible for discharge into public sewer. The anaerobic pre-treatment with faster conversion rate at shorter time with added advantage of biogas generation will be a better approach to opt before aerobic polishing treatment. In view of the promising results obtained in the above mentioned experiments more systematic and detailed evaluation of anaerobic pre-treatment of tannery waste-water followed by aerobic treatment is necessitated to develop necessary design criteria for full scale treatment facility. To overcome the difficulties of fixed-film reactors, the feasibility of anaerobic-aerobic concept of tannery waste treatment was further studied by the use of an anaerobic contact reactor in series with an aerobic activated sludge. Anaerobic-aerobic reactor system exhibited the ability to provide 98 per cent removal of suspended solids, 86 per cent removal of COD and 70 per cent removal of sulphides from raw combined tannery beamhouse waste-water without pre-treatment.

Treatment of Waste-water

Primary treatment

The primary treatment units principally comprise coarse screens, two numbers of equalisation-cum-settling tanks (used alternately) and sludge drying beds. The settling tanks, are of about 1–2 days capacity each, thereby acting also as equalisation tanks. Alternatively separate equalisation tank and settling tank/clarifier have been provided by a few tanneries. Depending on the quality of composite effluent, addition of neutralising chemicals and coagulants like lime, alum, ferric chloride, etc. would be required for effective precipitation of chromium and removal of suspended solids in the sedimentation process.

The sludge from the settling tank/clarifier is removed and dried on sludge drying beds made up of filtering media like gravel, sand with supporting masonry structure. For operational reasons, sludge drying beds are divided into four or more compartments. The sludge drying period varies from 4 to 8 days depending upon the type of sludge, atmospheric conditions, etc. The dried sludge from the sludge drying beds can be used as manure or for landfill if it is from vegetable tannery waste. In case of chrome tannery wastes, the dried sludge should be buried or disposed off suitably as per the directions of regulatory and local bodies.

Secondary biological treatment

The pretreated effluent needs suitable secondary biological treatment to meet the pollution control standards. The biological treatment units generally adopted by the Indian tanneries are anaerobic lagoon, aerated lagoon, extended aeration systems like oxidation ditch, etc.

Anaerobic lagoon is a simple anaerobic treatment unit adopted by some tanneries. The system is suitable for tanneries located outside town limits and having sufficient land. The depth of the lagoon may be 3–5 metres and the detention time may be 10–20 days depending upon the pollutional load and atmospheric conditions. The lagoon is provided impervious lining to prevent any subsurface infilteration of waste-water. The anaerobic lagoon is an open type digester with no provision for mixing and gas collection. No power is required for this system and its performance has proved to be efficient. pH control, sulphate reduction and atmospheric temperature are important factors in the performance of the system. pH in the range of 7.0 to 8.5, sulphates — amounts less than 500 mg/l and atmospheric temperature of 25°–40°C are favourable for anaerobic lagoon treatment.

Anaerobic contact filter and upflow anaerobic sludge blanket (UASB) are other types of anaerobic treatment system for tannery wastes under pilot scale study. These are closed type units made up of RCC or steel. These units occupy less land area since the detention time is about 1–2 days in case of contact filter and about 8 to 10 hours in case of UASB. These systems are found to be efficient for treating tannery effluent combined with domestic sewage. Though the capital cost would be high as compared to anaerobic lagoon, this system can be adopted for partial treatment of tannery wastes when adequate admixture of sewage is possible.

About 60 per cent reduction in BOD is reported to be achieved by this system. Aerated lagoon is a shallow water tight pond of about 2–3 metres depth with a detention time of about 4–6 days. Fixed or floating type surface aerators are provided to transfer oxygen from atmospheric air to the effluent for biological treatment using micro-organisms under aerobic conditions. Many tanneries have provided aerated lagoons as a second stage biological treatment unit and the system is found suitable for treating low organic loads.

The typical treatment system adopted by the tanneries processing raw hides and skins into finished leather is shown in Fig. 17.1. The typical treatment system adopted by the tanneries processing semi-finished leather into finished leather is shown in Fig. 17.2. The primary treatment units are mostly similar in all the treatment combinations and the secondary biological treatment combinations vary depending upon the mode of disposal and other local environmental conditions.

Waste Control Measures

Good housekeeping and water conservation

As in all waste control programs, good housekeeping is the first step to prevent wastage of water and materials in tannery. Economical use and reuse of water are necessary to reduce the volume of the waste-water. The water usage in tanneries can be considerably reduced by:

1. Better housekeeping.
2. Alteration of processes and low float systems to use less water.
3. Segregation of cleaner fractions of the waste-water for direct reuse without treatment.
4. Recycle after complete or partial treatment.

The adoption of batch washing as an alternative to continuous rinsing using the lattice door can reduce water consumption. It has been observed that paddles may have some advantages over drums for certain types of production from the point of view of water usage. Direct reuse is possible in a tannery since some wash waters are relatively clean, such as from washing after bating, pickling, neutralising and dyeing which can be used for less important tasks such as washing after soaking and liming and floor washing.

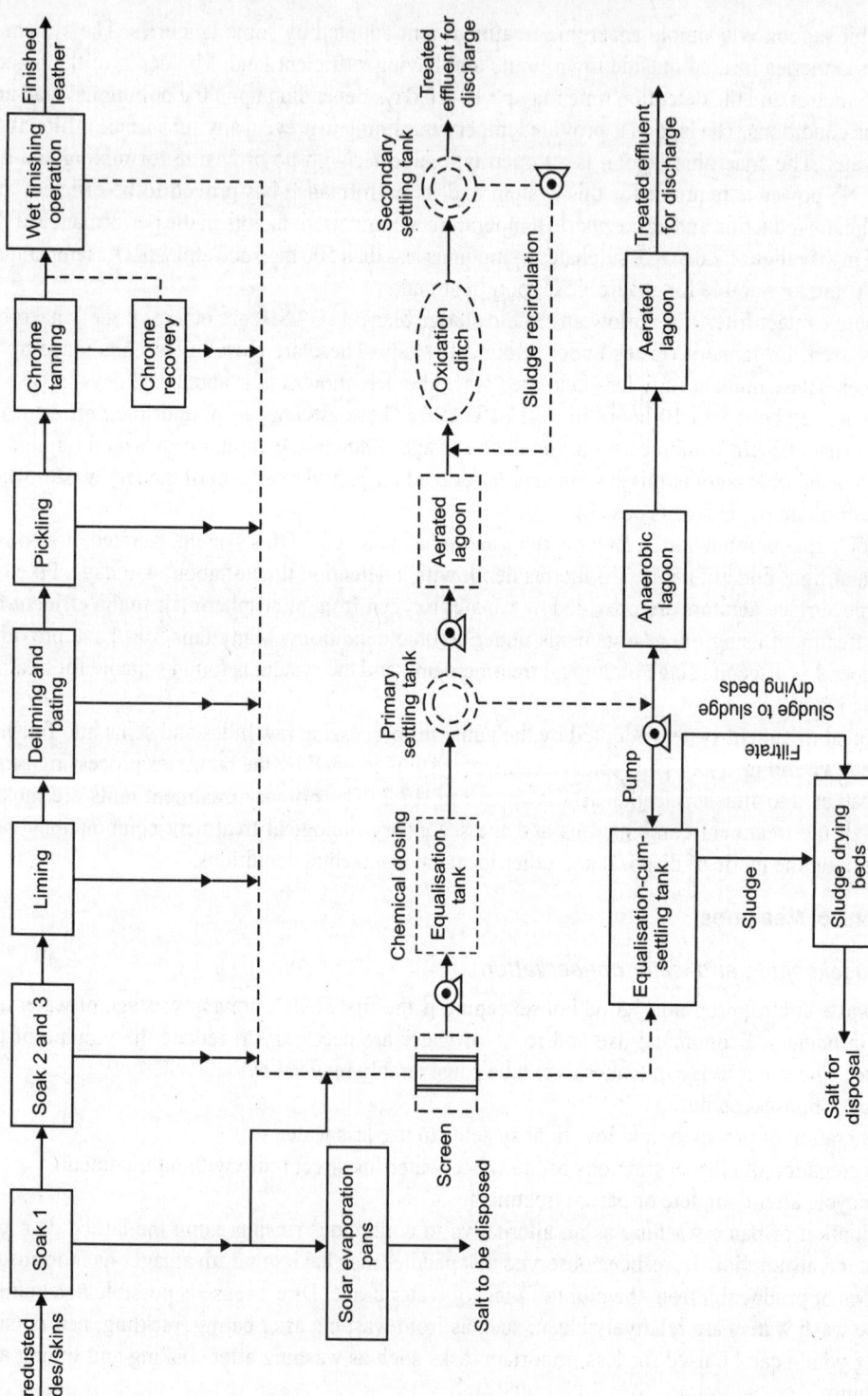

Fig. 17.1. Typical treatment system for raw to finishing tannery waste-water.

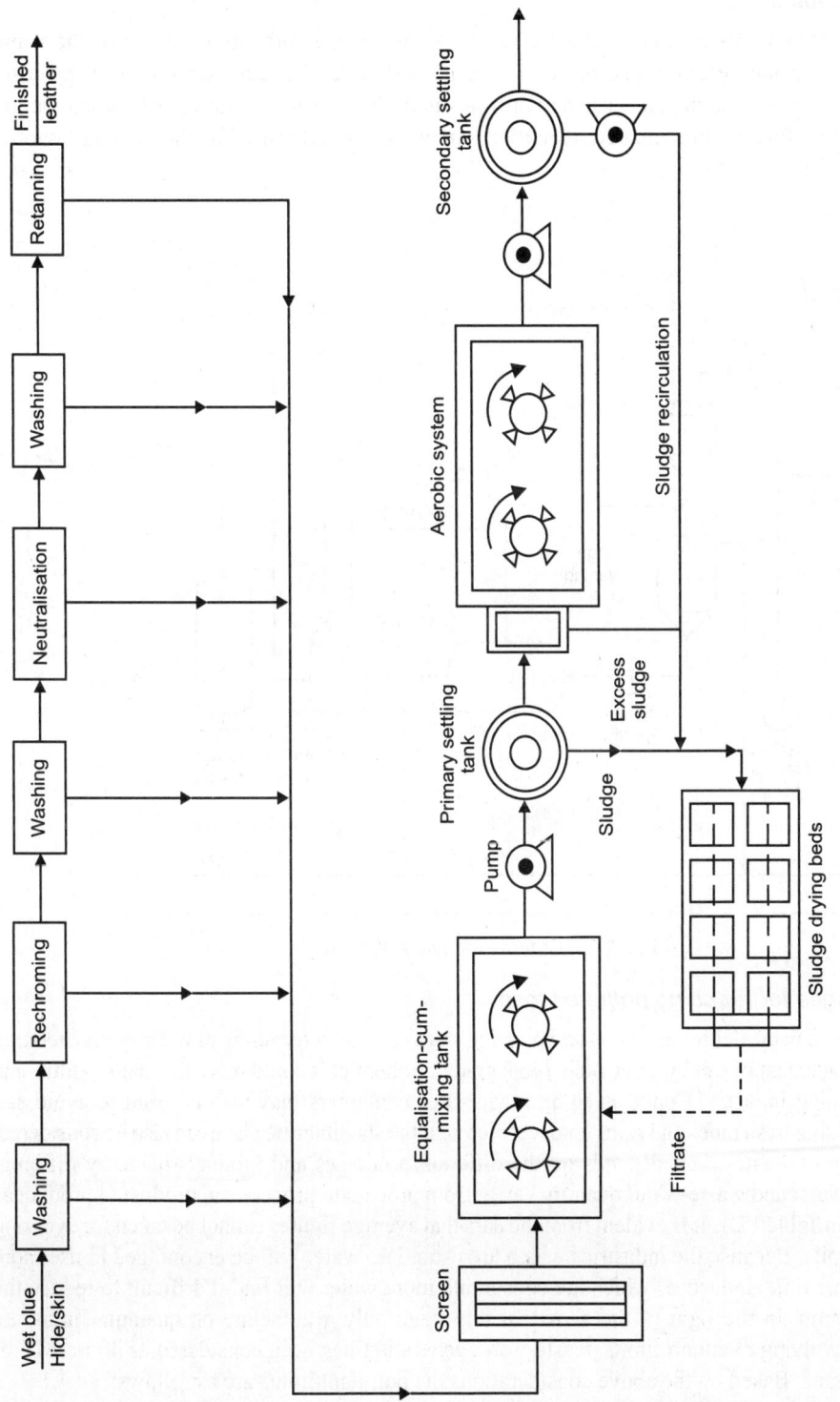

Fig. 17.2. Typical treatment system for wet blue to finishing tannery waste-water.

Recovery and reuse

Segregation of chrome waste stream, chrome recovery and reuse is proved to be one of the viable procedures. The chrome recovery and reuse in medium and large size tanneries should be practised because it makes good economical sense and prevents pollution due to chromium. The spent chrome liquor is captured, filtered, precipitated, reactified, strengthened and reused as the tanning liquor as shown in Fig. 17.3.

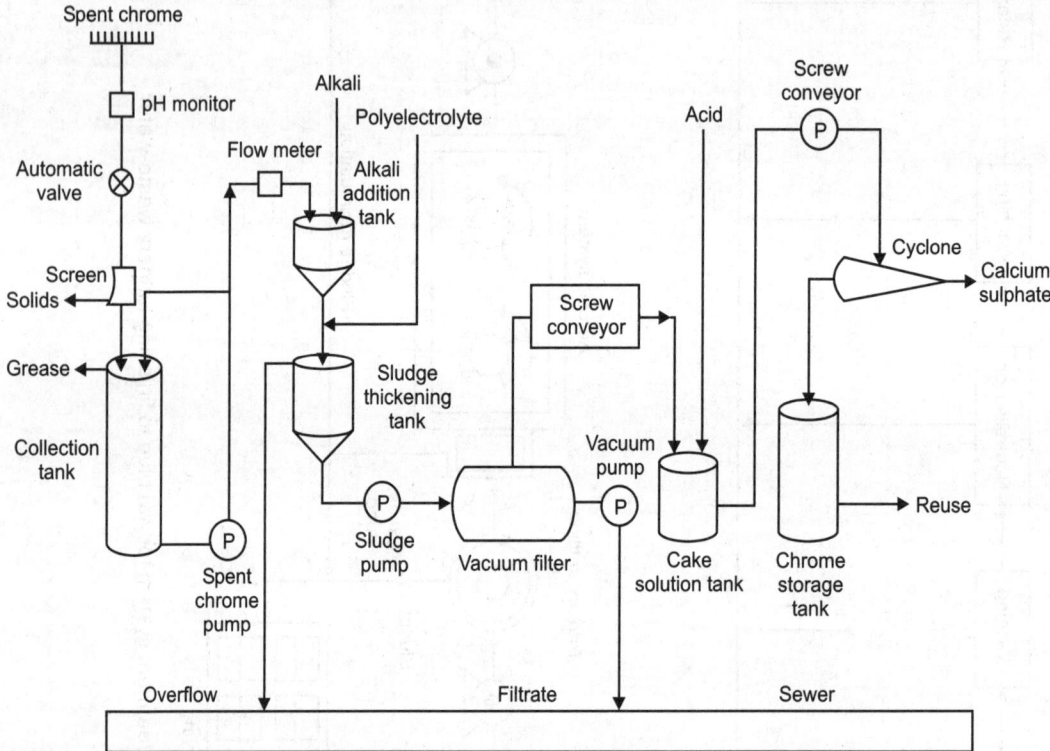

Fig. 17.3. Chrome recovery system.

Process changes for reducing pollution load

Water can be saved in the deliming, bating and pickling stages either by recycling of water or regeneration of liquors. Replacement of salt by some other biodegradable chemicals could solve the major pollutional problem of tanning industry. Conveyance by refrigerator containers may also be tried to avoid salt problem. Processing fresh hides and skins upto wet blue near the slaughtering place can also be considered.

　　The quantity of waste-water depends on the different processes and products made by different tanneries. The water and waste-water quantity varies from process to process and product to product as already shown in Table 17.1. It is evident from the data that average figures cannot be taken for evolving the quantum limits. Because the industries which are using less water will be encouraged to use more water at the same time, industries which are consuming more water will find it difficult to reduce the water consumption. In the light of the above, at this stage only a guideline on quantum limits are prescribed. In evolving quantum limits, waste-water generation has been considered as 40 m³/MT of raw hide processed. Based on the above considerations the pollutant limits are as follows:

Parametre	Limit (kg/MT of hide/skin processed)
BOD	4.0
Suspended solids	4.0
Total chromium	0.08
Sulphides	0.08
Oil and grease	0.40

The experience gained during the implementation of quantum limits, could be reviewed to firmly establish the quantum limits.

Waste-water in Leather Industry

As already discussed leather making involves the stabilisation of a putrescible collagenous matrix skin, against degradation by micro-organisms and thermomechanical stress. This involves a series of operations as demonstrated with the help of a flow chart Fig. 17.4. Traditionally the leather processing has been grouped as pretanning, tanning and post-tanning—aiming at preparation to tanning, the stabilisation against putrefaction and giving an aesthetic appeal.

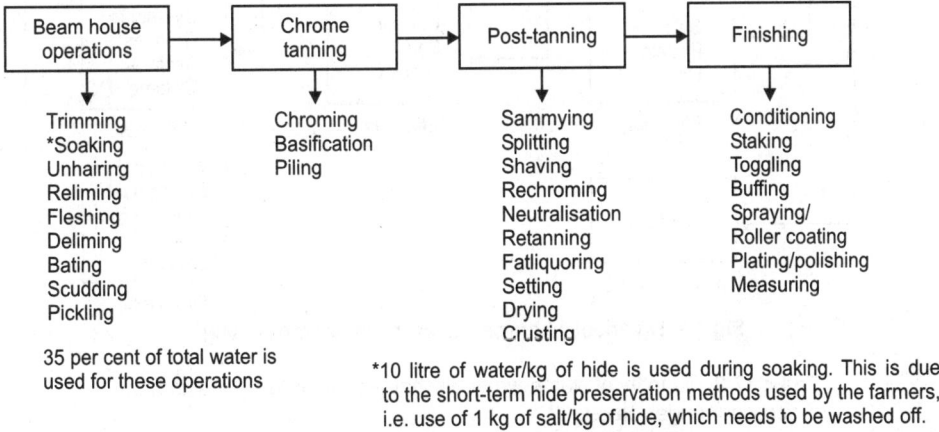

Fig. 17.4. Process sequence in leather processing.

The various chemical inputs into leather processing are demonstrated in the Fig. 17.5. Water forms the main medium of transport for the chemicals. Much of the leather processing steps depend on a large use of water. This explains in part the development of the industry along the river basins. The leather industry employs about 35–55 litres of water per kilogram of hide processed. With the present annual processing capacity of 0.9 billion kilograms of hides and skins in India, it could be estimated that nearly 30–50 billion litres of liquid effluent is generated annually. This gives rise to two major problems for the leather industry, viz. the availability for good quality water and the need for treatment of such large quantities of effluents. This leads to major investments in effluent treatment plants.

The pretanning operations consume nearly 15–20 litres of water per kilogram of hide processed, while the tanning operation consumes 1.5–2.5 litres of water and post-tanning 6–8 litres per kilogram of hide processed. The operation-wise break-up of water usage is given in Table 17.3. Each of these operations give rise to characteristic pollutant loads. The various pollutant loads from each operation is presented in Table 17.4. Treatment of the large volumes of effluents and their subsequent discharge into sewers, rivers, etc. is only a short-term solution, which adds to compounded problems of investment and recurring expenditure. Other technological options to contain the usage of water, their reuse are to be considered for evolving long-term solutions for the leather industry.

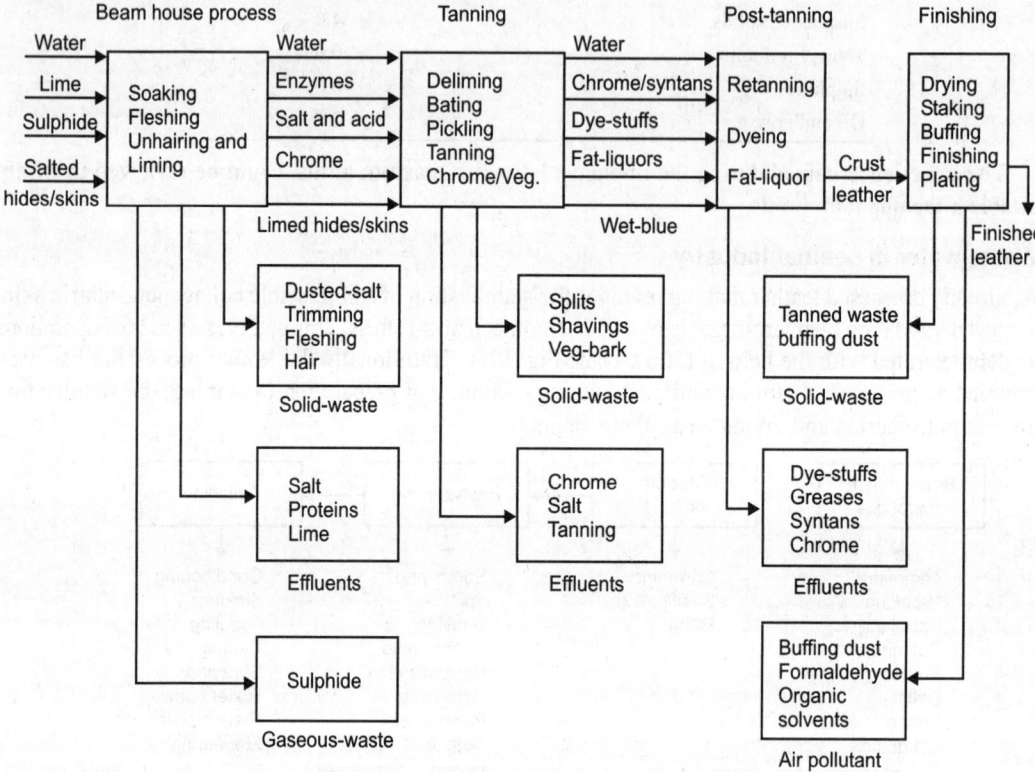

Fig. 17.5. Various chemicals used in leather processing.

Table 17.3. Quantity of waste-water discharged for 1000 kg of skin/hide processed/range values.

Operation	Quantity cu.m/day
Soaking I	4.0–5.0
Soaking II and III	5.0–6.0
Liming	4.0–6.0
Deliming	2.5–3.0
Pickling	2.0–3.0
Chrome tanning	1.5–2.5
Vegetable tanning	1.5–2.0
Myrobing	0.5–1.0
Stripping	3.0–3.5
Bleaching	2.5–3.5
Rechroming	1.5–2.2
Neutralisation	2.0–3.0
Dyeing and Fatliquoring	2.5–3.5
Washings	10.0–13.0

Table 17.4. Characteristics of tannery waste-waters.

Parameter	Soaking	Liming	Deliming	Pickling	Chrome tanning fatliquoring	Dyeing (including washing)	Composite
Vol. of effluent in m³/T of hides/skins	6–9	3–4	1–2	0.5–1.0	1–2	1–2	30–40
pH	7.5–8.0	10.0–12.8	7.0–9.0	2.0–3.0	2.5–3.0	3.5–4.5	7.0–9.0
BOD	1100–2500	5000–10000	1000–3000	400–700	350–800	1000–2000	1000–3000
COD	3000–6000	10000–25000	2500–7000	1000–3000	1000–2500	2500–7000	2500–8000
Total solids	35000–55000	30000–50000	4000–10000	35000–70000	30000–60000	4000–10000	15000–25000
Dissolved solids	32000–48000	24000–30000	2500–6000	34000–67000	29000–57500	3400–9000	13000–21000
Suspended solids	3000–7000	6000–20000	1500–4000	1000–3000	1000–2500	600–1000	2000–4000
Chlorides as Cl	15000–30000	4000–8000	1000–2000	20000–30000	15000–25000	500–1000	6000–9500
Total chromium as Cr	–	–	–	–	2000–5000	40–100	100–250

Note: All values except pH are in mg/l

PULP AND PAPER

Pulp and paper mill industry is one of the major industries which contributes a lot towards the pollution of our water environment. The pulp and paper mill wastes characteristically contain very high COD and colour; the presence of lignin in the waste, which is derived from the raw cellulosic materials and is not easily biodegradable, makes the COD/BOD ratio of the waste very high. It may be noted that the pollution potential of paper mills are negligible compared to that of the pulp mills. As such, it is the pulp-making process that is mainly responsible for the pollution problems associated with integrated pulp and paper mills.

Water Usage

Government policy on water is greatly emphasising for decreased use and increased discharge quality. High water consumption in the mills is due to old/obsolete machinery leading to higher water usage. The lack of chemical recovery at smaller mills on one side increases the chemical costs but on the other side it increases the water use, poor discharge quality and large effluent discharge. During the paper making process, the water is added or removed at a number of places.

The steps involve multistage dilution, squeezing, drainage and evaporation during paper making, thus water reuse is essential for water conservation. The flow sheet of the processes involving straw, rag and waste paper is given in Fig. 17.6.

Treatment for the Waste-water

For mills using agricultural residues as raw materials

The unit process involved in the treatment of waste-water from agricultural residue-based paper mills are:
1. Equalisation of flow from pulp wash section.
2. Primary clarification for combined waste-water.
3. Secondary biological treatment.
4. Sludge drying beds or lagoons for primary sludge depending on the availability of land.

Equalisation for pulp wash waste-water

Pulp washing section accounts for about 20–25 per cent of total waste-water and contributes around 70–80 per cent of pollution load from small paper mills using agricultural residues as raw materials. These waste-waters are discharged intermittently since, in most of the mills, washing of pulp is done in batches using pouchers. Normally two pouchers are employed and operated in series and thus generate two washes.

The period of washing is more or less same in both but the quantity of water used varies appreciably and thus the first wash water is more concentrated than the second wash water. The flow variation has been observed to be fairly wide since the minimum and maximum values, respectively are 0.15 and 2.3 times the average flow.

Discharge of these washes to the main waste-water stream (on an intermittent basis), as is being practised will alter the composition of the combined waste-water appreciably. This necessitates provision of equalisation to the pulp washes and discharge at a constant rate into the main sewer. Two alternatives can be considered for this purpose.

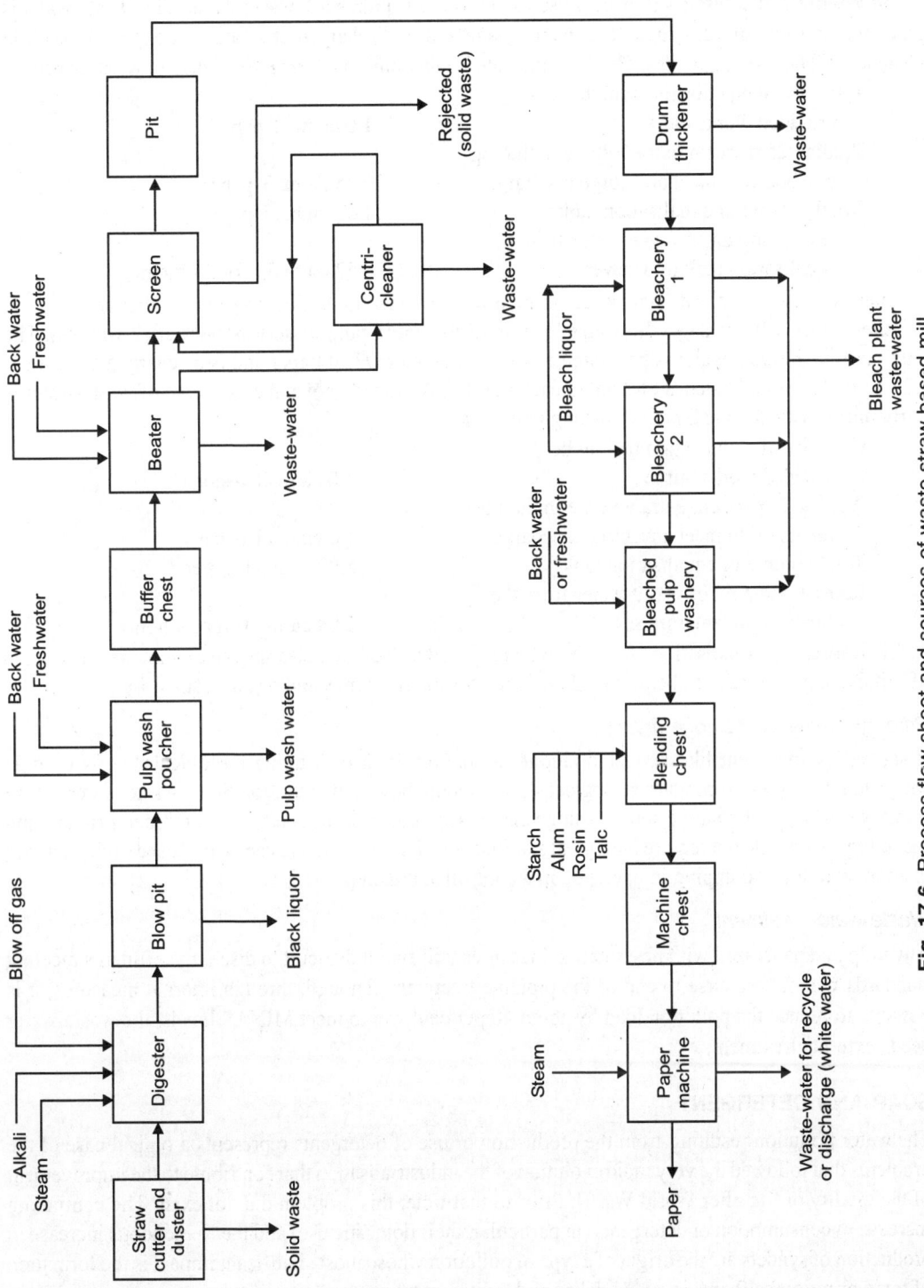

Fig. 17.6. Process flow sheet and sources of waste straw based mill.

Alternative 1: It envisages flow equalisation for the first pulp wash waste-water. This waste-water is generated at a rate of 12.5 cu.m./T of paper made and is discharged at a rate of 1.56 cu.m./batch in 8 batches of 2 hours each per day. The discharge rate is fairly uniform during the 2 hours of washing period.

Capacity of equalisation tank to be provided works out to	1.0 cu.m./T paper
Adding 25 per cent extra volume including free board to meet any surge discharge	0.25 cu.m./T paper
Total volume of equalisation tank	1.25 cu.m./T paper
Rate of pumping of waste-water from the equalisation tank into sewer	0.52 cu m./hr for 24 hours

Therefore, two equalisation tanks are to be provided to facilitate cleaning and maintenance.

Alternative 2: It envisages flow equalisation of the entire pulp wash waste-water (2 washes). The total pulp wash waste-water is generated at a rate of 50 cu.m./T of paper and is discharged at a rate of 6.25 cu.m./batch in 8 batches of 2 hours each per day. As stated above, the flow rate of waste-water is fairly uniform during the 2 hours washing period.

Capacity of equalisation tank to be provided works out to	4.0 cu.m./T paper
Adding 25 per cent extra volume including free board to meet any surge discharge	1.0 cu.m./T paper
Total volume of equalisation tank	5.00 cu.m./T paper
Rate of pumping of waste-water from the equalisation tank into sewer	2.08 cu.m./hr for 24 hours

Therefore, two equalisation tanks are to be provided to facilitate cleaning and maintenance. Pumps of suitable capacity have to be provided one each for the two tanks and one as a stand-by.

Primary clarifier and sludge drying

If secondary treatment like activated sludge, oxidation ditch or rotating biological disc is used as suggested, the excess secondary biological sludge should be added to the combined waste-water before primary settling such that the settled sludge can be filtered on drying bed. The combined primary and secondary sludges do not require biological stabilisation. The dried sludge can be disposed off by burning in an incinerator or dumping in open pits in a controlled manner.

Waste-water treatment

Any pulp and paper mill with best internal measures will find it difficult to discharge effluents meeting standards without recourse to end of the pipeline treatment. Though, through internal measures, it is possible to reduce the pollution load by about 40 per cent, yet to meet MINAS levels, the waste-water needs external treatment.

SOAP AND DETERGENT

The water pollution resulting from the production or use of detergents represents a typical case of the problems that followed the very rapid evolution of the industrialisation that contribute to the improvement of the quality of life after World War II. Prior to that time, this problem did not exist. The continuing increase in consumption of detergents (in particular, their domestic use) and the tremendous increase in production of syndets are the origin of a type of pollution whose most significant impact is the formation of toxic or nuisance foams in rivers, lakes and treatment plants.

Soaps and detergents are formulated products designed to meet various cost and performance standards. The formulated products contain many components, such as surfactants to tie up unwanted materials (commercial detergents usually contain only 10–30 per cent surfactants), builders or polyphosphate salts to improve surfactant processes and remove calcium and magnesium ions and bleaches to increase reflectance of visible light. They also contain various additives designed to remove stains (enzymes), prevent soil redeposition, regulate foam, reduce washing machine corrosion, brighten colours, give an agreeable odour, prevent caking and help processing of the formulated detergent.

Sources of Detergents in Water and Waste-water

The detergent concentrations that actually find their way into waste-waters and surface water bodies have quite diverse origins: (i) soaps and detergents, as well as their component compounds, are introduced into waste-waters and water bodies at the point of their manufacture, at storage facilities and distribution warehouses and at points of accidental spills on their routes of transportation, (ii) the additional industrial origin of detergent pollution notably results from the use of surfactants in various industries, e.g. textiles, cosmetics, leather tanning and products, paper, metals, dyes and paints, production of domestic soaps and detergents and from the use of detergents in commercial/industrial laundries and dry cleaners, (iii) the contribution from agricultural activities is due to the surface runoff transporting of surfactants that are included in the formulation of insecticides and fungicides, and (iv) the origin with the most rapid growth since the 1950s comprises the waste-waters from urban areas and it is due to the increased domestic usage of detergents and, equally important, their use in cleaning public spaces.

Impacts on Waste-water Treatment Processes

The effect of surfactants on waste-water oils and greases depends on the nature of the latter, as well as on the structure of the lipophilic group of the detergent that assists solubilisation. As is the case, emulsification could be more or less complete. This results in a more or less significant impact on the efficiency of physical treatment designed for their removal. On the other hand, the emulsifying surfactants play a role in protecting the oil and grease molecules from attacking bacteria in a biological unit process.

In water treatment plants, the coagulation/flocculation process was found early to be affected by the presence of surfactants in the raw water supply. In general, the anionic detergents stabilise colloidal particle suspensions or turbidity solids that, most times, are negatively charged. Langelier reported problems with water clarification due to syndets, although according to Nichols and Koepp and Todd concentrations of surfactants on the order of 4 to 5 ppm interfered with flocculation. The floc, instead of settling to the bottom, floats to the surface of sedimentation tanks. Other studies, such as those conducted by Smith and Cohen, indicated that this interference could be not so much due to the surfactants themselves, but to the additives included in their formulation, i.e. phosphate complexes. Such interference was observed both for alum and ferric sulphate coagulant, but the use of certain organic polymer flocculants was shown to overcome this problem.

Surfactants are only partially biodegraded in a sewage treatment plant, so that a considerable proportion may be discharged into surface water bodies with the final effluent. The shorter the overall detention time of the treatment plant, the higher the surfactant concentration in the discharged effluent.

Waste-water Characteristics

Waste-waters from the manufacturing, processing and formulation of organic chemicals such as soaps and detergents cannot be exactly characterised. The waste-water streams are usually expected to contain trace or larger concentrations of all raw materials used in the plant, all intermediate compounds produced during manufacture, all final products, coproducts and by-products and the auxiliary or processing chemicals employed. It is desirable, from the viewpoint of economics, that these substances not be lost, but some losses and spills appear unavoidable and some intentional dumping does take place during housecleaning and vessel emptying and preparation operations.

According to a study by Zimmerman presenting estimates of industrial waste-water generation as well as related pollution parameter concentrations, the waste-water volume discharged from soap and detergent manufacturing facilities per unit of production ranges from 0.3 to 2.8 gal/lb. (2.5–23.4 l/kg) of product. The reported ranges of concentration (mg/l) for BOD, suspended solids, COD and grease were 500 to 1200, 400 to 2100, 400 to 1800, and about 300, respectively.

These data were based on a study of the literature and the field experience of governmental and private organisations. The values represent plant operating experience for several plants consisting of 24 hours composite samples taken at frequent intervals. The ranges for flow and other parametres generally represent variations in the level of plant technology or variations in flow and quality parametres from different subprocesses. In particular, the more advanced and modern the level of production technology, the smaller the volume of waste-water discharged per unit of product. The large variability (up to one order of magnitude) in the ranges is generally due to the heterogeneity of products and processes in the soap and detergent industry.

Waste-water Control and Treatment

In-plant control and recycle

Significant in-plant control of both waste quantity and quality is possible particularly in the soap manufacturing subcategories where maximum flows may be 100 times the minimum. Considerably less in-plant water conservation and recycle are possible in the detergent industry, where flows per unit of product are smaller.

The largest in-plant modification that can be made is the changing or replacement of the barometric condensers. The waste-water quantity discharged from these processes can be significantly reduced by recycling the barometric cooling water through fat skimmers, from which valuable fats and oils can be recovered and then through the cooling towers.

The only waste with this type of cooling would be the continuous small blowdown from the skimmer. Replacement with surface condensers has been used in several plants to reduce both the waste flow and quantity of organics wasted.

Significant reduction of water usage is possible in the manufacture of liquid detergents by the installation of water recycle piping and tankage and by the use of air rather than water to blowdown filling lines. In the production of bar soaps the volume of discharge and the level of contamination can be reduced materially by installation of an atmospheric flash evaporator ahead of the vacuum drier. Finally, pollutant carry-over from distillation columns such as those used in glycerine concentration or fatty acid separation can be reduced by the use of two additional special trays.

In a recent document presenting techniques adopted by the French for pollution prevention, a new process of detergent manufacturing effluent recycle is described in which washout effluents from reaction and/or mixing vessels and washwater leaks from the paste preparation and pulverisation pump operations are collected and recycled for use in the paste preparation process. The claim was that pollution generation at such a plant is significantly reduced and although the savings on water and raw materials are small, the capital and operating costs are less than those for building a waste-water treatment facility. Besselievre reported in a review of water reuse and recycling by the industry that soap and detergent manufacturing facilities showed an average ratio of reused and recycled water to total waste-water effluent of about 2:1. That is, over two-thirds of the generated waste-water stream in an average plant was being reused and recycled. Of this volume, about 66 per cent was used as cooling water and the remaining 34 per cent for the process or other purposes.

Waste-water treatment methods

The soap and detergent manufacturing industry makes routine use of various physico-chemical and biological pre-treatment methods to control the quality of its discharges. A survey of these treatment processes is presented in Table 17.5 which also shows the usual removal efficiencies of each unit process on the various pollutants of concern. According to Nemerow the origin of major wastes is in washing and purifying soaps and detergents and the resulting major pollutants are high BOD and saponified soaps (oily and greasy, alkali and high-temperature wastes), which are removed primarily through air-flotation and skimming and precipitation with the use of $CaCl_2$ as a coagulant.

Table 17.5. Treatment methods in the soap and detergent industry.

Pollutant and method	Efficiency (percentage of pollutant removed)
Oil and grease	
API-type separation	Up to 90% of free oils and greases.
	Variable on emulsified oil.
Carbon adsorption	Up to 95% of both free and emulsified oils.
Flotation	Without the addition of solid phase, alum or iron, 70-80% of both
Mixed-media filtration	Up to 95% of free oils. Efficiency in removing emulsified oils unknown.
Coagulation/sedimentation with iron, alum or solid phase (bentonite, etc.)	Up to 95% of free oil. Up to 90% of emulsified oil.
Suspended solids	
Mixed-media filtration	70–80%
Coagulation/sedimentation	50–80%
BOD and COD	
Bioconversions (with final clarifier)	60–95% or more
Carbon adsorption	Up to 90%
Residual suspended solids	
Sand or mixed-media filtration	50–95%
Dissolved solids	
Ion exchange or reverse osmosis	Up to 90%

Figure 17.7 presents a composite flow diagram describing a complete treatment train of the unit processes that may be used in a large soap and detergent manufacturing plant to treat its wastes. As a minimum requirement, flow equalisation to smooth out peak discharges should be utilised even at a production facility that has a small-volume batch operation. Larger plants with integrated product lines may require additional treatment of their waste-waters for both suspended solids and organic materials reduction.

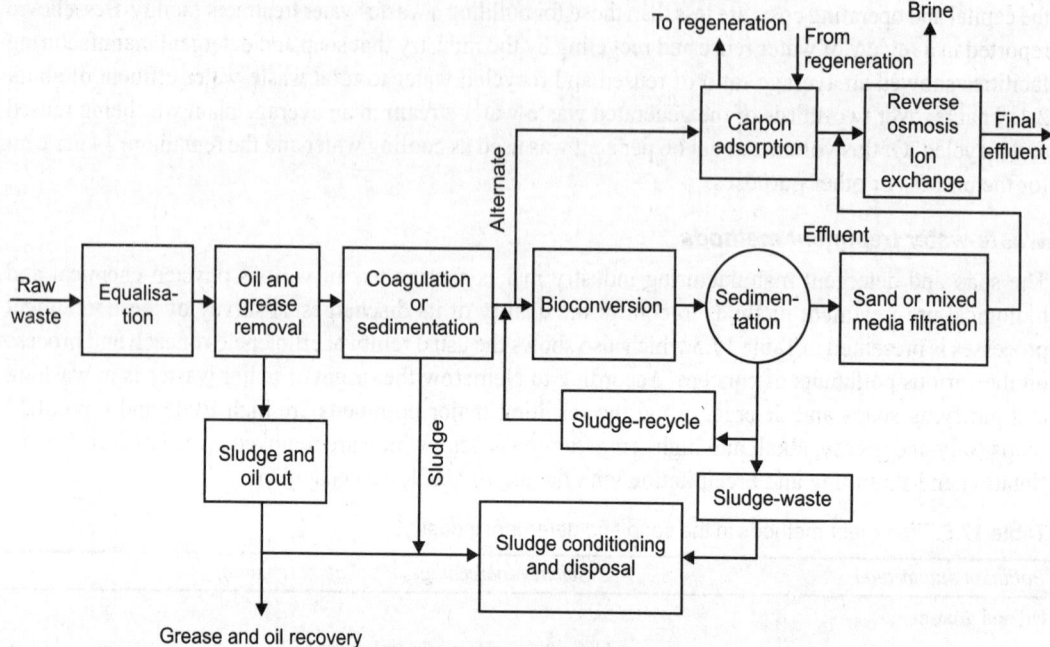

Fig. 17.7. Composite flowsheet of waste treatment in soap and detergent industry.

Coagulation and sedimentation are used by the industry for removing the greater portion of the large solid particles in its waste. On the other hand, sand or mixed-bed filters used after biological treatment can be utilised to eliminate fine particles. One of the biological treatment processes or, alternatively, granular or powdered activated carbon is the usual method employed for the removal of particulate or soluble organics from the waste streams. Finally, as a tertiary step for removing particular ionised pollutants or total dissolved solids (TDS), a few manufacturing facilities have employed either ion exchange or the reverse osmosis process.

Flotation or foam fractionation

One of the principal applications of vacuum and pressure (air) flotation is in commercial installations with colloidal wastes from soap and detergent factories. Waste-waters from soap production are collected in traps on skimming tanks, with subsequent recovery floating of fatty acids. Foam separation or fractionation can be used to an extra advantage. Not only do surfactants congregate at the air/liquid interfaces, but other colloidal materials and ionised compounds that form a complex with the surfactants tend to also be concentrated by this method. An incidental, but often important, advantage of air flotation processes is the aerobic condition developed, which tends to stabilise the sludge and skimmings so that they are less likely to turn septic.

However, disposal means for the foamate can be a serious problem in the use of this procedure. It has been reported that foam separation has been able to remove 70 to 80 per cent of synthetic detergents, at a wide range of costs. Gibbs reported the successful use of fine bubble flotation and 40 minutes detention in treating soap manufacture wastes, where the skimmed sludge was periodically returned to the soap factory for reprocessing. According to Wang the dissolved air flotation process is also feasible for the removal of detergents and soaps from water.

Activated carbon adsorption

Colloidal and soluble organic materials can be removed from solution through adsorption onto granular or powdered activated carbon, such as the particularly troublesome hard surfactants. Refractory substances resistant to biodegradation, such as ABS, are difficult or impossible to remove by conventional biological treatment and so they are frequently removed by activated carbon adsorption. The activated carbon application is made either in mixed-batch contact tanks with subsequent settling or filtration or in flow-through GAC columns or contact beds. Obviously, because it is an expensive process, adsorption is being used as a polishing step of pretreated waste effluents. Nevertheless, according to Kucharski much better results of surfactant removal have been achieved with adsorption than coagulation/settling.

Coagulation/flocculation/settling

The coagulation/flocculation process was found to be affected by the presence of surfactants in the raw water or waste-water. Such interference was observed for both alum and ferric sulphate coagulant, but the use of certain organic polymer flocculants was shown to overcome this problem. However, chemical coagulation and flocculation for settling may not prove to be very efficient for such waste-waters. Wastes containing emulsified oils can be clarified by coagulation, if the emulsion is broken through the addition of salts such as $CaCl_2$ the coagulant of choice for soap and detergent manufacture waste-waters. Also lime or other calcium chemicals have been used in the treatment of such wastes whose soapy constituents are precipitated as insoluble calcium soaps of fairly satisfactory flocculating ('hardness' scales) and settling properties. Treatment with each can be used to remove practically all grease and suspended solids and a major part of the suspended BOD. Using carbon dioxide (carbonation) as an auxiliary precipitant reduces the amount of calcium chloride required and improves treatment efficiency. The sludge from $CaCl_2$ treatment can be removed either by sedimentation or by air or vacuum flotation.

Ion exchange and exclusion

The ion-exchange process has been used effectively in the field of waste disposal. The use of continuous ion-exchange and resin regeneration systems has further improved the economic feasibility of the applications over the fixed-bed systems. One of the reported special applications of the ion-exchange resins has been the removal of ABS by the use porous anion exchanger that is a strong base and depends on a chloride cycle. This resin system is regenerated by removing a great part of the ABS absorbed on the resin beads with the help of a mixture of HCl and acetone. Other organic pollutants can also be removed by ion-exchange resins and the main problem is whether the organic material can be eluted from the resin using normal regeneration or it is economically advisable to simply discard the used resin. Wang and Wood successfully used the ion-exchange process for the removal of cationic surfactant from water.

The separation of ionic from non-ionic substances can be effected by the use of ion exclusion. Ion-exchange can be used to purify glycerine for the final product of chemically pure glycerine and reduce losses to waste, but the concentration of dissolved ionisable solids or salts (ash) largely impacts on the overall operating costs. Economically, when the crude or sweet water contains under 1.5 per cent ash, straight ion exchange using a cation-anion mixed bed can be used, whereas for higher percentages of dissolved solids, it is economically feasible to follow the ion-exchange with an ion-exclusion system. For instance, waste streams containing 0.2 to 0.5 per cent ash and 3 to 5 per cent glycerine may be economically treated by straight ion-exchange, while waste streams containing 5 to 10 per cent ash and 3 to 5 per cent glycerine have to be treated by the combined ion-exchange and ion-exclusion processes.

Biological treatment

Regarding biological destruction, as mentioned previously, surfactants are known to cause a great deal of trouble due to foaming and toxicity in municipal treatment plants. The behaviour of these substances depends on their type, i.e. anionic and non-ionic detergents increase the amount of activated sludge, whereas cationic detergents reduce it and also the various compounds decompose to a different degree. The activated sludge process is feasible for the treatment of soap and detergent industry wastes but, in general, not as satisfactory as trickling filters. The turbulence in the aeration tank induces frothing to occur and also the presence of soaps and detergents reduces the absorption efficiency from air bubbles to liquid aeration by increasing the resistance of the liquid film.

On the other hand, detergent production waste-waters have been treated with appreciable success on fixed-film process units such as trickling filters. Also, processes such as lagoons, oxidation or stabilisation ponds and aerated lagoons have all been used successfully in treating soap and detergent manufacturing waste-waters. Finally, Vath demonstrated that both linear anionic and nonionic ethoxylated surfactants underwent degradation, as shown by a loss of surfactant properties, under anaerobic treatment.

PESTICIDE

Pesticide industry has been identified as one of the highly polluting industries needing pollution control on top priority. Pesticides include insecticides, fungicides, herbicides, rodenticides, nematocides, etc. and are manufactured from chemicals. Recycling and reuse of waste-water within the industry will help minimise fresh water requirements and the simultaneous reduction in waste-water volume for final treatment and discharge thereby reducing related costs. The environment is being polluted by industrial wastes, animal and plant wastes, automobile exhausts and agricultural chemicals, which include pesticides and fertilisers. The major source of environmental contamination by pesticides is the deposits resulting from the application of these chemicals to control agricultural pests and pest causing public health problems. What concerns the society the most is the biological effects of these pesticide derivatives, which at various stages of environmental alteration can come in contact with many biological systems.

Fate of pesticides in the environment. Pesticides are generally applied to the soil, plant, water bodies and human settlements by man-mounted equipment, by tractors or by aircrafts, either as liquids, dusts or granules. If these pesticides hit only the target species, there would be least pollution. Often, as little as 25–50 per cent of the pesticide formulations land in the crop area when applied by aircraft and the remaining drift in the atmosphere to contaminate very remote areas of the abiotic environment. The migration of pesticides and their fate, and the routes for the loss of pesticides in the environment are given in Figs 17.8 and 17.9.

Effects on water. The factors which influence the persistence of pesticides in water are solubility, bottom mud, organic matter, temperature and pH.

The pesticides that are of importance in connection with water quality include chlorinated hydrocarbons and their derivatives, persistent herbicides, soil insecticides, pesticides that are easily leached out from soil, and pesticides that are systematically added to water supplies for disease vector control for other purpose. Of these compounds, only the chlorinated hydrocarbon persist in the environment and are known to have drifted over thousands of kilometers and from water melted from Antarctic snow. Traces of chlorinated hydrocarbon pesticides in water may accumulate progressively in different steps of a food chain; for example, DDT can accumulate in fish to levels more than 10,000 times the concentration present in the surrounding water.

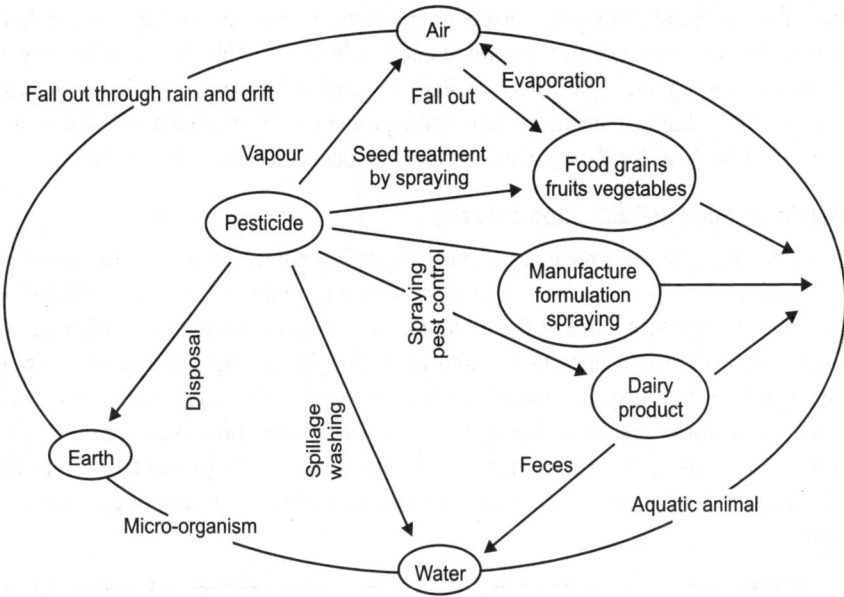

Fig. 17.8. Pesticide cycle in the environment.

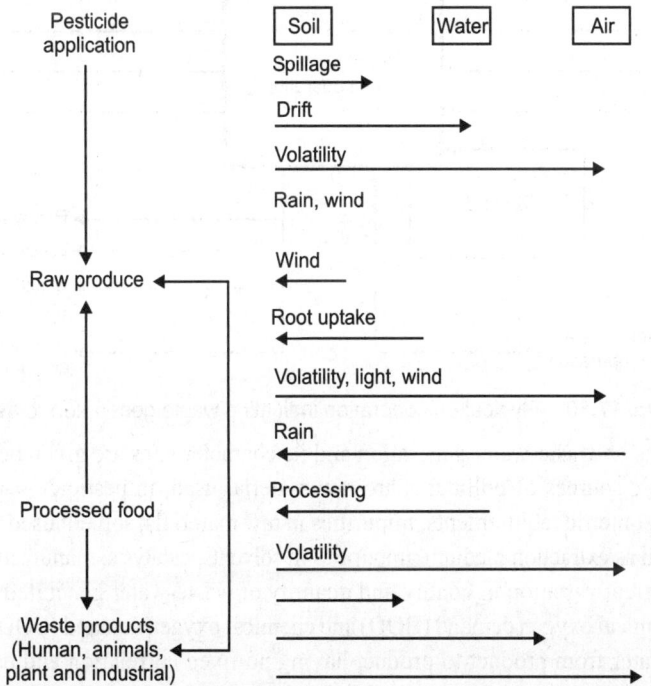

Fig. 17.9. Routes for the loss of pesticides to the environment.

Guideline values are recommended for several chlorinated hydrocarbon pesticides because of the possibility of their presence in water. These guideline values are derived from the acceptable daily intake (ADI) values with the assumption that is not more than 1 per cent of the ADI would be derived from drinking water. The presence of these toxic chemicals in water has also significance as they are picked up by unicellular aquatic organisms like plankton, and in the process pesticides get accumulated in body tissues manifolds higher than that of surrounding water by a phenomenon called bioconcentration. This plankton serves as food for fishes. Thus, residues of pesticides enter the food chain.

Waste-water Generation in Pesticides Industry

The entire manufacturing process for a particular product is a combination of various unit operations. The unit operations, where water is used and waste-water is generated, are to be identified so as to accurately identify the characteristics of the waste-water and the quantum of pollution load generated. A schematic diagram of an unit operation is shown in Fig. 17.10. The waste-water flow lines from process to treatment to final disposal are to be identified so that the entire status of the waste-water generation from the factory is depicted at one place. The waste-water flow lines in a typical industry are shown in Fig. 17.11. The network for monitoring waste-water may be fixed so as to analyse the characteristics of each of the waste-water streams and to evaluate the performance of the pollution control systems.

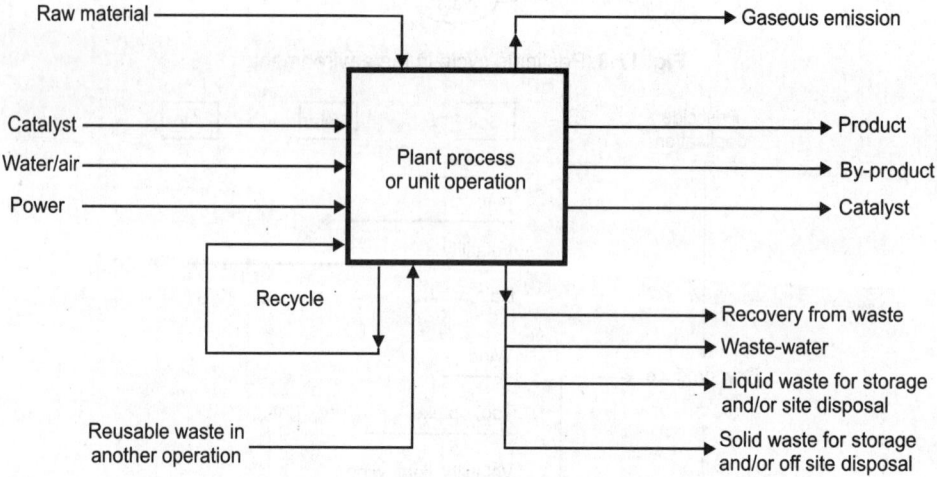

Fig. 17.10. A typical unit operation indicating waste generation routes.

The observations on waste-water generation and its characteristics are given below:

1. The possible sources of pollutants are raw material used, in pesticide synthesis in excess of their stoichiometric requirements, impurities in raw materials, solvent used as a carrier medium, solvent used as extraction medium, impurities in solvents, catalysts, manufacturing products, etc.

2. There is a great variation in quality and quantity of waste-water generated per unit of product. The biochemical oxygen demand (BOD) and chemical oxygen demand (COD) are widely varying in waste-water from product to product having no fixed correlation and has to be determined only on case to case basis. In general, BOD is in the range of 1000–5000 mg/l. The variation in the waste characteristics is hourly, daily and seasonal. The observations on manufacturing processes mentioned in above, also contribute to these variations. In general, the waste-water is toxic and not easily biodegradable in nature.

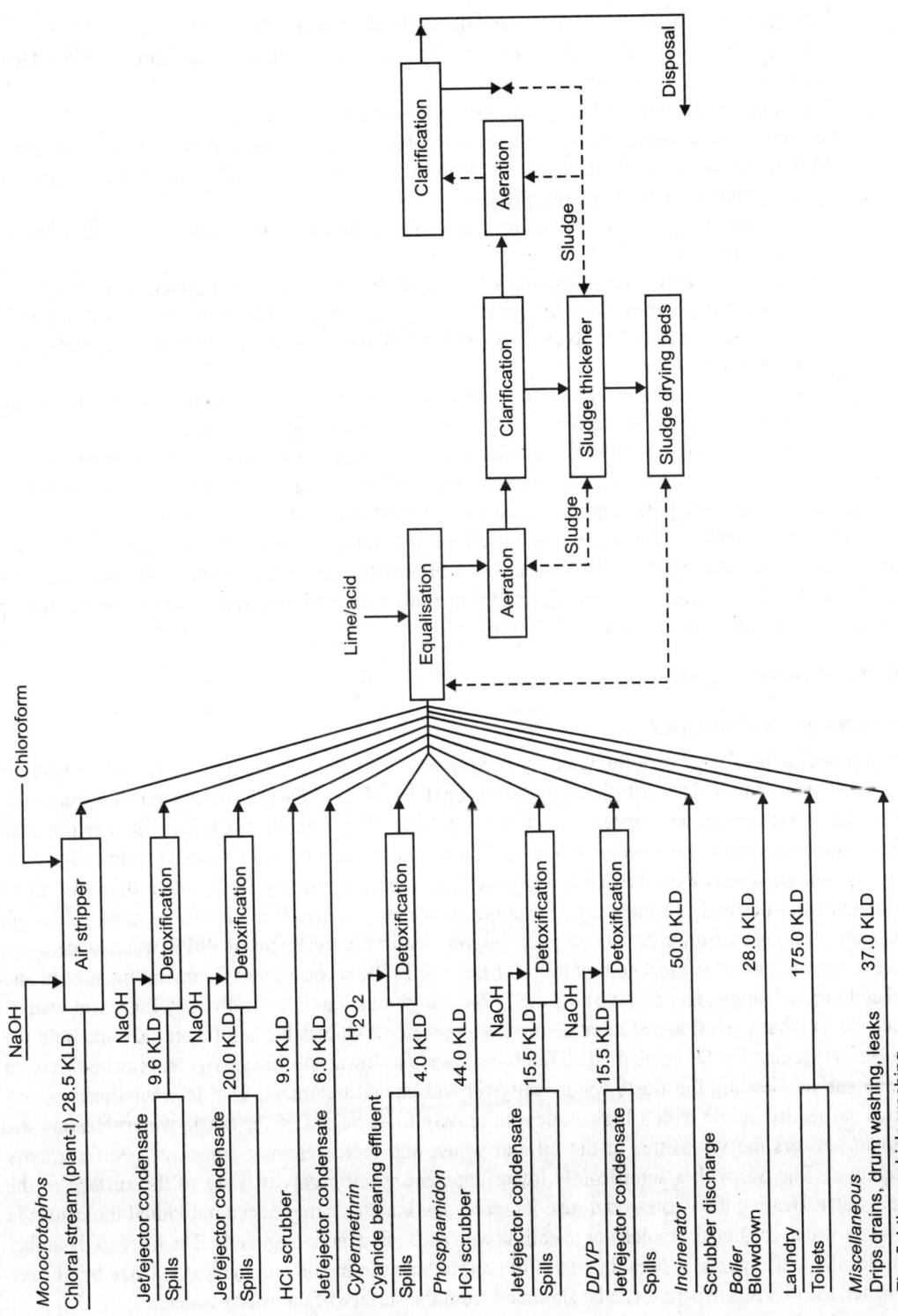

Fig. 17.11. Waste-water flow-lines in a typical industry.

3. Some process streams are inorganic and do not need biological treatment.
4. Some streams are organic and/or inorganic and have very high dissolved solids (TDS). These cannot be treated biologically.
5. Details of various other pollutants in the waste-water are given below:
   (a) Volatile organics: Benzene, toluene, chlorobenzene, etc. which are used as raw materials.
   (b) Halomethanes: Methylchloride, chloroform, carbon tetrachloride, etc. which are used as raw materials and extraction solvents.
   (c) Cyanides: Raw materials favouring cyanide formation are cyanamides, cyanates, thiocyanates and cyanuric chloride.
   (d) Phenols: Phenols are compounds having hydroxyl group (OH) attached directly to an aromatic ring. These may be found throughout the pesticides industry as raw materials, impurities in raw materials or as by-products of reactions utilising related compounds, such as chloro-benzenes.
   (e) Heavy metals: These are used as catalysts or as raw materials which are incorporated into the active ingredient (technical) as in case of metallo-organic pesticides.
   (f) Pesticides: Traces of the finished product, i.e. the pesticides themselves are present.

In view of the wide variety of process technology options, the quantity of water used and waste-water generation are widely varying from product to product and industry to industry.

Hence, it is difficult to summarise a specific limit for water use and waste-water generation as is done for other industries such as distilleries, sugar, breweries, etc. However, the pesticides industries may realise that there is wide scope in reducing the quantities of water usage and waste-water generation which may effect in reduction in cost of treatment.

Treatment Technologies

Pretreatment technologies

Many pesticide products are formulated by mixing active pesticide ingredients with inert materials (e.g. surfactants, emulsifiers, petroleum hydrocarbons), to achieve specific application characteristics. When these 'inerts' mix with water, emulsions may form. These emulsions reduce the performance efficiency of many treatment unit operations, such as chemical oxidation and activated carbon adsorption. In many situations, emulsion breaking is a necessary pretreatment step to facilitate the removal of pollutants from pesticide formulating, packaging and repackaging (PFPR) waste-waters. Although emulsion breaking is a pretreatment step, its importance in the treatment of PFPR waste-waters can make it a major part of the technology train for treating PFPR waste-waters. Temperature control and acid addition are simple, inexpensive methods of breaking emulsions in a variety of PFPR waste-waters. Acid (e.g. sulphuric acid) added to emulsified waste-water dissolves the solid materials that hold the emulsions together. The de-emulsified oil floats because of its lower specific gravity and can be skimmed off the surface, leaving the waste-water ready for subsequent treatment. The de-emulsification also causes suspended solids with a higher specific gravity to settle out of the waste-water. Heating the emulsion lowers the viscosities of the oil and water, and increases their apparent specific gravity differential. The oil, with a significantly lower apparent specific gravity, rises to the surface of the waste-water. Heating the waste-water also increases the kinetic energy of the individual molecules in the waste-water, causing the molecules to collide with each other more frequently. The increased number of molecule collisions aid in breaking the film present between the oil and the water. Once freed from the water, the oil rises, where it can be skimmed from the surface of the waste-water.

Secondary treatment

At secondary treatment plants, the primary effluent is further treated in a biological process, such as the activated sludge process. The activated sludge process and its variants are the most commonly used secondary processes for medium-to-large treatment plants. In the process, a special consortium of bacteria and other micro-organisms are grown. They metabolise the pollutants as they grow. This removes the BOD and TSS and can detoxify many pollutants and adsorb others. The secondary process removes at least 85 per cent of the BOD and TSS, and well-designed and operated plants may remove 95 per cent of the BOD and TSS. The effluent is clear with only a few pinhead sized suspended solids. It is not potable and still contains pathogens. It may contain trace metals and hard-to-treat organics compounds, such as pesticides. Nevertheless, it is usually suitable to be discharged into many receiving waters.

Tertiary treatment technologies

Activated carbon adsorption

Activated carbon effectively removes organic constituents from waste-water through the process of adsorption. The term 'activated carbon' refers to carbon materials, such as coal or wood, that are processed through dehydration, carbonisation, and oxidation to yield a material that is highly adsorbent, due to a large surface area and high number of internal pores per unit mass. As waste-water flows through a bed of carbon materials, molecules that are dissolved in the water may become trapped in these pores. In general, organic constituents (including many pesticide active ingredients) with certain chemical structures (such as aromatic functional groups), high molecular weights, and low water solubility are amenable to activated carbon adsorption. These constituents adhere to the stationary carbon material, so the waste-water leaving the carbon bed has a lower concentration of pesticide than the waste-water entering the carbon bed. Eventually, as the pore spaces in the carbon become filled, the carbon becomes exhausted and ceases to adsorb contaminants. Spent carbon may be regenerated or disposed of; cost and/or other regulatory factors (e.g. Resource Conservation and Recovery Act or RCRA) generally determine the choice. Carbon adsorption depends on process conditions, such as temperature and pH and process design factors such as carbon/waste-water contact time and the number of the carbon columns. If performed under the right conditions, activated carbon adsorption can be an effective treatment technology for PFPR industry waste-waters. Carbon adsorption capacity depends on the characteristics of the adsorbed compounds, the types of compounds competing for adsorption, and the characteristics of the carbon itself. If several constituents that are amenable to activated carbon adsorption are present in the waste-water, they may compete with each other for carbon adsorption capacity. This competition may result in low adsorption or even desorption of some constituents.

Activated carbon comes in two sizes: powdered carbon which has a diameter of less than 200 mesh, and granular carbon which has a diameter greater than 0.1 millimetre. While granular carbon is more commonly used in waste-water treatment; powdered carbon is used less frequently because the small particle size creates regeneration and design problems. Activate carbon is obtained from vendors in bulk or in a variety of container sizes. At smaller facilities, the container in which the carbon is sold is intended to be used as the carbon bed, with influent waste-water passing into one end of the container and treated effluent water passing out of the opposite end. At larger facilities, carbon is purchased and added to a column that is installed at the facility.

Carbon is regenerated by removing the adsorbed organic compounds through steam, thermal, or physical/chemical methods. Thermal and steam regeneration are the most common methods to regenerate carbon used for waste-water treatment. These methods volatilise the organic compounds that have adsorbed onto the carbon. Afterburners are required to ensure the destruction of the organic vapours; a scrubber may also be necessary to remove particulates from the air stream. Physical/chemical regeneration uses a solvent, which can be a water solution, to remove the organic compounds. Carbon is usually shipped back to the vendor for regeneration, although some facilities with larger carbon beds may find it economical to regenerate carbon on-site.

Chemical oxidation

Chemical oxidation modifies the structure of pollutants in waste-water to similar, but less harmful, compounds, through the addition of an oxidising agent. During chemical oxidation, one or more electrons transfer from the oxidant to the targeted pollutant, causing its destruction. One common method of chemical oxidation, referred to as alkaline chlorination, uses chlorine (usually in the form of sodium hypochlorite) under alkaline conditions to destroy pollutants, such as cyanide and some pesticide active ingredients. However, facilities treating waste-water, using alkaline chlorination should be aware that the chemical oxidation reaction may generate toxic chlorinated organic compounds, including chloroform, bromodichloromethane, and dibromochloromethane, as by-products. Adjustments to the design and operating parameters may alleviate this problem, or an additional treatment step (e.g. steam stripping, air stripping, or activated carbon adsorption) may be required to remove these by-products.

Chemical oxidation can also be performed with other oxidants (e.g. hydrogen peroxide, ozone, and potassium permanganate) or with the use of ultraviolet light. Although these other methods of chemical oxidation can effectively treat PFPR waste-waters, they typically entail higher capital and/or operating and maintenance costs, greater operator expertise, and/or more extensive waste-water pretreatment than alkaline chlorination.

Chemical precipitation

Chemical precipitation is a treatment technology in which chemicals (e.g. sulphides, hydroxides, and carbonates) react with organic and inorganic pollutants present in waste-water to form insoluble precipitates. This separation treatment technology is generally carried out in the following four phases:
1. Addition of the chemical to the waste-water;
2. Rapid (flash) mixing to distribute the chemical homogeneously throughout the waste-water.
3. Slow mixing to encourage flocculation (formation of the insoluble solid precipitate).
4. Filtration, settling, or decanting to remove the flocculated solid particles.

These four steps can be performed at ambient conditions and are well suited to automat control. Hydrogen sulphide or soluble sulphide salts (e.g. sodium sulphate) are chemicals commonly used in the PFPR industry during chemical precipitation. These sulphides are particularly effective in removing complexed and heavy metals (e.g. mercury, lead, and silver) from industrial waste-waters. Hydroxide and carbonate precipitation can also be used to remove metals from PFPR waste-waters, but these technologies tend to be effective on a narrower range of contaminants.

Hydrolysis

Hydrolysis is a chemical reaction in which organic constituents react with water and break into smaller (and less toxic) compounds. Basically, hydrolysis is a destructive technology in which the original molecule forms two or more new molecules. In some cases, the reaction continues and other products are formed. Because some pesticide and active ingredients react through this mechanism, hydrolysis can be an effective treatment technology for PFPR waste-water. The primary design parameter considered for hydrolysis is the half-life, which is the time required to react 50 per cent of the original compound. The half-life of a reaction generally depends on the reaction, pH and temperature and the reactant molecule (e.g. the pesticide active ingredient). Hydrolysis reactions can be catalysed at low pH, high pH or both, depending on the reactant molecule. In general, increasing the temperature increases the rate of hydrolysis. Identifying the best conditions for the hydrolysis reaction results in a shorter half-life, thereby reducing both the size of the reaction vessel required and the treatment time required.

FERTILISER

In Fertiliser plants, a variety of wastes are discharged as water pollutants in the form of processing chemicals like sulphuric acid; process intermediates like ammonia, phosphoric acid; final products like urea, ammonium sulphate, ammonium phosphate, etc. In addition, oil bearing waste from compressor houses of ammonia and urea plants, some portion of the cooling water and the wash water from the scrubbing towers for the purification of gases, also come as waste.

Wash water from, the scrubbing towers may contain toxic substances like arsenic, potassium carbonate, etc. in a nitrogenous fertiliser plant, while that in a phosphatic fertiliser plant may contain a mixture of carbonic acid, hydrofluoric acid and fluorosilicic acid. Both alkaline and acidic wastes are also expected from the boiler feed water treatment plant, the wastes being generated during the regeneration of anion and cation exchanger units. Fertilisers can be classified as nitrogenous fertilisers, phosphatic fertilisers and complex fertilisers. Plants may be producing nitrogenous fertilisers, like urea, ammonium sulphate, ammonium nitrate and ammonium chloride, or phosphatic fertilisers like super- phosphates; there are plants where complex fertilisers containing both nitrogen and phosphates, like ammonium phosphate and ammonium sulphate phosphate are produced. Also, some fertiliser units are only involved in mining and undertake no other activity.

Waste-water from Fertiliser Plants

A variety of waste streams are discharged from fertiliser plants in the form of:
1. Processing chemicals like sulphuric acid.
2. Process intermediates like ammonia, phosphoric acid, etc.
3. Final products like urea, ammonium sulphate, ammonium phosphate, etc.

In addition, oil bearing wastes from compressor houses of ammonia and urea plants, and some portion of cooling water and wash water from the scrubbing towers for the purification of gases, also come as waste. Wash water from the scrubbing towers may contain toxic substances like arsenic, monoethanolamine, potassium carbonate, etc. in a nitrogenous fertiliser plant, while that in a phosphatic fertiliser plant may contain a mixture of carbonic acid, hydrofluoric acid and fluosilicic acid. Both alkaline and acidic wastes are also expected from the boiler feed water treatment plant, the wastes being generated during the regeneration of anion and cation exchanger units. Previously, waste cooling water contained toxic elements like chromates, zinc, etc. which were used for corrosion control. The development of non-chromate technology using quaternary ammonium compounds has eliminated these toxic substances. Additional pollutants like phenol and cyanide will be introduced in the list of pollutants in a fertiliser plant where ammonia is derived from the waste ammoniacal liquor of coke ovens.

Effects of wastes on receiving streams

All the components of waste from the fertiliser plants induce adverse effects in streams. Acids and alkalies can destroy normal aquatic life. Arsenic, fluorides, and ammonium salts are found to be toxic to fish. Amines are not only toxic to fish but also exert a high oxygen and chlorine demand. Presence of different types of salts renders the stream unfit for use as a source of drinking water in the downstream side. Nitrogen and other nutrient content of the waste encourages growth of aquatic plants in the stream.

Treatment of fertiliser waste-water

Major pollutants in fertiliser waste-water for which treatment is necessary include oil, arsenic, ammonia, urea, phosphate and fluoride. Oil is removed in a gravity separator. Arsenic containing waste is segregated and after its concentration, the solid waste is disposed of in a safe place. Phosphate and fluoride bearing wastes are also segregated and chemically coagulated by lime; clarified effluent, which still contains some amount of phosphate and fluoride, is diluted by mixing with other wastes. Several alternatives are there for the treatment of ammonia bearing wastes, including:

1. Steam stripping.
2. Air stripping in towers.
3. Lagooning after pH adjustment.
4. Biological nitrification and denitrification.

For all practical purposes, 'steam stripping' for ammonia removal from fertiliser wastes has been found to be uneconomical. Removal of ammonia gas from the solution in an air stripping tower, packed with red wood stakes, is found to be a very efficient method. Very encouraging results are obtained from some laboratory and pilot plant studies conducted by national environmental engineering research institute (NEERI) in the removal of ammonia by simply lagooning the waste. It was found that considerable reduction in the ammonia content can be accomplished just by retaining the ammoniacal waste in an earthen tank, about 1 metre deep, for a day or two, after pretreatment of the waste by lime to increase the pH to 11.0. However, no reduction in urea content was observed within this period in wastes containing urea; thus waste containing both urea and ammonia required to be retained in the lagoon for a longer period, to allow urea to decompose to ammonia first. Biological nitrification involves oxidation of ammonia to nitrate, via nitrite under aerobic conditions; this is followed by denitrification of the nitrified effluent under anaerobic condition, in which gaseous N_2 and N_2O are the end products and are released into the atmosphere. The denitrification requires addition of some quantity of carbonaceous matter in the reactor. In all the ammonia removal methods described above, urea remains untouched. If urea removal is required, wastes containing urea must be retained for a sufficiently long time in an earthen lagoon to allow it to decompose first to ammonia. The effluent treatment of a complex fertiliser plant are given in Fig. 17.12. Pollution control aspects related to single superphosphate (SSP) fertiliser, straight nitrogeneous fertiliser and complex fertilisers are discussed here.

Straight Nitrogenous Fertilisers

Urea and salts of ammonia are referred to as straight fertilisers and if combined with other nutrients such as phosphates and potash, they are called complex/mixed fertilisers. Nitrogenous fertiliser plants use a large quantity of water mainly for process cooling, steam generation and process use, resulting in waste-water generation at various points in the manufacturing process. In an ammonia plant, if partial oxidation process is used, then the carbon of hydrocarbon feedstock is not completely combusted.

Disposal of the final treated effluent and monitoring

Disposal of the final treated effluent to the receiving water system is an important aspect in the pollution control system.

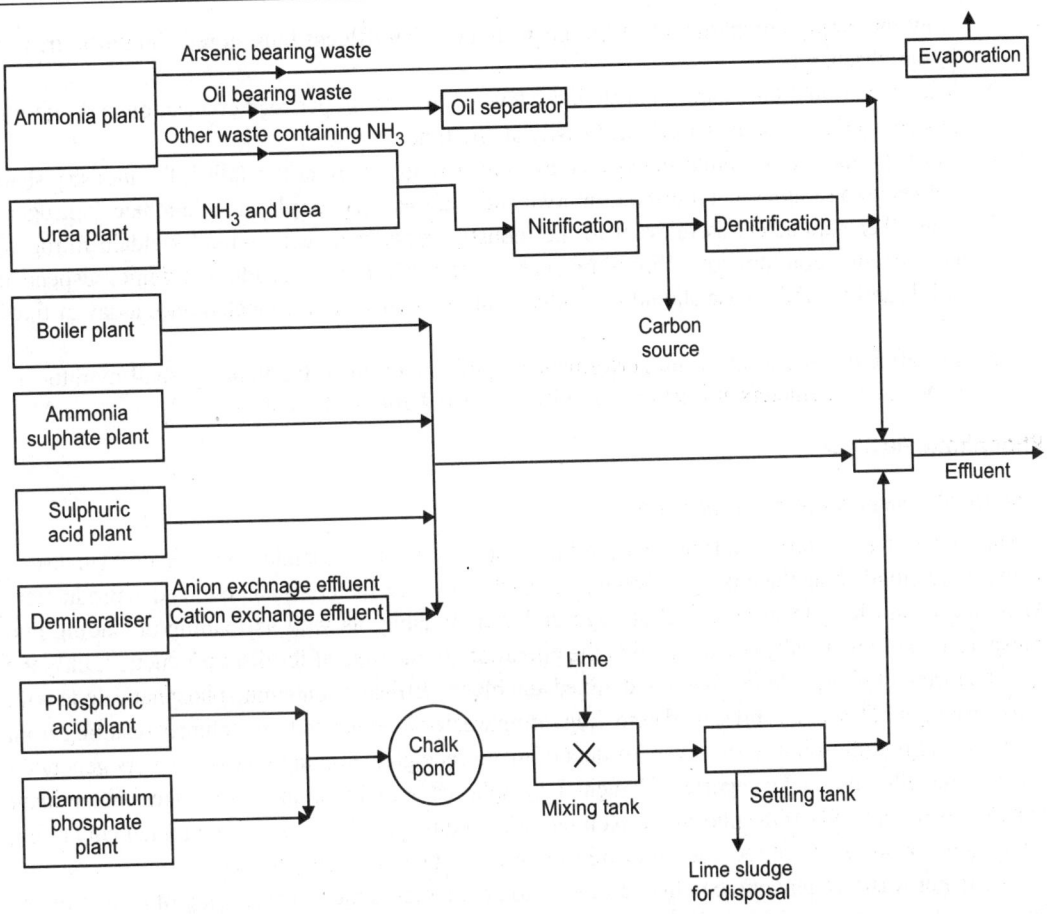

Fig. 17.12. Flow diagram for effluent treatment of a complex fertiliser plant.

The detailed procedure to be adopted for disposal system is given as under:

1. It is desired that all the effluents after treatment shall be routed to a properly lined guard pond for equalisation and final control.

2. The guard pond should have two compartments, each of at least four hours capacity. All the effluent streams shall be connected to these compartments by a parallel connection system. One compartment of the guard pond shall be used for the routine disposal of effluent, while the other compartment will remain empty and be utilised when effluent do not conform the limits.

3. In the guard pond, an automatic monitoring system for flow and the relevant pollutants shall be provided with high-level alarm system. The parameters necessary for automatic monitoring are pH, ammoniacal nitrogen, nitrate nitrogen and hexavalent chromium. Monitoring of nitrate nitrogen is applicable to the industries where nitric acid is produced and used for production of fertilisers, and the monitoring of hexavalent chromium is applicable to the industries where chromate-based inhibitors are used for cooling water conditioning.

4. When pollutants in the final effluent exceed the stipulated figures as indicated by the alarm system and the effluent stream responsible cannot be identified, all the effluent streams shall

flow the empty compartment of the guard pond. The effluent thus stored should be treated before discharge.

5. The area around the guard pond shall be developed with proper road connection and lighting system so that it can be approached easily at any time.

6. Till the continuous monitoring system as indicated above are not installed, the industry shall collect grab samples at a four-hour interval and analyse these for pH and ammoniacal nitrogen.

7. The other parameters as relevant to the industry concerned such as total kjeldahl nitrogen, hexavalent chromium, total chromium cyanide, nitrate nitrogen, vanadium, arsenic, suspended solids and oil and grease should be analysed in the grab sample collected once a day at fixed hours.

8. For effective appraisal of the performance of treatment units, the industry shall monitor the concerned parameters at least once a shift before and after treatment.

Phosphate Fertiliser

Industrial operation and waste-water

The phosphate manufacturing and phosphate fertiliser industry is a basic chemical manufacturing industry, in which essentially both the mixing and chemical reactions of raw materials are involved in production. Also, short- and long-term chemical storage and warehousing, as well as loading/unloading and transportation of chemicals, are involved in the operation. In the case of fertiliser production, only the manufacturing of phosphate fertilisers and mixed and blend fertilisers containing phosphate along with nitrogen and/or potassium is presented here. Regarding waste-water generation, volumes resulting from the production of phosphorus are several orders of magnitude greater than the waste-waters generated in any of the other product categories. Elemental phosphorus is an important waste-water contaminant common to all segments of the phosphate manufacturing industry, if the phossy water (water containing colloidal phosphorus) is not recycled to the phosphorus production facility for reuse.

The major waste-water source in the defluorination processes is the wet scrubbing of contaminants from the gaseous effluent streams. However, process conditions normally permit the use of recirculated contaminated water for this service, thereby effectively reducing the discharged waste-water volume.

Waste-water characteristics and sources

Waste-waters from the manufacturing, processing and formulation of inorganic chemicals such as phosphorus compounds, phosphates and phosphate fertilisers cannot be exactly characterised. The waste-water streams are usually expected to contain trace or large concentrations of all raw materials used in the plant; all intermediate compounds produced during manufacture; all final products, coproducts and by-products; and the auxiliary or processing chemicals employed. It is desirable from the viewpoint of economics that these substances not be lost, but some losses and spills appear unavoidable and some intentional dumping does take place during housecleaning, vessel emptying and preparation operations.

Few fertiliser plants discharge waste-waters to municipal treatment systems. Most use ponds for the collection and storage of waste-waters, pH control, chemical treatment and settling of suspended solids. Whenever available retention pond capacities in the phosphate fertiliser industry are exceeded, the waste-water overflows are treated and discharged to nearby surface water bodies. The range of waste-water characteristics and concentrations for typical retention ponds used by the phosphate fertiliser industry are given in Table 17.6.

Table 17.6. Raw waste-water characteristics of phosphate fertiliser industry retention ponds.

Quality parameter	Phosphate
Suspended solids (mg/l)	800–1200
pH	1–2
Ammonia (mg/l)	450–500
Sulphate (mg/l)	4000
Chloride (mg/l)	58
Total phosphate (mg/l)	3–5 M
Fluoride (mg/l)	6–8.5 M
Aluminium (mg/l)	110
Iron (mg/l)	85
Radium 226 (picocuries/l)	60–100

M = Thousand.

The specific types of waste-water sources in the phosphate fertiliser industry are: (i) water treatment plant wastes from raw water filtration, clarification, softening and deionisation, which principally consist of only the impurities removed from the raw water (such as carbonates, hydroxides, bicarbonates and silica) plus minor quantities of treatment chemicals, (ii) closed-loop cooling tower blowdown, the quality of which varies with the makeup of water impurities and inhibitor chemicals used (the only cooling water contamination from process liquids is through mechanical leaks in heat exchanger equipment. Table 17.7 highlights the normal range of contaminants that may be found in cooling water blowdown systems), (iii) boiler blowdown, which is similar to cooling tower blowdown but the quality differs as given in Table 17.8, (iv) contaminated water or gypsum pond water, which is the impounded and reused water that accumulates sizable concentrations of many cations and anions, but mainly fluorine and phosphorus [concentrations of 8500 mg/l F and in excess of 5000 mg/l P are not unusual; concentrations of radium 226 in recycled gypsum pond water are 60 to 100 picocuries/l and its acidity reaches extremely high levels (pH 1–2)], (v) waste-water from spills and leaks that, when possible, is reintroduced directly to the process or into the contaminated water system; and (vi) nonpoint source discharges that originate from the dry fertiliser dust covering the general plant area and then dissolving in rain water and snowmelt that become contaminated.

Table 17.7. Range of concentrations of contaminants in cooling water.

Cooling water contaminant	Concentration (mg/l)
Chromate	0–250
Sulphate	500–3000
Chloride	35–160
Phosphate	10–50
Zinc	0–30
TDS	500–10000
TSS	0–50
Biocides	0–100

Table 17.8. Range of concentrations of contaminants in boiler blowdown waste.

Boiler blowdown contaminant	Concentration (mg/l)
Phosphate	5–50
Sulphite	0–100
TDS	500–3500
Zinc	0–10
Alkalinity	50–700
Hardness	50–500
Silica (SiO_2)	25–80

In the specific case of waste-water generated from the condenser water bleedoff in the production of elemental phosphorus from phosphate rock in an electric furnace, Horton reported that the flow varies from 10 to 100 gpm (2.3–23 m^3/hr), depending on the particular installation. The most important contaminants in this waste are elemental phosphorus, which is colloidally dispersed and may ignite if allowed to dry out and fluorine that is also present in the furnace gases. The general characteristics of this type of waste-water (if no soda ash or ammonia were added to the condenser water) are given in Table 17.9.

Table 17.9. Range of concentrations of contaminants in condenser waste from electric furnace production of phosphorus.

Quality parameter	Concentration or value
pH	1.5–2.0
Temperature	120°–150°F
Elemental phosphorus	400–2500 ppm
Total suspended solids	1000–5000 ppm
Fluorine	500–2000 ppm
Silica	300–700 ppm
P_2O_5	600–900 pp
Reducing substances as (I_2)	40–50 ppm
Ionic charge of particles	Predominantly positive (+)

As previously mentioned, fertiliser manufacturing may create problems within all environmental media, i.e. air pollution, water pollution and solid wastes disposal difficulties. In particular, the liquid waste effluents generated from phosphate and mixed and blend fertiliser production streams originate from a variety of sources and may be summarised as follows: (i) ammonia-bearing wastes from ammonia production, (ii) ammonium salts such as ammonium phosphate, (iii) phosphates and fluoride wastes from phosphate and superphosphate production, (iv) acidic spillages from sulphuric acid and phosphoric acid production, (v) spent solutions from the regeneration of ion-exchange units, (vi) phosphate, chromate, copper sulphate and zinc wastes from cooling tower blowdown, (vii) salts of metals such as iron, copper, manganese, molybdenum and cobalt, (viii) sludge discharged from clarifiers and backwash water from sand filters, and (ix) scrubber wastes from gas purification processes.

Considerable variation, therefore, is observed in quantities and waste-water characteristics at different plants. The most important factors that contribute to excessive in-plant materials losses and therefore, probable subsequent pollution are the age of the facilities (low efficiency, poor process control), the state of maintenance and repair (especially of control equipment), variations in feedstock and difficulties in adjusting processes to cope and an operational management philosophy such as consideration for pollution control and prevention of materials loss. Because of process cooling requirements, fertiliser manufacturing facilities may have an overall large water demand, with the waste-water effluent discharge largely dependent on the extent of in-plant recirculation. Facilities designed on a once-through process cooling flowstream generally discharge from 1000 to over 10,000 m^3/hr waste-water effluents that are primarily cooling water.

According to research results reported by Fuller, the removal of semicolloidal matter in settling areas or ponds seems to be one of the primary problems concerning water pollution control. The results of dissolved oxygen (DO) and BOD surveys indicated that receiving streams were actually improved in this respect by the effluents from phosphate operations. On the other hand, no detrimental effects on fish were found, but there is the possibility of destruction of fish food aquatic micro-organisms and plankton under certain conditions.

The waste-water characteristics vary from one production facility to the next and even the particular flow magnitude and location of discharge will significantly influence its aquatic environmental impact. The degree to which a receiving surface water body dilutes a waste-water effluent at the point of discharge is important, as are the minor contaminants that may occasionally have significant impacts. Fertiliser manufacturing wastes, in general, affect water quality primarily through the contribution of nitrogen and phosphorus, whose impacts have been extensively documented in the literature. Significant levels of phosphates assist in inducing eutrophication and in many receiving waters they may be more important (growth-limiting agent) than nitrogenous compounds. Under such circumstances, programs to control eutrophication have generally attempted to reduce phosphate concentrations in order to prevent excessive algal and macrophyte growth.

In addition to the above major contaminants, pollution from the discharge of fertiliser manufacturing wastes may be caused by such secondary pollutants as oil and grease, hexavalent chromium, arsenic and fluoride. As reported by Beg in certain cases, the presence of one or more of these pollutants may have adverse impacts on the quality of a receiving water, due primarily to toxic properties or can be inhibitory to the nitrification process. Finally, oil and grease concentrations may have a significant detrimental effect on the oxygen transfer characteristics of the receiving surface water body.

Waste-water control and treatment

The pollution control and treatment methods and unit processes used are discussed in more detail in the following section.

In-plant control, recycle and process modification

The primary consideration for in-plant control of pollutants that enter waste streams through random accidental occurrences, such as leaks, spills and process upsets, is establishing loss prevention and recovery systems. In the case of fertiliser manufacture, a significant portion of contaminants may be separated at the source from process wastes by dedicated recovery systems, improved plant operations, retention of spilled liquids and the installation of localised interceptors of leaks such as oil drip trays for pumps and compressors. Also, certain treatment systems installed (i.e. ion-exchange, oil recovery and hydrolyser-stripper systems) may, in effect, be recovery systems for direct or indirect reuse of effluent constituents. Finally, the use of effluent gas scrubbers to improve in-plant operations by preventing gaseous product losses may also prevent the airborne deposition of various pollutants within the general plant area, from where they end up as surface drainage runoff contaminants.

Cooling water

Cooling water constitutes a major portion of the total in-plant wastes in fertiliser manufacturing and it includes water coming into direct contact with the gases processed (largest percentage) and water that has no such contact. The latter stream can be readily used in a closed-cycle system, but sometimes the direct contact cooling water is also recycled (after treatment to remove dissolved gases and other contaminants and clarification). By recycling, the amount of these waste-waters can be reduced by 80 to 90 per cent, with a corresponding reduction in gas content and suspended solids in the wastes discharged to sewers or surface water.

Process modifications

The following are possible process modifications and plant arrangements that could help reduce waste-water volumes, contaminant quantities and treatment costs: (i) in ammonium phosphate production and mixed and blend fertiliser manufacturing, one possibility is the integration of an ammonia process condensate steam stripping column into the condensate boiler feedwater systems of an ammonia plant, with or without further stripper bottoms treatment depending on the boiler quality makeup needed, (ii) contaminated waste-water collection systems designed so that common contaminant streams can be segregated and treated in minor quantities for improved efficiencies and reduced treatment costs, (iii) in ammonium phosphate and mixed and blend fertiliser production, another possibility is to design for a lower-pressure steam level (i.e. 42–62 atm) in the ammonia plant to make process condensate recovery easier and less costly; and (iv) when possible, the installation of air-cooled vapour condensers and heat exchangers would minimise cooling water circulation and subsequent blowdown.

Recently new techniques have been adopted by French company for pollution prevention, for a new process modification for steam segregation and recycle in phosphoric acid production in which, raw water from the sludge/fluorine separation system is recycled to the heat-exchange system of the sulphuric acid dilution unit and the waste-water used in plaster manufacture. Furthermore, decanted supernatant from the phosphogypsum deposit pond is recycled for treatment in the water filtration unit. This process modification permits an important reduction in pollution by fluorine and that it makes the treatment of effluents easier and in some cases allows specific recycling. Finally, the new process produced a small reduction in water consumption, either by recycle or discharging a small volume of polluted process water downstream and required no particular equipment and very few alterations in the mainstream lines of the old process.

Waste-water treatment methods

Phosphate manufacturing

Nemerow summarised the major characteristics of wastes from phosphate and phosphorus compounds production (i.e. clays, limes and tall oils, low pH, high suspended solids, phosphorus, silica and fluoride) and suggested the major treatment and disposal methods such as lagooning, mechanical clarification, coagulation and settling of refined waste-waters.

Phosphate fertiliser production

Contaminated water from the phosphate fertiliser is collected in gypsum ponds and treated for pH adjustment and control of phosphorus and fluorides. Treatment is achieved by 'double liming' or a two-stage neutralisation procedure, in which phosphates and fluorides precipitate. The first treatment stage provides sufficient neutralisation to raise the pH from 1 to 2 to a pH level of at least 8. The resultant effectiveness of the treatment depends on the point of mixing of lime addition and on the constancy of pH control. Fluosilisic acid reacts with lime and precipitates calcium fluoride in this step of the treatment.

The waste-water is again treated with a second lime addition to raise the pH level from 8 to at least 9 (where phosphate removal rates of 95 per cent may be achieved), although two-stage dosing to pH 11 may be employed. Concentrations of phosphorus and fluoride with a magnitude of 6500 and 9000 mg/l, respectively, can be reduced to 5 to 500 mg/l P and 30 to 60 mg/l F. Soluble orthophosphate and lime react to form an insoluble precipitate, calcium hydroxy apatite. Sludges formed by lime addition to phosphate wastes from phosphate manufacturing or fertiliser production are generally compact and possess good settling and dewatering characteristics and removal rates of 80 to 90 per cent for both phosphate and fluoride may be readily achieved. The seepage collection of contaminated water from phosphogypsum ponds and reimpoundment is accomplished by the construction of a seepage collection ditch around the perimeter of the diked storage area and the erection of a secondary dike surrounding the first. The base of these dikes is usually natural soil from the immediate area and these combined earth/gypsum dikes tend to have continuous seepage through them. The seepage collection ditch between the two dikes needs to be of sufficient depth and size to not only collect contaminated water seepage, but also to permit collection of seeping surface runoff from the immediate outer perimeter of the seepage ditch. This is accomplished by the erection of the small secondary dike, which also serves as a backup or reserve dike in the event of a failure of the primary major dike.

The sulphuric acid plant has boiler blowdown and cooling tower blowdown waste streams, which are uncontaminated. However, accidental spills of acid can and do occur and when they do, the spills contaminate the blowdown streams. Therefore, neutralisation facilities should be supplied for the blowdown waste streams, which involves the installation of a reliable pH or conductivity continuous-monitoring unit on the plant effluent stream. The second part of the system is a retaining area through which noncontaminated effluent normally flows. The detection and alarm system, when activated, causes a plant shutdown that allows location of the failure and initiation of necessary repairs. Such a system, therefore, provides the continuous protection of natural drainage waters, as well as the means to correct a process disruption. Mixed fertiliser treatment technology consists of a closed loop contaminated water system, which includes a retention pond to settle suspended solids. The water is then recycled back to the system. There are no liquid waste streams associated with the blend fertiliser process, except when liquid air scrubbers are used to prevent air pollution. Dry removals of air pollutants prevent a waste-water stream from being formed.

Phosphate and fluoride removal

Phosphates may be removed from waste-waters by the use of chemical precipitation as insoluble calcium phosphate, aluminium phosphate and iron phosphate. The liming process has been discussed previously, lime being typically added as a slurry and the system used is designed as either a single or two-stage one. Polyelectrolytes have been employed in some plants to improve overall settling and clariflocculators or sludge-blanket clarifiers are used in a number of facilities. Alternatively, the dissolved air flotation process is also feasible for phosphate and fluoride removal. A number of aluminium compounds, such as alum and sodium aluminate, have also been used as phosphate precipitants at an optimum pH range of 5.5 to 6.5, as have iron compounds such as ferrous sulphate, ferric sulphate, ferric chloride and spent pickle liquor. The optimum pH range for the ferric salts is 4.5 to 5 and for the ferrous salts it is 7 to 8, although both aluminium and iron salts have a tendency to form hydroxyl and phosphate complexes. As reported by Ghokas, sludge solids produced by aluminium and iron salts precipitation of phosphates are generally less settleable and more voluminous than those produced by lime treatment. According to Sprecht, in the two-step process to remove fluorides and phosphoric acid, water entering the first step may contain about 1700 mg/l F and 5000 mg/l P_2O_5 and it is treated with lime slurry or ground limestone to a pH of 3.2 to 3.8. Insoluble calcium fluorides settle out and the fluoride concentration is lowered to about 50 mg/l F, whereas the P_2O_5 content is reduced only slightly. The clarified supernatant is transferred to another collection area where lime slurry is added to bring the solution to pH 7 and the resultant precipitate of P is removed by settling. The final clear water, which contains only 3 to 5 mg/l F and practically no P_2O_5, is either returned to the plant for reuse or discharged to surface waters. The two-step process is required to reduce fluorides in the water below 25 mg/l F, because a single-step treatment to pH 7 lowers the fluoride content only to 25 to 40 mg/l F. In the process where the triple phosphate is to be granulated or nodulised, the material is transferred directly from the reaction mixer to a rotary dryer and the fluorides in the dryer gases are scrubbed with water. In making defluorinated phosphate by heating phosphate rock, one method of fluoride recovery consists of absorption in a tower of lump limestone at temperatures above the dewpoint of the stack gas, where the reaction product separates from the limestone lumps in the form of fines. A second method of recovery consists of passing the gases through a series of water sprays in three separate spray chambers, of which the first one is used primarily as a cooling chamber for the hot exit gases of the furnace. In the second chamber, the acidic water is recycled to bring its concentration to about 5 per cent equivalence of hydrofluoric acid in the effluent, by withdrawing acid and adding freshwater to the system. In the final chamber, scrubbing is supplemented by adding finely ground limestone blown into the chamber with the entering gases. Hydrochloric acid is sometimes formed as a by-product from the fluoride recovery in the spray chambers and this is neutralised with NaOH and lime slurry before being transferred to settling areas.

PAINTS AND DYES

Paint industry uses varied raw material such as resins, solvents, drying oils, pigments and extenders. The major waste generated by the paint manufacturing industry are empty raw material packages containing trace elements, equipment cleaning wastes and spills. Empty raw material packages are generated during unloading of materials to high speed mixers or mixing tanks. Water solvents are generated from equipment cleaning. Even after distillation of waste solvents for reuse, a residual paint sludge remains. The paint sludge contains solvents and residual toxic metals such as mercury, lead and chromium. Waste rinse water is generated from equipment cleaning with water and for caustic solutions. Wastes containing undispersed pigments are contained in waste filter cartridges.

The equipment cleaning wastes can be minimised by employing more efficient cleaning methods, like reduction in the frequency of equipment cleaning, use of rubber wipers to reduce the amount of paint left on the walls of a mix tank, use of teflon-lined tanks to reduce adhesion and use of plastic or foam to clean pipes to improve drainage.

Waste-water Generation

Waste-water quality

On close examination of manufacturing process of various paints, it is observed that water is not a constituent in any of the production processes of resins, varnishes, lacquers, etc. However, water is required for the manufacture of stiff or water-based paints and cooling of the ball mills or sand mills in manufacture of oil paints. Water is used for washing of floors which are littered by spillage of powders, solvents, etc. when a particular batch is finished. In cases where there is vast difference in the colour shade of next batch, the mills and containers, etc. have to be cleaned. This cleaning is done by xylene which is recovered and reused.

Once in a while the vessels are cleaned with caustic soda. This cleaning constitutes the major part of the waste-water generated which is highly alkaline in nature. Other source is the cooling tower blow down. The mills used for grinding powders and the solvents used in mixing have to be maintained at room temperature. They are cooled by circulating cooling water. Cooling tower blow down generally enters the effluent stream. In summary, it can be stated that waste-waters from the paint manufacturing industries generally tend to be alkaline, contain some oil and grease and biochemical oxygen demand (BOD), chemical oxygen demand (COD) and suspended solids (SS). The waste-waters can be assumed to contain small amounts of the products.

The BOD and COD values give only a gross measure of organics in the wastes. Some of the organic and inorganic compounds used in the manufacturing operations are classified as toxic and hazardous. Chromium, copper, lead, zinc, ethyl benzene, di-(2-ethylhexyl) phthalate, tetrachloroethylene and toluene were found in high concentrations.

Waste-water characteristics from main sections

Waste-waters from caustic cleaning

It was observed during survey that caustic cleaning resulted in maximum production of waste-water in the paint industry. Vessels used in manufacturing processes have to be cleaned before change of product. This is done using water and alkali. Thus, the waste-water pH is in range of 8.5 to 13.5. This waste-water consists of suspended solids in the range of 200 to 600 mg/l. BOD varies from 475 to 2400 mg/l and the COD is between 1100–3800 mg/l. The oil and grease is within a range of 32–150 mg/l. The phenolic concentration of this waste-water is 12.5 mg/l.

Waste-water from resin house

The waste-water from resin house is that which accumulates from the condenser into the separator. The water layer is disposed off in the waste-water stream and the upper layer of solvents is reused. The water from separator is generally acidic with pH range from 3.2 to 6.3. The suspended solids range from 240 to 400 mg/l. The BOD and COD of waste-water vary from 225 to 60,000 mg/l and 240 to as high as 78,000 mg/l respectively. The phenolics are between 6 and 86 mg/l. The oil and grease is within a range of 14–25 mg/l.

Waste-water from stiff paints

Stiff paints are water base paints and the main product is distemper. Cleaning of hoppers, grinders and containers consume large amount of water and contributes 400 to 700 mg/l of suspended solids, a large fraction of which is settleable. Waste-water contributed by stiff paint section is very less in quantity because of less production in this section and is not generated continuously. As observed during the survey, characteristics of the waste-water stream are COD 1215 to 6000 mg/l, BOD 380 to 980 mg/l and phenolic content varies from 6.4 to 100 mg/l. The oil and grease content of the waste-water is 252 mg/l.

Combined waste-water

The characteristics of combined waste-water is presented in Table 17.10.

Table 17.10. Combined waste-water characteristics.

Parameters mg/l, except pH	Range
pH	6.5–10.5
SS	220–1200
COD	300–5700
BOD	1700–3100
Oil and Grease	22–138
Phenolics	18–55

Reduction of waste-water

Schemes for reducing the generation of waste-water at source should be practised. This is to reduce the effluent load rather than finding methods to treat it. Unnecessary use of water not only adds to the quantity of effluent and the cost of treating it, but also increases the wastage of heat, power and/or product in the effluent. Steps that can be taken to reduce the generation of waste-water are discussed below:

Good housekeeping

Good housekeeping reduces generation of both waste-waters and solid wastes. The cooling water is usually uncontaminated and thus, should be collected and reused. It could be used for floor washing or discharged separately into the receiving water bodies rather than mixing with polluted water and discharging into the treatment plant. Accidental spills and leakages should be reduced to a minimum through proper maintenance of equipment and training of personnel. In case of caustic cleaning, instead of washing away the caustic solution, it can be collected, stored and used for further cleaning. In case of stiff paints, water from first cleaning should be collected and used later as process water for a similar type of batch. Waste-water volume can also be reduced through reuse of rinse water for preparation of alkali solution. The above procedures can reduce the quantity of generation of caustic cleaning water significantly.

Recovery of wastes

A large number of solvents are used in paint manufacturing and a majority of them are recovered and therefore not lost in the waste-water streams. In case of oil paints, solvents are added in grinders which are closed units, therefore, loss of solvents through evaporation is considerably reduced. High temperature is maintained in resin and varnish manufacture, resulting in evaporation of solvents added. These solvent vapours, along with the water vapours generated through chemical reactions are condensed and collected in a separator. The solvent layer is removed and reused in the next batch. Figure 17.13 shows the paint production process along with sources of liquid and solid waste generated.

Treatability aspects of waste-water

Combined effluent from paint industries can be satisfactorily treated using the usually physico-chemical and/or biological treatment methods. The treatment consists of coagulant addition and adjustment of pH to an optimum level for maximum precipitation, the precipitated material is removed by gravity separation, either on batch basis or in a continuous flow tank. Ultra filtration and activated carbon adsorption have also been tried but have not been found to be cost effective. Using physico-chemical processes, removals of 90 per cent or greater can be achieved for some significant toxic pollutants. Even after this treatment if some toxic pollutants remain, biological treatment can reduce their concentration. Some of the organics which remain in the supernatant after physico-chemical treatment are biodegradable and can be removed through biological treatment.

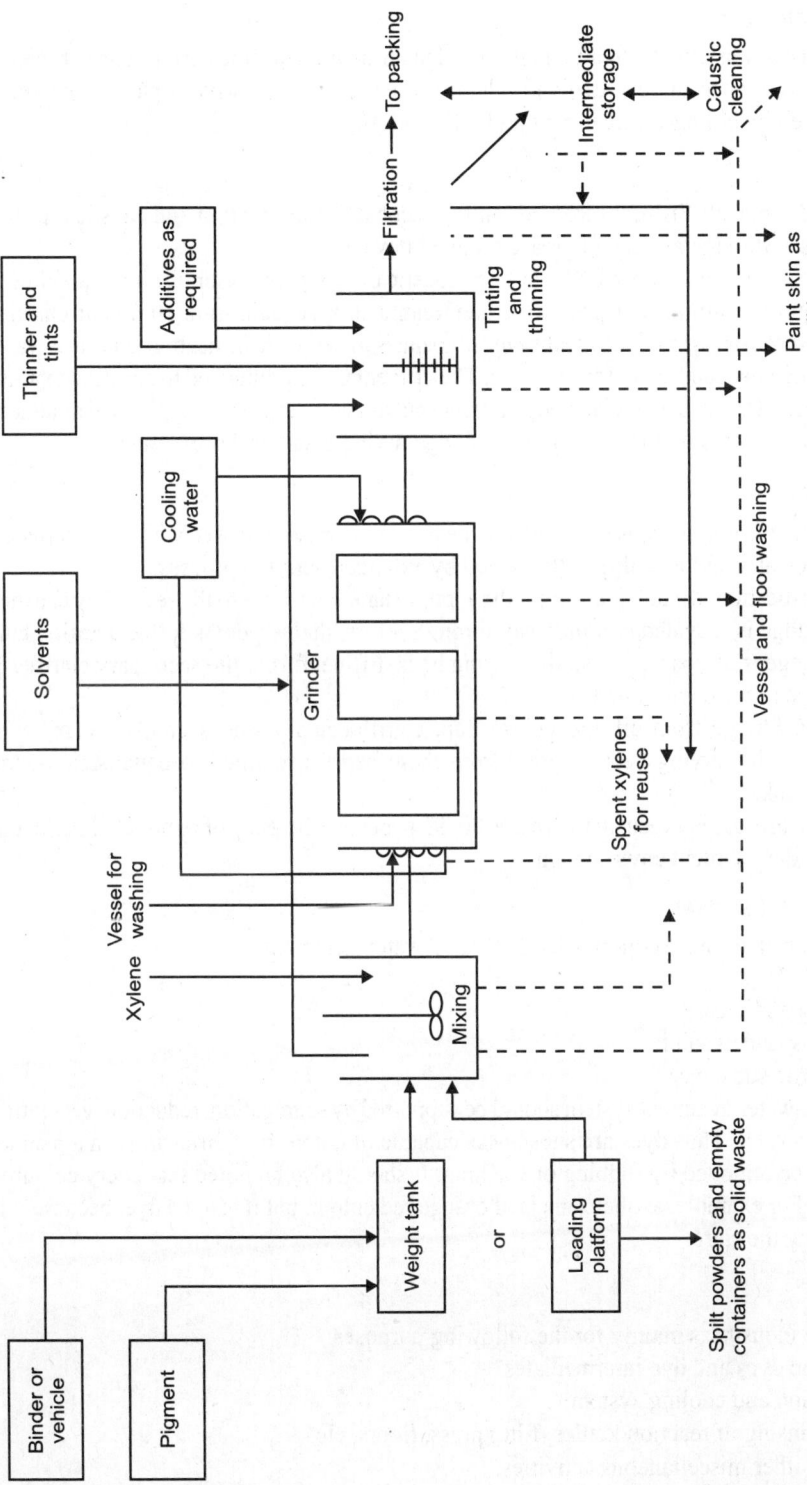

Fig. 17.13. Flow diagram for manufacturing paints with general sources of solid and liquid waste generation and possible solvent recovery steps.

Waste-water treatment

Considering the practice of waste-water treatment followed in the country and the performance efficiency of each operation, the best alternative is evolved. The treatment system consists of physico-chemical treatment units followed by biological treatment units (Fig. 17.14).

Primary treatment

1. Oil and grease removal: Effluents from all units except stiff paint section and caustic cleaning waste are passed through an oil and grease removal device.
2. Equalisation-cum-neutralisation: Effluent from caustic cleaning operation is highly alkaline in nature and requires neutralisation prior to further treatment. An equalisation-cum-neutralisation tank is provided with an agitator. Effluent from stiff paint is mixed with the neutralised waste-water, dosed with a coagulant and sent to flash mixer. The effluent is then subjected to clariflocculation.
3. Clariflocculation: The effluent is clarified in clariflocculator and subjected to biological treatment. Sludge generated in this unit is carried to the sludge drying beds for dewatering.

Secondary treatment

1. Extended aeration: Domestic waste-water from the factory premises is mixed with the supernatant from clariflocculator and is biologically treated by extended aeration process.
2. Secondary clarification: Mixed liquor from the aeration tank overflows to the secondary clarifier. The settled sludge is recycled continuously through return sludge pumps to the aeration tank and excess sludge is discharged to sludge drying beds. Effluent from the secondary clarifier is fit for discharge to the environment.
3. Sludge drying: Sludge from oil and grease trap, clariflocculator and secondary clarifier is dewatered on sludge drying beds. Filtrate from these beds is returned to equalisation-cum-neutralisation tank.

The above treatment process is expected to achieve 90–95 per cent efficiency of removal of pollutants and thus acceptable to the recipient environment.

Recommended treatment system

The following minimum steps are recommended for waste-water treatment:
1. Adjustment of pH.
2. Removal of oil and grease.
3. Removal of suspended solid.
4. Removal of toxic substance.

The proposed waste-water treatment system should be supported by segregation, reduction, generation and recycling of waste-water. Thus dyes are substances capable of colouring fabrics in such a manner that the colour cannot be removed by rubbing or washing. It should also be noted that every coloured substance is not a dye. For example, azobenzene is of orange red colour, but it is not a dye, because it is not capable of colouring the fibre.

Waste Generation

The water usage in the industry is mainly for the following purposes:
1. Synthesis of the dyes and dye intermediates.
2. Steam generation and cooling system.
3. Washing and rinsing of reaction kettles, filter press, floors, etc.
4. Domestic and other miscellaneous activities.

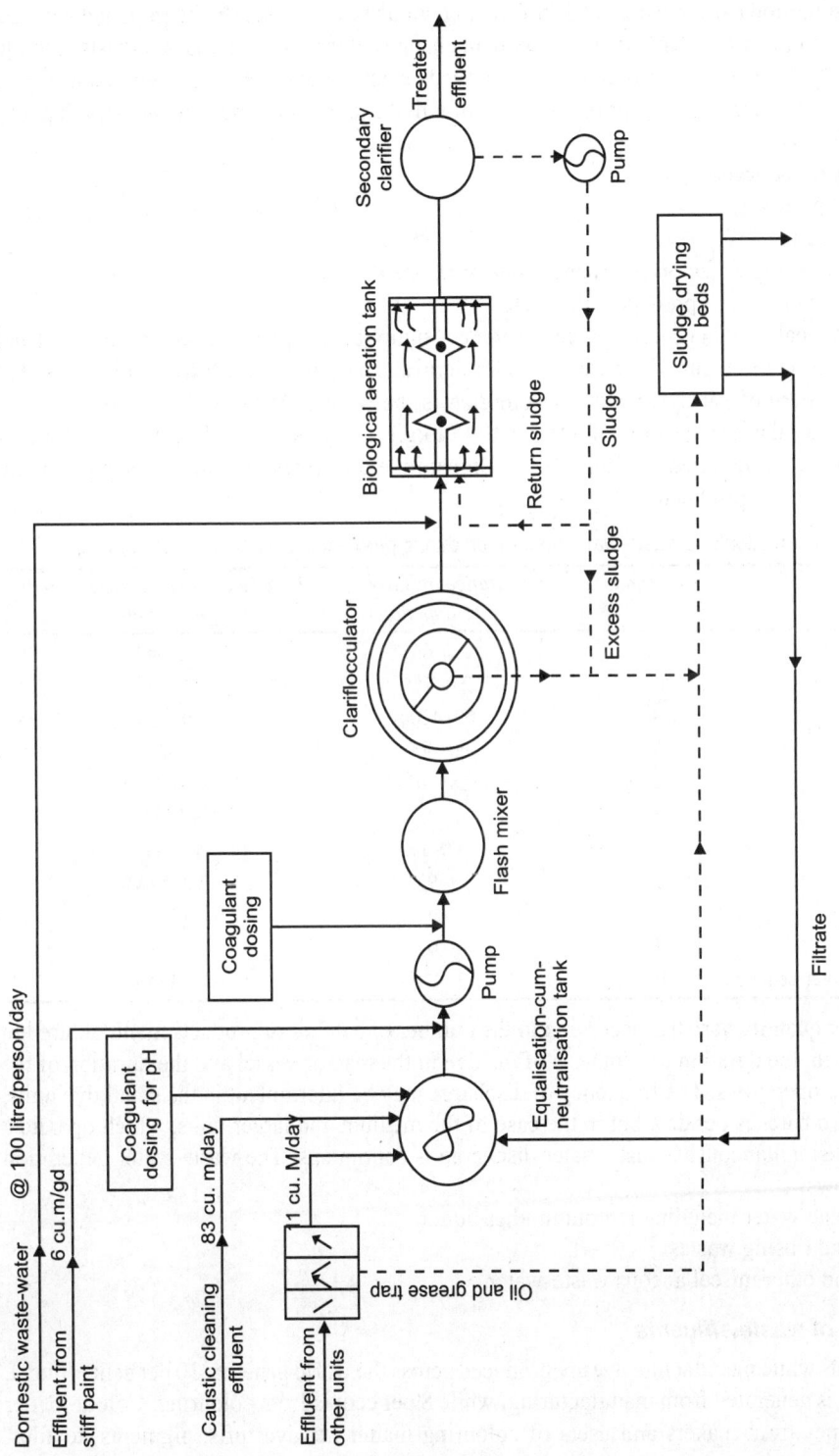

Fig. 17.14. Typical waste scheme for a factory producing 100 m³/day industrial waste-water and domestic waste-water @ 100 l/person/day.

The water consumption pattern varies widely from one industry to another. In the same industry the rate of water consumption often changes due to frequent changes of the feed material synthesis reaction and desired products. The change of product pattern needs cleaning and washing which consumes a substantial quantity of water. Thus water requirement of a dye and dye intermediate industry depends on the following factors:

1. Type of dye produced.
2. Number of products.
3. Gross production.
4. Pattern of working of factory, i.e. continuous or in one shift only.
5. Frequency of change of product pattern, etc.

In-depth study reveals that in general, process water consumption is highest, next to it is the cool and boiler make-up water requirement. The water needs for domestic purposes is the lowest. It is found that for production of one kg of vat dye, water consumption is the largest (1528 to 10345 l/kg) whereas naphthol dye consumes the least quantity of water (6 to 17 l/kg.). The generation of waste-water follows the trend of consumption of water. In Table 17.11 comparative quantities of water consumption and waste-water generation are provided.

Table 17.11. Water consumption and waste-water generation during production of various types of dye products.

Type of product	No. of units	Range of water consumption	Litre/kg of product waste-water generation
Direct dyes	15	2.5–667	1.0–644
Reactive dyes	11	2.0–186	2.0–157
Basic dyes	4	60–4200	50–200
Azo dyes	8	90–400	8.0–213
Vat dyes	2	1528–10345	1389–7980
Dye intermediates	6	36–230	9.0–74
Naphthol dye	2	6.0–17	5.0–8.0
Pigments	3	93–923	7.0–7.85
Indigosol colours	1	529	429
Disperse dyes	1	70	12–42.5
All varieties of dyes/intermediates	10	13–2300	11–1146

The waste-water quantity varieties according to the number of batches of products manufactured in a day, week or month, the duration of synthesis of the dye in the reactor vessel and the duration of the washing and rinsing operations. The frequency of discharge may be intermittent in the small dye units, operating one or two batches per day, but in the case of the medium and larger units, which operate a number of reactors simultaneously, waste-water discharge is continuous. The waste-water generation sources are as follows:

1. Process waste-water including left out mother liquor.
2. Washing and rinsing wastes.
3. Sanitary and other miscellaneous waste-water.

Characterisation of waste effluents

Some of the effluents while manufacture dye are produced across the world generate 10 per cent effluent, in which 2 per cent is generated from manufacturing, while 8 per cent is from colouring. Colour matter in waste water comes from makers and users of colouring matter i.e. dyestuffs, pigments, textiles, dyeing units and tanneries. The pulp and paper sector and distilleries which use raw materials with colour as a by-product, also discharge colour matter into waste water. Effluent left after dyeing contains unused dyes in the form of organic and inorganic compounds, toxic metals, suspended and dissolved solids.

Liquid effluents

Liquid effluents from dye manufacturing are mainly waste-water (biodegradable and non-biodegradable) or high COD waste-water. About 15 per cent of the dye that is manufactured and used by the industry is discharged into the water. Of this, approximately 2–3 per cent is discharged by the dye manufacturing industry and about 14–15 per cent by the textile dyeing industry.

Waste-water characteristics

The process waste-water is mainly the mother liquor left over after the product is isolated and separated by filter press. This waste-water is of smaller volume and highly concentrated in terms of pollutants. The vessel washings also contain similar type of pollutants but with lower concentration. It has been identified that the waste-waters of the industries have the following characteristics:

1. High levels of BOD and COD.
2. High acidity.
3. High TDS.
4. Deep colour of different shades.
5. High levels of chlorides and sulphates.
6. Presence of phenolic compounds.
7. Presence of heavy metals, e.g. copper, cadmium, chromium, lead, manganese, mercury, nickel and zinc.
8. Presence of oil and grease.

The dye industry waste-waters, if derived from naphthalene and anthracene bases are resistant to biodegradation. The colour removal is also not adequate by the conventional chemical and biological treatment. The characteristics of the combined waste-water of a dye industry is presented in Table 17.12. In Tables 17.13 and 17.14 the characteristics of combined waste-water of two dye industries engaged in the production of other products are given.

Table 17.12. Waste-water treatment in dyestuff Industries.

Characteristics, mg/l except pH, temperature and colour	Waste-water Untreated	Waste-water Treated	Efficiency % removal
pH	3.4	7.2	–
Temperature°C	45	30	–
Colour	Dark brown	Dark brown	–
TDS	5455	5658	–
SS	96	65	32
BOD	503	49	90
COD	1518	824	44
Chloride	1952	1927	–
Sulphate	1675	1028	–
Cyanide	0.075	0.04	47
Oil and grease	19.3	7.2	62.5

Note: Treatment system: Equalisation, neutralisation, flocculation, sedimentation, extended aeration and clarification. The sludges are dried in sludge drying bed.

Table 17.13. Waste-water treatment in dye industry.

Characteristics, mg/l except pH	Waste-water		Efficiency % removal
	Untreated	Treated	
pH	1.9	6.0–8.5	–
TDS	3670	6650	–
SS	370	200	46
BOD	350	275	22
COD	930	500	46
Chloride	1125	800	–
Sulphate	1985	–	–
Phosphate	23	–	–
Total Nitrogen	17	–	–
Oil and grease	20	10	50
Phenol	10.8	4.0	66

Note: Treatment system: Equalisation and neutralisation by aeration adopting batch process.

Table 17.14. Waste-water treatment in dye-intermediate(s) industry.

Characteristics, mg/l except pH, temperature and colour	Waste-water		Efficiency % removal
	Untreated	Treated	
pH	3.0	7.5	–
Temperature °C	29	29	–
Colour	Brown yellow	Pale yellow	–
TDS	1000	500	–
SS	200	50	75
BOD	1000	100	90
COD	1500	400	73
Cyanide	3	Nil	100
Oil and grease	10	1.0	90

Note: Treatment system: Oil and grease removal, homogenisation, neutralisation, primary clarification, biological treatment and secondary clarification.

Reduction of waste-water

In-plant control

It is essential to have proper in-plant control measures before going for waste-water treatment. Some of the relevant measures in-plant are summarised below.

Reduction of waste

The volume of waste-water can be reduced by proper control of freshwater consumption. The cooling water blow-down may be reduced by raising the concentration factor. Timely maintenance of the units may be done to prevent leakage, spillage, etc. Minimal usage of water for washing and rinsing may be practised. The last wash water may be recycled for first washing. Application of counter-current washing may also reduce waste-water generation. Dry cleaning of the floor is preferred. When floor washing is absolutely necessary, the treated waste-waters may be used. The pollution load can be reduced by recovery of chemicals and solvents as far as practicable. Spills, leakages, overflows, etc. may not be allowed to join the waste-water stream.

Segregation

Storm waters need segregation to reduce the volume of waste-water. Similarly the uncontaminated and less contaminated waste-water streams like cooling water blow-down, boiler blow-down, condensate, etc. are to be segregated and should not be permitted to join the process waste-water stream. The highly contaminated and coloured mother liquors should be segregated and collected separately. Sometimes strong wastes of the reaction vessels are required to be discharged. These unforeseen strong wastes are to be collected in separate holding tanks and drawn into the waste-water treatment plant at a regulated rate.

Process modification

The production process equipments should be modified so as to generate less wastes. The raw materials used in the synthesis may be substituted by choice of more readily biodegradable chemicals.

Effluent control

The waste-water streams originating as mother liquor and strong waste-waters as mentioned earlier should be segregated and collected in two separate collection tanks. The rest of the process waste-waters may be diverted to the equalisation pond.

Oil and grease and the floating matters present in the waste-water streams are to be arrested by putting up appropriate control measures at the boundary limit of each plant. The waste-water collected in equalisation pond will be monitored to ascertain the concentration of pollutants. Based on the data obtained, the waste-waters of holdings, ponds are to be diverted to the equalisation pond at regulated rates so that the concentration of pollutants are acceptable to the biological treatment system.

The equalisation of waste-waters is followed by pH adjustment and clarification in primary clarifier. Clarifier may be of sludge blanket type. Ferrous sulphate and lime may be used for precipitation of heavy metals. The overflow of the clarifier after pH adjustment will flow to the extended aeration type biological treatment unit. The effluent of biological treatment unit will be clarified in convenient type clarifier using coagulant. The reduction of colour may be achieved by segregating and controlled discharge of the mother liquor, which also contains most concentrated form of the chemicals left after dye synthesis. It is envisaged that substantial colour reduction of the waste streams may be achieved by the three-stage treatment system as proposed above. However, if the desired limit of colour (400 Hazen unit) is not achieved by the above treatment then one or both of the following steps are to be adopted:

1. Evaporation of the segregated mother liquor and washings of vessels by solar evaporation or by indirect heating using steam.
2. Powered activated carbon adsorption of the treated waste-water.

DRUGS AND PHARMACEUTICALS

The pharmaceutical industry, although a strong and important entity in itself, is frequently considered to be part of the chemical industry. The pollution problems of the pharmaceutical industry, particularly those of the major companies, parallel those of the chemical industry. Because the problems of the industry are not unique, this section will be concerned with emphasising the problems, where they should be looked for, and some of the more interesting solutions to solve these. The pharmaceutical companies facing major problems in connection with both air and water pollution are the ones that operate fermentation plants, principally in the production of antibiotics (penicillin, streptomycin, etc.).

Waste Generation

Waste generation from the basic drug manufacturing houses are higher than the formulating units. The nature and the quantity of waste generated vary depending upon the processes involved.

In general, the waste and emissions generated during manufacturing of pharmaceuticals depend on the raw materials and equipment used, as well as the manufacturing, compounding and formulation process employed. In designing bulk manufacturing processes, consideration is given to the availability of the starting materials and their toxicity, as well as the wastes (e.g. mother liquors, filter residues, and other by-products) and the emissions generated. When bulk manufacturing reactions are complete, the solvents are physically separated from the resulting product. Due to purity concerns, solvents are not often reused in a pharmaceutical process. They may be sold for non-pharmaceutical uses, utilised for fuel blending operations, recycled, or destroyed through incineration.

Effluent Generation

Pharmaceutical manufacturers use water for process operations, as well as for other non-process purposes. However, the use and discharge practices and the characteristics of the waste-water will vary, depending on the operations conducted at the facility. Additionally, in some cases, water may be formed as part of a chemical reaction.

Chemical methods

Cyanide destruction

Several cyanide destruction treatment technologies are currently used in the pharmaceutical manufacturing industry, including alkaline chlorination, hydrogen peroxide oxidation, and basic hydrolysis.

1. The alkaline chlorination treatment process involves reacting free cyanide with hypochlorite (formed by reacting chlorine gas with an aqueous sodium hydroxide solution) to form nitrogen and carbon dioxide. The reaction is a two-step process and is normally performed separately in two reactor vessels. Because treatment is normally performed in batches, it is necessary to use an additional equalisation tank to store accumulated waste-water during treatment. The reactors need to be equipped with agitators, and both reaction steps require close monitoring of pH and oxidation reduction potential (ORP). These reactions are normally performed at ambient temperatures.

2. Hydrogen peroxide treatment involves adding hydrogen peroxide to cyanide-bearing waste-water to convert free cyanide to ammonia and carbonate ions. This treatment is normally performed batch-wise, in a reaction vessel. The treatment process consists of heating the waste-water to approximately 125°C and adjusting the pH in the reaction vessel to approximately 11. Hydrogen peroxide is added to the vessel and is allowed to react for approximately one hour. Equipment required for this process includes reaction vessels, storage vessels for hydrogen peroxide and a pH adjustment compound (typically sodium hydroxide), an equalisation tank, and feed systems for hydrogen peroxide and sodium hydroxide.

3. Hydrolysis treatment involves reacting free cyanide with water under basic conditions to produce ammonia. This process requires approximately one hour and is typically performed at a temperature between 170° and 250°C, at a pH of 9 and 12. Hydrolysis is normally performed in a reactor vessel equipped with a heat exchanger and a system to store and deliver sodium hydroxide (or other basic compounds).

Oxidising agents like ozone, UV rays and hydrogen peroxide

Ozone gas is highly reactive and is used in the treatment of waste-water. Ozone is effective in oxidising antibiotics, betablockers, antiphlogistics, lipid regulator metabolites, the antiepileptic drug, carbamazepine and the natural estrogen, estrone from STP. Also, the combined treatment of hydrogen peroxide and UV radiations reduces the amount of diclofenac present in waste-water.

Fermentation

Waste-waters from a fermentation unit emanate mainly from the recovery and purification of the final product and also from washings of floor, vessels, equipments, etc. About 5 m^3/kg of final product is the waste-water generated. The main sources of wastes generation from a penicillin fermentation plant are listed in Table 17.15.

Table 17.15. Waste-water generation from an antibiotic plant.

Source of waste-water generation	Nature of waste-water and solid waste	Ranges for average characteristics of combined effluent in mg/l except pH	
Fermentation block	Floor and equipment washings,	pH	4.0 to 8.0
	leakages of valves/ machines, contaminated batches,	TSS	100 to 1000
	cooling waters, laboratory and utility wastes	BOD	500 to 6,000
Filtration/centrifugation	Mycellium cakes, filter washings, floor washings	BOD	Upto 10,000
Recovery and purification block	Faecal wastes, acid and alkali wastes (from regeneration of ion-exchangers), floor washings laboratory wastes	CN^- Heavy metals Oil and grease	0.1 to 1 1 to 5 20 to 50
Style and finishing block	Floor and equipments, washings and other utility wastes	Phenol (after mycellium cake separation)	1 to 5

WASTE-WATER TREATMENT

A general flow diagram for the treatment of waste-waters from various types of pharmaceutical products is presented in Fig. 17.15. Depending upon the nature of the waste-water, selection/omission of the specific treatment units can be made.

Before selection of a particular treatment system for effluents of pharmaceutical industries the following aspects are required to be considered:
1. Good housekeeping practices.
2. Segregation of certain waste-water streams.
3. Process and equipment modifications.
4. Recovery of by-products and recycle possibilities.

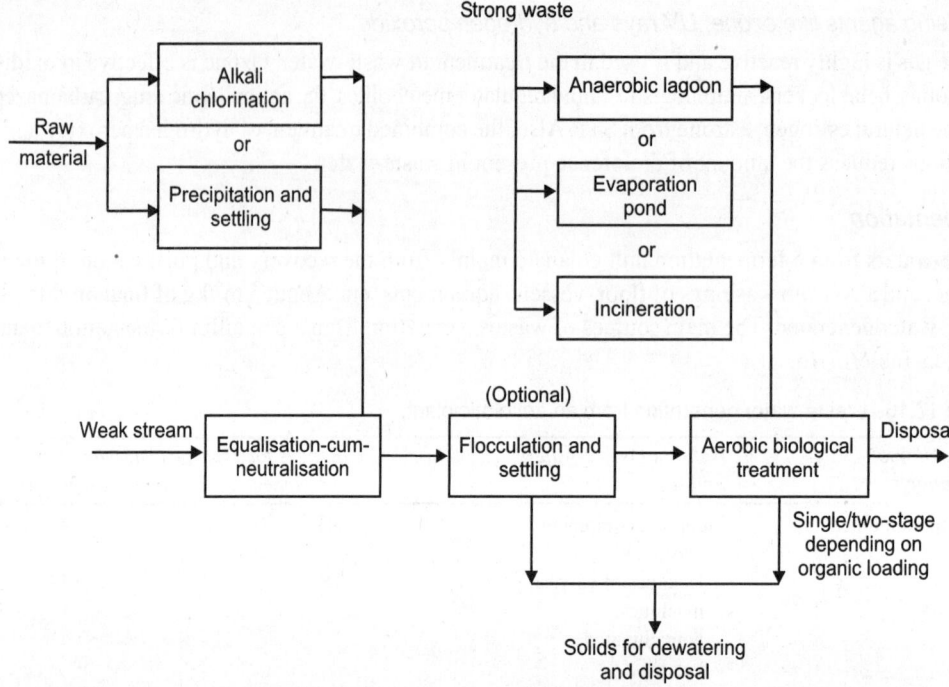

Fig. 17.15. A general flow diagram for the treatment of waste-waters from pharmaceutical industries.

Certain waste-waters containing pathogenic micro-organisms from fermentation particularly vaccine section need autoclaving before discharge into common treatment plant. Various treatment processes are discussed below.

Physical treatment

The common physical treatments are plain settling, dissolved air flotation, adsorption, solar evaporation. These methods can be adopted individually or in combination depending upon the waste-water quality and quantity.

Plain settling

Water insoluble compounds can be removed from the waste-water by plain settling. The settling characteristics can be improved with the addition of suitable coagulants. The system consists of a settling basin with adequate volume to provide sufficient settling time for the impurities.

Dissolved air flotation

Suspended, colloidal and emulsified impurities can be removed from waste-water by air flotation technique. Air dissolved in the waste-water under pressure when released in the flotation tank forms bubbles which trap the lighter impurities and lift them to the surface. The materials form a floating layer which is removed by a skimmer mechanism.

Primary treatment/chemical treatment

Segregation of different waste streams is an important step for economic design of a treatment plant.

The following streams may be segregated for this purpose:

1. Strong process liquors.
2. Streams containing cyanide, heavy metals, toxic chemicals.
3. Condensate and cooling waters.
4. Acidic and alkaline streams.

Various treatment alternatives for strong process liquors are incineration, solar evaporation and treatment by anaerobic filter. The toxic effluents either can be incinerated or treated by suitable technologies like carbon adsorption, ion-exchange, chemical precipitation, reverse osmosis, etc. Condensate and cooling waters can be recycled and reused. The acidic and alkaline waste streams can be either treated separately with acid/alkali for pH correction or may be combined suitably with other waste streams.

The effluent containing chemical sludge, settleable solids and high oil concentration (over 50 mg/l) can be treated by coagulation, flocculation and settling after neutralisation. Coagulants like alum, $FeSO_4$, $FeCl_3$, etc. with/without polyelectrolytes can be used. The coagulation process also breaks oil emulsions and nullifying the zeta potential.

Pre-aeration for 2 to 3 hours by means of diffused air may help to bring down the BOD load of about 30 to 40 per cent. Diffused aeration is known to bring oily and fatty matters in suspension form in the waste-water.

Secondary treatment

Various factors are responsible to select a suitable treatment system, i.e. quality and quantity of influent to be treated, desired degree of treatment, site conditions, change of products and overall economics. Secondary biological treatments employed are mainly aerobic and in some cases anaerobic followed by aerobic treatment.

Trickling filter, extended aeration and conventional activated sludge systems are generally practised. Anaerobic filter and anaerobic lagooning are also being used for treatment of pharmaceutical industry waste-waters.

Tertiary treatment

Desired effluent quality is the basis for the selection of any treatment scheme. In most cases, pharmaceutical industry effluents are not suitable for land disposal for farming due to the presence of high concentration of dissolved salts. Tertiary treatments are required to kill virus and bacteria and to remove other impurities like colour, bad smell, etc. Chlorination and sand filtration may generally be practised for tertiary treatment.

Biological treatment

Various factors are responsible for the selection of biological treatment system, i.e. quality and quantity of waste-water to be treated, desired degree of treatment, site conditions and overall economics. Biological treatment systems employed for pharmaceutical wastes are mainly aerobic in nature. In some cases where the organic loading is high, anaerobic treatment units are used prior to aerobic system.

The widely used anaerobic treatment unit for pharmaceutical wastes is anaerobic lagoon and for aerobic treatment, trickling filter, aerated lagoon or aeration tank followed by clarifier are used.

SUGAR AND DISTILLERY

Sugar

In sugar mills, the waste generated includes water used as splashes to extract the maximum amount of juice, and those used to cool roller bearings. As such, mill house waste contains high BOD due to the presence of sugar and oil from the machineries.

The filter cloths, used for filtering the juice, need occasional cleaning. The wash water thus produced, though small in volume, contains high BOD and suspended solids.

It is well known that sugarcane is the basic raw material for the extraction of sugar in India. After crushing of sugarcane, the juice is squeezed and the left-over cellulosic bagasse, which otherwise is a solid waste for the industry, is rich in fibre contents. Since, such a fibrous substance is a good raw material for the pulp and paper manufacturing industries, it can easily be sold to these industries.

After the purification of the juice, it is transferred to evaporator pans for concentration prior to crystallisation of sugar. In the evaporator pans, the juice is vacuum evaporated so that its water content is reduced. These water vapours are, in fact, one of the purest forms of water when condensed, without any carry over of sugar, and can be used as boiler-feed water.

In this regard, the 'polybaffle entrainment catcher' in pan evaporation for sugar recovery provides an alternative over the conventional pan evaporation technique. Such water will not only reduce the scaling in the boiler tubes, but also conserve large quantities of freshwater to be used. In addition, this hot condensate needs far lesser amount of energy (when compared with the freshwater) to be transformed into steam for obvious reasons, resulting in money saving to the industry. More recently, it is emerging that if hot condensate is used to extract the last traces of sugar from the crushed sugarcane, not only does this leads to increased recovery, but it also saves large quantities of freshwater that is used otherwise. It also requires less heat energy for the sulphitation process. According to an estimate by the Central Pollution Control Board, in general, approximately 1500 to 2000 litres of water is used per ton of cane crushed. The recycling and reuse of hot condensate water can reduce this consumption to as low as 100 to 200 litres per ton of cane crushed. Proper housekeeping, periodic checking and maintenance of pipe joints, valves and glands can further reduce this. Ideally speaking, the industry can be brought down to zero effluent discharging; if not, the small quantities of waste-water can easily be treated in the end-of-pipe treatment, requiring least amount of energy and resources.

Sources of waste-water and the characteristics of the wastes

Wastes from the mill house include water used as splashes to extract maximum amount of juice, and is used to cool the roller bearings. As such, mill house waste contains high BOD due to the presence of sugar, and oil from the machineries. The filter cloth, used for filtering the juice, needs occasional cleaning. The wash water thus produced, though small in volume, contains high BOD and suspended solids. A large volume of water is required in the barometric condensers of the multiple effect evaporators and vacuum pans. The water is usually partially or fully recirculated, after cooling through a spray pond. This cooling water gets polluted as it picks up some organic substances from the vapour of the boiling syrup in the evaporators and vacuum pans. The water from spray pond when it overflows, becomes a part of the waste-water, and is usually of low BOD in a properly operating sugar mill. But because of poor maintenance and bad operating conditions, a substantial amount of sugar may entrain in the condenser water; this polluted water, instead of being recirculated, is discarded. These discharges contribute substantially to the waste volume and moderately to BOD in many sugar mills.

Additional waste originates due to the leakages and spillages of juice, syrup and molasses in different sections, and also due to the handling of molasses. The periodical washings of the floor also contribute a lot to the pollution load. Though these wastes are small in volume and are discharged intermittently, they have a very high BOD. The periodic blow off of the boilers produce another intermittent waste discharge. This waste is high in suspended solids, low in BOD and is usually alkaline.

Characteristics of wastes from the different sections of common sugar mills and that of the combined waste are given in Table 17.16. Composition of the waste and its volume varies widely from mill to mill, depending upon the availability of water.

Table 17.16. Characteristics of sugar mill wastes.

Characteristics	Mill-house waste	Filter cloth washings	Condenser water	Boiler house and floor washings	Combined waste (excluding condenser water)
Rate of flow litres per ton of cane crushed	730	360	1640	230	–
pH	6.7	9.5	–	7.2	4.6–7.1
COD, mg/l	–	–	–	–	600–4380
BOD, mg/l (5-day 20°C)	210	1765	–	5150	300–2000
Total solids, mg/l	1760	6970	–	5130	870–3500
Total volatile solids, mg/l	–	–	–	–	400–2200
Total suspended solids, mg/l	910	4000	–	120	220–800
Total nitrogen mg/l	–	–	–	–	10–40
COD/BOD ratio	–	–	–	–	1.3–2.0

Effects of the waste on receiving water

The fresh effluent from a sugar-mill decomposes rapidly after few hours of stagnation. It has been found to cause considerable difficulties when their effluent gets an access to water courses, particularly the small and non-perennial streams in the rural areas. The rapid depletion of oxygen due to biological oxidation followed by anaerobic stabilisation of the waste causes a secondary pollution of offensive odour, black colour, and leads to fish mortality. No question of the discharge of this waste into the sewers arises, as most sugar mills are situated in the unsewered rural areas.

Treatment of the wastes

Like any other industry, the pollution load in sugar mills can be reduced with a better water and material economy practised in the plant. Judicious use of water in various plant practices and its recycling, wherever practicable, will reduce the volume of waste to a great extent. Volume of mill house waste can be reduced by recycling the water used for splashing. Dry cleaning of floors or floor washings using controlled quantity of water will also reduce the volume of waste to certain extent. The organic load of the waste can only be reduced by a proper control of the operations.

Distillery Industry

The distillery industry uses sugar cane molasses, cereal and other agro products to produce alcoholic beverages. Typical primary products of a grain alcohol distillery include the beverages: whiskey, gin and vodka. Cooked slurries of ground whole grains, called 'mash', are fermented, without prior filtration, using a yeast that operates rapidly at about 95°F. Ratios of the various grains are varied within specified limits and malted barley furnishes saccharifying enzymes. Complete conversion of the grain starch to alcohol is sought in a fermentation time of 2–3 days. The alcohol content of a finished fermentation is about 7 per cent.

Treatment of distilling industry effluents

Most of the distilleries are facing the environmental issue of treatment and disposal of its spent wash. A wide range of technologies have been tried for the treatment of spent wash, however, none of these methods have been found to be effective and economically viable, in order to achieve the standard norms set by the Central Pollution Control Board, Government of India.

Primary treatment

Primary treatment methods selectively remove materials, which could interfere with the physical operation and subsequent treatment processes. The processes are based on exploitation of the physical properties of the contaminants, and are generally used at the initial stages of effluent treatment. It helps in improving the treatment efficiencies by reduction of surface area. Screening, flow equalisation, comminution, mixing, flotation, flocculation and sedimentation are the methods used during the initial stages of effluent treatment. Chemicals are also used in conjugation with physical treatment, in which sedimentation is performed, by the addition of coagulants and other additives such as alum, lime, ferric chloride, activated charcoal, etc. Different chemical treatment methods were investigated, based on chemical properties of the pollutants, and as a result, chemical treatment with Fe(III) and anionic polymer, followed by H_2O_2 +Fenton reagent was selected as the best option because of its performance and viability.

Strong oxidants like ozone and hydrogen peroxide were used. Though they result in decolourisation, they are not accepted at present, because of economic reasons. It has been observed that inorganic flocculants such as iron sulphate, iron chloride and aluminium sulphate do aid the separation of colouring matter. Oxidation by ozone combined with hydrogen peroxide is found to be very effective in the degradation of some aromatics present in waste-water and removal of colour. Due to generation of sludge, accumulation of metals, cost, and related contaminants, chemical methods are not successful.

Secondary treatment

Secondary treatment is also known as biological treatment, in which soluble and colloidal form of organic matter is removed. Both aerobic and anaerobic micro-organism is used in secondary treatment. Anaerobic digestion is the most suitable option for the treatment of high strength organic effluents. The presence of biodegradable components in the effluents, coupled with the advantages of the anaerobic process over other treatment methods, makes it an attractive option. In addition, a modification in the existing reactor designs for improving the efficiency of digestion has also been suggested.

Anaerobic and aerobic treatment systems can remove most of the biologically removable organics, CODs and colour. Biologically treated distillery waste-water is nonbiodegradable. Anaerobic treatment of effluents from different alcohol-producing industries over a long-term period is performed. Since most of the COD and colour in biologically treated distillery waste-water is nonbiodegradable, identification and optimisation of biotechnological treatment methods are a necessary today.

Tertiary or alternative treatment

A variety of treatment methods and strategies, like thermal pre-treatment, wet air oxidation, concentration-incineration, anaerobic treatment, etc. have been suggested or tested for the treatment of distillery waste-water. As individual schemes, these are either incomplete, impractical or unviable. Thus, there is an urgent need to assess the possibility of combining the available partial treatment schemes, for the complete treatment.

TEXTILE INDUSTRY

The textile industry actually represents a range of industries with operations and processes as diverse as its products. It is almost impossible to describe a typical textile effluent because of such diversity. After fabrics are manufactured, they are subjected to several wet processes collectively known as 'finishing' and it is in these finishing operations that the major waste effluents are produced. These finishing processes are complex and ever-changing. This is a fact of life that is reflected in the variety of chemicals that find their way into textile finishing waste-waters.

Water Pollution from Boilers

Boilers meant for raising steam by combustion of fuels like coal, fuel oil, etc. contribute substantially in causing pollution in two ways:
1. Water pollution caused due to discharge of boiler blowdown water.
2. Air pollution caused due to emission of chimney gases into the atmosphere.

In order to ensure that concentration of dissolved solids in the boiler water does not exceed the permissible limit, some amount of boiler water has to be discharged from the boilers either continuously or intermittently. The boiler blowdown water so discharged contains very high amount of dissolved solids ranging from as low as 3500 mg/l for water tube boilers to as high as 10,000 mg/l for Lancashire boilers.

In fact in some cases, the blowdown water from the Lancashire boilers are known to contain as much as 60,000 mg/l of dissolved solids. Whereas raw water with high carbonate alkalinity is fed into the boilers without softening, hardness precipitates inside the boilers giving rise to high level of suspended solids in the blowdown.

The alkalinity of the blowdown is also very high and is generally about 15 per cent of the dissolved solids. Likewise, pH of the blowdown is also about 10.5 to 11.0. Thus the blow-down adds to the water pollution in terms of dissolved solids, suspended solids, alkalinity and pH. Table 17.17 gives the characteristics of boiler blowdown water.

Table 17.17. Characteristics of boiler blowdown water.

pH	10.5 to 11.5
TDS	18000 mg/l
Suspended solids	1500 mg/l
Temperature	95° to 100°C
Chlorides as Cl	4500 mg/l
Sulphate as SO_4	1700 mg/l
Total alkalinity as $CaCO_3$	8000 mg/l

Water Pollution from Water-Treatment Plants

It is required that the water used for processing of textiles and in generation of steam in the boilers should not be hard. The type of softening treatment generally imparted to the hard water is ion-exchange softening process. There is no water pollution from the softening process as such. However, when the ion-exchange resin is regenerated with concentrated solution of common salt, its effluent will contain very high amount of sodium ions apart from high dissolved solids. In case of partial demineralisation process, the effluent contains weak acid. Some mills have water treatment plant of precipitation type. Here lime is added to precipitate carbonate hardness and soda ash is added to remove noncarbonate hardness. Some coagulants like alum or sodium aluminate, etc. are added to accelerate the precipitation process. In this type of treatment, some amount of water is drained out from the bottom of the reaction vessel to get rid of the precipitates. This waste-water contains high amount of suspended solids and adds to the water pollution when it joins the composite stream.

Characteristics of Textile Waste-waters

From the preceding sections, it is apparent that the characteristics of the waste-waters from a textile plant will depend on the specific operations in the plant. It is misleading to speak of a typical textile effluent. The type of fibre involved and machinery employed are the main factors determining the type and quantities of chemicals present in the textile waste-waters.

Finishing processes discussed earlier may be either batch or continuous. For batch processes, the discharge is intermittent, with the interval between discharged depending on the operations. All the waste-waters from a batch process are likely to come from the same operation, the first being the most heavily contaminated and the last rinse the most dilute. For continuous processes, a steady flow of effluents with moderate concentrations is expected.

Developing a database from sampling the textile effluent discharges at frequent intervals should lead to establishing reliable average values. In addition to frequent sampling, the other approach to determining the characteristics of textile waste-waters is to study the process, its waste components and volumes. For example, the amount of organic matter that is removed from a fabric in the course of normal textile processing can be visualised when one considers that about 10 per cent of the gross weight of a cotton fabric consists of natural impurities that may be removed in processing. For a firm that manufactures 20 tons of fabric per week, about 2 tons per week of impurities are discharged to the sewer. With known discharge volumes, the concentrations of the impurities in the effluent may be estimated.

Kremer and others reviewed the pollutants generated by the textile industry. They divided the pollutants into four general groups, i.e. sizes, detergents, dyes and priority pollutants. They reported that most of the priority pollutants contained in textile industry effluents are aromatics, halogenated hydrocarbons and heavy metals.

Cotton and linen contribute organic matter from the noncellulosic materials that are present in the natural fibres, whereas wool contains sand and grease that are removed during scouring. Synthetic fibres may contain spinning oils and antistatic dressings. Textile wastes are generally coloured, highly alkaline and high in BOD, suspended solids and temperature. The raw waste-water (pH = 9) of a bleachery had 660 mg/l of BOD, 2080 mg/l of COD, 34 mg/l of oil and 2700 mg/l of TDS. Randall and King reported that waste-waters from textile dyeing and finishing operations may be characterised as high in organic matter, both biodegradable and nonbiodegradable. In addition, they tend to be high in surfactants and contain potentially significant concentrations of oils, phenols and heavy metals such as chromium, zinc and copper. Randall and King reported the characteristics of raw waste-waters from three plants. The BODs varied between 260 and 560 mg/l. The colour (APHA units) varied between 1000 and 1335. The hue was brown, red to black or yellow to black. Kertell and Hill reported the characteristics of the waste-water from a dye and finishing company. The average BOD was 371 mg/l and the average colour was 113. About 70 mg/l of oil and grease were present. The total solids concentration was about 480 mg/l. Troxler and Hopkins reported that the introduction of continuous dyeing machines had significantly increased the strength of carpet finishing waste-water. This is due to the use of natural bean gum thickeners as a viscosity modifier for the dye solutions. The average BODs and COD from 'beck dyeing' waste-water were 232 and 943 mg/l, respectively. The average BODs and COD from continuous dyeing waste-water were 930 and 2912 mg/l, respectively.

Davis reported that the BODs of a dyehouse waste-water varied between 20 and 1250 mg/l with an average of 634 mg/l. The colour (APHA units) varied between 7,700 and 13,100 mg/l. The average flow of this plant was 1.58 mgd.

A wide variety of methods have been used for reporting results related to colour-removal processes. These include the use of American public health association (APHA) colour units, transmittance, hue, intensity, etc. The current EPA standards are, however, based on a colour analysis procedure developed by the American dye manufacturer's institute.

Treatment of Textile Waste-waters

To solve the problems of treatment of waste-waters from a textile plant, several alternatives should be included. The alternatives are the following:
1. In-plant control for waste reduction.
2. Treatment to 'reuse standard' on an external (end-of-line) basis or by closed-loop recycle systems.
3. Direct discharge to municipal waste treatment systems (i.e. POTWs).
4. 'On-site treatment' of textile waste-waters at POTWs before combining with municipal waste-waters.
5. 'Pretreatment' of textile waste-waters at the plant before discharging to sewer.

In-plant control

A major portion of the waste load is inherent in the methods of textile processing and is independent of the efficiency of the processing plants. For example, the chemicals used for sizing to wrap yarns must be taken off before subsequent bleaching and dyeing. In the industry the normal practice is washing the material with a good flow of running water. This washing operation is the most water-intensive in any textile mills and considerable quantity of water is consumed in this process. Since the concentration level of pollutants in the combined waste-water are not generally high, at the cost of increase in concentration at the inlet level of the treatment plant, water can be saved by judicious use of water. The biological treatment units are designed on the basis of organic loading and hence increase in concentration will not affect the size of the biological unit. On the contrary size of the units designed on hydraulic loading can be reduced apart from saving precious water.

Recently suitable enzymes have been developed which can act upon the starch applied during sizing rapidly at high temperatures ensuring faster degradation of the starch without affecting the material to be treated. With this the desizing operation can be made continuous using only a fraction of water originally required.

Chemicking is a treatment with a weak solution of sodium hypochlorite. Often excessive alkalinity of the bath detracts from bleaching. For satisfactory treatment, the bath alkalinity should not be more than 10 per cent of the available chlorine. The spent bleach can be further replenished at the required time thus saving about 2 to 3 litres of water per kg of fabric for each chemicking operation. Similarly the scouring bath also can be used 8 to 10 times after replenishing.

In mercerising 20 per cent caustic soda is used for treating the fabric and the treatment period ranges from 30 seconds for fabrics to 2 minutes for yarns. Caustic soda on cellulose material is a very tenacious chemical to wash and industries use larger quantity of water than required. This causes not only loss of water but also requires unduly long time to recover caustic from wash waters. The measures by which water can be conserved in mercerising are:

1. Use higher temperature for the wash water.
2. Use counter-counter system of washing whereby caustic soda content in the fabric is progressively reduced.
3. Reduce the concentration of the caustic soda to the optimal limit of 20 per cent strength.
4. Increase the time of washing and the force of the wash water.

Most of the caustic used in textile mills comes from the kiering and mercerising sections. Dialysis and evaporation methods have been employed to purify and reuse the caustic soda from these waste streams.

Treatment to reuse

Groves and Buckley evaluated membrane separation technology for the reuse of textile effluents. They studied two pilot-plant applications: (i) high-temperature ultrafiltration of desizing effluent for polymer size recovery and water reuse, and (ii) hyperfiltration (reverse osmosis) of mixed cotton/synthetic fibre dyehouse effluents for water reuse. The membrane separation processes may offer potential for the recovery of various chemicals like sizing agents. Groves and Buckley concluded that the use of closed-loop recycle systems is technically feasible for textile waste-waters. They also discussed the 'fouling' problems and requirements of cleanings for the restoration of the design flux.

Davis reported that ultrafiltration has several applications for the recovery of textile sizes. Also, latex recovery can be accomplished using ultrafiltration membranes. Tinghul showed how the science of reverse osmosis offers a basis for the choice of membrane materials for use in reverse osmosis applications involving the separation of dyes in aqueous solutions.

The reuse of a textile effluent may be economic only if the plant faces an acute shortage of water. Complete reuse will probably be unrealistic under any circumstances for many more years.

Direct discharge to POTWs

For many textile plants, direct discharge to POTWs may be the best alternative. A mill's waste-water may be clean enough or low enough in volume to be treated by the POTW at little or no extra cost. Even if preliminary or primary treatment is required, the cost to the mill may be much less than if a complete treatment facility were required. Jones listed three advantages of combined treatment, i.e. the direct discharge of textile waters to POTWs. Potential economy of operation is the first advantage. Textile waste-waters may not contain enough nutrients (nitrogens and phosphorus) required for biological treatment. Hence, combining such waste-waters with nutrient-rich domestic waste-waters appears to be another advantage. Finally, in combined treatment, the dilution of highly concentrated textile wastes can be achieved, which prevents shock loads of toxic materials from killing the bacteria in the treatment plant. Newlin studied the economic feasibility of treating textile wastes in municipal treatment plants. The general conclusion, based on 26 municipalities serving some 100 textile mills, was that problems and conditions were so diverse that each case must be given individual attention. Three of the six textile mills covered by the study would have saved money by direct discharges, as compared with the costs of treating their own wastes. However, three plants were saving money by treating their own waste-waters. In general, the mills with small amounts of waste-waters were better off paying their service charges to the POTWs. Jones concluded that the findings regarding the cost of waste-water treatment in relation to total textile manufacturing costs in the southeastern United States raise doubts about the significance of waste-water treatment requirements as an important factor in a competitive strategy as long as the effluent standards do not change drastically. On the other hand, some old plants may face treatment costs that are high enough to create financial problems for them.

On-site treatment of textile waste-waters

Figure 17.16 shows schematic diagrams for alternatives (a) and (b). It is important to note how 'on-site treatment' and 'pretreatment' are defined for this discussion.

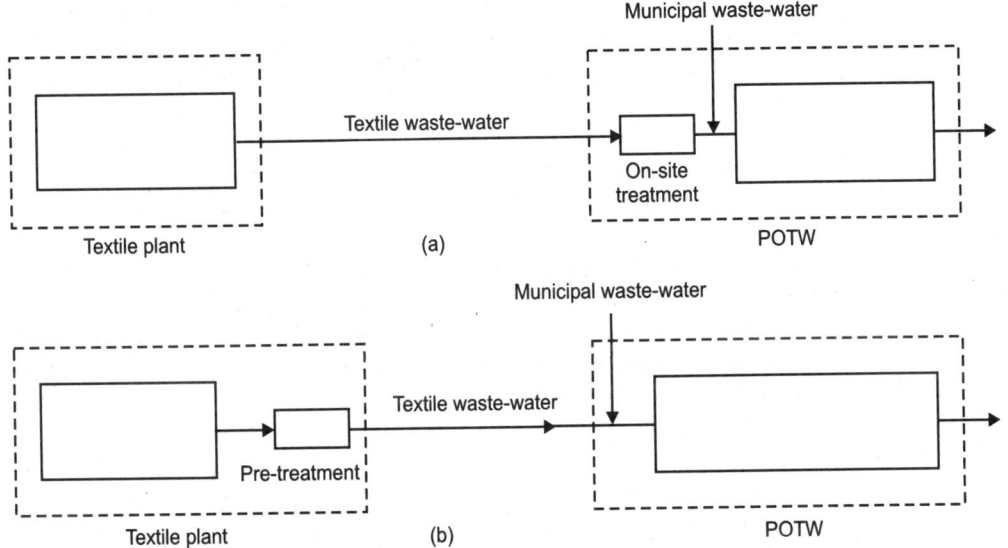

Fig. 17.16. Systemic flow diagrams for: (a) on-site treatment; and (b) pretreatment of textile waste-waters.

On-site treatment refers to any additional treatment at the POTW before combining the textile effluents with the municipal waste-waters for the subsequent treatment. Such additional treatment of the textile waste-waters at the POTW may be physico-chemical or biological. On-site treatment appears to be feasible when several small, closely located textile plants discharge the waste-waters to the same POTW.

Pretreatment of textile waste-water

Pretreatment refers to the treatment of waste-waters by the textile plants before discharging to the sewer (Fig. 17.16). Again, such treatments may be physico-chemical or biological. A combination of physico-chemical and biological (both anaerobic and aerobic) processes may also be feasible. Such pre-treatment appears to be feasible for large textile plants.

Physical/chemical treatment

Textile waste-waters may be treated using physical/chemical processes either at the POTW (on-site) or at the plant (pre-treatment). Experience has demonstrated that chemical processes remove biodegradable organic matter. Some of the physical/chemical processes are coagulation clarification, multimedia filtration, granular carbon adoption, dissolved air flotation and ozonation.

Coagulation/clarification

Coagulation/clarification is an effective process for textile waste-water treatment. This method may be, especially effective for colour removal. Typical coagulants are alum, ferrous sulphate, ferric sulphate and ferric chloride with lime or sulphuric acid for pH control. Other widely used coagulants are cationic, anionic and nonionic organic polymers. For effective coagulation, the experimental determination of the optimum dosage is required. Some waste-waters may require very high coagulant dosage. Chemical dosages in the range of 500 to 1000 mg/l are not uncommon for textile waste-waters. The addition of large amounts of chemicals result in the production of significant quantities of waste solids. The ultimate disposal of these wastes may be very expensive. Hence, the cost calculations of coagulation/clarification should include the additional costs for ultimate disposal. Stahr reported that the most economical approach to colour removal by coagulation appears to involve the use of a cationic polymer coagulant aid with alum as the primary coagulant. The resulting sludges from this process are reported to be more easily dewatered and conditioned than the sludges produced through the use of alum alone. Stahr also reported that the critical parametres defining the optimum polymer dose for colour removal include the waste solution pH, the concentration and types of dyes present and the charge density of the polymer being added. Abo-Elela reported that coagulation using a lime-ferrous sulphate combination was effective in removing the organic contaminants of textile waste-waters. Davis reported that the optimum ferric chloride coagulant dose for a composite dyehouse waste-water was 400 mg/l. Brower and Reed studied the treatment of textile dye wastes with sodium hydroxide and ferric chloride. Sodium hydroxide was used to minimise additional sludge production at the design pH of 7. Adding sodium hydroxide before coagulation produced a weak floc with little colour removed, whereas adding sodium hydroxide after coagulation removed more than 85 per cent of the colour.

Granular carbon

Granular carbon adsorption worked well for some textile waste-waters, whereas for others it was found that a portion of the organic removal occurred from physical filtering rather than an adsorption mechanism. McKay evaluated a model to explain the adsorption of selected dyes on activated carbon. The feasibility of activated carbon treatment of dye house waste-waters frequently depends on costs associated with the regeneration of spent carbon. Thermal regeneration has been the primary means of regenerating granular activated carbon. Posey and Kim studied the feasibility of solvent regeneration of exhausted activated carbon using methanol as the organic solvent. They found that for the three dye compounds tested, solvent regeneration was not cost-competitive with thermal regeneration because of the large amounts of methanol required.

Land Treatment of Textile Waste-waters

Overcash and Rendall studied the feasibility of land treatment of textile waste-waters. They concluded that from an investment perspective, wool scouring with dyeing and finishing and woven fabric dyeing and finishing require the greatest expenditure for land treatment. At the other extreme, knit fabric dyeing and finishing involve the smallest investment. To sum up due to the variability of textile waste-water characteristics, treatability studies should be conducted on a case-by-case basis in order to identify and confirm the required design parametres. For textile waste-water treatment, physical/chemical, biological or a combination of both processes may be suitable. A combined aerobic/anaerobic treatment may be shown to be a feasible process for the treatment of dye waste-waters. Land treatment may be suitable for some dye wastes. After identifying the technically feasible processes, the final selection should be based on the economics of the processes.

FOOD PROCESSING INDUSTRY

Agro-industries or food-processing industries, are concerned primarily with the production of edible goods for human and animal consumption from raw agricultural products. These industries include: (i) canning, (ii) dairy, (iii) brewing and distilling, (iv) meat packing and rendering (including poultry and feedlots), (v) sugar refining, (vi) soft drinks and beverages, and (vii) miscellaneous, including coffee, seafood, rice, grains, and bakeries.

Variety is the one word to describe water use in the agro-industries. Not only are there thousands of products, but agro-industries are well established in many countries with different availabilities of water resources. The industry is not as heavily concentrated in the developed countries as other industries since food processing is nearly a universal need.

Sources and characteristics of waste-water. Water is in contact with raw or finished products in most processes in the agro-industries. This contact results in wastes containing organic matter, in dissolved or colloidal states and in varying degrees of concentration. The sources of these waste-waters are water that contacts spoiled raw material or finished products, rinsing or washing water, transporting water, process water, cooling water, spills and water used for cleaning equipment.

The waste streams exhibit extreme ranges among the industries. The effluent BOD can range from 100 to 1,00,000 ppm. Suspended solids can range from nearly zero to 1,20,000 ppm and pH values range from 3.5 to 11.0. The volumes of the waste also vary greatly. Table 17.18 shows data that provide a brief summary of waste-water problems facing the food industries.

Table 17.18. Characteristic water pollution control problems of the agro-industries.

Industry	Characteristics attending water pollution control	Industry	Characteristics attending water pollution control

Canning, frozen and dehydrated fruits and vegetables, soup, potato chips, speciality items, baby food, etc.	Large seasonal volumes
	  Variation in effluent
	  strength and volume
	Highly biodegradable effluents
	Some soluble organics
	  difficult to remove chemically
	
	Water colouration by strong
	  pigments in some raw
	  products
	Liquid wastes highly
	  putrescible and cannot be
	  stored for long periods of time
		Coffee	Evaporation and other effluents

		Chocolate	Suspended fats in effluents
		Fish and seafood	Liquid wastes highly putrescible and cannot be stored for long periods of time Wastes have water-colouring properties
Edible oils	High concentrations of fats, oils and greases; BOD$_5$; suspended solids, dispersed organics; and dissolved solids Fats, oils and greases difficult to remove to acceptable level for direct discharge to waterways	Red meat	Highly biodegradable effluents Liquid wastes highly putrescible and cannot be stored for long periods of time Relatively large volumes of waste-water
		Poultry	Highly organic effluents, high in suspended solids and floating materials such as grease

(Contd ...)

Industry	Characteristics attending water pollution control	Industry	Characteristics attending water pollution control
	Highly biodegradable effluents Relatively large volumes of waste-water		Highly biodegradable effluents Relatively large volumes of waste-water
Dairy	Highly biodegradable effluents Variation in flow rates and characteristics		Fats and grease in high concentration
	Whey from cheese production	Milling	Highly organic effluents, high in suspended solids
Pickle	Brine, high dissolved solids in effluents		Highly biodegradable effluents Large volumes of waste-water
Tea	Evaporation effluent	Sugar	Liquid wastes highly putrescible and cannot be stored for long periods of time

Waste-water treatment practices: Since the primary component of agro-industries' effluents is organic wastes, biological forms of treatment are effective. The alternatives for the treatment of agro-industries' waste are in-plant treatment with the effluent discharged directly to the water environment or the effluent discharge to a municipal waste treatment plant with or without in-plant pretreatment. Discharge to municipal treatment facilities usually requires pretreatment owing to higher concentrations of organics. In-plant treatment for direct discharge or as pretreatment will most likely use aerobic or anaerobic biological treatment, with the most effective methods being activated sludge, biological filtration, anaerobic digestion, oxidation ponds, lagoons and spray irrigation. The selection of a method will depend upon the degree of treatment required, nature of the organic wastes, concentration of organic matter, volume of wastes and local costs.

Table 17.19 suggests treatment methods for various types of waste that may be encountered in the agro-industries.

Table 17.19. Waste-water treatment methods of removing various categories of pollutants in the effluents of the agro-industries.

Pollutant	Treatment method
Free and emulsified oils and greases	Gravity separation
	Coagulation and sedimentation
	Carbon adsorption
	Granular media filtration
	Flotation
	Impressed current
Suspended solids	Sedimentation
	Coagulation and sedimentation
	Granular media filtration
Dispersed organics	Biological conversion
	Carbon adsorption

(Contd ...)

Pollutant	Treatment method
Dissolved solids (inorganic)	Evaporation
	Ion exchange
	Reverse osmosis
	Electrodialysis
Acidity and alkalinity	Neutralisation
Sludge from processes	Anaerobic digestion
	Aerobic digestion
	Centrifuging
	Incineration
	Stabilisation pond
	Thickening
	Vacuum filtration
	Wet oxidation
	Landfill
	Fertiliser or soil conditioner
	Animal feed

A new system has been developed that can treat the caustic waste separately installed. The system is composed of land treatment for the noncaustic waste and an aerobic/anaerobic pond system with a sand filter and pH adjustment system (Fig. 17.17).

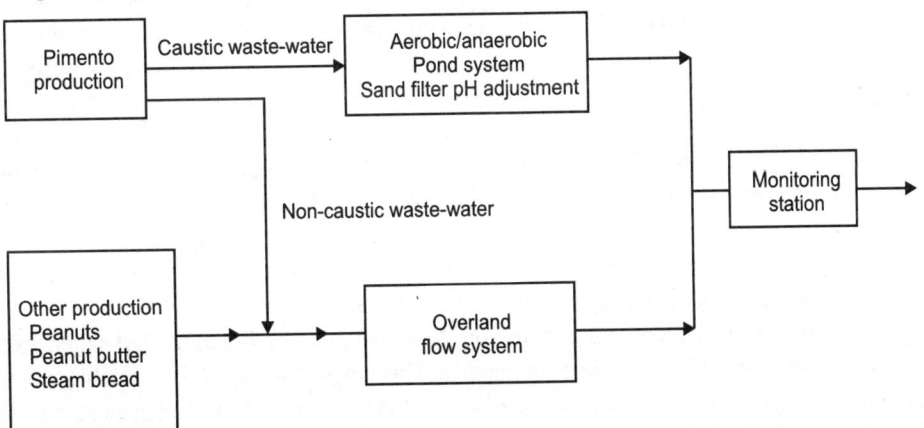

Fig. 17.17. Different treatment for different food-processing waste-water.

The performance of combined system is given in Table 17.20.

Table 17.20. Combined treatment performance of a food-processing plant.

Nature of waste-water	BOD_5 in (mg/l)	BOD_5 out (mg/l)	Fraction of BOD_5 removed (%)	Total suspended solids in (mg/l)	Total suspended solids out (mg/l)	Fraction of suspended solids removed (%)
Caustic	13640	85	99	3680	300	94
Noncaustic	3600	200	94	600	40	93

Dairy

The rapid expansion of dairy industry is one of the significant developments in the food processing field during this period. The subject dairy and food engineering occupies a major importance in the food science curriculum and this emphasis is likely to continue. In dairy industry, water has a multipurpose use. Water used for the purpose of processing, cleaning and other general uses should be of potable standard and absolutely free from microbial contaminants. Use of contaminated water in the plant, results in milk and milk products to be unsafe for human consumption.

Waste prevention

Waste disposal in the milk industry may be divided into two programs, first, waste prevention or saving, and second, waste treatment. The utilisation of by-products and a waste-saving program will materially reduce the loss of milk solids and simplify the requirements for treatment. Such a program should always precede the design of treatment facilities.

The first step in the program is to segregate all possible clean water from the water containing milk solids. Segregation necessitates changes in the drain system of the plant in order to provide a separate line for cooling water, ice machine water, boiler blow-down, roof drains and vacuum pan water. The condenser water from the vacuum pan will contain entrained solids, but because of its large volume it must be segregated from the plant wastes.

Methods of treatment

All floor wastes should pass through a simple combination sand trap, fat trap, screen tank. This is desirable for all plants even if the waste is discharged to a large stream or a municipal sewer. The simplest means of waste disposal for the milk plant is the discharge of the waste to a municipal sewer system. However, a sewer service charge can be expected which should be adequate to permit the municipality to construct the additional waste treatment facilities and pay for their operation.

Some of the effluent can be used directly for the irrigation or to spread them over an area of grassland or cultivated land. The composition of the effluent does, however, impose limits of a maximum of 200 litres per m^2 per year. Furthermore, the irrigation must be done at least 200 metres away from an inhabited area. Return of by-products waste to farmers is the most economical method of disposal. Farmers use such by-products as whey for feeding purposes. Effluent may be treated by mechanical, chemical and biological means. A combination of these methods is often used.

Mechanical treatments are screening, filtration, sedimentation, or flotation. The latter process is aided by gas bubbles to which the particles of impurities becomes attached. Only insoluble constituents can be removed by these methods. Reverse osmosis would also remove dissolved constituents, but this method is not suitable for large scale use at the present time because large volume of effluent produced makes the process too costly to run.

Chemical treatment consists of the precipitation of dissolved substances by means of suitable precipitating agents such as iron sulphate, iron chloride, aluminium sulphate, and lime, etc. A sedimentable coagulum is formed which also contains suspended solids. The coagulum may be subsequently separated from the water by mechanical means. Chemical purification is not sufficient for dairy effluents because it does not remove the dissolved lactose.

The most suitable method of treating dairy effluents is biological purification. Dissolved or colloidally suspended organic compounds are decomposed by oxidation with the aid of aerobic bacteria. Oxygen must be supplied by means of artificial ventilation. Organic substances can also be decomposed by reduction of anaerobic bacteria in a septic tank.

Bakery

The bakery industry is one of the world's major food industries and varies widely in terms of production scale and process. Traditionally, bakery products may be categorised as bread and bread roll products, pastry products (e.g. pies and pasties) and speciality products (e.g. cake, biscuits, donuts and speciality breads). The major equipment includes miller, mixer/kneading machine, bun and bread former, fermentor, bake ovens, cold stage and boilers. The main processes are milling, mixing, fermentation, baking and storage. Fermentation and baking are normally operated at 40°C and 160°–260°C, respectively. Depending on logistics and the market, the products can be stored at 4°–20°C.

Waste-water

Waste-water in bakeries is primarily generated from cleaning operations including equipment cleaning and floor washing. It can be characterised as high loading, fluctuating flow contains rich oil and grease. Flour, sugar, oil, grease and yeast are the major components in the waste.

The ratio of water consumed to products is about 10 in common food industry, much higher than that of 5 in the chemical industry and 2 in the paper and textiles industry. Normally, half of the water is used in the process, while the remainder is used for washing purposes (e.g. of equipment, floor and containers).

Typical values for waste-water production are summarised in Tables 17.21–17.22. Different products can lead to different amounts of waste-water produced. As shown in Table 17.21, pastry production can result in much more waste-water than the others. The values of each item can vary significantly as demonstrated in Table 17.22. The waste-water from cake plants has higher strength than that from bread plants. The pH is in acidic to neutral ranges, while the 5-day biochemical oxygen demand (BOD_5) is from a few hundred to a few thousand mg/l, which is much higher than that from the domestic waste-water. The suspended solids (SS) from cake plants is very high. Grease from the bakery industry is generally high, which results from the production operations.

Table 17.21. Summary of waste production from the bakery industry.

Manufacturer	Products	Waste-water production (litre/ton production)	COD (kg/ton production)	Contribution to total COD loading (%)
Bread and bread roll	Bread and bread roll	230	1.5	63
Pastry	Pies and sausage rolls	6000	18	29
Speciality	Cake, biscuits, donuts and Persian breads	74	–	–

The waste strength and flow rate are very much dependent on the operations, the size of the plants and the number of workers. Generally speaking, in the plants with products of bread, bun and roll, which are termed as dry baking, production equipment (e.g. mixing vats and baking pans) are cleaned dry and floors are swept before washing down.

The waste-water from cleanup has low strength and mainly contains flour and grease (Table 17.22). On the other hand, cake production generates higher strength waste, which contains grease, sugar, flour, filling ingredients and detergents.

Due to the nature of the operation, the waste-water strength changes at different operational times. As demonstrated in Table 17.22, higher BOD_5, SS, total solids (TS) and grease are observed from 1 to 3 am which results from lower waste-water flow rate after midnight.

Table 17.22. Waste-water characteristics in the bakery industry.

Type of bakery	pH	BOD₅ (mg/l)	SS (mg/l)	TS (mg/l)	Grease (mg/l)
Bread plant	6.9–7.8	155–620	130–150	708	60–68
Cake plant	4.7–8.4	2240–8500	963–5700	4238–5700	400–1200
Variety plant	5.6	1600	1700	–	630
Unspecified	4.7–5.1	1160–8200	650–13430	–	1070–4490

Bakery waste-water lacks nutrients; the low nutrient value gives BOD_5:N:P of 284:1:2. This indicates that to obtain better biological treatment results, extra nutrients must be added to the system. The existence of oil and grease also retards the mass transfer of oxygen. The toxicity of excess detergent used in cleaning operations can decrease the biological treatment efficiency. Therefore, the pretreatment of waste-water is always needed.

Bakery waste treatment

Generally, bakery industry waste is nontoxic. It can be divided into liquid waste, solid waste and gaseous waste. In the liquid phase, there are high contents of organic pollutants including chemical oxygen demand (COD), BOD_5, as well as fats, oils and greases (FOG) and SS. Waste-water is normally treated by physical, chemical and biological processes.

Pretreatment systems

Pretreatment or primary treatment is a series of physical and chemical operations, which precondition the waste-water as well as remove some of the wastes. The treatment is normally arranged in the following order: screening, flow equalisation and neutralisation, optional FOG separation, optional acidification, coagulation-sedimentation and dissolved air flotation. The pretreatment of bakery waste-water is presented in Fig. 17.18. In the bakery industry, pretreatment is always required because the waste contains high SS and floatable FOG. Pre-treatment can reduce the pollutant loading in the subsequent biological and/or chemical treatment processes; it can also protect process equipment. In addition, pretreatment is economically preferable in the total process view as compared to biological and chemical treatment.

Biological treatment

The objective of biological treatment is to remove the dissolved and particulate biodegradable components in the waste-water. It is a core part of the secondary biological treatment system. Micro-organisms are used to decompose the organic wastes. With regard to different growth types, biological systems can be classified as suspended growth or attached growth systems. Biological treatment can also be classified by oxygen utilisation: aerobic, anaerobic and facultative. In an aerobic system, the organic matter is decomposed to carbon dioxide, water and a series of simple compounds. If the system is anaerobic, the final products are carbon dioxide and methane.

Compared to anaerobic treatment, the aerobic biological process has better quality effluent, easier operation, shorter solid retention time, but higher cost for aeration and more excess sludge. When treating high-load influent (COD > 4000 mg/l), the aerobic biological treatment becomes less economic than the anaerobic system. To maintain good system performance, the anaerobic biological system requires more complex operations. In most cases, the anaerobic system is used as a pretreatment process. Suspended growth systems (e.g. activated sludge process) and attached growth systems (e.g. trickling filter) are two of the main biological waste-water treatment processes. The activated sludge process is most commonly used in treatment of waste-water.

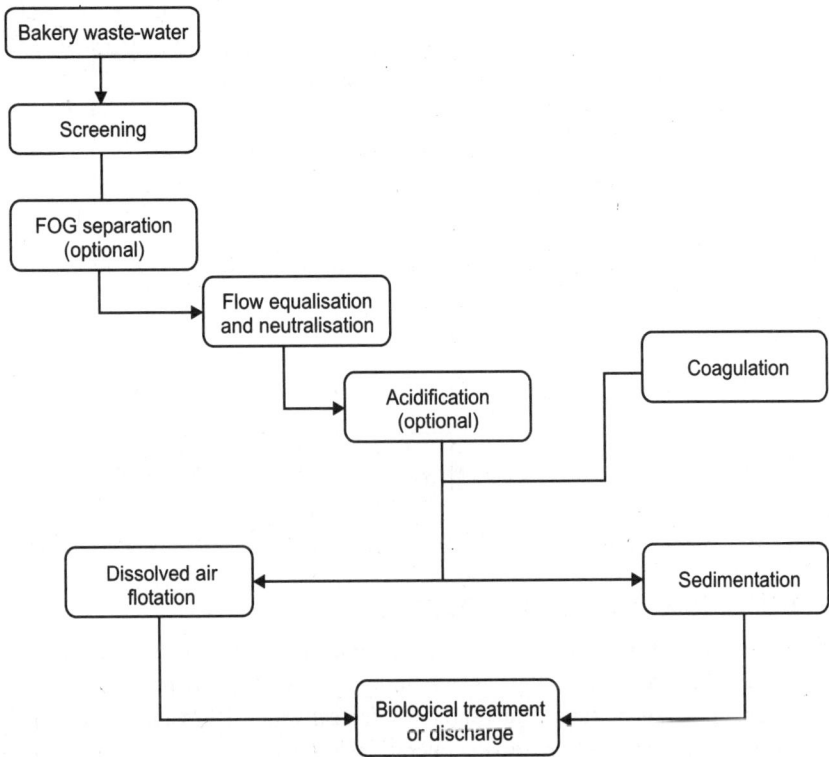

Fig. 17.18. Bakery waste-water pre-treatment system process flow diagram.

The trickling filter is easy to control and has less excess sludge. It has higher resistance loading and low energy cost. However, high operational cost is its major disadvantage. In addition, it is more sensitive to temperature and has odour problems. Comprehensive considerations must be taken into account when selecting a suitable system.

Aerobic Treatment

Activated sludge process

In the activated sludge process, suspended growth micro-organisms are employed. A typical activated sludge process consists of a pretreatment process (mainly screening and clarification), aeration tank (bioreactor), final sedimentation and excess sludge treatment (anaerobic treatment and dewatering process). The final sedimentation separates micro-organisms from the water solution. In order to enhance the performance result, most of the sludge from the sedimentation is recycled back to the aeration tank(s), while the remaining is sent to anaerobic sludge treatment. A recommended complete activated sludge process is given in Fig. 17.19.

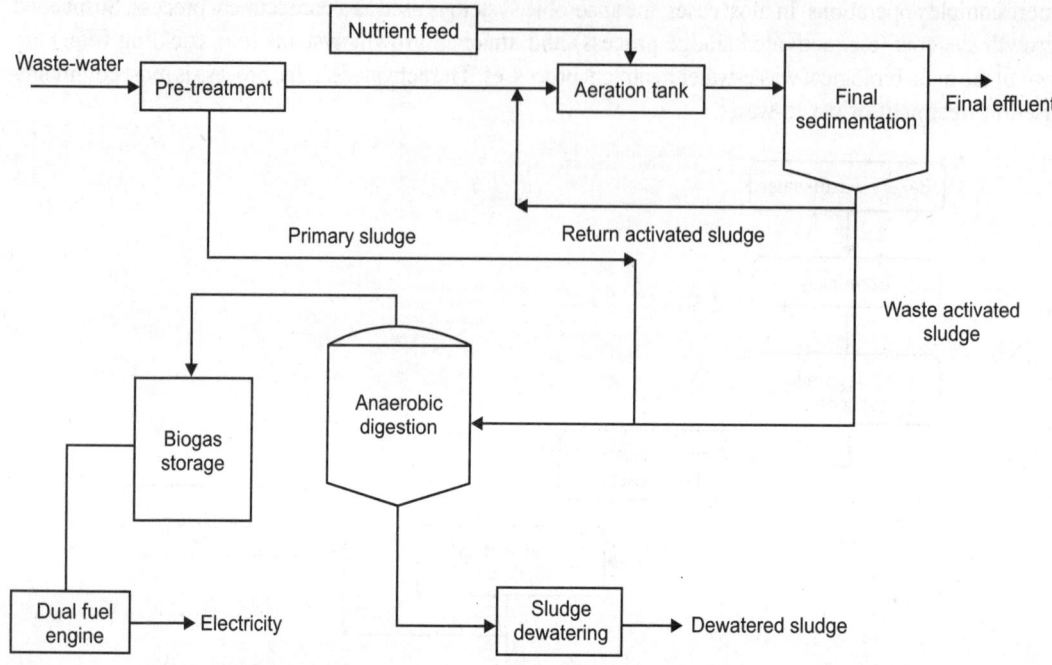

Fig. 17.19. Process flow diagram of activated sludge treatment of bakery waste-water.

SEAFOOD PROCESSING AND MEAT PRODUCTS

The seafood industry consists primarily of many small processing plants, with a number of larger plants located near industry and population centers. Numerous types of seafood are processed, such as mollusks (oysters, clams, scallops), crustaceans (crabs and lobsters), saltwater fishes and freshwater fishes. As in most processing industries, seafood-processing operations produce waste-water containing substantial contaminants in soluble, colloidal and particulate forms. The degree of the contamination depends on the particular operation; it may be small (e.g. washing operations), mild (e.g. fish filleting) or heavy (e.g. blood water drained from fish storage tanks).

Treatment of Seafood Processing Wastes

Waste-water from seafood-processing operations can be very high in biochemical oxygen demand (BOD), fat, oil and grease (FOG) and nitrogen content. Literature data for seafood processing operations showed a BOD production of 1–72.5 kg of BOD per ton of product. White fish filleting processes typically produce 12.5–37.5 kg of BOD for every ton of product. BOD is derived mainly from the butchering process and general cleaning and nitrogen originates predominantly from blood in the waste-water stream.

Seafood-processing waste-water characterisation

Seafood-processing waste-water characteristics that raise concern include pollutant parameters, sources of process waste and types of wastes. In general, the waste-water of seafood-processing waste-water can be characterised by its physico-chemical parameters, organics, nitrogen and phosphorus contents. Important pollutant parameters of the waste-water are five-day biochemical oxygen demand (BOD_5), chemical oxygen demand (COD), total suspended solids (TSS), fats, oil and grease (FOG) and water usage. As in most industrial waste-waters, the contaminants present in seafood-processing waste-waters are an undefined mixture of substances, mostly organic in nature.

Primary treatment

In the treatment of seafood-processing waste-water, one should be cognizant of the important constituents in the waste stream. This waste-water contains considerable amounts of insoluble suspended matter, which can be removed from the waste stream by chemical and physical means. For optimum waste removal, primary treatment is recommended prior to a biological treatment process or land application. A major consideration in the design of a treatment system is that the solids should be removed as quickly as possible. It has been found that the longer the detention time between waste generation and solids removal, the greater the soluble BOD_5 and COD with corresponding reduction in by-product recovery. For seafood-processing waste-water, the primary treatment processes are screening, sedimentation, flow equalisation and dissolved air flotation. These unit operations will generally remove up to 85 per cent of the total suspended solids and 65 per cent of the BOD_5 and COD per cent in the waste-water.

Screening

The removal of relatively large solids (0.7 mm or larger) can be achieved by screening. This is one of the most popular treatment systems used by food-processing plants, because it can reduce the amount of solids being discharged quickly. Usually, the simplest configuration is that of flow-through static screens, which have openings of about 1 mm. Sometimes a scrapping mechanism may be required to minimise the clogging problem in this process.

Fish solids dissolve in water with time; therefore, immediate screening of the waste streams is highly recommended. Likewise, high-intensity agitation of waste streams should be minimised before screening or even settling, because they may cause breakdown of solids rendering them more difficult to separate. In small-scale fish-processing plants, screening is often used with simple settling tanks.

Sedimentation

Sedimentation separates solids from water using gravity settling of the heavier solid particles. In the simplest form of sedimentation, particles that are heavier than water settle to the bottom of a tank or basin. Sedimentation basins are used extensively in the waste-water treatment industry and are commonly found in many flow-through aquatic animal production facilities. This operation is conducted not only as part of the primary treatment, but also in the secondary treatment for separation of solids generated in biological treatments, such as activated sludge or trickling filters.

Flow equalisation

A flow equalisation step follows the screening and sedimentation processes and precedes the dissolved air flotation (DAF) unit. Flow equalisation is important in reducing hydraulic loading in the waste stream. Equalisation facilities consist of a holding tank and pumping equipment designed to reduce the fluctuations of the waste streams. The equalising tank will store excessive hydraulic flow surges and stabilise the flow rate to a uniform discharge rate over a 24-hour day. The tank is characterised by a varying flow into the tank and a constant flow out.

Separation of oil and grease

Seafood-processing waste-waters contain variable amounts of oil and grease, which depend on the process used, the species processed and the operational procedure. Gravitational separation may be used to remove oil and grease, provided that the oil particles are large enough to float towards the surface and are not emulsified; otherwise, the emulsion must be first broken by pH adjustment. Heat may also be used for breaking the emulsion but it may not be economical unless there is excess steam available.

Flotation

Flotation is one of the most effective removal systems for suspensions that contain oil and grease. The most common procedure is that of dissolved air flotation (DAF), which is a waste treatment process in which oil, grease and other suspended matter are removed from a waste stream.

Biological treatment

To complete the treatment of the seafood-processing waste-waters, the waste stream must be further processed by biological treatment. Biological treatment involves the use of micro-organisms to remove dissolved nutrients from a discharge. Organic and nitrogenous compounds in the discharge can serve as nutrients for rapid microbial growth under aerobic, anaerobic or facultative conditions. The three conditions differ in the way they use oxygen. Aerobic micro-organisms require oxygen for their metabolism, whereas anaerobic micro-organisms grow in absence of oxygen; the facultative micro-organism can proliferate either in absence or presence of oxygen although using different metabolic processes. The biological treatment processes used for waste-water treatment are broadly classified as aerobic and anaerobic treatments.

Aerobic and facultative micro-organisms predominate in aerobic treatments, while only anaerobic micro-organisms are used for the anaerobic treatments. If micro-organisms are suspended in the waste-water during biological operation, this is known as a 'suspended growth process', whereas the micro-organisms that are attached to a surface over which they grow are said to undergo an 'attached growth process'. Biological treatment systems are most effective when operating continuously 24 hr/day and 365 days/year. Systems that are not operated continuously have reduced efficiency because of changes in nutrient loads to the microbial biomass. Biological treatment systems also generate a consolidated waste stream consisting of excess microbial biomass, which must be properly disposed. Operation and maintenance costs vary with the process used. The principles and main characteristics of the most common processes used in seafood-processing waste-water treatment are explained in this section.

Aerobic process

In seafood processing waste-waters, the need for adding nutrients (the most common being nitrogen and phosphorus) seldom occurs, but an adequate provision of oxygen is essential for successful operation. The most common aerobic processes are activated sludge systems, lagoons, trickling filters and rotating disc contactors.

Activated sludge systems

In an activated sludge treatment system, an acclimatised, mixed, biological growth of micro-organisms (sludge) interacts with organic materials in the waste-water in the presence of excess dissolved oxygen and nutrients (nitrogen and phosphorus). The micro-organisms convert the soluble organic compounds to carbon dioxide and cellular materials. Oxygen is obtained from applied air, which also maintains adequate mixing. The effluent is settled to separate biological solids and a portion of the sludge is recycled; the excess is wasted for further treatment such as dewatering.

Anaerobic treatment

Anaerobic biological treatment has been applied to high BOD or COD waste solutions in a variety of ways. Treatment proceeds with degradation of the organic matter, in suspension or in a solution of continuous flow of gaseous products, mainly methane and carbon dioxide, which constitute most of the reaction products and biomass. Its efficient performance makes it a valuable mechanism for achieving compliance with regulations for contamination of recreational and seafood-producing wastes. Anaerobic treatment is the result of several reactions: the organic load present in the waste-water is first converted to soluble organic material, which in turn is consumed by acid-producing bacteria to produce volatile fatty acids, plus carbon dioxide and hydrogen. The methane-producing bacteria consume these products to produce methane and carbon dioxide. Typical micro-organisms used in this methanogenic process are *metanobacterium, methanobacillus, metanococcus*, and *methanosarcina*.

Physico-chemical treatments

Coagulation/flocculation

Coagulation or flocculation tanks are used to improve the treatability of waste-water and to remove grease and scum from waste-water. In coagulation operations, a chemical substance is added to an organic colloidal suspension to destabilise it by reducing forces that keep them apart, that is, to reduce the surface charges responsible for particle repulsions. This reduction in charges is essential for flocculation, which has the purpose of clustering fine matter to facilitate its removal. Particles of larger size are then settled and clarified effluent is obtained.

In seafood processing waste-waters, the colloids present are of an organic nature and are stabilised by layers of ions that result in particles with the same surface charge, thereby increasing their mutual repulsion and stabilisation of the colloidal suspension. This kind of waste-water may contain appreciable amounts of proteins and micro-organisms, which become charged due to the ionisation of carboxyl and amino groups or their constituent amino acids. The oil and grease particles, normally neutral in charge, become charged due to preferential absorption of anions, which are mainly hydroxyl ions.

Several steps are involved in the coagulation process. First, coagulant is added to the effluent and mixing proceeds rapidly with high intensity. The purpose is to obtain intimate mixing of the coagulant with the waste-water, thereby increasing the effectiveness of destabilisation of particles and initiating coagulation. A second stage follows in which flocculation occurs for a period of up to 30 minutes. In the latter case, the suspension is stirred slowly to increase the possibility of contact between coagulating particles and to facilitate the development of large flocs. These flocs are then transferred to a clarification basin in which they settle and are removed from the bottom while the clarified effluent overflows.

Several substances may be used as coagulants. The pH of several waste-waters of the proteinaceous nature can be adjusted by adding acid or alkali. The addition of acid is more common, resulting in coagulation of the proteins by denaturing them, changing their structural conformation due to the change in their surface charge distribution. Thermal denaturation of proteins can also be used, but due to its high energy demand, it is only advisable if excess steam is available. In fact, the 'cooking' of the blood-water in fishmeal plants is basically a thermal coagulation process.

Another commonly used coagulant is polyelectrolyte, which may be further categorised as cationic and anionic coagulants. Cationic polyelectrolytes act as a coagulant by lowering the charge of the waste-water particles, because waste-water particles are negatively charged. Anionic or neutral polyelectrolyte are used as bridges between the already formed particles that interact during the flocculation process, resulting in an increase of floc size.

Since the recovered sludges from coagulation/flocculation processes may sometimes be added to animal feeds, it is advisable to ensure that the coagulant or flocculant used is not toxic.

Electrocoagulation

Electrocoagulation (EC) has also been investigated as a possible means to reduce soluble BOD. It has been demonstrated to reduce organic levels in various food- and fish-processing waste streams. During testing, an electric charge was passed through a spent solution in order to destabilise and coagulate contaminants for easy separation.

Disinfection

Disinfection of seafood-processing waste-water is a process by which disease-causing organisms are destroyed or rendered inactive. Most disinfection systems work in one of the following four ways: (i) damage to the cell wall, (ii) alteration of cell permeability, (iii) alteration of the colloidal nature of protoplasm, and (iv) inhibition of enzyme activity. Disinfection is often accomplished using bactericidal agents. The most common agents are chlorine, ozone (O_3) and ultraviolet (UV) radiation.

Waste-water flow

Water is used in the slaughterhouse for carcass washing after hide removal from cattle, calves and sheep and after hair removal from hogs. It is also used to clean the inside of the carcass after evisceration and for cleaning and sanitising equipment and facilities both during and after the killing operation. Associated facilities such as stockyards, animal pens, the steam plant, refrigeration equipment, compressed air, boiler rooms and vacuum equipment will also produce some waste-water, as will sanitary and service facilities for staff employed on site: these may include toilets, shower rooms, cafeteria kitchens and laboratory facilities. The proportions of water used for each purpose can be variable.

The quantity of waste-water will depend very much on the slaughterhouse design, operational practise and the cleaning methods employed. Waste-water generation rates are usually expressed as a volume per unit of product or per animal slaughtered and there is a reasonable degree of consistency between some of the values reported from reliable sources for different animal types.

Waste-water characteristics

Effluents from slaughterhouses and packing houses are usually heavily loaded with solids, floatable matter (fat), blood, manure and a variety of organic compounds originating from proteins. As already stated the composition of effluents depends very much on the type of production and facilities. The main sources of water contamination are from lairage, slaughtering, hide or hair removal, paunch handling, carcass washing, rendering, trimming and cleanup operations. These contain a variety of readily biodegradable organic compounds, primarily fats and proteins, present in both particulate and dissolved forms. The waste-water has a high strength, in terms of biochemical oxygen demand (BOD), chemical oxygen demand (COD), suspended solids (SS), nitrogen and phosphorus, compared to domestic waste-waters. The actual concentration will depend on in-plant control of water use, by-products recovery, waste separation source and plant management. In general, blood and intestinal contents arising from the killing floor and the gut room, together with manure from stockyard and holding pens, are separated, as best as possible, from the aqueous stream and treated as solid wastes. This can never be 100 per cent successful, however and these components are the major contributors to the organic load in the waste-water, together with solubilised fat and meat trimmings.

The aqueous pollution load of a slaughterhouse can be expressed in a number of ways. Within the literature reports can be found giving the concentration in waste-water of parameters such as BOD, COD and SS. These, however, are only useful if the corresponding waste-water flow rates are also given. Even then it is often difficult to relate these to a meaningful figure for general design, as the unit of productivity is often omitted or unclear. These reports do, however, give some indication as to the strength of waste-waters typically encountered and some of their particular characteristics, which can be useful in making a preliminary assessment of the type of treatment process most applicable. At best it can be concluded that slaughterhouse waste-waters have a pH around neutral, an intermediate strength in terms of COD and BOD, are heavily loaded with solids and are nutrient-rich.

The waste-water contains a high density of total coliform, fecal coliform and fecal streptococcus groups of bacteria due to the presence of manure material and gut contents. Numbers are usually in the range of several million colony forming units (CFU) per 100 ml. It is also likely that the waste-water will contain bacterial pathogens of enteric origin such as *Salmonella* sp., and *Campylobacter jejuni*, gastrointestinal parasites including *Ascaris* sp., *Giardia lamblia* and *Cryptosporidium parvum* and enteric viruses. It is, therefore, essential that slaughterhouse design ensures the complete segregation of process wash water and strict hygiene procedures to prevent cross-contamination. The mineral chemistry of the waste-water is influenced by the chemical composition of the slaughterhouse's treated water supply, waste additions such as blood and manure, which can contribute to the heavy metal load in the form of copper, iron, manganese, arsenic and zinc and process plant and pipework, which can contribute to the load of copper, chromium, molybdenum, nickel, titanium and vanadium.

Waste-water minimisation

As indicated previously, the overall waste load arising from a slaughterhouse is determined principally by the type and number of animals slaughtered. The partitioning of this load between the solid and aqueous phases will depend very much upon the operational practices adopted, however, and there are measures that can be taken to minimise waste-water generation and the aqueous pollution load.

Minimisation can start in the holding pens by reducing the time that the animals remain in these areas through scheduling of delivery times. The incorporation of slatted concrete floors laid to falls of 1 in 60 with drainage to a slurry tank below the floor in the design of the holding pens can also reduce the amount of washdown water required. Alternatively, it is good practice to remove manure and lairage from the holding pens or stockyard in solid form before washing down. In the slaughterhouse itself, cleaning and carcass washing typically account for over 80 per cent of total water use and effluent volumes in the first processing stages. One of the major contributors to organic load is blood, which has a COD of about 4,00,000 mg/l and washing down of dispersed blood can be a major cause of high effluent strength. Minimisation can be achieved by having efficient blood collection troughs allowing collection from the carcass over several minutes. Likewise the trough should be designed to allow separate drainage to a collection tank of the blood and the first flush of wash water. Only residual blood should enter a second drain for collection of the main portion of the wash water. An efficient blood recovery system could reduce the aqueous pollution load by as much as 40 per cent compared to a plant of similar size that allows the blood to flow to waste. The second area where high organic loads into the waste-water system can arise is in the gut room. Most cattle and sheep abattoirs clean the paunch (rumen), manyplies (omasum) and reed (abomasum) for tripe production. A common method of preparation is to flush out the gut manure from the punctured organs over a mechanical screen and allow water to transport the gut manure to the effluent treatment system.

Typically the gut manure has a COD of over 1,00,000 mg/l, of which 80 per cent dissolves in the wash water. Significant reductions in waste-water strength can be made by adopting a 'dry' system for removing and transporting these gut manures. The paunch manure in its undiluted state has enough water present to allow pneumatic transport to a 'dry' storage area where a compactor can be used to reduce the volume further if required. The tripe material requires washing before further processing, but with a much reduced volume of water and resulting pollution load.

The small and large intestines are usually squeezed and washed for use in casings. To reduce water, washing can be carried out in two stages: a primary wash in a water bath with continuous water filtration and recirculation, followed by a final rinse in clean potable water. Other measures that can be taken in the gut room to minimise water use and organic loadings to the aqueous stream include ensuring that mechanical equipment, such as the hasher machine, are in good order and maintained regularly.

Other methods can also be employed to minimise water usage. These will not in themselves reduce the organic load entering the waste-water treatment system, but will reduce the volume requiring treatment and possibly influence the choice of treatment system to be employed. For example, high-strength, low-volume waste-waters may be more suited to anaerobic rather than aerobic biological treatment methods. Water use minimisation methods include:

1. The use of directional spray nozzles in carcass washing, which can reduce water consumption by as much as 20 per cent.
2. Use of steam condensation systems in place of scald tanks for hair and nail removal.
3. Fitting washdown hoses with trigger grips.
4. Appropriate choice of cleaning agents.
5. Reuse of clear water (e.g. chiller water) for the primary washdown of holding pens.

Waste-water treatment processes

The degree of waste-water treatment required will depend on the proposed type of discharge. Waste-waters received into the sewer system are likely to need less treatment than those having direct discharge into a watercourse. In the European countries, direct discharges have to comply with the urban waste water treatment directive and other water quality directives. In the United States the EPA has proposed effluent limitations guidelines and standards (ELGs) for the meat and poultry products industries with direct discharge. These proposed ELGs will apply to existing and new meat and poultry products (MPP) facilities and are based on the well-tested concepts of 'best practicable control technology currently available' (BPT), the 'best conventional pollutant control technology' (BCT), the 'best available technology economically achievable' (BAT) and the 'best available demonstrated control technology for new source performance standards' (NSPS). In summary, the technologies proposed to meet these requirements use, in the main, a system based on a treatment series comprising flow equalisation, dissolved air flotation and secondary biological treatment for all slaughterhouses; and require nitrification for small installations and additional denitrification for complex slaughterhouses.

There is some potential, however, for segregation of waste-waters allowing specific individual pre-treatments to be undertaken or in some cases, bypass of less contaminated streams. Depending on local conditions and regulations, water from boiler houses and refrigerating systems may be segregated and discharged directly or used for outside cleaning operations.

Primary and secondary treatment

Primary treatment

Grease removal is a common first stage in slaughterhouse waste-water treatment, with grease traps in some situations being an integral part of the drainage system from the processing areas. Where the option is taken to have a single point of removal, this can be accomplished in one of two ways: by using a baffled tank or by DAF. A typical grease trap has a minimum detention period of about 30 minutes, but the period need not to be greater than 1 hour. Within the tank, coagulation of fats is brought about by cooling, followed by separation of solid material in baffled chambers through natural flotation of the less dense material, which is then removed by skimming.

Secondary treatment

Secondary treatment aims to reduce the BOD of the waste-water by removing the organic matter that remains after primary treatment. This is primarily in a soluble form. Secondary treatment can utilise physical and chemical unit processes, but for the treatment of meat wastes biological treatment is usually favoured.

Physico-chemical secondary treatment

Chemical treatment of meat-plant wastes is not a common practice due to the high chemical costs involved and difficulties in disposing of the large volumes of sludge produced.

Biological secondary treatment

Using biological treatment, more than 90 per cent efficiency can be achieved in pollutant removal from slaughterhouse wastes. Commonly used systems include lagoons (aerobic and anaerobic), conventional activated sludge, extended aeration, oxidation ditches, sequencing batch reactors and anaerobic digestion. A series of anaerobic biological processes followed by aerobic biological processes is often useful for sequential reduction of the BOD load in the most economic manner, although either process can be used separately. As noted above, slaughterhouse waste-waters vary in strength considerably depending on a number of factors. For a given type of animal, however, this variation is primarily due to the quantity of water used within the abattoir, as the pollution load (as expressed as BOD) is relatively constant on the basis of live weight slaughtered. Hence, the more economical an abattoir is in its use of water, the stronger the effluent will be and *vice versa*. The strength of the organic degradable matter in the waste-water is an important consideration in the choice of treatment system. To remove BOD using an aerobic biological process involves supplying oxygen (usually as a component in air) in proportion to the quantity of BOD that has to be removed, an increasingly expensive process as the BOD increases. On the other hand an anaerobic process does not require oxygen in order to remove BOD as the biodegradable fraction is fermented and then transformed to gaseous endproducts in the form of carbon dioxide (CO_2) and methane (CH_4).

Anaerobic treatment

Anaerobic digestion is a popular method for treating meat industry wastes. Anaerobic processes operate in the absence of oxygen and the final products are mixed gases of methane and carbon dioxide and a stabilised sludge. Anaerobic digestion of organic materials to methane and carbon dioxide is a complicated biological and chemical process that involves three stages: hydrolysis, acetogenesis, and finally methanogenesis. During the first stage, complex compounds are hydrolysed to smaller chain intermediates. In the second stage acetogenic bacteria convert these intermediates to organic acids and then ultimately to methane and carbon dioxide via the methanogenesis phase (Fig. 17.20).

In most of the countries, anaerobic systems using simple lagoons are by far the most common method of treating abattoir waste-water. These are not particularly suitable for use in the heavily populated regions of western Europe due to the land area required and also because of the difficulties of controlling odours in the urban areas where abattoirs are usually located. The extensive use of anaerobic lagoons demonstrates the amenability of abattoir waste-waters to anaerobic stabilisation, however, with significant reductions in the BOD at a minimal cost.

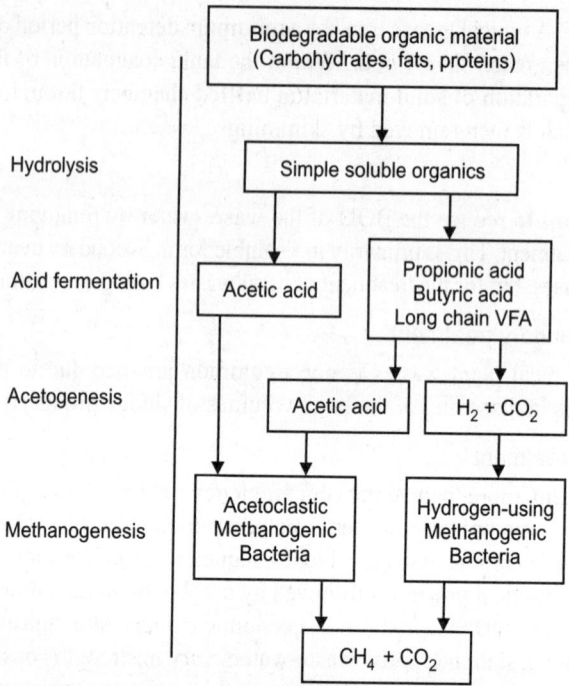

Fig. 17.20. The microbial phases of anaerobic digestion.

The anaerobic lagoon consists of an excavation in the ground, giving a water depth of between 10 and 17 ft (3–5 metres), with a retention time of 5–15 days. Common practice is to provide two ponds in series or parallel and sometimes linking these to a third aerobic pond. The pond has no mechanical equipment installed and is unmixed except for some natural mixing brought about by internal gas generation and surface agitation; the latter is minimised where possible to prevent odour formation and re-aeration. Influent waste-water enters near the bottom of the pond and exits near the surface to minimise the chance of short-circuiting. Anaerobic ponds can provide an economic alternative for purification. The BOD reductions vary widely, although excellent performance has been reported in some cases, with reductions of up to 97 per cent in BOD, up to 95 per cent in SS and up to 96 per cent in COD from the influent values.

Anaerobic lagoons are not without potential problems, relating to both their gaseous and aqueous emissions. As a result of breakdown of the waste-water, methane and carbon dioxide are both produced. These escape to the atmosphere, thus contributing to greenhouse gas emissions, with methane being 25 times more potent than carbon dioxide in this respect. Gaseous emissions also include the odouriferous gases, hydrogen sulphide and ammonia. The lagoons generally operate with a layer of grease and scum on the top, which restricts the transfer of oxygen through the liquid surface, retains some of the heat and helps prevent the emission of odour.

The use of fabricated anaerobic reactors for abattoir waste-water treatment is also well established. To work efficiently these are designed to operate either at mesophilic (around 95°F or 35°C) or thermophilic (around 130°F or 55°C) temperatures. Anaerobic filters have also been applied to the treatment of slaughterhouse waste-waters. These maintain a long SRT by providing the micro-organisms with a medium that they can colonise as a biofilm. Unlike conventional aerobic filters, the anaerobic filter is operated with the support medium submerged in an upflow mode of operation. Because anaerobic filters contain a support medium, there is potential for the interstitial spaces within the medium to become blocked and effective pre-treatment is essential to remove suspended solids as well as solidifiable oils, fats and grease.

The third type of high-rate anaerobic system that can be applied to slaughterhouse waste-waters is the upflow anaerobic sludge blanket reactor (UASB). This is basically an expanded-bed reactor in which the bed comprises anaerobic micro-organisms, including methanogens, which have formed dense granules. The mechanisms by which these granules form are still poorly understood, but they are intrinsic to the proper operation of the process. The influent waste-water flows upward through a sludge blanket of these granules, which remain within the reactor as their settling velocity is greater than the up flow velocity of the waste-water. The reactor therefore exhibits a long sludge retention time, high biomass density per unit reactor and can operate at a short HRT.

Aerobic treatment

Aerobic biological treatment for the treatment of biodegradable wastes has been established for over a hundred years and is accepted as producing a good-quality effluent, reliably reducing influent BOD by 95 per cent or more. Aerobic processes can roughly be divided into two basic types: those that maintain the biomass in suspension (activated sludge and its variants) and those that retain the biomass on a support medium (biological filters and its variants). There is no doubt that either basic type is suitable for the treatment of slaughterhouse waste-water and their use is well documented in works such as Broils and Broughton, where aerobic processes are compared with anaerobic ones. In selecting an aerobic process a number of factors need to be taken into account. These include the land area available, the head of water available, known difficulties associated with certain waste-water types (such as bulking and stable foam formation), energy efficiency and excess biomass production. It is important to realise that the energy costs of conventional aerobic biological treatment can be substantial due to the requirement to supply air to the process.

It is, therefore, usual to only treat to the standard required, as treatment to a higher standard will incur additional cost. For example, in order to convert ammonia to nitrate requires 4.5 moles of oxygen for every mole of ammonia converted. In effect this means that a 1 mg/l concentration of ammonia has an equivalent BOD of 4.5 mg/l. It is, therefore, only usual to aim for the conversion of ammonia to nitrate when this is required.

The most common aerobic biological processes used for the treatment of meat industry wastes are biological filtration, activated sludge plants, waste stabilisation ponds and aerated lagoons.

Activated sludge

The activated sludge process has been successfully used for the treatment of waste-waters from the meat industry for many decades. It generally has a lower capital cost than standard-rate percolating filters and occupies substantially less space than lagoon or pond systems. In the activated sludge process the waste-waters are mixed with a suspension of aerobic micro-organisms (activated sludge) and aerated. After aeration, the mixed liquor passes to a settlement tank where the activated sludge settles and is returned to the plant inlet to treat the incoming waste. The supernatant liquid in the settlement tank is discharged as plant effluent. Air can be supplied to the plant by a variety of means, including blowing air into the mixed liquor through diffusers; mechanical surface aeration; and floor-mounted sparge pipes. All the methods are satisfactory provided that they are properly designed to meet the required concentration of dissolved oxygen in the mixed liquor (greater than 0.5 mg/l) and to maintain the sludge in suspension; for nitrification to occur it may be necessary to maintain dissolved oxygen concentrations above 2.0 mg/l.

The activated sludge process can be designed to meet a number of different requirements, including the available land area, the technical expertise of the operator, the availability of sludge disposal routes and capital available for construction. The first step in the design of an activated sludge system is to select the loading rate, which is usually defined as the mass ratio of substrate inflow to the mass of activated sludge (on a dry weight basis); this is commonly referred to as the food to micro-organism (F:M) ratio and is usually reported as lb BOD/lb MLSS day (kg BOD/kg MLSS day). For conventional operation the range is 0.2–0.6; the use of higher values tends to produce a dispersed or nonflocculent sludge and lower values require additional oxygen input due to high endogenous respiration rates. Systems with F:M ratios above 0.6 are sometimes referred to as high rate, while those below 0.2 are known as extended aeration systems. The latter, despite their higher capital and operating costs are commonly chosen for small installations because of their stability, low sludge production and reliable nitrification. Because of the stoichiometric relationship between F:M ratio and mean cell residence time (MCRT), high-rate plants will have an MCRT of less than 4 days and extended aeration plants of greater than 13 days. Because of the low growth rates of the nitrifying bacteria, which are also influenced markedly by temperature, the oxidation of ammonia to nitrates (nitrification) will only occur at F:M ratios less than 0.1. It is also sometimes useful to consider the nitrogen loading rate, which for effective nitrification should be in the range 0.03–0.08 lb N/lb MLSS-day (kg N/kg MLSS day).

Conventional plants can be used where nitrification is not critical, for example, as a pre-treatment before sewer discharge. One of the main drawbacks of the conventional activated sludge process, however, is its poor buffering capability when dealing with shock loads. This problem can be overcome by the installation of an equalisation tank upstream of the process, or by using an extended aeration activated sludge system. In the extended aeration process, the aeration basin provides a 24–30 hours (or even longer) retention time with complete mixing of tank contents by mechanical or diffused aeration. The large volume combined with a high air input results in a stable process that can accept intermittent loadings. A further disadvantage of using a conventional activated sludge process is the generation of a considerable amount of surplus sludge, which usually requires further treatment before disposal. Some early work suggested the possible recovery of the biomass as a source of protein, but concerns over the possible transmission of exotic animal diseases would make this unacceptable in Europe. The use of extended aeration activated sludge or aerated lagoons minimises biosolids production because of the endogenous nature of the reactions. The size of the plant and the additional aeration required for sludge stabilisation does, however, lead to increased capital and operating costs. Considering the high concentrations of nitrogen present in slaughterhouse waste-water, ammonia removal is often regarded as essential from a regulatory standpoint for direct discharge and increasingly there is a requirement for nutrient removal. It is, therefore, not surprising that most modern day designs are of an extended aeration type so as to promote reliable nitrification as well as to minimise sludge production. Efficient designs will also attempt to recover the chemically bound oxygen in nitrate through the process of denitrification, thus reducing treatment costs and lowering nitrate concentrations in the effluent.

Design criteria and loadings for activated sludge treatment have been widely reported and reliable data can be found in a number of reports.

In recent years, a great deal of interest has been shown in the use of sequencing batch reactors (SBRs) for food-processing waste-waters, as these provide a minimum guaranteed retention time and produce a high-quality effluent. A batch process also often fits well with the intermittent discharge of an industrial process working on one or two shifts. Advantages are an ideal plug flow that maximises reaction rates, ideal quiescent sedimentation and flow equalisation inherent in the design. Decanting can be achieved using floating outlets and adjustable weirs, floating aerators are commonly employed and an anoxic fill overcomes problems of effluent turbidity as well as providing ideal conditions for denitrification reactions.

BIOLOGICAL METHODS OF WASTE-WATER TREATMENT

Biological treatment — the use of bacteria and other micro-organisms to remove contaminants by assimilating them — has long been a mainstay of waste-water treatment in the chemical process industries (CPI). Because they are effective and widely used, many biological-treatment options are available today. They are, however, not all created equal, and the decision to install a biological-treatment system requires ample thought. When considering biological waste-water treatment for a particular application, it is important to understand the sources of the waste-water generated, typical waste-water composition, discharge requirements, events and practices within a facility that can affect the quantity and quality of the waste-water, and pretreatment ramifications. Consideration of these factors will allow you to maximise the benefits your plant gains from effective biological treatment. Those benefits can include:

1. Low capital and operating costs compared to those of chemical-oxidation processes.
2. True destruction of organics, versus mere phase separation, such as with air stripping or carbon adsorption.
3. Oxidation of a wide variety of organic compounds.
4. Removal of reduced inorganic compounds, such as sulphides and ammonia, and total nitrogen removal possible through denitrification.
5. Operational flexibility to handle a wide range of flows and waste-water characteristics.
6. Reduction of aquatic toxicity.

All biological-treatment processes take advantage of bacteria's remarkable ability to use diverse waste-water constituents to provide the energy for microbial metabolism and the building blocks for cell synthesis. This metabolic activity can remove contaminants that are as varied as the raw materials, by-products and products generated by the CPI.

Selection Criteria

Biological-treatment technologies vary greatly in their strengths and weaknesses. The following are application criteria, which are normally relevant in evaluating various biological-treatment options for the chemical process industries (CPI):

1. Bioassay/toxicity control: The ability to control and minimise the impact of toxic constituents in waste-water on indicating organisms when the treated water is released.
2. BOD removal efficiency: The ability to remove biodegradable, organic compounds.
3. COD removal efficiency: The ability to remove chemically oxidisable substances that may or may not be biodegradable.
4. O&M costs: The cost to operate and maintain the treatment method.
5. Sludge production: The amount of residual biological solids generated by the biological-treatment process.
6. Sludge disposal costs: The cost to collect, dewater and dispose of residual sludge from the treatment method, either on-site or off-site.

7. Performance in winter and summer: The degree in which high or low ambient temperatures will affect biological treatment.

8. Performance on high—and low-temperature water: The degree in which high and low waste-water temperature will affect biological treatment.

9. Operator attention: The relative amount of time required to operate the biological treatment system.

10. Upset recovery: The amount of time it takes for a treatment method to recover from upset conditions. Upset conditions are defined as abnormal variations in the flow or characteristics of the waste-water, which can detrimentally affect a biological treatment system.

11. Expandability: The ease of expanding the treatment capacity to accommodate either an overall plant expansion or an increase in loading.

12. Nitrification efficiency: The relative ease of converting ammonia contained in waste-water to nitrates.

13. VOC containment: The relative ease with which the biological-treatment equipment can be enclosed to contain and collect VOC emissions.

14. VOC stripping potential: The relative ease with which the biological-treatment system will strip volatile organic compounds from the waste-water.

15. Ease of installation: The total amount of time and labour required to install the treatment method.

16. Energy efficiency: The amount of energy used by a treatment method.

17. Ease of secondary containment: The ability and ease with which the treatment system can be provided with secondary containment in case of overflow, spills or leaks.

18. Space requirements: The area required by the treatment method.

Knowing the composition of the water to be handled is essential for planning a treatment process. In petroleum refineries, for example, excessive amounts of spent caustic can quickly overwhelm a waste-water treatment system due to the normally high chemical oxygen demand (COD) of the spent caustic. Another issue can be a significant increase in ammonia and sulphide loads that result from upsets in the operation of sour-water strippers. These loads can, in turn, upset a biological-treatment system if it is not designed to handle ammonia and sulphide.

In addition to understanding the source and composition of the waste-water, one must also recognise when pretreatment steps are needed to provide adequate protection for a biological treatment system. In most petroleum and petrochemical facilities, for example, raw waste-water normally contains free oil, which can have serious, detrimental effects. Oil can coat and kill bacteria, causing the micro-organisms to float out of the system, and can interfere with oxygen-transfer efficiency. Another source for concern at refineries is a potential upset in desalter operations that can lead to significant oil/water emulsions in the waste-water and thereby negatively impact the biological-treatment system. To prevent these types of problems in petroleum-industry systems, process steps prior to biological treatment are normally included. Pretreatment for this industry typically includes the use of oil/water separators, an equalisation tank to moderate spikes in waste-water composition, and off-spec waste-water storage. Figure 17.21 shows a typical waste-water-treatment system for a petroleum facility. Even properly pretreated waste-water can still contain a wide variety of compounds which may or may not be biodegradable. There can also be significant concentrations of sulphides, ammonia, amines, mercaptans and other compounds that require modifications to the treatment process in order to meet discharge objectives. Vendors can be helpful in setting up pilot-plant or bench-scale tests to assist in determining if biotreatment is a viable option for a particular waste-water composition. The types of compounds present, the concentration of each and the ultimate discharge requirements are key to selecting a proper biological-treatment system. This is true whether the waste-water is being discharged directly to the environment or to a publicly owned treatment works (POTW) or if it is to be reused within the facility.

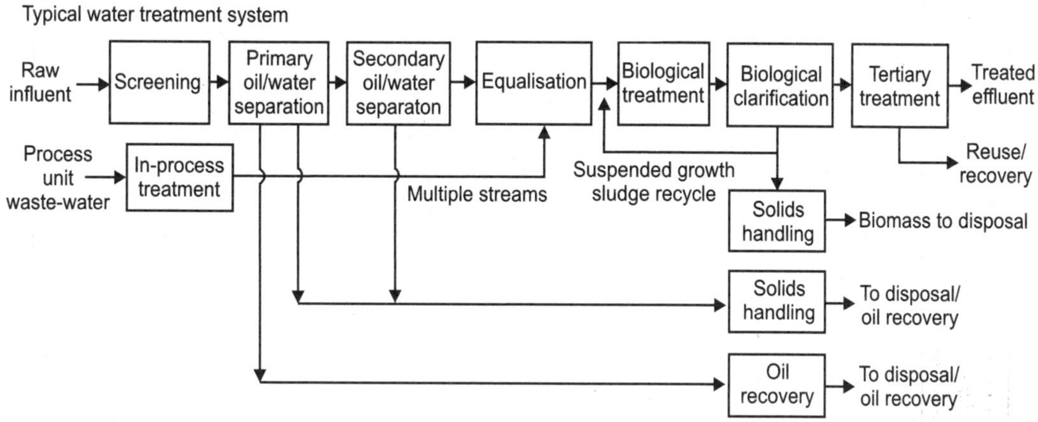

Typical water treatment system

Fig. 17.21. In order to protect the micro-organisms, some biological-treatment processes include pretreatment steps such as screening, oil/water separation and equalisation to moderate fluctuations in waste-water composition. This figure shows a schematic of a waste-water treatment system which is typical for hydrocarbon related industries.

Once the factors discussed above have been resolved, selection from the many available options can begin. Biological treatment methods vary widely, ranging from fixed-film technologies like rotating and submerged biological contactors to technologies like sequencing batch reactors and continuous flow activated-sludge systems. When evaluating the options, one needs to consider their effectiveness in the presence of constraints such as toxicity, COD, biochemical oxygen demand (BOD), and levels of nitrogen and sulphur compounds. Perhaps less obvious, but equally important, is to answer questions such as these:

1. How will the treatment method operate in cold or warm climates?
2. Can the system treat low- and high-temperature waste-waters?
3. How much sludge will the treatment method produce?
4. Can the system recover from up-sets?
5. How much will the system cost to operate and maintain, and does it require extensive operator attention?

All of these factors affect a company's bottom line and the quality of its end product. When evaluating treatment methods, it is essential to examine all of these criteria before making a technology selection.

Evaluation Guidelines

Table 17.23 is an application guide to help determine which biological-treatment technologies are most applicable for typical waste-water applications in the CPI. This table will assist in identifying the top two or three biological treatment technologies to investigate first. Subsequent detailed applications engineering is still critical to the process, but the guide is meant to prioritise potential technologies for further analysis. The parameters listed in Table 17.23 and explained in the box on selection criteria are those typically encountered when evaluating the addition or upgrade of a waste-water biological-treatment system in the CPI.

Table 17.23. Evaluation of biological treatment and aeration technologies for the CPI.

Evaluation parameter	Trickling filter	Rotating biological contactor	Submerged biological contactor	Disc aeration	Surface aeration	Fine bubble aeration	Coarse bubble aeration	Jet aeration	Sequencing batch reactor *	Membrane bio-reactor *	PACT*
Effective bioassay/ toxicity control				✓	✓	✓	✓	✓	✓	✓	✓
Effective BOD removal efficiency	✓	✓		✓	✓	✓	✓	✓	✓	✓	✓
Effective COD removal efficiency				✓	✓	✓	✓	✓	✓	✓	
Low O&M costs		✓	✓								
Low sludge production	✓	✓	✓							✓	
Low sludge disposal costs	✓	✓	✓								
Good operability: winter	✓	✓	✓	✓		✓		✓	✓	✓	✓
Good operability: summer	✓			✓	✓		✓	✓	✓	✓	✓
Good performance: high water temperature	✓				✓				✓	✓	✓
Good performance: low water temperature			✓	✓				✓			
Minimal operator attention		✓	✓								
Quick upset recovery	✓	✓.	✓						✓		
Easy expandability		✓	✓								
Efficient nitrification	✓	✓				✓	✓	✓	✓	✓	
Easy to enclose for VOC containment			✓								✓

(Contd ...)

Evaluation parameter	Trickling filter	Rotating biological contactor	Submerged biological contactor	Disc aeration	Surface aeration	Fine bubble aeration	Coarse bubble aeration	Jet aeration	Sequencing batch reactor *	Membrane bio-reactor*	PACT*
Low VOC stripping potential	✓					✓			✓	✓	✓
Easy installation		✓	✓		✓						
Energy efficient	✓	✓	✓			✓					
Ease to secondary containment		✓	✓								
Minimal space requirements						✓	✓	✓	✓	✓	✓

Comparisons are made based on waste-water with a COD of 600 mg/l and BOD of 250 mg/l.

* All listed technologies are products, except for sequencing batch reactors (SBRs), membrane bioreactors (MBRs) and powdered activated-carbon treatment (PACT), which are processes that can incorporate the remaining listed products. SBRs can incorporate fine-bubble, coarse-bubble and jet aeration. PACT can potentially incorporate all listed products except for trickling filters. MBRs can incorporate all listed products except for fixed-film treatment technologies like trickling filters, RBCs and SBCs.

Influent conditions used in the product/process evaluation are 600 mg/l COD and 250 mg/l BOD, amounts most commonly found in CPI waste-waters after appropriate pretreatment. For applications in which the influent conditions fall substantially outside of these criteria, the results in the table may not apply. In such cases, other biological processes such as anaerobic treatment and a fixed-film, fluidised-bed of activated carbon, should also be considered.

Biological Treatment Options

There are three basic categories of biological treatment: aerobic, anaerobic and anoxic. Aerobic biological treatment, which may follow some form of pretreatment such as oil removal, involves contacting waste-water with microbes and oxygen in a reactor to optimise the growth and efficiency of the biomass. The micro-organisms act to catalyse the oxidation of biodegradable organics and other contaminants such as ammonia, generating innocuous by-products such as carbon dioxide, water, and excess biomass (sludge).

Anaerobic (without oxygen) and anoxic (oxygen deficient) treatments are similar to aerobic treatment, but use micro-organisms that do not require the addition of oxygen. These micro-organisms use the compounds other than oxygen to catalyse the oxidation of biodegradable organics and other contaminants, resulting in innocuous by-products. The three individual types of biological-treatment technologies — aerobic, anaerobic or anoxic — can be run in combination or in sequence to offer greater levels of treatment. Regardless of the type of system selected, one of the keys to effective biological treatment is to develop and maintain an acclimated, healthy biomass, sufficient in quantity to handle maximum flows and the organic loads to be treated.

Maintaining the required population of 'workers' in a bioreactor is accomplished in one of two general ways:

1. Fixed film processes — micro-organisms are held on a surface, the fixed film, which may be mobile or stationary with waste-water flowing past the surface/media. These processes are designed to actively contact the biofilm with the waste-water and with oxygen, when needed.
2. Suspended growth processes — biomass is freely suspended in the waste-water and is mixed and can be aerated by a variety of devices that transfer oxygen to the bioreactor contents.

It is also possible to combine both methods in a single reactor for more effective treatment.

Fixed-film Options

Biotowers (trickling filters)

Biotowers, or trickling filters as they are often called, consist of a layer of media in a tank. Waste-water flowing into the biotower may have gone through an earlier treatment step to remove oil and coarse or settleable solids. Rotary distributor arms or fixed nozzles are used to spray the pretreated waste-water over the surface of the media.

The water then trickles downward through the bed. Air circulates upward through the media as treated water is removed by an underdrain system. As the waste-water trickles downward through the bed, a biological slime of microbes develops on the surface of the media. Continuous flow provides the needed contact between the microbes and the organics. As the slime layer gets thicker, it occasionally sloughs off of the media surface, requiring settling to remove the sloughed biosolids.

While biotowers generally are less efficient at removal of BOD and COD than other technologies, they do generate very little sludge and have a very low potential for stripping volatile organic compounds (VOC). Low VOC stripping potential can be an advantage for environmental reasons.

Rotating biological contactors

Rotating biological contactors (RBCs) consist of vertically arranged, plastic media on a horizontal, rotating shaft. The biomass-coated media are alternately exposed to waste-water and atmospheric oxygen as the shaft slowly rotates at 1–1.5 rpm, with about 40 per cent of the media submerged. High surface area allows a large, stable biomass population to develop, with excess growth continuously and automatically shed and removed in a downstream clarifier.

RBC systems have been installed in many petroleum facilities because of their ability to quickly recover from upset conditions. The RBC system is easily expandable should the need arise, and RBCs are also very easy to enclose should VOC containment become necessary.

Submerged biological contactors

Submerged biological contactors (SBCs), big brothers of the RBC, operate at nearly 90 per cent submergence with coarse-bubble diffused aeration providing a means of both aeration and motive force for rotation. Because of greater submergence, the load on the shaft is significantly less than that of an RBC. The SBC also provides nearly three times the surface area of a conventional RBC per foot of shaft length. With its compact design, the SBC is very easy to cover for VOC and odour containment. Unlike the RBC, the SBC system is driven completely by air, making it one of lowest maintenance and lowest operation-intensive, biological-treatment systems available. Like the RBC, the SBC is modular and can easily be expanded.

Suspended-Growth Options

Diffused aeration

Diffused aerators add air to waste-water, increasing dissolved oxygen content and supplying micro-organisms with oxygen necessary for aerobic biological treatment. Fine-bubble diffused-aeration systems are available in various types including ceramic and membranes, and are highly efficient. More reliable, but less efficient, coarse-bubble aeration systems are also available, and are normally manufactured of corrosion-resistant, stainless-steel components. Both systems are compatible with new installations and replacement of existing gas-aeration equipment. Fine-bubble aerators offer very low VOC stripping potential, and both fine and coarse diffusers provide good BOD and COD removal efficiency.

Jet aeration

The jet-aeration system is designed to provide required aeration as well as maintain suspension of biological solids, with the flexibility to either aerate or mix independently without the need for additional equipment. Air flowrates to the system can be varied. When aeration requirements decrease and air is completely shut off, pumps provide the required mixing action to enhance process control and save energy. The subsurface discharge leads to smooth and quiet operation, with no misting, splashing or spray from the basin. This also translates to low VOC release to the atmosphere. Since jet aeration requires no moving parts in the basin, the system offers long life with no in-basin routine maintenance required.

Surface aeration

For efficient surface aeration, high- and low-speed floating aerators provide pumping action that transfers oxygen by breaking up the waste-water into a spray of droplets. The large surface area of the spray allows oxygen to enter the waste-water from the atmosphere. At the same time, the oxygen-enriched water is dispersed and mixed, resulting in effective oxygen delivery. High- and low-speed surface aerators offer excellent oxygen transfer and low operating costs. They are able to handle environmental extremes such as high temperatures.

Another alternative for surface aeration is the use of horizontally mounted aeration discs or rotors. These disc or rotor aerators can be used in oxidation ditches known as looped, 'race track' reactor configurations. They provide stable operation with resulting high-quality effluent. The aerators are above water for easy maintenance and are energy efficient. Other multichannel processes use a concentric arrangement of looped reactors, which is particularly energy efficient and designed to achieve total nitrogen removal through simultaneous nitrification/denitrification. Disc and rotor surface aerators offer good BOD and COD removal efficiencies, and are very easy to replace if necessary.

Reactors in a vertical-loop configuration are also available for surface aeration. They are essentially oxidation ditches flipped on their sides. Upper and lower compartments separated by a horizontal baffle run the length of the tank. Surface-mounted discs or rotors provide mixing and deliver oxygen. Typically, two or more basins make up the system. The first basin operates as an aerated anoxic reactor and the second basin is operated under aerobic conditions. These types of reactors also have high BOD/COD removal efficiency.

Biological Treatment Processes

All of the previously noted aeration technologies, both fixed film and suspended growth, can be considered treatment products. However, there are some technologies that are actually processes because they can incorporate a number of different aeration technologies in their design. These processes include sequencing batch reactors, membrane bioreactors and powdered activated-carbon treatment (PACT) systems.

Sequencing batch reactors

A variation of the conventional activated-sludge system (in such systems, a clarifier is used to settle and recycle biomass back to an aeration basin) is the sequencing batch reactor (SBR). The SBR is a fill-and-draw, non-steady-state, activated-sludge process in which one or more reactor basins are filled with waste-water during a discrete time period and then operated in batch mode. In a single reactor basin, the SBR accomplishes equalisation, aeration and clarification in a timed sequence. Depending upon desired treatment objectives, the SBR can be operated in aerobic, anoxic or anaerobic conditions to encourage the growth of desirable micro-organisms. Aeration in this system is typically achieved with jet aeration, fine-bubble diffused aeration or coarse-bubble diffused aeration.

One of the advantages of SBRs is good operability in the winter, making them well suited for installations in colder climates. SBRs also take up little space because all of the treatment steps take place in a single reactor basin. Additionally, since the process is controlled by microprocessors, the plant operator is given tremendous flexibility to modify the treatment scheme to match changes in influent flow and loading characteristics.

PACT systems

When conventional biological treatment alone does not meet desired treatment requirements, powdered activated carbon can be added to enhance treatment efficiency. Activated carbon can be used in most suspended-growth, biological-treatment systems. The addition of carbon allows both physical adsorption and biological assimilation to occur simultaneously. PACT systems can be operated either aerobically or anaerobically.

Using powdered activated carbon in conjunction with traditional biological treatment provides excellent effluent bioassay results, provides for toxicity control within the bioreactor, and promotes higher nitrification efficiency than that of a conventional activated-sludge system. PACT systems also provide a buffering effect to shock or upset conditions, allowing the treatment system to recover quickly or even continue treatment with little or no detrimental effects. The use of activated carbon also decreases VOC emissions and improves COD removal efficiency.

Membrane bioreactors

In addition to the traditional types of biological treatment, speciality products have also been introduced to perform more than just one treatment step. Membrane bioreactor (MBR) systems are unique processes, which combine anoxic- and aerobic-biological treatment with an integrated membrane system that can be used with most suspended-growth, biological waste-water-treatment systems.

In the MBR, waste-water is screened before entering the biological treatment tank. Aeration within the aerobic-reactor zone provides oxygen for biological respiration and maintains solids in suspension. To retain active biomass in the process, the MBR relies on submerged membranes rather than clarifiers, eliminating sludge-settling issues. This allows the biological process to operate at longer than normal sludge ages (typically 20–100 days for a MBR) and to increase mixed-liquor, suspended-solids (MLSS) concentrations (typically 8000–15,000 mg/l in a MBR) for more effective removal of pollutants.

High MLSS concentrations and long-solids retention time promote numerous process benefits including stable operation, complete nitrification and reduced biosolids production. High MLSS concentrations also reduce biological-volume requirements and the associated space needed to only 20–30 per cent of conventional biological processes.

Chapter 18

Coal and Metal Mining

INTRODUCTION

The water pollution problems of coal mining have received great attention. Most of these water pollution problems are not completely unique to the coal industry. The mining of metals such as iron, lead, zinc and copper sometimes has acid drainage problems similar to those of coal. Gold mining, phosphate mining, sand and gravel washing and other mineral dressing operations have the problem of suspended solids in their effluents. These suspended solids differ from coal's problem principally in the colour of their effluent.

Wherever thermal energy is required in large quantities, coal is one fuel which is always considered and most frequently used to produce this energy. Coal is the most valuable mineral mined, and coal mining is the largest segment of the mining industry. As bituminous coal is the principal type mined, this chapter will be devoted primarily to water pollution control from the mining and processing of bituminous coal.

CHARACTER OF WASTE DISCHARGES

The water discharged from bituminous coal mines cannot be readily characterised. It is improper to call this water an industrial waste as it has not been used in the mining process. In fact, it is an unwanted and costly intruder to the mining operation and in an operating mine it must necessarily be handled and disposed of so that mining can continue. Mine drainage also issues from many abandoned mines, exceeding in quantity and pollutional effect the drainage from active mines. Because water pollution from acid mine drainage is the largest pollution problem of the coal mining industry, it is proper to consider it here even though it is not an industrial waste in the usual sense.

The drainage from bituminous coal mines today may vary from highly acid waters to water of drinking quality, and good quality drainages are known even in normally acid regions. Acid mine drainage must be understood rather than characterised and usually prevented rather than treated. To understand acid mine drainage, a working knowledge of the geology, hydrology, chemistry and bacteriology involved is necessary. The acid constituents of almost all acid mine drainages originate from oxidative destruction of iron disulphide. Iron disulphide in the crystalline form of pyrite is frequently found in the thin layers or partings between layers of coal in a coal bed. Additionally, pyrite may be found in the coal substance itself, both in the form of small nodules dispersed throughout the coal and in large masses or lens-shaped concretions occurring at random in certain coal seams. Pyrite is often found in layers of slate, sandstone and other rock overlying the coal.

The pyrite in and near coal seams may appear as bright, sparkling, yellow crystals characteristic of fool's gold or as a gray-black, hard, pyritic mass. Early investigators, on the basis of appearance and oxidation rate, hypothesised that the iron sulphides in coal consisted of pyrite, marcasite and pyrrhotite. However, pyrrhotite is iron monosulphide, FeS and chemical analyses quickly demonstrated that the mineral was a disulphide compound. The pyrite crystal is isometric while marcasite is orthorhombic. X-ray diffraction patterns of these two crystals are easily distinguished and studies of the iron disulphide in several coal seams have readily identified the presence of pyrite, whereas marcasite has not been found. It is concluded that the iron disulphide mineral associated with and near coal seams is pyrite rather than marcasite. The high rate of oxidation of this material is presumed to result from the small particle size and the intimate association with carbonaceous materials.

Pyrite reacts with dry atmospheric oxygen to form ferrous sulphate and sulphur dioxide. In the presence of moisture the reaction proceeds more rapidly and ferrous sulphate and sulphuric acid are formed. When the pyritic material is exposed to atmospheric oxygen under wet conditions, bacterial action also may take place. The bacterial action may somewhat accelerate the rate of oxidation of pyrite; however, the degree of acceleration is a subject of controversy among the various researchers who have investigated it. The acidity of mine drainage is a function of the degree of contact between flowing water and oxidised acid-producing materials and the amount of alkaline material which the water may contact and dissolve either before or after its contact with acid-producing materials. The pH of mine drainage cannot be correlated with the acidity of this drainage and hence is not a suitable yardstick for determining the effect which a given mine drainage might have when discharged into a particular stream. The lack of correlation between pH and acidity is often confusing and has led many people, both mine operators and water pollution control officials, to erroneous conclusions about the quality of mine drainage water. The lack of correlation between pH and acidity, however, can be understood if we remember that mine drainage is a mixture of unstable chemical compounds that are continuously changing under ambient conditions. Some of these unstable compounds include unreacted alkali materials, ferrous sulphate, ferric sulphate, a variety of ferric hydroxide sulphate complexes and sulphuric acid that may be occluded and mechanically held in the precipitates.

Most acid mine drainages contain large amounts of sulphate, calcium, magnesium and iron and may also contain aluminium, manganese and other heavy metals. The actual amount of each component and the ratio between components vary over extremely wide ranges, depending upon the peculiarities of each specific location. An average acid mine discharge would contain perhaps 500–1000 mg/l acidity as $CaCO_3$, 100–300 mg/l iron, approximately 2000 mg/l sulphate and total dissolved solids of approximately 3000 mg/l. The pH might vary from about 6.0 down to 2.8. Acid mine drainages rarely contain a high concentration of chloride and almost never contain oxidisable organic materials.

Water discharged from the preparation of coal is not greatly dissimilar to the water discharged from sand and gravel washing or other mineral preparation operations. The principal water pollutant in these waste-waters is suspended solids. Additionally, the water may contain an increased but not usually troublesome amount of dissolved inorganic solids such as calcium or magnesium sulphates and iron. Waste-water from coal preparation plants is sometimes acidic in character, but usually it is deliberately kept alkaline to minimise corrosion of the processing equipment. This water has a negligible content of oxidisable organic materials. Historically, chlorides were at one time dissolved in the water of certain preparation processes to give the solution a high specific gravity and to assist in the separation of extraneous mineral matter from coal. These processes have largely disappeared and chlorides are rarely a problem in waste-waters from coal preparation plants.

Waste Disposal and Pollution Prevention

The process through which acid mine drainage is formed and discharged to cause water pollution may be illustrated as a 6-link chain. Figure 18.1 shows this chain and the part each link plays in the formation and discharge of acid mine drainage. Any treatment or abatement procedure intended to lessen the pollutional effect from acid mine drainage must necessarily break this chain at one or more of its links.

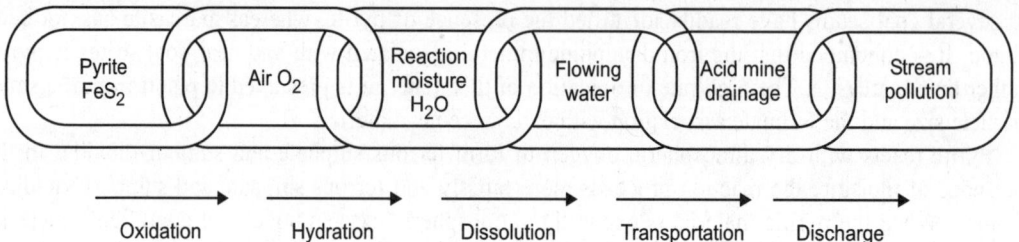

Fig. 18.1. The chain of the formation and water pollution of acid mine drainage.

Transport water

The acid oxidation products of pyrite must be dissolved and transported to the streams before they can cause water pollution. If these products are not dissolved and carried away by flowing water, they remain in the mine and cause no pollution. Control of water flow, in both underground and surface mining operations, is probably the most effective method available to minimise the amount of acid discharged from a mine. The principle, as stated for application to mining situations, is that the contact between water and acid-producing materials shall be minimised.

In surface mining, the contact between water and acid-producing materials can be minimised by diverting the flow of surface runoff and flowing stream waters around and away from the mining operation. This may take the form of high wall diversion ditches, drains and conduits to carry flowing water through or around the mining operation and re-channelling or diverting of streams away from the mine. Contact between water which does gain entry to the mine and acid-forming materials can be minimised by removing this water from the mine as quickly as possible after it accumulates.

In underground mining, contact between flowing water and acid-producing materials can be minimised by sealing off the surface of the earth above the mine to close cracks, fissures, sink holes and other openings when they can be detected, by picking up water as close as possible to its points of entry in the mine and by conducting it through and out of the mine either in closed conduits or in ditches or sewers that prevent further contact of the water with acid-producing materials.

Acid mine drainage

Acid mine drainage as it discharges from a mine is a complex solution of ferrous and ferric iron salts, calcium, magnesium, sometimes manganese and other sulphate salts. The flow of mine drainage may vary from only a few gallons per hour to millions of gallons per day, depending on the specific drainage point being considered.

In some of the drier areas of the country, where annual evaporation equals or exceeds annual rainfall, it is possible to impound low flows of acid drainage and hold them permanently in a pond so they do not flow into any stream. Particularly in surface mines this may be accomplished by proper handling of the overburden. No cases are known where this has proved practicable for underground mining.

Where extremely low flows of acid drainage are encountered, they can sometimes be neutralised through the use of lime or related alkalies. Because of the costs and other factors involved, neutralisation is usually uneconomical and completely impractical. In those situations where it can be used, it is a difficult operation, especially on continuous flows. Special equipment is required, which will compel intimate contact between the individual particles of alkali and the acid water. As acid mine water containing dissolved iron is neutralised, the iron precipitates and tends to coat individual particles of alkali, rendering unavailable and useless, the remaining alkaline material in the particle.

Under some conditions it may be practicable to neutralise small standing pools of acid mine drainage before these pools are discharged to the streams. Such neutralisation can be accomplished by the proper application of lime to the acid water. However, neutralisation again is difficult because of the coating effect of the precipitating iron hydroxide.

The suspended ferric hydroxide carried with acid mine drainage can sometimes be reduced by permitting the water to stand in a lagoon for many hours or days before releasing it to the stream. Additionally, lagooning of acid mine drainages that contain large amounts of ferrous sulphate may permit the sulphate to oxidise and deposit some of its iron content. When these procedures are applicable and where space and property availability permit, they are considered and used. The precipitated and settled iron hydroxide progressively fills lagoons of this type and the sludge must be removed or a new pond constructed after prolonged use.

Stream pollution

Stream pollution by coal mine drainage occurs only after the natural alkalinity of the receiving water has been used up and the concentration of undesirable elements has reached such a point that it renders the receiving waters unsuitable for their proposed downstream uses. The ultimate downstream effect of mine drainage frequently can be minimised by discharging these waters as uniformly as possible from the mine. This may be accomplished through an elaborate system of pumping controls to hold the water level in the mine at a particular point and to discharge water from the mine in direct relationship to rate of inflow. Additionally, surface impoundments can be built and controlled so that water pumped intermittently from a mine is released as uniformly as practicable to the receiving streams. Under these conditions the effect on the receiving stream is relatively uniform and the downstream effects are minimised. Control of the discharge of black solids from a coal cleaning operation is quite similar to the control of suspended solids from any other mineral preparation operation. The black solid material is finely divided clay, black shale and other minerals, along with coal. Normally, the first step in clarification of coal washery waste-waters is to concentrate the solid material through the use of hydraulic cyclones or thickeners. From these operations the clarified water is returned to the coal washing circuit while the high-solids water is further concentrated. The underflow from the final clarification stage, usually a thickener, may be passed through filtration equipment such as a drum type continuous filter or more normally, may be pumped to a settling pond where the solids are permitted to settle.

The clarified water from the settling pond may be either returned to the preparation circuit or discharged into a receiving stream. As new preparation plants are designed and built, more and more effort is being expended to make these plants closed-circuit units so there is normally no process water discharge into the streams.

Most coal preparation plants have suspended solids removal equipment installed and operating; however, there are frequent complaints from water pollution control officials that this equipment does not perform satisfactorily. Probably the most common problem other than normal maintenance is overloading of the clarification system so it cannot function properly. This condition could be overcome by designing and installing larger equipment and clarification systems are often unwisely designed at absolute minimum safety factors. A more immediate and economic solution usually can be attained by properly balancing the water circuits in the coal washing operations so the minimum amount of water necessary for proper coal preparation is used and discharged to the clarification system. Proper and constant attention to this detail may forestall the requirement of installing larger clarification equipment and may at the same time yield better water pollution control.

There have been specific instances where the fine coals recovered from washery water clarification have had economic value; in a few cases their value has been great enough to defray a considerable portion of the expense of water clarification. Normally, however, the solids recovered from water clarification operations are of no economic value and must be disposed of by permanent impoundment, by disposal with other solid refuse where it will be permanently stored, or by disposal in specially constructed pits, so it does not again become waterborne and pollute the streams.

METAL MINING

This section covers industrial waste-waters resulting from the removal of ores from the ground and from their subsequent treatment to produce an ore concentrate. Emphasis in this section is on those metallic ores that are mined in substantial amount in various parts of the world. These include iron, copper, zinc, lead, molybdenum and uranium. Of less significance are metal mining operations involving aluminium, tungsten, vanadium, gold, manganese, magnesium, mercury and lithium. Vanadium concentrate is chiefly obtained as a by-product from uranium ore milling operations and may eventually be recovered in substantial quantity from wastes of the phosphorus industry.

The metal mining industry is not unique with respect to control of industrial waste-waters. There is similarity between the mining procedures and subsequent ore upgrading steps in coal mining, metal mining and nonmetal mining. The metal mining industry produces substantial mine drainage, as does the coal mining industry and contends with a paucity of water in arid areas, as do many nonmetal mining industries. Most ore processing steps, require water. Many of them produce waste-waters. Although some metal mining operations bring in water to alleviate dust, it is usual that water is drained or pumped from the mine or pit in order to avoid flooding. Often this results in an acid discharge from the operation. Water is, of course, necessary in ore washing operations, in wet grinding, jigging, tabling, aqueous classification and chemical processing. It is axiomatic that water is also used for power generation, cooling and for sanitary purposes.

Although the metal mining industry requires much water, it also produces much water by virtue of its operations penetrating the water table. Moreover, the relative geographic isolation of the industry and the geologic circumstances which seem to concentrate its effort in water-scarce areas have combined to render the industry sensitive to its water problems. Because of such factors, the industry is knowledgeable with respect to waste-water control, water recovery and water reuse and has been able to maintain its water resources ample for growth. Simultaneously, pollutional aspects of its waterborne wastes have been thoughtfully controlled.

Industrial Wastes

The principal characterisation of metal mining waste-waters is their settleable solids. In various forms, as mud and slimes washed from the ore during processing, as gangue from wet gravity separation operations or froth flotation systems and as undissolved residues from chemical leaching procedures, these suspended solids represent the industry's major waterborne waste. This is not surprising when it is borne in mind that most domestic ores contain from 1 per cent to 10 per cent of the desired concentrate, hence the bulk of the ore fed to the mill must be discarded. Another waste, acid drainage from both active and abandoned metal mines, may, as a tonnage flow, dwarf the production of waterborne waste solids. However, such mine flows are relatively dilute and their dissolved mineral content, converted to potential solids, is minor compared with the waterborne undissolved solids in the waste-water of the industry. Another source of wastes is the wide variety of reagents used in froth flotation processes. Most of these reagents appear in the effluent from the mill. Some, such as slime depressors, adhere to the waterborne gangue; others remain entrained or dissolved in the process waters.

Finally, the metal mining industry utilises some, though relatively few, leaching operations, usually employing soda ash, caustic soda or acids, chiefly sulphuric. These reagents, or water-soluble products thereof, further add to the waste-water streams of the industry.

Water Pollution Potential

The pollutional potential of waste-waters from the metal mining industry is relatively low. There have been and are exceptions to this, such as waterborne wastes from cyanide leaching circuits or waterborne radio-activity from uranium milling operations. However, cyanide leaching circuits are so few and so isolated as to preclude their further consideration. Water-borne radio-active waste discharged present some problems in uranium milling districts until prompt and decisive action on the part of all concerned corrected the situation. Except for the relatively negligible input of sanitary wastes (where sanitary sewers enter mine and mill waste streams), wastes from the metal mining industry have no significant biochemical oxygen demand. Moreover, these wastes, on the average, have practically no toxicity. This is because the processing steps in the industry consist essentially of water washing, gravity separations in a water medium, flotation using small dosages of nontoxic reagents and leaching involving either sodium alkalies or sulphuric acid; and because the potentially toxic metal ions are efficiently scavenged from wastes as part of the milling operation.

Control of Industrial Wastes

The principal control of waste effluents by the metal mining industry is usually accomplished by a settling pond or ponds. Few operating mills may be found without their nearby tailings ponds, settling ponds, impounding dams, lagooning areas or similar devices.

Sometimes these structures take advantage of the local topography, but more often they must be constructed by damming, excavating, diking or a combination of the three. In most metal mining localities, substantial land area is available and the use of much land area for waste disposal is fortunately both technically and economically feasible. The common tailings pond serves many purposes. It is a primary settler for the gangue, a clarifier for the water, a treatment tank for pH adjustment or chemical precipitation as desired, a water storage area, a surge tank for controlling discharge into public waters and an investment storage because, in many instances, the mill tailings contain secondary values which can be reclaimed at a future time. It is of particular significance that the reservoir of water in the pond (the pond being necessary *per se* to settle out solids) is, by virtue of its relatively low burden of dissolved process chemicals, at least a tempting and usually a vital source of water for reuse. Depending upon the ratio of atmospheric evaporation to precipitation and the ground and effluent characteristics affecting seepage, many tailings ponds never overflow their confines and some may go nearly dry during certain periods of the year. This is especially true in the arid and semi-arid areas of the country. On the other hand, this situation may be reversed in normal and heavy rainfall areas. In either situation, tailings ponds at least retain substantially all, if not all, of the waste solids from the metal mining operation.

In addition to tailings pond techniques, the industry utilises many innovations for industrial waste-water control primarily dictated by local circumstances. If the effluent to the tailings pond is alkaline, acid mine drainage may also be run into the tailings pond to effect a neutralisation and a better grade of water for reuse. Conversely, when the mill effluent is acid and particularly when acid mine drainage is also entering the tailings pond, pulverised limestone or lime is frequently added to the effluent ahead of its discharge to the tailings pond to effect pH control of the clarified effluent in the settling area. In some instances, acid effluent from copper mills containing dissolved copper, or copper-pregnant acid drainage from the mines, is first passed over scrap iron to precipitate the copper, while simultaneously dissolving iron. The value of the recovered 'cement' copper is considerably greater than the cost of lime subsequently used for pH control and iron precipitation in the settling pond. Where flotation reagents, untreated sanitary sewage or other chemicals that might adversely affect the reuse of tailings pond waters are present, activated carbon may be added to the pond influent for the purpose of adsorbing undesired organic matter into the common precipitate. The impounding of radio-active wastes has produced a problem as to legal responsibility when operations are abandoned and ownership of the property changes hands. Sometimes tailings pond waters are suitable for livestock watering, irrigation and sanitary facilities. Possibly no other industrial waste control device produces so much reusable water as the tailings pond and at a price that the industry can afford. Conversely, probably no industry is better situated than the metal mining industry, in terms of geographic and technological considerations, to utilise tailings ponds to their fullest advantage.

Specific Industry Controls

Copper and uranium are the most significant of the metal mining industries in which pollution potential could be serious, as a result of dissolved substances, if control measures were not being practiced. Iron, zinc, lead, complex ore operations, molybdenum and other segments of the industry have no pollution problems as far as dissolved wastes are concerned and of course, solid tailings are impounded by practically all industry units.

The iron industry, by the very nature of its operations, adds little to its process water other than hardness. It uses water primarily for washing its ores and avidly reuses water from its settling ponds. In some plants, where a heavy medium of finely divided ferrosilicon is involved, a slight solubility of iron in plant water may result. However, clarified water from the iron mining industry can generally be discharged into public waters without detrimental effect.

In zinc production, the usual processing involves mining, crushing, grinding, flotation, thickening and filtering. This industry produces considerable mine drainage, a typical composition of which is 1450 mg/l Zn, 800 mg/l Fe and pH 2.3.

Water from the settling and impounding of zinc mill process tailings tends to be alkaline, so the combination of mine water and clarified tailings pond overflow results in nothing more than a hard water with a pH in the vicinity of 7.0. The zinc content of drainage from both active and abandoned mines presents a challenging problem with respect to development of an economic process for zinc recovery, but is not considered to be a pollution problem. Essentially, the foregoing statements on the zinc industry apply also to those industries processing lead-zinc, copper-zinc and complex lead-zinc-silver-gold-copper ores. Mine drainage from these operations has a pH near 7.0 and clarified mill process water from the tailings ponds generally ranges from neutral to slightly alkaline because of the slightly alkaline nature of the mill concentration systems usually employed. It is not uncommon to find the waste-water from these operations in demand for irrigation and livestock watering, particularly in arid areas. There are a few abandoned mines in which ageing processes have been undisturbed for a long period and the sulphur bodies present have produced so much sulphuric acid that the drainage water has a pH as low as 2.5 or 3.0. However, it is general practice not to pump water of this acidity; and when such water does overflow the mine, it is usually diluted to a harmless acid strength by the freshet producing the overflow.

The mining of gold, involving cyanide leaching, could beget a water pollution problem, were it not for the fact that gold operations are so limited and their locales so isolated. In one gold operation, using a cyanidation process, waste-water overflows the settling pond at a pH of approximately 9.0 and is almost immediately lost into the ground of the isolated area. In another plant, mill tailings are impounded in a dry mountain canyon, apparently capable of serving this purpose for some generations ahead.

Summarising the foregoing observations, it may be concluded that the mining industry enjoys relatively high standards of industrial waste-water control. This is largely the result of the inherent technology of the industry and the practical necessity of water reuse, but it is no less a tribute to the awareness of the industry of the value of protecting and respecting natural water resources.

Industry trends

An assessment of the long range aspects of industrial waste-water control in the metal mining industry must be based on the premises that the industry already has the situation well in hand, it is not an expanding industry as a whole, it is essentially wed to its ore reserves and metal mining industry processes (statistics notwithstanding) may well consume less rather than more water even though production of some metal concentrates does increase.

Chapter 19

Iron and Steel Industry

INTRODUCTION

The iron and steel industry, as treated here, includes pig iron production, steel making, rolling operations and those finishing operations common in steel mills, i.e. cold reduction, tin plating and galvanising. Most steel firms operate iron ore mines, ore beneficiation plants, coal mines, coal cleaning plants and coke plants; many have fabricating plants or produce a variety of speciality steel products.

IRON AND STEEL INDUSTRY

Manufacturing Operations

Manufacturing operations of the iron and steel industry may be grouped as pig iron manufacture, steelmaking processes, rolling mill operations and finishing operations. A single mill is not likely to incorporate all of the many combinations and variations of these operations that are possible. Most mills specialise in the production of broad categories of steel products; in a large mill, however, the product list is long.

The manufacture of pig iron is accomplished in the blast furnace. Steel-making processes include pneumatic processes, open hearth processes and electric furnace processes. Rolling mill operations include rolling of blooms, slabs and billets; scarfing and other preparations of semi-finished steel; rolling of shapes, bars, strip and plates; wire drawing; tube drawing and pipe forming; and pickling or other oxide removal operations. Finishing operations include tin plating, galvanising, cold reduction and coating.

Blast furnaces

The blast furnace process consists essentially of charging iron ore, limestone and coke into the top of the furnace and blowing heated air into the bottom. Combustion of the coke provides the heat necessary to attain the temperatures at which the metallurgical reducing reactions take place. The incandescent carbon of the coke accounts for about 20 per cent of the reduction of the iron oxides; the carbon monoxide formed between the coke and the oxygen of the blast accounts for the remaining reduction accomplished. The function of the limestone is to form a slag, fluid at the furnace temperature, which combines with unwanted impurities in the ore. Two tons of ore, 1 ton of coke, ½ ton of limestone and 3½ tons of air produce approximately 1 tonne of iron, ½ ton of slag and 5 tons of blast furnace gas containing the fines of the burden carried out by the blast; these fines are referred to as flue dust.

Characteristics of Steel Mill Wastes

Wastes from the various operations in steelmaking vary widely in characteristics and in volume water pollutant in a typical steel mill complex are shown in Table 19.1. These wastes generally have physical and chemical effects on receiving streams different from the oxygen-consuming characteristics of municipal sewage and organic industrial wastes. Because the waste streams vary so widely and are usually separated by the distances between the several operations, composite effects are of little significance; treatment and disposal generally must be considered for the separate wastes (Fig. 19.1).

Table 19.1. Water pollutants in a steel mill complex.

Description	Source	Disposition
Oil		
Rolling oils	Rolling mills, cold and hot-rolling	Part adheres to scale; free oil collected for incineration
Lubricants	Motor power: steam engines, forge and hammer pistons, gear drives, electric motors. Fabrication machinery; machining, forging, drawing, etc.	Floor spills soaked up on adsorption compounds; cutting oils and other emulsions segregated for incineration; oily water skimmed at source, and free oil collected for incineration
Hydraulic oils	Motive power pumps for positioning devices or for press operation	Spills segregated for pickup and incineration
Quenching oil	Heat treatment	Seldom replaced; can be incinerated
Fuel oil	Boiler plant, furnaces, soaking pits	Spills segregated by dikes around storage tanks; incinerated
Solvents	Paint shops, degreasing operation	Collected for incineration
Tars and pitch	Coke plant and by-products recovery	Collected for incineration
Suspended solids		
Scale	Rolling mills	Sinter plant for recovery as sinter
Sand	Foundry	Slag pile, landfill
Burden fines	Air washers at sinter plant and skip hoist charging area; Blast furnace gas washers	Sinter plant for recovery as sinter
	Open hearth and BOF fines	Sinter plant if Zn is low; otherwise buried in slag pile
Fly ash	Coal fired furnaces	Cement or cement block additive
Coal and coke	Coke plant	Collected and burned
Chemicals		
Pickle liquor	Acid pickling	Regenerated or neutralised
Acid sludge	By-products plant	Regenerated or burned
Caustic wash	By-products plant	Incinerated
Lime	Mould or stool coating; water softener sludge	Recovered for pickle liquor treatment; controlled release to waste-water
Brine	Zeolite regeneration	Reclaimed or controlled release to waste-water
Cleaners	Surface treatment, degreasing	Segregated for oil breakout or incineration

(Contd...)

Description	Source	Disposition
Toxic chemicals	Coke plant, gas line drip legs, metal treatment	Chemical or biological destruction; incineration
Heat		
Cooling water	Furnaces, heat treatment, roll cooling, air conditioning, heat exchangers	Cooling towers
Boiler blowdown	Steam plant	Recover to heat feedwater
Sanity wastes		
Domestic water	Change rooms, toilets, cafeterias, etc.	Segregate and treat by standard methods

Water Use in the Industry

The water requirements of steel plants vary widely, depending primarily upon the quantity and quality of the available supply. The use of as little as 1500 gal of water per ton of product has received much attention in one instance where recirculation is extensively practiced, due primarily to short supply. A figure of 65,000 gal per ton of product has also been widely quoted and has been valid in certain installations that have had practically unlimited water supply. The use of 30,000–40,000 gal of water per ton of product has been typical of many large plants; actual consumptive use of water, i.e. water withdrawn but not returned, is probably less than 1000 gal per ton of product. A recent industry survey indicated a maximum water use of 49,000 gal per ton of product and an average use of 17,000 gal per ton. Water use in the various departments of a typical integrated mill is approximately as shown in Table 19.2. Most of the water required by a steel plant is used for indirect cooling and needs no treatment, provided it is not excessively hard; chlorination is often desirable to prevent slime formation.

Table 19.2. Water use in an integrated steel mill.

| Department | Volume | |
	Gallon per ton of finished steel	Per cent of total
Blast furnace	10000	25
Open hearths	5000	12½
Coke plants	5000	12½
Hot mills	10000	25
Finishing mills	8000	20
Sanitary, boiler and other uses	2000	5
Total	40000	100

The water used in Blast furnace gas washing and in hot mills for roll cooling and scale transport is not necessarily of high quality; it is usually used as pumped. In the various finishing operations such as cold reduction, stainless strip rolling, electrolytic tin lines and galvanising, purer water is required and treated water is often used.

Waste-waters

The various waste-waters from a typical steel plant are considered here individually, roughly segregated according to the operations from which they result. It must be remembered, however, from the previous descriptions of the various operations that no such clearcut segregation exists in actual practice. Indeed one of the major problems in installing waste treatment facilities in older mills is the segregation of waste streams from integrated operations.

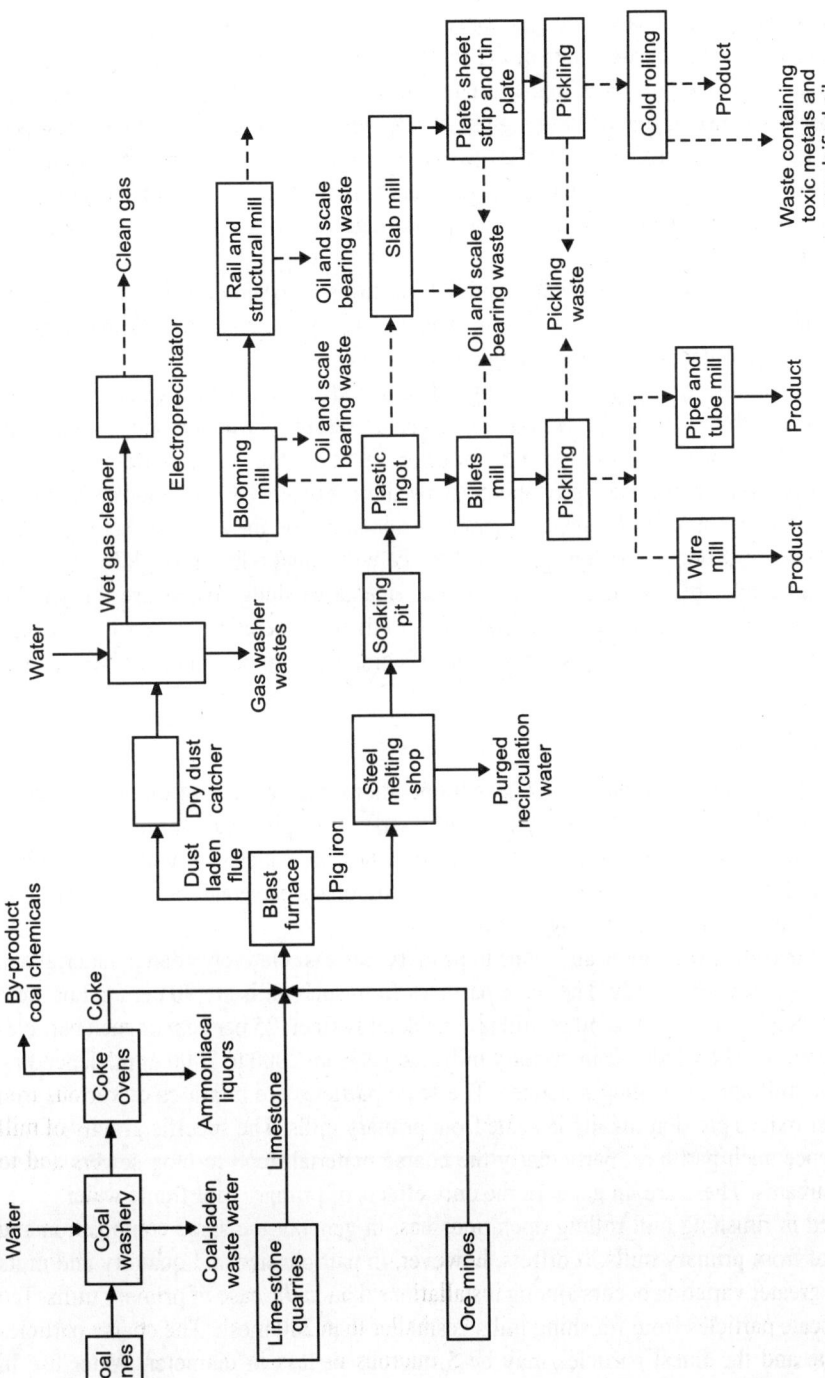

Fig. 19.1. Typical flow sheet showing iron and steel processes, and pollution status.

Gas-washer waters

The water used in washing Blast furnace flue gas contains from 1000 to 10,000 mg/l of suspended solids, depending upon the furnace burden, size of the furnace, operating methods employed and type of gas-washing equipment. Following a 'slip' in the furnace, the concentration of solids in washer water may exceed 30,000 mg/l. The use of fine ore and high blast rates result in the highest concentrations of solids in the washer water; the top pressure used in the furnace is also an important factor. The efficiencies of dry dust catchers and wet washers vary considerably and account for many of the differences found in various installations. Conventional wet washers use an average of about 3000 gpm of water; the newer venturi scrubbers use 600–1000.

The wash water from electrostatic precipitators adds little to the washer flow, but increases the concentration of the finest particles in the waste stream. Blast furnaces producing ferromanganese have a high percentage of semicolloidal dust particles in the washer water.

The fume from pneumatic steelmaking processes, open hearth and electric furnaces and hot scarfing operations is often eliminated by electrostatic precipitators or venturi scrubbers which produce waterborne wastes. These suspensions are generally similar to gas-washer water, but the particles are much finer.

The flue dust particles in washer water are probably 50 per cent finer than 10 microns and approximate the composition of the furnace burden; the specific gravity is about 3.5 on the average. Effects on the receiving streams include objectionable colour and interference with aquatic life through formation of bottom deposits and impedance of light transmission; in extreme cases sludge banks are formed that interfere with navigation. Gas-washer waters, especially from furnaces operating on ferromanganese, contain appreciable though highly variable concentrations of complex cyanides and may have a toxic effect on aquatic life.

Scale-bearing waters

Scale-bearing water originates in the various rolling mill operations and consists of the water used to dislodge scale and to cool the rolled product, plus the water used to transport scale through the flumes beneath the mill line. The characteristics and quantities of scale-bearing water vary widely depending upon the particular rolling operations. The total iron loss in the form of scale averages about 2½ per cent, from the blooming mill through the final rolling operation.

Scale produced in the rolling of blooms and slabs in primary mills is relatively coarse material and most of it settles out of suspension readily. The scale particles from such mills are 90 per cent or more coarser than 200 mesh. Scale produced in a billet mill is considerably finer; 25 per cent of such particles may be finer than 200 mesh. The water use in primary mills ranges from 2000 to 7000 gpm, depending upon the design of the mill and the rolling practices. The scale particles are mixtures of various iron oxides, with the higher oxides pre-dominating in scale from primary mills. The specific gravity of mill scale is about 5.0, hence such particles, particularly the coarse material, tend to clog sewers and to deposit in receiving streams. These are, in general, the only effects of primary mill flume water.

The scale produced in finishing mill rolling operations has, in general, the same composition and specific gravity as that from primary mills. It differs, however, in particle size and quantity and in its effects. Considerably greater variation occurs among installations than in the case of primary mills. Ten to 20 per cent of the scale particles from finishing mills is smaller than 200 mesh. The coarse particles are still relatively fine and the finest particles may be 5 microns or less in diameter. Water use in finishing mills ranges from 5000 gpm or less in bar mills and cold reduction mills to 25,000 gpm or more in the newest hot strip mills. Finishing mill flume water may settle in the receiving streams and form bottom deposits or sludge banks and may increase the turbidity of the stream or impart an objectionable colour if the scale particles are extremely fine.

Acid waters

Spent pickling solutions and acid rinsewaters differ widely in quantity, composition and concentration, depending upon the manner of pickling, production rate, type of steel being cleaned and the degree to which control over the operation is practiced. Spent pickling solutions of various types may be produced in different, separated operations at a large mill. Acid rinsewaters have the same relative proportions of iron salts and free acid as pickling solutions, but are much more dilute; 10–15 per cent of the acid used in pickling is discharged in rinsewaters. Spent pickling, solution discharges may inhibit bio-oxidation processes in streams and may be injurious to aquatic life if the quantity released in relation to the stream flow is sufficient to lower the pH of the stream significantly.

Sulphuric acid comprises about 90 per cent of the total of acids used in pickling steel. Spent sulphuric acid pickling solutions contain free acid, ferrous sulphate, undissolved scale and dirt and the various inhibitors and wetting agents, as well as dissolved trace metals. The spent solutions from continuous strip picklers contain 5–9 per cent free acid and 13–16 per cent ferrous sulphate; from batch operations the spent solutions may contain 0.5–2.0 per cent free acid and 15–22 per cent ferrous sulphate. Ten to 15 per cent of the acid used in pickling is discharged in the rinsewater as highly diluted free and combined acid. Spent sulphate pickling solutions are discharged at 170°F–190°F and can amount to 1,00,000 gal per day in a large mill.

Hydrochloric, nitric, phosphoric and hydrofluoric acids are used in pickling stainless steels. These acids may be used alone, in various combinations, or in combination with sulphuric acid. Stainless steel pickling practices vary widely, in the industry. A typical pickling operation for stainless steel plates consists of a 10 per cent sulphuric acid bath at 160°F, followed by a 10 per cent nitric acid, 4 per cent hydrofluoric acid bath at 150°F. A typical continuous pickling line for stainless steel strip consists of a 15 per cent hydrochloric acid bath at 160°F followed by a 4 per cent hydrofluoric acid, 10 per cent nitric acid bath at 170°F. Phosphoric acid is often used in pickling when a phosphate coating is desired. Hydrochloric acid is being used increasingly for mild steels in the new vertical tower pickling installations. These spent solutions contain free acids and the various iron salts, as well as undissolved scale and dirt, inhibitors, wetting agents and trace metals. Compositions vary widely according to the specific operation and plant practice.

Oil-bearing waters

Rolling oils, lubricants and hydraulic oils are present in the effluents from many operations in a steel mill and occur as both free and emulsified oils. Volumes of the waste streams and the concentrations of oil vary widely according to operating practices and housekeeping methods. Emulsified oil in an effluent can be esthetically objectionable and may add a significant BOD. Free oil is particularly objectionable in a stream because very small quantities can result in widespread surface films; larger quantities foul boats and docks and result in unsightly accumulations along stream banks. Severe oil pollution can have serious adverse effects on aquatic life, birds and land animals.

So-called soluble oils are present in the waste discharges from cold reduction mills, electrolytic tin lines and a variety of machine shop operations. Natural palm oil and synthetic proprietary substitutes are used in these operations and form stable emulsions when mixed with water at elevated temperatures, especially when kerosene and various detergent cleaning compounds are used. Concentrations of soluble oils vary according to the degree of recirculation practices; volumes of the waste streams likewise vary widely. Typically, the effluent from once-through use in a cold reduction mill will contain 200 mg/l oil, 25 per cent of which is a stable emulsion.

Lubricating oils and hydraulic fluids are present in the effluents from all rolling operations and most other machine operations. These oils exist mostly as free, floating films and the quantities depend primarily upon machine maintenance and manual lubrication practices, i.e. upon housekeeping.

Miscellaneous

Other waste-waters from steel mills include alkaline cleaning solutions, water used in granulating slag and cooling water. Alkaline cleaning solutions are used to remove rolling oils prior to finishing operations. Caustic soda, soda ash, silicates and phosphates are common cleaning agents. Spent cleaning solutions contain saponified oils and dirt and have substantial residual alkalinity. Total volumes are small, ranging from 1000 to 10000 gal per week for individual operations; these quantities are usually dumped batchwise. The effects on the receiving stream are probably not adverse, especially if the volumes of acid wastes are relatively large, as is usual.

The quenching of blast furnace slag produces small quantities of water containing slag particles. Effluent from a slag pit may range from 100 to 200 gpm and is usually of a clear appearance. The highly abrasive nature of the suspended slag particles is the principal objection to such effluents; the bulk gravity of slag particles ranges from 0.8 to 1.5 because of expansion in the granulating process.

Cooling water discharges comprise the largest percentage of steel mill effluents and are usually 10°–15°F warmer than the water withdrawn from the source of supply. The rise in temperature is the only change in water used for indirect cooling and is usually not significant if the effluent is discharged into a reasonably large stream. Discharged cooling water can have an adverse effect at certain plants where the receiving stream is small and supports a temperature-sensitive fish population. More often than not, the cooling water discharge is of better quality than the water withdrawn from the stream, because of the treatment used for corrosion control.

Disposal of Steel Mill Wastes

Methods of waste disposal in the steel industry vary widely from plant to plant. The age of the mill is probably the most important single factor accounting for these variations. In older mills, space for large treatment facilities is often not available; the space required may well be of more potential value for production facilities than the total direct costs of a waste treatment plant. Other factors influencing the variability in methods include the effluent standards that are applicable, the attitudes of management and the competitive position of the particular operations involved, as well as the characteristics of the waste streams.

Gas-washer water

Blast furnace gas-washer water is usually treated by plain gravity sedimentation in mechanically cleaned circular clarifiers or in simple rectangular sedimentation basins. Circular clarifiers are used almost exclusively in newer installations.

A typical modern installation may consist of a 75 ft diameter clarifier, handling gas-washer water at 6.9 mgd from a blast furnace rated at 1200 tons per day. The effluent would contain approximately 80 mg/l suspended solids; the underflow of 180 tons per day of wet dust would be pumped to the sinter plant for additional thickening and filtration on leaftype vacuum filters.

Older installations might be typified by two 13 ft × 111 ft rectangular sedimentation basins handling gas-washer water at 3.5 mgd from a Blast furnace rated at 880 tons per day. The effluent concentration might average 250 mg/l suspended solids. Sludge would be dredged from the basins by clamshell buckets at the rate of about 22 tons of wet dust per day and hauled to the sinter plant in railroad cars.

The clarified gas-washer water may be recirculated either wholly or in part. Recirculation is practiced when supply conditions dictate water economy and usually requires secondary treatment such as chemical flocculation. The effluent from simple rectangular basins is not ordinarily suitable for recirculation without such extra treatment.

The use of separate clarifiers for each Blast furnace or pair of furnaces may result in the discharge of untreated wastes whenever clarifier operation is interrupted. A more satisfactory arrangement consists of collecting the gas-washer water from all Blast furnaces, with treatment in centrally located clarifiers. Interconnection of the clarifiers insures continuous treatment even if the operation of one clarifier is interrupted for an extended period.

Some plants operate the washer water clarifiers in series, the underflow of each being added to the influent of the next. A single line to the sinter plant and the agglomerating effect of added sludge are possible benefits of this scheme.

Where effluent requirements are stringent or where existing equipment is called upon to handle greater than design flows, chemical flocculation or the various polyelectrolytes may be used, usually in secondary treatment units. Polyelectrolytes alone have usually not resulted, in plant practice, in the rather spectacular improvements indicated by laboratory experiments. With improved methods of determining and controlling optimum dosages and with probable price reductions, these materials will doubtless become more commonly used.

The design of circular clarifiers and rectangular sedimentation basins for gas-washer water requires specialised techniques; the conventional criteria for sanitary wastes are not satisfactory. The methods outlined in the following section on scale-bearing waters are generally applicable for gas-washer water clarification.

Scale-bearing waters

Rolling mill flume water has long been partially clarified in small, simple sedimentation basins known as scale pits, in order to prevent sewer clogging. These pits are usually small in relation to the water flow and the deposited scale is cleaned out periodically with clamshell buckets. In newer mills, scale pits are larger and are designed with the objective of water pollution control; continuous mechanical cleaning is often incorporated.

Flume water clarification

A scale pit typical of older practice was 18 ft wide, 30 ft long and 8 ft deep, to handle flume water at the rate of 3500 gpm from the slab rolling section of a hot strip mill. Effluent concentration averaged 200 mg/l of suspended solids; there was no provision for removing oil. The pit effluent went directly to a river. Flume water from the finishing end of the mill contained only line scale, not likely to clog the sewer and went to the river untreated.

When the mill cited above was rebuilt, flume water treatment was improved to provide more effective pollution control. The scale pit was tripled in size and handles all water from the mill; oil is removed continuously through split-pipe skimmers. The scale pit effluent goes to a 35 ft diameter clarifier for additional solids removal and oil separation and the clarifier effluent is returned to the mill for reuse. Little or no waste-water from this mill is now discharged.

Treatment following once-through use in newer mills usually consists of primary clarification in a scale pit and secondary clarification and oil removal in relatively larger rectangular sedimentation basins. Scale pits are typically cleaned by dredging with clamshell buckets and secondary clarifiers are usually cleaned continuously by scrapers on endless chains. Many variations of this basic scheme are found in various rolling mills; in fact, few installations are identical. Often the secondary clarification includes chemical treatment, typically with additions of lime, ferric sulphate and polyelectrolyte coagulating agents. Chemical treatment is most often used with circular clarifiers as the secondary basins. When chemical treatment is used, the water is generally reused in the mill; it may be passed through cooling towers, especially in the warm weather months. The recovered scale is sintered for use in the Blast furnace or open hearth furnaces. Mill scale is comparable to high grade iron ore and is thus a salvaged material of considerable value. Generally speaking, the recovery of mill scale from primary scale pits shows an economic return; more than 90 per cent of recovered scale is obtained from these pits. Scale removal from secondary pits must be justified on the basis of pollution control or as necessary for water reuse.

Steel mill waste-water sedimentation

Research sponsored by the American Iron and Steel Institute has resulted in design procedures that are, applicable for steel mill sedimentation equipment, including scale pits, secondary basins for scale removal and gas-washer water clarifiers. These procedures predict basin performance in terms of an empirical measure known as the sedimentation index (SI), expressed in minutes. The sedimentation index may be interpreted as the settling time, under specified laboratory conditions, that will result in sedimentation equal to that of a particular basin at a specified flow rate. Values of SI are approximately 0.10 for simple scale pits, 1.0 for secondary mill scale basins, 10.0 for small gas-washer water clarifiers and 30.0 for large washer water clarifiers. This work has shown that there are optimum ratios of width and depth to length for rectangular basins and that large circular clarifiers have less volumetric efficiency than smaller clarifiers at comparable flow rates. Overflow rates and superficial linear velocities are not adequate criteria for the design of steel mill sedimentation equipment.

Rolling mill flume water and Blast furnace gas-washer water should be sampled with care when such samples are to be used as the basis of basin design for required effluent concentrations. Composite samples should be taken over periods of typical operation; samples should be randomised so as not to coincide with process cycles such as slab rollings or Blast furnace chargings. Settling rate tests should be made soon after collection because many of these suspensions cannot be effectively reconstituted after settling has occurred. Existing installations similar to contemplated new installations can often be used as sample sources for design purposes, but differences in raw water quality due to location and season of the year should be borne in mind.

Pickling solutions

Few industrial wastes have received as much attention as spent pickling solutions in terms of research and process development effort. The recovery of by-products from waste treatment processes seems attractive, but has not proved economically sound. Relatively dilute solutions of cheap bulk chemicals are involved and the quantities are large in comparison with most possible markets for by-products.

Other waste-waters

Other steel mill waste-waters such as alkaline cleaning solutions, slag pit effluents and sanitary wastes present few special problems, but are important in planning pollution control comprehensively. Alkaline cleaning solutions may be used as additional alkaline agents in spent pickling solution neutralisation, or may simply be diluted prior to discharge if the quantities are relatively small. Slag pit effluents are treated by rotary screening if discharge is to a navigable stream or a recreational stretch of a stream. Sanitary wastes may be conventionally treated in a mill-operated facility or sent to a municipal sewage treatment plant. The greatest problem encountered with sanitary wastes is in segregating them from process waste streams, especially in older mills; the cost of sewer segregation is usually the greatest cost of treating these wastes.

Water reuse

Reuse of water in the steel industry will increase in the future. Some of this increased reuse will be for the purpose of conservation in localised situations of water shortages, as in circumstances where the low flow period reduces surface streams to a critical point or where the groundwater faces serious depletion. The principal factor influencing reuse will probably be the increasing requirement for high effluent quality and the criterion for the extent of reuse will be economics. The completely closed process water system is, of course, the final answer in industrial waste-water treatment. Even under conditions of abundant supply, complete recirculation can become economical when effluent quality requirements become sufficiently high. Such system will probably provide the solution to waste-water control problems in the steel industry increasingly in the future.

NONFERROUS METALS

Current usage divides all metals into three groups: iron and steel including alloy steels; ferroalloys; and nonferrous metals. Waste-water from the first two groups is already discussed in Iron and Steel. This section deals with waste-water control in the nonferrous production industry. The subject matter of this section is confined to the processes and related operations intermediate between production of metal ores and the finished product. These intermediate processes are usually those involved in the extraction or refining of commercially pure metal from ores and the fabrication of the metal into usable shapes. The four major nonferrous metals are aluminium, copper, zinc and lead. Most of this section is related to these four metals, as they constitute well over 70 per cent of all nonferrous metals produced.

The processes described are concerned with either the extraction of pure metal from the ore or fabrication of the metal. Extraction of pure metal includes a variety of purification methods, such as dissolving metal compounds by leaching, production of oxides, reduction of the oxides to metal by smelting and refining by electrolysis. Smelters and refiners are primary or secondary, depending on whether they use natural ores or scrap as their principal source for metals. Fabrication of metals includes such operations as alloying, casting, extrusion, forging, rolling, wire-drawing and heat treating and provides sheets, wire, tubing and other industrial shapes.

Major Nonferrous Metals

Production of aluminium from bauxite ore includes an aqueous extraction of aluminium oxide (alumina), followed by electrolytic reduction of molten alumina. Almost 50 per cent of the aluminium produced is made into sheets, including plates and foil; about 25 per cent into extruded tubing; and about 20 per cent into castings. Much of the rest is made into rolled shapes.

Copper is extracted from sulphide concentrates and from 'cement' copper by smelting in a reverberatory furnace to produce anodes of about 98 per cent copper, followed by electrolytic refining using aqueous sulphate solutions. Most of the ores, used are sulphides. Copper oxides are converted to 'cement' copper by leaching the ores with sulphuric acid and precipitating the copper with scrap iron. About 55 per cent of the copper is fabricated in wire mills and about 40 per cent in brass mills.

In the rolling and drawing of tubes and wires in several steps, the metal tends to become hard and annealing is required after every two or three steps. The oxide scale formed in annealing is removed by dipping the metal products in sulphuric acid baths, followed by rinsing in water.

The major waste-waters in the copper and brass industry are these rinses; they contain a considerable amount of dissolved copper, zinc, chromium and sulphuric acid. The acid or 'pickle' baths, although they are dumped only infrequently, provide waste-waters containing the same toxic compounds; spent liquor wastes may be considered related to rinse-water wastes, but are of higher concentration and much lower volume.

Oil-bearing waste-waters are formed from lubrication, similar to those formed in the aluminium industry. These are frequently discharged into municipal sewers or rivers with little or no treatment. Characteristics and disposal of this type of waste are discussed under aluminium. Other wastes of the copper and brass industry are the solid scrap, almost all of which is recovered for reuse and zinc fume from the electrolytic melting furnaces, most of which is discharged in stack gases without treatment.

Pickle rinsewaters from fabrication

Rinsewater and acid bath dumps are discussed together because both contain the same noxious compounds—copper, zinc, chromium and acid—and are related in other ways. Of these two wastes, rinsewaters contain the larger mass of contaminant—90 per cent of the total in one study. Although acid bath dumps are more concentrated, the flow rates of the relatively dilute rinsewaters are large, averaging 200–1000 gal/ton of product.

Rinsewater concentrations vary with time and with the individual plant, but some not at typical values are indicative. Pickle baths are batch vessels of about 1000 gal capacity, filled with a 5–10 per cent sulphuric acid solution. During the time they are used for pickling, the acid content becomes depleted and the metal content accumulates. When spent, the pickle liquor is discarded and a new batch of acid is prepared. Dumping cycles vary, but are frequently once a month.

Bright dip baths, used to remove stains on the finished tube or wire, operate similarly to pickle baths except that 3–8 per cent sodium dichromate is added to fresh batches and dumping cycles are usually every week or every few days. Typical compositions of pickle baths contain in mg/l: 80,000 sulphuric acid, 10,000 copper and 10,000 zinc, with maximum values 2–4 times these concentrations. Bright dip baths, when dumped, have similar copper content, somewhat lower acid and zinc content and substantial chromium content (20,000 mg/l).

Discharge of these wastes without treatment is toxic to aquatic life and harmful to sewers and sewage plants; dilution is seldom adequate. Acidity of water below a pH of 6 is often lethal to the aquatic life that forms food for fish. The presence of copper, zinc or chromium above 2 mg/l is lethal to fish; furthermore, natural purification of a stream is inhibited. In a Japanese study, the lethal concentration for salmon was reported as 0.05 mg/l copper or 0.6 mg/l zinc. Permissive metal concentrations are usually set between 0.02 and 1.0 mg/l. These metal sulphate wastes are acidic and corrosive, so they reduce the life of municipal sewers and corrode sewage treatment plant equipment. These wastes also interfere somewhat with the biological treatment of sewage and are not completely removed by municipal sewage treatment processes.

Pickle washwaters from the copper industry have most of the undesirable qualities possessed by steel industry pickle wastes and copper, zinc and chromium are considerably more toxic than iron. The most objectionable liquid wastes in the production and fabrication of copper are the iron sulphate solutions from leaching of oxide ores and the pickle rinsewaters from wire and brass mills. Another, but minor, waste-water is the process water effluent containing entrained solid.

Zinc and lead

Zinc and lead wastes are similar to wastes from the copper industry; the major ores are sulphides and the primary production of metal is by smelting and electrolysis, except that most of the zinc is refined by distillation. Ores of zinc and lead often occur in the same deposit, sometimes with copper deposits.

Other metals

Magnesium

In the production of magnesium hydroxide, the water effluent contains fewer salts than the influent sea water used as raw material. Furthermore, magnesium salts are not toxic in the usual concentrations found in water. Little waste-water is formed during production of the metal by electrolysis; recovery of the metal and of the chlorine gas evolved is almost complete.

Gold

Gold is extracted from gold ores by cyanidation, amalgamation and rest from placers or base metal ores, but only which produces a potential waste-water hazard. This hazard is due to the high toxicity of cyanide and arises not only from the sodium and calcium cyanides used to form complexes with gold, but also from the cyanide sometimes used in the flotation circuits of mineral beneficiation plants.

After the gold cyanide complexes are split or electrolysed to remove the gold, the cyanides are frequently recycled, thus minimising waste discharge. Waste cyanide solutions can be oxidised to nitrogen and carbon dioxide for electroplating wastes.

Thus, the major potentially objectionable waste-waters arising from primary production and fabrication of nonferrous metals are the mud slurries from bauxite, fluorine solutions from aluminium refining, oil-bearing wastes from lubrication in fabrication, iron sulphate solutions from copper oxide leaching, pickle washings from brass mills, entrained solids from many operations and cyanides from gold extraction. Of all these, pickle washwaters from brass mills are probably the most objectionable in quantity and toxicity.

Waste Disposal

From a long-range view, the best solution to waste problems is waste elimination by process changes. Wastes may be eliminated in existing plants by reuse of the noxious material or by recovery of by-products for sale; in new plants, choice of an alternate process may eliminate production of the undesirable material. Examples of these approaches in the nonferrous metal industries have been mentioned. Even where these approaches are not used, process changes to minimise waste formation have reduced treatment costs and decreased the amount of waste discharged to the surroundings. Treatment and disposal should be considered only where process alternatives are not available or where a temporary expedient is needed.

Where wastes are noxious materials that can be converted to harmless form by chemical reactions, treatment may involve such decomposition. In the nonferrous industries, however, noxious materials usually persist through treatment and ultimate disposal becomes significant. Treatment of such waste-waters often involves separation of the deleterious substances from water by chemical reaction and phase change to gas or solid form, usually designed so that more concentrated mixtures of the noxious material are produced. Less frequently, treatment may involve concentration to another liquid phase; examples of these concentration treatment methods include ion-exchange and the separation of immiscible liquids. On the other hand, disposal of these materials involves the permanent or semipermanent relocation of the waste. Frequently called 'ultimate disposal', this includes dispersion of wastes by dilution and storage of solid materials or slurries on land. Therefore, for these persistent materials, treatment in itself is not adequate, but must be followed by some form of disposal. Waste disposal may be accomplished, in one or more steps, with or without treatment—either of which may be satisfactory or unsatisfactory, depending upon such conditions as nature of the surroundings and concentration and amount of waste. An example of unsatisfactory treatment, indicative of some parts of the nonferrous metal industry today, occurs when a gaseous waste is converted into a noxious liquid waste, as in the scrubbing of fluoride gases in the aluminium industry. This treatment is unsatisfactory at many locations because it is incomplete; satisfactory disposal in these cases requires subsequent treatment and disposal of the new form of the waste. Before considering disposal practices for specific wastes, three aspects common to most of the nonferrous metal industry should be examined: effect of surrounding, plant size and waste concentration.

The primary production of these metals often occurs in arid, sparsely populated areas, whereas much of the secondary production and fabrication is located in or near heavily populated industrial centers where water is often more plentiful. Much of the primary aluminium, however, is produced in sparsely settled areas having plentiful water. A popular method of treatment and disposal in sparsely populated areas is lagooning, because of low land values. This method, however, is often considered too expensive within heavily populated industrial areas, even though smaller plants are the rule. Partly because of the more favourable attitude on the part of large companies, which usually operate larger facilities, it is generally true that treatment of wastes from primary production is better accepted than adequate treatment for wastes at small fabrication plants. It also is generally true that waste disposal in these industries is more of a problem in industrial centers where water is plentiful, but where treatment costs are higher and the accumulation of wastes from several industries is more likely to occur. These, of course, are generalisations for which notable exceptions occur and which will probably be changed in time. Waste-waters contaminated by metal ions in large concentrations offer the best possibility of economic extraction of the metals for reuse or sale as by-products; treatment costs are often low, as in precipitation of iron sulphate by evaporation in the copper industry. Dilute solutions, because of their large water volume, are so costly to treat that the practice of discharge of these solutions to large waterways is widespread. This dilution method is possible only if sufficiently abundant water flows are near and only as long as it is condoned by the public and the government. Where valuable metals are involved, as in electrolytic copper wastes, it is usually economical to concentrate the solutions and to recover the metals. If dilute solutions cannot be stored and reused, the alternative is expensive treatment; dilute wastes arising from the washing of pickled metal are a current example of this problem. Other rinsewaters and cooling waters used in fabrication also pose the problems of dilute solutions.

The treatment method most frequently used in the nonferrous metal industries is sedimentation of solids; it is used for entrained particulate matter as well as for precipitates formed by evaporation and by treatment with alkaline chemicals. Disposal methods most frequently used for waste-waters of this industry are discharged into rivers and oceans and discharge of solids onto land areas.

Copper

Iron sulphate solutions from leaching

Leaching wastes are formed in the iron launders used to extract copper from oxide ore and to recover copper from tailings of sulphide ore, low grade ore and mine waters. Because these operations are located mainly in the arid rocky area, treatment of this waste is necessary to prevent making streams unpotable and unfit for agricultural or recreational use.

Most of the ferrous sulphate wastes are treated with lime to make them alkaline. The solutions are then transferred to lagoons where soluble ferrous hydroxide oxidises to ferric hydroxide, which precipitates, aided somewhat by water evaporation. In this way, iron is disposed of by land storage and clean effluents are produced. At some locations, the effluents are reused. The copper sulphate leach solution is electrolysed to deposit copper at the cathodes and the resultant regenerated sulphuric acid liquid is recycled. In this way, production of iron sulphate waste is avoided.

Integrated waste treatment for primary production

Where more than one waste-water is produced, the possibility of combining them in some way should be examined. An example of integrated treatment, using leaching wastes and ore tailings, is used by various companies. The essential feature of this joint treatment is a combination of the alkaline tailing waste with the acidic leaching waste to form a neutral iron-free effluent for discharge. Two tailing waste sources are used: sand and slime sediment formed from lagooning tailing wastes in previous years and waste-water from currently used tailing disposal lagoons; both contain residual lime. The leaching waste-water, having a pH of 4, first passes through the old tailings lagoon where it is partially neutralised by contact with the residual lime. By thus passing over the area, it controls dusting which otherwise presents an air pollution problem. After collection, the water flows through a ditch where clear water from the active tailing lagoon, of pH 11, is added to produce a pH of about 7. At times of high leach flow or cold weather, milk of lime is added to complete neutralisation. This combined stream flows into a 400 acre settling pond which permits oxidisation of the ferrous hydroxide to ferric hydroxide. In settling, ferric hydroxide carries down with it any other solids suspended in the stream.

Pickle rinsewaters from fabrication

At plants where pickle bath rinsewaters and dumps are segregated from other wastes, rinsewaters have been treated successfully by neutralisation and precipitation. A good example of segregated wastes treated in this way, where rinsewaters contain 10–20 mg/l each of copper and zinc and have a pH of about 2.5. To even out large fluctuations in flow rate that occur (0–800 gpm), these dilute wastes are first collected in an equalisation lagoon; based on the average flow rate of 400 gpm, this lagoon has an 8-hour capacity.

A steady 400 gpm flow-from the equalisation lagoon is pumped to a 1000 gal mixing tank where slaked lime is added to neutralise the acid and to raise the pH to 11–12. The spent pickle liquor, when dumped, is sent to a separate storage tank from which it is pumped slowly, over several days, to the mixing tank where it is combined with rinsewaters. These strong wastes are dumped when the mill is shut down, using the same pump and parts of the piping used for rinsewaters. Provision is made for a pre-treatment reduction to the trivalent form of the hexavalent chromium in dichromate pickle liquors; this is done by addition of sodium bisulphite and acid to the liquors in the storage tank.

The high pH liquid from the mixing tank is separated in a clariflocculator, composed of a flocculator in the center of an annular clarifier; at average flowrates, these two parts have detention times of 20 and

170 minutes, respectively. The slurry underflow is discharged to one of two sludge lagoons, of 250 day combined capacity; top water from the sludge lagoon is returned to the equalisation lagoon and compacted sludge is removed periodically. The clear effluent of pH 11–12 and containing 1–2 mg/l each of copper and zinc is diluted fourfold by mixing with other process water effluent and discharged to the nearby river. These treatment facilities have for several years produced a final effluent containing acceptable levels of copper, zinc and pH.

Attempts to reduce the size of treatment facilities by process changes in the mills themselves have been successful. In one approach, by changes in cycle time and drain angles, more of the pickle liquor is allowed to drain back into the acid bath before rinsing. This improvement in rinse procedure results in a decrease in the amount of acid waste produced and a reduction in the flow rate of rinsewater required, both of which allowed construction of a smaller waste treatment plant than would otherwise have been possible. The other approach also involved rinse procedures; countercurrent rinses were used and the rinsewater flow rate was controlled by pH. As a result, the wastes produced are fairly consistent in concentration, so smaller variations in lime addition are required; fluctuations in flow are decreased by the equalising lagoon.

WASTE-WATER TREATMENT IN METAL FABRICATING PLANTS

Generally the waste-waters originating in metal fabricating plants are similar regardless of whether they originate from small shops or large production facilities. For the most part the waste-waters from such plants contain small amounts of metal particles, free and soluble oils, and various cleaning compounds used in cleaning either the product or the shop itself. Some plants have waste-waters from air pollution control devices for painting operations. These residues are treated in a manner similar to free oils. Normally, the wastes are slightly acid or alkaline, and they are usually opaque, milk-coloured, and contain some free (nonsoluble) oil.

The simplest form of treating such wastes is by means of gravity skimming tanks. In many cases such skimming will be sufficient to permit discharge of the industrial waste to a sewer. Optimum gravity skimming tanks are generally designed along standards set forth by the American Petroleum Institute which relate tank dimensions to particle size, oil rise rates and detention time as well as to the nature of the type of oil to be skimmed. Such tanks operate best when supplied with a limited, uniform influent flow rate. As a result it is often wise to optimise the skimming tank design and then utilise large holding or equalising tanks upstream of the actual skimming tanks to permit pumps to deliver the waste-water to the gravity skimmers at a fixed design rate. The gravity skimming will remove free oils. The metal particles will settle in the bottom of the tank and can be removed manually at infrequent intervals or automatically by means of drag conveyors if such metal particles are deposited in significant quantities.

If the metal fabricating plant waste-water contains considerable soluble oils, generally used as a machining coolant, further treatment will be required in addition to gravity skimming. Usually waste-waters from a metal fabricating plant will be contaminated with soluble oils, since even if soluble oils are not used in plant processes, they may well develop because of the mixing of free oils and metal cleaners or emulsifying agents when brought together in the waste-water collection system.

Batch Treatment

Under these conditions a basic decision must be made when considering such waste treatment: whether or not the qualities and quantities of the waste-waters are such that a batch system is in order of whether a constant flow system utilising continuous skimming and other oil removal methods would be preferred.

Generally speaking, small quantities of less than 2000 gpd can best be handled by batch type systems, with larger quantities being treated by a continuous system. However, in the event there is a possibility of toxic contaminants such as plating waste-waters, a batch treatment system becomes essential. A batch type system more readily permits testing and additional treatment to be applied to the waste-waters if this becomes necessary. Certain cleaning compounds can mix with the soluble oil waste and convert it into a jellylike material which will defy treatment by usual methods. Such complications can be accommodated only by a batch system. These treatment systems consist of parallel batch collection and skimming tanks such as those indicated in Fig. 19.2. When sufficient waste-waters have been accumulated in one batch tank, the free oil is skimmed and conveyed to a waste oil collection tank. Miscellaneous grit and metal particles can be allowed to accumulate in the batch tanks until manual cleaning is required. The remaining waste-waters containing soluble oils are then cracked in either the batch tank or a separate retention tank to break the oil emulsion. This is usually done by means of adding acid or alum until the pH has been lowered sufficiently to cause the emulsion to break down. The free oil is then skimmed or decanted in a separator and the remaining waters are neutralised by means of caustic soda or lime. If sodium hydroxide (caustic soda) is utilised the clarified waste-water may be high in dissolved solids consisting mainly of the soluble sodium salts resulting from neutralisation. Frequently such waste cannot be discharged directly to a stream, but it is usually acceptable in municipal sewer systems.

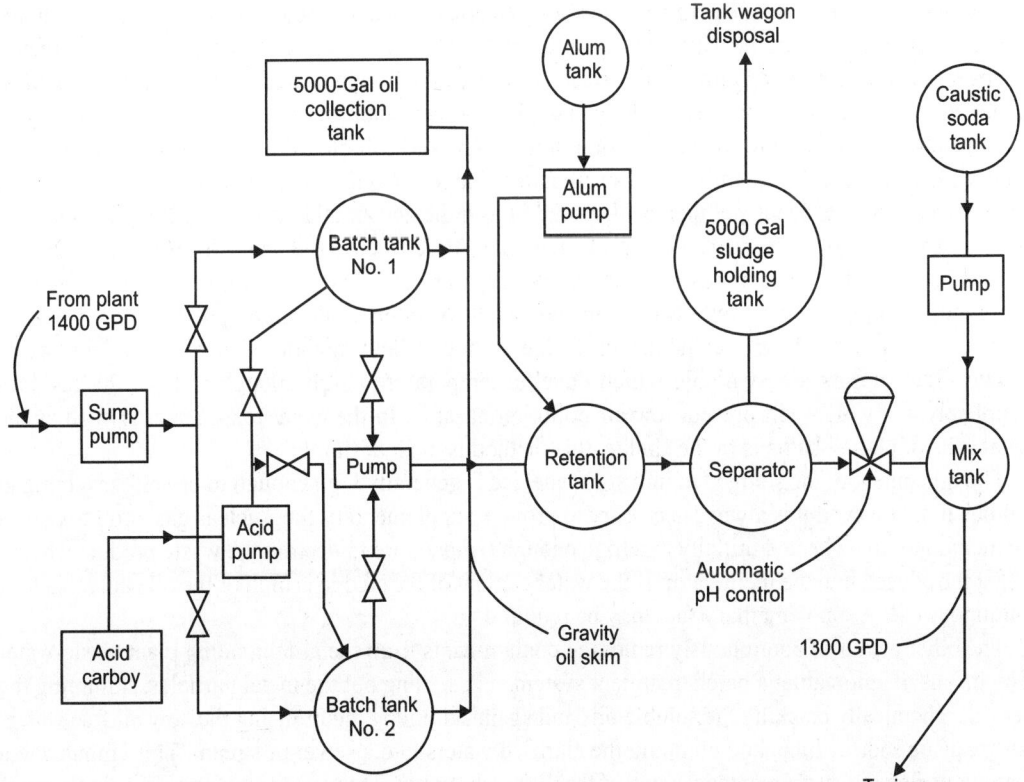

Fig. 19.2. A small soluble oil and alkaline waste treatment system.

If lime is used for neutralisation the waste-waters will contain few dissolved solids because of the relative insolubility of the resulting calcium salts. However, neutralisation with lime will precipitate a sludge consisting of the calcium salts of the acid. This presents the problem of disposal of the resulting sludge. This sludge can be collected and stored in a holding tank for ultimate disposal. Not infrequently it is wise to dry this sludge on vacuum filters, since with its high water content it can present a difficult problem with respect to storage and/or disposal of the precipitate. Removal of most of the water will reduce the volume of the precipitate or sludge to a more easily handled quantity. While lime neutralisation of acid waste-waters is less expensive than caustic soda neutralisation, the cost of sludge handling or drying may far outweigh the savings achieved by the use of lime. The volume and concentrations of waste-waters must be known and the resulting sludge volumes calculated before a comparative economic study can be made.

Several new proprietary compounds have recently been marketed which will crack oil emulsions. Such chemicals usually will not require subsequent neutralisation with possible resulting sludge complications.

Continuous Treatment

Perhaps the most successful of all metal fabricating plant continuous waste treatment systems are those involving air flotation. In this process the soluble oil contaminated waste-waters are collected in large hold of equalising tanks. This permits one-shift operation of the air flotation unit at fixed input rates. Operation for three shifts would permit smaller hold tanks to be used but might not provide the quantitative or qualitative equalising of the waste-waters. After collection the waste-waters are conveyed by transfer pumps to the retention tank. Acid, alkali and soda ash are added as indicated, and the waste is detained in the retention tank under air pressure long enough to absorb quantities of air. Upon release through a pressure regulator valve into the air flotation unit the cracked soluble oil is floated to the top surface along with the minute floc quantities and swept into a sludge hold tank by scrappers or flight conveyors. The relatively clear effluent can then be disposed of as indicated and the heavy oil filled floc disposed of by incineration or tank wagon disposal methods. The effluent may require further filtration or sedimentation in detention ponds to meet stream requirements. The less stringent requirements for discharges to sanitary sewers will usually obviate needs for filters or detention ponds.

Incineration of resulting precipitates or sludge is an excellent method of disposing of this waste product. Incinerators are available which develop temperatures high enough to burn the residues completely and which will present few air pollution hazards. In the event toxic material such as the metal salts of plating baths is in the sludge, this method is most desirable.

The oil content of the sludge fed into such a device is generally high enough to be self-sustaining in combustion: however, it is always necessary to provide supplementary fuel such as gas or oil to permit the incinerator to be heated initially to a high enough temperature to destroy the waste product when it is first introduced into the incinerator. If the water content of the sludge is high or the oil content is low, continuous use of supplementary fuel may be required.

The other means of continuously removing contaminants from metal fabricating plant waste-water is by means of automating a batch treatment system, e.g. settling out the metal particles, skimming the free oils, chemically cracking the soluble oils and again skimming, neutralising the now oil free waters using caustic soda or lime, and dumping the clarified waters into a sewer or stream. This amounts to a continuous flowing, instrumented version of the the batch system. This would require pH sensing devices in the batch tanks to regulate the acid pump. Level switches in the batch tanks would be needed to activate the pump to the retention tank and the alum pump. Also, automatic monitoring and recording of the final effluent properties would be required. There are many points in this type of system which must be instrumented to achieve the required automatic control using pH and level sensing devices, and it may therefore be difficult to achieve reliability. The clinging of residual oils and/or sludge to the electrodes of pH sensing devices and the possibilities of unexplained materials finding their way into the system and fouling the chemical treatment and sensing devices almost preclude a trouble free automatic system.

As a result, for most plants the most practical approach to an automatic system is the previously mentioned air flotation system. Several waste-water treatment equipment manufacturers construct such devices, and in general most of them are satisfactory. However, care must be taken not to overload an air flotation machine. A slight overload of the machine will cause substantial reductions in the pollutant removal efficiencies. A 10 per cent hydraulic overload may cause as much as 50 per cent reduction in oil removal efficiencies and result in carryover of oil bearing floc with the clarified effluent. All automatic or batch systems must be examined to be sure that valves, piping materials, tanks, coatings and linings are suitable for the oils, pH variations and chemicals used in the waste treatment process. Such considerations will be important to the cost of the facilities.

System Sizing

As in all industrial waste treatment problems, the question of anticipating waste quantities and strengths in conjunction with the design of a new metal fabricating waste-water treatment facility is much more difficult than measuring the known quantities of properties of a waste that occurs in existing facilities. As a result, the waste-water treatment facility for a new proposed metal fabricating plant must be designed much more conservatively than one for an existing plant in which the effluent quantities and properties can be measured.

A possible solution, if it can be achieved, is to go into production on a new facility and, once waste quantities are determined, design the waste-water treatment equipment. This takes cooperation on the part of state and regulatory agencies and may require tank wagon disposal or scavenger service for the first few months of operation.

Since waste-water treatment is usually an overhead cost, all possible means to defray expenses must be examined. In the case of metal fabricating plant waste-waters, several possibilities exist. The recovered waste oil may be reused or at least sold for road oil, and the treated waste-water may be reused as previously discussed to satisfy plant process, cooling, or other non-potable water demands.

Carried to the ultimate end, such reuse may virtually eliminate plant waste-water discharges, thereby negating the involvement with waste-water regulatory agencies.

Thus, to make greater reuse of water possible, more stringent requirements on quality of the effluent to the nation's waterways will be enforced on cities and industries. Some relief will be accomplished by better water planning, for instance, by storing of water from periods of high precipitation. Research will be required to develop improved methods of treating sanitary wastes. Economical means are required to remove a higher percentage of pollutants. However, if the techniques now available were applied, a vast improvement in the nation's waterways would result.

Chapter 20
Coal Based Thermal Power Plants

INTRODUCTION

In India, Thermal power constitutes about 2/3rd of total installed capacity, mostly in the form of coal based thermal power plants. A coal based thermal power plant utilises a huge quantity of natural water resources to generate steam, to cool condenser and to dispose of ash produced by the coal burning. A typical 210 MW unit boiler which produces about 700 T of steam from dimineralised (DM) water to run turbine with 1.5 per cent make-up, requires 30,000 T of cooling water and 500 T of ash transport water in an hour. In the water-steam cycle of a unit, different chemicals are added at different stages of cycle and blow-downs are performed to maintain chemical regime in view of preventing the costly components and equipment of the plants from corrosion, erosion, scale formation, etc. Waste-water streams produced in water-stream cycle are: boiler blow down, circulatory water blow down, cooling tower blow down and condenser blow down, etc. A large quantity of water is also used to maintain moisture to avoid fugitive emissions at different stages of coal handling and to transport huge quantity of ash in wet disposal system. These two systems generate coal handling effluents and ash pond overflow. In addition, different run-offs of the plants at oil handling systems, etc. meet together forming a main plant effluent.

This chapter describes characteristics of each effluent measured adopting standard sampling and analytical techniques for the key parameters such as pH, TSS, BOD, TDS, Cl, SO_4, O and G and heavy metals, etc. discussing the importance of right choice of sampling and analytical methods. Keeping in view the typical characteristics of waste-water streams of a power plant, various reuse and recycle schemes are conceptualised. Visualised approaches are described in this chapter to manage the waste-water of a coal fired thermal power plant in order to achieve an optimum water conservation and to meet the Minimum National Standards (MINAS).

In coal fired thermal power stations, to maintain the chemistry of water-stream cycle, chemicals are fed at various stages and blowdowns are carried out at other stages resulting in a number of waste streams. In order to meet regulatory restrictions and to manage waste-water, meaningful physico-chemical measurements need to be carried out. This physico-chemical analysis involves: Collection and preservation of the samples; preparation and pre-treatment of samples; analytical evaluation and presentation of the data in meaningful form.

WASTE-WATER MANAGEMENT

Waste-water management at a power plant is necessary for both in water-short regions and from water conservation point of view to minimise the discharges meeting environmental regulations.

Presently no specific water treatment are in use except the following:

1. DM plant wastes are equalised in two tanks namely neutralisation pits where necessary pH corrections are made before letting into the drains.

2. Ash slurries are transported to ash ponds/dykes where after settling of solids, water is allowed to seep through ash to the ground to evaporate and any overflow is discharged to receiving water without treatments.

3. Hot water drainage from the once through cooling system are allowed to pass through a canal before discharge to surface water bodies, i.e. river or lake, etc.

4. In recirculating cooling system, cooling water is chlorinated. In addition it is treated with corrosion and scale inhibitors.

It has been observed that most of the waste-water streams are produced intermittently on a batch basis except the ash transportation slurries and cooling water discharges. These intermittent waste-water streams can be collected in 'equalisation reservoirs' in batches as and where generated. From these equalisation reservoirs, waste-water can be discharged in a continuous stream eliminating flow surges. This process may help in smoothening the pollutants loading at the discharge ruling out the possibility of upsets in the quality of waste stream due to sudden variation in quality and quantity of the influent intermittent waste-stream.

The waste-water management system may consist of either several individual specified treatment of each waste-water stream at or near its source or an integrated system treating all waste-water streams at a single location using common equipments for various treatments or combining all waste-water streams with reuse and recycling possibilities of most voluminous stream while low volume waste streams can be handled as guided by USEPA (Table 20.1).

Table 20.1. Ways of handling, treating, disposing of low-volume wastes.

Waste-stream	Treatment	Predominant disposal method
Boiler waterside cleaning waste	If organic chelating agents used, waste can be incinerated	Co-disposal with high volume waste in pond or landfill after treatment
	If acids used, waste is often neutralised and metals are precipitated with lime and flocculants	
Boiler fireside cleaning waste	Sometimes neutralised and precipitated waste can be diverted to ash ponds with high-volume wastes without treatment	Co-disposal in pond, no treatment
	If metal contents are high, chemical coagulation and settling is used	Ponding after treatment
Air-heater cleaning	Settling in ash pond, neutralised and coagulated if combined with other streams before treatment	Co-disposal in pond, no treatment Ponding with treatment
Coal-pile run-offs	Neutralised by diverting to alkaline ash pond. Fine coal dust caught in perimeter ditch often diverted back to coal pile	Co-disposal of sludge in landfill after treatment
Make-up water treatment wastes	Usually codisposed in pond	Co-disposal in pond

(Contd...)

Waste-stream	Treatment	Predominant disposal method
Cooling tower sludge	Limited information—sometimes combined with waste-water treatment sludge	Landfilling
Demineraliser water sludge	Equalised in tanks than combined in ash ponds	Ponding
Pyrites wastes	Disposed of in landfills with disposed of in landfills with bottom ash or diverted to ash pond	Ponding Landfilling
Waste-water treatment	Usually ponded with ash or as a separate waste. Solids may be disposed of in landfill with ash	Ponding Landfilling

However, at present authorities are insisting for recycling of ash transport water for further sluicing of ash, treatment of any part of the ash transport water discharged onto land surface water and reuse to meet the prevailing standards set by 'Pollution Control Boards'. Recycling and treatment of waste-water for reuse can be complementary approaches to waste-water management and pollution control as recycling reduces the volume of water released to the water course and amount of treatment is minimised.

Reuse and Recycle Possibilities

There are various possible ways to reuse ash transport water within power plants. The possibilities of water reuse at a particular plant will depend on the presence of the systems in which overflow could be reused, such as limestone or lime scrubbers for removal of SO_2 emissions or cooling towers for reducing waste heat discharges and the need to reuse the transport water for fly ash, bottom ash or both to meet the stipulated limits and to conserve water. In additions to possible reuse of sluice water, there are also the possibilities of recycling the used water in a closed loop ash sluicing system.

Assuming that major sources of waste (ash ponds receiving bottom ash or fly-ash or both) can be combined with anyone of the reuse or recycling possibilities, a number of combinations can be drawn. Three schemes which can be of help in Indian context are discussed below.

Complete reuse of bottom ash sluice water

All small waste streams, such as chemical cleaning waste, floor drains and treated sanitary waste would be discharged into the closed loop bottom ash sluicing system. Also cooling tower blowdown could be used, without treatment as the main source of water for replenishing evaporation losses and other losses of water from the ash sluicing system.

In the closed loop system, scaling problems could occur after a few cycles because of the supersaturation of carbonate and non-carbonate hardness. Treatment of sidestream by some processes such as lime soda softening, ion-exchange or reverse osmosis would be required to control scaling. Lime soda process appeal to be most economical for removing suspended solid, hardness and silica. The effluent from the sidestream treatment process would be reused to dilute the recycled water stream to help avoid exceeding the solubility limits within the ash sluicing system. The sludge from the first stage of the lime soda treatment, the excess lime softening process is primarily calcium carbonate, magnesium hydroxide, calcium sulphate and other solids. The sludge from the second stage, the soda ash process is calcium carbonate. Both of these sludges would be dewatered for recovery or ultimate disposal. These sludges would contain trace metals that could affect decisions on their ultimate disposal.

The proportion of sidestream to total flow from the ash pond that would need treatment to avoid exceeding the solubility limits can be determined with the equation:

$$Qs/Qt = \{[(1+ki)(CiT-CiL)]/[(1+ki)(CiT-CiS)]\}$$

where,

Qs	= sidestream flow rate through softening.	
Qt	= bottom ash pond overflow rate.	
ki	= concentration increase factor of species in the pipe after each ash sluicing.	
CiT	= concentration of species in the ash pond effluent.	
CiS	= concentration of species i in the sidestream.	
CiL	= concentration of species i in the sluice water just below the solubility limit.	

The control limits of, Ci are given in Table 20.2. The relationship among the variables are described in Fig. 20.1a.

Table 20.2. Control limits for ash pond recirculating water composition.

Characteristics	Constraint	
pH and hardness	Langeleir saturation index	pH = measured pH
	Isat = pH-pHs = 0.0–1.0	pHs = pH at saturation with $CaCO_3$
	Ryznar stability index	
	Istab = 2pHs-pH = 6.0–7.0	Is = Ionic strength
Sulphate and calcium		
	CSO_4 x CCa = 200000 when Is = 0.01	
	CSO_4 x CCa = 460000 when Is = 0.05	
	CSO_4 x CCa = 730000 when Is = 0.10	
Magnesium and silica		
	CMg x CSi = 8505	CSO_4 = Concentration of SO_4 in mg/l
	$CSiO_2$ = 150	CCa = Concentration of Ca in mg/l
		CMg = Concentration of Mg in mg/l
		$CSiO_2$ = Concentration of SiO_2 in mg/l

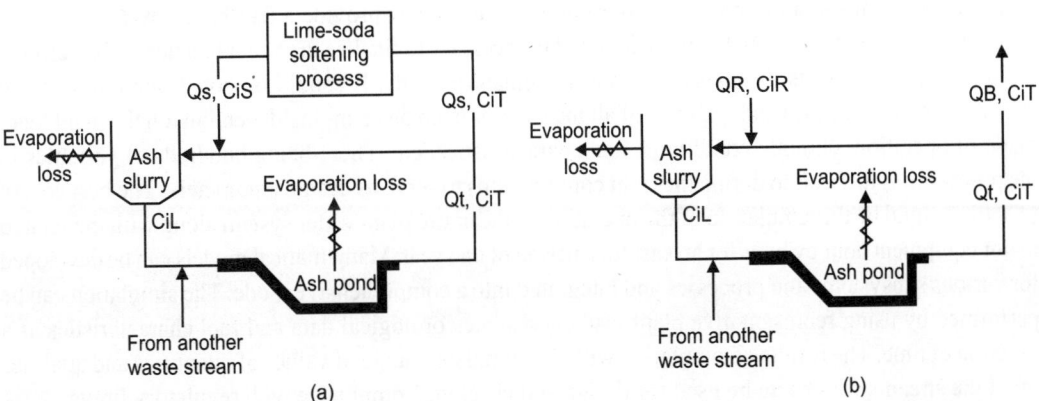

Fig. 20.1. Flow and mass balance diagram for (a) Complete closed-loop system; and (b) Partial closed-loop system.

Complete reuse of ash sluice water

It would involve closed loop recycling of the ash transport water for both bottom and fly ash. It would be applicable where the effluent limitations guidelines could not be achieved with once through settling, where there is a shortage of water or where trace metal concentration are at undesirable levels.

The scheme is similar to previous except that fly ash sluice water is also recycled. Thus, above water balance model discussed also would be applicable to this case.

Effluent from the sidestream treatment process possibly could be used for boiler make up water and in house service purposes. This effluent could also be used as make up for the cooling tower and blowdown from the cooling tower could be used as make up water for ash transport.

Recycle of combined ash sluicing water with treatment of blowdown for discharge

In this blowdown would be treated to meet applicable effluent guidelines before discharge to a natural water way. This scheme would be useful when there is a shortage of water, where the once through ash sluicing system does not meet effluent guidelines for suspended solids and possibly pH, particularly for acidic effluents. Treatment of only the blowdown would reduce scaling in the treatment facility. Concentrations of trace metal would increase in the blowdown in proportion to the number of times the sluicing water is recycled, however, the total mass load of metals discharged to the receiving stream should not increase and therefore the concentration of these metals in the receiving stream should remain essentially unchanged (assuming the blowdown is discharged to the stream from which make-up water is taken).

In this partial closed loop ash sluicing system [Fig. 20.1(b)], scaling problems can be avoided by determining the relationship among the blowdown rate, the make-up water rate and ash sluicing rate in the equation:

$$\{(QT - QB)/QR\} \ \{[CiL - (1 + ki)CiR]\}/\{[(1 + ki)CiT - CiL]\}$$

where,

QB = ash pond blow down rate.

QR = make up water rate in the sluicing loop.

CiR = concentration of species in the make up water and other varibles as defined earlier.

These applications for reuse and recycling of ash pond overflow require detailed evaluation. Bench scale studies, demonstration projects and economic studies may provide a feasible answer.

Once the scheme is reasonably worked out, it is necessary to finalise the water balance and determine the qualities of effluents discharged to satisfy regulatory limits. It should be noted that during plant operation, flow quantities and qualities of all the streams keep on changing depending on the plant load, made of operation, climatic conditions, fuel characteristics, etc. There being hundreds of variables in the process, it is difficult to define the right combinations to serve as design parameters. The best course of action would be the simulation of the operation of the entire plant water system along with the related major equipment hour by hour for at least for a period of one year. Mathematical models can be developed for various subsystems and processes and integrated into a comprehensive mode. The simulation can be performed by using representative plant load curves, meteorological data and fuel characteristics as a function of time. The results of simulation will give simulated range of values of quantities and qualities of all the streams which can be used for design and checking compliance with regulatory limits.

The most common method of final waste-water treatment may be evaporation ponds and mechanical evaporation. In India, high solar radiation and net evaporation rates may favour application of evaporation ponds in most of the areas. In fact evaporation ponds may be used wherever practical for final disposal of waste streams is necessary. The evaporation ponds may need to be lined by natural clay or synthetic material to provide an impervious barrier for the concentrated salts.

CONCLUSION

A systematic measurements of all the waste-water streams in operating units of various sizes is required to generate realistic data on the quality and quantity of waste-stream. This may also include experience record of various methods adopted. Characterisation of various waste streams should be carried out with the aims:

1. To find out various possible ways to reuse and recycle the waste streams within coal fired thermal power plant and to evaluate all the possible ways depending on:
   (a) Characteristics of ash pond overflow and other waste-streams such as cooling tower blow downs, coal pile drainage, clariflocculator sludge, etc.
   (b) Bench scale studies pertaining to: (i) identification of concentrations of heavy trace metals resulting from present system and repeated contact cycles of recirculated sluice water with ash, and (ii) determination of solubility limits of alkaline compounds.

2. To study the applicability of the various schemes in view of:
   (a) Water quality problem.
   (b) Water flow and mass balance between the waste-water-streams and other power plant operations which require water.
   (c) The number of allowable cycle of sluicing water without causing problems with plant operation.
   (d) The economics of waste-water reuse and recycling by carrying out further pilot plant studies.

Chapter 21

Cement and Ceramic Industry

INTRODUCTION

Portland cement is a powder that, when mixed with water, will bind sand and stone into a hardened mass called concrete. Portland cement concrete is an attractive construction product due to its low cost, high compressive strength and durability. Concrete is an important ingredient in economic development, especially the development of infrastructure and large public projects for the development of natural or human resources, such as dams, bridges, railroads, schools, airports and the like. A ceramic is an inorganic, nonmetallic solid prepared by the action of heat and subsequent cooling. Ceramic materials may have a crystalline or partly crystalline structure or may be amorphous (e.g. a glass). Because most common ceramics are crystalline, the definition of ceramic is often restricted to inorganic crystalline materials, as opposed to the noncrystalline glasses. The earliest ceramics were pottery objects made from clay, either by itself or mixed with other materials, hardened in fire. Later ceramics were glazed and fired to create a coloured, smooth surface. Ceramics now include domestic, industrial and building products and art objects.

CEMENT

The basic raw material for the production of cement is lime. Lime is obtained from a variety of sources, primarily limestone, cement rock, oyster shell marl or chalk, all of which are primarily calcium carbonate. In addition, silica, alumina and iron ore are needed. These are obtained from sand, clay, shale, iron ore and blast-furnace slag. The selection and amount of additional ingredients is a function of the desired properties of the cement produced. The raw material grinding and kiln steps can be performed by either of two production processes, wet or dry. The choice between the two depends on the water and chemical content of the raw-materials, the availability of process water and the cost and availability of fuel. In the wet process the raw materials are ground with water and are fed into the kiln as a slurry. In the dry process the raw materials are dried, dry ground and fed into the kiln pneumatically. The remaining steps are identical in the wet and dry processes. Figure 21.1 shows cement production and associated water pollutants.

Water Use

The cement industry in terms of total water use does play an important role in most industrial economics. Since the cement industry is found in so many geographical locations the water use characteristics vary quite a bit from country to country. Water in the cement production process has three basic uses: cooling, process and service and sanitary.

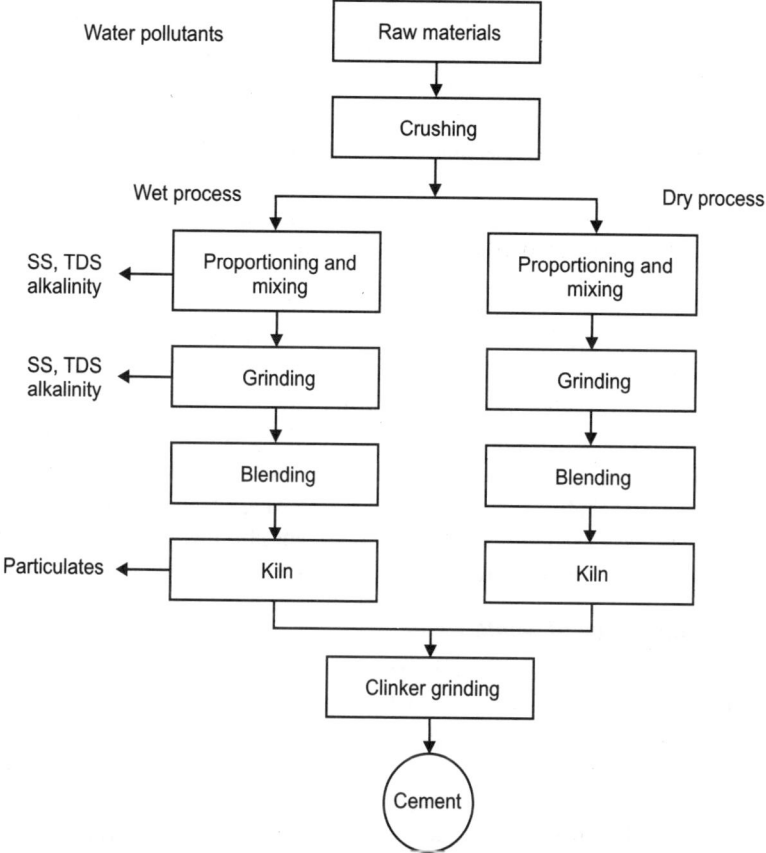

Fig. 21.1. Cement production and associated water pollutants.

Cooling water

The major water use at most cement plants is for cooling. Cooling water is used for bearings on the kiln and grinding equipment, air compressors, burner pipes and finished cement. Most cooling is non-contact. Cooling use is approximately the same for both the wet and dry processes.

Process water

Process water is needed only for the wet process. Here water is added to the raw materials to aid in grinding and to make a slurry for feeding the ground material to the kiln. The process water enters the kiln as part of the slurry and is evaporated, providing no liquid wastes.

Service and sanitary water

Water is needed in certain plants for the preparation of raw material, either washing or beneficiation. Water is also used in the disposal of collected kiln dust. In the wet process, kiln dust is leached of soluble alkalis to recover raw materials by mixing the dry dust with water to form a slurry. The slurry is then put into a clarifier. The treated dust slurry is returned to be used in the raw material slurry preparation. For both the wet and dry processes, collected dust can be mixed with water and fed to a settling pond where the settled solids are not reused and the clarified water discharged. Although this method is practised, it is not recommended to discharge untreated waste-water. Large quantities of cement dust are generated due to grinding and handling of cement. Water is used to control cement dust by being sprayed on roadways and parking lots and used for washing trucks.

One possible use for water in the cement industry is for the prevention of air pollution. This use is not frequently employed now, but may become more important in the future. The air pollution from the kiln stacks can be treated by wet gas-scrubbers. The scrubbers will use large quantities of water (on the order of 10 times that of the total water use). Therefore, a former air pollution problem becomes a water pollution problem and the scrubber effluent must be treated.

Recycling and reuse

There is some potential for recycling and reuse of waste-water within the cement production process. Cooling water can easily be recycled by installing cooling towers or cooling ponds for the dissipation of waste heat. For the wet process, cooling towers or pond blow-down can be used in raw material slurry preparation. In a wet or dry-process plant, if kiln-dust leaching is being used, the blow-down can be used in the preparation of kiln-dust slurry. In a wet-process plant, all other waste-water can be reused in the kiln process except the leachate effluent from the clarifier, which must be disposed of in a containment pond. In the dry process, there are limited number of reuse possibilities. However, leachate waste could be treated to produce water suitable for reuse by means of electrodialysis.

Sources and characteristics of waste-water

Similar wastes are produced by the wet and dry process. However, in the wet process the wastes enter the waste-water stream while in the dry process the wastes enter the atmosphere. In the wet process, spillage and overflow from slurry formation produce suspended solids, dissolved solids and alkalinity in the preparation and grinding stage. However, the major sources of waste-waters are generally produced when addressing the dust or air pollution problem, i.e. the collection of kiln dust and its disposal. The amount of waste generated by the wet process is slightly more than the dry process due to more use of leaching. Table 21.1 lists the basic pollutants for an average cement plant using either wet or dry process.

Table 21.1. Comparison of water pollution loads from wet and dry-process cement plants.

	Wet-process			Dry-process		
	Average load per unit of product		Percentage of plants reporting loads less than 0.005 kg per ton of product	Average load per unit or product		Percentage of plants reporting loads less than 0.005 kg ton of product
Pollutant	(lb/T)	(kg/T)		(lb/T)	(kg/T)	
Alkalinity	0.394	0.79	50	0.096	0.19	75
Total dissolved solids	1.723	3.45	36	0.611	1.22	32
Total suspended solids	–	–	38	–	–	74
Sulphate	0.535	1.07	50	–	–	67
Potassium	1.075	2.15	46	0.040	0.08	50

Waste-water treatment practices

Few steps in the manufacture of cement directly produce liquid wastes. In non-leaching plants, contact of raw material or final product with water provides the major source of the waste load. These waste sources can be reduced through good cleaning and maintenance practices or by collecting these waste sources for treatment. One important area of waste is the storage of raw material and finished products.

Protection of these materials from precipitation and spillage on the plant grounds will reduce significantly pollutants entering the waste-water stream. For leaching plants, settling ponds, containment ponds and clarifying are used as pre-treatment processes. The discharge of the containment pond is then treated by electrodialysis for final discharge or recycled for slurry formation. The cooling water can be recycled through cooling towers or ponds and the blow-down reused.

CERAMIC INDUSTRY

Manufacture of ceramic is an ancient art. In general ceramic can be defined as 'the art and science of making solid articles by the action of heat on earthen materials, which have inorganic nonmetallic materials as their essential component'. This definition includes not only materials, such as pottery, porcelain, refractories, structural clay products, abrasives, porcelain enamels, cement and glass but also non-metallic magnetic materials, ferrotrics manufactured single crystals, glass ceramic and variety of products which were not in existence a few year ago. Ceramic products are wide ranging. Each of these products need different composition of raw materials different glazing materials and also to be fired to a definite maximum temperature, which differ for each product, and to a definite firing and cooling time temperature schedule.

Thus manufacturing processes for production of ceramics are equally wide ranging. However, the basic manufacturing process remains same. In general the manufacturing process can be divided in following steps. The steps are also shown in Fig. 21.2.

Recommended Waste-water Treatment Options for Units Manufacturing Various Ceramic Products

Due to different nature of industrial (non-biodegradable) and domestic (biodegradable) waste-water from such units, segregation of the two is recommended. After the initial step of segregation, the following treatment systems are recommended for these waste-waters.

For industrial waste-water

The waste-water from ball mill washing and propylene drum washing should be taken to independent settling tanks. The settled sludge should be periodically removed and dispose off to commensurate with applicable disposal practices, because the sludge from glaze preparation section would contain heavy metals and thus need to be disposed off in a secure manner.

The overflow from these settling tanks should be taken to the effluent treatment plant for final treatment. Such a system would help in reducing the load on effluent treatment plant. A schematic diagram of the recommended system is shown in Fig. 21.3.

Waste-water Characteristics

The results of the samples collected at the inlet and outlet of settling tank are presented in Table 21.2.

Table 21.2. Characteristics of waste-water at inlet and outlet of settling tank.

Source	pH	TSS mg/l	BOD mg/l	COD mg/l	Oil and Grease mg/l
Inlet	7.8	1817	< 10	840	12
Outlet	8.0	163	< 10	< 15	12

The characteristics of the effluent show that the unit can dispose its water in municipal sewer. However, in the absence of any such facility, physico-chemical treatment of waste-water to remove high TSS and O/G concentration is required before its final disposal.

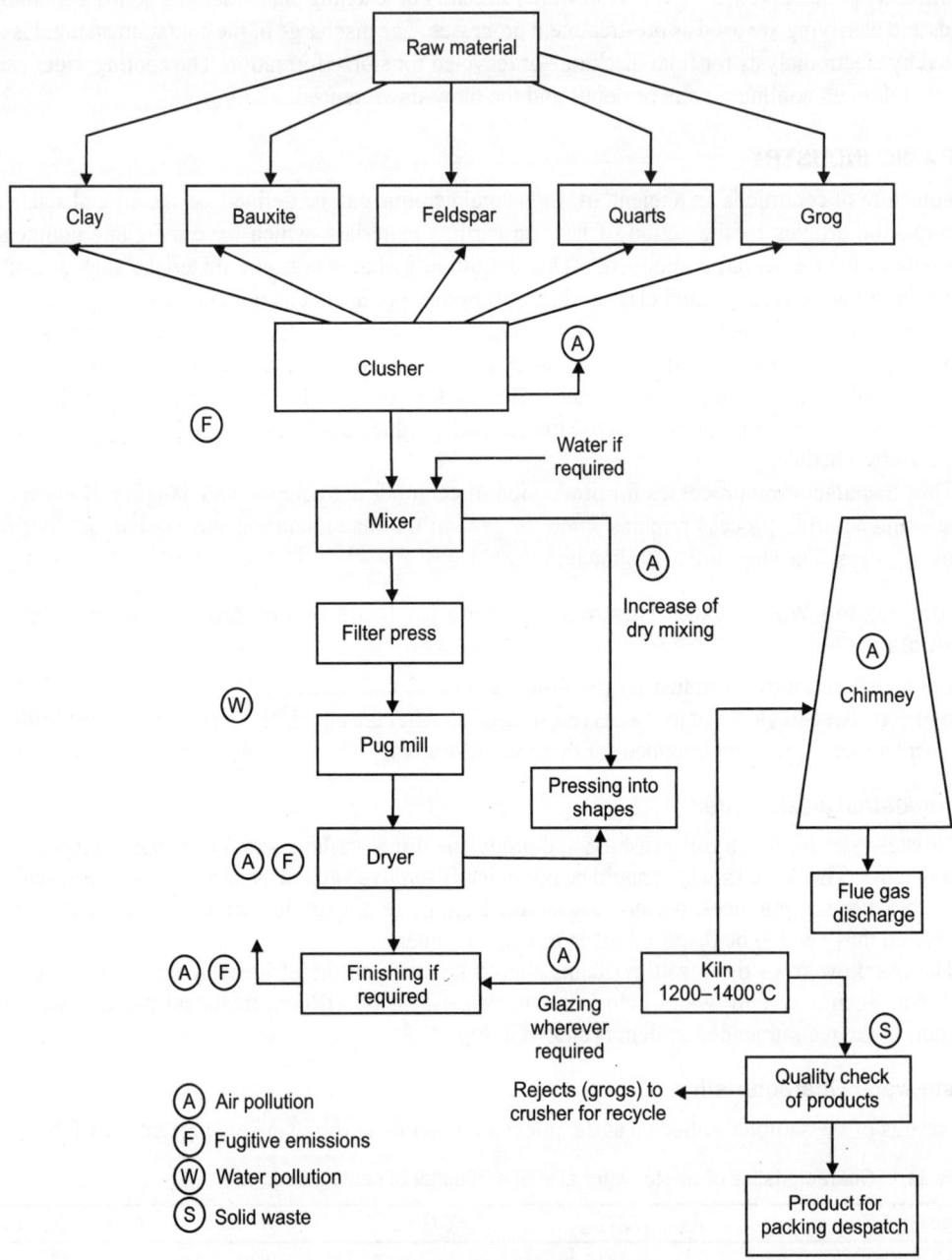

Fig. 21.2. Various operations in ceramic manufacturing.

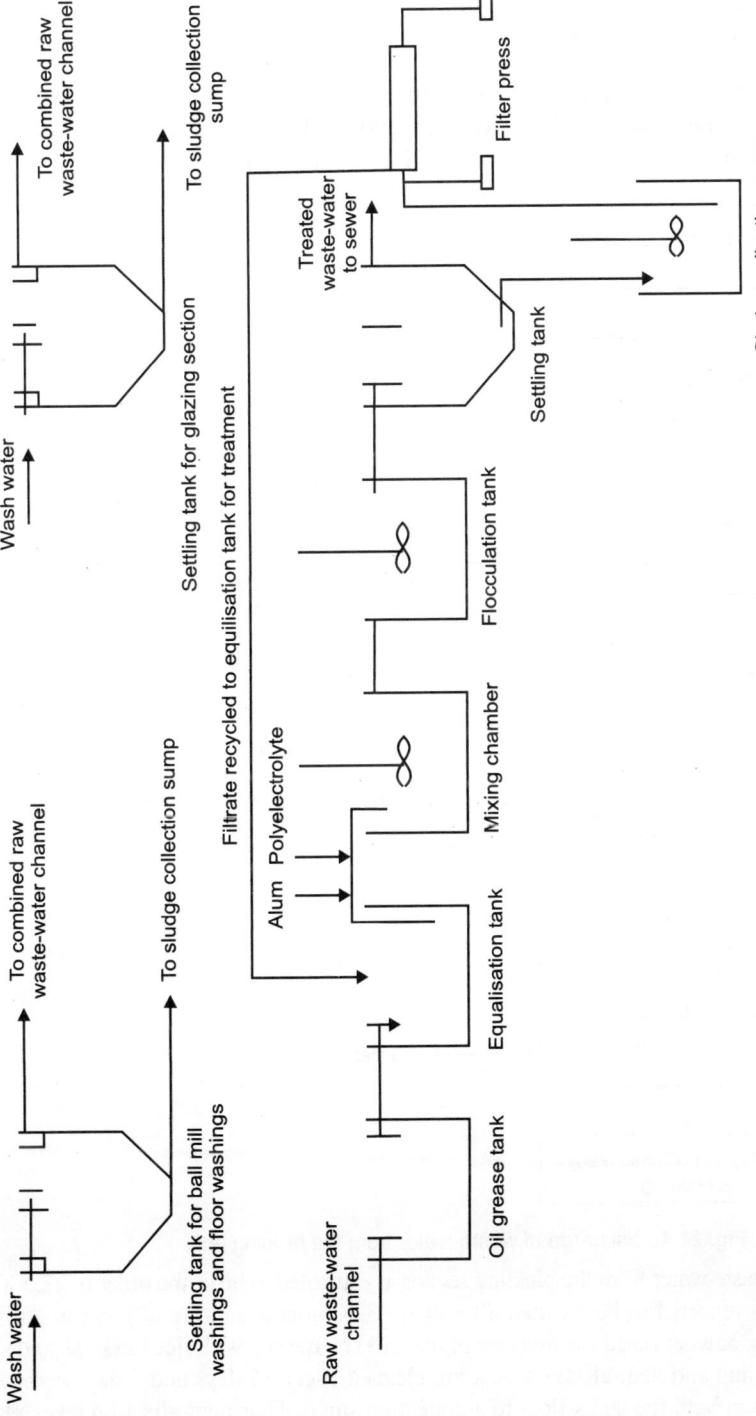

Wash water

To combined raw
waste-water channel

To sludge collection
sump

Wash water

Settling tank for glazing section

Filtrate recycled to equilisation tank for treatment

Treated
waste-water
to sewer

Filter press

Flocculation tank

Settling tank

Alum Polyelectrolyte

Mixing chamber

Sludge collection sump

Equalisation tank

To combined raw
waste-water channel

To sludge collection sump

Settling tank for ball mill
washings and floor washings

Raw waste-water
channel

Oil grease tank

Fig. 21.3. Schematic flow diagram of a continuous waste-water treatment system for a large scale ceramic unit.

Sources of waste-water generation

The source of waste-water generation in the unit are as follows:

Industrial

About 52.5 m³/day (estimated) of industrial waste-water is discharged to the municipal sewage everyday. Major industrial usage of raw water is in the pickling section and in ball mills section. Pickling unit: The process and source of waste-water generation are given in Fig. 21.4.

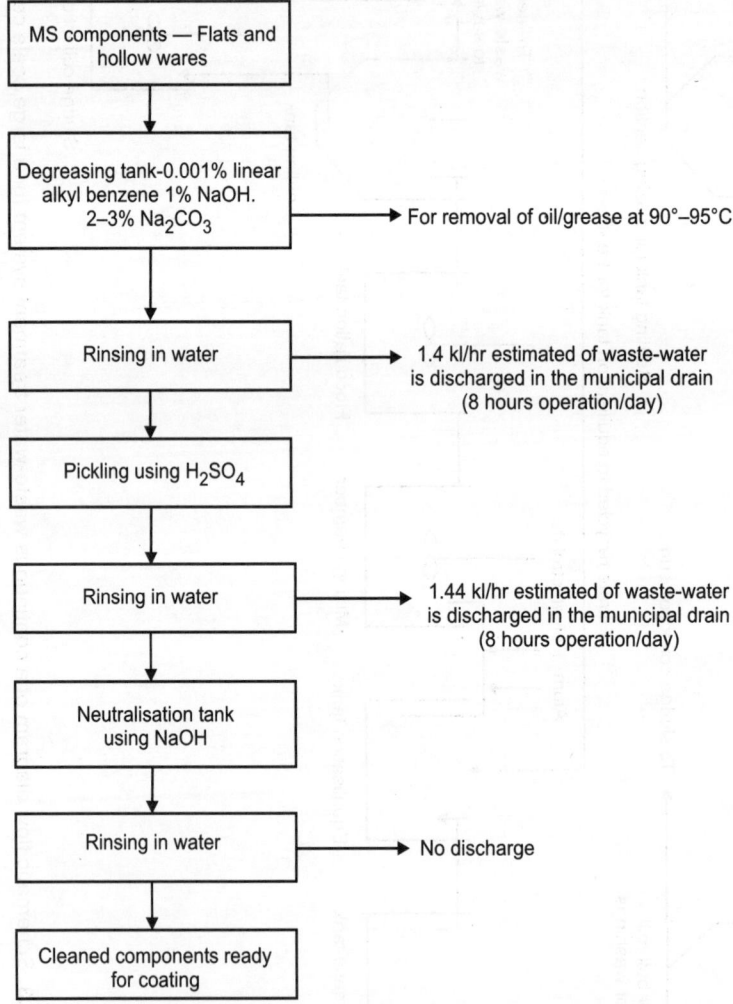

Fig. 21.4. Discharge of waste-water from the pickling unit.

Total discharge of waste-water from the pickling section is estimated to be of the order of 22.5 Kl/day. The degreasing tank is reported to be cleaned after every 3–4 months and the effluent is utilised in manufacturing scouring powder (used for washing of utensils) by mixing with rice husk ash and sold to local market. The pickling and neutralisation tank are cleaned every 15 days and 7 days respectively and the waste-water from both the tanks flow to a collection sump. Thus neutralisation takes place in the collection sump in course of time and the neutralised effluent is discharged to the municipal drain. Expected pollution parameters in the waste-water are pH TSS, and grease and Fe.

Domestic

Domestic water is used in the toilets and for drinking purpose. It is reported that about 18 m³/day of domestic waste-water is discharged to the municipal sewer.

Waste-water treatment system: The unit does not have any waste-water treatment system either for industrial or for domestic waste-water, and discharges their effluent waste-water (of the order of 45 m³/day) to the municipal sewer.

Waste-water characteristics: Spot samples of water have been collected from the pickling unit and the ball mill section and analysed in government approved laboratories, for pH, TSS, oil and greases, metals, results of which are summarised in Table 21.3.

Table 21.3. Waste-water characteristics from different sources.

Source of Sample	BOD	pH	TSS mg/l	CO	Ni	Fe	Heavy metals mg/l Na	Ca	K	Oil and grease mg/l
Post alkali treatment rinsing	< 30	7.2	114	–	–	–	–	–	–	37
Post acid treatment rinsing	< 30	7.1	714	–	–	2.82	–	–	–	27
Ball mill washing	< 30	7.0	48	0.1	0.6	–	255	148	7	–
Final discharge	< 30	7.3	81	–	–	0.88	–	–	–	–

Water pollution prevention techniques

In ceramic industries water is used mainly in following areas:

1. Wet grinding of raw materials.
2. Preparation of slip for moulding the required shape.
3. Glaze preparation.
4. Glazing (in spray glazing).
5. Flour washing and equipment/container washing.
6. Domestic use.

The water pollution prevention techniques can be divided in two types of measures as discussed below:

Housekeeping measures

It is a common practice that hose pipes used for floor washing, equipment washing are kept open, resulting increase in total volume to be treated in effluent treatment plant, i.e. higher capital investment and recurring cost. A simple measure, i.e. usage of self-closing type water hose pipes will reduce the avoidable capital and recurring cost of the effluent treatment plant. Careful handling of fuel oil to prevent spillage will help in bringing down the oil and grease in waste-water.

Process modifications operational

Wet grinding is done mostly in ball mills, from where the slurry is sent to underground blungers for blungering. In production of potteries, stone wares, sanitary wares, porcelain fire bricks, etc. after blungering, the excess water is removed in filter presses. In many cases, the tank capacity does not commensurate the wash water hence part of it overflows down the drain. Sometimes part of the water is intentionally drained for quality reasons. Efforts should be made to maximise utilisation of wash water. At least the filter press wash water could be completely recycled without any adverse effect on quality.

Prepared glaze is normally stored in PVC or metallic containers. In case of dip-glazing there is no water discharge, however in case of automatic spraying, the unutilised glaze from spray which normally is washed down the drawing can be recycled. If not fully, at least the first wash from the drain could be recycled for use.

In ball mill and other equipment washing (in raw material handling sections), it is a common practice to discharge the whole wash water, which is quite substantial. It is recommended that, the equipment should first be rinsed with a little quantity of water and wash water should be collected separately and recycled. If process permits then subsequent wash water can also be recycled.

Otherwise it can be taken to a settling tank and overflow to the effluent treatment plant. It may be noted that the wash water is the major source of TSS and heavy metals. The data from one such unit shows that the TSS concentration from such washing is more than 25,000 mg/l. Avoidance of wash water will not only reduce the pollution load, but also reduce the capital and recurring cost of effluent treatment plant.

Water pollution

The water discharged from potteries, porcelain, small scale sanitary ware manufacturing units, decoration wares, fire bricks, stone wares is very low. However, in big units and specially sanitary wares, tiles manufacturing units the water discharged is substantial. The characteristics of the waste-water analysis shows the presence of high TSS concentration along with heavy metal depending upon the glaze. The waste-water should be treated before its final disposal. For proper design of waste-water treatment system, it is recommended to segregate non-biodegrade industrial waste-water from bio-degradable domestic waste-water of the unit. The segregated industrial waste-water is then recommended to be treated by dosing alum singly or in combination with polyelectrolytes, followed by sedimentation. For domestic waste-water anaerobic treatment in a septic tank is recommended before its final disposal. However, if sewer facilities exist the domestic water can be directly discharged into sewer without any treatment.

Chapter 22

Electrical and Electronic Industries

INTRODUCTION

The manufacture of microelectronic equipment generates liquid and gaseous pollutants and the treatment and removal of these waste products are essential to provide safe operating conditions in the factory and in the external environment. This chapter primarily deals with the identification, collection and treatment of liquid and airborne waste products so that purification occurs to fulfil the commitment to the community and to comply with governmental standards. The opening portion of this chapter deals with the sources of liquid waste, including the wastes from water softening, deionising, plating and etching processes. The liquid treatment process is then described beginning with the collection and segregation of the liquid wastes into separate sewer systems.

LIQUID WASTE CONTROL

Preliminary Planning

Sources

Many of the chemicals used for microelectronic processes eventually end up as a waterborne waste problem. There is a large demand for soft water and deionised water in microelectronics. At present time zeolite softening techniques are employed using NaCl to regenerate the zeolite material. The waste from the regeneration process contains high concentrations of soluble salts requiring proper disposal. The deionisation process utilises both strong alkali and strong acid in the regeneration of the ion-exchange resins. This regeneration process is relatively inefficient and high concentrations of soluble salts result in the effluent.

Etching processes produce waste that presents unique waste treatment problems. Etching consists of the removal of silicon, gold, copper and other metals from the various substrates and the etchant employed must be capable of dissolving the metals and maintaining the metals in solution. The removal of the metal from the waste stream becomes very difficult because the primary purpose of the etchant is to maintain the metal in solution. Conversely the treatment process must be capable of reversing or neutralising the basic action of the etchant.

There are two waste streams generated from these processes—concentrated spent etchant and dilute rinsewater containing a low concentration of the etchant. In some cases the concentrated waste can be considered for recovery of both etchants and metals to avoid a waste disposal problem. The dilute waste is acidic and contains a variable quantity of metal and acid depending on the individual operations.

Preceding many plating operations are metal cleaning processes which produce toxic wastes. The cleaning or metal preparation waste consists of both acid and alkali materials and cyanide compounds. The spent concentrated cleaning bath must be disposed of periodically as well as the continuous flow of dilute rinsewater following each cleaning operation.

The plating or metal finishing techniques employed by a specific shop can be numerous and widely different in the toxic wastes produced. Cyanide is used in many plating operations (copper, zinc, cadmium, gold, etc.). Nickel and chrome plating are also widely used in microelectronics shops. The waterborne waste from these operations is limited to the diluted rinsewater following the plating operations and the volume is variable because of the different rinsing techniques employed by each shop.

Collection of waste

The various process wastes must be segregated to allow for treatment by the different disposal techniques. For instance, deadly poisonous cyanide gas would be liberated if cyanide waste and acid wastes were mixed in the same sewer system. Thus, the treatment techniques and the effluents dictate the complexity of the sewer collection system as well as the type of construction materials required for corrosion resistance. For a typical microelectronics factory separate sewer systems may be installed for: (i) concentrated acids, (ii) concentrated alkali, (iii) diluted acid-alkali, (iv) concentrated cyanide, (v) diluted cyanide, and (vi) chrome-wastes. The construction material may vary from steel pipe for certain types of cyanide effluents to various reinforced plastics for more corrosive wastes. Unplasticised polyvinyl chloride (PVC) is probably the most common construction material used for corrosive wastes at temperatures less than 150°F.

Effluent requirements

The first step in the design of the waste treatment plant is the determination of the requirements established for the receiving stream or municipal sewer. These requirements are established by the regulatory agency having jurisdiction at the point of discharge of the waste and are set by a municipal sewer ordinance or the stream standards of a state agency.

The Water Quality Control Act of 1965 has required each state to establish stream standards for all interstate waters. These standards must be established to protect the present use of the stream as well as anticipated future use. In some streams a zoning type approach has been applied where the upstream water quality requirement are extremely high and where a certain degree of pollution is to be tolerated in the lower or tidal reaches of the stream.

This necessitates that a high degree of treatments be employed. Stream standards also establish the effluent requirement for the municipal sewage treatment systems; thus, municipalities must enact appropriate sewer use codes to comply with these standards. Table 22.1 presents typical limits for waste discharge to a municipal sewer system. Biological treatment processes are not capable of complete removal of the metals and therefore, a municipality must limit the concentration in order to comply with stream standards.

The regulatory agency having jurisdiction may have some specific requirement in regard to installation of industrial waste treatment systems. In some areas the agency may require a batch treatment in lieu of a flow through type system, back-up chemical feed equipment or duplicate instrumentation.

In addition, there may be regulations regarding detention time or the furnishing of metering and sampling facilities that must be determined prior to preparation of plans and specifications.

Table 22.1. Typical limits or standards for waste discharged in a municipal sewer ordinance.

Constituent	Limit
Free acid	None
pH	5 to 10
Fat and oil	100 ppm
Cadmium	2 ppm
Chromium (total)	3 ppm
Copper	1 ppm
Lead	0.3 ppm
Nickel	1 ppm
Zinc	2 ppm
Cyanide	1 ppm
Suspended solids	500 ppm
Dissolved solids	2000 ppm

Waste characteristics

The next step is the determination of the actual quantity and quality of the waste by a complete inventory of all operations. This inventory includes an estimate of work flow through the shop, concentration of cleaning and plating baths, estimated rinsewater flow, estimated drag-out, dump schedules of concentrated plating and cleaning tanks and other data. This information will determine the sewer system required to adequately handle the waste.

At this point consideration should be given to possible recovery of the chemicals, metals and water to reduce the amount of wastes to be treated. Many recovery procedures are well established and have recovered the associated capital investment in a relatively short time period. Ion-exchange techniques for the recovery of specific metals such as chrome or gold and the use of evaporation facilities for recovery of chrome or cyanide solutions have been employed for many years. Similarly, recovery techniques for waste etchants from printed circuit board operations can be instituted to recover the large quantity of metal and etchant that often are wasted. A separate recovery system would be required for each specific process to avoid contamination by different process effluents being served by the same recovery unit. Nonetheless, a waste treatment system is usually required because recovery systems will service only a part of the factory effluents.

Treatment Systems

Design concept compliance with affluent code mandates the need for automatic controls with each treatment system. These controls pace the addition of chemicals to ensure complete treatment in the most economical manner. Due to the variability of the waste a manual control system would cause fluctuation between over-treatment and incomplete treatment, with over-treatment usually predominating. Automatic control systems are of proven quality to produce the desired effluent characteristics.

The selection of either a batch type or a flow-through system is dependent on the waste flow unless the regulatory agency so specifies a required system. If the waste flow is greater than 50 gpm a batch treatment system is usually not economical since the tankage becomes excessive. A batch treatment system requires a minimum of three duplicate systems to provide maximum flexibility of operation;

one tank being filled, one tank being treated and the third tank being emptied. The flow-through or continuous system is used for the large flows but must always be preceded by a equalisation and surge tank.

pH adjustment

This process is basically concerned with acidic and alkaline solutions and elimination of most of their acidic or alkaline characteristics. Acids and alkalies are the converse of each other; the hydrogen ions (low pH) of an acid will react with the hydroxyl ions (high pH) of an alkaline solution to form water. The acid anion (mostly the chloride or sulphate ions) and the alkali cation (mostly the sodium ion) form a salt which normally remains in solution. Table 22.2 shows the basic equations involved in the pH adjustment process.

Table 22.2. pH adjustment.

Self neutralisation	$H^+ + OH^- \rightarrow H_2O$
Neutralisation with caustic soda	$HCl + NaOH \rightarrow NaCl + H_2O$
Neutralisation with lime	$2HCl + Ca(OH)_2 \rightarrow CaCl_2 + 2H_2O$

It is desirable to collect the strong acid waste and the strong alkali waste separate from the dilute waste in order to make the pH adjustment process controllable. These strong wastes are then pumped at a constant rate to the treatment system. The best features of the neutralisation process with optimum pH adjustment control are achieved with a continuous flow-through system (Fig. 22.1).

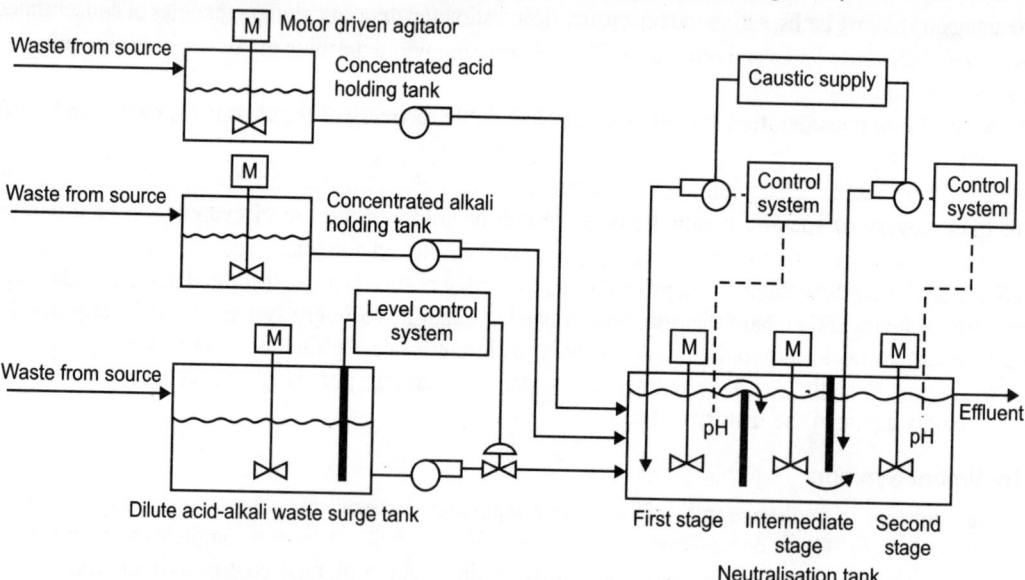

Fig. 22.1. Typical waste treatment systems—acid-alkali wastes: flow-through neutralisation system.

Three tanks are used with each tank furnished with agitation capability. The initial pH adjustment should be made in the first tank while the final adjustment would be made in the third tank. Figure 22.1 shows that the strong acid and strong alkali wastes are mixed with the dilute waste just ahead of the first neutralisation tank. (For a batch type treatment system the neutralisation tanks should be designed to mix the waste and neutralising agent effectively to make the process controllable.) The system should be as fully automatic as is practical and be provided with chemical feed equipment and the instrumentation needed to continually control additions of the neutralising agent.

The treatment of acid waste is recommended to normally maintain a pH of zero mineral acidity in the first stage of neutralisation and a pH in the second stage of neutralisation as required in the subsequent combined waste treatment system.

The choice of the alkali to be used for neutralisation is dependent on local conditions and the actual requirements of the alkali. Being low in cost, lime is preferred for waste which requires extensive treatment. Caustic soda is employed for waste having a relatively low treatment requirement because of its relative ease in handling. Lime is the currently accepted alkali used for removal of fluoride bearing wastes. The partially soluble CaF_2 is formed and removed as a sludge, its solubility in water being 7 mg/l. A better method of fluoride removal is an important need of the microelectronics industry, since hydrogen fluoride (HF) is a commonly used etchant.

Figure 22.2 presents a schematic flow diagram of a continuous flow-through type system. The system consists of a reaction tank furnished with agitation and facilities to control the addition of acid and the reducing agent followed by a retention chamber to ensure the complete reduction of chrome. Preferably the trivalent chromium should not be precipitated in this system since the effluent will be discharged directly to the combined waste treatment system for further treatment. The reduction of chrome can be accomplished using either sulphur dioxide gas, one of the sodium salts of sulphur dioxide or ferrous sulphate as the reducing agent.

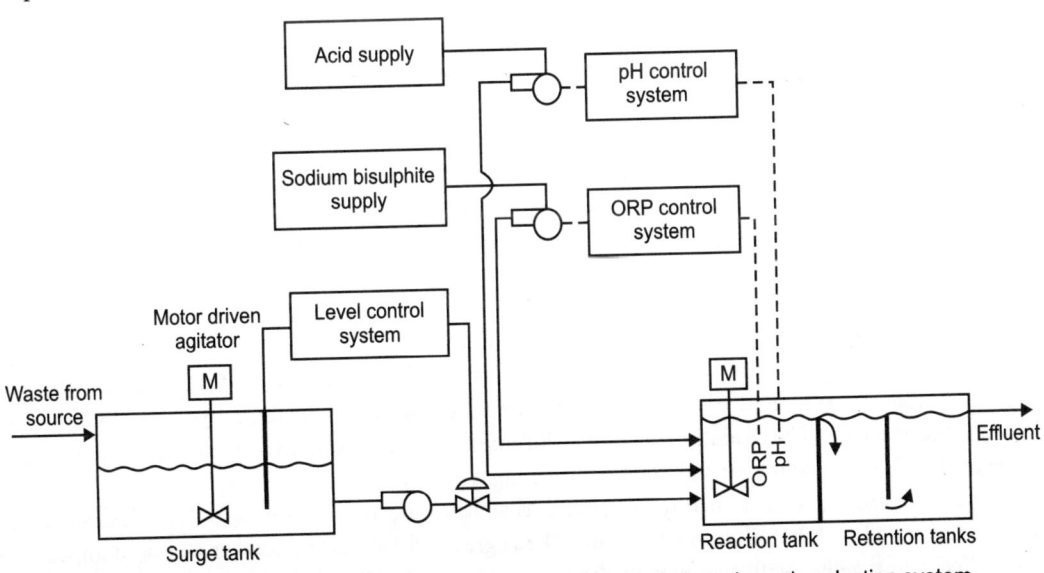

Fig. 22.2. Typical waste treatment systems — chrome wastes: flow-through reduction system.

The pH of the waste can be maintained at the proper level with acids usually sulphuric acid. The reduction with ferrous sulphate can be accomplished over a wide pH range, but this process produces an excessive quantity of an iron bearing sludge.

The use of sulphur dioxide or one of its sodium salts reacts at a pH range of approximately 2.0 to 3.0 and produces a waste containing a minimum quantity of sludge.

The system should be provided with chemical feed equipment and instrumentation to continually control the process. The addition of the reducing agent is controlled to maintain a sulphite-trivalent chromium oxidation-reduction potential level, i.e. a small sulphite residual that ensures the completion of the reduction reaction. The instrumentation also controls the rate of acid additions to maintain a pH suitable for the reaction.

Cyanide treatment

There are several oxidation methods that have been employed for cyanide destruction: (i) biological, (ii) chemical, (iii) electrolytic, (iv) incineration, and (v) radiation. The chemical method employing the alkaline chlorination procedure is the most widely employed because: (i) the method lends itself to automatic control, (ii) the process can be controlled to stop the reaction at the cyanate level, and (iii) the process ensures 100 per cent destruction of the cyanide. This process is basically concerned with the oxidation of cyanide to carbon dioxide and nitrogen in an alkaline environment using chlorine as the oxidising agent in two stages of treatment. In the first stage, the chlorine oxidises the cyanide to cyanogen chloride and this is then converted to cyanate. In the second stage the chlorine oxidises the cyanate to nitrogen and carbon dioxide. Table 22.3 presents the basic equation of the oxidation process.

Table 22.3. Cyanide treatment.

Chlorine solution	$Cl_2 + H_2O$	$\rightarrow HOCl + HCl$
Hypochlorite	$HOCl + HCl + 2NaOH$	$\rightarrow NaOCl + NaCl + 2H_2O$
1st stage oxidation	$NaOCl + NaCN + H_2O$	$\rightarrow CNCl + 2NaOH$
Hydrolysis reaction	$CNCl + 2NaOH$	$\rightarrow NaCNO + NaCl + H_2O$
2nd stage oxidation	$3NaOCl + 2NaCNO + H_2O$	$\rightarrow 2NaHCO_3 + N_2 + 3NaCl$

In the first stage of treatment chlorine reacts with the cyanide radical (CN) at any pH to form cyanogen chloride (CNCl). This compound is so volatile and toxic that it is essential that it be converted by being hydrolysed to the non-volatile and less toxic cyanate as quickly as possible. It is important to complete such conversion before the second stage reaction occurs because the release of the end-product gases, carbon dioxide and nitrogen, in the second stage reaction tends to increase the liberation of cyanogen chloride. The rate of conversion of the cyanogen chloride to cyanate is dependent upon the pH of the waste, its conversion or hydrolysis being practically nil at pH 7.5 or less, fairly rapid at pH 8.0–8.5 quite rapid at pH 9.0–9.5 and exceedingly rapid (a matter of minutes) at a pH of 10 or above.

In the second stage of treatment chlorine reacts with the cyanate radical (CNO) to produce nitrogen gas and carbon dioxide. Part of the carbon dioxide reacts with carbonate or hydroxyl alkalinity to produce bicarbonates. The oxidation reaction is infinitely slow at a pH above 10, becomes increasingly rapid as the pH drops below 10 and reaches an optimum at pH 7.5.

Figures 22.3 and 22.4 illustrate flow-through and batch type treatment systems for cyanide destruction. The concentrated and the dilute wastes should be segregated so that the former can be pumped at a controlled rate together with the later to the treatment system. The oxidation of the cyanide can be accomplished using either liquid chlorine or a hypochlorite solution as the source of chlorine. The pH of the waste must be maintained at the proper level in each stage of the process by the use of either acid or alkali.

Figure 22.3 illustrates a recirculated flow-through system designed to perform the oxidation in two separate stages and consists of two sets of reaction tanks placed in series. The first tank will be employed for the first stage of oxidation, cyanide to cyanate, while the second series of tanks provides for the second stage of oxidation, cyanate to nitrogen and carbon dioxide.

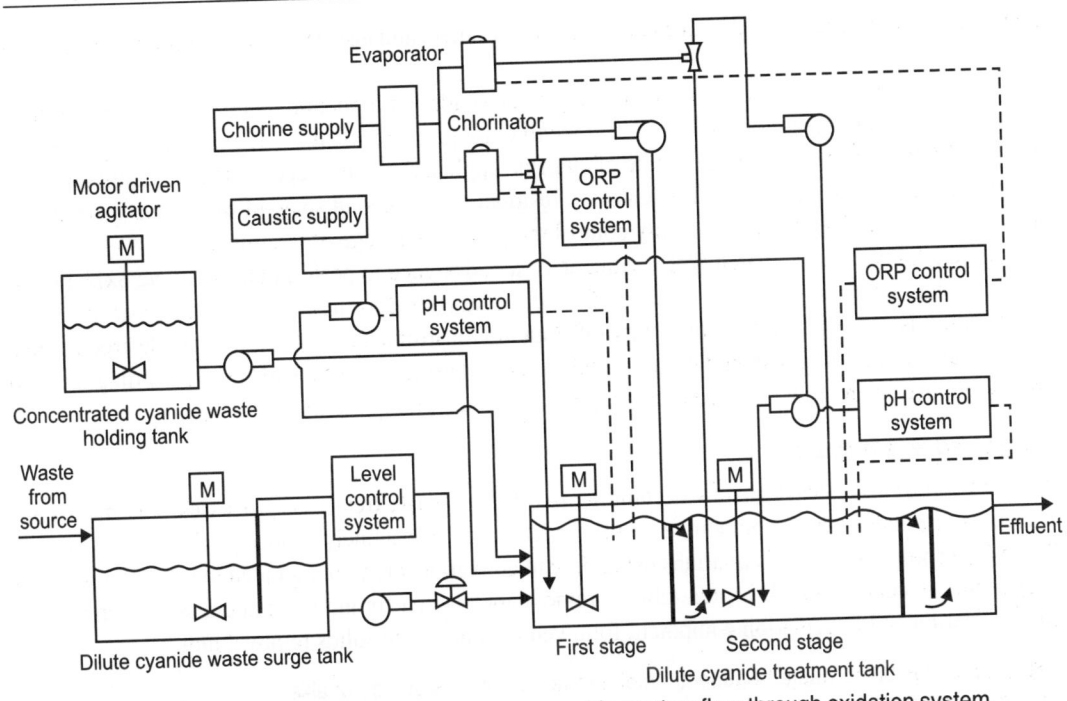

Fig. 22.3. Typical waste treatment systems—cyanide wastes: flow-through oxidation system.

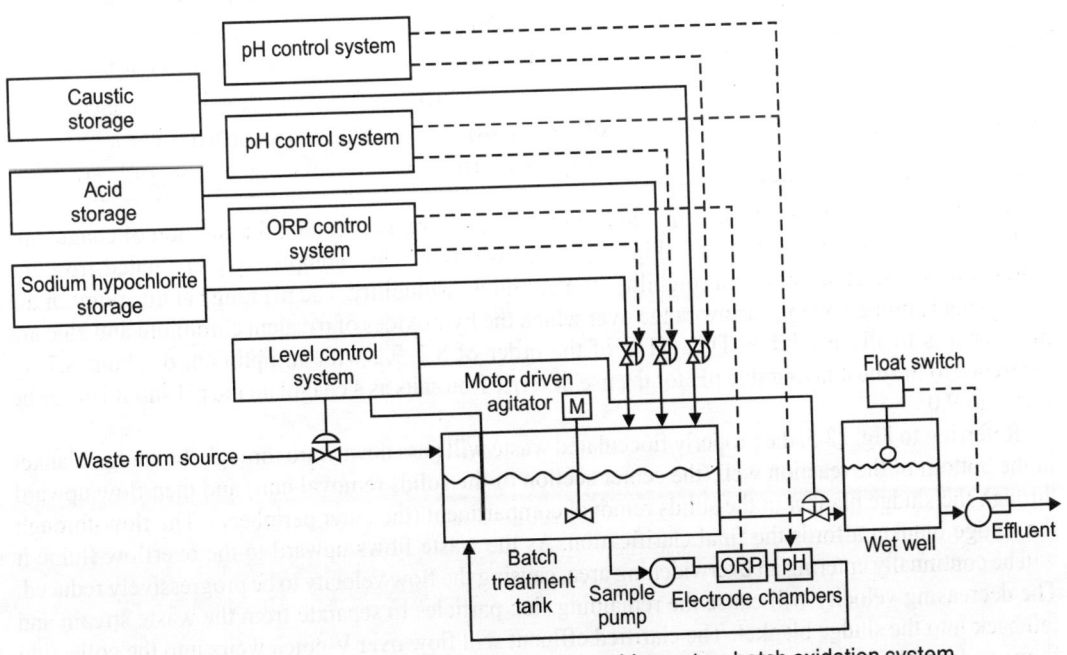

Fig. 22.4. Typical waste treatment systems—cyanide wastes: batch oxidation system.

Positive displacement or retention tanks should be installed following each stage of treatment to ensure the completion of the reactions prior to the discharge of the waste to the next stage of treatment.

The system should be provided with chemical feed equipment and instrumentation to control the system automatically.

The rate of chlorine application is controlled by an oxidation reduction potential (ORP) instrument to maintain a chloramine-cyanate level, i.e. a small chloramine residual in the first stage of treatment. Another ORP instrument controls the rate of chlorine application to the second stage to maintain a free chlorine (ORP) level, i.e. a small free chlorine residual in the second stage of treatment. Instruments should also be provided to automatically control the pH in each stage of treatment.

Figure 22.4 illustrates a batch type treatment system designed to perform the complete oxidation of cyanide. The system is a stepwise titration of the cyanide with chlorine to produce the desired results. The titration is controlled automatically by ORP and pH instrumentation to attain the desired results. The first and second stages should be separated by the proper control of the pH. The flow diagram also shows the instrumentation required to automatically fill and empty the treatment tank.

Combined waste treatment

The removal of toxic heavy metals is common to all three waste treatment processes. The metals are present as insoluble hydrous compounds, oxides and hydroxides in a finely divided particulate suspension. Table 22.4 shows the general equations using lime to convert any remaining metal ions into hydroxides preparatory to coagulation. It is preferable to remove metals in a common treatment system consisting of coagulation and flocculation equipment followed by a high rate solids removal unit.

Table 22.4. Combined waste treatment—metal removal (lime as source of alkali).

Iron	$2FeCl_3 + 3Ca(OH)_2$	$\rightarrow 2Fe(OH)_3 + 3CaCl_2$
Zinc	$ZnCl_2 + Ca(OH)_2$	$\rightarrow Zn(OH)_2 + CaCl_2$
Copper	$CuCl_2 + Ca(OH)_2$	$\rightarrow Cu(OH)_2 + CaCl_2$
Chromium	$Cr_2(SO_4)_3 + 3Ca(OH)_2$	$\rightarrow 2Cr(OH)_3 + 3CaSO_4$
Nickel	$NiCl_2 + Ca(OH)_2$	$\rightarrow Ni(OH)_2 + CaCl_2$
Tin	$SnCl_4 + 2Ca(OH)_2$	$\rightarrow SnO_2 + 2CaCl_2 + 2H_2O$

The three treated waste streams should be combined in a rapid mix tank for addition of coagulants and final pH adjustment as shown in Fig. 22.5. The metals are then completely precipitated from the solution to the point of minimum solubility or maximum insolubility. The pH range at this point in the processing is limited by the narrow range over which the hydroxides of trivalent chromium and zinc are more or less totally insoluble. This pH is of the order of 8.2–9.2 with an optimum of about 8.7. In deference to the most favourable pH for the use of ferric iron salts as a coagulant the pH should never be less than 9.0.

Referring to Fig. 22.5, the properly flocculated waste will pass downward through the sludge blanket in the bottom of the reaction well (the center section of the solids removal unit) and then flow upward through the sludge blanket in the solids removal compartment (the outer periphery). The flow-through the sludge blanket affords the final clarification. As the waste flows upward to the overflow flume it will be continually entering an everwidening area, causing the flow velocity to be progressively reduced. The decreasing velocity will cause the remaining floc particles to separate from the waste stream and fall back into the sludge blanket. The clarified effluent will flow over V-notch weirs into the collecting flume and then by gravity to the effluent meter pit for flow measurement and final disposal. The sludge drying beds, centrifuges and rotary drum vacuum filters have been utilised to concentrate the sludge. Each of those can produce a relatively dry sludge having the consistency of wet clay suitable for landfill.

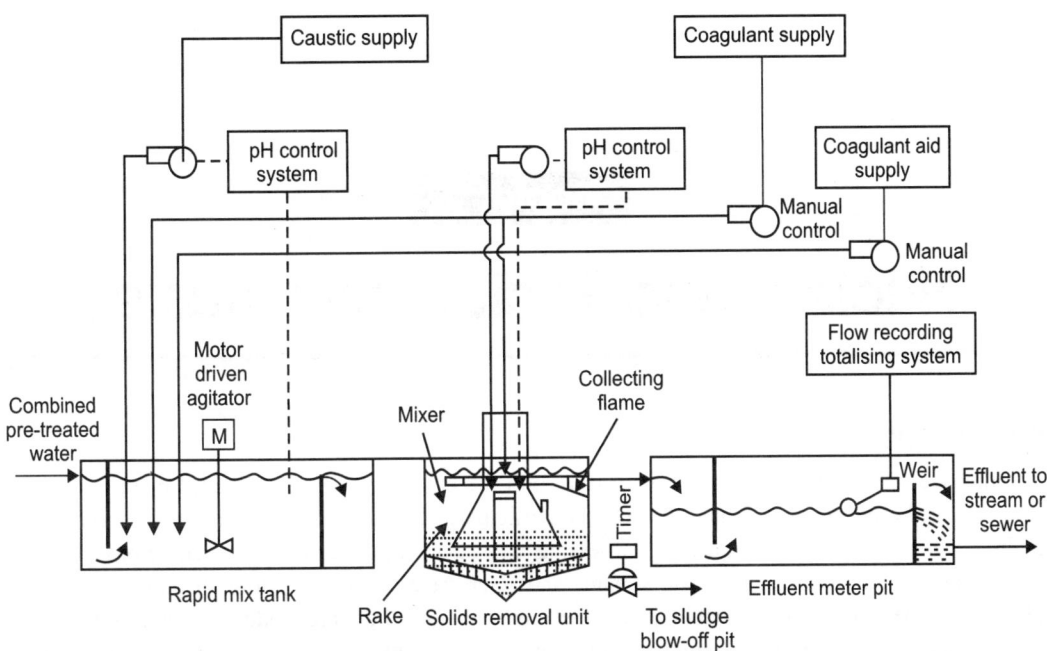

Fig. 22.5. Typical waste treatment system—combined wastes: flow-through solids removal system.

Effectiveness of Treatment

The installation of the properly designed treatment processes and the proper operation will provide an effluent having the following characteristics:

1. One that is free of hexavalent and trivalent chromium.
2. One that contains no cyanides except the iron cyanide complexes (the alkaline chlorination process does not destroy the less toxic iron cyanide complexes; it converts the ferrocyanide to ferricyanide).
3. One that has a pH of about 9.0 required for the efficient removal of the metal hydroxides (a final pH adjustment tank can be installed to maintain any required pH in the effluent).
4. One that will contain on the average not more than 5 mg/l, of suspended solids, mostly insoluble metallic compounds whose total metallic content is about 2.5 mg/l and consisting of a mixture of all the various metals initially present in the raw wastes.

Chapter 23

Nuclear and Radioactive Wastes

INTRODUCTION

The menace of radioactive pollution spreading into the environment has increased extensively as a result of the discovery of artificial radioactivity, particularly due to the development of the atom bomb, hydrogen bomb and of techniques of harnessing nuclear energy. Actually, this dangerous pollution enters into the environment in waste streams and stack gases from operations of power processing plants. From neutron bombardment of atomic fuel, heavy radionuclides are produced which are extremely toxic. Once these radioactive elements find access into the environment, they enter the ecocycling processes and ultimately into food chain and metabolic pathways.

Radioactive pollution is a physical type of environmental pollution that arrived suddenly in the process of industrial revolution. It differs from other pollutions of air, water, soil or land in the respect that it not only critically affects individuals but also brings physiological changes in the subsequent generations. The adverse effects due to pesticides, fertilisers, detergents, drugs, oils and plastics, etc. are primarily on the environment while radioactive pollutants directly or indirectly hit the target–humans.

Radionuclides contaminate air, water, soil and dangerously deteriorate man's vital life-support system on earth. Any radioisotope, with a sufficient long half-life introduced into the environment anywhere in the biosphere is reported to find its way into man's body. There is no threshold or safe dose for radiation effects, so even the smallest increase in radiation above the natural background radiations is reported to cause risks and great human misery. Radiation stress leads to reduction in species diversity and can also alter key population interactions, disrupting the equilibria.

The increasing installation of power plants, nuclear testing, X-ray fluoroscopy, radars, luminous dials of clocks and watching colour TV, etc. pose new dimensions of radiation pollution and a serious threat to the future generation.

Radioactive waste is generated during the production of electricity by nuclear power plants, by the eventual disposal of those facilities, and during the manufacturing and disposal of nuclear weapons and machines used in medical diagnosis and treatments, academic and industrial research, and certain industrial applications. Radioactive waste produces ionising radiation, which can damage or destroy living tissues. Ionising radiation transfers energy when it encounters biochemicals, causing them to become electrically charged or ionised, which can damage their essential metabolic function.

Unlike conventionally toxic chemicals, the degree of danger from radioactive waste decreases over time. The half-life of a radioactive substance (or radioisotope) is the time required for one-half of an initial quantity to decay to other isotopes. Each radioisotope has a unique half-life, which can be only fractions of a second long, or as great as billions of years. The longer the half-life of a radioisotope, the longer is the period for which it must be safely stored or disposed until it is no longer hazardous.

The regulations for transporting radioactive waste are stringent due to the possibility of a transportation accident. Various containers are used for transporting specific kinds of waste. High-level waste has the most rigorous standards, and the containers in which it is shipped must be capable of withstanding tremendous pressure, impact, and heat, and are waterproof. There have been accidents in North America involving trucks and trains carrying radioactive waste, but no significant amount of radioactivity has ever been released to the environment as a result.

Storage can be defined as 'a method of containment with a provision for retrieval'. High-level and transuranic wastes are typically stored in on-site, deep-water storage ponds with thick, stainless steel-lined concrete walls. After about five years, the spent fuel has lost much of its radioactivity and can be moved into dry storage facilities. These are usually on-site, above-ground facilities in which the waste is stored in thick, concrete canisters.

Low-level waste is stored in concrete cylinders in shallow burial sites at nuclear plants or at designated waste sites. Since these wastes are not as much of a concern as high-level wastes, the regulations for their storage are not as strict. Basically, the waste must be covered and stored so that contact with ambient water is minimal.

RECENT METHODS TO DISPOSE CRITICALLY DANGEROUS RADIO-WASTES

In nuclear waste management terminology, radioactive waste can be disposed according to the following categories, depending on the rate of disintegration: (i) method to dispose low-level waste, (ii) method to dispose intermediate-level waste, and (iii) method to dispose high-level waste.

Method to Dispose Low-level Waste

Low-level radioactive wastes range from 0 to 1 micro Ci per litre, where Curie (Ci) is defined as the quantity of radioactive isotope which decays at the rate of 3.7×10^{10} dis./sec. One micro curie (μ Ci) is one millionth of a curie. Low-level radio-waste, containing very short living and weak beta radioactivity, can be extremely diluted and discharged in water bodies like other domestic, chemical and industrial wastes. These wastes may be released into the aquatic environment only after necessary preliminary treatment so that the radioactivity reaches a lower level. Before disposal, the low-level radio-wastes should be fixed in bitumen and stored for a sufficient time to reduce radioactivity. These wastes were so far being stored in the liquid form in storage tanks deep underground to protect the environment from radiation hazard. After this the waste may be disposed of into sewer system or river or sea without harming any aquatic organism. The technique adopted in the management of low-level wastes is to pretreat it in such a way that the concentration of radionuclides ultimately disposed of is much less than the maximum permissible concentration (MPC) as recommended by the ICRP, Vienna.

Generally, radio-waste that comes from hospitals belongs to this category. However, permissible discharge depends on dilution and dispersion capability of the receiving medium.

In the nuclear industry, because of the reconcentration process, it is necessary to pretreat radio-effluents contained in the radio-waste residues and discharge the effluents with minimum radioactivity.

Method to Dispose Intermediate-level Waste

(1 μ Ci to 100 Ci/L)—The intermediate level-wastes originate as sludges from the decontamination process, in fuel reprocessing, in the later stages of solvent extraction and in the form of ion exchange

resins. Waste from research laboratories and nuclear reactor processes have radioactivity a little more energetic and longer lasting. Such wastes are safely contained in concrete-lined tanks till their radioactivity reaches a minimum.

The amount of radioactive waste generated by nuclear power reactors is not much. Thus, the waste produced in larger quantities can be treated by well-accepted methods before their disposal.

The intermediate level radioactive waste may be packed in sturdy concrete boxes covered by steel casing that is not easily corroded by chemicals. The waste can also be sealed in concrete filled steel drums and then discharged into a depth of 1000 fathoms (a measure of six feet). The waste may be stacked in deepest shafts of old mines not in use. Even then, access to such mines is guarded to prevent any pilferage. However, steel casings can be safely buried in salt mines in remote areas after one to six years to allow for dissipation of heat.

Method to Dispose High-level Waste

Highly radioactive waste (more than 100 Ci/L) can not be discharged directly. For the remnants of highly radioactive fuel used in nuclear power reactors and nuclear bombs, the disposal has to be made with much caution, since these wastes can remain radioactive even after ten generations.

Chemical method of disposal

Radionuclides from high-level wastes are segregated by coagulation, precipitation or by the ion-exchange method. The concentrated material, which is in the solid form, is then stored or buried underground. After decontamination, the waste-water, which is highly diluted, may be discharged into the sea. Deep sea water has stable matrix with pH 8.2 and has a steady distribution of its chemical constituents. As the radio-effluents from nuclear industry meet sea water, various chemical and biological interactions take place. The dissolved inorganic species containing the multivalent elements hydrolyse in the stream and form suspension of radionuclides.

The high concentration of radioisotopes and their accumulation in aquatic plants and animals has shown that this technique is not so adequate.

CONVERTING RADIOACTIVE WASTE INTO SOLID FORM

Currently, the method to dispose high-level radio-waste is to dump them into sea or store them underground for several years. Nowadays, small quantities of highly active longer lasting wastes are converted into solid masses such as concrete and buried under sea or ground. Such longer life wastes, moulded into insoluble solid masses, can be stored in salt mines without any accidental leakage.

The radioactive waste may be concentrated by evaporation and the condensate with extremely low radioactivity may be discharged. The concentrated bottoms are then once again stored in special types of durable tanks.

This method is also not so adequate for high-level wastes as there have been reports of leakage of radiations from the tanks. After all, no metallic tank can be expected to last as long as the long-life radio-nuclide.

OTHER RECENT DISPOSAL METHODS

Following methods have been recently devised to dispose of high-level radioactive wastes: (i) reprocessing method, (ii) immobilisation techniques, and (iii) vitrification.

Reprocessing Method

Under this technology, these wastes are first reprocessed to separate the non-usable wastes from the reusable uranium and plutonium. The non-usable liquid wastes are then solidified by passing them through an evaporator, which avoids any possibility of radiation leakage to nearby soil or underground water.

Immobilisation Techniques

Waste contaminated by radionuclides can be collected and immobilised in the following ways:

1. By heating: The radioactive waste may be heated with sulphur and the pitch solidified or can be mixed with cement and water, which is then allowed to settle. This method, however, involves a mere entrapment of radioisotopes and is mainly used for intermediate radio-wastes.

2. In tank-solidification: In this method, the liquid waste is evaporated to a solid cake and left in underground containers. The main drawback of this process is that the nitre cake is freely soluble in water and may be washed away when it comes in contact with underground water.

3. Using zeolites: Certain natural and synthetic zeolites may fix a given radio-nuclide by base exchange or by accommodating them in appropriate channels in their structure. However, the method is extremely selective and is sensitive to salt content and pH of the solution. It requires final sintering and vitrification for permanent fixation of radioactive substances.

4. Calcination: High-level radioactive wastes may be dried and calcined either alone or with additives. Calcination involves the heating up of the waste materials with or without additives to a temperature where little or no melting occurs and ingredients react with each other to form an inert material. Experimental results indicate that besides loss of water and oxides of nitrogen, little or no reaction takes place amongst the additives.

However, there is a serious problem of volatilisation of caesium, ruthenium and other radionuclides polluting the environment.

By Vitrification

The ultimate and safe disposal of high-level dangerous nuclear waste produced after recovery of plutonium and unburnt uranium from irradiated spent nuclear fuels is a growing world problem. The radioactive waste, in the form of aqueous nitric acid stream, contains numerous fission fragments and has a radioactivity of 5 to 10 curie per litre of nuclear waste.

In all the developed countries, including India, this radioactive waste is stored in underground stainless steel tanks with strict surveillance over radiation leakage, off gas release and temperature increase. Radionuclides have half-lives varying from a second to a number of years. Because of long half lives of toxic fission products, these wastes have to be stored safely in the tanks for period extending from 1600 to 2000 years. But the main worry is that the tank life may not be exceed a few decades.

Advantages of the vitrification method

Disposal experts all over the world believe that the best way to permanently dispose of high-level waste is to immobilise the fission products by incorporating them in a solid matrix. Immobilisation in a glassy matrix appears to be most desirable solution to radioactive waste disposal problems. Immobilisation of fission products in a glassy or ceramic matrix offers the following advantages:

1. Glass can accommodate in its structure various types of cations and anions like nitrate and sulphate.

2. The fission products form a part and parcel of the structural block of glass.
3. Fixation in glass is essentially irreversible, which means that fission products will not come out of the glass easily when they come in contact with various attacking agents.
4. Glass may have very high leach resistance, that is it does not dissolve easily in water.
5. Almost all the radioactive elements occupy a definite site in the glass matrix.
6. Volatile fission products like ruthenium (Ru-106) are accommodated in the glass structure either surrounded with six oxygen atoms having one oxygen atom in common between two octahedrons or in tetrahedral holes, depending upon the alkali content of the glass.

The radioactive waste after vitrification (that is conversion into glassy or ceramic material) is then calcined with alumina or zirconia and leached. Clay can also be used to take up the leached radionuclides through absorption or ion-exchange after which it is fired. Firing causes the exchange in crystal structure of clay and the radioisotopes cannot be leached or desorbed.

ENVIRONMENTAL PROTECTION IN NUCLEAR INDUSTRY

The main steps in a nuclear power development programme are referred to as nuclear fuel cycle operations and consists of: (i) mining and milling of radioactive ores, (ii) fuel fabrication, (iii) reactor operations, and (iv) irradiated fuel reprocessing.

All these operations like any other industrial operations, generate wastes. The unique characteristic that distinguishes from other industrial operations is the radioactive nature of some of the waste produced.

The wide spectrum of nuclear fuel cycle operations results in the generation of radioactive as well as non-radioactive wastes of different types, quantities and constituents having widely different physico-chemical characteristics, toxicity and environmental behaviour. In view of the radioactive nature of the waste products generated, a comprehensive and sound environmental management policy has been evolved for management of these wastes and to minimise its environmental impact. The primary objectives of the environmental protection policy adopted for this purpose are:

1. The operations in nuclear power plants/installations shall not interfere in any manner with proper utilisation of environmental resources in the area outside its control.
2. No deleterious effects, either of acute or chronic nature, shall accrue from nuclear operations and disturb the ecological balance of all life forms, including man.
3. Radioactive and non-radioactive pollutants released to the environment shall be at such concentration levels and quantities that the resultant accumulation of radioactivity and other toxins in any component of the environment, including life forms, will not affect them in a manner detrimental to the ecosystem.

These objectives are realised by proper management of the wastes generated in different fuel cycle operations and control of environmental releases conforming to stringent standards.

RADIOACTIVE WASTE MANAGEMENT

The radioactive wastes generated during the various nuclear fuel cycle operations are highly variable in nature, composition, volume, radioactivity levels and half-lives of radionuclides contained. They are classified into solid, liquid and gaseous wastes. Solid and liquid wastes are nominally labelled as low, medium and high-level wastes, depending upon the surface dose and or radio-nuclide content. An elaborate segregation and collection system is necessary before choosing the treatment and disposal methods. The basic scheme for management of radioactive wastes is given in Fig. 23.1. The widely accepted principles for management of radioactive wastes are: (i) dilute and disperse, (ii) delay and decay, and (iii) concentrate and contain.

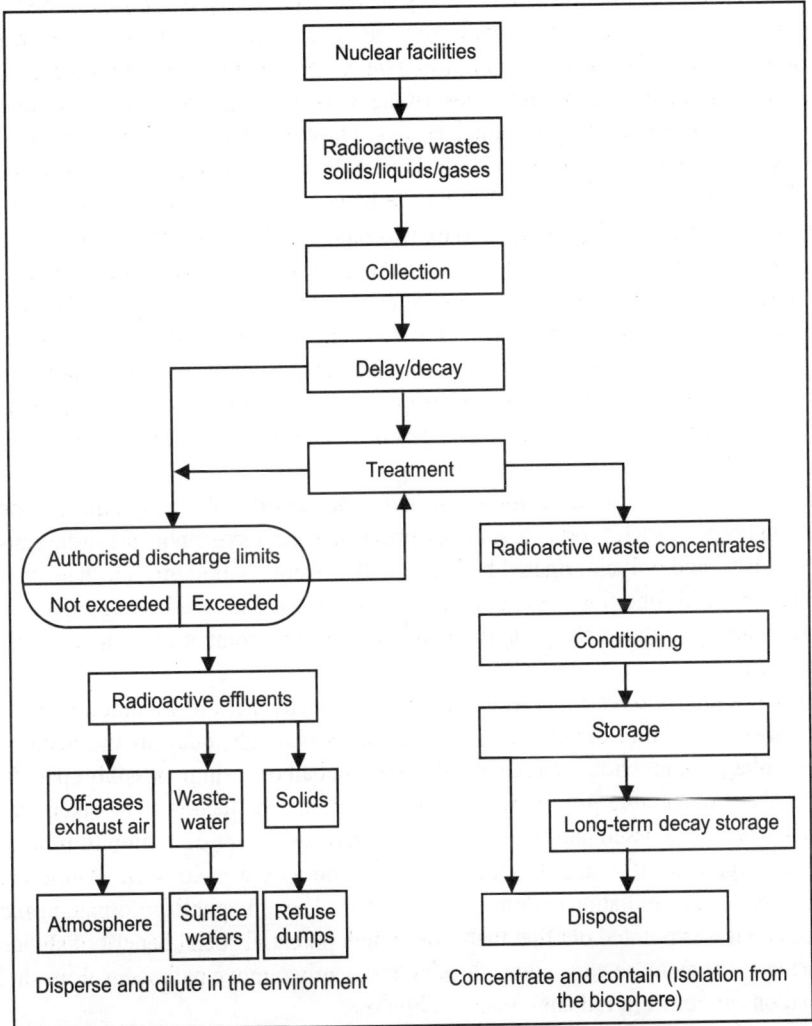

Fig. 23.1. Basic scheme for the management of radioactive wastes.

WASTE TREATMENT BY RADIATION

It is apparent that no waste-water treatment employed at present can guarantee an effluent without human and/or animal pathogens and that in some cases treated sludges contain pathogens in higher concentration than in the untreated raw sludge, resulting in a threat to human and/or animal health through dietary practice. The relative importance of treated but still polluted water and sludges, to say nothing of untreated ones, has to be established in comparison to other pathways of infection before any drastic measures should be taken to improve the hygienic quality of current practice. However, there is a need for improved treatment and radiation might provide an effective means to achieve it. Unfortunately, essential information is still lacking on radiation effect on pathogens, especially on parasites under practical conditions.

Radiosensitivity of micro-organisms: A survey of radiosensitivity of viruses in respect to waste-water treatment showed that radiosensitivity is affected by temperature and suspending medium. It was dressed that assessment of the quantity of micro-organisms present in the water is essential for effective radiation treatment. In the case of viruses, most of the work has been done with bacteriophages which may not be a reasonable indicator of animal viruses. Only two types of viruses, which are directly relevant to waste-water have so far been examined, i.e. polio virus and adenovirus, the latter being more resistant than the former, and considered to be the better indicator. The existence of radioresistant asporogenic bacteria in natural high radioactivity was reported. This indicates an occurrence of natural selection (or adaptation) of micro-organisms as a result of long, continuous exposure to ionising radiation, and is a warning signal that bacteria may occur which cannot be destroyed by irradiation.

Disinfection and microbiological control: One study strongly emphasised, from the hygienic point of view, the need to interrupt the transmission cycle of pathogenic micro-organisms starting from sewage sludge, through plants and animals to man, and proposed effective sanitation of sludge or sewage water as most essential. If radiation is to be used as a practical method of microbiological control, the cost-effectiveness must be improved.

Physical and chemical modification of aqueous pollutants: Various aqueous pollutants of environmental importance such as organic solvents, phenols, linear alkylsulphonate surfactants, pesticides, anthraquinonic dyes, and polychlorinated biphenyls either from a laboratory mechanistic approach or under simulated practical conditions. While some of the data indicated the advantage of ionising radiation in decomposing these pollutants, the yield is mostly too low to be competitive with other methods such as ozone treatment.

Technological and economic aspects: Economy and effectiveness of radiation treatment of waste depend primarily on the choice of radiation source. Sources available today are the electron accelerator (low and high voltage), gamma-ray source of radioactive cobalt or cesium, possibly spent fuel elements and fission products obtainable from power reactor operation. The factors to be taken into account in making a choice were discussed and the importance of having a clear definition of process objectives and a critical comparison of available radiation energy options was stressed. Public acceptance of municipal sewage sludge irradiation systems, including social benefit, public information and safeguards were outlined and it was pointed out that there was a high potential social benefit of sludge irradiation in view of the fact that many sewage plants do not consistently remove pathogens from sludge and that sludge irradiation can result in reducing major public risk.

Pilot-plant design and operating experience: It was reported that in spite of problems left unsolved at the experimental stage, several pilot plants have already been operating to accumulate considerable data in the Fed. Rep. of Germany, the USA and the USSR. Some others are at the planning stage in Hungary and other countries. Although it is noted that we are at the stage where necessary work of an experimental nature is being done by national agencies, but not at the expense of national authorities in charge of water treatment, it is felt that it may not be long before the first truly commercial plant will be built by such authorities, based on the operating experiences obtained by this pilot plant study. A comprehensive report was made on the pilot plant of the Abwasserverband Ampergruppe at Geiselbullach, FRG, for the irradiation of sewage sludge; this plant has accumulated operating experience for about one year. Starting with the cost calculation, all the aspects involved were analysed and evaluated, covering dosimetry, bacteriology, virology, dewatering properties of irradiated sludge, effect of treated sludge on plant and soil, and chemical analysis of irradiated sludge. It was quite apparent that this kind of comprehensive study is essential in promoting the introduction of radiation treatment to be accepted both by the scientific community and by the public in general, and that it could be a model for other pilot plants, operating or in the planning stage, to follow, so that data could be compared for proper assessment.

There exists no one perfect process, conventional or otherwise, which can solve every problem in all situations. The ionising radiation alone or in combination with other methods, has the potential utility of contributing towards the solution of certain problems of waste treatment and reuse of spent resources. Take sewage sludge for example radiation treatment showed better sedimentation and dewatering, and provided improved sludges as fertiliser or as possible animal feed additives. Cost-benefit considerations are still obscure. The choice of radiation source to be employed is still controversial. It was shown, however, that given a clearly-cut requirement in terms of type and nature of effluent, throughput rate and dose to be applied, technological solutions exist and cost estimates can be made for the specific technology. For the evaluation of performance, especially in comparison to other options, co-operative efforts from scientists, engineers and administrative authorities of all relevant disciplines are essential.

To elaborate recommendations regarding the activities necessary to promote this particular application of ionising radiation. Excerpts of the recommendations are as follows:

1. An attempt should be made to understand surface properties of suspended sewage particles and their reaction with radiolytic species.
2. End products, as well as their pollutional effects, should be identified when pollutants of public health and environmental concern are reported as being destroyed by radiation.
3. Present gaps in knowledge on the radiation resistance of bacteria, viruses and parasites in their natural environment should be filled, especially with regard to dose rate effects. The character of radiation resistance in naturally occurring highly resistant microbes should be investigated.
4. Experimental protocol for chemists and microbiologists working in radiation treatment of sludge and waste-water should be developed.
5. Synergistic effects of radiation with chemicals (chlorine, ozone, air, etc.) and physical properties (heat, vibration, etc.) should be pursued.
6. Exchange of information and experiences should be facilitated between pilot plants presently in operation or to be commissioned in the near future.

Chapter 24

Common Effluent Treatment Plant

INTRODUCTION

Chemical process industry has attracted a great deal of opprobrium in many developing countries in the recent past mainly on account of its rather poor record in treatment and management of its liquid and solid wastes. A sudden expansion of this industry in the past few decades, when many of these countries did not have appropriate and effective environment protection laws, has been responsible for increased pollution from this industry in these countries. However, there has been a dramatic change in the situation in the past few years in many countries as public awareness has grown. National laws are in place; and enforcement thereof is becoming more effective day by day. International opinion against pollution causing industries such as leather, sugar, textiles, etc. in developing countries is threatening the booming export of products of such industries taking place from these countries. All these developments have brought to the centre stage the topic of dealing with the problem of pollution in South East Asia. This has become critical for the survival and growth of the industrial sector, particularly in the developing world.

Quality and quantity of raw effluents change not only from industry to industry but for the same type of industry also. Slaughter-house waste-waters are quality-wise different from those from textile industries, but there is no surprise that effluents from cotton textiles industry differ from those of synthetic textiles. Surprisingly, however, effluents from one nylon textile factory may not be same, in quantity and quality, compared to those from another nylon factory. Even for the same industry, the quantity and quality fluctuations take place from day to day and even from hour to hour on the same day.

Design of effluent treatment plant (ETP) is, therefore, to be tailored to suit that industry and there is no unique design of treatment even for industries of one type. This can be well understood by taking an example from the industry, say tannery.

There are many similarities in the structure of the tanning industry in countries of South East Asia. Though organised medium and large scale tanneries are coming up fast in these countries, the preponderance of small scale tanneries in all these countries, generally found in clusters, is a prominent feature. Invariably the medium and large tanneries take measures to treat the wastes generated by them. However, this is not the case with the small scale tanneries. Small scale tanneries do not admit universal definition. There are small tanners having one or two drums in India, Indonesia, Nepal, Bangladesh, China and Sri Lanka. At the same time there are small tanners having four to six drums too. The common features of such small tanners in these countries are primitive work methods, limited work space generally laid out haphazardly, severe financial constraints leading to virtual hand to mouth existence, non-availability of vacant land near the workshed, one-man operations with no support from technically qualified personnel and lack of awareness of damage to environment caused by solid and liquid wastes discharged. When the governments in these countries and the public raised their voices against such unacceptable practices, the tanners had to find a solution. The one redeeming feature of these small scale units is that these are found in clusters. This provided the opportunity of a group of contiguous tanneries joining together to treat the wastes generated by them. This gave birth to the concept of common effluent treatment plants, referred to as CETP, in India.

A CETP may be described as an effluent treatment plant designed to receive the liquid waste generated by a group of contiguous industries and treat this by application of an appropriate technology in a cost-effective manner to achieve discharge standards prescribed by the Pollution Control Authority of the country or region.

PRINCIPLES OF EFFLUENT TREATMENT PLANT (ETP)

The creation of the CETP first requires all industries in contiguity to agree to join together. There has to be a strong motive for doing this. In countries like India, the tremendous pressure exerted by the Pollution Control Authorities, the public and the courts of law has been the main driving force. Similar pressure is applied in varying degrees in different countries. The pattern of organisation that has become popular in India is for a group of industries to form themselves into a company by contributing towards its share capital. The company is charged with the responsibility of designing the plant, arranging finance, engaging contractors and implementing the project. Subsequently the company is responsible for operation and maintenance of the project. In some parts of India, public sector agencies came forward to help the industry.

Need for Effluent Treatment

Effluent treatment is needed to achieve one of the following objectives.
1. To satisfy legal requirements, laid down by pollution control authorities or local municipal authorities. These authorities prescribe limits (upper limits for most of the parameters except dissolved oxygen content and lower and upper limits for pH) for parameters of treated effluent relevant to the final mode of disposal or end-use. The final disposal permitted may be on land or into a nala, river or sea.
2. To recycle the same in industry or factory in the same process. For example, an industry manufacturing C_2H_2 (Acetylene gas) uses water and calcium carbide as reactants to obtain C_2H_2 and lime slurry, as effluent. After clarifying lime slurry effluent and removing lime, the same can be recycled as (alkaline) water to react on calcium carbide.
3. To reuse the same in industry/factory or outside the factory for a different purpose, say as cooling water, or for irrigation or even as boiler feed water.
4. Quality improvement expected of treated effluent will depend on statutory requirement as above or quality required for recycle or reuse.

Common Quality Parameters

Common quality parameters are pH (indicative of acidity and alkalinity), BOD (Biochemical Oxygen Demand indicative of oxygen needs of biodegradable organic matters for their stabilisation through aerobic process) and suspended solids. The upper limits for the above are given in Table 24.1. The above values vary, within reasonable limits, from authorities to authorities that enforce the pollution control acts.

Table 24.1. Common quality parameters

Upper limits for parameters for disposal	pH	20°C 5D BOD mg/l	Suspended solids mg/l
Into river nala	5.5 - 9.0	20	30
Into creek waters/sea	5.5 - 9.0	100	100
On land for irrigation	5.5 - 9.0	150	200
Into municipal sewers	5.5 - 9.0	350	500

The quality parameters, however, may vary substantially and may be very strict to suit the end-use. For example, for reuse in high pressure boilers, the quality parameters will be (for boiler pressure of 250–400 psig) i.e. 17 to 27 kg/cm^2) as follows:
1. Turbidity 5.
2. Minimum pH value 9.0.
3. Oxygen consumed 4 mg/l.

The turbidity is indirectly related to suspended solids and O$_2$ consumed is related to BOD.

Common Problems

CETPs for treatment of effluent are operating in India for some years. During this period valuable experience has been gained in operation and maintenance of CETPs. Based on the visit to the CETPs, discussion with the managers and operators and others concerned with the enforcement of environment law in the country, the following have been identified as common problems:
1. Frequent disruption in the conveyance system resulting in overflow of effluent.
2. Stock loading and overloading of the CETP.
3. Corrosion of machinery and equipment; break-downs; below-rated-capacity performance of pumps, aerators, etc.
4. Inefficient performance of various treatment units of a CETP—much below the designed capacity.
5. Inconsistent results in the effluent discharged after treatment.
6. Difficulty in management and disposal of sludge generated.
7. Higher cost of treatment than anticipated.

Common Causes

A CETP generally consists of the following units—pretreatment systems; collection and conveyance system; screens; receiving sump; screens and grit remover; equalisation tank; chemical dosing (flash mixer); primary clarifier; anaerobic/aerobic treatment or two stage aerobic treatment; secondary clarifier; sludge thickener; sludge drying (mechanical or solar drying beds); discharge of effluent.

The efficiency of a CETP depends on the efficiency of each individual unit. If any of these units does not work properly it will be reflected in the operation of other units and the overall efficiency of the plant. Having stated this general principle, let us now look at the common causes that are responsible for the common problems seen in the operation of the CETPs which are disussed below.

Design of the Plant

The key to successful operation of a CETP is proper design. Proper design of the CETP calls for accurate data with regard to the quality and quantity of effluent generated in the cluster. Invariably it has been found that the tanners have either understated or overstated such figures for their own reasons. The result has been a CETP of either under or over capacity. This has been responsible for overloading of the CETP in some instances.

Management Structure

Management of CETPs by companies is under the control of Board of Directors, elected in terms of the provisions of the country's company law. The Board appoints a technically or otherwise qualified person as the General Manager of the plant who is responsible for the day to day operation and management of the plant. The general manager thereafter engages key technical personnel. Some CETP companies have been operating their plants with the help of contractors. In such cases, the contractors bring in their expertise and personnel and only very few key technical personnel are engaged by the CETP company.

System of management

While some CETP companies directly manage the CETP with their own staff, others engage contractors. It has been generally seen that the contractors do not show needed commitment. Nor are they very well versed in the management of the CETP. As the CETP companies attempt to keep the contract fee as low as possible, the contractors have no compunction in cutting corners. Though the contractors may find short-term solutions, in the long-term they cause great damage to the machinery and equipment of the CETP. Contract management of CETPs does not appear as a desirable solution.

Where the CETPs are managed by the companies themselves too, unfortunately properly qualified and minimum required technical personnel are not often engaged. Absence of qualified and trained personnel has been found to be responsible for poor maintenance, frequency breakdowns and higher cost of operation of CETPs.

Training

On the one hand the CETP companies generally engage not properly qualified personnel. On the other hand, even these people are not put through proper training. Over a period of time, many of such personnel consider themselves 'experts' not needing training. This is a dangerous situation. Training of all personnel, operators, supervisors and managers, of the CETP is a must. Every CETP company should insist on this and arrange for this to achieve better results. Such training should be organised on an annual basis.

Maintenance

Regarding maintenance of machinery and equipment, the less said the better. No regular maintenance is resorted to. As the tannery effluent is very corrosive in nature constant painting of the machinery and equipment is required to prevent corrosion. Lack of preventive maintenance results in frequent breakdowns and shorter life of machinery and equipment. Besides these are not able to operate at rated capacity.

Purchases

There have been instances of purchase of cheaper machinery resulting in serious problems of operation. But more significantly many CETPs do not purchase the chemicals of the right quality. This leads to poor operation results on the one hand; on the other the quantity of sludge generated too increases. Purchasing poor quality chemicals, in the hope of effecting economy, is being penny wise and pound foolish.

A CETP has two basic units—collection and conveyance system and treatment plant. Technical management of a CETP includes effective management of both these segments.

Collection and Conveyance System

The collection and conveyance system includes treatment of segregated effluent stream, pre-treatment and collection of effluent in collection, pumping stations. Proper management of a collection and conveyance system is an essential prerequisite to good performance of the CETP.

Management of Segregated Effluent Streams

A CETP is suppose to segregate the effluent streams from main effluent which is admitted into the collection lines of CETP. Data need for design of effluent treatment plants is as follows:

ETP shall have to be designed keeping in view the expansion program of the industry—say over next 10 years at least.

Effluent Quantity

The quantity of effluent is say m³/day (or litres/day) along with the fluctuations in rate of flow over every hour, during the week, are to be obtained. Effluent quantity will have a relation with water consumption in the industry, which may vary from industry to industry and even for industries of the same nature, and as such, data of water consumption from water metre readings is helpful in assessing quantity of effluent. Water consumption and quantity of effluent have also a bearing with the production of that industry and therefore production figures also help to assess the effluent.

For an existing industry, quantity of effluent per day and quantity of fluctuations can actually be measured by installation of V-notch in the conduits that carry the effluent streams. Data of water consumption and effluent generated in relation with the production is given in *Annexure I* of this chapter. This data is of some use for assessing the water needs and effluent generated for a proposed industry.

The actual quantities of water consumption and therefore of effluent generated may substantially vary from the data in *Annexure A* because the industry may use different methods of production which may need lesser quantities of water, the industry may follow water saving techniques like counter-washings or first washings with waters of inferior quality or the industry may reuse/recycle the waste-waters after treatment partly or fully. There is one asbestos cement company that settles their effluents in sedimentation basins, reuses the entire settled effluent as curing water for their products and the 'asbestos-cement sludge' from the sedimentation basins as make-up material with normal batches of raw material. Data of the type as in *Annexure I* will not be applicable for such industries.

Quantity Variations

ETP, for proper functioning, shall be designed for average flow rate and it shall be seen that effluent reaches treatment plant at average rate. For high variations in flow rates, it would be convenient to have equalisation tank as the first unit that would absorb the variations in the flow rate. Pumps will pump effluent at constant average rate from equalisation tank and feed the same to the ETP.

Effluent Quality

Apart from three common parameters, i.e. pH, BOD and suspended solids mentioned above, there are many other quality parameters that need to be determined for raw effluents, along with their range of variation.

They can be listed as follows:

1. COD (chemical oxygen demand).
2. Oil and grease in floating form or in emulsified form.
3. Detergents.
4. Heavy metals such as chromium, mercury, zinc, lead, etc. or toxic materials such as CN, phenols. Toxicity is determined by a bioassay test that gives an upper limit of the toxic substance that will not kill the fish for a contact period of 24 to 48 hours.
5. All types of solids, say dissolved, settleable, organic (or volatile), inorganic and total.

Quality parameters

1. pH: If the effluent is not within the desired pH range, neutralisation tank will be needed to neutralise acidity with say lime or caustic and alkalinity with say hydrochloric acid. Acid and/ or alkali-proof lining needs to be provided for all treatment units/channels, etc. till pH is brought within the range of 5 to 9. If equalisation tank is very first unit, it also needs acid-alkali resistant lining.
2. Suspended solids: If suspended solids are marginal, provision of settling unit becomes unnecessary. If suspended solids are present, then they may settle in equalisation tank which is not meant to work as settling tank. Therefore, compressed air agitation needs to be provided in equalisation tank.

Volatile solids (organic solids)

If organic solids are not high, biological treatment units are not needed. COD and BOD are two parameters that determine total oxidisable matter and biologically oxidisable matter. The ratio of COD to BOD is about 2 for domestic sewage which is amenable to biological treatment. High value of COD/BOD above and away from 2 is indicative of reduced biodegradability.

Bioassay test

Bioassay test is fish mortality test and it indicates toxicity of the effluent. If the combined effluent of the industry indicates toxic nature of the combined effluent, it would be necessary to determine toxicity of every stream, segregate by laying separate sewers in the toxic stream, give preparatory or separate treatment to toxic stream to detoxify it and treat the balance bulk of the effluent with detoxified stream.

Example of the above will be cyanide and chromium effluent streams from electroplating industries. If these toxic streams are mixed with remaining effluent streams, biological treatment of the combined streams will not be possible.

PRETREATMENT OF WASTE-WATERS

It is often necessary to subject certain industrial waste-waters to adequate pretreatment so as to render them amenable to biological treatment. Typical examples of such wastes are those discharged by the electroplating industry (containing cyanides, chromium, cadmium, nickel, etc.), vegetable oil industry (containing vegetable oil and grease) engineering industry (mineral oil and grease). Removal of these contaminants from the waste streams becomes necessary on account of their ability to inhibit or to completely kill the biomass which is essential for biological treatment of the effluents.

Thus, removal of cyanides from electroplating waste-water is conventionally done by alkaline-chlorination method.

The probable chemical reaction with excess chlorine in the presence of caustic soda is:

$$2NaCN + 5Cl_2 + 12\ NaOH \rightarrow N_2 + 2Na_2CO_3 + 10NaCl + 6H_2O$$

About 6 kg each of caustic soda and chlorine are required to oxidise 1 kg of cyanide ion to nitrogen. Similarly, hexavalent chromium (Cr^{6+}) from chromium plating bath is removed from the spent bath using the process of reducing toxic hexavalent chromium to its non-toxic trivalent state under acid conditions, followed by raising pH value by adding an alkali such as lime and precipitating trivalent chromium as chromium hydroxide, $Cr(OH)_3$. The reactions are:

$$H_2Cr_2O_7 + 6FeSO_4 + 6H_2SO_4 \rightarrow Cr_2\ (SO_4)_3 + 3Fe_2\ (SO_4)_3 + 7H_2O.$$

$$Cr_2(SO_4)_3 + 3Ca(OH_2) \rightarrow 2\ Cr\ (OH)_3 + 3CaSO_4$$

$$Fe_2(SO_4)_3 + 3Ca(OH)_2 \rightarrow 2Fe(OH)_3 + 3CaSO_4$$

For every 1 mg/l of chromium, about 16 mg/l of copper as ($FeSO_4$) 6 mg/l of sulphuric acid and 9.5 mg/l of lime are required. These reactions result in the formation of about 2 mg/l of chromium hydroxide $Cr(OH)_3$, 0.4 mg/l of ferric hydroxide $Fe(OH)_3$ sludge, and 2 mg/l of calcium sulphate ($CaSO_4$).

COMMON ETP FACILITY AND DESIGN THUMB RULES

Common ETP will have the following treatment units. Thumb rules for their design are mentioned.

Raw Effluent Pump House

Normally, raw effluent will reach the pump house through the last or outfall sewer say at a depth of 2000–4000 mm. The depth of sump, which is around 1500–2500 mm, is to be provided below this depth. It, therefore, becomes uneconomic, to provide equalisation capacity in the receiving sump itself, as the sewer bringing effluent is at depth. Therefore, raw effluent sump is designed for about 10 minutes average flow for large flows (exceeding 3000 m³/d or so) to 30 minutes average flows for small flows.

Number of pumps, including standbys provided are peak flow/average flow + 1 or 2. For small flows only 2 pumping sets (1 working and 1 standby) are provided. 2 pumps (1 working 1 standby) of average flow and 2 pumps (1 working and 1 standby) of peak flow would be a practical combination.

If the flow reaching the pump house is acidic or alkaline and if neutralisation cannot be achieved in the channels or conduits or ahead of them before effluent reaches the raw effluent pump house, it would be necessary to provide acid/alkali proof lining to sump and to use acid/alkali resistant pump sets.

Equalisation Tank

For fluctuating flow rates, there is a need of this unit to absorb the peaks and troughs in the flow rates. The capacity of equalisation tank is decided from the cumulative mass curve of cumulative flow (on y-axis) against hours (on x-axis) or from the cumulative largest continued excess of flow over uniform cumulative pumping at the average rate over the same period.

Contents of equalisation tank need to be agitated continuously to prevent settlement of suspended solids. Such agitation by compressed air is more satisfactory as the level of effluent in equalisation tank is variable and may be zero once in a day. Pumping at an uniform average rate from equalisation tank is a part of equalisation arrangement.

Equalisation tank achieves complete equalisation of flow, as it is meant for the same. It additionally helps to suppress the peak and troughs in quality. If parameters such as BOD, suspended solids, COD have extreme variations, equalisation tank would flatten these variations.

Neutralisation Tank

If neutralisation is not achieved in equalisation tank, neutralisation facility in the form of a tank and agitating device may be provided. Retention period in the tank varies from 10 to 20 minutes and agitation power may be calculated at 5 w to 15 w per m^3 capacity of the tank.

Settling Tank

Settling tank is also known as settling basin, sedimentation tank or basin, clarifier. Settling tank surface area is decided for a surface loading of 10–20 m^3/m^2/day. Minimum side water depth of 2000 mm is provided. Detention time is from 90 minutes to 120 minutes. If settling tank is provided with hoppers, they are given a slope of 50° to 60° with horizontal. Such steep hoppers will not need sludge scraping mechanisms. If circular sedimentation tanks with scraping mechanism are provided, they are given a slope of 7 per cent at the bottom. The process of settlement of suspended solids can be hastened with the addition of coagulants like alum, iron chlorides, etc. and/or patented special polyelectrolytes. If chemicals are to be added, then a separate chemical dosing unit, chemical reaction tank and a flocculator will be needed. Chemical reaction tank with a flash agitator has retention time of 1 to 2 minutes. Settling of relatively high concentration of suspended solids follows zone settling or being in physical contact with each, they settle at reduced rate of subsidence. Under these circumstances, the clarifier works as a sludge thickener and apart from hydraulic loading, solids surface load rate needs to be checked. These rates vary for types of sludges. They are in the following range:

Type of sludge	Solids loading rate (kg/m²/day)
Primary or primary with lime addition	20–30
Primary and biological sludge or only biological sludge	6–10

Aerobic Biological Treatment

In aerobic biological treatment, micro-organisms use the waste's organic biodegradable component, in the presence of oxygen, to produce their cell growth and end products of carbon dioxide and water. Biological treatment processes available are:

1. Activated sludge (AS) process and its modifications like extended aeration including oxidation ditch process and Aerated Lagoon. Guidelines for design of activated sludge process are given in Table 24.2.

Table 24.2. Design of Activated sludge process.

Design parameter	Value for AS process (conventional)	Value for AS rocess (extended aeration)
Volumetric loading kg of BOD$_5$/M³/d	0.4–0.8	0.15–0.2
MLVSS mg/l	1500–4000	2000–6000
F/M ratio kg of BOD$_5$/day/kg of MLVSS	0.2–0.5	0.05–0.15
Detention time hrs.	6–10	18–36
Recycle of sludge %	50–100	75–150
O$_2$ supply kg of O$_2$ per kg of BOD$_5$ removed	0.6–0.8	1–1.5
Sludge retention time (days)	5–10	20–30
Waste sludge kg per kg of BOD$_5$ removed	0.4–0.6	0.15–0.3

Aerated Lagoon process is like activated sludge system without recirculation. Typical design thumb rules are:

Organic loading (per hectare)	30–300 kg of BOD_5/ha/day
Depth	2 M–5 M
Detention time	5–10 days
BOD_5 removed	80 per cent

Pond Systems

Aerobic stabilisation pond systems have been used to treat domestic and industrial waste-waters. Aerobic and facultative ponds use oxygen from photosynthesis of algae in the presence of sunlight to stabilise organic matter. Design rules in ponds are given below:

1. Bottom and dike material shall be of compacted clay with liners to minimise leakage and seepage. Dike top width shall be 2.5–3 M minimum and side slopes of 3 horizontal to 1 vertical. Erosion control at waste-water level due to wind-wave action is necessary. Free board of 1 M is provided.
2. Ponds shall be minimum two or three in number. They are operated in series normally expected for larger systems. Area shall be properly lighted.
3. For aerobic pond design, depth is shallow say upto 0.6 1 M, detention time 5–20 days and organic loading of 100–200 kg BOD_5/ha/day. BOD removal is 85–90 per cent.
4. For facultative pond, depth is from 1 M to 2.5–3 M, detention time 30–90 days, organic loading 25–60 kg of BOD_5/h/day. BOD removal of 75–95 per cent is expected.
5. Trickling filters: The trickling filter process employs an attached growth biological system based on passing organic waste-water over the biological slime growth on solid media. The slime layer increases in thickness and loses its ability to cling to the media. Losing the slime layer is called sloughing and is a function of organic as well as hydraulic loading.

Trickling filters are classified as low rate, high rate and roughing with respect to application rate of organic and hydraulic loading. Design data is given in Table 24.3.

Table 24.3. Design data of trickling filters.

Parameter	Low rate TF	High rate TF	Roughing TF
Organic loading BOD_5/day/M^3	0.08–0.4	0.4–1.6	1.6–8
Hydraulic loading M^3/day/M^2	1–4	10–40	30–120
Recirculation ratio	None	0.5–3	0.5–3
Media Bed depth M	1.5–3	1.0–2.5	6–9
Sloughing	Intermittent	Continuous	Continuous
BOD removal efficiency after sedimentation	80–90	65–85	40–70

Primary function of a roughing filter is to reduce high organic loading by its use as an intermediate treatment process upstream of an actuated sludge process. Physical properties of trickling filter media are given in Table 24.4.

Sludge Collection and Disposal

Except for the pond system and aerated lagoon, settlement facility in the form of settling tank is provided after biological treatment. This tank is called secondary clarifier while the one preceding the biological treatment is called primary clarifier.

Table 24.4. Physical properties of trickling filter media.

Media	Nominal size in mms	Void space %
Granite	100	50
Slag	50–75	50
Plastic	10 × 10 × 20	95

In case of activated sludge type biological treatment, settled biological sludge-cell mass from secondary clarifier is recirculated to aeration tank. In trickling filter system, the clarified effluent from secondary clarifier is recirculated to trickling filter to dilute the raw effluent settled in primary clarifier.

Water sludge from primary clarifier and excess sludge from secondary clarifier may contain from 0.1 per cent to 4 per cent of solids and balance as water and needs to be collected, treated and disposed. The combined sludge is collected in a sump and pumped to an anaerobic digester and economically treated at temperature of 30°–33°C. Common design basis is volatile solids loading of 1 kg/d per m^3 of digester capacity, with a total retention time of 20–40 days.

ANAEROBIC DIGESTION

Anaerobic digestion is the biological decomposition of organic matter in the absence of oxygen. The process occurs in two interdependent stages; the acid stage and methane stage. In the acid stage facultative organisms called 'acid formers' degrade the complex organics of waste-water to volatile organic acids, primarily acetic acid. In the second step, these volatile acids are fermented to methane and carbon dioxide by anaerobes called 'methane bacteria'.

Of these two phases methane fermentation phase is important because:

1. The only mechanism of COD/BOD removal is the production of methane. Acid production solubilises the complex organics, does not accomplish stabilisation.
2. This step has been found to be the rate limiting step in the reaction sequence. Volatile acid (VA)/alkalinity ratio gives the indication of progress and balance between the two stages of acid fermentation and methane fermentation. The normal ratio is less than 0.1. Any increase in VA/alkalinity ratio indicates either excessive feeding of raw sludge to the digester or removal of too much digested sludge. The detention time is dependent on the volatile solids content of the sludge pumped. Usually it varies from 30–60 days. It can be expressed as days per filling.

Reduction of per cent volatile solids may go up to 70 per cent. The gas production is dependent on the organic content of the sludge depending upon the composition of the waste. The supernatant liquor to be removed from digester may have a high BOD and is returned to the inlet of treatment plant or into the raw effluent sump for complete treatment.

Digested sludge is spread on shallow rectangular sand beds with an underdrain arrangement. Sludge is run on the beds to a depth of 200 mm. Sludge may dry to 25 per cent solids after 2–4 weeks. A normal area requirement is 1 M^2 of bed area per 1 kg of BOD destroyed. Digested sludge (and chemical sludge from pretreatment) is also separated by vacuum filtration on a revolving drum, covered with filter cloth, and partially submerged in sludge.

A vacuum of 80–90 kN/M^2 is applied and a layer of sludge 3–6 mm thick is built on cloth. Filter yields of 800–1000 kg of dry solids per day per M^2 of filter cloth are possible. Solids are 25 per cent to 30 per cent by weight in dried cake. The underdrainage from sand filters and filtrate from vacuum filters is led back at the inlet of treatment plant for complete treatment.

Land Requirement of Effluent Treatment Plant

Land requirement varies from 0.5 to 1 M^2 per M^3/d of effluent depending upon the extent of treatment provided. This land requirement is only for treatment units, with just adequate interspaces and approach road strips. It is not inclusive of roads, garden, laboratory and administrative office and quarters for operating staff.

Cost of Treatment Plant

The break-up of cost of treatment plant will be:

Civil cost	30–50 per cent of total cost
Mechanical equipment	30–50 per cent of total cost
Balance for piping, valves, electrification, approach strips, etc.	20 per cent

Cost of civil work can be conveniently calculated by summing up of cubic contents of all treatment units, inclusive of free board and calculating the cost on the basis of total cubic contents.

REPAIR, MAINTENANCE AND OPERATION OF TREATMENT (RMO) PLANT

Proper design and installation of ETP is the first step; proper operation and maintenance of ETP is equally important step if one wishes to achieve the desired quality of final effluent as per design.

It is observed that upto 15 per cent of the capital cost will be maintained and operation cost of ETP per year. Cost involved depends on following factors:

1. Cost of consumables and chemicals.
2. Cost of power.
3. Salary of operation staff.
4. Maintenance, repair and replacement of civil/electrical/mechanical components of ETP.

The above 15 per cent does not include the depreciation on the cost of ETP and interest on the invested capital. Further, while calculating the capital cost, not the book value but the present cost of ETP after allowing for the inflation and increase in labour cost is to be taken. An ETP built in 1998 at Rs. 40 lakhs may cost Rs. 1 crore in 2011. RMO charges in 2011 will be calculated on Rs. 1 crore and not on Rs. 40 lakhs.

Annexure-I. Indian standards related to disposal of treated effluents. (Ref. Latest IS 2490 Part D).

Parameters		Inland surface waters	Discharged into public sewers	On land for irrigation	Marine disposal
pH		5.5 to 9.0	5.5 to 9.0	5.5 to 9.0	5.5 to 9.0
Temperate max.		(Not to exceed 40°C)	45°C	40°C	45°C
Colour (platinum cobalt scale)	1	100 units	–	–	100 units
Total suspended solids	mg/1	100	(Relaxable 750 by local unit)	200	100
Total dissolved solids (inorganic)	mg/1	2100	2100	2100	–

(Contd ...)

Parameters		Inland surface waters	Discharged into public sewers	On land for irrigation	Marine disposal
BOD (5 days 20°C)	mg/l	30	350	100	100
COD	mg/l	100	–	–	250
Oil and Grease	mg/l	10	20	10	20
Chlorides (as Cl)	mg/l	1000	1000	600	–
Phenolic compounds	mg/l	1.0	5.0	–	5.0
Cyanides (as CN)	mg/l	0.2	0.2	0.2	0.2
Sulphate (as SO_4)	mg/l	1000	1000	1000	–
Sulphides (as S)	mg/l	2.0	–	–	5.0
Pesticides	mg/l	–	–	–	–
Phosphate (Dissolved as P)	mg/l	5.0	–	–	5.0
Total residual chlorine	mg/l	1.0	–	–	1.0
Fluoride (as F)	mg/l	1.5	15	–	10
Boron (as B)	mg/l	2.0	2.0	0.75	–
Arsenic (as As)	mg/l	0.2	0.2	0.10	0.2
Sodium percentage	mg/l	–	–	60	–
Cadmium (as Cd)	mg/l	1.0	2.0	0.01	2.0
Copper (as Cu)	mg/l	3.0	3.0	0.2	3.0
Lead (as Pb)	mg/l	0.1	1.0	5.0	1.0
Hexavalent Cr. (as Cr^{+6})	mg/l	0.1	2.0	0.1	1.0
Total chromium (as Cr)	mg/l	2.0	2.0	1.0	2.0
Mercury (as Hg)	mg/l	0.01	0.01	0.01	0.01
Nickel (as Ni)	mg/l	3.0	3.0	0.2	4.0
Selenium (as Se)	mg/l	0.05	0.05	0.02	0.05
Radioactive materials					
Alpha emitters	UC/ml	10^{-7}	10^{-7}	10^{-8}	10^{-7}
Beta emitters	UC/ml	10^{-6}	10^{-6}	10^{-7}	10^{-6}
Ammoniacal nitrogen (as N) total kjeldahl	mg/l	50	50	–	50
Nitrogen (as N)	mg/l	100	–	–	100
Free ammonia (as NH_3)	mg/l	5.0	–	–	5.0
Zinc	mg/l	5.0	15.0	2.0	15.0

Annexure–A. Water required by certain important industries of the world.

Industry and products	Unit of production	Water required in M^3 per unit of production
Chemicals		
Alcohol	Kilo-litres	50–150
Ammonia	Ton of NH_3	150
Ammonium sulphate	Ton	900
Calcium carbide	Ton	150

(Contd ...)

Industry and products	Unit of production	Water required in M^3 per unit of production
Calcium metaphosphate	Ton of Ca $(PO_3)_2$	10
Carbon dioxide	Ton	100
Caustic soda	Ton of NaOH	100
Cellulose nitrate	Ton	50
Charcoal and wood chemicals	Ton of $CaAc_2$	300
Corn refining	Ton of starch	1.5
Gasoline	Kilo litre	10–40
Gun powder	Ton	750
Hydrochloric acid	Ton of 20% Be°HCl	130
Hydrogen	Ton of H_2	3000
Lactose	Ton	900
Oxygen	100 cu. m.	10
Soap	Ton	2
Soda ash	Ton	75
Sulphuric acid	Ton of 100% H_2SO_4	5–25
Sulphur	Ton	10
Food and beverages		
Beans, green	Ton	75
Beer	Kilo litres	10–20
Bread	Ton	2–4
Butter	Ton	20
Canned fish	Ton	1–60
Canned fruits and vegetables	Ton	10–50
Cheese	Ton	2
Gelatine	Ton	50–100
Milk powder	Ton	50–200
Potato, flour	Ton of potato	10–20
Poultry	Ton	5–50
Sugar	Ton of sugar beets	10–100
Sugar	Ton of sugarcane	1–5
Whisky	Kilo litre	15
Wheat milling	Ton	80
Petroleum products		
Aviation	Kilo litres	25
Gasoline	Kilo litres	400
Kerosene	Kilo litres	40
Oil refinery	Ton of crude petroleum	10–30

(Contd ...)

Industry and products	Unit of production	Water required in M^3 per unit of production
Pulp and paper		
Average industry demand	Ton of dry pulp	90–150
Bleached pulp	Ton of dry pulp	175–500
Fine paper	Ton of dry pulp	900
Kraft pulp	Ton of dry pulp	400
Sulphite pulp	Ton of dry pulp	300–800
Soda pulp	Ton of dry pulp	400
Paper	Ton of dry pulp	175
Paperboard	Ton of dry pulp	75–400
Strawboard	Ton of dry pulp	100
Textiles		
Artificial silk	Ton	2000
Cotton	Ton	10–300
Cotton bleaching	Ton	250–350
Cotton dyeing	Ton	35–70
Linen	Ton	750
Rayon	Ton of yarn	400–900
Wool	Ton of yarn	150–900
Metal and metal products		
Aluminium	Ton	1400
Rolled steel	Ton	350
Finished steel	Ton	300
Fabricated steel	Ton	200
Steel sheets	Ton	60
Average of all products	Ton	75–130
Mining and quarrying		
Bauxite	Ton of ore	10
Gold	Ton of ore	1
Iron ore	Ton of ore	5
Miscellaneous		
Automobiles	Per vehicle	40
Coal	Ton	5–15
Electric power	Kw	0.3–0.7
Explosives	Ton	850
Fertilisers	Ton	300
Glass	Ton	75
Leather tanning	Ton of raw hide	75
Synthetic rubber	Ton	900–3000
Laundering	Ton	30

SECTION V

Case Studies Related to Water Pollution

Chapter 25

Case Studies

INTRODUCTION

This chapter discusses the case studies of Hindustant Petroleum Corporation Limited (HPCL)—Vishakhapatnam; JK Rayon and Synthetics—Kanpur; and Radioactive waste treatment technology at Czech Nuclear Power Plant.

CASE STUDY-1 WATER POLLUTION CONTROL AND DISPOSAL OF WASTE-WATER IN HPCL REFINERY VISHAKHAPATNAM (ANDHRA PRADESH)

Hindustan Petroleum Corporation Limited (HPCL) is a global fortune 500 company in the energy sector. HPCL has two refineries located in Mumbai (West Coast) and Visakh/Vizag (East Coast) with capacities of 6.5 MMTPA and 8.3 MMTPA respectively, churning out a wide range of petroleum products, bitumen, etc. and various grades of lubricants, specialities and greases as per BIS and international standards. Over the years HPCL's capacity of production has expanded massively through various upgradation initiatives. The refineries known for the full utilisation of capacity and world class performance are the foundations of HPCL's successful journey towards meeting India's energy requirements. Mumbai refinery is a Lube based refinery with the highest lube production capacity in India. The offsite product handling facilities of refineries at Mumbai and Visakhapatnam (Vizag) has been automated and facilities upgraded to produce green fuels like unleaded petrol and low sulphur diesel. HPCL is committed to upheld India's position in the global energy scenario as a useful contributor.

The refineries are operated efficiently to comply with international quality standard. Both Refineries have been operated at their capacity utilisation above 100 per cent. The consistent maintenance of standard has fetched the two refineries numerous awards. The refineries can claim the lion's share of HPCL's contribution in the field of energy conservation, environment and safety. For HPCL, success is never an end in itself and hence the refineries will go through further upgradation in future. The basic operations of refinery are shown in Fig. 25.1. In this case study the two effluent treatment units of Visakhapatnam (Vizag) refinery, old waste-water treatment plant (OWWTP) and central waste-water treatment plant (CWWTP) with biological system for treating process effluent waters and associated cooling water treatment facilities at each unit are studied in view of Vizag refinery expansion, toxic effects of refinery effluent water to biological life and finally to see whether the characteristics of treated effluent waters are in conformance with MINAS standards prescribed by Central Board for Prevention and Control of Water Pollution (CBCWP), New Delhi (India) or not.

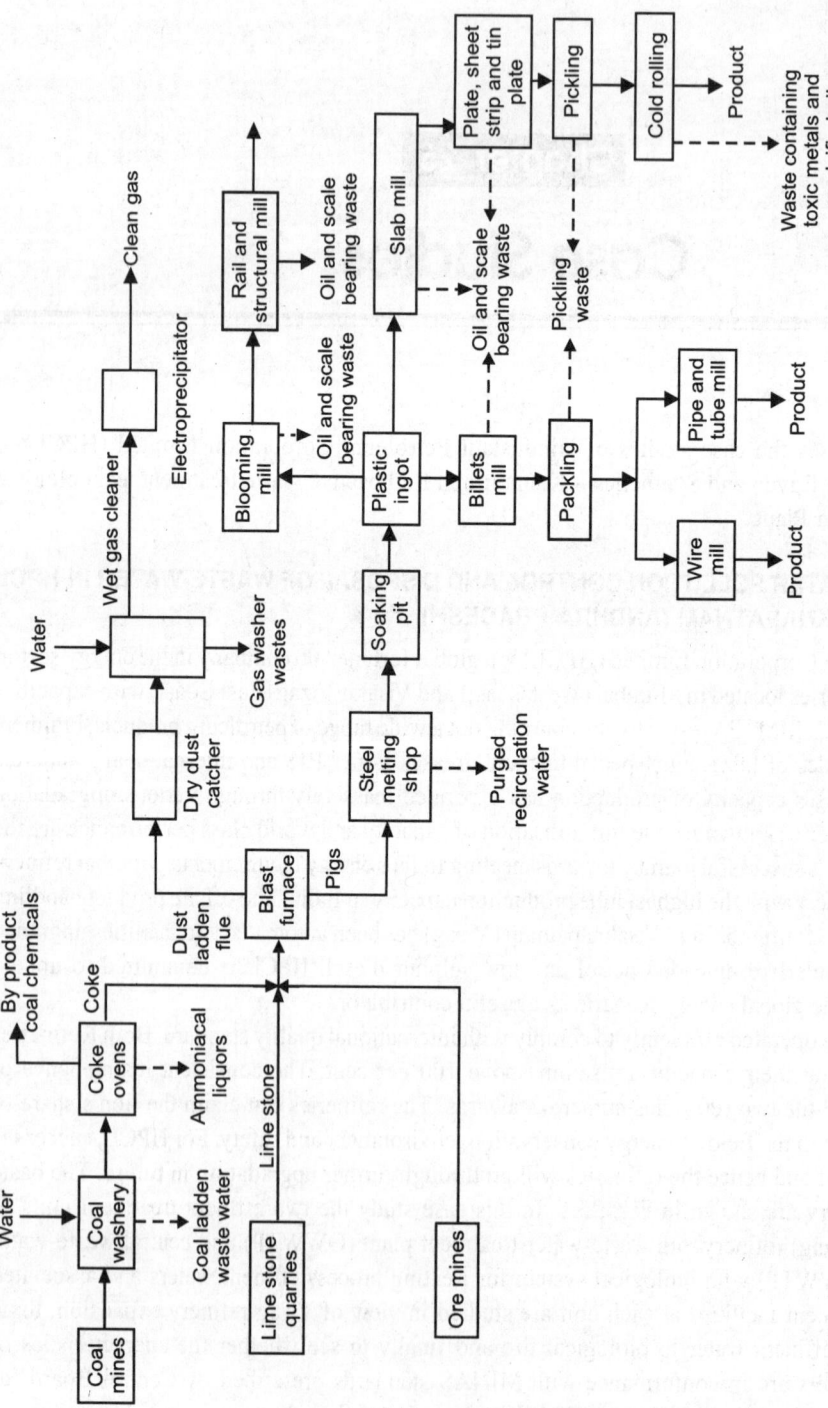

Fig. 25.1. Basic operations of refinery.

The segregation of waste-waters, the in-plant pretreatment facilities for removing highly contaminated wastes, the different influent streams of ETP-l, CWWTP and treated effluent characteristics from different units namely TPI total free oil separator, ammonia stripping, total sulphide removal, emulsified oil removal, two stage biological waste treatment, DMF unit of CWWTP are studied. It is observed that the DMF outlet of ETP-l, DMF outlet of CWWTP and MHF outlet effluent characteristics are meeting MINAS standards.

Due to their nature of operation, all process and manufacturing units are liable to pollute the environment. The level of operation varies from industry to industry. However, there are various ways and means by which water pollution can be controlled and kept below permissible levels. A petroleum refinery has to take special steps for its safe operation and environmental management. In view of stringent standards for refinery effluents known as Minimal National Standards (MINAS) prescribed by CBCWP, New Delhi, the Vizag refinery has modified the existing treatment facilities and introduced, emulsified oil removal and biological systems for treatment process effluent water. The two effluent water treatment units namely old waste-water treatment plant (OWWTP) and central waste-water treatment plant (CWWTP) with biological system for treating process effluent water and associated water treatment facilities at each unit are shown in Fig. 25.2 and 25.3 respectively. This study is related to treatment aspects and treated effluent characteristics.

Waste-water Segregation

The plant waste-water are segregated into five systems: (i) sour condensate, (ii) oily water, (iii) cooling water returns, (iv) boiler blow down waters, and (v) sanitary sewer waters. This approach permits handling the most highly contaminated wastes through the complete treatment sequence and selectively by passing cleaner effluents around some treatment operations.

In plant pretreatment maximum use of in-plant pretreatment was to remove pollutants from highly contaminated wastes before they are routed to main waste-water treatment facilities. It consists of sour water strippers to remove sulphide, ammonia, from the various process units, neutralisation pits for collecting and neutralising the acidic and alkaline effluents and the blow down drums to recover maximum oil from process units, thus reducing the loading on the treatment plant. After the pretreatment the process effluent is routed to main treating units. The segregated process effluent stream characteristics after in-plant pretreatment of Vizag refinery are given in Table 25.1.

Central Waste-water Treatment Plant (CWWTP)

The CWWTP (Fig. 25.1) receives the treated process effluent from existing new (VREP) units and old units, crude tank bottom drains, merox chemical waste and sour water from all units. These waste-water streams are subjected to free oil removal, ammonia stripping, sulphide precipitation, emulsified oil removal, biological treatment and final treatment before disposal.

Process description

The effluent streams are initially subjected to free oil removal in different units. The removal efficiency is more than 95 per cent. Free oil removal is accomplished in tilted plate separators using gravity. The suspended solids settle down on the plate and slide down the trough of the corrugated plate to the sludge compartment.

The deoiled water is sent to the ammonia fixing tank. The deoiled sour water streams over flowing from TPI units gravitates to a stripper feed pump.

Table 25.1. Effluent characteristics (after pretreatment) of waste-water streams of Vizag refinery, Vishakhapatnam

Parameter	Samples									
	A		B		C		D a		D b	
	Range	Avg	Range	Avg	Range	Avg	Range	Avg	Range	Avg
Flow (m³/hr)	50–70	65	100–140	133	0.88–1.4	1.1	45.60	50.0	13.17	15.5
TDS	2600–4000	3400	4500–6000	5300	11000–18000	16,000	200–300	260	40–70	60
SS	48–150	100	30–54	40	300–600	500	—	—	36–58	46
Ammonia (as NH_3)	100–900	700	0.12–0.18	0.16	560–700	600	—	—	—	—
Cyanide (as CN)	16–40	28	0.4–0.7	0.5	0.26–0.32	0.28	—	—	0.5–1.2	0.8
Sulphides (as S)	30–70	56	8.4–12.6	10.8	2800–9000	7900	6–10.0	8.0	4.0–8.0	6.0
Mercaptans (as R'SH)	—	—	7.0–9.0	8.0	1980–2300	2160	—	—	—	—
Thiophenol (as RSH)	—	—	0.02–0.06	0.05	20–30	25	—	—	—	—
Phenolic compounds	36–80	68	1.20–2.00	1.50	2230–5600	4600	7–12.0	9.0	200–320	240
Oil and Grease	260–400	348	800–900	860	3000–3900	3500	290–450	360	7500–11000	9000
BOD, 5 days	1400–1800	1590	140–190	170	22000–34000	29,000	90–150	110	200–370	290
COD	2200–7000	6080	190–360	300	70000–90000	82,000	200–360	290	850–1200	1010

Sampling points:

A = Process sour water from old and new units

B = General effluent like floor washings, drips/leaks, crude tank bottom drain waters, etc.

C = Combined spent caustic/merox wastes

D: a = Contaminated rain waters from surge tank   b = Effluent from bitumen blowing units.

TDS = Total dissolved solids   SS = Suspended solids

Note: All parameters are expressed in mg/l,

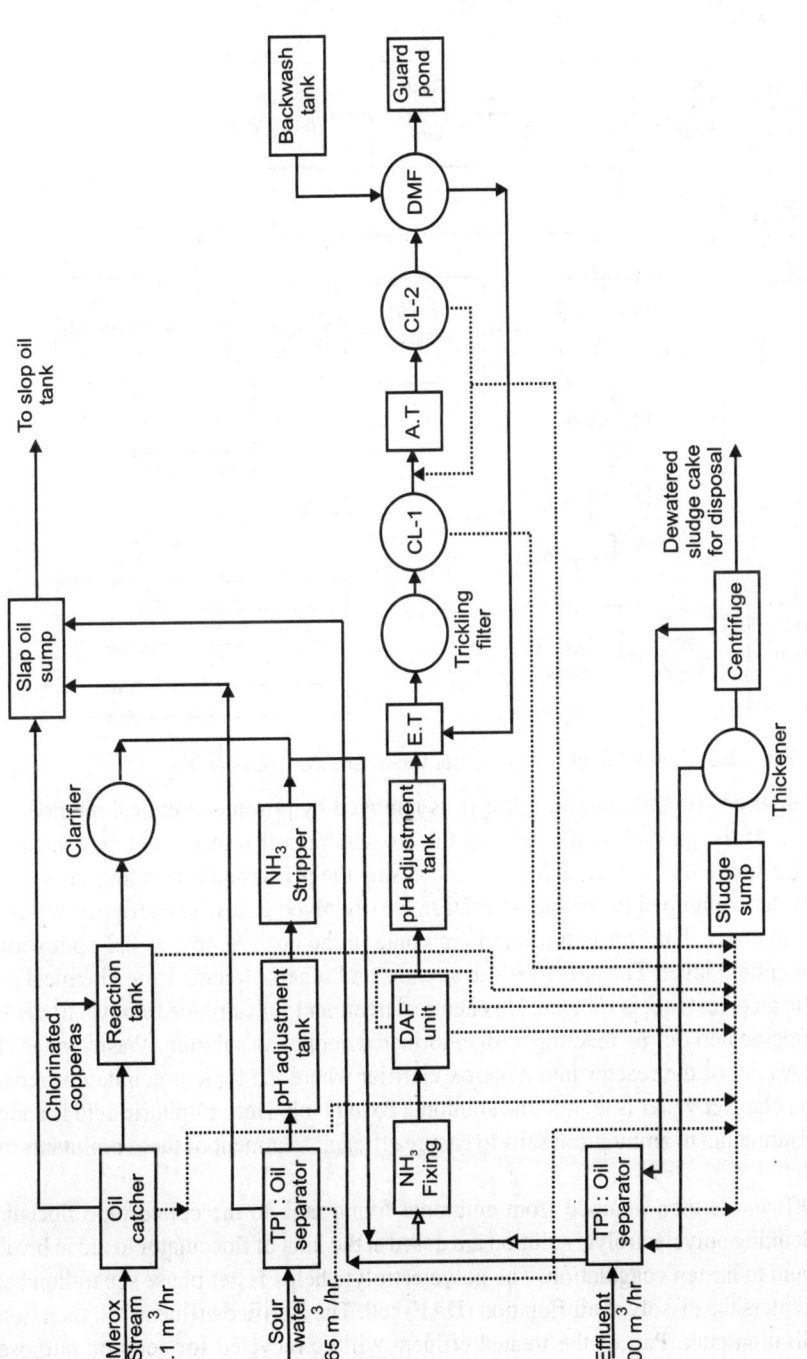

Fig. 25.2. Flow sheet of old waste-water treatment plant (CWWTP).

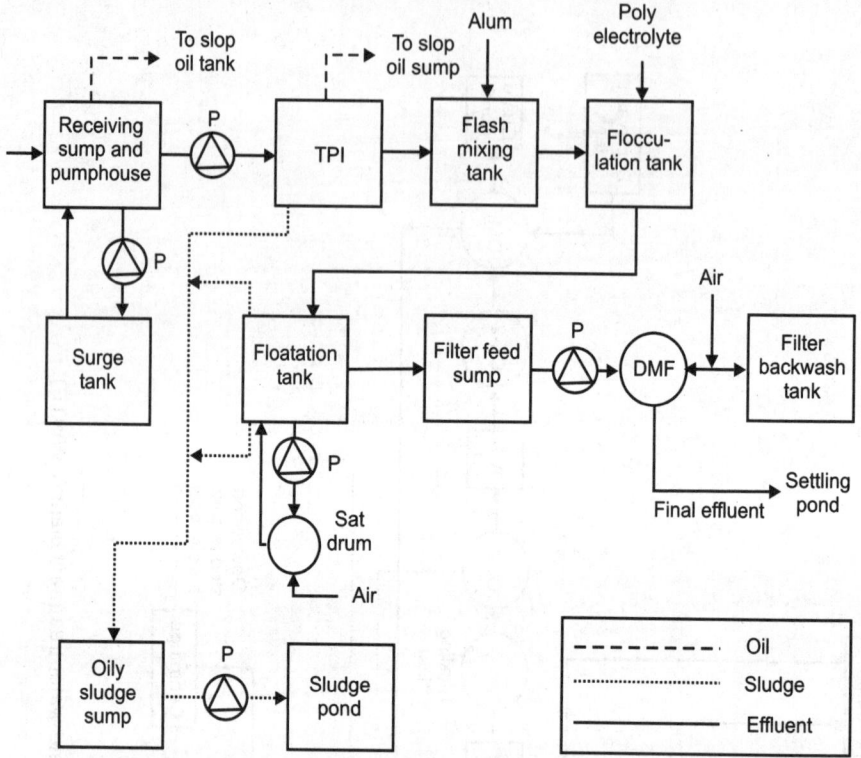

Fig. 25.3. Effluent of old waste-water treatment plant (OWWTP).

Here high concentrations of ammoniacal nitrogen is removed by physico-chemical method which involves volatilisation of the gaseous ammonia into the air. The rate of transfer can be enhanced by converting most of the ammonia to gaseous form by increasing the pH through the addition of caustic. The ammonia rich water is pumped to stripper to release the dissolved gases. The stripped water falls into a pump below the tower. The concentration of ammonia in the air is as low as 0.2 ppm which is much below the perceptible level. The spent caustic streams and others streams from chemical sewer are initially deoiled in an oil catcher is subjected to chemical treatment for sulphide removal of pH >7.0. The sulphides are precipitated out by reacting with chlorinated copper as solution. Waste-water along with precipitates flows out of the reactor into a merox clarifier where the toxic precipitate is received for safe disposal. The clarifier water is let into the ammonia fixing tank. Here sulphuric acid is added to convert any residual ammonia to ammonium salts to ensure efficient treatment of these pollutants in the down stream units.

The combined effluent is then pumped from ammonia fixing tank to the coiled pipe flocculator. Alum solution and deoiling polyelectrolyte solution are dosed at the inlet of flocculator to aid in breaking the emulsion of oil and to hasten coagulation. The polyelectrolyte helps faster phase separation before the flocculated feed enters the dissolved air flotation (DAF) cell. The clarified effluent will then flow by gravity to the equalisation tank. Part of the treated effluent will be recycled for aeration purpose. In equalisation tank the quality and quantity of the feed will be equalised for biological system. The residence time is 8 hours and the overflow from the tank is of uniform quality.

Biological treatment

The equalised flow is then subjected to a two stage biological treatment. The first stage comprises at trickling filter, a concrete tank packed with a bed of tone media to which micro-organisms are attached as a film. The feed is continuously spread over the media by means of rotating distribution pipes. Continuous availability of food matter and aerobic conditions are conducive for the growth of micro-organisms and capable of breaking the specific pollution in the feed. The waste-water along with the sheared off solids emerges from the filter through a drain to an underground sump. From the sump the effluent is pumped by recirculation pumps for recycling. The quantity of recycling will vary with the feed characteristics, BOD removal efficiency, etc. and is set during plant operation. Then the effluent goes to clarifier - I and the overflow flows to aeration tank where it is subjected to activated sludge treatment, a second stage biological treatment. For the proper oxidation of organic matter contained in nutrient deficient waste, nutrients mainly nitrogen and phosphorous are added to feed chamber of biofilter in the form of urea and phosphoric acid. Ferric sulphate is also added to inlet of aeration tank to take care of traces of hydrocarbons. The aeration tank has a retention time of 7 hours sufficient to degrade the residual organic content in waste-water to the desired level. Under optimum conditions this concentration of biomass will range between 3000 to 5000 mg/l^{-1}. The reacted mass from aeration tank flows to clarifier-II from where the clarified effluent flows to the filter feed pump. The clarifier under flow is recycled back to the bioreactor by means of biosludge recirculation pumps to maintain the desired level of active biomass. A portion of this sludge is periodically send to the combined sludge symptom maintain biomass balance in aeration tank. The treated effluent is then passed through dual media filter (DMF) to remove traces of pollutants, suspended solids, etc. This is a polishing stage. The pressure filter consist of a sand layer supported by gravel layer with an anthracite layer above. The treated effluent is pumped from the top of the filter. The anthracite layer removes traces of pollutants left behind while the sand removes fine solids. The solids accumulated in the DMF are removed by back washing with water after scouring it with air. The filtered water flows into the guard pond from where it flows to the equalisation pond. Here the treated process effluent is joined by cooling (sea) water which return from skimming pond. After equalisation, the combined effluent passes through hay filters into effluent channel which joins the main effluent channel near VR-WWTP.

Waste-water treatment plant

It receives and treats waste-waters from pump drips/floor washings/ contaminated rain water from old units and product tank bottom drains and cooling water returns. Under normal operations these streams do not contain phenols, sulphides or any other dissolved organic impurities. The treatment steps include free oil removal, emulsified oil removal, and filtration through hay filters. The effluent characteristics from ETP-1 are shown in Table 25.2.

Sludge Handling System

Oily sludge from TPI separator, chemical sludge from merox clarifier and biosludge from clarifier I and II are collected in combined sludge pump. This combined sludge is taken to the sludge thickener for partial dewatering.

Slop Oil Recovery System

The slop oil recovered from various free oil removal stages is collected in a slop oil sump and will be pumped to existing wet slop oil tanks.

Table 25.2. Effluent water characteristics at different stages of central wastewater treatment plant (CWWTP).

Parameter	Samples, concentration, mg/l^{-1}											
	S_1	S_2	S_3	S_4	S_5	S_6	S_7	S_8	S_9	S_{10}	S_{11}	S_{12}
pH	9.5	9.6	9.4	7.6	10.8	7.68	7.46	7.6	7.5	7.38	0.45	6–8.5
Suspended solids	250	40	60	45	50	56	44	30	20	20	15	<20
Ammonia (NH$_4^+$–N)	1.0	0.58	7.80	650	110	26	24	8.0	7.0	5.8	5.2	–
Sulphide (S^{2-})	7900	880	76	10.5	56	24	18	0.5	0.5	0.45	5.2	0.5
Cyanide (CN)	0.6	0.25	22.0	0.6	20.8	3.6	3.0	0.15	0.15	0.12	0.10	–
Phenolic compounds	4600	520	74.8	19.4	71.0	48	46	1.0	1.0	0.8	0.76	1.0
Oil and grease	1420	256	164	670	152	12.0	10	7.0	5.0	8.6	6.0	10.0
Biochemical oxygen demand (BOD)	29,540	31,360	1680	770	690	278	260	14.6	14.8	14.8	14.0	15
Chemical oxygen demand (COD)	82,600	9060	6340	1600	2200	850	830	118	106	98	90	250

Sampling points

S_1: Effluent water of merox waste sample collected after oil catcher of CWWTP.
S_2: Effluent water sample collected from clarifier (after chlorinated copperas tank).
S_3: Sour water sample collected after TPI oil separator-I.
S_4: Bitumen and rain water effluent collected after TPI oil separator-II.
S_5: Sour water sample collected after ammonia stripping unit.
S_6: Water sample collected after DAF unit.
S_7: Water sample collected before entering into secondary waste-water treatment (Equalisation tank).
S_8: Water sample collected in clarifier-II.
S_9: DMF outlet (Guard pond).
S_{10}: ETP-I outlet.
S_{11}: MHF outlet.
S_{12}: MINAS standards for oil refineries.
Note: All parameters are expressed in mg/l, except pH.

Through Cooling Water

Sea water is used as the cooling medium through system at Vizag refinery oil and grease is the major contaminant due to leaks in the heat exchangers. Only free oil removal is carried out before disposal. Sanitary waste-water from the refinery is treated through septic tanks before discharge.

Thus, it is evident from Table 25.2, at different stages of treatment units (from S_1– S_9) the dominant pollutants were reduced at each treatment stage, finally the effluent values of DMF outlet (S_9), ETP-1 outlet (S_{10}) and MHF outlet (S_{11}) are meeting MINAS standards given in S_{12}. The free oil removal was 95 per cent from API and TPI oil separators. The emulsified oil was removed by alum, polyelectrolytes in dissolved air flotation (DAF) units.

Phenolic compounds, sulphides and oxygen demand (mainly due to the presence of soluble hydrocarbons and sulphides) were reduced by chemical treatment with chlorinated copperas and by two stage biological treatment. The provision of two stage biological system enhances the bio-oxidation and provides sufficient inbuilt safety and protection against any possible shock loads or failure of the system.

To produce maximum results it is necessary that the treatment system be well designed and properly operated. After going through different parameters at different sampling points (after different treatment units) it can be concluded that almost all the treatment units are performing properly and also the overall performance is better than the individual performance.

However, the operating conditions namely flow, maintenance of toxic element concentration, recycling of sludge for maintaining the MLSS concentration in the bioreactor are not maintained properly, the effluent water quality has never been satisfactory and one cannot expect an improperly operated treatment plant to produce satisfactory results.

This means the effluent water quality does not meet stringent MINAS, the techno-economically acceptable standards, particularly for sulphide and BOD, prescribed by Central Board for Prevention and Control of Water Pollution (CBCWP) for oil refineries. Unlike Mathura refinery, the Vizag refinery is letting their effluents into sea watercourses, so the MINAS standards may be slightly relaxed particularly for sulphide and BOD.

CASE STUDY-2 JK RAYON INDUSTRY, KANPUR

The present investigation is a case study on JK Rayon, Kanpur describing the mechanism of zinc, which is used in the regeneration process of cellulose, escapes through various routes. Based on the study, of effluent to MINAS, are recommended. Besides this Ganga water quality of a short stretch in the vicinity of JK Rayon is also discussed.

The JK Rayon Industry, Kanpur is one of the old units producing Viscose Filament Yarn (VFY) and commissioned plant in 1966 for treating zinc bearing effluent. So far known this could be the first plant treating zinc bearing effluent from VFY producing industrial unit.

The present study is undertaken to assess the pollutional load generated and the performance of effluent treatment plant in the Industry.

Characterisation of the waste-waters emanating from the various sources as well as of the total combined effluent is carried out.

Measure to ensure the Minimal National Standards (MINAS) for man-made fibre industry are recommended. The Minimal National Standards applicable to this specific type of industry are furnished in Table 25.3.

Table 25.3. Minimal national standard for man-made fibre Industries.

Parameter	Concentration not to exceed
pH	5.5 to 9
Suspended solids	100 mg/l
Biochemical Oxygen demand (5 day, 20°C)	30 mg/l
Zinc	1 mg/l

The industry produces viscose filament yarn, sulphuric acid and carbon disulphide; the latter two produced for captive use. The capacity for the products are listed below:

Product	Capacity (Tons/day)
Viscose Filament Yarn	9
Sulphuric acid	20
Carbon disulphide	3

Water Use and Waste-water Generation

The source of water used in the industry is the river Ganga. The quantity abstracted is 9900 kilolitre per day (KLD) out of which 660 KLD is for domestic use in the colony. The waste-water generated is reported at 9320 KLD and the measurements carried out during the present study indicated a final waste-water flow of 10,735 KLD. The two types of waste-water streams generated in the industry are the zinc bearing and the other non-zinc bearing. The sources of waste-water generation in the industry and the present collection system are shown in Figure 25.4.

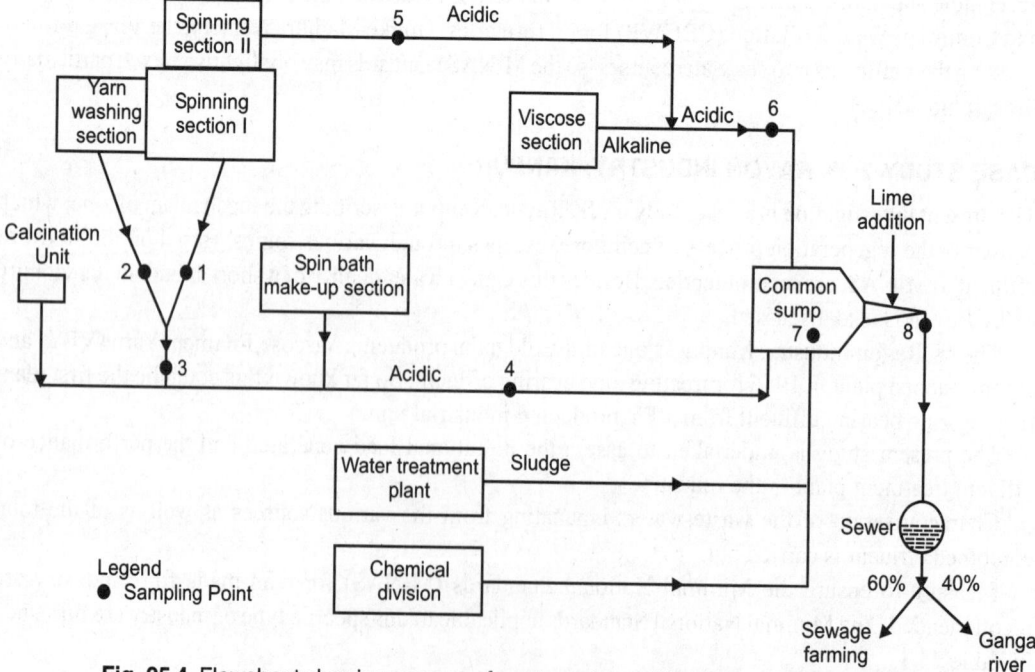

Fig. 25.4. Flowsheet showing sources of waste-water and location of sampling points.

The important pollutant discharged from the industry is zinc which, at 18.20 kg per ton of viscose filament yarn produced, is utilised in the regeneration process of cellulose.

The major discharges of zinc is from continuous and intermittent spin bucket wash:

1. From the spinning section.
2. From the washing of yarn in the after treatment section. The other minor source of zinc discharge is the spin bath section also known as make-up section. All the three waste-water streams are acidic in character. The sources of zinc bearing effluents are discussed below.

Spinning Section

The ripened viscose liquid of viscosity upto 70 poise is pumped in the spinning section through spinnerettes which are submerged in the spinning bath containing 13 grams per litre (g/l) of zinc sulphate and 137 g/l of sulphuric acid. Viscose cellulose as it comes out of the fine holes of the spinnerettes into the spinning bath gets solidified and regenerated into filaments having about 300 degrees of polymerisation. The filaments are wound in a spinning bucket rotating at about 8000 rpm. The filaments are intermittently washed in the spinning bucket to remove impurities. The spinning wash water thus generated contain very high amount of zinc which is mostly taken to the spin bath make-up section. However during such washing the waste-water containing significant amount of zinc also enter drains through leakages from the 120 positions provided in any one spinning machine. The spent spin bath solution is taken to calcination unit where sodium sulphate is recovered. The calcination unit was not in operation in the industry during the reported study.

After treatment section

The bobbins on which filaments are wound referred as cake are transferred to the after treatment section where these are subjected to a number of washing with hot and cold water in a counter current manner before sulphurising, softening agent washing and bleaching. The washing is done to remove coagulated liquor and sulphur, etc. The bobbins are washed in batches on racks continuously for 25 minutes with 5 minutes interval between any two batches. It is reported that the maximum quantity of zinc is lost during after treatment operation if the spinning bucket wash is intermittent. The most concentrated zinc wash water from after treatment section is reportedly taken to the spin bath section for spinning solution make-up.

Spin bath section

The solution for the coagulation and regeneration of cellulose is prepared in this section. The requisite amounts of zinc and sulphuric acid are maintained at 13 g/l and 137 g/l, respectively. The most concentrated zinc wash water from the after treatment section and from the continuous of intermittent spinning bucket wash are sent to this section to be used in the preparation of the solution. The waste-water containing zinc emanating from this section is due to the washing of floor, filterback wash and cooling water blow down.

Non-zinc bearing effluents

The other sources of waste-water generation are viscose section, dialyser unit, chemical section and water treatment plant. The viscose liquid after being fed into spinnerettes is filtered to remove suspended impurities. The filters are washed. The wash water containing fibrous material is drained. In the dialyser unit the viscose liquid containing beta and gamma-cellulose as 4.5 and 2.3 per cent as well as lignin are removed and drained. The waste-water emanating from viscose and dialyser units are alkaline in nature.

Assessment of Pollution Loads and the Present Treatment

Two hourly composited samples are collected from the locations identified in Fig. 25.4. Grab samples are also collected once at the final outlet and the spinning section. Two types of waste-waters are generated, one acidic and the other alkaline, in the manufacture of viscose filament yarn. The acidic waste-water contains zinc, the concentration of which is less in continuous washing compared to intermittent washing. Apparently the volume of effluent should be more in continuous washing.

It is reported that if the washing are continuous the loss of zinc from spinning section would be about 80–90 per cent of the total loss and the balance from after treatment section. In intermittent spray wash more loss from after treatment and less from spinning section may be expected. It is not possible to measure the rate of waste-water flow at the various sampling locations because of the old drainage system which is inaccessible. Therefore, estimation of quantity of zinc discharged from the two sources could not be done.

Results of analyses of samples collected from the locations indicated in Fig. 25.4 are furnished in Table 25.4. The final effluent had a zinc concentration of 20 mg/l which is very high when compared with the desired limit of 1 mg/l. The two important zinc bearing streams, acidic in nature, are from the spinning sections I and II having concentrations of 40.55 mg/l and 57.25 mg/l, respectively (sampling locations 1 and 5, Table 25.4 and Fig. 25.4). The stream from spinning section II inspite of having lesser number of machines compared to section I showed higher concentration of zinc at 57.25 mg/l because the zinc bearing effluent emanating from this section is not diluted with wash water from after treatment section as in the case of the effluent from section I. The stream from the after treatment section had zinc at concentration of 8.1 mg/l and pH value of 6.0 (sampling location 2). The alkaline stream from the viscose section mixes with the effluent of spinning section II and even after the mixing the resultant pH and zinc concentration were found to be 2.6; and 36.58 mg/l, respectively (sampling location 6). Samples from location 4 showed a zinc concentration of 32.63 mg/l which is higher than that of location 3. This is due to the contribution of zinc from the spin bath section. Moreover in spin bath, filter back wash is carried out in batches causing occasional discharges. The two main drain carrying waste-water from different sources run parallely and terminate in a sump (location 7, Fig. 25.4) where the waste-waters from the chemical division and water treatment plant also enter. The waste-water from the sump flows through a channel and over a V-notch and then over a bucket containing lime, obviously, it is no treatment at all. The samples collected before and after lime addition show zinc concentrations of 15.84 mg/l and 20.03 mg/l respectively and pH is found to be almost same at 2.7 and 2.9 indicating no treatment whatsoever is taking place (sampling locations 7 and 8). The pH should range between 9.5 and 10.0 for zinc precipitation. Additional of lime or keeping a bucket of lime at the final outlet for neutralisation is highly improper. A grab sample collected on the second day of the investigation from the final discharge point flowed a zinc concentration of 19.8 mg/l which tallied with the previous day result of the composited sample at 20.30 mg/l. The management of the industry mentioned that at sampling location 8 they could obtain zinc concentration as low as 0.8 mg/l and BOD value as 52.5 mg/l.

A grab sample is collected on the second day of the investigation from the basement of the spinning section I which showed high zinc content at 82.62 mg/l. The flow from spinning section I is also measured as 1,426 KLD. According to this flow, loss of zinc from this section is calculated to be 118 kg/day amounting to 70 per cent of the zinc consumption reported at 180 kg/day. The exact loss of zinc from spinning section II could not be estimated because of difficulty in measuring flow of spinning waste-water from this section. Heavy losses are expected from spinning section I and II because the washings were continuous. The total discharge of zinc from the industry to Ganges river and agricultural land is calculated to be 215 kg/day which is close to the reported consumption of 180 kg/day. A grab sample from continuous spray bucket showed a zinc concentration of 151 mg/l. As reported by this industry this wash water is taken to spin bath section for the preparation of spinning solution. However through leakages some of the wash water enter the drains.

Table 25.4. Characteristics of the waste-water stream.

Sampling location	Sampling location number and description	pH	Zn
1.	Spinning section I (outside)	2.0	40.55
2.	After treatment section (washing of yarn)	6.0	8.09
3.	Combined effluent (1 and 2)		
4.	Combined effluent (3, boiler) house spin bath, cooling water	2.4	16.31
	blow down	2.4	32.63
5.	Spinning section II	2.0	57.25
6.	Combined (5 and viscose section)	2.6	36.58
7*.	Total combined effluent (common sump before lime addition)	2.7	15.84
8**.	Total combined effluent (after lime addition)	2.9	20.02 (flow 10735 m³/day), Zn 215 kg/day
+	Spinning section I (basement drain)	2.1	82.62 (flow 1425.7 m³/day), Zn 118 kg/day
++	Continuous spin bucket was wash grab	1.85	151.0
+++	Zinc in sludge in the settling pit	-	2537.0 mg/kg of dry sludge

Sampling location 1 to 8 are marked in the flowsheet.
* Location 7 - BOD 115, COD-248, SO_4^- 1200, SS-120
** Location 8 - BOD 105, COD-240, SO_4^- 1325, SS-200

Location +, ++ and +++ are not shown in the flowsheet. All the parameters are expressed in ppm except pH.

Achievement of 1 mg/l of zinc is not difficult provided control is done properly in two stage precipitation. The two stage precipitation would help the industry to recover zinc and to avoid problems of sludge disposal.

As already mentioned, the industry provided data to show that by proper operation only by lime treatment the level of zinc in the treated effluent is brought to 0.8 mg/l and Biochemical Oxygen Demand (BOD) to 52.5 mg/l. This suggests that the existing treatment systems should be rectified and operated properly.

Sulphate and Biochemical Oxygen Demand

During regeneration process sodium sulphate is formed which is recovered. However, sodium sulphate is also carried away along with the filaments, which enter the drains during spinning and the washing of fibres and from spin bath make-up. Although sulphate, an important parameter in rayon effluent is not included in the MINAS, it is thought expedient to estimate this to know the actual concentration being discharged. The concentration of sulphate is determined in the final effluent and found to be quite high at 1325 mg/l whereas the suggested permissible concentration is discharges into inland surface waters is 1000 mg/l (IS 2296–1982). The Biochemical Oxygen Demand (BOD) of the final effluent is found to be more than 100 mg/l compared to the permissible limit of 30 mg/l (Table 25.3). There is a sludge settling pit which is full of sludge and is not in operation. Zinc sludge stored in the settling tank is collected for zinc analysis. The presence of zinc is found to be 2537 mg/kg of sludge on dry basis.

Effect on the River Ganga

A number of industries, textiles, jute, tanneries, explosive, chemicals and metallurgical, are discharging their effluents into the river Ganga near Kanpur. There are a number of drains discharging domestic waste-waters into the river. Grab samples from locations X and Y (Fig. 25.5) were collected and the results of analysis are furnished in Table 25.5.

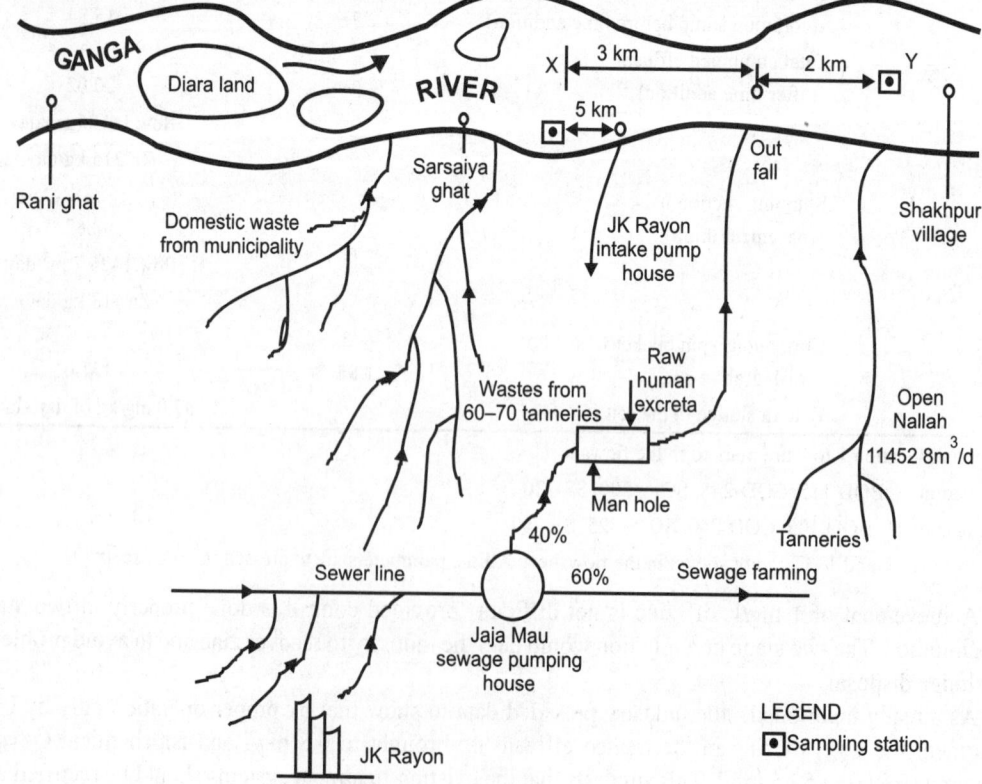

Fig. 25.5. Kanpur stretch—Ganga river.

The objective is to see the level of zinc in river water and bottom sediment, upstream and down stream of the outfall of the waste-water from Jajmou sewage pumping station, which includes waste-water from JK Rayon (location Z).

The waste-water from JK Rayon is discharged into a municipal sewer which carries other industrial and domestic waste-water. About 60 per cent of the waste-water is pumped for irrigational use. The other 40 per cent is letout through a manhole, into which human excreta collected from the town is also dumped, to the river Ganga. The 40 per cent flow is estimated to be 1,4,652 KLD, and the total therefore will be approximately 2,86,632 KLD.

Based on the very scanty data obtained, some possible interpretation regarding water quality in terms of organic matter as well as heavy metals is given. The value of zinc in river water upstream of outfall (location Z) is 0.14 mg/l (location X) and at down stream of the outfall is 0.22 mg/l (location Y). The build-up is not considered significant. The reason could be that zinc before it reaches downstream is mostly trapped along with the heavy sludge and settle in the vicinity of the outfall. It is seen that the concentration of zinc in bottom sediment down stream is lower than that at upstream (Table 25.5). The possible reason for such observation could be the presence of high concentration of sulphates 1325 mg/l discharged from rayon effluent. This will result in more accumulation of zinc in the bottom sediment in the river at the outfall due to sulphate reducing bacteria which converts sulphate to sulphide and thus more zinc sulphide will precipitate and settle. With regard to the concentration of total chromium at these locations a striking difference in river water samples and bottom mud samples were observed. The concentration of chromium upstream of outfall is 20 mg/l and at downstream is 0.57 mg/l. This is due to the discharges from the tanneries as shown in Fig. 25.5. Chromium in the bottom mud at locations X and Y are 0.58 mg/kg and 26.35 mg/kg respectively on dry basis exhibiting considerable build-up of chromium down stream which may affect benthos.

JK Rayon industry also discharges organic matter and its level is reflected by the measure of BOD (Table 25.5) which is insignificant as compared to other sources like tanneries and municipal wastes, but more than 30 mg/l. It has been reported that presence of heavy metal at high level suppresses BOD. The presence of zinc to 1 mg/l in watercourse suppresses BOD by about 8 per cent and other metals like copper and chromium suppress upto 50 per cent at that concentration. At upstream the BOD is 64 mg/l as compared to 23 mg/l downstream (Table 25.5). Of course, values of BOD at both the location were very high for a natural watercourse.

Recommendation

Restructuring of the drain system should be done for easy assessment of the effluent, with flow measuring devices. This may be elaborated as follows:

1. The drains are narrow and deep. They need to be widened with proper gradient for easy flow and accessibility. The drainage system carrying zinc-laden waste-water should be segregated and be brought to one common outlet for appropriate treatment.
2. The sludge disposal site should be free from percolation to avoid ground water pollution due to zinc.
3. BOD is found more than what is prescribed in MINAS and therefore the waste-water should be treated biologically to reduce its level to 30 mg/l.
4. Sulphate in the final effluent is found to be quite high. Hence recovery of sodium sulphate from the spent spin bath solution in the calcination unit should be effectively done to minimise sulphate in the discharge.

Table 25.5. Analysis of Ganga water collected from upstream and downstream from outfall of Jajman sewage pumping station.

Location	Parameters						
	pH	COD	BOD	SO$_4$ (mg/l)	Zn	Cr (Total)	Total Coliform (MPN/100 ml)
X (300 m upstream from outfall)	8.5	128	64	350	0.14 (42.94 mg/kg*)	20.0 (0.58 mg/kg*)	2,40,000
Y (200 m down-stream from outfall)	2.0	56	23	31	0.22 (26.07 mg/kg*)	0.57 (26.35 mg/kg*)	4,60,000

* Figure given in parenthesis are the contents of metal in the bottom sediment on dry basis.

CASE STUDY-3 RADIOACTIVE WASTE TREATMENT TECHNOLOGY AT CZECH NUCLEAR POWER PLANTS

This case study describes the main technologies for the treatment and conditions of radioactive wastes of Czech nuclear power plants. The main technologies are bituminisation for liquid radioactive wastes and supercompaction for solid radioactive wastes

By the term of waste we usually understand a thing, of which its owner (waste producer) wants to dispose. Waste might also be some movable material, the disposal of which is necessary to protect human health and the environment, even when the producer of the waste is unknown.

Radioactive waste from nuclear power plants (NPPs) include materials and equipment, which are not possible to be introduced into the environment due to radionuclide content or unremovable surface radioactive contamination. For that reason, their handling is controlled by special procedures (processing, treatment, transport). The boundary between radioactive waste and other waste is not precise. Discharge limits for release of radioactive materials into the environment, have been adopted by the competent authority. Radioactive waste arises during maintenance and operational activities in radioactive environments. Radioactive waste arising in a NPP may have specific activity up to 10^8 Bq/kg with about 30 year half-life.

Objective of Radioactive Waste Handling at Dukovany NPP

The main objective of waste management is to isolate radioactive waste from the environment. Radwastes from normal Dukovany NPP operation are disposed of in a surface disposal facility on Dukovany NPP site. The basic objective is to isolate waste from environment for about 300 years, which correspond to 10 half-lives of the dominant radionuclide (^{137}Cs). This decay period results in an activity reduction by 3 orders of magnitude (i.e. from 10^8 to 10^5 Bq/kg). Waste has to be in a suitable form for disposal prior its deposition in the disposal facility.

Minimisation principle

An important principle of radioactive waste treatment is its minimisation, which is a process leading to the smallest practical amounts of treated waste. Keeping waste quantities low has economic, environmental and political advantages. It is also very important for public acceptance.

For that reason, it is useful to give examples of possible additional waste reduction:

1. Recycling boric acid from liquid waste.
2. Treating low activity waste to reduce chemical content (water is recycled).
3. Release of decontaminated metal (the volume of treated decontamination solutions must not exceed metal volume for recycling).
4. Optimisation and selection of chemicals used in controlled area to minimise foaming in evaporator (shampoo, soap, washing powder, cleansing powder, floor detergents, etc.). Foaming in evaporator can cause premature saturation of demineraliser resin capacity.
5. Avoid bringing unnecessary objects into the controlled area (packages).
6. Avoid unnecessary entry of persons.
7. Usage of thinner (lighter) foils as protection against contamination.
8. Usage of optimum concentrations of drained media (i.e. usage of 12.8 per cent boric acid instead of 12 per cent increase volume of chemicals in waste-water above the level necessary during maintenance).
9. Replacement of service water by condensate or demineralised water in areas with leakage (reduction of salt volume in condensates).

Categorisation of Waste—Radwaste Catalogue

Categories of wastes from the Dukovany NPP classified from the technical point of view are given in Table 25.6.

Table 25.6. Categories according to which wastes from Dukovany NPP are classified.

Category	Waste characteristics	Source	Technology of treatment	One-year production (t/y)	Activity (MBq/kg)
Compactible/ Combustible	Condemned personal protective aids, decontamination and cleaning clothes, packing materials, paper, PE sheets	The biggest part originates during unit inspections and repairs	High-pressure compaction, combustion	20	1–2
Non-combustible	Glass, wires, cans, metal particles, ceramics, filters	Mainly during inspections and repairs	High-pressure compaction	3	1–2
Wood	Wooden transport packages, pallets, scaffold flooring, planks	Contingent origination, air-conditioning filters replacement	High-pressure compaction, combustion	1	0.1
Flammable but unfit for combustion	PVC, PTFE (teflon) - foils, sealing materials	Previously extensively used materials in RCA	High-pressure compaction	1	0–1
Large metal objects	Structural material of carbon and stainless steel	Extensive reconstructions	Disposal without treatment or decontamination and recycling	10	0.01

(Contd ...)

Category	Waste characteristics	Source	Technology of treatment	One-year production (t/y)	Activity (MBq/kg)
Resins	Condemned purification station fillings	Regular substance replacement, contingent leakage during technological operations	Insertion into HIC[a], bitumenisation	30	100
Other sorbents	Active coal, vapex (perlite), zeolites	–	Insertion into HIC, cementing	1	not given
Sludge	Sediments in tanks, mixture of organic and inorganic substances of nonstandard composition	Floor washing and cleaning, dust from material separation and abrasion, crystallisation beyond design basis	Insertion into HIC, cementing	5	10
Waste-water	Usually diluted solutions of chemical inorganic substances containing impurities	Uncontrolled leakage, sampling, laboratory water, spilling of liquids	Concentrate bitumenisation	350	1
Oils and solvents	Depreciated lubricants, solution residues and scintillators	Filling exchange, laboratories, elimination of non-applicable and contaminated liquids	Washing by demineralised water, combustion	2	0.001
Ash, fly ash, slag	Residues after combustion and melting	External incinerator, melting furnace	Insertion into HIC, cementing	0	not given

[a]HIC means high integrity container.

Methodology of Radwaste Processing and Treatment

Processing and treatment of liquid radwastes

Origination and composition of liquid radwastes

The following main sources of liquid radwastes at Dukovany NPP need to be considered:
1. Recovery solutions and purification plant flushing.
2. Liquids collected in plant systems.
3. Decontaminating solutions.
4. Waste-water from laundry and water from changing rooms if it is not possible to discharge it to sewage drainage from the radiation point of view.
5. Alkaline agents for radioactive concentrate stabilisation.

According to purification plant type, recovery solutions and purification plant flushings contain nitrates (from cation resin recovery by nitric acid), borates (boric acid washed out from highly basic anion exchanger), potassium hydroxide, sodium hydroxide and ammonia (from anion exchanger recovery). In highly reducing conditions nitrates also arise.

Liquids collected in drain systems contain salts from primary coolant circuit and salts originating in service and cooling water, with sulphate, chloride and calcium content. A substantial part of this kind of waste-water originates in the course of unit outages during drainage of pipelines and tanks. Intakes to the drainage system are then a significant source of waste-water organic pollution (cleaning and others).

Decontamination solutions contain particularly salts of manganese (reduction of $KMnO_4$ used in alkaline decontamination solution), sodium and potassium and residues of citric acid and oxalic acid which are, however, in the given environment, rather unstable.

Laundry waste-water and changing room water usually do not represent a significant part of waste-water because they are discharged, if their activity is low, to the inactive sewage system of Dukovany NPP. If it is necessary to treat them with radioactive water, they introduce surfactants, phosphates and increased organic pollution with to the treatment systems. Alkaline agents are added before entering the evaporator prior to waste-water concentration.

Collection, concentration and storage of liquid radwastes

The primary section of Dukovany NPP is equipped with a unified sewage system. The main disadvantage of this system is mixing of all sorts of waste-water and therefore recycling of separate waste-water is complicated and difficult. This unified sewage system is routed from the twin unit 1 to radioactive drain sump tanks and waste-water is pumped from there to a sedimentation tank located in the auxiliary service building and then through an overflow tank to waste-water hold-up tanks. Waste-water from laundry, laboratories and changing rooms of operational building 1 is collected in control tanks and if it is not possible to discharge it, it is pumped to the radioactive drain sump tanks.

Regenerative and flushing water are routed through a tank to the radioactive drain sump tank, and from there to the overflow tank and then by the same path as the rest of water to the waste-water hold-up tanks. The situation of the twin unit 2 is similar. Waste-water in the waste-water hold-up tanks is analysed to determine the need for further concentration in evaporators. Usually it contains 0.5–2 g of salt/l with pH ca. 8. Crystallisation of borates with relatively low solubility might occur during cool down, storage or other handling in neutral and acid media. Prior to concentration, it is necessary to alkalise this waste-water by addition of sodium hydroxide to pH at least 11 to reach solubility 60 g of boric acid/l. Evaporation is used to concentrate the salt concentration in waste liquid about 50 times. The concentrated residue is pumped to concentrate tanks. Both auxiliary service buildings are interconnected through three pipelines for liquid radwastes and they are surrounded by a protective trough. Composition of concentrate changes both with time and between individual NPP units depending particularly on unit mode. In the Table 25.7 is a typical composition of RA concentrates produced at Dukovany NPP.

Table 25.7. Typical composition of wastes at Dukovany NPP.

pH		11.4
Specific weight	kg/m^3	1110
Salt contents	g/l	147
Boric acid	g/l	65
Nitrates	g/l	30
Nitrites	g/l	2
Sulphates	g/l	3
Chlorides	g/l	3
Oxalates	g/l	2
Citranes	g/l	1
Sodium	g/l	35

(Contd ...)

Potassium	g/l	6
134-Cs	kBq/l	300
137-Cs	kBq/l	400
58-Co	kBq/l	10
60-Co	kBq/l	60
90-Sr	Bq/l	10
239+240 Pu	mBq/l	200

Since saturated solution is introduced into the condensate storage tanks, it means that boric acid salts crystallise out on the bottom, vessel internals and the like. When the temperature or pH value increases, they gradually go back into solution. The aim of chemical operation control during concentration is to reach purposely maximum concentration (to minimise volume) but without exceeding the solubility limit of salts present, particularly borates. This problem is not simple. Disposal of crystallised salts is very demanding from the technical and economical point of view.

Liquid concentrate treatment

The objective of liquid radioactive concentrate treatment is to transform it to a form suitable for disposal. Such a form has to be a solid state and therefore we often talk about liquid waste solidification. Two basic technologies are mainly used for this purpose in the case of nuclear power plant wastes:

1. Cementation.
2. Bituminisation.

In the future, it might be that technology used to date for high-level radioactive waste solidification from reprocessing plants could be applied to NPP wastes-vitrification.

Cementation is a relatively simple technology is based on chemical reaction of mineral components, contained in cement, with water. This reaction is called hydration. Water is chemically combined in the final product—with a certain fixation ability for radionuclides contained in charge water. It is evident that the mass ratio of water to cement is not unlimited and the maximum for solid product is usually about 0.4. As a result, waste volume and weight increase significantly during this process. Another disadvantage is relatively high leachability, that is, release of fixed components from the product in the case of contact with water.

Bituminisation consists in fixing of contaminated salts into a water resistant matrix, using asphalt. During the process of liquid radwaste bituminisation water evaporates with temperatures above 120°C and contained salts are homogenised with asphalt. During vitrification, the process, the waste material is melted at temperatures about 1000°C with glass-forming material, resulting in a product with very low leachability. The volume reduction is very good.

Bituminisation at Dukovany NPP

Liquid radwaste bituminisation is the process used in the Dukovany radioactive waste treatment facility. The basic equipment is a layer rotor evaporator with a vertical double-shell drum of inside diameter 600 mm, made from molybdenum steel, for 1.1 MPa steam heating. Fitted in the axis is a shaft with swinging blades that in radial direction nearly touch the internal surface of the heated shell. Asphalt and liquid concentrate are tangentially sprayed to the upper part of evaporator. The function of the blades is to spread the mixture on the internal heated surface during the rotor revolution (300 rev./hour) and so create a thin layer from which water evaporates and residual salts are mixed with asphalt. Drained bitumen product, containing ca. 40 per cent of mass salts, flows down the evaporator wall and continues through heated piping to drums where it is lidded automatically and then transported to the disposal facility. Theoretical capacity of the evaporator is 240 litres per hour.

Concentrate is transported to the radioactive waste treatment facility building from the tanks located in the auxiliary service building jusses or directly by a submersible sump pump from the storage tank. The tanks, each of 7 m^3 effective volume, are filled through this pump. There is also the possibility of pH regulation by nitric acid or sodium hydroxide addition in these tanks. Required pH values, according to technical specifications, is 11–11.5. The lower value is to avoid crystallisation and the upper value prevents so called asphalt alkaline pyrolysis, which is exothermic reaction of organic components at higher temperature in asphalt with high pH values and with multivalent metal catalysis. Manganese content is limited to 15 g/l for the same reason.

Asphalt is transported to Dukovany NPP by trailer and is stored with temperature of ca 120°C in a tank located outside the radioactive waste treatment facility building. It is very important to select an asphalt type carefully because product quality depends on it (softening point, penetration, leachability). Some asphalt types are inclined to carbon and hard substances insoluble in common organic solvents and non-fusible by on site temperatures. Then pipeline fouling and evaporator vibration occur. To eliminate these problems, various additives are added to asphalt at Dukovany NPP.

Asphalt is pumped from the storage tank through steam heated pipelines to a tank and from there it is sprayed by heated gear feed pumps to the evaporator. Feed is regulated in such a way to reach 40 mass per cent of salt contents in the resulting product, which means that a feed rate of ca 50–60 kg of asphalt/hr.

Solid radwaste processing and treatment

Solid radwastes at Dukovany NPP are produced principally during refuelling and maintenance. Solid radwaste processing involves collecting, sorting, bagging and storing of radioactive waste. Bags of waste are classified according to radioactivity content.

Waste bags with surface dose rate lower than 1 μGy/hr are sorted in a storage carousel. These bags are assumed to be inactive and can be released as normal trash. Before disposal, they are additionally sorted with respect to materials to paper (scrap), other combustible waste (clothes and plastics) and metal. Waste bags with surface dose rate higher than 100 μGy/hr are more radioactive. Noncompactible objects are removed and they are stored for several years.

Waste bags with dose rate 1–100 μGy/hr are low active. Since their activity might be caused only by one object it is worthwhile to sort these bags according to their activity. Sorting is performed in a semiautomatic sorting box and it enables operating personnel reliable separation of radwastes and inactive wastes. Solid radwastes produced by this sorting are stored together with radwastes of higher dose rates and inactive wastes are removed again.

Dukovany NPP does not have equipment for further routine solid waste treatment. The sorted radwaste bags are stored in reinforced concrete cells pending periodic treatment campaigns using high pressure compaction.

During the campaigns performed about every ten years, the stored radwastes are removed, then treated by compaction and placed into radioactive waste disposal. Treatment is performed in two stages. Bags with compactible waste are gradually inserted into drums and low-pressure compacted. Low-pressure non-compactible wastes (small metal, glass, etc.) are inserted into drums without compacting. Low-pressure compaction of up to 15 waste bags (50 liltre each) in one 200 litres drum is possible.

When the drums are full they are transported to a high-pressure compactor operated by a contractor (IAEA technology). About 2500 filled drums were re-compacted with super press AEA (Great Britain) in 2004. Moulded cakes were inserted into 820 larger drums and they were disposed in radioactive waste disposal.

The following objects are not suitable for high-pressure compacting in AEA device:

1. Drums with weight above 300 kg.
2. Drums with dose rate above 3 mSv/h.
3. Drums dimensionally unfit.
4. Wet or highly oiled up wastes.
5. Toxic wastes.
6. Metal waste longer than 550 mm and 60–250 mm in diameter.
7. Pulverised waste.
8. Wastes containing significant volume of plutonium, uranium, americium and tritium.

Combustion is often used for solid radwaste treatment in foreign countries. The higher volume reduction is reached by combustion. Organic substance destruction prevents subsequent microbial decomposition and gas production. Disadvantages of radwaste combustion are both inaccessibility in the Czech Republic, high price and also necessity of ash treatment (by cementing, sintering, vitrification). It is usually effective to use low-active waste technologies for metal decontamination (half-dry decontamination). If the metal is too highly contaminated, fragmented metal is stored in the drums without treatment. It is possible to place large metal parts in the radioactive waste disposal facility as nonstandard radwaste, but extra permission is necessary. Contaminated metal melting is used during NPP decommissioning in foreign countries but this process is very expensive considering the negligible quantity of contaminated metals at Dukovany NPP.

Treatment of other radwastes

Moist radwaste

Exhausted contaminated ion exchangers are taken out of service from purification plants at Dukovany NPP. Ion exchange resins are usually in the form of small beads (smaller than 1 mm) with capacity to capture ions and impurities from solutions. Filters might contain other media such as activated charcoal, perlite or even sand. Contaminated sludges are radioactive wastes of valuable composition. They originate from impurities, from condensation of chemical substances or crystallisation of saturated solutions. They contain variable water percentage. They are called, together with ion exchange resins and other sorbents, as 'moist radwastes'. Moist radwastes are usually transported by water flow with the help of pumps. Their treatment for disposal is only being planned so far. They might be cemented, calcined, bituminised, vitrified, or they might be disposed of without a matrix. This particular method has been chosen for application at Dukovany NPP. The wastes are pumped into a special container without sorting or any other treatment and dehydrated.

This leak proof container, known as a HIC (high integrity container) represents a multi-layer barrier against radionuclide release. Waste dried by vacuum pump is of minimum volume because it is not mixed with binder (cement, bitumen). Water resistance is provided by the container maintaining its leak tightness for about 400 years. Moist radwastes might attain high activity (up to 10^9 Bq/kg). High activity and atypical size of HIC containers might represent certain problems with manipulation. Nonstandard composition, which might cause difficulties for some technologies due to changeable additive composition, is not a problem for HIC technology.

Oils and solvents

Oily substances, that are also in stable form classified as hazardous wastes, are involved as a rule. Radioactive oils are usually burnt in foreign countries. This method will presumably be used for radwaste disposal at Dukovany NPP and this activity will be provided by contractor (i.e. incinerator). Oils have been treated in the past at Dukovany NPP by liquid-liquid extraction. The method consists in extraction of radioactive substances by demineralised water. A mixture of oil and water is sucked by a pump into a closed vessel. After some hours the phase equilibrium is established and water discharged from the lower part of the vessel. The water contains radioactive substances and it is treated in purification system. After repeated extraction, oil is recycled as inactive waste. During oil collection it is necessary to segregate certain oils. The most important consideration is not to mix the most toxic oil substances (containing PCB and the like) with the rest of oils. It is very important to store non-chlorinated oils separately from chlorinated substances because burning of mixtures is very expensive in the case of chlorinated contents. Individual sorts of oils are also not mixed, if it is possible, due to viscosity changes.

It is possible to decontaminate solvents through extraction but also through distillation. Only distillation of perchlorethylene used in the radioactive waste treatment bituminisation line has been performed at Dukovany NPP.

Ash, fly ash and slag

These radwastes may arise if Dukovany NPP contracts for incineration or melting of wastes in the foreign countries in the future. However, the most probable case is that the contractor will treat these wastes with the use of cement technology and return them in our 200 litres drums or put them into our HIC casks.

Glossary

Absorption field	:	A system of properly sized and constructed narrow trenches partially filled with a bed of washed gravel or crushed stone into which perforated or open joint pipe is placed. The discharge from the septic tank is distributed through these pipes into trenches and surrounding soil. While seepage pits normally require less land area to install, they should be used only where absorption fields are not suitable and well-water supplies are not endangered.
Acid soil	:	Soil with a pH value <6.6.
Activated sludge	:	Sludge particles produced in raw or settled waste-water (primary effluent) by the growth of organisms (including zoogleal bacteria) in aeration tanks in the presence of dissolved oxygen. The term 'activated' comes from the fact that the particles are teeming with fungi, bacteria and protozoa. Activated sludge is different from primary sludge in that the sludge particles contain many living organisms which can feed on the incoming waste-water.
Advanced waste treatment	:	A process intendedlit to remove the BOD and suspended solids as well as nitrogen or phosphorus that has not been eliminated at the primary stage. Advanced waste-water treatment techniques include filtration, carbon adsorption, ion exchange, distillation, reverse osmosis, electrodialysis and microstraining. Full-scale physical/chemical treatment plants use such techniques as filtration and carbon adsorption in lieu of secondary treatment.
Aeration tank	:	The tank where raw or settled waste-water is mixed with return sludge and aerated; this is the same as an aeration bay, aerator or reactor.
Aerobe	:	An organism that requires free oxygen for growth.
Aerobic	:	(i) Having molecular oxygen as a part of the environment, (ii) growing only in the presence of molecular oxygen, as in aerobic organisms, and (iii) occurring only in the presence of molecular oxygen, as in certain chemical or biochemical processes such as aerobic respiration.
Aerotolerant anaerobes	:	Microbes that grow under both aerobic and anaerobic conditions, but do not shift from one mode of metabolism to another as conditions change. They obtain energy exclusively by fermentation.
Agar	:	Complex polysaccharide derived from certain marine algae that is a gelling agent for solid or semisolid microbiological media. Agar consists of about 70 per cent agarose and 30 per cent agaropectin. Agar can be melted at temperature above 100°C; gelling temperature is 40°–50°C.
Agarose	:	Non-sulphated linear polymer consisting of alternating residues of D-galactose and 3,6-anhydro-L-galactose. Agarose is extracted from seaweed and agarose gels are often used as the resolving medium in electrophoresis.

597

Alkaline substance	:	Chemical compounds in which the basic hydroxide (OH–) ion is united with a metallic ion, such as sodium hydroxide (NaOH) or potassium hydroxide (KOH). These substances impart alkalinity to water and are employed for neutralisation of acids. Lime is the most commonly used alkaline material in waste-water treatment.
Alga (plural, algae)	:	Phototrophic eukaryotic micro-organism. Algae could be unicellular or multicellular. Blue-green algae are not true algae; they belong to a group of bacteria called cyanobacteria.
Backflow	:	The flow of water or other liquids, mixtures, or substances into the distributing pipes of a potable supply of water from any source or sources other than its intended source.
Alkaline soil	:	Soil having a pH value >7.3.
Alkalophile	:	Organism that grows best under alkaline conditions (up to a pH of 10.5).
Allochthonous flora	:	Organisms that are not indigenous to the soil but that enter soil by precipitation, diseased tissues, manure and sewage. They may persist for some time but do not contribute in a significant way to ecologically significant transformations or interactions.
Anabolism	:	Metabolic processes involved in the synthesis of cell constituents from simpler molecules. An anabolic process usually requires energy.
Anaerobe	:	An organism that lives and reproduces in the absence of dissolved oxygen, instead deriving oxygen from the breakdown of complex substances.
Anaerobic	:	(i) Absence of molecular oxygen, (ii) growing in the absence of molecular oxygen, such as anaerobic bacteria, and (iii) occurring in the absence of molecular oxygen, as a biochemical process.
Anaerobic respiration	:	Metabolic process whereby electrons are transferred from an organic or in some cases, inorganic compounds to an inorganic acceptor molecule other than oxygen. The most common acceptors are nitrate, sulphate and carbonate.
Anion	:	A negatively charged ion in an electrolyte solution, attracted to the anode under the influence of a difference in electrical potential. Chloride is an anion.
Anion exchange capacity	:	Sum total of exchangeable anions that a soil can adsorb. Expressed as centimoles of negative charge per kilogram of soil.
Anoxic	:	Literally 'without oxygen'. An adjective describing a microbial habitat devoid of oxygen.
Antagonist	:	Biological agent that reduces the number or disease-producing activities of a pathogen.
API separator	:	A facility developed by the Committee on Disposal or Refinery Wastes of the American Petroleum Institute for separation of oil from waste-water in a gravity differential and equipped with means for recovering the separated oil and removing sludge.
Aseptic	:	Free from living germs of disease, fermentation or putrefaction.
Associative symbiosis	:	Close but relatively casual interaction between two dissimilar organisms or biological systems. The association may be mutually beneficial but is not required for accomplishment of a particular function.
Attached growth processes	:	Waste-water treatment processes in which the micro-organisms and bacteria treating the wastes are attached to the media in the reactor. The wastes being treated flow over the media. Trickling filters, biotowers and RBCs are attached growth reactors. These reactors can be used for removal of BOD, nitrification and denitrification.

Attenuation	:	Reduction of the signal power of field strength as a function of distance through a material. Also refers to shielding effectiveness.
Attenuation	:	Reduction of the signal power of field strength as a function of distance through a material. Also refers to shielding effectiveness.
Autoradiography	:	Detecting radioactivity in a sample, such as a cell or gel, by placing it in contact with a photographic film.
Batch process	:	A treatment process in which a tank or reactor is filled, the waste-water (or solution) is treated or a chemical solution is prepared and the tank is emptied. The tank may then be filled and the process repeated. Batch processes are also used to cleanse, stabilise or condition chemical solutions for use in industrial manufacturing and treatment processes.
Beneficial uses (of water)	:	The waters of the state that may be protected against quality degradation include, but are not necessarily limited to, domestic, municipal, agricultural and industrial supply; power generation; recreation, aesthetic enjoyment; navigation; preservation and enhancement of fish, wildlife, and other aquatic resources or preserves.
Biochemical oxygen demand (BOD)	:	The measure of decomposable organic material in domestic or industrial waste-waters, as represented by the oxygen utilised over a period of 5 days at 20°C and as determined by the appropriate procedure in 'Standard Methods'.
Biodegradable	:	Substance capable of being decomposed by biological processes.
Biofilm	:	A slime layer which naturally develops when bacteria attach to an inert support that is made of a material such as stone, metal or wood. There are also non-filamentous bacteria that will produce an extracellular polysaccharide that acts as a natural glue to immobilise the cells. In nature, non-filament-forming micro-organisms will stick to the biofilm surface, locating within an area of the biofilm that provides an optimal growth environment (i.e., pH, dissolved oxygen, nutrients). Since nutrients tend to concentrate on solid surfaces, a micro-organism saves energy through cell adhesion to a solid surface rather than by growing unattached and obtaining nutrients randomly from the medium. *Pseudomonas* and *Nitrosomonas* strains are especially well known for their ability to form a strong biofilm.
Bioflocculation	:	The clumping together of fine, dispersed organic particles by the action of certain bacteria and algae.
Bioremediation	:	Use of micro-organisms to remove or detoxify toxic or unwanted chemicals from an environment.
Biosolid	:	The residue of waste-water treatment. Formerly called sewage sludge.
Biostimulation	:	Any process that increases the rates of biological degradation, usually by the addition of nutrient, oxygen or other electron donors and acceptors so as to increase the number of indigenous micro-organisms available for degradation of contaminants.
BOD test	:	A procedure that measures the rate of oxygen use under controlled conditions of time and temperature. Standard test conditions include dark incubation at 20°C for a specified time (usually 5 days).
Bonded sewer house connection sewer	:	Any house connection sewer from a lot or part of a lot, which does not have a public sewer directly in front, at the rear or on its sides of and which has not been directly assessed for a public sewer.

Bulking sludge	:	Clouds of billowing sludge that occur throughout secondary clarifiers and sludge thickeners when sludge becomes too light and will not settle properly. In the activated sludge process, bulking is usually caused by filamentous bacteria.
Cess pool	:	A lined, excavation in the ground which receives the discharge of a drainage system or part thereof, so designed to retain the organic matter and solids discharging therein, but permitting liquid to seep through.
Chemical oxygen demand (COD)	:	The measure of chemically decomposable material in domestic or industrial waste-water as represented by the oxygen utilised as determined by the appropriate procedure described in 'Standard Methods'.
Chemical precipitation	:	Precipitation induced by addition of chemicals; the process of softening water by the addition of lime and soda ash as the precipitants.
Chlorination	:	The application of chlorine to water or waste-water, generally for the purpose of disinfection, but frequently for accomplishing other biological or chemical results.
Chlorine demand	:	The difference between the amount of chlorine added to a waste-water sample and the amount remaining at the end of a 30-minute period as determined by the procedures given in 'Standard Methods'.
Clarification	:	A process in which suspended material is removed from a waste-water. This may be accomplished by sedimentation, with or without chemicals or filtration.
Clarifier	:	Settling tank, sedimentation basin. A tank or basin in which waste-water is held for a period of time, during which the heavier solids settle to the bottom and the lighter material will float to the water surface.
Coagulants	:	Chemicals which cause very fine particles to clump (floc) together into larger particles. This makes it easier to separate the solids from the water by settling, skimming, draining or filtering.
Coliform bacteria	:	Non-pathogenic microbes found in fecal matter that indicate the presence of water pollution; are thereby a guide to the suitability for potable use.
Colloids	:	Very small, finely divided solids (particles that do not dissolve) that remain dispersed in a liquid for a long time due to their small size and electrical charge.
Combination waste and vent system	:	A specially designed system of waste piping, embodying the horizontal wet venting of one or more sinks or floor drains by means of a common waste and vent piping adequately sized to provide free movement of air above the flow line of the drain.
Combined sewer	:	A sewer designed to carry both sanitary waste-water and storm or surface-water run-off.
Contact stabilisation	:	Contact stabilisation is a modification of the conventional activated sludge process. In contact stabilisation, two aeration tanks are used. One tank is for separate re-aeration of the return sludge for at least four hours before it is permitted to flow into the other aeration tank to be mixed with the primary effluent requiring treatment.
Contamination	:	An impairment of the quality of the waters by waste to a degree which creates a hazard to public health through the spread of disease.
Continuous vent	:	A vertical vent that is a continuation of the drain to which it connects.
Continuous waste	:	A drain connecting the compartments of a set of fixtures to a trap or connecting other permitted fixtures to a common trap.
Conventional treatment	:	The preliminary treatment, sedimentation, floatation, trickling filter, rotating biological contactor, activated sludge and chlorination of waste-water.

Degradation	:	A growth phase in which the availability of food begins to limit cell growth.
Deionised water	:	Water that goes through an ion exchange process in which all positive and negative ions are removed.
Denitrification	:	An anaerobic biological reduction of nitrate nitrogen to nitrogen gas, the removal of total nitrogen from a system and or an anaerobic process that occurs when nitrite ions are reduced to nitrogen gas and bubbles are formed as a result of this process. The bubbles attach to the biological floc in the activated sludge process and float the floc to the surface of the secondary clarifiers. This condition is often the cause of rising sludge observed in secondary clarifiers or gravity thickeners.
Disinfection	:	The process designed to kill most micro-organisms in waste-water, including essentially all pathogenic (disease-causing) bacteria. There are several ways to disinfect, with chlorine being the most frequently used in water and waste-water treatment plants.
Dissolved air floatation: (DAF)	:	Dissolved air floatation is one of many designs for waste treatment.
Dissolved oxygen (DO)	:	A measure of the oxygen dissolved in water expressed in milligrams per litre.
Dissolved solids	:	The solid matter in solution in waste-water which can be obtained by evaporation of a sample from which all suspended matter has been removed by filtration as determined by the procedures in 'Standard Methods'.
Distributor	:	The rotating mechanism that distributes the waste-water evenly over the surface of a trickling filter or other process unit.
Domestic sewage	:	The water-borne wastes derived from ordinary living processes and of such character as to permit satisfactory disposal without special treatment into the sanitary sewer system.
Domestic waste-water	:	The water-carried wastes produced from non-commercial or non-industrial activities, which result from normal human living processes.
Drainage system	:	All the piping within public or private premises, which conveys sewage or other liquid wastes to a legal point of disposal, but not including the mains of a public sewer system, or a public sewage treatment or disposal plant.
DOUR	:	Dissolved oxygen uptake ratio.
Downstream side	:	The side of a product stream that has already passed through a given filter system; portion located after the filtration unit.
Dual chamber test method	:	Measures near field shielding effectiveness by indicating the signal attenuation caused by passage through test material.
Ecosystem	:	Groupings of various organisms interacting with each other and their environment.
Effluent	:	Literally anything which flows out or is discharged. Usually applied to the discharge of a waste material into a water body or the atmosphere.
Effective stack height	:	The sum of the stack height and the plume rise.
Emulsion	:	A liquid mixture of two or more liquid substances not normally dissolved in one another, one liquid held in suspension in the other.
Equalising basin	:	A holding basin in which variations in flow and composition of liquid are averaged. Such basins are used to provide a flow of reasonably uniform volume and composition to a treatment unit. Also called a balancing reservoir.
Estuaries	:	Bodies of water which are located at the lower end of a river and are subject to tidal fluctuations.

Eurythermal	:	Bodies of water which are located at the lower end of a river and are subject to tidal fluctuations.
Extractables	:	Substances that can be leached from a filter during the filtration process or under other specified conditions.
Facultative anaerobe	:	A bacterium capable of growing under aerobic conditions or anaerobic conditions in the presence of an inorganic ion i.e., SO_4, NO_3.
Facultative pond	:	The most common type of pond in current use. The upper portion (supernatant) is aerobic, while the bottom layer is anaerobic. Algae supply most of the oxygen to the supernatant.
Flocculation	:	The bonding together of coagulated particles to form settleable or filterable solids.
Fermentation	:	A type of heterotrophic metabolism in which an organic compound rather than oxygen is the terminal electron (or hydrogen) acceptor. Less energy is generated from this incomplete form of glucose oxidation than is generated by respiration, but the process supports anaerobic growth.
Filter aid	:	A chemical (usually a polymer) added to water to help remove fine colloidal suspended solids.
Filter medium	:	The permeable portion of a filtration system that provides the liquid-solid separation, such as screens, papers non-wovens, granular beds and other porous media.
Grit	:	The heavy material present in waste-water, such as sand coffee grounds, eggshells, gravel and cinders.
Headworks	:	The facilities where waste-water enters a waste-water treatment plant. The headworks may consist of bar screens, comminutors, a wet well and pumps.
Humus	:	The dark organic material in soils, produced by the decomposition of soils. The matter that remains after the bulk of detritus has been consumed (leaves, roots). Humus mixes with top layers of soil (rock particles), supplies some of the nutrient needed by plants—increases acidity of soil; inorganic nutrients more soluble under acidic conditions, become more available, for example wheat grows best at pH 5.5–7.0. Humus modifies soil texture, creates loose, crumbly texture, that allows water to soak in and nutrients retained; permits air to be incorporated into soil.
Incineration	:	The conversion of dewatered waste-water solids by combustion (burning) to ash, carbon dioxide and water vapour.
Indirect waste pipe	:	A pipe that does not connect directly with the drainage system but conveys liquid wastes by discharging them into a plumbing fixture, interceptor or receptacle, which is directly connected to the drainage system.
Industrial waste-water	:	All water-carried wastes and waste-water of the community, excluding domestic waste-water and uncontaminated water; and all waste-water from any producing, manufacturing, processing, institutional, commercial, agricultural, or other operation where the waste-water discharged includes significant quantities of waste of non-human origin.
Infiltration	:	The seepage of groundwater into a sewer system, including service connections. Seepage frequently occurs through defective or cracked pipes, pipe joints, connections or manhole walls.
Interceptor (gravity separation)	:	Any facility designed, constructed and operated for the purpose of removing dangerous, deleterious or prohibited constituents from waste-water by differential gravity separation before discharge into the public sewer.

Interceptor sewer	:	A collecting sewer that intercepts and collects sewage from a number of lateral or local public sewers.
Interface	:	The common boundary layer between two substances such as between water and a solid (metal) or between water and a gas (air) or between a liquid (water) and another liquid (oil).
Media	:	The material in the trickling filter on which slime accumulates and organisms grow. As settled waste-water trickles over the media, organisms in the slime remove certain types of wastes thereby partially treating the waste-water. Also the material in a rotating biological contactor (RBC) or in a gravity or pressure filter.
Metabolism	:	All of the processes or chemical changes in an organism or a single cell by which food is built up (anabolism) into living protoplasm and by which protoplasm is broken down (catabolism) into simpler compounds with the exchange of energy.
MLSS	:	Mixed liquor suspended solids—the volume of suspended solids in the mixed liquor of an aeration tank.
MLVSS	:	Mixed liquor volatile suspended solids—the volume of organic solids that can evaporate at relatively low temperatures (550°C) from the mixed liquor of an aeration tank. This volatile portion is used as a measure or indication of micro-organisms present. Volatile substances can also be partially removed by air stripping.
MPN index	:	Most Probable Number of coliform-group organisms per unit volume of sample water. Expressed as a density or population of organisms per 100 ml of sample water.
Nitrification	:	An aerobic process in which bacteria change the ammonia and organic nitrogen in waste-water into oxidised nitrogen (usually nitrate). The second-stage BOD is sometimes referred to as the 'nitrification stage' (first-stage BOD is called the 'carbonaceous stage').
Nitrifying bacteria	:	Bacteria that change the ammonia and organic nitrogen in waste-water into oxidised nitrogen (usually nitrate).
Oil retention boom	:	A floating baffle used to contain and prevent the spread of floating oil on a water surface.
Organic waste	:	Waste material which comes—mainly from animal or plant sources. Organic waste generally can be consumed by bacteria and other small organisms. Inorganic wastes are chemical substances of mineral origin.
Ozonation	:	The application of ozone to water, waste-water or air, generally for the purposes of disinfection or odour control.
Parts per million (ppm)	:	The unit commonly used to designate the concentration of a substance in a waste-water in terms of weight, i.e. one pound per million pounds, etc. ppm is synonymous with the more commonly used term mg/l (milligrams per litre).
Potable water	:	Water that does not contain objectionable pollution, contamination, minerals or infective agents and is considered satisfactory for drinking.
Preliminary treatment	:	The removal of metal, rocks, rags, sand, eggshells and similar materials which may hinder the operation of a treatment plant. Preliminary treatment is accomplished by using equipment such as racks, bar screens, comminutors and grit removal systems.
Pretreatment	:	Treatment by using screens, degritters, degreasers and scum removal devices to eliminate settleable solids, such as floating debris, sand and grit.

Primary sewage treatment	:	Treatment of raw sewage to remove the largest impurities (suspended solids or liquids, such as clays, bits of organic wastes, oil droplets). In domestic wastes, BOD is reduced about 30 to 40 per cent through this. This system is mostly physical, employing sedimentation tanks, flotation tanks, flocculation systems and occasional screening.
Private sewage disposal system	:	A septic tank with the effluent discharging into a subsurface disposal field, one or more seepage pits, a combination of subsurface disposal field and seepage pits or such other facilities as may be permitted under the local governing agency.
Putrefaction	:	Biological decomposition of organic matter with the production of ill-smelling products associated with anaerobic conditions.
RAS	:	Return activated sludge—settled activated sludge that is collected in the secondary clarifier and returned to the aeration basin to mix with incoming raw settled waste-water.
RASVSS	:	Return activated sludge volatile suspended solids.
Reclaimed water	:	Water which as the result of treatment of waste is suitable for direct beneficial use or a controlled use that would not otherwise occur.
Recycle	:	The use of water or waste-water within (internally) a facility before it is discharged to a treatment system.
REDOX	:	Biological reductions/oxidations. These reactions usually require enzymes to mediate the electron transfer. The sediment in the bottom of a lake, sludge in a sewerage works or septic tank will have a very low redox potential and will likely be devoid of any oxygen. This sludge or waste-water will have a very high concentration of reductive anaerobic bacteria, indeed the bulk of the organic matter may in fact be bacteria. As the concentration of oxygen increases the oxidation potential of the water will increase. A low redox potential or small amount of oxygen is toxic to anaerobic bacteria, therefore as the concentration of oxygen and redox potential increases the bacterial population changes from reductive anaerobic bacteria to oxidative aerobic bacteria. Measurement of REDOX potential is also referred to as ORP.
Refractory materials	:	Material difficult to remove entirely from waste-water such as nutrients, colour, taste and odour-producing substances and some toxic materials.
Reverse osmosis	:	The principle involving diffusion of water through a semipermeable membrane from a dilute to a concentrated solution. By applying pressure greater than osmotic pressure, to the concentrated solution the diffusion of water through the membrane is reversed. Pure water passes from the concentrated into dilute solution. This results in a separation of dissolved solids.
R/O unit	:	Reverse osmosis unit for water purification in small aquariums and miniature yard-ponds, utilises a membrane under pressure to filter dissolved solids and pollutants from the water. Two different filter membranes can be used: the CTA (cellulose triacetate) membrane is less expensive, but only works with chlorinated water and removes 50–70 per cent nitrates and the TFC membrane, which is more expensive, removes 95 per cent of nitrates, but is ruined by chlorine.
Run-off	:	Water running down slopes rather than sinking in (again, result of poor humus content). Example: erosion due to deforestation.
Sanitary sewage	:	(i) Domestic sewage, excluding with storm or surface water, (ii) sewage discharging from sanitary conveniences, (iii) the water supply of a community after it has been used and discharged into a sewer.

Secondary sewage treatment	:	Treatment operations designed to remove dissolved and colloidal organic compounds such as proteins, sugars, starches and phenols. These organics may themselves be harmful, but additionally they consume oxygen and increase BOD. Typical processes include activated sludge, trickling filters, stabilisation ponds and aeration lagoons. All work on the same basic principle: the accelerated biological degradation or consumption of organic compounds.
Sedimentation	:	The process of subsidence and deposition of suspended matter from a waste-water by gravity.
Septicity	:	Septicity is the condition in which organic matter decomposes to form foul-smelling products associated with the absence of free oxygen. If severe, the waste-water turns black, gives off foul-odours, contains little or no dissolved oxygen and creates a heavy oxygen demand.
Septic tank	:	A watertight receptacle which receives the discharge of a drainage system or part thereof, designed and constructed so as to retain solids, digest organic matter through a period of detention and allow the liquids to discharge into the soil outside of the tank through a system of open joint piping or a seepage pit, meeting the requirements of the local governing authority.
Settleable solids	:	Those solids in suspension which will pass through a 2000 micron sieve and settle in one hour under the influence of gravity.
Sewerage	:	Any and all facilities used for collection, conveying, pumping, treating and disposing of waste-water.
Sewerage system	:	A network of waste-water collection, conveyance, treatment and disposal facilities interconnected by sewers.
Sloughings	:	Trickling-filter slimes that have been washed off the filter media. They are generally quite high in BOD and will lower effluent quality unless removed.
Sludge	:	The settleable solids separated from liquids during processing; the deposits of foreign materials on the bottoms of streams or other bodies of water.
Solid wastes	:	Non-liquid-carried wastes normally considered to be suitable for disposal with refuse at sanitary landfill.
Soluble BOD	:	Soluble BOD is the BOD of water that has been filtered in the standard suspended solids test.
Surface media	:	Captures particles on the upstream surface with efficiencies in excess of depth media, sometimes close to 100 per cent with minimal or no off-loading. Commonly rated according to the smallest particle the media can repeatedly capture. Examples of surface media include ceramic media, microporous membranes, synthetic woven screening media and in certain cases, wire cloth. The media characteristically has a narrow pore size distribution.
TDS	:	Total dissolved solids is commonly estimated from the electrical conductivity of the water. Pure water is a poor conductor of electricity. Impurities dissolved in the water cause an increase in the ability of the water to conduct electricity. Conductivity, usually expressed in units of microsimens, formerly micromhos or in mg/l, thus becomes an indirect measure of the level of impurities in the water.
Tertiary sewage treatment	:	Treatment operations designed to eliminate residual dissolved organic and inorganic compounds. Inorganics are usually removed in electrodialysis, reverse osmosis or iron exchange. In large concentrations, distillation, freezing or other desalination-type techniques are more appropriate for removing inorganics. One

		of the most effective means of treating organics is by absorption on activated carbon.
Tidal waters	:	All coastal ocean waters of a state, including bays and estuaries upstream to the inland limit of tidal action.
TOC	:	Total organic carbon—a measure of the amount of organic carbon in water.
Transpiration	:	The process by which water vapour is released to the atmosphere by living plants, a process similar to people sweating.
Trickling filter	:	An attached culture, waste-water treatment system. A large tank generally filled with rock or rings. Waste-water is sprayed over the top of the media, providing the opportunity for the formation of slimes or biomass to remove wastes from the waste-water, through revolving arms which have spray nozzles. Water is pumped from the bottom of a trickle filter to a secondary clarifier.
TSS	:	Total suspended solids.
Turbidity	:	The amount of suspended matter in waste-water, obtained by measuring its light scattering ability.
WAS	:	Waste activated sludge, mg/l. The excess growth of micro-organisms which must be removed from the process to keep the biological system in balance.
Waste-water	:	The used water and solids from a community that flow to a treatment plant. Storm water, surface water and groundwater infiltration may also be included in the waste-water that enters a waste-water treatment plant. The term 'sewage' usually refers to household wastes, but this word is being replaced by the term 'waste-water'.
Water content	:	Water contained in a material expressed as the mass of water per unit mass of oven-dry material.
Water-retention curve	:	Graph showing soil-water content as a function of increasingly negative soil water potential.
Xerophile	:	Organism adapted to grow at low water potential, i.e. very dry habitats.

References

Adams, T.K. and Smith, D.J., *Water Pollution*, Addison Wesley Longman, Harlow, Essex.

Bloomfield, A., Jonassen and T. Schneider, *Industrial Waste-water Treatment*, Butterworth, London.

Bryan Bergeron, *Desalination by Reverse Osmosis*, Pearson Education, Singapore.

Creighton, M.O. and Schwartz, R.M., *Chemistry for Environmental Engineering*, Academic Press, London.

Dagley, R., *Reverse Osmosis Technology*, Academic Press, London.

Fersht, R. and Kri, L., *Water Filtration Technology*, John Wiley and Sons, New York.

Gorodkin, J., *Water Disinfection—Chemical and Analytical Aspects*, Applied Science Publishers, London.

Harwood, S. and Benner, S.A., *Disinfection: Water and Waste-water*, Academic Press, London.

Johnson, C., *Unit Processes in Drinking Water Treatment*, Tata McGraw Hill, New York.

K.A. De Jong, *Membrane Separation Technologies*, Chapman and Hall, London.

Kanehisa, M., *Advance Waste-water Treatment Technologies*, Pergamon Press, Oxford, London.

Kari, L. and Paun, G., *Microbial Aspects of Waste-water*, Chilton Book Co., USA.

Morrison, D.W., *Industrial Waste-Water*, Cold Spring Horbour Press, UK.

Munn, R.F., *Physical Methods of Treatment of Water*, John Wiley & Sons, New York.

Pigman, D.E., *Biological Methods of Treatment of Water*, Reston Publishing Co., Reston, Virginia.

Palmer, P.A., *Quality and Tests of Water*, Chapman and Hall, New York.

Ricci, F., *Water Recycling Criteria*, Johny Wiley and Sons, New York.

Richard Dybowshi and Stephen Roberts, *Sewage and Its Disposal*, Springer-Verlog, London.

Richard M. Twyman, *Handbook of Hazardous Waste Treatment and Disposal*, Bio Scientific Publishers, Oxford.

Richardson, D. and Coffee, L., *Kinetics of Water and Waste-water Technology*, University Press, Cambridge.

Schrowebel, J., *Water Resources Engineering*, Pergamon Press, New York.

Segel, T.F. and Waterman, M.S., *Water Pollution*, Prentice Hall, London.

Snell, I.D. and Snell, C.T., *Hydrology and Hydraulic System*, D. Van Nostrand, New York.

Stephen Misener and Stephen A. Krawetz, *Water Supply and Pollution Control*, Human Press Inc., New Jersey.

Whistler, W.J. and Lipman, D.J., *Chemical Inactivation of Viruses in Water*, Heinemann, London.

Wolfrom, K. and Snell, J., *Water Structure and Behaviour*, Heinemann, London.

Index